Lecture Notes in Computer Sci

Commenced Publication in 1973
Founding and Former Series Editors:
Gerhard Goos, Juris Hartmanis, and Jan van Leeuwen

Palash Sarkar   Tetsu Iwata (Eds.)

Advances in Cryptology – ASIACRYPT 2014

20th International Conference on the Theory
and Application of Cryptology and Information Security
Kaoshiung, Taiwan, December 7-11, 2014
Proceedings, Part I

 Springer

Volume Editors

Palash Sarkar
Indian Statistical Institute
Applied Statistics Unit
203, B.T. Road, Kolkata 700108, India
E-mail: palash@isical.ac.in

Tetsu Iwata
Nagoya University
Department of Computer Science and Engineering
Furo-cho, Chikusa-ku, Nagoya 464-8603, Japan
E-mail: iwata@cse.nagoya-u.ac.jp

ISSN 0302-9743   e-ISSN 1611-3349
ISBN 978-3-662-45610-1   e-ISBN 978-3-662-45611-8
DOI 10.1007/978-3-662-45611-8
Springer Heidelberg New York Dordrecht London

Library of Congress Control Number: 2014954246

LNCS Sublibrary: SL 4 – Security and Cryptology

Typesetting: Camera-ready by author, data conversion by Scientific Publishing Services, Chennai, India

Printed on acid-free paper

Springer is part of Springer Science+Business Media (www.springer.com)

Preface

It is with great pleasure that we present the proceedings of Asiacrypt 2014 in two volumes of *Lecture Notes in Computer Science* published by Springer. The year 2014 marked the 20th edition of the International Conference on Theory and Application of Cryptology and Information Security held annually in Asia by the International Association for Cryptologic Research (IACR). The conference was sponsored by the IACR and was jointly organized by the following consortium of universities and government departments of the Republic of China (Taiwan): National Sun Yat-sen University; Academia Sinica; Ministry of Science and Technology; Ministry of Education; and Ministry of Economic Affairs. The conference was held in Kaohsiung, Republic of China (Taiwan), during December 7-11, 2014.

An international Program Committee (PC) consisting of 48 scientists was formed approximately one year earlier with the objective of determining the scientific content of the conference. As for previous editions, Asiacrypt 2014 also stimulated great interest among the scientific community of cryptologists. A total of 255 technical papers were submitted for possible presentations approximately six months prior to the conference. Authors of the submitted papers are spread all over the world. Each PC member could submit at most two co-authored papers or at most one single-authored paper, and the PC co-chairs did not submit any paper. All the submissions were screened by the PC members and 55 papers were finally selected for presentation at the conference. These proceedings contain the revised versions of the papers that were selected. The revisions were not checked and the responsibility of the papers rest with the authors and not the PC members.

The selection of papers for presentations was made through a double-blind review process. Each paper was assigned four reviewers and submissions by PC members were assigned five reviewers. Apart from the PC members, the selection process was assisted by a total of 397 external reviewers. The total number of reviews for all the papers was more than 1,000. In addition to the reviews, the selection process involved an extensive discussion phase. This phase allowed PC members to express opinion on all the submissions. The final selection of 55 papers was the result of this extensive and rigorous selection procedure.

The decision of the best paper award was based on a vote among the PC members, and it was conferred upon the paper "Solving LPN Using Covering Codes" authored by Qian Guo, Thomas Johansson, and Carl Löndahl. In addition to the best paper, three other papers were recommended for solicitations by the Editor-in-Chief of the *Journal of Cryptology* to submit expanded versions to the journal. These papers are "Secret-Sharing for NP" authored by Ilan Komargodski, Moni Naor, and Eylon Yogev; "Mersenne Factorization Factory" authored by Thorsten Kleinjung, Joppe W. Bos, and Arjen K. Lenstra; and

"Jacobian Coordinates on Genus 2 Curves" authored by Huseyin Hisil and Craig Costello.

In addition to the regular presentations, the conference featured two invited talks. The invited speakers were decided through an extensive multi-round discussion among the PC members. This resulted in very interesting talks on two different aspects of the subject. Kennth G. Paterson spoke on "Big Bias Hunting in Amazonia: Large-Scale Computation and Exploitation of RC4 Biases," a topic of importance to practical cryptography, while Helaine Leggat spoke on "The Legal Infrastructure Around Information Security in Asia," which had an appeal to a wide audience.

Along with the regular presentations and the invited talks, a rump session was organized. This session contained short presentations on latest research results, announcements of future events, and other topics of interest to the audience.

Many people contributed to Asiacrypt 2014. We would like to thank the authors of all papers for submitting their research works to the conference. Thanks are due to the PC members for their enthusiastic and continued participation for over a year in different aspects of selecting the technical program. The selection of the papers was made possible by the timely reviews from external reviewers, and thanks are due to them. A list of external reviewers is provided in these proceedings. We have tried to ensure that the list is complete. Any omission is inadvertent and if there is an omission, we apologize to that person.

Special thanks are due to D. J. Guan, the general chair of the conference, for working closely with us and ensuring that the PC co-chairs were insulated from the organizational work. This work was carried out by the Organizing Committee and they deserve thanks from all the participants for the wonderful experience. We thank Daniel J. Bernstein and Tanja Lange for expertly organizing and chairing the rump session.

We thank Shai Halevi for developing the IACR conference management software, which was used for the whole process of submission, reviewing, discussions, and preparing these proceedings. We thank Josh Benaloh, our IACR liaison, and San Ling, Asiacrypt Steering Committee Representative, for guidance and advice on several issues. Springer published the volumes and made these available before the conference. We thank Alfred Hofmann, Anna Kramer, Christine Reiss and their team for the professional and efficient handling of the production process.

December 2014
Palash Sarkar
Tetsu Iwata

Asiacrypt 2014
The 20th Annual International Conference on Theory and Application of Cryptology and Information Security

Sponsored by the *International Association for Cryptologic Research (IACR)*

December 7–11, 2014, Kaohsiung, Taiwan (R.O.C.)

General Chair

D. J. Guan — National Sun Yat-sen University, Taiwan, and National Chung Hsing University, Taiwan

Program Co-chairs

Palash Sarkar — Indian Statistical Institute, India
Tetsu Iwata — Nagoya University, Japan

Program Committee

Masayuki Abe — NTT Secure Platform Laboratories, Japan
Elena Andreeva — K.U. Leuven, Belgium
Paulo S. L. M. Barreto — University of Sao Paulo, Brazil
Daniel J. Bernstein — University of Illinois at Chicago, USA, and Technische Universiteit Eindhoven, The Netherlands
Guido Bertoni — STMicroelectronics, Italy
Jean-Luc Beuchat — ELCA, Switzerland
Debrup Chakraborty — CINVESTAV-IPN, Mexico
Chen-Mou Cheng — National Taiwan University, Taiwan
Jung Hee Cheon — Seoul National University, Korea
Ashish Choudhury — IIIT Bangalore, India
Sherman S.M. Chow — Chinese University of Hong Kong, Hong Kong SAR
Kai-Min Chung — Academia Sinica, Taiwan
Carlos Cid — Royal Holloway, University of London, UK
Jean-Sébastien Coron — University of Luxembourg, Luxembourg

Aurélie Bauer
Carsten Baum
Anja Becker
Amos Beimel
Rishiraj Bhattacharya
Begül Bilgin
Olivier Billet
Elia Bisi
Nir Bitansky
Olivier Blazy
Céline Blondeau
Andrej Bogdanov
Alexandra Boldyreva
Joppe W. Bos
Elette Boyle
Zvika Brakerski
Nicolas Bruneau
Christina Brzuska
Sébastien Canard
Anne Canteaut
Claude Carlet
Angelo De Caro
David Cash
Dario Catalano
André Chailloux
Donghoon Chang
Pascale Charpin
Sanjit Chatterjee
Jie Chen
Wei-Han Chen
Yu-Chi Chen
Ray Cheung
Céline Chevalier
Dong Pyo Chi
Ji-Jian Chin
Alessandro Chisea
Chongwon Cho
Kim-Kwang Raymond
   Choo
HeeWon Chung
Craig Costello
Giovanni Di Crescenzo
Dana Dachman-Soled
Ivan Damgård
Jean-Luc Danger

Bernardo David
Patrick Derbez
David Derler
Srinivas Devadas
Sandra Diaz-Santiago
Vassil Dimitrov
Ning Ding
Yi Ding
Christoph Dobraunig
Matthew Dodd
Nico Döttling
Rafael Dowsley
Frédéric Dupuis
Stefan Dziembowski
Maria Eichlseder
Martianus Frederic
   Ezerman
Liming Fang
Xiwen Fang
Pooya Farshim
Sebastian Faust
Omar Fawzi
Serge Fehr
Victoria Fehr
Matthieu Finiasz
Dario Fiore
Rob Fitzpatrick
Pierre-Alain Fouque
Tore Kasper Frederiksen
Georg Fuchsbauer
Eiichiro Fujisaki
Philippe Gaborit
Tommaso Gagliardoni
David Galindo
Wei Gao
Pierrick Gaudry
Peter Gaži
Laurie Genelle
Irene Giacomelli
Sergey Gorbunov
Dov Gordon
Samuel Dov Gordon
Robert Granger
Jens Groth
Felix Guenther

Nicolas Guillermin
Sylvain Guilley
Siyao Guo
Divya Gupta
Patrick Haddad
Nguyen Manh Ha
Iftach Haitner
Shai Halevi
Fabrice Ben Hamouda
Shuai Han
Christian Hanser
Mitsuhiro Hattori
Carmit Hazay
Qiongyi He
Brett Hemenway
Jens Hermans
Takato Hirano
Jeffrey Hoffstein
Dennis Hofheinz
Deukjo Hong
Hyunsook Hong
Wei-Chih Hong
Sebastiaan de Hoogh
Jialin Huang
Kyle Huang
Qiong Huang
Yan Huang
Yun Huang
Zhengan Huang
Andreas Hülsing
Michael Hutter
Jung Yeon Hwang
Malika Izabachene
Abhishek Jain
Dirmanto Jap
Stanislaw Jarecki
Eliane Jaulmes
Jérémy Jean
Mahabir Jhanwar
Guo Jian
Shaoquan Jiang
Pascal Junod
Chethan Kamath
Pierre Karpman
Aniket Kate

Jonathan Katz
Elif Bilge Kavun
Akinori Kawachi
Yutaka Kawai
Sriram Keelveedhi
Dakshita Khurana
Franziskus Kiefer
Eike Kiltz
Jihye Kim
Jinsu Kim
Minkyu Kim
Miran Kim
Myungsun Kim
Sungwook Kim
Taechan Kim
Mehmet Sabir Kiraz
Susumu Kiyoshima
Ilya Kizhvatov
Markulf Kohlweiss
Ilan Komargodski
Takeshi Koshiba
Simon Kramer
Ranjit Kumaresan
Po-Chun Kuo
Thijs Laarhoven
Fabien Laguillaumie
Russell W.F. Lai
Tanja Lange
Adeline Langlois
Martin M. Laurisden
Rasmus Winther
   Lauritsen
Changmin Lee
Hyung Tae Lee
Kwangsu Lee
Moon Sung Lee
Younho Lee
Wang Lei
Tancrède Lepoint
Gaëtan Leurent
Kevin Lewi
Allison Lewko
Liangze Li
Wen-Ding Li
Guanfeng Liang

Kaitai Liang
Benoît Libert
Changlu Lin
Huijia (Rachel) Lin
Tingting Lin
Yannis Linge
Helger Lipmaa
Feng-Hao Liu
Joseph Liu
Zhen Liu
Daniel Loebenberger
Victor Lomné
Yu Long
Patrick Longa
Cuauhtemoc
   Mancillas-López
Atul Luykx
Vadim Lyubashevsky
Houssem Maghrebi
Mohammad Mahmoody
Alex Malozemoff
Mark Manulis
Xianping Mao
Joana Treger Marim
Giorgia Azzurra Marson
Ben Martin
Daniel Martin
Takahiro Matsuda
Mitsuru Matsui
Ingo von Maurich
Filippo Melzani
Florian Mendel
Bart Mennink
Sihem Mesnager
Arno Mittelbach
Payman Mohassel
Amir Moradi
Tomoyuki Morimae
Kirill Morozov
Nicky Mouha
Pratyay Mukherjee
Gregory Neven
Khoa Nguyen
Phon Nguyen
Ivica Nikolić

Ventzislav Nikov
Svetla Nikova
Ryo Nishimaki
Adam O'Neill
Miyako Ohkubo
Tatsuaki Okamoto
Cristina Onete
Claudio Orlandi
David Oswald
Elisabeth Oswald
Khaled Ouafi
Carles Padro
Jiaxin Pan
Omer Paneth
Anat Paskin
Rafael Pass
Kenneth G. Paterson
Arpita Patra
Roel Peeters
Chris Peikert
Geovandro
   C. C. F. Pereira
Olivier Pereira
Ludovic Perret
Edoardo Persichetti
Krzysztof Pietrzak
Bertram Poettering
Geong-Sen Poh
David Pointcheval
Antigoni Polychroniadou
Raluca Ada Popa
Manoj Prabhakaran
Baodong Qin
Somindu C. Ramanna
Samuel Ranellucci
C. Pandu Rangan
Vanishree Rao
Jean-René Reinhard
Ling Ren
Oscar Reparaz
Alfredo Rial
Jefferson E. Ricardini
Silas Richelson
Ben Riva
Matthieu Rivain

Thomas Roche
Francisco
    Rodríguez-Henríquez
Lil María
    Rodríguez-Henríquez
Mike Rosulek
Arnab Roy
Hansol Ryu
Minoru Saeki
Amit Sahai
Yusuke Sakai
Olivier Sanders
Fabrizio De Santis
Yu Sasaki
Alessandra Scafuro
Christian Schaffner
John Schanck
Tobias Schneider
Peter Schwabe
Gil Segev
Nicolas Sendrier
Jae Hong Seo
Karn Seth
Yannick Seurin
Ronen Shaltiel
Elaine Shi
Koichi Shimizu
Ji Sun Shin
Naoyuki Shinohara
Joseph Silverman
Marcos A. Simplicio Jr
Boris Skoric
Daniel Slamanig
Nigel Smart
Fang Song
Douglas Stebila
Damien Stehlé
Rainer Steinwandt
Marc Stottinger

Mario Strefler
Takeshi Sugawara
Ruggero Susella
Koutarou Suzuki
Alan Szepieniec
Björn Tackmann
Katsuyuki Takashima
Syh-Yuan Tan
Xiao Tan
Qiang Tang
Christophe Tartary
Yannick Teglia
Sidharth Telang
Isamu Teranishi
Adrian Thillard
Aishwarya
    Thiruvengadam
Enrico Thomae
Susan Thomson
Mehdi Tibouchi
Tyge Tiessen
Elmar Tischhauser
Arnaud Tisserand
Yosuke Todo
Jacques Traoré
Roberto Trifiletti
Viet Cuong Trinh
Raylin Tso
Toyohiro Tsurumaru
Hoang Viet Tung
Yu-Hsiu Tung
Dominique Unruh
Berkant Ustaoglu
Meilof Veeningen
Muthuramakrishnan
    Venkitasubramaniam
Daniele Venturi
Frederik Vercauteren
Damien Vergnaud

Andrea Visconti
Ivan Visconti
Niels de Vreede
Mingqiang Wang
Wei Wang
Yanfeng Wang
Yuntao Wang
Hoeteck Wee
Puwen Wei
Qiaoyan Wen
Erich Wenger
Qianhong Wu
Keita Xagawa
Hong Xu
Weijia Xue
Takashi Yamakawa
Bo-Yin Yang
Guomin Yang
Wun-She Yap
Scott Yilek
Eylon Yogev
Kazuki Yoneyama
Ching-Hua Yu
Yu Yu
Tsz Hon Yuen
Aaram Yun
Mark Zhandry
Cong Zhang
Guoyan Zhang
Liang Feng Zhang
Tao Zhang
Wei Zhang
Ye Zhang
Yun Zhang
Zongyang Zhang
Yongjun Zhao
Yunlei Zhao
Vassilis Zikas

Organizing Committee

Advisors

Lynn Batten	Deakin University, Australia
Eiji Okamoto	Tsukuba University, Japan
San Ling	Nanyang Technological University, Singapore
Kwangjo Kim	Korea Advanced Institute of Science and Technology, Korea
Xuejia Lai	Shanghai Jiaotong University, China
Der-Tsai Lee	National Chung Hsing University, Taiwan, and Academia Sinica, Taiwan
Tzong-ChenWu	National Taiwan University of Science and Technology, Taiwan

Secretary

Chun-I Fan	National Sun Yat-sen University, Taiwan

Treasurer

Chia-Mei Chen	National Sun Yat-sen University, Taiwan

Local Committee Members

Shiuhpyng Shieh	National Chiao Tung University, Taiwan
Ching-Long Lei	National Taiwan University, Taiwan
Wen-Guey Tzeng	National Chiao Tung University, Taiwan
Hung-Min Sun	National Tsing Hua University, Taiwan
Chen-Mou Cheng	National Taiwan University, Taiwan
Bo-Yin Yang	Institute of Information Science, Academia Sinica, Taiwan

Sponsors

National Sun Yat-sen University
Academia Sinica
Ministry of Science and Technology
Ministry of Education
Ministry of Economic Affairs

Table of Contents – Part I

Cryptology and Coding Theory

Solving LPN Using Covering Codes 1
  Qian Guo, Thomas Johansson, and Carl Löndahl

Algebraic Attack against Variants of McEliece with Goppa Polynomial
of a Special Form 21
  Jean-Charles Faugère, Ludovic Perret, and Frédéric de Portzamparc

New Proposals

Bivariate Polynomials Modulo Composites and Their Applications 42
  Dan Boneh and Henry Corrigan-Gibbs

Cryptographic Schemes Based on the ASASA Structure: Black-box,
White-box, and Public-key (Extended Abstract) 63
  Alex Biryukov, Charles Bouillaguet, and Dmitry Khovratovich

Authenticated Encryption

Beyond $2^{c/2}$ Security in Sponge-Based Authenticated
Encryption Modes 85
  Philipp Jovanovic, Atul Luykx, and Bart Mennink

How to Securely Release Unverified Plaintext in Authenticated
Encryption 105
  Elena Andreeva, Andrey Bogdanov, Atul Luykx, Bart Mennink,
  Nicky Mouha, and Kan Yasuda

Forging Attacks on Two Authenticated Encryption Schemes COBRA
and POET 126
  Mridul Nandi

Symmetric Key Cryptanalysis

Low Probability Differentials and the Cryptanalysis of Full-Round
CLEFIA-128 141
  Sareh Emami, San Ling, Ivica Nikolić, Josef Pieprzyk, and
  Huaxiong Wang

Automatic Security Evaluation and (Related-key) Differential
Characteristic Search: Application to SIMON, PRESENT, LBlock,
DES(L) and Other Bit-Oriented Block Ciphers 158
    Siwei Sun, Lei Hu, Peng Wang, Kexin Qiao, Xiaoshuang Ma, and
    Ling Song

Scrutinizing and Improving Impossible Differential Attacks:
Applications to CLEFIA, Camellia, LBlock and SIMON 179
    Christina Boura, María Naya-Plasencia, and Valentin Suder

A Simplified Representation of AES 200
    Henri Gilbert

Side Channel Analysis I

Simulatable Leakage: Analysis, Pitfalls, and New Constructions 223
    Jake Longo, Daniel P. Martin, Elisabeth Oswald,
    Daniel Page, Martijin Stam, and Michael J. Tunstall

Multi-target DPA Attacks: Pushing DPA Beyond the Limits of a
Desktop Computer 243
    Luke Mather, Elisabeth Oswald, and Carolyn Whitnall

GLV/GLS Decomposition, Power Analysis, and Attacks on ECDSA
Signatures with Single-Bit Nonce Bias 262
    Diego F. Aranha, Pierre-Alain Fouque, Benoît Gérard,
    Jean-Gabriel Kammerer, Mehdi Tibouchi,
    and Jean-Christophe Zapalowicz

Soft Analytical Side-Channel Attacks 282
    Nicolas Veyrat-Charvillon, Benoît Gérard,
    and François-Xavier Standaert

Hyperelliptic Curve Cryptography

On the Enumeration of Double-Base Chains with Applications to
Elliptic Curve Cryptography 297
    Christophe Doche

Kummer Strikes Back: New DH Speed Records 317
    Daniel J. Bernstein, Chitchanok Chuengsatiansup, Tanja Lange,
    and Peter Schwabe

Jacobian Coordinates on Genus 2 Curves 338
    Huseyin Hisil and Craig Costello

Factoring and Discrete Log

Mersenne Factorization Factory 358
    Thorsten Kleinjung, Joppe W. Bos, and Arjen K. Lenstra

Improving the Polynomial time Precomputation of Frobenius
Representation Discrete Logarithm Algorithms: Simplified Setting for
Small Characteristic Finite Fields 378
    Antoine Joux and Cécile Pierrot

Invited Talk I

Big Bias Hunting in Amazonia: Large-Scale Computation and
Exploitation of RC4 Biases (Invited Paper)......................... 398
    Kenneth G. Paterson, Bertram Poettering, and Jacob C.N. Schuldt

Cryptanalysis

Multi-user Collisions: Applications to Discrete Logarithm,
Even-Mansour and PRINCE 420
    Pierre-Alain Fouque, Antoine Joux, and Chrysanthi Mavromati

Cryptanalysis of Iterated Even-Mansour Schemes with Two Keys 439
    Itai Dinur, Orr Dunkelman, Nathan Keller, and Adi Shamir

Meet-in-the-Middle Attacks on Generic Feistel Constructions 458
    Jian Guo, Jérémy Jean, Ivica Nikolić, and Yu Sasaki

XLS is Not a Strong Pseudorandom Permutation 478
    Mridul Nandi

Signatures

Structure-Preserving Signatures on Equivalence Classes and Their
Application to Anonymous Credentials 491
    Christian Hanser and Daniel Slamanig

On Tight Security Proofs for Schnorr Signatures 512
    Nils Fleischhacker, Tibor Jager, and Dominique Schröder

Zero-Knowledge

Square Span Programs with Applications to Succinct NIZK
Arguments 532
    George Danezis, Cédric Fournet, Jens Groth, and Markulf Kohlweiss

Better Zero-Knowledge Proofs for Lattice Encryption and Their
Application to Group Signatures   551
    Fabrice Benhamouda, Jan Camenisch, Stephan Krenn,
    Vadim Lyubashevsky, and Gregory Neven

Author Index   573

Table of Contents – Part II

Encryption Schemes

Concise Multi-challenge CCA-Secure Encryption and Signatures with
Almost Tight Security 1
    Benoît Libert, Marc Joye, Moti Yung, and Thomas Peters

Efficient Identity-Based Encryption over NTRU Lattices 22
    Léo Ducas, Vadim Lyubashevsky, and Thomas Prest

Order-Preserving Encryption Secure Beyond One-Wayness 42
    Isamu Teranishi, Moti Yung, and Tal Malkin

Outsourcing and Delegation

Statistically-secure ORAM with $\tilde{O}(\log^2 n)$ Overhead 62
    Kai-Min Chung, Zhenming Liu, and Rafael Pass

Adaptive Security of Constrained PRFs 82
    *Georg Fuchsbauer, Momchil Konstantinov, Krzysztof Pietrzak, and
    Vanishree Rao*

Obfuscation

Poly-Many Hardcore Bits for Any One-Way Function and a Framework
for Differing-Inputs Obfuscation 102
    Mihir Bellare, Igors Stepanovs, and Stefano Tessaro

Using Indistinguishability Obfuscation via UCEs 122
    Christina Brzuska and Arno Mittelbach

Indistinguishability Obfuscation versus Multi-bit Point Obfuscation
with Auxiliary Input 142
    Christina Brzuska and Arno Mittelbach

Bootstrapping Obfuscators via Fast Pseudorandom Functions 162
    Benny Applebaum

Homomorphic Cryptography

Homomorphic Authenticated Encryption Secure against
Chosen-Ciphertext Attack 173
    Chihong Joo and Aaram Yun

Authenticating Computation on Groups: New Homomorphic Primitives
and Applications...   193
    Dario Catalano, Antonio Marcedone, and Orazio Puglisi

Compact VSS and Efficient Homomorphic UC Commitments   213
    Ivan Damgård, Bernardo David, Irene Giacomelli,
    and Jesper Buus Nielsen

Secret Sharing

Round-Optimal Password-Protected Secret Sharing and T-PAKE
in the Password-Only Model   233
    Stanislaw Jarecki, Aggelos Kiayias, and Hugo Krawczyk

Secret-Sharing for NP   254
    Ilan Komargodski, Moni Naor, and Eylon Yogev

Block Ciphers and Passwords

Tweaks and Keys for Block Ciphers: The TWEAKEY Framework.....   274
    Jérémy Jean, Ivica Nikolić, and Thomas Peyrin

Memory-Demanding Password Scrambling   289
    Christian Forler, Stefan Lucks, and Jakob Wenzel

Side Channel Analysis II

Side-Channel Analysis of Multiplications in $GF(2^{128})$:
Application to AES-GCM.......................................   306
    Sonia Belaïd, Pierre-Alain Fouque, and Benoît Gérard

Higher-Order Threshold Implementations   326
    Begül Bilgin, Benedikt Gierlichs, Svetla Nikova, Ventzislav Nikov,
    and Vincent Rijmen

Masks Will Fall Off: Higher-Order Optimal Distinguishers   344
    Nicolas Bruneau, Sylvain Guilley, Annelie Heuser, and Olivier Rioul

Black-Box Separation

Black-Box Separations for One-More (Static) CDH
and Its Generalization   366
    Jiang Zhang, Zhenfeng Zhang, Yu Chen, Yanfei Guo,
    and Zongyang Zhang

Black-Box Separations for Differentially Private Protocols   386
    Dakshita Khurana, Hemanta K. Maji, and Amit Sahai

Composability

Composable Security of Delegated Quantum Computation   406
    Vedran Dunjko, Joseph F. Fitzsimons, Christopher Portmann,
    and Renato Renner

All-But-Many Encryption: A New Framework for Fully-Equipped UC
Commitments   426
    Eiichiro Fujisaki

Multi-Party Computation

Multi-valued Byzantine Broadcast: The $t < n$ Case   448
    Martin Hirt and Pavel Raykov

Fairness versus Guaranteed Output Delivery in Secure Multiparty
Computation   466
    Ran Cohen and Yehuda Lindell

Actively Secure Private Function Evaluation........................   486
    Payman Mohassel, Saeed Sadeghian, and Nigel P. Smart

Efficient, Oblivious Data Structures for MPC   506
    Marcel Keller and Peter Scholl

Author Index   527

Solving LPN Using Covering Codes

Qian Guo[1,*], Thomas Johansson[2], and Carl Löndahl[2,**]

[1] Dept. of Electrical and Information Technology, Lund University, Lund, Sweden
and School of Computer Science, Fudan University, Shanghai, China
qian.guo@eit.lth.se
[2] Dept. of Electrical and Information Technology, Lund University, Lund, Sweden
{thomas.johansson,carl.londahl}@eit.lth.se

Abstract. We present a new algorithm for solving the LPN problem. The algorithm has a similar form as some previous methods, but includes a new key step that makes use of approximations of random words to a nearest codeword in a linear code. It outperforms previous methods for many parameter choices. In particular, we can now solve instances suggested for 80-bit security in cryptographic schemes like HB variants, LPN-C and Lapin, in less than 2^{80} operations.

1 Introduction

In recent years of modern cryptography, much effort has been devoted to finding efficient and secure low-cost cryptographic primitives targeting applications in very constrained hardware environments (such as RFID tags and low-power devices). Many proposals rely on the hardness assumption of *Learning Parity with Noise* (LPN), a fundamental problem in learning theory, which recently has also gained a lot of attention within the cryptographic society. The LPN problem is well-studied and it is intimately related to the problem of decoding random linear codes, which is one of the most important problems in coding theory. Being a supposedly hard problem[1], the LPN problem is a good candidate for post-quantum cryptography, where other classically hard problems such as factoring and the discrete log problem fall short. The inherent properties of LPN also makes it ideal for lightweight cryptography.

The first time the LPN problem was employed in a cryptographic construction was in the Hopper-Blum (HB) identification protocol [17]. HB is a minimalistic protocol that is secure in a *passive* attack model. Aiming to secure the HB scheme also in an *active* attack model, Juels and Weis [18], and Katz and Shin [19] proposed a modified scheme. The modified scheme, which was given the name HB$^+$, extends HB with one extra round. It was later shown by Gilbert et al. [14] that the HB$^+$ protocol is vulnerable to active attacks, i.e. *man-in-the-middle attacks*,

---

* Supported in part by the National Natural Science Foundations of China (Grants No. 61170208) and Shanghai Key Program of Basic Research (Grant No. 12JC1401400).
** Supported by the Swedish Research Council (Grants No. 621-2012-4259).
[1] LPN with adversarial error is $\mathcal{NP}$-hard.

P. Sarkar and T. Iwata (Eds.): ASIACRYPT 2014, PART I, LNCS 8873, pp. 1–20, 2014.

where the adversary is allowed to intercept and attack an ongoing authentication session to learn the secret. Gilbert et al. [12] subsequently proposed a variant of the Hopper-Blum protocol called HB$^\#$. Apart from repairing the protocol, the constructors of HB$^\#$ introduced a more efficient key representation using a variant of LPN called TOEPLITZ-LPN.

In [13], Gilbert et al. proposed a way to use LPN in encryption of messages, which resulted in the cryptosystem LPN-C. Kiltz et al. [22] and Dodis et al. [9] showed how to construct message authentication codes (MACs) using LPN. The existence of MACs allows one to construct identification schemes that are provably secure against active attacks. The most recent contribution to LPN-based constructions is a two-round identification protocol called Lapin, proposed by Heyse et al. [16], and an LPN-based encryption scheme called HELEN, proposed by Duc and Vaudenay [10]. The Lapin protocol is based on an LPN variant called RING-LPN, where the samples are elements of a polynomial ring.

The two major threats against LPN-based cryptographic constructions are generic algorithms that decode random linear codes (information set decoding (ISD)) and variants of the BKW algorithm, originally proposed by Blum et al. [3]. Being the asymptotically most efficient[2] approach, the BKW algorithm employs an iterated collision procedure on the queries. In each iteration, colliding entries sum together to produce a new entry with smaller dependency on the information bits but with an increased noise level. Once the dependency from sufficiently many information bits are removed, the remaining are exhausted to find the secret. Although the collision procedure is the main reason for the efficiency of the BKW algorithm, it leads to a requirement of an immense amount of queries compared to ISD. Notably, for some cases, e.g., when the noise is very low, ISD yields the most efficient attack.

Levieil and Fouque [26] proposed to use Fast Walsh-Hadamard Transform in the BKW algorithm when searching for the secret. In an unpublished paper, Kirchner [23] suggested to transform the problem into systematic form, where each information (key) bit then appears as an observed symbol, pertubated by noise. This requires the adversary to only exhaust the biased noise variables rather than the key bits. When the error rate is low, the noise variable search space is very small and this technique decreases the attack complexity. Building on the work by Kirchner [23], Bernstein and Lange [5] showed that the ring structure of RING-LPN can be exploited in matrix inversion, further reducing the complexity of attacks on for example Lapin. None of the known algorithms manage to break the 80 bit security of Lapin. Nor do they break the parameters proposed in [26], which were suggested as design parameters of LPN-C [13] for 80-bit security.

In this paper, we propose a new algorithm for solving the LPN problem based on [23,5]. We employ a new technique that we call subspace distinguishing, which exploits coding theory to decrease the dimension of the secret. The trade-off is a small increase in the sample noise. Our novel algorithm performs favorably in comparison to »state-of-the-art« algorithms and we manage to break previously

---

[2] For a fixed error rate.

Table 1. Comparison of different algorithms for solving LPN with parameters $(512, 1/8)$

Algorithm	Complexity ($\log_2$)		
	Queries	Time	Memory
Levieil-Fouque [26]	75.7	87.5	84.8
Bernstein-Lange [5]	68.6	85.7	77.6
New algorithm	66.3	79.9	75.3

unbroken parameters of HB variants, Lapin and LPN-C. As an example, we attack the common $(512, 1/8)$-instance of LPN and break its 80-bit security barrier. A comparision of complexity of different algorithms[3] is shown in Table 1.

The organization of the paper is as follows. In Section 2, we give some preliminaries and introduce the LPN problem in detail. Moreover, in Section 3 we give a short description of the BKW algorithm. We briefly describe the general idea of our new attack in Section 4 and more formally in Section 5. In Section 6, we analyze its complexity. The results when the algorithm is applied on various LPN-based cryptosystems are given in Section 7 and in Section 8, we describe some aspects of the covering-coding technique. Section 9 concludes the paper.

2   The LPN Problem

We will now give a more thorough description of the LPN problem. Let Ber_η be the Bernoulli distribution and let $X \sim \mathsf{Ber}_\eta$ be a random variable with alphabet $\mathcal{X} = \{0, 1\}$. Then, $\mathbf{Pr}[X = 1] = \eta$ and $\mathbf{Pr}[X = 0] = 1 - \mathbf{Pr}[X = 1] = 1 - \eta$. The bias ϵ of X is given from $\mathbf{Pr}[X = 0] = 1/2(1 + \epsilon)$. Let k be a security parameter and let $\mathbf{x}$ be a binary vector of length k.

Definition 1 (LPN oracle). *An* LPN *oracle* Π_{LPN} *for an unkown vector* $\mathbf{x} \in \{0, 1\}^k$ *with* $\eta \in (0, \frac{1}{2})$ *returns pairs of the form*

$$\left(\mathbf{g} \xleftarrow{\$} \{0, 1\}^k, \langle \mathbf{x}, \mathbf{g} \rangle + e\right),$$

where $e \leftarrow \mathsf{Ber}_\eta$. *Here,* $\langle \mathbf{x}, \mathbf{g} \rangle$ *denotes the scalar product of vectors* $\mathbf{x}$ *and* $\mathbf{g}$.

We also write $\langle \mathbf{x}, \mathbf{g} \rangle$ as $\mathbf{x} \cdot \mathbf{g}^{\mathrm{T}}$, where $\mathbf{g}^{\mathrm{T}}$ is the transpose of the row vector $\mathbf{g}$. We receive a number n of noisy versions of scalar products of $\mathbf{x}$ from the oracle Π_{LPN}, and our task is to recover $\mathbf{x}$.

---

[3] The Bernstein-Lange algorithm is originally proposed for RING-LPN, and by a slight modification [5], one can apply it to the LPN instances as well. It shares the beginning steps (i.e., the steps of Gaussian elimination and the collision procedure) with the new algorithm, so for a fair comparison, we use the same implementation of these steps when computing their complexity.

Problem 1 (LPN). *Given an LPN oracle Π_{LPN}, the (k, η)-LPN problem consists of finding the vector $\mathbf{x}$. An algorithm $\mathcal{A}_{\mathrm{LPN}}(t, n, \delta)$ using time at most t with at most n oracles queries solves (k, η)-LPN if*

$$\mathbf{Pr}\left[\mathcal{A}_{\mathrm{LPN}}(t, n, \delta) = \mathbf{x} : \mathbf{x} \xleftarrow{\$} \{0, 1\}^k\right] \geq \delta.$$

Let $\mathbf{y}$ be a vector of length n and let $y_i = \langle \mathbf{x}, \mathbf{g}_i \rangle$. For known random vectors $\mathbf{g}_1, \mathbf{g}_2, \ldots, \mathbf{g}_n$, we can easily reconstruct an unknown $\mathbf{x}$ from $\mathbf{y}$ using linear algebra. In the LPN problem, however, we receive instead noisy versions of $y_i, i = 1, 2, \ldots, n$. Writing the noise in position i as $e_i, i = 1, 2, \ldots, n$ we obtain $z_i = y_i + e_i = \langle \mathbf{x}, \mathbf{g}_i \rangle + e_i$. In matrix form, the same is written as $\mathbf{z} = \mathbf{xG} + \mathbf{e}$, where $\mathbf{z} = \begin{bmatrix} z_1 & z_2 & \cdots & z_n \end{bmatrix}$, and the matrix $\mathbf{G}$ is formed as $\mathbf{G} = \begin{bmatrix} \mathbf{g}_1^{\mathrm{T}} & \mathbf{g}_2^{\mathrm{T}} & \cdots & \mathbf{g}_n^{\mathrm{T}} \end{bmatrix}$. This shows that the LPN problem is simply a decoding problem, where $\mathbf{G}$ is a random $k \times n$ generator matrix, $\mathbf{x}$ is the information vector and $\mathbf{z}$ is the received vector after transmission of a codeword on the binary symmetric channel with error probability η.

2.1   Piling-up Lemma

We recall the piling-up lemma, which is frequently used in analysis of the LPN problem.

Lemma 1 (**Piling-up lemma**). *Let $X_1, X_2, \ldots X_n$ be independent binary random variables where each $\mathbf{Pr}[X_i = 0] = \frac{1}{2}(1 + \epsilon_i)$, for $1 \leq i \leq n$. Then,*

$$\mathbf{Pr}[X_1 + X_2 + \cdots + X_n = 0] = \frac{1}{2}\left(1 + \prod_{i=1}^{n} \epsilon_i\right).$$

3   The BKW Algorithm

The BKW algorithm is due to Blum, Kalai and Wasserman [3]. In the spirit of generalized birthday algorithms, their approach uses an iterative sort-and-match procedure on the columns of the generator matrix $\mathbf{G}$, which iteratively reduces the dimension of $\mathbf{G}$.

Initially, one searches for all combinations of two columns in $\mathbf{G}$ that add to zero in the last b entries. Assume that one finds two columns $\mathbf{g}_{i_1}^{\mathrm{T}}, \mathbf{g}_{i_2}^{\mathrm{T}}$ such that

$$\mathbf{g}_{i_1} + \mathbf{g}_{i_2} = [* * \cdots * \underbrace{0\ 0 \cdots 0}_{b \text{ symbols}}],$$

where $*$ means any value. Then a new vector $\mathbf{g}_1^{(2)} = \mathbf{g}_{i_1} + \mathbf{g}_{i_2}$ is formed. An "observed symbol" is also formed, corresponding to this new column by forming $z_1^{(2)} = z_{i_1} + z_{i_2}$. If $y_1^{(2)} = \langle \mathbf{x}, \mathbf{g}_1^{(2)} \rangle$, then $z_1^{(2)} = y_1^{(2)} + e_1^{(2)}$, where now $e_1^{(2)} = e_{i_1} + e_{i_2}$. It can be verified that $\mathbf{Pr}\left[e_1^{(2)} = 0\right] = 1/2(1 + \epsilon^2)$.

There are two approaches to realize the above merging procedure. One, raised by Blum et al. [3], called *LF1* type by Levieil and Fouque [26], and later adopted by Bernstein and Lange [5], is choosing one sample in each partition with the same last b entries, and then adding it to the remaining samples in the same partition. Thus, the number of samples reduces by about 2^b after this operation. The other method is a heuristic called *LF2* in [26], which computes any pair with the same last b entries. It produces more samples at the cost of increased dependency, thereby gaining more efficiency in practice but losing rigorous analysis in theory. We will use the *LF1* setting throughout the remaining part of the paper.

Put all such new columns in a matrix $\mathbf{G}_2$,

$$\mathbf{G}_2 = \begin{bmatrix} \mathbf{g}_1^{(2)\mathrm{T}} & \mathbf{g}_2^{(2)\mathrm{T}} & \cdots & \mathbf{g}_{n-2^b}^{(2)\mathrm{T}} \end{bmatrix}.$$

If n is the number of columns in $\mathbf{G}$, then the number of columns in $\mathbf{G}_2$ will be $n - 2^b$. Note that the last b entries of every column in $\mathbf{G}_2$ are all zero. In connection to this matrix, the vector of observed symbols is

$$\mathbf{z}_2 = \begin{bmatrix} z_1^{(2)} & z_2^{(2)} & \cdots & z_{n-2^b}^{(2)} \end{bmatrix},$$

where $\mathbf{Pr}\left[z_i^{(2)} = y_i^{(2)}\right] = 1/2(1 + \epsilon^2)$, for $1 \leq i \leq n - 2^b$.

We now iterate the same, picking one column and then adding it to another suited column in $\mathbf{G}_i$ giving a sum with an additional b entries being zero, forming the columns of $\mathbf{G}_{i+1}$. Repeating the same procedure an additional $t - 2$ times will reduce the $\mathbf{G}_i$ of unknown variables to $k - bt$ in the remaining problem.

For each iteration the noise level is squared. By the piling-up lemma we have that

$$\mathbf{Pr}\left[\sum_{j=1}^{2^t} e_i = 0\right] = \frac{1}{2}\left(1 + \epsilon^{2^t}\right).$$

Hence, the bias decreases quickly to low levels. The remaining unknown key variables are guessed and for each guess we check whether the bias is present or not. The procedure is summarized in Algorithm 1.

4   Essential Idea

In this section we try to give a very basic description of the idea used to give a new and more efficient algorithm for solving the LPN problem. A more detailed analysis will be provided in later sections, and a graphical interpretation of the key step is given in Appendix A.

Assume that we have an initial LPN problem described by $\mathbf{G} = \begin{bmatrix} \mathbf{g}_1^\mathrm{T} & \mathbf{g}_2^\mathrm{T} & \cdots & \mathbf{g}_n^\mathrm{T} \end{bmatrix}$ and $\mathbf{z} = \mathbf{x}\mathbf{G} + \mathbf{e}$, where $\mathbf{z} = \begin{bmatrix} z_1 & z_2 & \cdots & z_n \end{bmatrix}$, where $z_i = y_i + e_i = \langle \mathbf{x}, \mathbf{g}_i \rangle + e_i$.

As previously shown in [23] and [5], we may through Gaussian elimination transform $\mathbf{G}$ into systematic form. Assume that the first k columns are linearly

---

Algorithm 1. BKW Algorithm

Input: Matrix $\mathbf{G}$ with k rows and n columns and received vector $\mathbf{z}$, algorithm parameters b, t

1 Put the received word as a first row in the matrix, $\mathbf{G}_1 \leftarrow \begin{bmatrix} \mathbf{z} \\ \mathbf{G} \end{bmatrix}$;

2 **for** $i = 1$ **to** t **do**

3 $\quad$ For $\mathbf{G}_i$, partition the columns by the last $b \cdot i$ bits;

4 $\quad$ Form pairs of columns from each partition and form $\mathbf{G}_{i+1}$;

5 **for** $\mathbf{x} \in \{0,1\}^{k-bt}$ **do**

6 $\quad$ Find the vector $\begin{bmatrix} 1 & \mathbf{x} & \mathbf{0} \end{bmatrix}$ such that $\begin{bmatrix} 1 & \mathbf{x} & \mathbf{0} \end{bmatrix} \mathbf{G}_{t+1}$ has minimal weight;

---

independent and forms the matrix $\mathbf{D}$. With a change of variables $\hat{\mathbf{x}} = \mathbf{x} \mathbf{D}^{-1}$ we get an equivalent problem description with $\hat{\mathbf{G}} = \begin{bmatrix} \mathbf{I} & \hat{\mathbf{g}}_{k+1}^{\mathrm{T}} & \hat{\mathbf{g}}_{k+2}^{\mathrm{T}} & \cdots & \hat{\mathbf{g}}_n^{\mathrm{T}} \end{bmatrix}$. We compute

$$\hat{\mathbf{z}} = \mathbf{z} + \begin{bmatrix} z_1, z_2, \ldots, z_k \end{bmatrix} \hat{\mathbf{G}} = \begin{bmatrix} \mathbf{0}, \hat{z}_{k+1}, \hat{z}_{k+2}, \ldots, \hat{z}_n \end{bmatrix}.$$

In this situation, one may start performing a number of BKW steps on columns $k+1$ to n, reducing the dimension k of the problem to something smaller. This will result in a new problem instance where noise in each position is larger, except for the first systematic positions. We may write the problem after performing t BKW steps in the form $\mathbf{G}' = \begin{bmatrix} \mathbf{I} & \mathbf{g}_1'^{\mathrm{T}} & \mathbf{g}_2'^{\mathrm{T}} & \cdots & \mathbf{g}_m'^{\mathrm{T}} \end{bmatrix}$ and $\mathbf{z}' = \begin{bmatrix} \mathbf{0}, z_1', z_2', \ldots z_m' \end{bmatrix}$, where now $\mathbf{G}'$ has dimension $k' \times m$ with $k' = k - bt$ and m is the number of columns remaining after the BKW step. We have $\mathbf{z}' = \mathbf{x}' \mathbf{G}' + \mathbf{e}'$, $\mathbf{Pr}\left[x_i' = 0 \right] = 1/2(1 + \epsilon)$ and $\mathbf{Pr}\left[\mathbf{x}' \cdot \mathbf{g}_i'^{\mathrm{T}} = z_i \right] = 1/2(1 + \epsilon^{2^t})$.

Now we will explain the basics of the new idea proposed in the paper. In a problem instance as above, we may look at the random variables $y_i' = \mathbf{x}' \cdot \mathbf{g}_i'^{\mathrm{T}}$. The bits in $\mathbf{x}'$ are mostly zero but a few are set to one. Let us assume that c bits are set to one. Furthermore, $\mathbf{x}'$ is fixed for all i. We usually assume that $\mathbf{g}_i'$ is generated according to a uniform distribution. However, assume that every column $\mathbf{g}_i'$ would be biased, i.e., every bit in a column position is zero with probability $1/2(1 + \epsilon')$. Then we observe that the variables y_i' will be biased, as

$$y_i' = \langle \mathbf{x}', \mathbf{g}_i' \rangle = \sum_{j=1}^{c} [\mathbf{g}_i']_{k_j},$$

where $k_1, k_2, \ldots k_c$ are the bit positions where $\mathbf{x}'$ has value one (here $[\mathbf{x}]_y$ denotes bit y of vector $\mathbf{x}$). In fact, variables y_i' will have bias $(\epsilon')^c$.

So how do we get the columns to be biased in the general case? We could simply hope for some of them to be biased, but if we need to use a larger number of columns, the bias would have to be small, giving a high complexity for an algorithm solving the problem. We propose instead to use a covering code to achieve something similar to what is described above. Vectors $\mathbf{g}_i'$ are of length

k', so we consider a code of length k' and some dimension l. Let us assume that the generator matrix of this code is denoted $\mathbf{F}$. For each vector $\mathbf{g}'_i$, we now find the codeword in the code spanned by $\mathbf{F}$ that is closest (in Hamming sense) to $\mathbf{g}'_i$. Assume that this codeword is denoted $\mathbf{c}_i$. Then we can write

$$\mathbf{g}'_i = \mathbf{c}_i + \mathbf{e}'_i,$$

where $\mathbf{e}'_i$ is a vector with biased bits. It remains to examine exactly how biased the bits in $\mathbf{e}'_i$ will be, but assume for the moment that the bias is ϵ'. Going back to our previous expressions we can write

$$y'_i = \langle \mathbf{x}', \mathbf{g}'_i \rangle = \mathbf{x}' \cdot (\mathbf{c}_i + \mathbf{e}'_i)^{\mathrm{T}}$$

and since $\mathbf{c}_i = \mathbf{u}_i \mathbf{F}$ for some $\mathbf{u}_i$, we can write

$$y'_i = \mathbf{x}' \mathbf{F}^{\mathrm{T}} \cdot \mathbf{u}_i^{\mathrm{T}} + \mathbf{x}' \cdot \mathbf{e}'^{\mathrm{T}}_i.$$

We may introduce $\mathbf{x}'' = \mathbf{x}' \mathbf{F}^{\mathrm{T}}$ as a length l vector of unknown bits (linear combinations of bits from $\mathbf{x}'$) and again

$$y'_i = \mathbf{x}'' \cdot \mathbf{u}_i^{\mathrm{T}} + \mathbf{x}' \cdot \mathbf{e}'^{\mathrm{T}}_i.$$

Since we have $\mathbf{Pr}\,[y'_i = z'_i] = 1/2(1 + \epsilon^{2^t})$, we get

$$\mathbf{Pr}\,[\mathbf{x}'' \cdot \mathbf{u}_i^{\mathrm{T}} = z'_i] = \frac{1}{2}(1 + \epsilon^{2^t}(\epsilon')^c),$$

where ϵ' is the bias determined by the expected distance between $\mathbf{g}'_i$ and the closest codeword in the code we are using, and c is the number of positions in $\mathbf{x}'$ set to one. The last step in the new algorithm now selects about $m = 1/(\epsilon^{2^t} \epsilon'^c)^2$ samples $z'_1, z'_2, \ldots, z'_m$ and for each guess of the 2^l possible values of $\mathbf{x}''$, we compute how many times $\mathbf{x}'' \cdot \mathbf{u}_i^{\mathrm{T}} = z'_i$ when $i = 1, 2, \ldots, m$. As this step is similar to a correlation attack scenario, we know that it can be efficiently computed using Fast Walsh-Hadamard Transform. After recovering $\mathbf{x}''$, it is an easy task to recover remaining unknown bits of $\mathbf{x}'$.

4.1   An Example Using Dimension $k = 160$

In order to illustrate the ideas and convince the reader that the proposed algorithm can be more efficient than previously known methods, we consider an example. We assume an LPN instance of dimension $k = 160$, where we allow at most 2^{24} received samples and we allow at most around 2^{24} vectors of length 160 to be stored in memory. Furthermore, the error probability is $\eta = 0.1$.

For this particular case, we propose the following algorithm. The first step is to compute the systematic form, $\hat{\mathbf{G}} = \begin{bmatrix} \mathbf{I} & \hat{\mathbf{g}}^{\mathrm{T}}_{k+1} & \hat{\mathbf{g}}^{\mathrm{T}}_{k+2} & \cdots & \hat{\mathbf{g}}^{\mathrm{T}}_n \end{bmatrix}$ and

$$\hat{\mathbf{z}} = \mathbf{z} + \begin{bmatrix} z_1 & z_2 & \cdots & z_k \end{bmatrix} \hat{\mathbf{G}} = \begin{bmatrix} \mathbf{0} & \hat{z}_{k+1} & \hat{z}_{k+2} & \cdots & \hat{z}_n \end{bmatrix}.$$

Here $\hat{\mathbf{G}}$ has dimension 160 and $\hat{\mathbf{z}}$ has length at most 2^{24}.

In the second step we perform $t = 4$ steps of BKW (using the *LF1* approach), the first step removing 22 bits and the remaining three each removing 21 bits. This results in $\mathbf{G}' = \begin{bmatrix} \mathbf{I} \ \mathbf{g_1'}^{\mathrm{T}} \ \mathbf{g_2'}^{\mathrm{T}} \cdots \mathbf{g_m'}^{\mathrm{T}} \end{bmatrix}$ and $\mathbf{z}' = \begin{bmatrix} \mathbf{0} \ z_1' \ z_2' \ldots z_m' \end{bmatrix}$, where now $\mathbf{G}'$ has dimension $75 \times m$ and m is about $3 \cdot 2^{21}$. We have $\mathbf{z}' = \mathbf{x}'\mathbf{G}'$, $\mathbf{Pr}\,[x_i' = 0] = 1/2(1 + \epsilon)$, where $\epsilon = 0.8$ and $\mathbf{Pr}\,\left[\mathbf{x}' \cdot \mathbf{g_i'}^{\mathrm{T}} = z_i\right] = 1/2(1 + \epsilon^{16})$. So the resulting problem has dimension 75 and the bias is $\epsilon^{2^t} = (0.8)^{16}$.

In the third step we then select a suitable code of length 75. In this example we choose a block code which is a direct sum of 25 $[3, 1, 3]$ repetition codes[4], i.e., the dimension is 25. We map every vector $\mathbf{g_i'}$ to the nearest codeword by simply selecting chunks of three consecutive bits and replace them by either 000 or 111. With probability $3/4$ we will change one position and with probability $1/4$ we will not have to change any position. In total we expect to change $(3/4 \cdot 1 + 1/4 \cdot 0) \cdot 25$ positions. The expected weight of the length 75 vector $\mathbf{e_i'}$ is $75/4$, so the expected bias is $\epsilon' = 1/2$. As $\mathbf{Pr}\,[x_i' = 1] = 0.1$, the expected number of nonzero positions in $\mathbf{x}'$ is 7.5. Assuming we have only $c = 6$ nonzero positions, we get

$$\mathbf{Pr}\,[\mathbf{x}'' \cdot \mathbf{u}_i^{\mathrm{T}} = z_i'] = \frac{1}{2}\left(1 + 0.8^{16}\left(\frac{1}{2}\right)^6\right) = \frac{1}{2}(1 + 2^{-11.15}).$$

In the last step we then run through 2^{25} values of $\mathbf{x}''$ and for each of them we compute how often $\mathbf{x}'' \cdot \mathbf{u}_i^{\mathrm{T}} = z_i'$ for $i = 1, \ldots, 3 \cdot 2^{21}$. Again since we use Fast Walsh-Hadamard Transform, the cost of this step is not much more than 2^{25} operations. The probability of having no more than 6 ones in $\mathbf{x}'$ is about 0.37, so we need to repeat the whole process a few times.

In comparison with other algorithms, the best approach we can find is the Kirchner, Bernstein, Lange approach [23,5], where one can do up to 5 BKW steps. Removing 21 bits in each step leaves 55 remaining bits. Using Fast Walsh-Hadamard Transform with $0.8^{-64} = 2^{20.6}$ samples, we can include another 21 bits in this step, but there are still 34 remaining variables that needs to be guessed.

Overall, the simple algorithm sketched above is outperforming the best previous algorithm using optimal parameter values[5].

Simulation. We have verified in simulation that the proposed algorithm works in practice. We use a rate $R = 1/3$ concatenated repetition code and query the oracle for 2^{24} samples. Simple pruning of the samples with too large distance from the codeword was used to approximate the behaviour of an optimal distinguisher.

---

[4] In the sequel, we denote this code construction as concatenated repetition code. For this $[75, 25, 3]$ linear code, the covering radius is 25, but we could see from this example that what matters is the average weight of the error vector, which is much smaller than 25.

[5] Adopting the same method to implement their overlapping steps, for the $(160, 1/10)$ LPN instance, the Bernstein-Lange algorithm and the new algorithm cost $2^{35.70}$ and $2^{33.83}$ bit operations, respectively. Thus, the latter offers an improvement with a factor roughly 4 to solve this small-scale instance.

Algorithm 2. New attacking algorithm

Input: Matrix $\mathbf{G}$ with k rows and n columns, received length n vector $\mathbf{z}$ and algorithm parameters t, b, k'', l, w_0, c

1 **repeat**
2   Pick random column permutation π;
3   Perform Gaussian elimination on $\pi(\mathbf{G})$ resulting in $\mathbf{G}_0 = [\mathbf{I}|\mathbf{L}_0]$;
4   **for** $i = 1$ **to** t **do**
5   Partition the columns of $\mathbf{L}_{i-1}$ by the last $b \cdot i$ bits;
6   Denote the set of columns in partition s by $\mathcal{L}_s$;
7   Pick a vector $\mathbf{a}_{is} \in \mathcal{L}_s$;
8   **for** $(\mathbf{a} \in \mathcal{L}_s)$ *and* $(\mathbf{a} \neq \mathbf{a}_{is})$ **do**
9           $\mathbf{L}_i \leftarrow [\mathbf{L}_i|(\mathbf{a} + \mathbf{a}_{is})]$;

10   Pick a $[k'', l]$ linear code with good covering property;
11   Partition the columns of $\mathbf{L}_t$ by the middle non-all-zero k'' bits and group them by their nearest codewords;
12   Set $k_1 = k - ab - k''$;
13   **for** $\mathbf{x}_2' \in \{0,1\}^{k_1}$ *with* $wt(\mathbf{x}_2') \leq w_0$ **do**
14   Update the observed samples;
15   **for** $\mathbf{y} \in \{0,1\}^l$ **do**
16   Use Fast Walsh-Hadamard Transform to compute the numbers of 1s and 0s observed respectively;
17   Perform hypothesis testing whose threshold is defined as a function of c;

18 **until** *acceptable hypothesis is found*

The average execution time is ~ 1.86 seconds on an Apple iMac 3.06 GHz Intel Core 2 Duo with 4 GB ram running OS X 10.9 (13A603).

5   Algorithm Description

Having introduced the key idea in a simplistic manner, we now formalize it by stating a new five-step LPN solving algorithm (see Algorithm 2) in detail. Its first three steps combine several well-known techniques on this problem, i.e., changing the distribution of secret vector [23], sorting and merging to make the length of samples shorter [3], and partial secret guessing [5], together. The efficiency improvement comes from a novel idea introduced in the last two subsections—if we employ a linear covering code and rearrange samples according to their nearest codewords, then the columns in the matrix subtracting their corresponding codewords lead to sparse vectors desired in the distinguishing process. We later propose a new distinguishing technique—subspace hypothesis testing, to remove the influence of the codeword part using Fast Walsh-Hadamard Transform. The algorithm consists of five steps, each described in separate subsections.

5.1   Gaussian Elimination

Recall that our LPN problem is given by $\mathbf{z} = \mathbf{x}\mathbf{G} + \mathbf{e}$, where $\mathbf{z}$ and $\mathbf{G}$ are known. We can apply an arbitrary column permutation π without changing the problem (but we change the error locations). A transformed problem is $\pi(\mathbf{z}) = \mathbf{x}\pi(\mathbf{G}) + \pi(\mathbf{e})$. This means that we can repeat the algorithm many times using different permutations.

Continuing, we multiply by a suitable $k \times k$ matrix $\mathbf{D}$ to bring the matrix $\mathbf{G}$ to a systematic form, $\hat{\mathbf{G}} = \mathbf{D}\mathbf{G}$. The problem remains the same, except that the unknowns are now given by the vector $\tilde{\mathbf{x}} = \mathbf{x}\mathbf{D}^{-1}$. This is just a change of variables. As a second step, we also add the codeword $\begin{bmatrix} z_1 & z_2 & \cdots & z_k \end{bmatrix} \hat{\mathbf{G}}$ to our known vector $\mathbf{z}$, resulting in a received vector starting with k zero entries. Altogether, this corresponds to the change $\hat{\mathbf{x}} = \mathbf{x}\mathbf{D}^{-1} + \begin{bmatrix} z_1 & z_2 & \cdots & z_k \end{bmatrix}$.

Our initial problem has been transformed and the problem is now written as

$$\hat{\mathbf{z}} = \begin{bmatrix} \mathbf{0} & \hat{z}_{k+1} & \hat{z}_{k+2} & \cdots & \hat{z}_n \end{bmatrix} = \hat{\mathbf{x}}\hat{\mathbf{G}} + \mathbf{e}, \tag{1}$$

where now $\hat{\mathbf{G}}$ is in systematic form. Note that these transformations do not affect the noise level. We still have a single noise variable added in every position.

Schoolbook implementation of the above Gaussian elimination procedure requires about $nk^2/2$ bit-operations; we propose however to reduce its complexity by using a more sophisticated space-time trade-off technique. We store intermediate results in tables, and then derive the final result by adding several items in the tables together. The detailed description is as follows.

For a fixed s, divide the matrix $\mathbf{D}$ in $a = \lceil k/s \rceil$ parts, i.e., $\mathbf{D} = \begin{bmatrix} \mathbf{D}_1, \mathbf{D}_2, \dots, \mathbf{D}_a \end{bmatrix}$, where $\mathbf{D}_i$ is a sub-matrix with s columns(except possibly the last matrix $\mathbf{D}_a$). Then store all possible values of $\mathbf{D}_i\mathbf{x}^{\mathrm{T}}$ for $\mathbf{x} \in \mathbb{F}_2^s$ in tables indexed by i, where $1 \leq i \leq a$. For a vector $\mathbf{g} = \begin{bmatrix} \mathbf{g}_1, \mathbf{g}_2, \dots, \mathbf{g}_a \end{bmatrix}$, the transformed vector is

$$\mathbf{D}\mathbf{g}^{\mathrm{T}} = \mathbf{D}_1\mathbf{g}_1^{\mathrm{T}} + \mathbf{D}_2\mathbf{g}_2^{\mathrm{T}} + \dots + \mathbf{D}_a\mathbf{g}_a^{\mathrm{T}},$$

where $\mathbf{D}_i\mathbf{g}_i^{\mathrm{T}}$ can be read directly from the table.

The cost of constructing the tables is about $\mathcal{O}(2^s)$, which can be negligible if memory in the BKW step is much larger. Furthermore, for each column, the transformation costs no more than $k \cdot a$ bit operations; so, this step requires

$$C_1 = (n - k) \cdot ka < nka$$

bit operations in total if 2^s is much smaller than n.

5.2   Collision Procedure

This next step contains the BKW part. The input to this step is $\hat{\mathbf{z}}$ and $\hat{\mathbf{G}}$.

We write $\hat{\mathbf{G}} = \begin{bmatrix} \mathbf{I} & \mathbf{L}_0 \end{bmatrix}$ and process only the matrix $\mathbf{L}_0$. As the length of $\mathbf{L}_0$ is typically much larger than the systematic part of $\hat{\mathbf{G}}$, this is roughly no restriction at all. We then use the a sort-and-match technique as in the BKW algorithm, operating on the matrix $\mathbf{L}_0$. This process will give us a sequence of matrices denoted $\mathbf{L}_0, \mathbf{L}_1, \mathbf{L}_2, \ldots, \mathbf{L}_t$.

Let us denote the number of columns of $\mathbf{L}_i$ by $r(i)$, with $r(0) = n-k$. Adopting the *LF1* type technique, every step operating on columns will reduce the number of samples by 2^b, yielding that $m = r(t) = r(0) - t2^b$. Apart from the process of creating the $\mathbf{L}_i$ matrices, we need to update the received vector in a similar fashion. A simple way is to put $\hat{\mathbf{z}}$ as a first row in the representation of $\hat{\mathbf{G}}$.

This procedure will end with a matrix $\begin{bmatrix} \mathbf{I} \ \mathbf{L}_t \end{bmatrix}$, where $\mathbf{L}_t$ will have all tb last entries in each column all zero. By discarding the last tb rows we have a given matrix of dimension $k - tb$ that can be written as $\mathbf{G}' = \begin{bmatrix} \mathbf{I} \ \mathbf{L}_t \end{bmatrix}$, and we have a corresponding received vector $\mathbf{z}' = \begin{bmatrix} \mathbf{0} \ z_1' \ z_2' \ \cdots \ z_m' \end{bmatrix}$. The first $k' = k - tb$ positions are only affected by a single noise variable, so we can write

$$[\mathbf{0}, \mathbf{z}'] = \mathbf{x}'\hat{\mathbf{G}} + \begin{bmatrix} e_1 \ e_2 \ \cdots \ e_{k'} \ \tilde{e}_1 \ \tilde{e}_2 \ \cdots \ \tilde{e}_m \end{bmatrix}, \tag{2}$$

for some unknown $\mathbf{x}'$ vector, where $\tilde{e}_i = \sum_{i_j \in \mathcal{T}_i, |\mathcal{T}_i| \le 2^t} e_{i_j}$ and $\mathcal{T}_i$ contains the positions that have been added up to form the $(k' + i)$th column of $\mathbf{G}'$. By the piling-up lemma, the bias for $\tilde{e}_i$ increases to ϵ^{2^t}.

We denote the complexity of this step C_2, where

$$C_2 = \sum_{i=1}^{t} (k + 1 - ib)(n - i2^b) \approx (k+1)tn.$$

5.3 Partial Secret Guessing Procedure

The previous procedure outputs $\mathbf{G}'$ with dimension $k' = k - tb$ and $m = n - k - t2^b$ columns. We removed the bottom tb bits of $\hat{\mathbf{x}}$ to form the length k' vector $\mathbf{x}'$, with $\mathbf{z}' = \mathbf{x}'\mathbf{G}' + \tilde{\mathbf{e}}$.

We now divide $\mathbf{x}'$ into two parts: $\mathbf{x}' = \begin{bmatrix} \mathbf{x}_1' \ \mathbf{x}_2' \end{bmatrix}$, where $\mathbf{x}_1'$ is of length k''. In this step, we simply guess all vectors $\mathbf{x}_2 \in \mathbb{F}_2^{k'-k''}$ such that $wt(\mathbf{x}_2) \le w_0$ for some w_0 and update the observed vector $\mathbf{z}'$ accordingly. This transforms the problem to that of attacking a new smaller LPN problem of dimension k'' with the same number of samples. Firstly, note that this will only work if $wt(\mathbf{x}_2) \le w_0$, and we denote this probability by $P(w_0, k' - k'')$. Secondly, we need to be able to distinguish a correct guess from incorrect ones and this is the task of the remaining steps. The complexity of this step is

$$C_3 = m \sum_{i=0}^{w_0} \binom{k' - k''}{i} i.$$

5.4 Covering-Coding Method

In this step, we use a $[k'', l]$ linear code $\mathcal{C}$ with covering radius d_C to group the columns. That is, we rewrite

$$\mathbf{g}_i' = \mathbf{c}_i + \mathbf{e}_i',$$

where $\mathbf{c}_i$ is the nearest codeword in $\mathcal{C}$, and $wt(\mathbf{e}_i') \leq d_C$. The employed linear code is characterized by a systematic generator matrix $\mathbf{F} = \begin{bmatrix} \mathbf{I} \ \mathbf{F}' \end{bmatrix}_{l \times k''}$; we thus obtain a corresponding parity-check matrix $\mathbf{H} = \begin{bmatrix} \mathbf{F}'^{\mathrm{T}} \ \mathbf{I} \end{bmatrix}_{(k''-l) \times k''}$.

There are several ways to select a code. One way of realizing the above grouping idea is by a table-based syndrome decoding technique. The procedure is as follows: 1) We construct a constant query time table containing $2^{k''-l}$ items, in each of which stores the syndrome and its corresponding minimum weight error vector. 2) If the syndrome $\mathbf{H}\mathbf{g}_i'^{\mathrm{T}}$ is computed, we then find its corresponding error vector $\mathbf{e}_i'$ by checking in the table; adding them together yields the nearest codeword $\mathbf{c}_i$.

The remaining task is to calculate the syndrome efficiently. We, according to the first l bits, sort the vectors $\mathbf{g}_i'$, where $0 \leq i \leq m$, and group them into 2^l partitions denoted by $\mathcal{P}_j$ for $1 \leq j \leq 2^l$. Starting from the partition $\mathcal{P}_1$ whose first l bits are all zero, we can derive the syndrome by reading its last $k'' - l$ bits without any additional computational cost. If we know one syndrome in $\mathcal{P}_j$, then we can compute another syndrome in the same partition within $2(k'' - l)$ bit operations, and another in a different partition whose first l-bit vector has Hamming distance 1 from that of $\mathcal{P}_j$ within $3(k'' - l)$ bit operations. Therefore, the complexity of this step is

$$C_4 = (k'' - l)(2m + 2^l).$$

Notice that the selected linear code determines the syndrome table, which can be pre-computed within complexity $\mathcal{O}(k'' 2^{k''-l})$. The optimal parameter suggests that this cost is acceptable compared with the total attacking complexity.

The expected distance to the nearest codeword determines the bias ϵ' in $\mathbf{e}_i'$. This plays important roles in the later hypothesis testing step: if we rearrange the columns $\mathbf{e}_i'$ as a matrix, then it is sparse; therefore, we can view the ith value in one column as a random variable R_i distributed according to $\mathsf{Ber}_{\frac{d}{k''}}$, where d is the expected distance. We can bound it by the covering radius[6]. Moreover, if the bias is large enough, then it is reasonable to consider R_i, for $1 \leq i \leq i_1$, as independent variables.

5.5   Subspace Hypothesis Testing

Group the samples $(\mathbf{g}_i', z_i')$ in sets $L(\mathbf{c}_i)$ according to their nearest codewords and define the function $f_L(\mathbf{c}_i)$ as

$$f_L(\mathbf{c}_i) = \sum_{(\mathbf{g}_i', z_i') \in L(\mathbf{c}_i)} (-1)^{z_i'}.$$

The employed systematic linear code $\mathcal{C}$ describes a bijection between the linear space $\mathbb{F}_2^l$ and the set of all codewords in $\mathbb{F}_2^{k''}$, and moreover, due to its systematic

---

[6] In the sequel, we replace the covering radius by the sphere-covering bound to estimate the expected distance d, i.e., d is the smallest integer, s.t. $\sum_{i=0}^{d} \binom{k''}{i} > 2^{k''-l}$. We give more explanation in Section 8.

feature, the corresponding information vector appears explicitly in their first l bits. We can thus define a new function

$$g(\mathbf{u}) = f_L(\mathbf{c}_i),$$

such that $\mathbf{u}$ represents the first l bits of $\mathbf{c}_i$ and exhausts all the points in $\mathbb{F}_2^l$.

The Walsh transform of g is defined as

$$G(\mathbf{y}) = \sum_{\mathbf{u} \in \mathbb{F}_2^l} g(\mathbf{u})(-1)^{\langle \mathbf{y}, \mathbf{u} \rangle}.$$

Here we exhaust all candidates of $\mathbf{y} \in \mathbb{F}_2^l$ by computing the Walsh transform.

The following lemma illustrates the reason why we can perform hypothesis testing on the subspace $\mathbb{F}_2^l$.

Lemma 2. *There exits a unique vector* $\mathbf{y} \in \mathbb{F}_2^l$ *s.t.,*

$$\langle \mathbf{y}, \mathbf{u} \rangle = \langle \mathbf{x}', \mathbf{c}_i \rangle.$$

Proof. As $\mathbf{c}_i = \mathbf{u}\mathbf{F}$, we obtain

$$\langle \mathbf{x}', \mathbf{c}_i \rangle = \mathbf{x}'\mathbf{F}^{\mathsf{T}}\mathbf{u}^{\mathsf{T}} = \langle \mathbf{x}'\mathbf{F}^{\mathsf{T}}, \mathbf{u} \rangle.$$

Thus, we construct the vector $\mathbf{y} = \mathbf{x}'\mathbf{F}^{\mathsf{T}}$ that fulfills the requirement. On the other hand, the uniqueness is obvious.

Given the candidate $\mathbf{y}$, $G(\mathbf{y})$ is the difference between the number of predicted 0 and the number of predicted 1 for the bit $z_i' + \langle \mathbf{x}', \mathbf{c}_i \rangle$. If $\mathbf{y}$ is the correct guess, then it is distributed according to $\mathsf{Ber}_{\frac{1}{2}(1 - \epsilon^{2^t} \cdot (\epsilon')^w)}$, where $\epsilon' = 1 - \frac{2d}{k''}$ and w is the weight of $\mathbf{x}'$; otherwise, it is considered random. Thus, the best candidate y_0 is the one that maximizes the absolute value of $G(\mathbf{y})$, i.e. $\mathbf{y}_0 = \arg\max_{\mathbf{y} \in F_2^l} |G(\mathbf{y})|$, and we need approximately $1/(\epsilon^{2^{t+1}} \cdot (\epsilon')^{2w})$ samples to distinguish these two cases. Note that false positives are quickly detected in an additional step and this does not significantly increase complexity.

Since the weight w is unknown, we assume that $w \leq c$ and then query for samples. If the assumption is valid, we can distinguish the two distributions correctly; otherwise, we obtain a false positive which can be recognized without much cost, and then choose another permutation to run the algorithm again. The procedure will continue until we find the secret vector $\mathbf{x}$.

We use the Fast Walsh-Hadamard Transform technique to accelerate the distinguishing step. As the hypothesis testing runs for every guess of $\mathbf{x}_2'$, the overall complexity of this step is

$$C_5 = l2^l \sum_{i=0}^{w_0} \binom{k' - k''}{i}.$$

6 Analysis

In the previous section we already indicated the complexity of each step. We now put it together in a single complexity estimate. We first formulate the formula for the possibility of having at most w errors in m positions $P(w, m)$ as follows,

$$P(w, m) = \sum_{i=0}^{w} (1 - \eta)^{m-i} \eta^i \binom{m}{i}.$$

Therefore, the success probability in one iteration is $P(w_0, k' - k'')P(c, k'')$. In each iteration, the complexity accumulates step by step, hence revealing the following theorem.

Theorem 1 (The complexity of Algorithm 2). *Let n be the number of samples required and a, t, b, w_0, c, l, k'' be algorithm parameters. For the* LPN *instance with parameter (k, η), the number of bit operations required for a successful run of the new attack is equal to $2^{f(k,n,a,t,b,w_0,c,l,k'',\eta)}$, where $f(k, n, a, t, b, w_0, c, l, k'', \eta)$ is a function[7] defined as follows,*

$$f(k, n, a, t, b, w_0, c, l, k'', \eta) =$$

$$\log_2 \left(ank + b2^b \frac{t(t+1)(2t+1)}{6} - ((k+1)2^b + nb)\binom{t}{2} + (k+1)tn \right.$$

$$+(k'' - l)(2(n - t2^b) + 2^l) + l2^l \sum_{i=0}^{w_0} \binom{k_1}{i} + (n - t2^b) \sum_{i=0}^{w_0} \binom{k_1}{i} i \right)$$

$$- \log_2 \left(\sum_{i=0}^{w_0} (1 - \eta)^{k_1 - i} \eta^i \binom{k_1}{i} \right) - \log_2 \left(\sum_{i=0}^{c} (1 - \eta)^{k'' - i} \eta^i \binom{k''}{i} \right) \quad (3)$$

under the condition that

$$n - t2^b > 1/(\epsilon^{2^{t+1}} \cdot (\epsilon')^{2c}), \quad (4)$$

where $\epsilon = 1 - 2\eta$, $\epsilon' = 1 - \frac{2d}{k''}$ and d is the smallest integer, s.t., $\sum_{i=0}^{d} \binom{k''}{i} > 2^{k''-l}$.

Proof. The complexity in one iteration is $C_1 + C_2 + C_3 + C_4 + C_5$, and the expected number of iterations is the inverse of $P(w_0, k_1)P(c, k'')$; the overall complexity, therefore, is C^*, where

$$C^* = \frac{C_1 + C_2 + C_3 + C_4 + C_5}{P(w_0, k_1)P(c, k'')}.$$

Substituting the detailed formulas into the above expression will end the proof. The condition (4) ensures that we have enough samples to determine the right guess with high probability.   □

---

[7] The symbol k_1 denotes $k - tb - k''$ for notational simplicity.

7   Results

We now present numerical results of the new algorithm attacking three key LPN instances, as shown in Table 2. All aiming for achieving 80-bit security, the first one is with parameter $(512, 1/8)$, widely accepted in various LPN-based cryptosystems (e.g., HB^+ [18], $HB^\#$ [12], LPN-C [13]) after the suggestion from Levieil and Fouque [26]; the second one is with increased length $(532, 1/8)$, adopted as the parameter of the irreducible RING-LPN instance employed in Lapin [16]; and the last one is a new design parameter[8] we recommend to use in the future. The attacking details on different protocols will be given later. We note that the new algorithm has significance not only on the above applications but also on some LPN-based cryptosystems without explicit parameter settings (e.g., [9,22]).

Table 2. The complexity for solving different LPN instances

LPN instance	Parameters								$\log_2 C^*$
	t	a	b	l	k''	w_0	c	$\log_2 n$	
$(512, 1/8)$	6	9	63	64	124	2	16	66.3	79.92
$(532, 1/8)$	6	9	65	66	130	2	17	68.0	81.82
$(592, 1/8)$	6	10	70	64	137	3	18	72.7	88.07

7.1 HB^+

In [26], Levieil and Fouque proposed an active attack on HB^+ by choosing the random vector **a** from the reader to be **0**. To achieve 80-bit security, they suggested to adjust the lengths of secret keys to 80 and 512, respectively, instead of being both 224. Its security is based on the assumption that the LPN instance with parameter $(512, 1/8)$ can resist attacks in 2^{80} bit operations. But we break it in $2^{79.9}$ bit operations, thereby yielding an active attack on 80-bit security of HB^+ authentication protocol straightforwardly.

7.2   LPN-C and $HB^\#$

Using similar structures, Gilbert et al. proposed two different cryptosystems, one for authentication ($HB^\#$) and the other for encryption (LPN-C). By setting the random vector from the reader and the message vector to be both **0**, we obtain an active attack on $HB^\#$ authentication protocol and a chosen-plaintext-attack on LPN-C, respectively. As their protocols consist of both secure version (RANDOM-$HB^\#$ and LPN-C) and efficient version ($HB^\#$ and Toeplitz LPN-C), we need to analyze separately.

---

[8] This instance requires $2^{82.3}$ bits memory using the new algorithm, and could withstand all existing attacks on the security level of 2^{80} bit operations.

Using Toeplitz Matrices. Toeplitz matrix is a matrix in which each ascending diagonal from left to right is a constant. Thus, when employing a Toeplitz matrix as the secret, if we attack its first column successively, then only one bit in its second column is unknown. So the problem is transformed to that of solving a new LPN instance with parameter $(1, 1/8)$. We then deduce the third column, the fourth column, and so forth. The typical parameter settings of the number of the columns (denoted by m) are 441 for HB$^\#$, and 80 (or 160) for Toeplitz LPN-C. In either case, the cost for determining the vectors except for the first column is bounded by 2^{40}, negligible compared with that of attacking one $(512, 1/8)$ LPN instance. Therefore, we break the 80-bit security of these »efficient« versions that use Toeplitz matrices.

Random Matrix Case. If the secret matrix is chosen totally at random, then there is no simple connection between different columns to exploit. One strategy is to attack column by column, thereby deriving an algorithm whose complexity is that of attacking a $(512, 1/8)$ LPN instance multiplied by the number of the columns. That is, if $m = 441$, then the overall complexity is about $2^{79.9} \times 441 \approx 2^{88.7}$. We may slightly improve the attack by exploiting that the different columns share the same random vector in each round.

7.3   Lapin with an Irreducible Polynomial

In [16], Heyse et al. use a $(532, 1/8)$ RING-LPN instance with an irreducible polynomial to achieve 80-bit security. We show here that this parameter setting is not secure enough for Lapin to thwart attacks on the level of 2^{80}. Although the new attack on a $(532, 1/8)$ LPN instance requires $2^{81.8}$ bit operations, larger than 2^{80}, there are two key issues to consider: 1) the RING-LPN problem is believed to be not harder than the standard LPN problem[9]; 2) we perform BKW steps using *LF1* setting in the new algorithm, but may obtain a more efficient attack in practice when adopting the *LF2* heuristic, whose effectiveness has been stated and proven in the implementation part of [26]. We suggest to increase the size of the employed irreducible polynomial in Lapin for 80-bit security.

8   More on the Covering-Coding Method

We in this section describe more aspects of the covering-coding technique, thus emphasizing the most novel and essential step in the new algorithm.

Sphere-Covering Bound. We use sphere-covering bound, for two reasons, to estimate the bias ϵ' contributed by the new technique. Firstly, there is a well-known conjecture [7] in coding theory, i.e., the covering density approaches 1

---

[9] For the instance in Lapin using a quotient ring modulo the irreducible polynomial $x^{532} + x + 1$, it is possible to optimize the procedure for inverting a ring element, thereby resulting in a more efficient attack than the generic one.

asymptotically if the code length goes to infinity. Thus, it is sensible to assume for a »good« code, when the code length k'' is relatively large. Secondly, we could see from the previous example that the key feature desired is a linear code with low average error weights, which is smaller than its covering radius. From this perspective, the covering bound brings us a good estimation.

By concatenating five $[23, 12]$ Golay codes, we construct a $[115, 60]$ linear code[10] with covering radius 15. Its expected weight of error vector is quite close to the sphere-covering bound for this parameter (with gap only 1). We believe in the existence of linear codes with length around 125, rate approximately $1/2$ and average error weight that reaches the sphere-covering bound. For explicit code construction, see [15] for details.

Using Soft Information. The weight of the error vector $\mathbf{e}'_i$ is different for different values of i, causing the confidence level to vary on different samples. However, the inherent assumption when using Fast Walsh-Hadamard Transform is a constant confidence level over all samples; thus, Fast Walsh-Hadamard Transform is not an optimal distinguishing method. For optimal distinguishing, soft information methods such as likelihood ratio tests are required. We show how to fully exploit soft distinguishing in the longer version of the paper[15].

Attacking Public-Key Cryptography. We know various decodable covering codes that could be employed in the new algorithm, e.g., rate about $1/2$ linear codes that are table-based syndrome decodable, concatenated codes built on Hamming codes, Golay codes and repetition codes, etc.. For the aimed cryptographic schemes in this paper, i.e., HB variants, LPN-C, and Lapin with an irreducible polynomial, the first three are efficient; but in the realm of public-key cryptography (e.g., schemes proposed by Alekhnovich [1], Damgård and Park [8], Duc and Vaudenay [10]), the situation alters. For these systems, their security is based on LPN instances with huge secret length (tens of thousands) and extremely low error probability (less than half a percent), so due to the competitive average weight of the error vector shown by the previous example in Section 4.1, the concatenation of repetition codes with much lower rate seems more applicable—by low-rate codes, we remove more bits when using the covering-coding method.

Alternative Collision Procedure. Although the covering-coding method is employed only once in the new algorithm, we could derive numerous variants, and among them, one may find a more efficient attack. For example, we could replace one or two steps in the later stage of the collision procedure by adding two vectors decoded to the same codeword together. This alternative technique is similar to that invented by Lamberger et al. in [24,25] for finding near-collisions of hash function. By this procedure, we could eliminate more bits in one step

---

[10] Using this code, we stand at the margin of breaking the 80-bit security of $(512, 1/8)$ LPN instances, with time complexity only $2^{80.5}$ and query complexity $2^{66.2}$.

at the cost of increasing the error rate; this is a trade-off, and the concrete parameter setting should be analyzed more thoroughly later.

9 Conclusions

In this paper we have described a new algorithm for solving the LPN problem that employs an approximation technique using covering codes together with a subspace hypothesis testing technique to determine the value of linear combinations of the secret bits. Complexity estimates show that the algorithm beats all the previous approaches, and in particular, we can present academic attacks on instances of LPN that has been suggested in different cryptographic primitives.

The new technique has only been described in a rather simplistic manner, due to space limitations. There are a few obvious improvements, one being the use of soft decoding techniques and another one being the use of more powerful constructions of good codes. There are also various modified versions that need to be further investigated. One such idea is to use the new technique inside a BKW step, thereby removing more bits in each step at the expense of introducing another contribution to the bias. An interesting open problem is whether these ideas can improve the asymptotic behavior of the BKW algorithm.

References

1. Alekhnovich, M.: More on Average Case vs Approximation Complexity. In: FOCS, pp. 298–307. IEEE Computer Society (2003)
2. Blum, A., Furst, M., Kearns, M., Lipton, R.: Cryptographic Primitives Based on Hard Learning Problems. In: Stinson, D.R. (ed.) CRYPTO 1993. LNCS, vol. 773, pp. 278–291. Springer, Heidelberg (1994)
3. Blum, A., Kalai, A., Wasserman, H.: Noise-Tolerant Learning, the Parity Problem, and the Statistical Query Model. Journal of the ACM 50(4), 506–519 (2003)
4. Berlekamp, E.R., McEliece, R.J., van Tilborg, H.C.A.: On the Inherent Intractability of Certain Coding Problems. IEEE Trans. Info. Theory 24, 384–386 (1978)
5. Bernstein, D.J., Lange, T.: Never trust a bunny. In: Hoepman, J.-H., Verbauwhede, I. (eds.) RFIDSec 2012. LNCS, vol. 7739, pp. 137–148. Springer, Heidelberg (2013)
6. Chose, P., Joux, A., Mitton, M.: Fast Correlation Attacks: An Algorithmic Point of View. In: Knudsen, L.R. (ed.) EUROCRYPT 2002. LNCS, vol. 2332, pp. 209–221. Springer, Heidelberg (2002)
7. Cohen, G., Honkala, I., Litsyn, S., Lobstein, A.: Covering codes. Elsevier (1997)
8. Damgård, I., Park, S.: Is Public-Key Encryption Based on LPN Practical? Cryptology ePrint Archive, Report 2012/699 (2012), http://eprint.iacr.org/
9. Dodis, Y., Kiltz, E., Pietrzak, K., Wichs, D.: Message Authentication, Revisited. In: Pointcheval, D., Johansson, T. (eds.) EUROCRYPT 2012. LNCS, vol. 7237, pp. 355–374. Springer, Heidelberg (2012)
10. Duc, A., Vaudenay, S.: HELEN: A Public-Key Cryptosystem Based on the LPN and the Decisional Minimal Distance Problems. In: Youssef, A., Nitaj, A., Hassanien, A.E. (eds.) AFRICACRYPT 2013. LNCS, vol. 7918, pp. 107–126. Springer, Heidelberg (2013)

11. Fossorier, M.P.C., Mihaljevic, M.J., Imai, H., Cui, Y., Matsuura, K.: A Novel Algorithm for Solving the LPN Problem and its Application to Security Evaluation of the HB Protocol for RFID Authentication. Cryptology ePrint archive, Report 2012/197 (2012), http://eprint.iacr.org/

12. Gilbert, H., Robshaw, M., Seurin, Y.: HB$^{\#}$: Increasing the Security and the Efficiency of HB$^+$. In: Smart, N.P. (ed.) EUROCRYPT 2008. LNCS, vol. 4965, pp. 361–378. Springer, Heidelberg (2008)

13. Gilbert, H., Robshaw, M.J.B., Seurin, Y.: How to encrypt with the LPN problem. In: Aceto, L., Damgård, I., Goldberg, L.A., Halldórsson, M.M., Ingólfsdóttir, A., Walukiewicz, I. (eds.) ICALP 2008, Part II. LNCS, vol. 5126, pp. 679–690. Springer, Heidelberg (2008)

14. Gilbert, H., Robshaw, M.J.B., Sibert, H.: An active attack against HB$^+$—a provably secure lightweight authentication protocol. Cryptology ePrint Archive, Report 2005/237 (2005), http://eprint.iacr.org/

15. Guo, Q., Johansson, T., Löndahl, C.: Solving LPN Using Covering Codes and Soft Information (in preparation)

16. Heyse, S., Kiltz, E., Lyubashevsky, V., Paar, C., Pietrzak, K.: Lapin: An Efficient Authentication Protocol Based on Ring-LPN. In: Canteaut, A. (ed.) FSE 2012. LNCS, vol. 7549, pp. 346–365. Springer, Heidelberg (2012)

17. Hopper, N.J., Blum, M.: Secure human identification protocols. In: Boyd, C. (ed.) ASIACRYPT 2001. LNCS, vol. 2248, pp. 52–66. Springer, Heidelberg (2001)

18. Juels, A., Weis, S.A.: Authenticating pervasive devices with human protocols. In: Shoup, V. (ed.) CRYPTO 2005. LNCS, vol. 3621, pp. 293–308. Springer, Heidelberg (2005)

19. Katz, J., Shin, J.S.: Parallel and concurrent security of the HB and HB$^+$ protocols. In: Vaudenay, S. (ed.) EUROCRYPT 2006. LNCS, vol. 4004, pp. 73–87. Springer, Heidelberg (2006)

20. Katz, J., Shin, J.S., Smith, A.: Parallel and concurrent security of the HB and HB$^+$ protocols. Journal of Cryptology 23(3), 402–421 (2010)

21. Kearns, M.: Effcient Noise-Tolerant Learning from Statistical Queries. J. ACM 45(6), 983–1006 (1998)

22. Kiltz, E., Pietrzak, K., Cash, D., Jain, A., Venturi, D.: Efficient Authentication from Hard Learning Problems. In: Paterson, K.G. (ed.) EUROCRYPT 2011. LNCS, vol. 6632, pp. 7–26. Springer, Heidelberg (2011)

23. Kirchner, P.: Improved Generalized Birthday Attack. Cryptology ePrint Archive, Report 2011/377 (2011), http://eprint.iacr.org/

24. Lamberger, M., Mendel, F., Rijmen, V., Simoens, K.: Memoryless near-collisions via coding theory. Designs, Codes and Cryptography 62(1), 1–18 (2012)

25. Lamberger, M., Teufl, E.: Memoryless near-collisions, revisited. Information Processing Letters 113(3), 60–66 (2013)

26. Levieil, É., Fouque, P.-A.: An Improved LPN Algorithm. In: De Prisco, R., Yung, M. (eds.) SCN 2006. LNCS, vol. 4116, pp. 348–359. Springer, Heidelberg (2006)

27. Lyubashevsky, V.: The Parity Problem in the Presence of Noise, Decoding Random Linear Codes, and the Subset Sum Problem. In: Chekuri, C., Jansen, K., Rolim, J.D.P., Trevisan, L. (eds.) APPROX 2005 and RANDOM 2005. LNCS, vol. 3624, pp. 378–389. Springer, Heidelberg (2005)

28. Mitzenmacher, M., Upfal, E.: Probability and computing - randomized algorithms and probabilistic analysis. Cambridge University Press (2005)

29. Munilla, J., Peinado, A.: HB-MP: A further step in the HB-family of lightweight authentication protocols. Computer Networks 51(9), 2262–2267 (2007)

30. Regev, O.: On Lattices, Learning with Errors, Random Linear Codes, and Cryptography. In: Gabow, H.N., Fagin, R. (eds.) Proceedings of 37th Annual ACM Symposium on Theory of Computing, pp. 84–93 (2005)
31. Stern, J.: A Method for Finding Codewords of Small Weight. In: Wolfmann, J., Cohen, G. (eds.) Coding Theory 1988. LNCS, vol. 388, pp. 106–113. Springer, Heidelberg (1989)
32. Stern, J.: A New Identification Scheme Based on Syndrome Decoding. In: Stinson, D.R. (ed.) CRYPTO 1993. LNCS, vol. 773, pp. 13–21. Springer, Heidelberg (1994)
33. Wagner, D.: A Generalized Birthday Problem. In: Yung, M. (ed.) CRYPTO 2002. LNCS, vol. 2442, pp. 288–304. Springer, Heidelberg (2002)

A   Illustrating the Procedure

In this section, we give an intuitive illustration of subspace hypothesis test performed as follows,

Rewrite $\mathbf{g}_i$ as codeword $\mathbf{c}_i = \mathbf{u}'\mathbf{F}$ and discrepancy $\mathbf{e}'_i$

$$
\begin{bmatrix} * \\ \vdots \\ * \\ \hline z'_i \\ \hline * \\ \vdots \end{bmatrix}
=
\underbrace{\begin{bmatrix} x_0 \\ \vdots \\ x_{k''} \\ \hline 0 \\ \vdots \end{bmatrix}^{\mathrm{T}}}_{\text{Secret } \mathbf{x}}
\underbrace{\begin{bmatrix} * & * & g_0 & * \\ \vdots & \vdots & \vdots & \vdots \\ * & * & g_{k''} & * \\ \hline & & 0 & \\ & & \vdots & \end{bmatrix}}_{\text{Query matrix}}
=
\begin{bmatrix} x_0 \\ \vdots \\ x_{k''} \\ \hline 0 \\ \vdots \end{bmatrix}^{\mathrm{T}}
\begin{bmatrix} * & * & (\mathbf{u}'\mathbf{F}+\mathbf{e}'_i)_0 & * \\ \vdots & \vdots & \vdots & \vdots \\ * & * & (\mathbf{u}'\mathbf{F}+\mathbf{e}'_i)_{k''} & * \\ \hline & & 0 & \\ & & \vdots & \end{bmatrix} .
$$

We can separate the discrepancy $\mathbf{e}'_i$ from $\mathbf{uF}$, which yields

$$
\begin{bmatrix} x_0 \\ \vdots \\ x_{k''} \\ \hline 0 \\ \vdots \end{bmatrix}^{\mathrm{T}}
\begin{bmatrix} * & * & (\mathbf{u}'\mathbf{F})_0 & * \\ \vdots & \vdots & \vdots & \vdots \\ * & * & (\mathbf{u}'\mathbf{F})_{k''} & * \\ \hline & & 0 & \\ & & \vdots & \end{bmatrix}
=
\begin{bmatrix} * \\ \vdots \\ * \\ \hline z'_i + \langle \mathbf{x}, \mathbf{e}'_i \rangle \\ \hline * \\ \vdots \end{bmatrix} .
$$

Finally, we note that $\mathbf{x}'_1\mathbf{F}^{\mathrm{T}} \in \mathbb{F}_2^l$, where $l < k''$. A simple transformation yields

$$
\begin{bmatrix} (\mathbf{x}'_1\mathbf{F}^{\mathrm{T}})_0 \\ \vdots \\ (\mathbf{x}'_1\mathbf{F}^{\mathrm{T}})_l \\ \hline 0 \\ \vdots \end{bmatrix}^{\mathrm{T}}
\begin{bmatrix} * & * & u'_0 & * \\ \vdots & \vdots & \vdots & \vdots \\ * & * & u'_l & * \\ \hline & & 0 & \\ & & \vdots & \end{bmatrix}
=
\begin{bmatrix} * \\ \vdots \\ * \\ \hline z'_i + \langle \mathbf{x}'_1, \mathbf{e}'_i \rangle \\ \hline * \\ \vdots \end{bmatrix} .
$$

Since $w_{\mathrm{H}}(\mathbf{e}'_i) \leq w$, the contribution from $\langle \mathbf{x}'_1, \mathbf{e}'_i \rangle$ is very small.

Algebraic Attack against Variants of McEliece with Goppa Polynomial of a Special Form

Jean-Charles Faugère[1,2,3], Ludovic Perret[1,2,3],
and Frédéric de Portzamparc[1,2,3,4]

[1] Inria, Équipe PolSys, Paris-Rocquencourt
[2] Sorbonne Universités, UPMC Univ Paris 06, Équipe PolSys,
LIP6, F-75005, Paris, France
[3] CNRS, UMR 7606, LIP6 UPMC, F-75005, Paris
[4] Gemalto, 6 rue de la Verrerie 92190, Meudon, France
jean-charles.faugere@inria.fr, ludovic.perret@lip6.fr,
frederic.urvoydeportzamparc@gemalto.com

Abstract. In this paper, we present a new algebraic attack against some special cases of Wild McEliece Incognito, a generalization of the original McEliece cryptosystem. This attack does not threaten the original McEliece cryptosystem. We prove that recovering the secret key for such schemes is equivalent to solving a system of polynomial equations whose solutions have the structure of a usual *vector space*. Consequently, to recover a basis of this vector space, we can greatly reduce the number of variables in the corresponding algebraic system. From these solutions, we can then deduce the basis of a GRS code. Finally, the last step of the cryptanalysis of those schemes corresponds to attacking a McEliece scheme instantiated with particular GRS codes (with a polynomial relation between the support and the multipliers) which can be done in polynomial-time thanks to a variant of the Sidelnikov-Shestakov attack. For Wild McEliece & Incognito, we also show that solving the corresponding algebraic system is notably easier in the case of a non-prime base field $\mathbb{F}_q$. To support our theoretical results, we have been able to practically break several parameters defined over a non-prime base field $q \in \{9, 16, 25, 27, 32\}$, $t \leqslant 6$, extension degrees $m \in \{2, 3\}$, security level up to 2^{129} against information set decoding in few minutes or hours.

Keywords: Public-key cryptography, McEliece cryptosystem, algebraic cryptanalysis.

1 Introduction

Algebraic cryptanalysis is a general attack technique which reduces the security of a cryptographic primitive to the difficulty of solving a non-linear system of equations. Although the efficiency of general polynomial system solvers such as Gröbner bases, SAT solvers ..., is constantly progressing such algorithms all face the intrinsic hardness of solving polynomial equations. As a consequence, the success of an algebraic attack relies crucially in the ability to find the best modelling in term of algebraic equations.

P. Sarkar and T. Iwata (Eds.): ASIACRYPT 2014, PART I, LNCS 8873, pp. 21–41, 2014.
© International Association for Cryptologic Research 2014

In [14,15], Faugère, Otmani, Perret and Tillich (FOPT) show – in particular – that the key-recovery of McEliece [20] can be reduced to the solving of a system of non-linear equations. This key-recovery system can be greatly simplified for so-called compact variants of McEliece, e.g. [4,21,2,16,23,1], leading to an efficient attack against various compact schemes [14,13]. However, it is not clear whether the attack of [14,15] could be efficient against non-compact variants of McEliece, the bottleneck being the huge number of variables and the high degree of the equations involved in the algebraic modelling.

We present a novel algebraic modelling that applies to the original McEliece system and to generalizations such as *Wild McEliece* [6] and *Wild McEliece Incognito* [8]. Note, however, that the resulting attack works only in some special cases, and in particular does not work for the original McEliece system. Wild McEliece uses *Wild Goppa codes*, that is Goppa codes over $\mathbb{F}_q, q \geqslant 2$, with a Goppa polynomial of the form Γ^{q-1} (Γ being an univariate polynomial of low degree). This form of the Goppa polynomial, generalizing the form used in the original McEliece system for $q = 2$, allows to increase the number of errors that can be added to a message (in comparison to a random Goppa polynomial of the same degree). In [8], Bernstein, Lange, and Peters generalized this idea by using Goppa polynomials of the form $f \, \Gamma^{q-1}$, with f another univariate polynomial. We shall call such Goppa codes *Masked Wild Goppa* codes. Like the authors of [8], we refer to this version as Wild McEliece *Incognito*. All in all, Wild McEliece/Wild McEliece Incognito allow the users to select parameters with a resistance to all known attacks, so in particular to the algebraic attack of [14,15], similar to that of binary Goppa codes but with much smaller keys. The security of Wild McEliece defined over quadratic extension has been recently investigated in [11], where the authors presented a polynomial time attack on the key when $t = \deg(\Gamma) > 1$.

1.1   Our Contributions

We present a completely new algebraic attack dedicated to Wild McEliece and Wild McEliece Incognito. To do so, we show that the key-recovery for such schemes is equivalent to finding the basis of a vector-space which is hidden in the zero-set of an algebraic system. To our knowledge, this is a new computational problem that never appeared in algebraic cryptanalysis before. Compared to the algebraic attack proposed in [14] for McEliece, our modelling intrinsically involves less variables. Informally, the multiplicity of the Goppa polynomial implies that the solutions of the algebraic system considered here have a structure of vector space. When the base field is $\mathbb{F}_q$ with $q > 2$, this simplifies its resolution. For instance, for a Wild McEliece Incognito scheme with parameters $q = 32, m = 2, n = 864, t = 2, deg(f) = 36$), we end up with an algebraic system having only 9 variables ([14] would require to consider algebraic equations with 1060 variables in the same situation). On a very high level, our attack proceeds in two main steps.

1. **Polynomial System Solving.** We have to solve a non-linear system of equations whose zero-set forms, unexpectedly, a vector space of some known

dimension d. Consequently, we can reduce the number of variables by fixing d variables in the initial and repeat several times the solving step to recover a basis of the vector space solution. This is the most computationally difficult part of the attack.

2. **Linear Algebra to Recover the Secret Key.** The second phase is the treatment of the solutions obtained at the first step so as to obtain a private description which allows to decode the public-key as efficiently as the private key. It involves computing intersections of vector spaces, solving linear systems, and polynomial interpolation. Thus, this part can be done efficiently, i.e. in polynomial time.

We detail below the main ingredients of our attack.

An Algebraic Modelling with a Vector Space Structure on the Zero Set. Let $\mathbf{G}_{pub} = (g_{i,j})_{\substack{0 \leqslant i \leqslant n-1 \\ 0 \leqslant j \leqslant k-1}} \in \mathbb{F}_q^{k \times n}$ be the public matrix of a Wild McEliece Incognito scheme. We denote by m its extension degree, and set $t = \deg(\Gamma)$. Our attack considers the system

$$\mathcal{W}_{q,a}(\mathbf{Z}) = \bigcup_{u \in \mathcal{P}_a} \left\{ \sum_{j=0}^{n-1} g_{i,j} Z_j^u = 0 \quad | \quad 0 \leqslant i \leqslant k-1 \right\}, \qquad (1)$$

with $\mathcal{P}_a = \{1, 2, \ldots, p^a - 1\} \cup \{p^a, p^{a+1}, \ldots, q\}$ being a subset of $\{1, \ldots, q\}$.

As a comparison, the modelling of Faugère, Otmani, Perret and Tillich [14] will necessarily introduce variables $\mathbf{X} = (X_0, \ldots, X_{n-1})$, $\mathbf{Y} = (Y_0, \ldots, Y_{n-1})$ and $\mathbf{W} = (W_0, \ldots, W_{n-1})$ for all the support and multipliers (that is, the vectors $\mathbf{y} = \Gamma(\mathbf{x})^{-1}$ and $\mathbf{w} = f(\mathbf{x})^{-1}$). In [14], the system is as follows:

$$\bigcup_{0 \leqslant u \leqslant t-1} \left\{ \sum_{j=0}^{n-1} g_{i,j} Y_j X_j^u = 0 \quad | \quad 0 \leqslant i \leqslant k-1 \right\}.$$

In our context, [14] would induce a system containing monomials of the forms $Y_i^{\ell_Y} X_i^{\ell_X}$ and even $Y_i^{\ell_Y} X_i^{\ell_X} W_i^{\ell_W}$ (for some ℓ_X, ℓ_Y, ℓ_W). Here, we use a single vector of variables $\mathbf{Z} = (Z_0, \ldots, Z_{n-1})$ and write very simple homogeneous equations. The secret-key $\mathbf{x}$, $\mathbf{y}$ and $\mathbf{w}$ will be recovered from $\mathbf{Z}$, but in a second step. The main advantage of this approach (Theorem 2) is that the solutions of $\mathcal{W}_{q,a}(\mathbf{Z})$ have a very unexpected property for a non-linear system: they form a vector space. This allows to reduce the number of "free" unknowns in $\mathcal{W}_{q,a}(\mathbf{Z})$ by the dimension of the solutions. For example, we end up with a system containing only 9 variables for an Incognito scheme with parameters $q = 32, m = 2, n = 864, t = 2, deg(f) = 36)$. The algebraic description of Goppa codes proposed in [14] would require to consider algebraic equations with 1060 variables for the same parameters.

To be more precise, the vector space underlying the solutions of (1) is closely related to Generalized Reed-Solomon (GRS) codes.

Definition 1 (Generalized Reed-Solomon codes). *Let* $\mathbf{x} = (x_0, \ldots, x_{n-1}) \in (\mathbb{F}_{q^m})^n$ *where all* x_i*'s are distinct and* $\mathbf{y} = (y_0, \ldots, y_{n-1}) \in (\mathbb{F}_{q^m}^*)^n$. *The*

Generalized Reed-Solomon code *of dimension t, denoted by* $\text{GRS}_t(\mathbf{x}, \mathbf{y})$, *is defined as follows*

$$\text{GRS}_t(\mathbf{x}, \mathbf{y}) \overset{def}{=} \big\{ (y_0 Q(x_0), \ldots, y_{n-1} Q(x_{n-1})) \mid Q \in \mathbb{F}_{q^m}[z], \deg(Q) \leqslant t - 1 \big\}.$$

We shall call $\mathbf{x}$ *the* support *of the code, and* $\mathbf{y}$ *the* multipliers.

Theorem 2 shows that the solutions of $\mathcal{W}_{q,a}(\mathbf{Z})$ contain a vector-space which is generated by sums of codewords of Generalized Reed-Solomon (GRS) codes $\text{GRS}_t(\mathbf{x}^\ell, \mathbf{y}^\ell)$ (where $(\mathbf{x}, \mathbf{y})$ is a key equivalent to the secret key). In Section 3.2, we explain more precisely how we can take advantage of this special structure for solving (1) and recover a basis of the vector subspace.

A Method to Isolate a GRS Code From a Sum of GRS. From a basis of this sum of GRS, we want to recover the basis of the code $\text{GRS}_t(\mathbf{x}, \mathbf{y})$. We refer to this phase as the *disentanglement*. We expose our solution in Section 4, which relies on a well-chosen intersection of codes. It is rigorously proved in characteristic 2 (Proposition 6). For other characteristics, we launched more than $100,000$ experiments and observed that Proposition 6 still held in all cases.

A Sidelnikov-Shestakov-Like Algorithm Recovering the Goppa Polynomial. Given a basis of a Generalized Reed-Solomon code $\text{GRS}_t(\mathbf{x}, \mathbf{y})$, the Sidelnikov-Shestakov attack [26] consists in recovering the secret pair of vectors $(\mathbf{x}, \mathbf{y})$. It is well-known that the Sidelnikov-Shestakov attack works in polynomial-time. In our case, we have to address a slight variant of this problem. There is a polynomial relation $\Gamma(z)$ linking $\mathbf{x}$ and $\mathbf{y}$ which is part of the private key. In Section 4.2, we provide an adaptation of [26] to obtain a key $(\mathbf{x}', \mathbf{y}', \Gamma')$ equivalent to the secret key, also in polynomial time. We are unaware of such an algorithm published so far.

A Weakness of Codes Defined Over Non-prime Base Fields. Independently of our algebraic attack, we prove a general result about Goppa codes defined over $\mathbb{F}_q$ (with $q = p^s$, p prime and $s > 0$) and whose polynomials have a factor $\Gamma(z)$ with multiplicity q. We show in Section 5 that the coordinate-vectors over $\mathbb{F}_p$ of the codewords of such a public code are codewords of a Wild Goppa code, defined over $\mathbb{F}_p$, with same secret support and Goppa polynomial $\Gamma(z)^p$ (Theorem 8). In other words, this construction gives access, from the public key, to a new code implying the same private elements. As a consequence, using non-prime base fields reveals more information on the secret key than expected by the designers. Any key-recovery attack can benefit from it. This is then an intrinsic weakness of Goppa codes defined over non-prime base fields. In our context, this property provides additional linear equations between the variables Z_j's of the system (1). We can reduce the number of variables from $(p^s - 1)mt$ to $(p - 1)mst$ essential variables, and make the codes defined over fields $\mathbb{F}_q$ with $q = p^s$ notably weaker (Corollary 10).

1.2   Impact of Our Work

In order to evaluate the efficiency of our attack, we considered various parameters for which [6] said that strength is "unclear" and that an attack would not be a "surprise" but for which no actual attack was known.

Information Set Decoding (ISD) is a generic decoding technique which allows message-recovery. This technique has been intensively studied since 1988 (e.g. [17,10,5,7,19,3]) and remains the reference to choose secure parameters in code-based cryptography. The latest results from [24] have been used to generate the parameters for Wild McEliece and Wild McEliece Incognito.

In [6, Table 7.1] numerous keys are presented which illustrate the key size reduction when the size of the field q grows. Another consequence of increasing q is pointed out by the authors of [6]: the low number of irreducible polynomials in $\mathbb{F}_{q^m}[z]$ entails a possible vulnerability against the SSA structural attack ([18,25]). Although the designers provide a protection (using non full-support codes) such that [18] is completely infeasible today, they warn that further progress in [18] may jeopardize the parameters with $q > 9$ and thus estimate that those parameters have unclear security. Our experiments reveal that, in the case of non-prime base fields, it is already possible to recover the secret key in some minutes with our attack using off-the-shelf tools (MAGMA [9] V2.19-1).

Getting around the alleged vulnerability against SSA was the main motivation for proposing Incognito: in [8, Table 5.1], they propose parameters considered fully secure, as all ISD-complexities are above 2^{128} and numbers of possible Goppa polynomials greater than 2^{256}. It turns out that, in the case of non-prime base fields, the extra-shield introduced in Incognito is not a protection against our attack. We can practically break the recommended parameters for $q \in \{16, 27, 32\}$. However, we could not solve (in less than two days) the algebraic systems involved for extension degrees $m \geqslant 4$ or $t \geqslant 7$, and for codes over $\mathbb{F}_p$, p prime. So, it does not threaten the original McEliece cryptosystem. To conclude, we highlight that Theorem 2 is valid for all Goppa codes whose Goppa polynomial has multiplicities and should be then taken into account by designers in the future. Figure 1 provides a diagram which recapitulates all the steps performed to solve the system (1) and recover the secret key.

2   Coding Theory Background

Let $\mathbb{F}_q$ be a finite field of $q = p^s$ elements (p prime, and $s > 0$). To define conveniently the various kinds of codes we will deal with, we introduce the following Vandermonde-like matrices:

$$
\mathbf{V}_t(\mathbf{x}, \mathbf{y}) \overset{\text{def}}{=} \begin{pmatrix} y_0 & \cdots & y_{n-1} \\ y_0 x_0 & \cdots & y_{n-1} x_{n-1} \\ \vdots & & \vdots \\ y_0 x_0^{t-1} & \cdots & y_{n-1} x_{n-1}^{t-1} \end{pmatrix}, \tag{2}
$$

where $\left(\mathbf{x} = (x_0, \ldots, x_{n-1}), \mathbf{y} = (y_0, \ldots, y_{n-1}) \right) \in \mathbb{F}_{q^m}^n \times \mathbb{F}_{q^m}^n$.

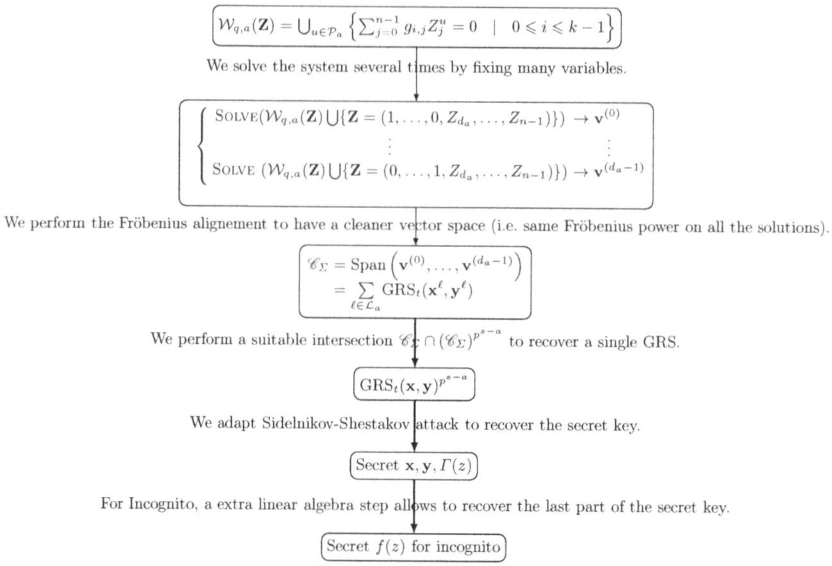

$$W_{q,a}(\mathbf{Z}) = \bigcup_{u \in \mathcal{P}_a} \left\{ \sum_{j=0}^{n-1} g_{i,j} Z_j^u = 0 \quad | \quad 0 \leqslant i \leqslant k-1 \right\}$$

We solve the system several times by fixing many variables.

$$\begin{cases} \text{SOLVE}(W_{q,a}(\mathbf{Z}) \bigcup \{\mathbf{Z} = (1,\dots,0,Z_{d_a},\dots,Z_{n-1})\}) \to \mathbf{v}^{(0)} \\ \quad\quad\quad \vdots \quad\quad\quad\quad\quad\quad\quad \vdots \\ \text{SOLVE }(W_{q,a}(\mathbf{Z}) \bigcup \{\mathbf{Z} = (0,\dots,1,Z_{d_a},\dots,Z_{n-1})\}) \to \mathbf{v}^{(d_a-1)} \end{cases}$$

We perform the Fröbenius alignement to have a cleaner vector space (i.e. same Fröbenius power on all the solutions).

$$\mathscr{C}_\Sigma = \text{Span}\left(\mathbf{v}^{(0)},\dots,\mathbf{v}^{(d_a-1)}\right)$$
$$= \sum_{\ell \in \mathcal{L}_a} \text{GRS}_t(\mathbf{x}^\ell, \mathbf{y}^\ell)$$

We perform a suitable intersection $\mathscr{C}_\Sigma \cap (\mathscr{C}_\Sigma)^{p^{s-a}}$ to recover a single GRS.

$$\text{GRS}_t(\mathbf{x}, \mathbf{y})^{p^{s-a}}$$

We adapt Sidelnikov-Shestakov attack to recover the secret key.

$$\text{Secret } \mathbf{x}, \mathbf{y}, \Gamma(z)$$

For Incognito, a extra linear algebra step allows to recover the last part of the secret key.

$$\text{Secret } f(z) \text{ for incognito}$$

Fig. 1. Overview of the attack

With suitable $\mathbf{x}$ and $\mathbf{y}$, the rows of such matrices $\mathbf{V}_t(\mathbf{x},\mathbf{y})$ define Generalized Reed-Solomon (GRS) codes (Definition 1). Alternant and Goppa codes can be viewed as the restriction of duals of GRS codes to the base field $\mathbb{F}_q$.

Definition 2 (Alternant/Goppa Codes). *Let* $\mathbf{x} = (x_0,\dots,x_{n-1}) \in (\mathbb{F}_{q^m})^n$ *where all* x_i's *are distinct and* $\mathbf{y} \in (\mathbb{F}_{q^m}^*)^n$. *The* alternant code *of order* t *is defined as* $\mathcal{A}_t(\mathbf{x},\mathbf{y}) \overset{def}{=} \left\{\mathbf{c} \in \mathbb{F}_q^n \mid \mathbf{V}_t(\mathbf{x},\mathbf{y})\mathbf{c}^T = \mathbf{0}\right\}$. *As for* GRS *codes,* $\mathbf{x}$ *is the* support, *and* $\mathbf{y}$ *the* multipliers. *Let* $g(z) \in \mathbb{F}_{q^m}[z]$ *be of degree* t *satisfying* $g(x_i) \neq 0$ *for all* $i, 0 \leqslant i \leqslant n-1$. *We define the* Goppa code *over* $\mathbb{F}_q$ *associated to* $g(z)$ *as the code* $\mathscr{G}_q(\mathbf{x}, g(z)) \overset{def}{=} \mathcal{A}_t(\mathbf{x},\mathbf{y})$, *with* $\mathbf{y} = g(\mathbf{x})^{-1}$. *The dimension* k *of* $\mathscr{G}_q(\mathbf{x},g(z))$ *satisfies* $k \geqslant n - tm$. *The polynomial* $g(z)$ *is called the* Goppa polynomial, *and* m *is the* extension degree. *Equivalently,* $\mathscr{G}_q(\mathbf{x}, g(z))$ *can be defined as:*

$$\mathscr{G}_q(\mathbf{x}, g(z)) \overset{def}{=} \left\{\mathbf{c} = (c_0,\dots,c_{n-1}) \in \mathbb{F}_q^n \mid \sum_{i=0}^{n-1} \frac{c_i}{z - x_i} \equiv 0 \bmod g(z)\right\}.$$

Goppa codes naturally inherit a decoding algorithm that corrects up to $\frac{t}{2}$ errors. This bound can be improved to correct more errors by using *Wild* Goppa codes, introduced by Bernstein, Lange, and Peters in [6]. We also recall the version of Wild Goppa code used in Wild McEliece Incognito [8]. We call such special version of Wild Goppa codes: *Masked Wild Goppa codes.*

Definition 3 (Wild Goppa/Masked Wild Goppa). *Let* $\mathbf{x}$ *be an* n-tuple $(x_0,\dots,x_{n-1})$ *of distinct elements of* $\mathbb{F}_{q^m}$. *Let* $\Gamma(z) \in \mathbb{F}_{q^m}[z]$ *(resp.* $f(z) \in$

$\mathbb{F}_{q^m}[z]$) be a squarefree polynomial of degree t (resp. u) satisfying $\Gamma(x_i) \neq 0$ (resp. $f(x_i) \neq 0$) for all $i, 0 \leqslant i \leqslant n-1$. A Wild Goppa code is a Goppa code whose Goppa polynomial is of the form $g(z) = \Gamma(z)^{q-1}$. A Masked Wild Goppa code is a Wild Goppa code whose Goppa polynomial is such that $g(z) = f(z)\Gamma(z)^{q-1}$.

The reason for using those Goppa polynomials lies in the following result.

Theorem 1. [6,8] Let the notations be as in Definition 3. It holds that

$$\mathscr{G}_q\big(\mathbf{x}, f(z)\Gamma^{q-1}(z)\big) = \mathscr{G}_q\big(\mathbf{x}, f(z)\Gamma^q(z)\big). \tag{3}$$

Thus, the code $\mathscr{G}_q\big(\mathbf{x}, f(z)\Gamma^q(z)\big)$ has dimension $\geqslant n - m\left((q-1)t + u\right)$.

This is a generalization of a well-known property for $q = 2$. The advantage of Wild Goppa codes (i.e. $f = 1$) compared to standard Goppa codes is that $\lfloor qt/2 \rfloor$ errors can be decoded efficiently (instead of $\lfloor (q-1)t/2 \rfloor$) for the same code dimension $(n - (q-1)mt$ in most cases). In fact, we can decode up to $\lfloor qt/2 \rfloor + 2$ using list decoding. This increases the difficulty of the syndrome decoding problem. Hence, for a given level of security, codes with smaller keys can be used (for details, see [6, Section 7] and [8, Section 5]).

3   An Algebraic Modelling with a Vector Space Structure on the Zero Set

The core idea of our attack is to construct, thanks to the public matrix, an algebraic system whose solution set $\mathcal{S}$ has a very surprising structure (Definition 4). It appears that $\mathcal{S}$ includes the union of several vector spaces. The vector spaces correspond in fact to sums of GRS codes (Definition 1) which have almost the same support $\mathbf{x}$ and multiplier vector $\mathbf{y}$ as the public-key of the attacked Wild McEliece Incognito scheme (Theorem 3). These vectors give a key-equivalent to the secret-key.

3.1   Description of the New Modelling

We consider the following algebraic equations:

Definition 4. Let $q = p^s$ (p prime and $s \geqslant 0$). Let $\mathbf{G}_{pub} = (g_{i,j})_{\substack{0 \leqslant i \leqslant n-1 \\ 0 \leqslant j \leqslant k-1}}$ be a generator matrix of a masked Wild Goppa code $\mathscr{C}_{pub} = \mathscr{G}_q(\mathbf{x}, f(z)\Gamma^{q-1}(z))$. For an integer $a, 0 < a \leqslant s$, we define the system $\mathcal{W}_{q,a}(\mathbf{Z})$ as follows :

$$\mathcal{W}_{q,a}(\mathbf{Z}) = \bigcup_{u \in \mathcal{P}_a} \left\{ \sum_{j=0}^{n-1} g_{i,j} Z_j^u = 0 \ \mid \ 0 \leqslant i \leqslant k-1 \right\} \tag{4}$$

with $\mathcal{P}_a = \{1, 2, \ldots, p^a - 1\} \cup \{p^a, p^{a+1}, \ldots, q\}$.

The parameter a in $\mathcal{P}_a$ determines the exponents considered for the Z_j's in the system (4). For $a = s$, we consider all the powers Z_j^u where u ranges in $\{1, \ldots, q\}$. Removing some exponents leads to a system with fewer equations and may seem counter-intuitive at first sight (the more equations, the better it is for solving a polynomial system). However, the situation is different here due to the specific structure of the solutions of $\mathcal{W}_{q,a}(\mathbf{Z})$, described in the following theorem.

Theorem 2. *Let the notations be as in Definition 4. Let* $\mathbf{y} = \Gamma(\mathbf{x})^{-1}$, $t = \deg(\Gamma)$ *and* $\mathcal{L}_a = \bigcup_{0 \leqslant r \leqslant s-1-a} \{p^r, 2p^r, \ldots, (p-1)p^r\} \cup \{p^{s-a}\}$. *The solutions* $\mathcal{S}$ *of* $\mathcal{W}_{q,a}(\mathbf{Z})$ *contain the union of* m *vector spaces which are sums of* GRS *codes:*

$$\bigcup_{0 \leqslant e \leqslant m-1} \left(\sum_{\ell \in \mathcal{L}_a} \mathrm{GRS}_t(\mathbf{x}^\ell, \mathbf{y}^\ell)^{q^e} \right) \subseteq \mathcal{S},$$

with $\mathrm{GRS}_t(\mathbf{x}^\ell, \mathbf{y}^\ell)^{q^e}$ *denoting all the elements of* $\mathrm{GRS}_t(\mathbf{x}^\ell, \mathbf{y}^\ell)$ *with coordinates raised to the power* q^e, *with* $0 \leqslant e \leqslant m - 1$.

Remark 1. When all the powers $\{1, \ldots, q\}$ are considered in the system, that is $a = s$, then $\mathcal{L}_a$ is reduced to $\{1\}$ and the solution set is a union of GRS codes. If $a < s$, the solution set is a bit more complex, but it has the great advantage of having a larger dimension; allowing then to solve the system (4) more efficiently. We will formalize this in Section 3.2.

Note that we state in Theorem 2 that we know a subset of the solutions. In practice, as the system is highly overdefined, we always observed that this subset was *all* the solutions.

Proof. The full proof of this result is postponed in Section A.3. We just give the global idea of the proof. The goal is to show the elements of $\sum_{\ell \in \mathcal{L}_a} \mathrm{GRS}_t(\mathbf{x}^\ell, \mathbf{y}^\ell)^{q^e}$ are solutions of $\mathcal{W}_{q,a}(\mathbf{Z})$. We can assume that $e = 0$ w.lo.g.

Let $\mathbf{z} = (z_1, \ldots, z_n) \in \sum_{\ell \in \mathcal{L}_a} \mathrm{GRS}_t(\mathbf{x}^\ell, \mathbf{y}^\ell)$. We write the coordinates of $\mathbf{z}$ as $z_j = \sum_{\ell \in \mathcal{L}_a} y_j^\ell Q_\ell(x_j^\ell)$, where the Q_ℓ's are polynomials of degree $\leqslant t - 1$ of $\mathbb{F}_{q^m}[z]$. We have to prove that

$$\sum_{j=0}^{n-1} g_{i,j} z_j^u = 0 \text{ for } u \in \mathcal{P}_1 \cup \mathcal{P}_2, \text{ where } \mathcal{P}_1 = \{1, 2, \ldots, p^a - 1\}, \mathcal{P}_2 = \{p^a, \ldots, p^s\}.$$

The idea is to develop $z_j^u = \left(\sum_{\ell \in \mathcal{L}_a} y_j^\ell Q_\ell(x_j^\ell) \right)^u$ with Newton multinomial. The development is performed slightly differently whether $u \in \mathcal{P}_1$ or $u \in \mathcal{P}_2$ (see Appendix A.3). In both cases, we end up with a result of the form $z_j^u = \sum_{u_x, u_y} \alpha_{u_x, u_y} y_j^{u_y} x_j^{u_x}$, so that our sum writes:

$$\sum_{j=0}^{n-1} g_{i,j} z_j^u = \sum_{u_x, u_y} \left(\alpha_{u_x, u_y} \sum_{j=0}^{n-1} g_{i,j} y_j^{u_y} x_j^{u_x} \right). \tag{5}$$

Then, we apply the next lemma (proved in Appendix A.2).

Lemma 3. *Let* $\mathbf{G}_{pub}$ *be a generator matrix of a masked Wild Goppa code* $\mathscr{C}_{pub} = \mathscr{G}_q(\mathbf{x}, f(z)\Gamma^{q-1}(z))$, $\mathbf{y} = \Gamma(\mathbf{x})^{-1}$, $\mathbf{w} = f(\mathbf{x})^{-1}$ *and* $t = \deg(\Gamma(z))$. *The values of* $\mathbf{x}$, $\mathbf{y}$, *and* $\mathbf{w}$ *satisfy the following set of equations for any value of* u_x, u_y, u, b *verifying the conditions* $0 \leqslant u_y \leqslant q, 0 \leqslant u_x \leqslant u_y t - 1, 0 \leqslant u \leqslant \deg(f) - 1, b \in \{0, 1\}$ *and* $(b, u_y) \neq (0, 0)$:

$$\left\{ \sum_{j=0}^{n-1} g_{i,j} \left(w_i x_i^u\right)^b y_j^{u_y} x_j^{u_x} = 0 \mid 0 \leqslant i \leqslant k - 1 \right\}.$$

We set $b = 0$ and obtain that $\sum_{j=0}^{n-1} g_{i,j} y_j^{u_y} x_j^{u_x} = 0$ for (u_y, u_x) such that $1 \leqslant u_y \leqslant t$ and $0 \leqslant u_x \leqslant u_y t - 1$. Thus to conclude that $\sum_{j=0}^{n-1} g_{i,j} z_j^u = 0$, we check that all the couples (u_x, u_y) appearing in the sum (5) satisfy those conditions.

3.2 Recovering a Basis of the Vector Subspace

We now explain more precisely how to use the particular structure of the solution set for solving the non-linear system (4). When looking for a vector in a subspace of $\mathbb{F}_{q^m}^n$ of dimension d, then you can safely fix d coordinates arbitrarily and complete the $n - d$ so as to obtain a vector of this subspace. This corresponds to computing intersections of your subspace with d hyperplanes. With this idea, we deduce the following corollary of Theorem 2.

Corollary 4. *Let* $\mathscr{C}_{pub} = \mathscr{G}_q(\mathbf{x}, f(z)\Gamma^{q-1}(z))$ *be a masked Wild Goppa code. Let* $t = \deg(\Gamma)$, $\mathcal{W}_{q,a}(\mathbf{Z})$, *and* $\mathcal{L}_a$ *be as defined in Theorem 2. Then, we can fix* $t \times \#\mathcal{L}_a$ *variables* $\mathbf{Z}_i$ *to arbitrary values in* $\mathcal{W}_{q,a}(\mathbf{Z})$. *The system obtained has* m *solutions (counted without multiplicities), one for each sum* $\sum_{\ell \in \mathcal{L}_a} \mathrm{GRS}_t(\mathbf{x}^\ell, \mathbf{y}^\ell)^{q^e}$.

In the rest of this article, we set $\lambda_{a,t} = t \times \#\mathcal{L}_a$. Our purpose is to find a basis of one of the vector spaces $\sum_{\ell \in \mathcal{L}_a} \mathrm{GRS}_t(\mathbf{x}^\ell, \mathbf{y}^\ell)^{q^e}$. To do so, we pick $\lambda_{a,t}$ independent solutions of $\mathcal{W}_{q,a}(\mathbf{Z})$ by fixing the variables $Z_0, Z_1, \ldots, Z_{\lambda_{a,t}-1}$ in $\mathcal{W}_{q,a,t}(\mathbf{Z})$ accordingly. Namely, for $0 \leqslant i \leqslant \lambda_{a,t} - 1$, we pick one solution $\mathbf{v}^{(i)}$ among the m solutions of the system

$$\mathcal{W}_{q,a}(\mathbf{Z}) \bigcup \{Z_i = 1, Z_j = 0 \mid 0 \leqslant j \neq i \leqslant \lambda_{a,t} - 1\}.$$

Thanks to Theorem 3 and Definition 1, we know that those solutions can be written as follows, for $Q_{i,\ell} \in \mathbb{F}_{q^m}[z]$ of degree lower than t and $0 \leqslant e_i \leqslant m - 1$:

$$\mathbf{v}^{(i)} = \left(0, \ldots, 1, \ldots, 0, \sum_{\ell \in \mathcal{L}_a} y_{\lambda_{a,t}}^{q^{e_i}} Q_{i,\ell}(x_{\lambda_{a,t}}^{q^{e_i}}), \ldots\right) \in \sum_{\ell \in \mathcal{L}_a} \mathrm{GRS}_t(\mathbf{x}^\ell, \mathbf{y}^\ell)^{q^{e_i}}. \quad (6)$$

After $\lambda_{a,t}$ resolutions of $\mathcal{W}_{q,a}(\mathbf{Z})$, the solutions $\mathbf{v}^{(i)}$ are not necessarily a basis of one of the vector spaces $\sum_{\ell \in \mathcal{L}_a} \mathrm{GRS}_t(\mathbf{x}^\ell, \mathbf{y}^\ell)^{q^e}$ because the Fröbenius exponents need not be identical for all $\mathbf{v}^{(i)}$'s. We explain in the next paragraph why this is not an issue in practice.

Simplication: Fröbenius Alignment. Let $\{\mathbf{v}^{(i)}\}_{0 \leqslant i \leqslant \lambda_{a,t}-1}$ be as defined in (6). We can suppose without loss of generality that $q_0 = q_1 = \ldots = q_{\lambda_{a,t}-1}$. This simplification requires less than $m^{(\lambda_{a,t}-1)}$ Fröbenius evaluations on the solutions. Indeed, $\mathbf{v} \in \sum_{\ell \in \mathcal{L}_a} \mathrm{GRS}_t(\mathbf{x}^\ell, \mathbf{y}^\ell)^{q^e}$, implies that $\mathbf{v}^q \in \sum_{\ell \in \mathcal{L}_a} \mathrm{GRS}_t(\mathbf{x}^\ell, \mathbf{y}^\ell)^{q^{e+1}}$. For the parameters considered in [6,8], m and t are rather small, making the cost of the Fröbenius alignment negligible. In the rest of this article, we assume that $q_0 = \ldots = q_{\lambda_{a,t}-1} = 0$, which is not a stronger assumption since the private elements of $\mathscr{C}_{pub}$ are already defined up to Fröbenius endomorphism.

Example 1. Pick for instance $q = 8$ and solve the system $\mathcal{W}_{q,a}$ with $a = 2$. Thanks to Theorem 2, after re-alignment of the Fröbenius exponents, we have a basis of the vector space $\mathrm{GRS}_t(\mathbf{x}, \mathbf{y}) + \mathrm{GRS}_t(\mathbf{x}^2, \mathbf{y}^2)$, that is:

$$\left\{ (y_i Q(x_i) + y_i^2 R(x_i^2))_{0 \leqslant i \leqslant n-1} | Q, R \in \mathbb{F}_{q^m}[z], \deg(Q), \deg(R) \leqslant t - 1 \right\}.$$

4   Recovering the Secret Key from a Sum of GRS – A Linear Algebra Step

Once we know a basis $(\mathbf{v}^{(i)})_{0 \leqslant i \leqslant \lambda_{a,t}-1}$ of $\sum_{\ell \in \mathcal{L}_a} \mathrm{GRS}_t(\mathbf{x}^\ell, \mathbf{y}^\ell)$, we aim at recovering the basis of a single GRS code. This *disentanglement* is done in Paragraph 4.1. Then, we show in 4.2 how to recover a private support $\mathbf{x}$ and Goppa polynomial $\Gamma(z)$ of the masked Wild Goppa code. This is the full description of a plain Wild Goppa code. In the Incognito case $(\deg(f) > 0)$, we explain in 4.3 that an extra linear step enables to find f. To sum up, the purpose of this section is to prove the following theorem.

Theorem 5. *Let $q = p^s$ (p prime and $s \geqslant 0$). Let $\mathbf{G}_{pub} = (g_{i,j})_{\substack{0 \leqslant i \leqslant n-1 \\ 0 \leqslant j \leqslant k-1}}$ be a generator matrix of a masked Wild Goppa code $\mathscr{C}_{pub} = \mathscr{G}_q(\mathbf{x}, f(z)\Gamma^{q-1}(z))$. Let $\mathbf{y} = \Gamma(\mathbf{x})^{-1}$, $t = \deg(\Gamma)$, and*

$$\mathcal{V} = \left(\sum_{\ell \in \mathcal{L}_a} \mathrm{GRS}_t(\mathbf{x}^\ell, \mathbf{y}^\ell) \right) \quad \text{where } \mathcal{L}_a = \bigcup_{r=0}^{s-1-a} \{p^r, 2p^r, \ldots, (p-1)p^r\} \cup \{p^{s-a}\}.$$

Once $\mathcal{V}$ is given, we can recover in polynomial-time a support $\mathbf{x}'$ and polynomials $f'(z), \Gamma'(z) \in \mathbb{F}_{q^m}[z]$ such that $\mathscr{C}_{pub} = \mathscr{G}_q(\mathbf{x}', f'(\Gamma')^{q-1})$. Stated differently, we can recover in polynomial-time a key $(\mathbf{x}', \Gamma', f')$ equivalent to the secret-key as soon as the system (4) *has been solved.*

4.1   Disentanglement of the System Solutions

The Sidelnikov-Shestakov [26] attack is a well known attack against McEliece schemes instantiated with GRS codes [22]. In our case, we can have a sum of GRS codes. In this situation, it seems not possible to apply directly [26] (because the vectors of $\sum_{\ell \in \mathcal{L}_a} \mathrm{GRS}_t(\mathbf{x}^\ell, \mathbf{y}^\ell)$ do not have the desired form; that is $(y_0 Q(x_0), \ldots, y_{n-1} Q(x_{n-1}))$). To overcome this issue, we propose to use well-chosen intersections to recover a basis suitable for Sidelnikov-Shestakov. To gain intuition, we provide a small example.

Example 2. We continue with the example 1. By squaring all the elements of $\mathrm{GRS}_t(\mathbf{x}, \mathbf{y}) + \mathrm{GRS}_t(\mathbf{x}^2, \mathbf{y}^2)$, we have a basis of $\mathrm{GRS}_t(\mathbf{x}^2, \mathbf{y}^2) + \mathrm{GRS}_t(\mathbf{x}^4, \mathbf{y}^4)$:

$$\left\{ (y_i^2 Q(x_i^2) + y_i^4 R(x_i^4))_{0 \leqslant i \leqslant n-1} \mid Q, R \in \mathbb{F}_{q^m}[z], \deg(Q), \deg(R) \leqslant t-1 \right\}.$$

We prove in Proposition 6 that, in charac. 2,

$$\left(\mathrm{GRS}_t(\mathbf{x}, \mathbf{y}) + \mathrm{GRS}_t(\mathbf{x}^2, \mathbf{y}^2)\right) \cap \left(\mathrm{GRS}_t(\mathbf{x}^2, \mathbf{y}^2) + \mathrm{GRS}_t(\mathbf{x}^4, \mathbf{y}^4)\right) = \mathrm{GRS}_t(\mathbf{x}^2, \mathbf{y}^2).$$

Hence, we have a basis of $\mathrm{GRS}_t(\mathbf{x}^2, \mathbf{y}^2)$.

Our general method to disentangle the solutions is proved in characteristic 2, but for other characteristics we need the following assumption:

Assumption 1. *Let $q = p^s$ with p prime. Let $\mathbf{x} \in \mathbb{F}_{q^m}^n$ be a support and $\mathbf{y} \in \mathbb{F}_{q^m}^n$ be defined by $\mathbf{y} = \Gamma(\mathbf{x})^{-1}$ for some polynomial $\Gamma(z) \in \mathbb{F}_{q^m}[z]$ of degree t. Let $\mathcal{L}$ and $\mathcal{L}'$ be two subsets of $\{1, \ldots, q\}$ with $(\#\mathcal{L} + \#\mathcal{L}')t < n$. Then, we have that:*

$$\left(\sum_{\ell \in \mathcal{L}} \mathrm{GRS}_t(\mathbf{x}^\ell, \mathbf{y}^\ell)\right) \cap \left(\sum_{\ell \in \mathcal{L}'} \mathrm{GRS}_t(\mathbf{x}^\ell, \mathbf{y}^\ell)\right) = \sum_{\ell \in \mathcal{L} \cap \mathcal{L}'} \mathrm{GRS}_t(\mathbf{x}^\ell, \mathbf{y}^\ell).$$

For the specific subsets $\mathcal{L}$ that we encountered, this assumption is rigorously proved in characteristic 2 (see Proposition 6). For bigger characteristics, though we could not find a formal proof, we launched more than $100,000$ experiments and found out that equality held in all cases. Now we generalize the method of intersection of codes proposed in Example 2.

Proposition 6. *Let $q = p^s$ (p prime and $s \geqslant 0$). Let also $a, 0 < a \leqslant s$, and $\mathcal{L}_a = \bigcup_{0 \leqslant r \leqslant s-1-a} \{p^r, 2p^r, \ldots, (p-1)p^r\} \cup \{p^{s-a}\}$. Then:*

$$\sum_{\ell \in \mathcal{L}_a} \mathrm{GRS}_t(\mathbf{x}^\ell, \mathbf{y}^\ell) \cap \left(\sum_{\ell \in \mathcal{L}_a} \mathrm{GRS}_t(\mathbf{x}^\ell, \mathbf{y}^\ell)\right)^{p^{(s-a)}} = \mathrm{GRS}_t(\mathbf{x}^{p^{s-a}}, \mathbf{y}^{p^{s-a}}).$$

Proof. Let $\Phi : (m_0, \ldots, m_{n-1}) \in \mathbb{F}_{q^m}^n \mapsto (m_0^{p^{s-a}}, \ldots, m_{n-1}^{p^{s-a}})$. First, remark that, as p^{s-a} is a power of the characteristic, it holds that $\Phi(\mathrm{GRS}_t(\mathbf{x}^\ell, \mathbf{y}^\ell)) = \mathrm{GRS}_t(\mathbf{x}^{p^{s-a}\ell}, \mathbf{y}^{p^{s-a}\ell})$ for all ℓ, and $\Phi(\sum_{\ell \in \mathcal{L}_a} \mathrm{GRS}_t(\mathbf{x}^\ell, \mathbf{y}^\ell)) = \sum_{\ell \in \Phi(\mathcal{L}_a)} \mathrm{GRS}_t(\mathbf{x}^\ell, \mathbf{y}^\ell)$. When $p = 2$, we fully prove the proposition in Appendix A.4. Otherwise (when $p > 2$), we rely on Assumption 1 with the sets $\mathcal{L}_a$ and

$$\Phi(\mathcal{L}_a) = \bigcup_{s-a \leqslant r \leqslant 2(s-a)-1} \{p^r, 2p^r, \ldots, (p-1)p^r\} \cup \left\{p^{2(s-a)}\right\}.$$

Then, we have $\mathcal{L}_a \cap \Phi(\mathcal{L}_a) = \{p^{s-a}\}$, and the desired equality. $\qquad\square$

Once a basis of $\mathrm{GRS}_t(\mathbf{x}^{p^{s-a}}, \mathbf{y}^{p^{s-a}})$ is known, we recover $\mathbf{x}, \mathbf{y}$ and $\Gamma(z)$ thanks to a variant of Sidelnikov-Shestakov described below.

4.2   Sidelnikov-Shestakov Adapted to Recover the Goppa Polynomial

In our attack, we have to adapt the classical Sidelnikov-Shestakov attack for special GRS codes, namely those for which there is an additional polynomial relation between the support and the multipliers.

Proposition 7. *Let $\mathbf{x}$ be an n-tuple $(x_0, \ldots, x_{n-1})$ of distinct elements of $\mathbb{F}_{q^m}$ and $\Gamma(z) \in \mathbb{F}_{q^m}[z]$ be a squarefree polynomial of degree t such that $\Gamma(x_i) \neq 0$, for all $i, 0 \leqslant i \leqslant n - 1$ Let $\mathbf{G}_{\mathrm{GRS}}$ be the generator matrix of a GRS code $\mathrm{GRS}_t(\mathbf{x}, \Gamma(\mathbf{x})^{-1})$. There is a polynomial-time algorithm which allows to recover a n-tuple $\mathbf{x}' = (x'_0, \ldots, x'_{n-1})$ of distinct elements of $\mathbb{F}_{q^m}$ and a squarefree polynomial $\Gamma'(z) \in \mathbb{F}_{q^m}[z]$ of degree t such that $\Gamma'(x'_i) \neq 0$, for all $i, 0 \leqslant i \leqslant n - 1$ and $\mathrm{GRS}_t(\mathbf{x}, \Gamma(\mathbf{x})^{-1}) = \mathrm{GRS}_t(\mathbf{x}', \Gamma'(\mathbf{x}')^{-1})$.*

This problem is very close to the one addressed in [26]. The only issue is that the homographic transformation on the support used in the original attack indeed preserves the GRS structure but not the polynomial link. Thus, polynomial interpolation over $\mathbf{x}$ and $\mathbf{y}^{-1}$ is not possible. We propose to avoid this homographic transformation by considering a well chosen *extended code*.

Definition 5. *Let $\mathscr{C}$ be a linear code of length n over $\mathbb{F}_q$. The extended code of $\mathscr{C}$, denoted by $\widetilde{\mathscr{C}}$, is a code of length $n + 1$ obtained by adding to each codeword $\mathbf{m} = (m_0, \ldots, m_{n-1})$ the coordinate $-\sum_{j=0}^{n-1} m_j$.*

Our algorithm, proved in the full version of this paper, is then the following.

Algorithm 1. Extended Version of Sidelnikov-Shestakov algorithm

---

INPUT : $\mathbf{G}_{GRS}$ generator matrix of $\mathscr{C}_{GRS} = \mathrm{GRS}_t(\mathbf{x}, \mathbf{y})$, with $\mathbf{y} = \Gamma(\mathbf{x})^{-1}$ $(\deg(\Gamma) = t)$
OUTPUT : Secret $\mathbf{x}$, $\mathbf{y}$, and $\Gamma(z)$

1: Build $\mathbf{P} = (p_{i,j})_{\substack{0 \leqslant i \leqslant n-t-1 \\ 0 \leqslant j \leqslant n-1}}$ a generator matrix of the dual of $\mathscr{C}_{GRS}$.

2: Deduce $\widetilde{\mathbf{P}}$ a matrix of the extended code (Definition 5) of the code spanned by $\mathbf{P}$

3: Build $(\mathbf{I}_t | \mathbf{U})$, with $\mathbf{U} = (u_{i,j})_{\substack{0 \leqslant i \leqslant t \\ t+1 \leqslant j \leqslant n}}$ a parity-check matrix of the code spanned by $\widetilde{\mathbf{P}}$ in systematic form

4: Solve the linear system with unknowns X_i's to find $\mathbf{x}$

$$\left\{ \frac{u_{i,j}}{u_{i',j}}(X_{i'} - X_j) = \frac{u_{i,n}}{u_{i',n}}(X_i - X_j) \mid 0 \leqslant i, i' \leqslant t, t+1 \leqslant j \leqslant n-1 \right\}.$$

5: Solve the linear system with unknowns Y_i's to find $\mathbf{y}$ (the x_i's were found at previous step)

$$\left\{ \sum_{i=0}^{n-1} p_{j,i} x_i^{\ell} Y_i = 0 \mid 0 \leqslant j \leqslant n-t-1, 0 \leqslant \ell \leqslant t-1 \right\}.$$

6: Interpolate $\Gamma(z)$ from $\mathbf{x}$ and $\mathbf{y}^{-1}$

4.3   Recovery of the Incognito Polynomial by Solving a Linear System

An extra step is necessary in the Incognito case to recover the other factor f of the Goppa polynomial. To do so, we recover the multipliers associated to f, that is the vector $\mathbf{w} = f(\mathbf{x})^{-1}$. Then, we perform polynomial interpolation. We note that once $\mathbf{x}$ and $\mathbf{y} = \Gamma(\mathbf{x})^{-1}$ are known, many of the equations of Lemma 3 become linear in $\mathbf{w}$. Namely,

$$\bigcup_{u_y=1}^{q} \left\{ \sum_{j=0}^{n-1} g_{i,j} w_i \left(y_j^{u_y} x_j^{u_x} \right) = 0 \mid 0 \leqslant i \leqslant k-1, 0 \leqslant u_x \leqslant u_y t + \deg(f) - 1 \right\}.$$

In practice, we observed that the linear system obtained has a rank defect and is not sufficient to find $\mathbf{w}$. However, we can also use the fact that $\mathscr{C}_{pub} \subset \mathscr{G}(\mathbf{x}, f(z))$ to prove that

$$\sum_{j=0}^{n-1} g_{i,j} w_i x_i^{\deg(f)} = \frac{1}{\mathrm{LC}(f)} \left(\sum_{j=0}^{n-1} g_{i,j} \right).$$

(This is rigorously done in the full version of this article.) Since $\mathbf{x}$ is known and setting $\mathrm{LC}(f) = 1$, we obtain new linear equations in the components of $\mathbf{w}$. Putting all the linear equations together, experiments show then that we obtain a unique solution $\mathbf{w}$, and f by polynomial interpolation.

5   Weakness of Non-prime Base Fields

The most (computationally) difficult part of our attack against Wild McEliece Incognito is to solve the algebraic system defined in Theorem 2. In this part, we aim at giving a better idea of the complexity of resolution by determining the exact number of "free variables" in the system. Namely, we show that we can eliminate many variables thanks to linear equations. The system $\mathcal{W}_{q,a}(\mathbf{Z}) = \bigcup_{u \in \mathcal{P}_a} \left\{ \sum_{j=0}^{n-1} g_{i,j} Z_j^u = 0 \mid 0 \leqslant i \leqslant k-1 \right\}$ of Theorem 2 obviously contains k linear equations by picking $u = 1$ ($1 \in \mathcal{P}_a$ by definition). We can easily derive other linear equations by applying the additive map $z \mapsto z^{(q^m/p^u)}$ to all the equations in degree p^u. As the solutions lie in $\mathbb{F}_{q^m}$, it holds that $(Z_j^{p^u})^{q^m/p^u} = Z_j$, and for $0 \leqslant i \leqslant k-1$

$$\left(\sum_{j=0}^{n-1} g_{i,j} Z_j^{p^u} \right)^{q^m/p^u} = \sum_{j=0}^{n-1} g_{i,j}^{q^m/p^u} Z_j = 0.$$

However, we observed that those linear equations were very redundant. To explain those linear dependencies, we found out a property of the masked Wild Goppa codes $\mathscr{G}_q(\mathbf{x}, f(z)\Gamma(z)^q)$ (Theorem 8). Namely, by simple operations on their generator matrices, we can build a generator matrix of the code $\mathscr{G}_p(\mathbf{x}, \Gamma(z)^p)$ over $\mathbb{F}_p$. This latter matrix allows to write many *independent* linear equations implying the private elements of $\mathscr{C}_{pub}$.

Theorem 8. *Let $q = p^s$ (p prime, and $s > 0$). Let $\mathbf{G}_{pub} = (g_{i,j})_{\substack{0 \leqslant i \leqslant n-1 \\ 0 \leqslant j \leqslant k-1}}$ be a generator matrix of a masked Wild Goppa code $\mathscr{C}_{pub} = \mathscr{G}_q(\mathbf{x}, f(z)\Gamma(z)^{q-1})$. We consider the scalar restriction of $\mathbf{m} \in \mathscr{C}_{pub} \subseteq \mathbb{F}_q^n$ into $\mathbb{F}_p^s$. This yields s components $\mathbf{m}^{(0)}, \ldots, \mathbf{m}^{(s-1)} \in \mathbb{F}_p^n$ (we write each $\mathbf{m} \in \mathbb{F}_q^n$ over a $\mathbb{F}_p$-basis, i.e. $\mathbf{m} = \mathbf{m}^{(0)}\theta_0 + \cdots + \mathbf{m}^{(s-1)}\theta_{s-1}$). Let $\mathscr{C}^{\mathbb{F}_p} \subseteq \mathbb{F}_p^n$ be the code generated by the coordinate vectors $\mathbf{m}^{(0)}, \ldots, \mathbf{m}^{(s-1)}$ for all the codewords $\mathbf{m} \in \mathscr{C}_{pub}$. Then, it holds that*

$$\mathscr{C}^{\mathbb{F}_p} \subseteq \mathscr{G}_p(\mathbf{x}, \Gamma(z)^p).$$

The proof can be found in the full version of this paper. In practice, we observed equality in the inclusion provided $s \dim(\mathscr{C}_{pub}) > \dim(\mathscr{G}_p(\mathbf{x}, \Gamma(z)^p))$. Note that $\mathscr{G}_p(\mathbf{x}, \Gamma(z)^p)$ is a Wild Goppa code with the same private elements $\mathbf{x}$ and $\mathbf{y} = \Gamma(\mathbf{x})^{-1}$ as $\mathscr{C}_{pub}$. This provides extra equations on the variables $\mathbf{Z}$ of $\mathcal{W}_{q,a}(\mathbf{Z})$ (proved in the full version):

Proposition 9. *Let $\mathscr{C}_{pub} = \mathscr{G}_q(\mathbf{x}, f(z)\Gamma^{q-1}(z))$ and $\mathcal{W}_{q,a}(\mathbf{Z})$ the associated system for $1 \leqslant a \leqslant s$. Let $\tilde{\mathbf{G}}_{\mathbb{F}_p} = (\tilde{g}_{i,j})_{\substack{0 \leqslant i \leqslant n-1 \\ 0 \leqslant j \leqslant k_p-1}}$ be a generator matrix of $\mathscr{G}_p(\mathbf{x}, \Gamma(z)^p)$ (with $k_p = \dim(\mathscr{G}_p(\mathbf{x}, \Gamma(z)^p))$). Then, the solutions of $\mathcal{W}_{q,a}(\mathbf{Z})$ satisfy:*

$$\bigcup_{\ell=0}^{p-1} \left\{ \sum_{j=0}^{n-1} \tilde{g}_{i,j} Z_j = 0 \mid 0 \leqslant u \leqslant t-1, 0 \leqslant i \leqslant k_p - 1 \right\}.$$

As $k_p \geqslant n - (p-1)mst$ (and in practice $k_p = n - (p-1)mst$), we have the following corollary.

Corollary 10. *The knowledge of $\mathbf{G}_{pub}$ gives access to $n-(p-1)mst$ independent linear relations between the Z_i's. The system $\mathcal{W}_{q,a}(\mathbf{Z})$ contains (at most) $(p-1)mst$ free variables.*

Remark 2. The number of "free" variables given in Corollary 10 is given without taking into account the vector space structure of the solutions. Thanks to Corollary 4, we know that $\lambda_{a,t}$ extra variables can be fixed to arbitrary values in $\mathcal{W}_{q,a}(\mathbf{Z})$.

For a Goppa polynomial of same degree, but without multiplicities, the number of free variables in the system would be $n-k \geqslant (p^s-1)mt$ instead of $(p-1)mst$. In particular, for a masked code, the number of variables describing it does not depend on the degree of the incognito polynomial f and the attack is not harder for masked codes. This explains why the codes defined over non-prime fields are the weakest ones.

6   Practical Experiments

We report below various experimental results performed with our attack on various parameters for which [6] said that strength is "unclear" and that an attack would not be a "surprise" but for which no actual attack was known. We also generated our own keys/parameters to see how the attack scales. We performed our experiments with off-the-shelf tools (MAGMA [9] V2.19-1) and using a 2.93 GHz Intel PC with 128 Gb. of RAM. As the polynomial system solving is by far the most costly step, we give timings only for this one. We performed it using the F_4 algorithm ([12]) of MAGMA. As explained in Section 4, it is necessary to solve the systems $\mathcal{W}_{q,a}(\mathbf{Z})$ a number of times equal to the dimension of the vector space of the solutions (Theorem 2). These resolutions are completely independent and can be executed in parallel. This is why we give the timings under the form (number of separate resolutions) × (time for one resolution). By #$\mathbf{Z}$, we denote the number of free variables remaining in the system after cleaning up the linear equations (Corollary 10) and fixing coordinates thanks to the vector space structure of the solutions (Corollary 4). The general formula is #$\mathbf{Z} = ((p-1)ms - \#\mathcal{L}_a)\,t$ for $q = p^s$ and $s > 1$.

In the experiments, we tried various parameters a for the systems $\mathcal{W}_{q,a}(\mathbf{Z})$. We give a comparison on some examples in Table 1 (the system $\mathcal{W}_{q,a}(\mathbf{Z})$ with $a = s$ can be solved in a reasonable amount of time in actually few cases).

Table 1. Comparison of the resolution times of $\mathcal{W}_{q,a}(\mathbf{Z})$ for various possible a's. The smallest possible a gives the best timings.

q	m	t	n	k	$\deg(f)$	Solving $\mathcal{W}_{q,a}(\mathbf{Z})$ with $a = s$	Solving $\mathcal{W}_{q,a}(\mathbf{Z})$, **optimal** a
32	2	2	678	554	0	2 × 12s (#$\mathbf{Z}$ = 18)	8× 0.08 s ($a = 2$, #$\mathbf{Z}$ = 9)
32	2	1	532	406	32	2 × 49s (#$\mathbf{Z}$ = 9)	4× 0.02 s ($a = 2$, #$\mathbf{Z}$ = 6)
32	2	3	852	621	24	3× (30 min 46s) (#$\mathbf{Z}$ = 37)	12× 0.6 s ($a = 2$, #$\mathbf{Z}$ = 18)
27	3	3	1312	1078	0	3× (3h 10 min) (#$\mathbf{Z}$ = 51)	15× 3.0 s ($a = 1$, #$\mathbf{Z}$ = 39)

It appeared that a should be chosen so as to maximize the dimension of the solution set (Theorem 2). This choice minimizes the number of variables. Namely, the best choice is to set $a = 1$ when $p > 2$. When $p = 2$, setting $a = 1$ would yield only "linear" equations (of degree $2^u, u \leqslant s$). So, we set $a = 2$ and the systems $\mathcal{W}_{2^u,2}(\mathbf{Z})$ contain only cubic equations. We recall that for $a = s$, Assumption 1 is not necessary, whereas we rely on it when $a < s$ and $p \neq 2$. In the rest of the experiments, we always pick the best choice for a.

In Table 2, we present experimental results performed with Wild McEliece (when $\deg(f) = 0$) and Incognito ($\deg(f) > 0$) parameters. For Wild McEliece, all the parameters in the scope of our attack were quoted in [6, Table 7.1] with the international biohazard symbol ☣. The reason is that, for those parameters, enumerating all the possible Goppa polynomials is computationaly feasible. In the current state of the art, to apply the SSA attack ([18]), one would not only have to enumerate the irreducible polynomials of $\mathbb{F}_{q^m}[z]$, but also all the possible support sets, as the support-splitting algorithm uses the support set as input. This introduces a factor $\binom{q^m}{n}$ in the cost of SSA, chosen by the designers in order to make the attack infeasible. However, the authors of [6] conclude that, even if no attack is known against those instances, algorithmical progress in support enumeration may be possible and therefor they do not recommend their use. In the case of non-prime base fields, experiments show that our attack represents a far more serious threat for the security of some of those instances: for $q \in \{32, 27, 16\}$ we could find the secret keys of parameters with high ISD complexity. We indicate, for each set of parameters, the ISD complexity (obtained thanks to Peters' software[1]), as it remains the reference to evaluate the security of a McEliece scheme. We also give the complexity of an SSA attack, which is in the current state-of-the-art $\binom{q^m}{n} \cdot q^{mt}/t$.

Regarding Wild McEliece Incognito, we broke the parameters indicated with a security of 2^{128} in [8, Table 5.1] for $q \in \{32, 27, 16\}$. For some other non-prime base fields, we give the hardest parameters in the scope of our attack in roughly one day of computation. Note that here, SSA complexity is given by $\binom{q^m}{n} \cdot (q^{m(t+s)}/(ts))$.

For the sake of completeness, we also include in Tables 2 Wild McEliece schemes with a quadratic extension. In [11], the authors already presented a poly-time attack in this particular case: it applies for the parameters with $q = 32$, but not for the other ones. We want to stress that our attack also works for $m = 2$ and any t ([11] does not work in the extreme case $t = 1$). Also, we emphasize that, whilst solving a non-linear system, our attack is actually faster than [11] in some cases. For $q = 32$ and $t = 4$, the attack of [11] requires 49.5 minutes (using a non-optimized MAGMA implementation according to the authors). We can mount our attack in several seconds with the techniques of this paper.

Table 2. Practical experiments with Wild McEliece & Incognito parameters. ISD complexity is obtained thanks to Peters' software[1]. SSA attack complexity is given under the form (support enumeration)·(Goppa polynomial enumeration).

q	m	t	n	k	$\deg(f)$	Key (kB)	ISD	SSA	Solving $\mathcal{W}_{q,a}(\mathbf{Z})$, optimal a
32	2	4	841	601	0	92	2^{128}	$2^{688} \cdot 2^{38}$	16× 10 s (#$\mathbf{Z} = 36$)
32	2	5	800	505	0	93	2^{136}	$2^{771} \cdot 2^{48}$	20× (2 min 45s) (#$\mathbf{Z} = 40$)
27	3	3	1312	1078	0	45	2^{113}	$2^{6947} \cdot 2^{41}$	15× 3.0 s (#$\mathbf{Z} = 39$)
27	3	4	1407	1095	0	203	2^{128}	$2^{7304} \cdot 2^{55}$	20× (6 min 34 s) (#$\mathbf{Z} = 52$)
27	3	5	1700	1310	0	304	2^{158}	$2^{8343} \cdot 2^{69}$	25× (1h 59 min) (#$\mathbf{Z} = 65$)
27	3	5	1800	1410	0	327	2^{160}	$2^{8679} \cdot 2^{69}$	25× (1h 37 min) (#$\mathbf{Z} = 65$)
16	3	6	1316	1046	0	141	2^{129}	$2^{3703} \cdot 2^{69}$	18× (36h 26 min) (#$\mathbf{Z} = 54$)
32	2	3	852	621	24	90	2^{130}	$2^{663} \cdot 2^{273}$	12× 0.6 s (#$\mathbf{Z} = 18$)
27	3	2	1500	1218	42	204	2^{128}	$2^{5253} \cdot 2^{225}$	10× 0.9 s (#$\mathbf{Z} = 26$)
25	3	3	1206	915	25	155	2^{117}	$2^{7643} \cdot 2^{632}$	15× (1h 2 min) (#$\mathbf{Z} = 57$)
16	3	6	1328	1010	16	160	2^{125}	$2^{3716} \cdot 2^{265}$	18× (36h 35 min) (#$\mathbf{Z} = 54$)
9	3	6	728	542	14	40	2^{81}	$2^{2759} \cdot 2^{191}$	18× (25h 13 min) (#$\mathbf{Z} = 54$)

---

[1] Available at http://christianepeters.wordpress.com/publications/tools/

7   Conclusion and Future Work

In practice, we could not solve (in less than two days) the algebraic systems involved when the number of free variables $\#\mathbf{Z}$ exceeds 65. We recall the relation $\#\mathbf{Z} = ((p-1)ms - \#\mathcal{L}_a)t$ (for $q = p^s$ and $s > 1$), which should help the designers to scale their parameters. An important remaining open question is to give a precise complexity estimates for the polynomial system solving phase in those cases.

Acknowledgements. This work was supported in part by the HPAC grant (ANR ANR-11-BS02-013) of the French National Research Agency. The authors would also like to thank (some of) the referees as well as PC chairs for their useful comments on a preliminary version of this paper.

References

1. Barbier, M., Barreto, P.S.L.M.: Key reduction of McEliece's cryptosystem using list decoding. In: Kuleshov, A., Blinovsky, V., Ephremides, A. (eds.) 2011 IEEE International Symposium on Information Theory Proceedings, ISIT 2011, St, St. Petersburg, Russia, July 31 - August 5, pp. 2681–2685. IEEE (2011)

2. Barreto, P.S.L.M., Lindner, R., Misoczki, R.: Monoidic codes in cryptography. In: Yang (ed.) [27], pp. 179–199

3. Becker, A., Joux, A., May, A., Meurer, A.: Decoding random binary linear codes in $2^{n/20}$: How $1 + 1 = 0$ improves information set decoding. In: Pointcheval, D., Johansson, T. (eds.) EUROCRYPT 2012. LNCS, vol. 7237, pp. 520–536. Springer, Heidelberg (2012)

4. Berger, T.P., Cayrel, P.-L., Gaborit, P., Otmani, A.: Reducing key length of the McEliece cryptosystem. In: Preneel, B. (ed.) AFRICACRYPT 2009. LNCS, vol. 5580, pp. 77–97. Springer, Heidelberg (2009)

5. Bernstein, D.J., Lange, T., Peters, C.: Attacking and defending the McEliece cryptosystem. In: Buchmann, J., Ding, J. (eds.) PQCrypto 2008. LNCS, vol. 5299, pp. 31–46. Springer, Heidelberg (2008)

6. Bernstein, D.J., Lange, T., Peters, C.: Wild McEliece. In: Biryukov, A., Gong, G., Stinson, D.R. (eds.) SAC 2010. LNCS, vol. 6544, pp. 143–158. Springer, Heidelberg (2011)

7. Bernstein, D.J., Lange, T., Peters, C.: Smaller decoding exponents: Ball-collision decoding. In: Rogaway, P. (ed.) CRYPTO 2011. LNCS, vol. 6841, pp. 743–760. Springer, Heidelberg (2011)

8. Bernstein, D.J., Lange, T., Peters, C.: Wild McEliece incognito. In: Yang (ed.) [27], pp. 244–254

9. Bosma, W., Cannon, J.J., Playoust, C.: The Magma algebra system I: The user language. Journal of Symbolic Computation 24(3-4), 235–265 (1997)

10. Canteaut, A., Chabaud, F.: A new algorithm for finding minimum-weight words in a linear code: Application to McEliece's cryptosystem and to narrow-sense BCH codes of length 511. IEEE Transactions on Information Theory 44(1), 367–378 (1998)

11. Couvreur, A., Otmani, A., Tillich, J.-P.: Polynomial time attack on wild McEliece over quadratic extensions. In: Nguyen, P.Q., Oswald, E. (eds.) EUROCRYPT 2014. LNCS, vol. 8441, pp. 17–39. Springer, Heidelberg (2014)
12. Faugère, J.-C.: A new efficient algorithm for computing gröbner bases (F4). Journal of Pure and Applied Algebra 139(1-3), 61–88 (1999)
13. Faugère, J.-C., Otmani, A., Perret, L., de Portzamparc, F., Tillich, J.-P.: Structural cryptanalysis of McEliece schemes with compact keys. IACR Cryptology ePrint Archive, 2014:210 (2014)
14. Faugère, J.-C., Otmani, A., Perret, L., Tillich, J.-P.: Algebraic cryptanalysis of Mceliece variants with compact keys. In: Gilbert, H. (ed.) EUROCRYPT 2010. LNCS, vol. 6110, pp. 279–298. Springer, Heidelberg (2010)
15. Faugère, J.-C., Otmani, A., Perret, L., Tillich, J.-P.: Algebraic Cryptanalysis of McEliece variants with compact keys – toward a complexity analysis. In: SCC 2010: Proceedings of the 2nd International Conference on Symbolic Computation and Cryptography, pp. 45–55. RHUL (June 2010)
16. Heyse, S.: Implementation of McEliece based on quasi-dyadic Goppa codes for embedded devices. In: Yang (ed.) [27], pp. 143–162
17. Leon, J.S.: A probabilistic algorithm for computing minimum weights of large error-correcting codes. IEEE Transactions on Information Theory 34(5), 1354–1359 (1988)
18. Loidreau, P., Sendrier, N.: Weak keys in the McEliece public-key cryptosystem. IEEE Transactions on Information Theory 47(3), 1207–1211 (2001)
19. May, A., Meurer, A., Thomae, E.: Decoding random linear codes in $\widetilde{O}(2^{0.054n})$. In: Lee, D.H., Wang, X. (eds.) ASIACRYPT 2011. LNCS, vol. 7073, pp. 107–124. Springer, Heidelberg (2011)
20. McEliece, R.J.: A Public-Key System Based on Algebraic Coding Theory, pp. 114–116. Jet Propulsion Lab (1978), DSN Progress Report 44
21. Misoczki, R., Barreto, P.S.L.M.: Compact McEliece keys from Goppa codes. In: Jacobson Jr., M.J., Rijmen, V., Safavi-Naini, R. (eds.) SAC 2009. LNCS, vol. 5867, pp. 376–392. Springer, Heidelberg (2009)
22. Niederreiter, H.: Knapsack-type cryptosystems and algebraic coding theory. Problems Control Inform. Theory 15(2), 159–166 (1986)
23. Persichetti, E.: Compact McEliece keys based on quasi-dyadic srivastava codes. J. Mathematical Cryptology 6(2), 149–169 (2012)
24. Peters, C.: Information-set decoding for linear codes over $\mathbb{F}_q$. In: Sendrier, N. (ed.) PQCrypto 2010. LNCS, vol. 6061, pp. 81–94. Springer, Heidelberg (2010)
25. Sendrier, N.: Finding the permutation between equivalent linear codes: The support splitting algorithm. IEEE Transactions on Information Theory 46(4), 1193–1203 (2000)
26. Sidelnikov, V., Shestakov, S.: On the insecurity of cryptosystems based on generalized Reed-Solomon codes. Discrete Mathematics and Applications 1(4), 439–444 (1992)
27. Yang, B.-Y. (ed.): PQCrypto 2011. LNCS, vol. 7071. Springer, Heidelberg (2011)

A   Appendix

A.1   A Technical Lemma

We prove a technical lemma which is useful for the proofs of Sections 3 and 4.

Lemma 1. *Let* $q = p^s$ *(p prime and $s > 0$), and* $Q = \gamma_t z^t + \cdots + \gamma_0 \in \mathbb{F}_{q^m}[z]$ *be a polynomial of degree t. For all j, it holds that:*

$$Q(z)^{p^j} = \gamma_t^{p^j}(z^t)^{p^j} + \cdots + \gamma_0^{p^j}$$
$$= \gamma_t^{p^j}(z^{p^j})^t + \cdots + \gamma_0^{p^j}$$
$$= F_{(j)}(Q)(z^{p^j}).$$

where $F_{(j)}(Q) = \gamma_t^{p^j} z^t + \cdots + \gamma_0^{p^j}$ *is the polynomial of same degree as Q obtained by raising all the coefficients to the p^j-power.*

A.2   Proof of Lemma 3

We want to prove that, under the conditions of Lemma 3 (that is $0 \leqslant u_y \leqslant q, 0 \leqslant u_x \leqslant u_y t - 1, 0 \leqslant u \leqslant \deg(f) - 1, b \in \{0, 1\}$ and $(b, u_y) \neq (0, 0)$), it holds that

$$\left\{ \sum_{j=0}^{n-1} g_{i,j} \left(w_i x_i^u \right)^b y_j^{u_y} x_j^{u_x} = 0 \mid 0 \leqslant i \leqslant k-1 \right\}.$$

Proof. The crucial remark is that, for any $\mathbf{c} \in \mathbb{F}_q^n$, $\sum_{i=0}^{n-1} \frac{c_i}{z - x_i} \equiv 0 \bmod f(z)\Gamma^q(z)$ implies $\sum_{i=0}^{n-1} \frac{c_i}{z - x_i} \equiv 0 \bmod f^b(z)\Gamma^{u_y}(z)$ for all $0 \leqslant u_y \leqslant q$ and $0 \leqslant b \leqslant 1$ (and $(u_y, b) \neq (0, 0)$). In other words, for those u_y, b, it holds that

$$\mathscr{C}_{pub} \subseteq \mathscr{G}_q(\mathbf{x}, f^b(z)\Gamma^{u_y}(z)).$$

As $\mathscr{G}_q(\mathbf{x}, f(z)^b \Gamma^{u_y}(z))$ has parity check matrix $\mathbf{V}_{d_{tot}}(\mathbf{x}, \mathbf{w}^b \mathbf{y}^{u_y})$ (with $d_{tot} = b \deg(f) + u_y t$), the matrix products $V_{d_{tot}}(\mathbf{x}, \mathbf{w}^b \mathbf{y}^{u_y}) \times \mathbf{G}_{pub}^{\mathrm{T}} = \mathbf{0}_{d_{tot} \times k}$ yield all the relations of the lemma. □

A.3   Proof of Theorem 2

Proof. We give the multinomial development of the $z_j^u = \left(\sum_{\ell \in \mathcal{L}_a} y_j^\ell Q_\ell(x_j^\ell) \right)^u$ under the form $z_j^u = \sum_{u_x, u_y} \alpha_{u_x, u_y} y_j^{u_y} x_j^{u_x}$ and show that u_x, u_y satisfy the conditions of Lemma 3. This is done separately for $u \in \mathcal{P}_1$ and $u \in \mathcal{P}_2$.

  Case $u \in \mathcal{P}_1$. We pick $u \in \{1, 2, \ldots, p^a - 1\}$ and use the multinomial formula to expand $\left(\sum_{\ell \in \mathcal{L}_a} y_j^\ell Q_\ell(x_j^\ell) \right)^u$. Namely, with $L_a = \#\mathcal{L}_a$, we have:

$$\left(\sum_{\ell \in \mathcal{L}_a} y_j^\ell Q_\ell(x_j^\ell) \right)^u = \sum_{\substack{0 \leqslant u_1, \ldots, u_L \leqslant u \\ u_1 + \ldots + u_L = u}} \left(\binom{u}{u_1, \ldots, u_L} y_j^{\left(\sum_{\ell \in \mathcal{L}_a} \ell u_\ell \right)} \prod_{\ell \in \mathcal{L}_a} Q_\ell(x_j^\ell)^{u_\ell} \right).$$

Let's look at each term $y_j^{u_y} x_j^{u_x}$ in the sum. For $u_1, \ldots, u_L$ non-negative integers with $u_1 + \ldots + u_L = u$, it holds that $u_y = \sum_{\ell \in \mathcal{L}_a} \ell u_\ell \leqslant \max(\mathcal{L}_a) \sum_{\ell \in \mathcal{L}_a} u_\ell \leqslant p^{s-a} u \leqslant p^s$. For each $y_j^{u_y}$, several terms $y_j^{u_y} x_j^{u_x}$ appear after expanding $\prod_{\ell \in \mathcal{L}_a} Q_\ell(x_j^\ell)^{u_\ell}$. In $Q_\ell(x_j^\ell)^{u_\ell}$ the maximal power u_x appearing is $\ell u_\ell(t-1)$ (as Q_ℓ has degree $t-1$). Thus, in $\prod_{\ell \in \mathcal{L}_a} Q_\ell(x_j^\ell)^{u_\ell}$, the maximal power is $(t-1) \sum_{\ell \in \mathcal{L}_a} \ell u_\ell = (t-1)u_y \leqslant t u_y - 1$.

Case $u \in \mathcal{P}_2$. We pick $b \in \{a, \ldots, s\}$. Then $z_j^{p^b} = \left(\sum_{\ell \in \mathcal{L}_a} y_j^\ell Q_\ell(x_j^\ell) \right)^{p^b} = \sum_{\ell \in \mathcal{L}_a} y_j^{\ell p^b} F_{(b)}(Q_\ell)(x_j^{\ell p^b})$ (Lemma 1). Pick $\ell \in \mathcal{L}_a$, it writes $\ell = \alpha p^c$ with $1 \leqslant \alpha < p$ and $0 \leqslant c \leqslant s - a$. Thus we have $\ell p^b = \alpha p^{c+b}$. The euclidian division of $c + b$ by s gives $c + b = ds + e$ with $0 \leqslant e < s$. The exponent ℓp^b then writes $\ell p^b = \alpha p^e p^{ds} = \alpha p^e q^d$. As $g_{i,j}^q = g_{i,j}$ it holds that $\left(\sum_{j=0}^{n-1} g_{i,j} y_j^{\alpha p^e q^d} F_{(b)}(Q_\ell)(x_j^{\alpha p^e q^d}) \right)^{q^m/q^d} = \sum_{j=0}^{n-1} g_{i,j} y_j^{\alpha p^e} F_{(b-ds)}(Q_\ell)(x_j^{\alpha p^e})$. As the $F_{(b-ds)}(R_\ell)$'s have degree lower than t, all the terms of the sum are of the form $y_j^{u_y} x_j^{u_x}$ with $u_y \leqslant q$ (since $\alpha p^e < p^s$) and $u_x \leqslant u_y t - 1$. $\square$

A.4   Proof of Proposition 6

When $p = 2$, we prove Proposition 6 without resorting to Assumption 1. We use the fact that the polynomial $\Gamma(z)$ linking $\mathbf{x}$ and $\mathbf{y}^{-1}$ is irreducible in the construction proposed in [6,8]. For $p = 2$, $\mathcal{L}_a$ is reduced to powers of 2, namely $\mathcal{L}_a = \{p^u\}_{0 \leqslant u \leqslant s-a}$. So the proof consists in showing that the intersection

$$\mathcal{I} = \left(\sum_{u=0}^{s-a} \mathrm{GRS}_t(\mathbf{x}^{p^u}, \mathbf{y}^{p^u}) \right) \cap \left(\sum_{u=s-a}^{2(s-a)} \mathrm{GRS}_t(\mathbf{x}^{p^u}, \mathbf{y}^{p^u}) \right)$$

is reduced to $\mathrm{GRS}_t(\mathbf{x}^{p^{s-a}}, \mathbf{y}^{p^{s-a}})$.

Proof. We pick $\mathbf{v} \in \mathcal{I}$. There exist polynomials $R_{p^u}, Q_{p^{s-a+u}} \in \mathbb{F}_{q^m}[z]$ (with $0 \leqslant u \leqslant s - a$) of degree lower than t such that

$$v_i = \sum_{u=0}^{s-a} y_i^{p^u} R_{p^u}(x_i^{p^u}) = \sum_{u=0}^{s-a} y_i^{p^{s-a+u}\ell} Q_{p^{s-a+u}}(x_i^{p^{s-a+u}\ell})$$

for all $0 \leqslant i \leqslant n - 1$. As $y_i = \Gamma(x_i)^{-1}$, we obtain polynomial relations in the x_i's by multiplying by $\Gamma(x_i)^{p^{2(s-a)}}$. This yields n relations,

$$\sum_{u=0}^{s-a} \Gamma(x_i)^{(p^{2(s-a)-u})} R_{p^u}(x_i^{p^u}) = \sum_{u=0}^{s-a} \Gamma(x_i)^{p^{2(s-a)-(s-a+u)}} Q_{p^{s-a+u}}(x_i^{p^{s-a+u}}).$$

We suppose here that the degree of this polynomial relation is lower than n, that is $(t-1)p^{2(s-a)} < n$, so that we can deduce the polynomial equality:

$$\sum_{u=0}^{s-a} \Gamma(z)^{(p^{2(s-a)-u})} R_{p^u}(z^{p^u}) = \sum_{u=0}^{s-a} \Gamma(z)^{p^{2(s-a)-(s-a+u)}} Q_{p^{s-a+u}}(z^{p^{s-a+u}}) \quad (7)$$

Modulo $\Gamma(z)$ all polynomials vanish but one, this yields $Q_{p^{2(s-a)}}(z^{p^{2(s-a)}}) \equiv 0 \mod \Gamma(z)$. Thanks to Lemma 1, we have $\Gamma(z)$ divides $Q_{p^{2(s-a)}}(z^{p^{2(s-a)}}) = \left(F_{(u)}(Q_{p^{2(s-a)}})(z)\right)^{p^{2(s-a)}}$ (for $u = ms - 2(s-a)$). As $\Gamma(z)$ is irreducible, this entails that $\Gamma(z)$ divides $F_{(u)}(Q_{p^{2(s-a)}})(z)$, but $F_{(u)}(Q_{p^{2(s-a)}})(z)$ has same degree as $Q_{p^{2(s-a)}}(z)$, which has degree lower than t (notations as in the proof of Theorem 2). Hence we deduce that $F_{(u)}(Q_{p^{2(s-a)}})(z) = 0$ and also its Fröbenius $Q_{p^{2(s-a)}} = 0$. Then, we look at the new relation of type (7) and start over with the polynomial $Q_{p^{2(s-a)}-1}(z^{p^{2(s-a)-1}})$. The proof that $Q_{p^{2(s-a)}-1} = 0$ is identical. One after the other, we prove that all the polynomials $R_{p^u}, Q_{p^{s-a+u}}$ are zero except the matching polynomials $R_{p^{s-a}}$ and $Q_{p^{s-a}}$ which are equal, so that $\mathbf{z} \in \mathrm{GRS}_t(\mathbf{x}^{p^{s-a}}, \mathbf{y}^{p^{s-a}})$. The problem when $p \neq 2$ is that the set $\mathcal{L}_a$ contains exponents which are not a pure power of p. $\qquad\square$

Bivariate Polynomials Modulo Composites and Their Applications

Dan Boneh and Henry Corrigan-Gibbs

Stanford University, Stanford CA 94305, USA

Abstract. We investigate the hardness of finding solutions to bivariate polynomial congruences modulo RSA composites. We establish necessary conditions for a bivariate polynomial to be one-way, second preimage resistant, and collision resistant based on arithmetic properties of the polynomial. From these conditions we deduce a new computational assumption that implies an efficient algebraic collision-resistant hash function. We explore the assumption and relate it to known computational problems. The assumption leads to (i) a new statistically hiding commitment scheme that composes well with Pedersen commitments, (ii) a conceptually simple cryptographic accumulator, and (iii) an efficient chameleon hash function.

Keywords: Algebraic curves, bivariate polynomials, cryptographic commitments, Merkle trees.

1   Introduction

In this paper, we investigate the cryptographic properties of bivariate polynomials modulo random RSA composites $N = pq$. We ask: for which integer polynomials $f \in \mathbb{Z}[x, y]$ does the function $f : \mathbb{Z}_N \times \mathbb{Z}_N \to \mathbb{Z}_N$ defined by f appear to be a one-way function, a second-preimage-resistant function, or a collision-resistant function? We say that a polynomial $f \in \mathbb{Z}[x, y]$ is one-way if the function $f : \mathbb{Z}_N \times \mathbb{Z}_N \to \mathbb{Z}_N$ defined by f is one-way (Section 3.1). We similarly define second-preimage-resistance (Section 3.2) and collision-resistance (Section 3.3) of polynomials $f \in \mathbb{Z}[x, y]$.

Using tools from algebraic geometry we develop a heuristic for deducing the cryptographic properties of a bivariate polynomial over $\mathbb{Z}_N$ from its arithmetic properties, namely from its properties as a polynomial over the rationals $\mathbb{Q}$. We give a number of necessary conditions for a bivariate polynomial to be one-way, second-preimage-resistant, or collision-resistant. We also provide examples of polynomials f that appear to satisfy each of these properties and we offer separations between these three classes.

Taking collision resistance as an example, we conjecture that a bivariate polynomial $f \in \mathbb{Z}[x, y]$ that defines an *injective* function $f : \mathbb{Q}^2 \to \mathbb{Q}$ gives a *collision resistant* function $f : \mathbb{Z}_N^2 \to \mathbb{Z}_N$ where N is a random RSA modulus of secret factorization (see Section 3.3). Constructing an explicit polynomial

P. Sarkar and T. Iwata (Eds.): ASIACRYPT 2014, PART I, LNCS 8873, pp. 42–62, 2014.

$f \in \mathbb{Z}[x, y]$ that is provably injective over the rationals is an open number theoretic problem [30]. However, even relatively simple polynomials appear to be
injective over $\mathbb{Q}^2$. For example, Don Zagier [13,34] conjectures that the polynomial $f_{\text{ZAG}}(x, y) := x^7 + 3y^7$, which we refer to as the *Zagier polynomial*, is
injective over the rationals. Since the only apparent efficient strategy for finding
collisions in f_{ZAG} over $\mathbb{Z}_N$ is to find rational collisions and reduce them modulo N,
we conjecture that f_{ZAG} is collision resistant over $\mathbb{Z}_N$. To build confidence in the
assumption that f_{ZAG} is collision resistant over $\mathbb{Z}_N$ we discuss potential collision-
finding strategies and relate them to existing number theoretic problems.

Applications. We demonstrate that the existence of low-degree collision-resistant
bivariate polynomials gives rise to very efficient instantiations of a number of
cryptographic primitives.

First, we derive a statistically hiding commitment scheme which is computationally inexpensive to evaluate and composes naturally with Pedersen commitments. By "nesting" these new commitments inside of Pedersen commitments,
we obtain an efficient zero-knowledge protocol for proving knowledge of an opening of a commitment which is nested inside of another commitment. Use of
nested commitments reduces the length of transactions in an anonymous e-cash
scheme [24] by roughly 70%.

Second, we demonstrate that the new commitment scheme, in conjunction
with Merkle trees, can serve as a simple replacement for one-way accumulators. Though the communication complexity of our accumulator construction
is asymptotically worse than that of strong-RSA accumulators [8]—$O(\log |S|)$
versus $O(1)$ for a set S being accumulated—our construction has the benefit of
being conceptually simple and easy to implement.

Third, from the same collision-resistant polynomial, we derive a new chameleon
hash function, signature scheme, claw-free permutation family, and a variable-
length algebraic hash function.

2   Related Work

Multivariate polynomials in $\mathbb{Z}_N$ have a long history in cryptography. For example, the security of the Ong-Schnorr-Shamir signature scheme [26] followed
from the hardness of finding solutions to a particular type of bivariate polynomial equation over $\mathbb{Z}_N$. Pollard and Schnorr later demonstrated a general attack
against the hardness of finding solutions to such equations [28].

Shamir related the hardness of factoring certain multivariate polynomials
modulo N to the problem of factoring the modulus N itself [33]. Schwenk and
Eisfeld proposed encryption and signature schemes reliant on the hardness of
finding roots of random *univariate* polynomials $f \in \mathbb{Z}[x]$ modulo a composite N,
and they prove that this problem is as hard as factoring N [31].

This work introduces a new statistically hiding commitment scheme based on
low-degree polynomials. Commitment schemes are used widely in cryptography.

Prior work has derived statistically hiding commitment schemes from the discrete log problem [27], the Paillier cryptosystem [12], and the RSA problem [3]. Verifying the correctness of opening a commitment in these existing schemes requires expensive modular exponentiations or elliptic curve scalar multiplications. Verifying an opening with our new commitment scheme requires just a few modular multiplications. By combining our new commitment scheme with traditional Pedersen commitments, we improve the communication efficiency of the Zerocoin decentralized e-cash construction [24].

Given a Pedersen commitment and a finite set of elements S, our commitment scheme leads to a simple zero-knowledge protocol for proving knowledge of an opening x of the commitment such that $x \in S$. The length of the proof is $\log |S|$. This technique, which uses Merkle trees [21], has applications to anonymous authentication and credential systems and it has the potential to replace traditional RSA one-way accumulators, introduced by Benaloh and De Mare [5] and revisited by Barić and Pfitzmann [4].

Camenisch and Lysyanskaya presented an efficient zero-knowledge protocol which proves that a value contained in a Pedersen commitment is also contained in a particular strong-RSA accumulator [8]. The Camenisch-Lysyanskaya accumulator produces a shorter proof of knowledge than ours does, but the conceptual simplicity and ease of implementation may make our Merkle-style proof more attractive for some applications.

The "zero-knowledge sets" of Micali, Rabin, and Kilian solve an orthogonal problem: a prover publishes a commitment to a set S and later can prove that $x \in S$ without leaking other information about S [23]. In contrast, we are interested in hiding the value x but allow the set of items S to be public.

3   Cryptographic Properties of Polynomials

We begin by surveying the cryptographic properties of integer polynomials modulo random RSA composites. Our goal is to relate the algebraic properties of polynomials to their cryptographic complexity. In particular, we identify families of integer polynomials that give rise to progressively stronger cryptographic primitives: one-way functions, second-preimage-resistant functions, and collision-resistant functions.

Notation. We write $x \xleftarrow{R} S$ to indicate that the variable x takes on a value sampled independently and uniformly at random from a finite set S. A function $f : \mathbb{Z} \to \mathbb{R}^+$ is *negligible* if it is smaller than $1/p(\lambda)$ for every polynomial $p()$ and all sufficiently large λ. We denote an arbitrary negligible function in λ as $\mathsf{negl}(\lambda)$. We use the notation $f(x) := x^2$ to indicate the definition of a new term.

In what follows, we let $\mathsf{RSAgen}(\lambda)$ denote a randomized algorithm that runs in time polynomial in λ. The algorithm generates two random $\mathsf{len}(\lambda)$-bit primes p and q and outputs $(p, q, N := p \cdot q)$. Here $\mathsf{len} : \mathbb{Z}^+ \to \mathbb{Z}^+$ is some fixed function that determines the size of the primes p and q as a function of λ.

Let $f \in \mathbb{Z}[x, y]$ be a bivariate polynomial. For $c \in \mathbb{Z}$ consider the curve $f(x, y) = c$. The *genus* of this curve is a standard measure of its "complexity:" conics have genus zero, elliptic curves have genus one, and so on (see, e.g. [2,18]). We define the genus of a polynomial f as follows:

Definition 1. *The genus of a polynomial $f \in \mathbb{Z}[x, y]$ is defined as*

$$\max_{c \in \mathbb{Q}} \left(\text{genus}(\ f(x, y) = c\ ) \right).$$

As we will see, the genus of a polynomial f has some relation to its cryptographic properties. While we focus on bivariate polynomials, most of the following discussion generalizes to multivariates.

We use the following terms throughout this section to describe relationships between curves. (For more precise definitions, see Hindry and Silverman [18, Sec. A.1.2].) A *rational map* from a curve C to another curve C' is a pair of rational functions g and h mapping points (x, y) on C to points $(g(x, y), h(x, y))$ on C'. A *birational map* from C to C' is a rational map which is a bijection between points on C and C' such that the map's inverse is also rational. Two curves C and C' and are *birationally equivalent* if there is a birational map from C to C'. An *automorphism* is a birational map from a curve to itself.

3.1   One-Way Polynomials

One-way functions are the basis of much of cryptography. A function $g : X \to Y$ is *one-way* if, given the image $c = f(x)$ of a random point $x \in X$, it is hard to find an x' such that $f(x') = c$. We first ask: what polynomials give rise to one-way functions?

Definition 2. *A polynomial f in $\mathbb{Z}[x_1, ..., x_\ell]$ is one-way if for every p.p.t. algorithm $\mathcal{A}$ the following advantage is a negligible function of λ:*

$$\mathsf{Adv}_{\mathcal{A}, f}(\lambda) := \Pr[N \leftarrow \mathsf{RSAgen}(\lambda),\ \bar{x} \xleftarrow{R} (\mathbb{Z}_N)^\ell,\ c \leftarrow f(\bar{x})\ :$$
$$f\big(\mathcal{A}(N, c)\big) = c \text{ in } \mathbb{Z}_N] \ .$$

Clearly linear polynomials are not one-way. A result of Pollard and Schnorr [28] shows that quadratic polynomials, indeed all genus zero polynomials, are not one-way.

Theorem 3. *A genus zero polynomial $f \in \mathbb{Z}[x, y]$ is not one-way.*

Proof sketch. For all $c \in \mathbb{Q}$ the curve $f(x, y) = c$ is of genus zero, or is a product of genus zero curves. A genus zero curve is birationally equivalent to a linear or quadratic curve $\tilde{f}(x, y) = 0$ [18, Theorem A.4.3.1]. If $\tilde{f}(x, y)$ is linear in one of the variables x or y then finding points on this curve is easy thereby breaking the one-wayness of f. This leaves the case where $\tilde{f}(x, y)$ is quadratic in both x and y. Let N be an output of $\mathsf{RSAgen}(\lambda)$. Let $\tilde{f} \in \mathbb{Z}[x, y]$ be a quadratic polynomial in x and y and let $c \in \mathbb{Z}_N$. There is an efficient algorithm that for most $c \in \mathbb{Z}_N$ finds an $(x_0, y_0) \in \mathbb{Z}_N^2$ such that $\tilde{f}(x_0, y_0) = c$ in $\mathbb{Z}_N$, breaking the one-wayness of f. See for example [6, Sec. 5.2] for a description of the algorithm. □

Theorem 3 played an important role in analyzing the security of the Ong-Schnorr-Shamir signature scheme [26]. The scheme depended on the difficulty of finding solutions (x, y) to the equation:

$$x^2 + ay^2 = b \quad \text{in } \mathbb{Z}_N$$

for known constants $a, b \in \mathbb{Z}_N$. Since this equation defines a genus-zero curve, Theorem 3 shows that it is possible to efficiently find solutions without knowledge of the factors of N. Pollard and Schnorr demonstrated an attack against the scheme soon after its publication [28,32].

One-way Polynomials. It is not known how to break the one-wayness of polynomials $f \in \mathbb{Z}[x, y]$ that are not genus zero. Thus, for example, even a simple polynomial such as $f(x, y) = y^2 - x^3$ may be one-way, although that would require further study.

3.2   Second Preimage Resistant Polynomials

A function $f : U \to V$ is *second preimage resistant* if, given $u \in U$, it is difficult to find a $u' \neq u \in U$ such that $f(u) = f(u')$. We define a similar notion for polynomials:

Definition 4. *A polynomial f in $\mathbb{Z}[x_1, ..., x_\ell]$ is second preimage resistant if, for every p.p.t. algorithm $\mathcal{A}$, the following advantage is a negligible function of λ:*

$$\mathsf{Adv}_{\mathcal{A}, f}(\lambda) := \Pr[N \leftarrow \mathsf{RSAgen}(\lambda), \ \bar{x} \xleftarrow{R} (\mathbb{Z}_N)^\ell, \ \bar{x}' \leftarrow \mathcal{A}(N, \bar{x}) :$$
$$f(\bar{x}) = f(\bar{x}') \text{ in } \mathbb{Z}_N \ \wedge \ \bar{x} \neq \bar{x}'] \ .$$

Since genus 0 polynomials are not one-way they are also not second preimage resistant. It is similarly straight-forward to show that no genus-one polynomial is second preimage resistant.

Proposition 5. *A genus one polynomial $f \in \mathbb{Z}[x, y]$ is not second preimage resistant.*

To see why, let $f \in \mathbb{Z}[x, y]$ be a polynomial such that $f(x, y) = c$ is a curve of genus one for all but finitely many $c \in \mathbb{Q}$. Then f is not second preimage resistant because of the group structure on elliptic curves. That is, let N be an output of $\mathsf{RSAgen}(\lambda)$. Choose a random pair $(x_0, y_0) \in \mathbb{Z}_N^2$ and set $c := f(x_0, y_0) \in \mathbb{Z}_N$. Then $P = (x_0, y_0)$ is a point on the curve $f(x, y) = c$ and so is the point $2P = P + P$ where addition refers to the elliptic curve group operation. With overwhelming probability $2P$ is not the point at infinity and therefore, given P as input, the adversary can output $2P$ as a second preimage for P. It follows that f is not second preimage resistant.

Even polynomials that give higher genus curves need not be second preimage resistant. For example, a hyperelliptic polynomial of genus $g \geq 2$ has the form

$f(x, y) = y^2 - h(x) \in \mathbb{Z}[x, y]$ where $h \in \mathbb{Z}[x]$ is a polynomial of degree $2g + 1$ or $2g + 2$. The simple fact that $f(x_0, y_0) = f(x_0, -y_0)$ immediately gives a second preimage attack on these polynomials: given (x_0, y_0) the attacker outputs $(x_0, -y_0)$ as a second preimage.

The fact that all curves of genus two are hyperelliptic [18, Theorem A.4.5.1] leads to the following proposition:

Proposition 6. *A genus two polynomial $f \in \mathbb{Z}[x, y]$ is not second preimage resistant.*

This proposition, in combination with Theorem 3 and Proposition 5 means that all second preimage resistant polynomials must have genus at least three.

As outlined above, elliptic (genus one) and hyperelliptic (genus two) polynomials are not second preimage resistant because there are non-trivial automorphisms on the associated curves. We say that a polynomial $f \in \mathbb{Z}[x, y]$ is *automorphism free* if, for all but finitely many $c \in \mathbb{Q}$, the curve $f(x, y) = c$ has no automorphisms over $\mathbb{Q}$, apart from the trivial map $(x, y) \mapsto (x, y)$. It is natural to conjecture that every automorphism-free polynomial $f \in \mathbb{Z}[x, y]$ is second preimage resistant.

Poonen constructs a large family of automorphism-free polynomials, in arbitrarily many variables and of arbitrarily large degree [29]. For example, he proves that the polynomial $f(x, y) = x^3 + xy^3 + y^4$ is automorphism-free over the rationals [29].

A Historical Aside: q-Way Preimage Resistance. A stronger notion of preimage resistance for a function $f : U \to V$, called *q-way preimage resistance*, states that given a random $v \in V$ and random points $u_1, \ldots, u_q$ in U such that $v = f(u_1) = \cdots = f(u_q)$, it is difficult to find a new point $u \in U \setminus \{u_1, \ldots, u_q\}$ such that $f(u) = v$.

As before, one can define a similar property for polynomials. That is, a polynomial f in $\mathbb{Z}[x, y]$ is *q-way preimage resistant* if, for a random RSA moduli N and a random $c \in \mathbb{Z}_N$, given q points on the curve $f(x, y) = c$ in $\mathbb{Z}_N$, it is hard to find another point on this curve.

Kilian and Petrank [19] proposed an authentication scheme whose security is based on the q-way preimage resistance of the polynomial $f_{\mathrm{KP}}(x, y) = x^e - y^e$, for some small odd e, say $e = 17$. In their scheme, q is the total number of users in the system. Naor [25] refers to the computational assumption that f_{KP} is q-way preimage resistant as the *Difference RSA Assumption*. We note that the polynomial f_{KP} is not even second preimage resistant because there is a non-trivial automorphism $(x, y) \mapsto (-y, -x)$ on the curve. In other words, for any point (x_0, y_0) we have that $f_{\mathrm{KP}}(x_0, y_0) = f_{\mathrm{KP}}(-y_0, -x_0)$. This bad symmetry appears to violate the security properties needed for the Kilian-Petrank identification scheme, but the scheme can be modified to resist such attacks.

Camenisch and Stadler [10, Sec. 6] used a similar assumption to construct group signatures. They need the polynomial $f_{\mathrm{CS}}(x, y) = x^{e_1} + ay^{e_2}$ to be q-way preimage resistant for some small e_1 and e_2. They propose using $e_1 = 5$ and

$e_2 = 3$. We observe in that next section that the polynomial $f(x, y) = x^5 + y^3$ is not collision resistant. Nevertheless, it may be second preimage resistant.

3.3   Collision-Resistant Polynomials

A function $f : U \to V$ is *collision resistant* if it is difficult to find a pair $u \neq u' \in U$ such that $f(u) = f(u')$. We define a similar notion for polynomials:

Definition 7. *For a polynomial f in $\mathbb{Z}[x_1, ..., x_\ell]$ and an integer N, we say that $\bar{x}, \bar{y} \in (\mathbb{Z}_N)^\ell$ are an N-collision for f if $f(\bar{x}) = f(\bar{y})$ in $\mathbb{Z}_N$ and $\bar{x} \neq \bar{y}$.*

Definition 8. *A polynomial f in $\mathbb{Z}[x_1, ..., x_\ell]$ is* collision resistant *if for every p.p.t. algorithm $\mathcal{A}$ the following advantage is a negligible function of λ:*

$$\mathsf{Adv}_{\mathcal{A}, f}(\lambda) := \Pr\left[\, N \leftarrow \mathsf{RSAgen}(\lambda) \quad : \quad \mathcal{A}(N) \text{ is an } N\text{-collision for } f \,\right].$$

In the previous two subsections, we observed that polynomials $f \in \mathbb{Z}[x, y]$ which are of genus $g \leq 2$ or which are hyperelliptic, are not second preimage resistant and thus are not collision resistant.

Even polynomials that *are* second preimage resistant are not necessarily collision resistant. For example, in Section 3.2 we suggested that the polynomial $f(x, y) = x^3 + xy^3 + y^4$ may be second preimage resistant. However, it is certainly not collision resistant, since for any $r \in \mathbb{Q}$, the points $(r^4, 0)$ and $(0, r^3)$ constitute a collision.

Attacking Collision Resistance over the Rationals. Suppose that a polynomial $f \in \mathbb{Z}[x_1, \ldots, x_\ell]$ has a *rational* collision. That is, there are rational points $\bar{x}_0 \neq \bar{x}_1$ in $\mathbb{Q}^\ell$ such that $f(\bar{x}_0) = f(\bar{x}_1)$. Then, for most[1] RSA moduli N, the points $\bar{x}_0$ and $\bar{x}_1$ give a collision for f in $\mathbb{Z}_N$. This breaks the collision resistance of f when the security parameter λ is sufficiently large. Indeed, for sufficiently large λ the attack algorithm can construct the fixed rational points $\bar{x}_0$ and $\bar{x}_1$ by exhaustive search and obtain collisions for f for most RSA moduli output by $\mathsf{RSAgen}(\lambda)$.

The discussion above shows that if a polynomial $f \in \mathbb{Z}[x_1, \ldots, x_\ell]$ has a *rational* collision then f is not collision resistant. We summarize this in the following proposition.

Proposition 9. *If a polynomial $f \in \mathbb{Z}[x_1, \ldots, x_\ell]$ is collision resistant then the function $f : \mathbb{Q}^\ell \to \mathbb{Q}$ must be injective.*

If $f \in \mathbb{Z}[x_1, \ldots, x_\ell]$ defines an injective function from $\mathbb{Q}^\ell$ to $\mathbb{Q}$ then f is said to be an *injective polynomial*. Proposition 9 shows that the search for collision-resistant polynomials must begin with the search for an injective polynomial over the rationals.

---

[1] The points $\bar{x}_0$ and $\bar{x}_1$ give a collision in $\mathbb{Z}_N$ whenever N is relatively prime to their denominators and $\bar{x}_0 \neq \bar{x}_1 \bmod N$. This holds with overwhelming probability for sufficiently large λ.

Injective Polynomials. Even the *existence* of bivariate injective polynomials is currently an open problem. Poonen [30] shows that they exist under certain number theoretic conjectures. Moreover, Poonen [30, Lemma 2.3] shows that if $f \in \mathbb{Z}[x,y]$ has only a *finite* number of rational collisions then one can use f to construct an injective polynomial $g \in \mathbb{Z}[x,y]$ by pre-composing f with a suitable polynomial map. In other words, an "almost" injective polynomial can be converted to an injective one.

Although proving that a particular polynomial is injective over $\mathbb{Q}$ is currently out of reach, there are simple polynomials that appear to have this property. In particular, Don Zagier[2] conjectures that the polynomial $f_{\text{ZAG}}(x,y) := x^7 + 3y^7$ (the "Zagier polynomial") defines an injective function from $\mathbb{Q}^2$ to $\mathbb{Q}$. As indirect evidence, Cornelissen [13, Remarque 10] and Poonen [30, Remark 1.7] remark that the four-variate generalization of the abc-conjecture [7] implies that $f(x,y) = x^e + 3y^e$ is injective over the rationals for "sufficiently large" odd integers e. Experimentally, we have confirmed that there are no rational collisions in f_{ZAG} for rationals with height less than 100.

ℓ-Variate Injective Polynomials over $\mathbb{Q}$ from Merkle-Damgård. Given a bivariate injective polynomial over $\mathbb{Q}$, it is possible to construct ℓ-variate injective polynomials over $\mathbb{Q}$ for every $\ell > 2$ using the Merkle-Damgård construction for collision-resistant hash functions [15,22]. For example, applying one step of Merkle-Damgård to f_{ZAG} shows that if f_{ZAG} is injective then so is the following three-variate polynomial:

$$g(x,y,z) = (x^7 + 3y^7)^7 + 3z^7 .$$

Injective Polynomials and Collision Resistance. Proposition 9 states that, for a polynomial f to be collision resistant over $\mathbb{Z}_N$, f must be injective over the rationals. The following conjecture asserts the converse: injectivity over the rationals is *sufficient* for collision resistance.

Conjecture 10. *If $f \in \mathbb{Z}[x_1, \ldots, x_\ell]$ is injective over $\mathbb{Q}$ then f is collision resistant.*

This conjecture is based on the intuition that the only *efficient* way to find collisions in f over $\mathbb{Z}_N$ is to find collisions in f over $\mathbb{Q}$. Since collisions over $\mathbb{Q}$ do not exist it may be difficult to find collisions over $\mathbb{Z}_N$.

We only state Conjecture 10 to stimulate further research on this topic. The conjecture is not needed for this paper. For the applications described in this paper, we only need the collision resistance of an explicit low-degree polynomial in $\mathbb{Z}[x,y]$. Nevertheless, if Conjecture 10 is true it would give a clean characterization of collision resistant polynomials in terms of their arithmetic properties.

For the applications in paper, the following assumption suffices.

---

[2] Gunther Corneliseen attributes to Don Zagier the suggestion that $f(x,y) = x^7 + 3y^7$ is collision-free over the rationals [13, Remarque 10].

Assumption 11. *The Zagier polynomial* $f_{\text{ZAG}}(x, y) = x^7 + 3y^7 \in \mathbb{Z}[x, y]$ *is collision resistant.*

We see that breaking Assumption 11 would either: (a) resolve a 15-year open number theoretic problem by showing that f_{ZAG} is non-injective, or (b) find $\mathbb{Z}_N$ collisions that are not rational collisions. We next review two potential avenues for attacks of type (b) and discuss why they do not apply.

Attack Strategy I: Related Non-injective Polynomials over $\mathbb{Q}$. One potential avenue for attacking the collision resistance of f_{ZAG} in $\mathbb{Z}_N$ is to look for a polynomial $h \in \mathbb{Z}[x, y]$ such that

$$g(x, y) := f(x, y) + N \cdot h(x, y)$$

is not injective over $\mathbb{Q}$. If (x_0, y_0) and (x_1, y_1) in $\mathbb{Q}^2$ are a rational collision for g then by reducing this pair modulo N we may[3] obtain a $\mathbb{Z}_N$ collision for $f(x, y)$. We say that h is "useful" if there exists a rational collision for $g(x, y)$ that gives a $\mathbb{Z}_N$ collision for $f(x, y)$. It is easy to show that there are many useful polynomials h: every $\mathbb{Z}_N$ collision for $f(x, y)$ gives a useful polynomial h. However, we do not know how to construct a useful h just given f and N. Furthermore, even if efficiently constructing a useful h is possible, the attack algorithm will need to find a rational collision on the resulting g and this may not be feasible in polynomial time.

Attack Strategy II: Algebraic Extensions. Another avenue for attacking the collision resistance of f_{ZAG} in $\mathbb{Z}_N$ is via algebraic extensions. Let g be an irreducible polynomial in $\mathbb{Z}[x]$ and consider the number field $\mathbb{K} = \mathbb{Q}[x]/(g)$. Suppose the adversary constructs g so that it knows an efficiently computable map $\rho : \mathbb{K} \to \mathbb{Z}_N$ (this can be done by choosing the polynomial g so that the adversary knows a zero of g in $\mathbb{Z}_N$). Now, even if f_{ZAG} is injective as a function $\mathbb{Q}^2 \to \mathbb{Q}$, it may not be injective as a function $\mathbb{K}^2 \to \mathbb{K}$. For example, f_{ZAG} is not injective over the extension $\mathbb{K} = \mathbb{Q}[\sqrt[7]{3}]$: the points $(\sqrt[7]{3}, 0)$ and $(0, 1)$ are a collision. If the adversary could find a collision of f_{ZAG} in $\mathbb{K}^2$ this collision may lead to a $\mathbb{Z}_N$ collision for f_{ZAG}. However, for a random RSA modulus N, it is not known how to efficiently construct an extension $\mathbb{K}$ such that (i) $f_{\text{ZAG}} : \mathbb{K}^2 \to \mathbb{K}$ is not injective, and (ii) the adversary has an efficiently computable map $\rho : \mathbb{K} \to \mathbb{Z}_N$.

Assumption 11 merits further analysis and we hope that this work will stimulate further research on this question.

Non-collision Resistant Polynomials. Simple variations of Zagier's polynomial are trivially not injective and therefore not collision resistant. For example, the polynomials

$$f_1(x, y) = x^7 + y^7 \quad \text{and} \quad f_2(x, y) = x^7 + 2y^7$$

---

[3] If (x_0, y_0) and (x_1, y_1) happens to reduce to the same point modulo N or if one of the denominators is not relatively prime to N then this rational collision for g does not give a $\mathbb{Z}_N$ collision for f.

in $\mathbb{Z}[x, y]$ are not collision resistant. The polynomial f_1 is not injective because for all $x_0 \neq y_0$ in $\mathbb{Z}$ the points (x_0, y_0) and (y_0, x_0) are a collision for f_1. The polynomial f_2 is not collision resistant because for all $t \neq 0$ in $\mathbb{Z}$ the points $(-t, 0)$ and $(t, -t)$ are a collision for f_2.

Similarly, polynomials of the form $f(x, y) = x^{e_1} + by^{e_2} \in \mathbb{Z}[x, y]$ for some $b \in \mathbb{Z}$ where $\gcd(e_1, e_2) = 1$ are not injective and therefore not collision resistant. To see why observe that if the equation $\alpha e_1 - \beta e_2 = 1$ has integer solutions (α_0, β_0) and (α_1, β_1) then $(t^{\alpha_0}, t^{\beta_1})$ and $(t^{\alpha_1}, t^{\beta_0})$ are a collision for f.

Random Self-reduction. Finally, we mention that the collision finding problem for the family of polynomials $\{x^e + ay^e\}_{a \in \mathbb{Z}_N}$ has a random self reduction. Given a collision-finding algorithm $\mathcal{A}(N, a)$ that outputs a $\mathbb{Z}_N$ collision in $x^e + ay^e$ for a non-negligible fraction of choices of $a \in \mathbb{Z}_N$, it is possible to construct a collision-finding algorithm $\mathcal{B}(N, a)$ that finds collisions for *every* choice of a with high probability. On input (N, a) Algorithm $\mathcal{B}$ chooses a random $r \leftarrow \mathbb{Z}_N$, and calls $\mathcal{A}(N, r^e a)$. When $\mathcal{A}$ outputs the collision $(x_0, y_0), (x_1, y_1)$, algorithm $\mathcal{B}$ obtains the following collision on the original curve: $(x_0, ry_0), (x_1, ry_1)$. If $\mathcal{A}$ fails then $\mathcal{B}$ can try again with a fresh random choice of $a \in \mathbb{Z}_N$. After an expected polynomial number of iterations algorithm $\mathcal{B}$ will find a collision for the given polynomial $x^e + ay^e$.

4   A Nestable Commitment Scheme from Polynomials over $\mathbb{Z}_N$

Having argued that it is infeasible to find collisions in the function $f_{\text{ZAG}}(x, y) = x^7 + 3y^7 \bmod N$ (Assumption 11), we now turn to the cryptographic applications of this new computational assumption. In this section, we demonstrate that the collision-resistance of f_{ZAG} leads to a commitment scheme where the procedure for verifying that a commitment was opened correctly uses only low-degree polynomials. The new commitment scheme is statistically hiding and its computational binding property is based on Assumption 11.

The commitment scheme composes naturally with zero-knowledge proofs of knowledge involving Pedersen commitments. In particular, given a Pedersen commitment C to one of our low-degree commitments, there is a succinct zero-knowledge protocol which proves knowledge of *an opening of an opening* of C. We call the inner commitment scheme *nestable*, since it can be efficiently nested inside of a Pedersen commitment. We discuss applications of nestable commitments in Sections 4.4 and 5.

4.1   Commitments

A commitment scheme is a tuple of efficient algorithms (Setup, Commit, Open), with the following functionalities:

Setup(λ) $\rightarrow$ pp. The Setup routine is a randomized algorithm that runs in time polynomial λ and returns public parameters pp. These parameters define a message space $\mathcal{M}$, a space of random blinding values $\mathcal{R}$, and a space of commitments $\mathcal{C}$. The following algorithms take the public parameters pp as an implicit argument.

Commit$(m) \to (c, r)$. Given a message $m \in \mathcal{M}$, return a commitment $c \in \mathcal{C}$ and a random blinding value $r \in \mathcal{R}$ used to open the commitment.

Open$(c, m, r) \to \{0, 1\}$. Given a commitment c, a message m, and a blinding value r, return "1" if (m, r) is a valid opening of c and "0" otherwise.

For correctness, we require that, for all $m \in \mathcal{M}$:

$$\Pr[\mathsf{pp} \leftarrow \mathsf{Setup}(\lambda); (c, r) \leftarrow \mathsf{Commit}(m) : \mathsf{Open}(c, m, r) = 1] \geq 1 - \mathsf{negl}(\lambda).$$

A statistically hiding commitment scheme must satisfy two security properties:

- **Statistically Hiding.** For any two messages m_0 and m_1 in $\mathcal{M}$, a commitment to m_0 is statistically indistinguishable from a commitment to m_1.
- **Computationally Binding.** For any p.p.t. adversary $\mathcal{A}$, the adversary has negligible advantage in producing two different valid openings of the same commitment. More precisely,

$$\Pr[\mathsf{pp} \leftarrow \mathsf{Setup}(\lambda); (c, m, r, m', r') \leftarrow \mathcal{A}(\mathsf{pp}) :$$
$$\mathsf{Open}(c, m, r) = 1 \wedge \mathsf{Open}(c, m', r') = 1 \wedge (m, r) \neq (m', r')] \leq \mathsf{negl}(\lambda).$$

4.2 Construction

The public parameters for our new commitment scheme consist only of an RSA modulus N, for which no one knows the factorization. To commit to a value $m \in \mathbb{Z}_N^*$, the committer samples a random blinding value r from $\mathbb{Z}_N^*$ and computes the value of f_{ZAG} at the point (m, r).

The construction of the new commitment scheme follows.

Setup$(\lambda) \to N$. The value N is an RSA modulus—the product of two random $\mathsf{len}(\lambda)$-bit primes p and q such that $\gcd(p-1, q-1, 7) = 1$. The commitment space $\mathcal{C}$ is $\mathbb{Z}_N$. The message space $\mathcal{M}$ and the space of blinding values $\mathcal{R}$ are $\mathbb{Z}_N^*$.

Commit$(m) \to (c, r)$. Choose a random blinding value $r \leftarrow \mathbb{Z}_N^*$ and set $c \leftarrow m^7 + 3r^7$ in $\mathbb{Z}_N$. Return r as the commitment secret.

Open$(c, m, r) \to \{0, 1\}$. Output "1" if $m, r \in \mathbb{Z}_N^*$ and if $c = m^7 + 3r^7$ in $\mathbb{Z}_N$. Output "0" otherwise.

Security Properties. The following theorem summarizes the security properties of the scheme.

Theorem 12. *The commitment scheme is statistically hiding and computationally binding under Assumption 11.*

Proof. Statistical hiding follows from a standard argument given in Appendix A. Computational binding follows directly from the collision resistance of f_{ZAG} over $\mathbb{Z}_N$. One issue is Setup algorithm generates a random N such that $\gcd(\phi(N), 7) =$

1 whereas Assumption 11 imposes no such restriction on N. Nevertheless, Assumption 11 implies the collision resistance of f_{ZAG} for this modified distribution of N: By way of contradiction, assume there were an algorithm $\mathcal{A}$ which finds collisions in f_{ZAG} with non-negligible probability ϵ when $\gcd(\phi(N), 7) = 1$. Since algorithm RSAgen in Assumption 11 generates such N with probability about $(5/6)^2 = 25/36$ it follows that $\mathcal{A}$ will find collisions in with probability at least $(25/36)\epsilon$ when N is sampled as in algorithm RSAgen, violating Assumption 11.

Efficiency. Generating and verifying standard Pedersen commitments requires two modular exponentiations (or elliptic curve scalar multiplications). In contrast, our scheme requires only a few modular *multiplications*. On a workstation with a 3.20 GHz processor, for example, computing 10,000 Pedersen commitments in a subgroup of order $\approx 2^{256}$ modulo a 2048-bit prime takes 16.54 seconds. Computing the same number of commitments using this new scheme takes 0.925 seconds—a factor of 17.9× speed-up.

4.3   Nestable Commitments

We say that a commitment scheme (Setup, Commit, Open) is *nestable* if, given Pedersen commitments to a message m, randomness r, and a commitment c, there is an succinct zero-knowledge proof of knowledge of values m, r, and c, such that $c = \text{Commit}(m, r)$. In other words, there is a succinct protocol for proving knowledge of an *opening of an opening* of a Pedersen commitment. For our purposes, a *succinct* zero-knowledge protocol is one in which proof length is $k|c|$ bits long, where k is a constant which does not depend on the security parameter.

We adopt the notation of Camenisch and Stadler [9] for specifying zero-knowledge proof-of-knowledge protocols. For example, $\text{PoK}\{x, y : X = g^x \lor Y = g^x\}$ indicates a protocol in which the prover and verifier share public values g, X, and Y, and the prover demonstrates knowledge of either a value x such that $X = g^x$ *or* a value y such that $Y = g^y$.

Given Pedersen commitments

$$C_m = g^m h^{s_m} \qquad C_r = g^r h^{s_r} \qquad C_c = g^c h^{s_c}$$

a *nestable* commitment scheme has a succinct zero-knowledge protocol which proves knowledge of the statement:

$$\text{PoK}\{m, r, c, s_m, s_r, s_c : C_m = g^m h^{s_m} \land C_r = g^r h^{s_r} \land C_c = g^{\text{Commit}(m,r)} h^{s_c}\}.$$

For the commitment scheme outlined above, $\text{Commit}(m, r) = m^7 + 3r^7 \bmod N$, so the proof of knowledge protocol is:

$$\text{PoK}\{m, r, c, s_m, s_r, s_c : C_m = g^m h^{s_m} \land C_r = g^r h^{s_r} \land C_c = g^{m^7 + 3r^7} h^{s_c}\}.$$

The group $\mathbb{G} = \langle g \rangle = \langle h \rangle$ used for the proof must be a group of *composite* order N, where N is the RSA modulus used in the commitment scheme. As usual for

Pedersen commitments, no one should know the discrete logarithm $\log_g h$ in $\mathbb{G}$. For example, $\mathbb{G}$ might be the order-N subgroup of the group $\mathbb{Z}_p^*$ for a prime $p = 2kN + 1$, where k is a small prime. Alternatively, $\mathbb{G}$ could be an elliptic curve group of order N.

The fact that the verification equation for our commitment scheme is a fixed low-degree polynomial means that this proof can be executed succinctly using standard techniques [10]. This proof requires only one challenge and 20 elements of $\mathbb{G}$. If N is a 2048-bit modulus, then the proof is roughly 5 KB in length.

In contrast, nesting Pedersen commitments inside of other Pedersen commitments does not lead to succinct proofs of knowledge. The shortest proofs of knowledge for nested Pedersen commitments require a number of group elements that is linear in the security parameter [11, Sec. 5.3.3], whereas our proof requires only a constant number of group elements.

Being able to prove knowledge of an opening of a commitment which is itself nested inside of a commitment proves useful in constructing distributed e-cash schemes (Section 4.4) and set membership proofs (Section 5).

4.4   Application Sketch: Anonymous Bitcoins

The Zerocoin scheme for anonymizing Bitcoin transactions requires a proof of knowledge of an opening of an opening of a commitment [24]. For this purpose, Zerocoin uses Pedersen commitments nested inside of Pedersen commitments, which requires a proof-of-knowledge of the form: $\mathsf{PoK}\{m, r, s : \hat{c} = \hat{g}^{(g^m h^r)} \hat{h}^s\}$. The number of group elements exchanged in this proof is linear in the security parameter, since the proof uses single-bit challenges.

By using our nestable commitment scheme for the "inner" commitment, we reduce the number of group elements from linear to constant in the security parameter. This reduces the length of anonymous coin transactions in the Zerocoin scheme by roughly 70% (down to 12.0 KiB from 39.4 KiB when using a 2048-bit RSA modulus). When instantiated with our nestable commitments, Zerocoin maintains its unconditional privacy property and maintains double-spending prevention under Assumption 11.

5   Succinct Set Membership Proofs

A *cryptographic accumulator*, first defined by Benaloh and De Mare [5], is a primitive which allows a prover to accumulate large set of values $S = \{x_1, \ldots, x_n\}$ into a single short value A. For every value x_i in the accumulator, there is an accompanying short witness w_i. By exhibiting a valid (x_i, w_i) pair, a prover can convince a verifier that the value x_i was actually accumulated into A. Informally, the security property of the accumulator requires that it be difficult to find a valid value-witness pair (x^*, w^*) such that $x^* \notin S$.

Benaloh and De Mare give one example application of this primitive: the administrator of a club can accumulate the names of the members of the club into an accumulator A, distribute a witness to each member, and publish the accumulator value A. The value A is a concise representation of the club's membership list. A person can prove membership in the club by revealing her name x_i and the witness w_i to a verifier.

Camenisch-Lysyanskaya extend the basic accumulator primitive to allow for zero-knowledge proofs of accumulator membership [8]. That is, a prover can convince a verifier that the prover "knows" a valid value-witness pair (x, w) for a particular accumulator A, without revealing x or w. This augmented primitive allows for privacy-preserving authentication: a club member can prove that she is *some* member of the club defined by a membership list A without revealing *which* member she is.

We provide a construction that offers the same functionality as the Camenisch-Lysyanskaya scheme with the cost of requiring slightly larger proofs—of length $O(\log |S|)$ instead of length $O(1)$. The benefit of our construction is its simplicity: compared with the Camenisch-Lysyanskaya proof, which requires a nuanced security analysis, ours is relatively straightforward.

5.1   Definitions

A *cryptographic accumulator* is a tuple of algorithms (Setup, Accumulate, Witness, Verify) with the following functionalities:

Setup(λ) $\rightarrow$ pp. Given a security parameter λ as input, output the public parameters pp. The other functions take pp as an implicit input. Setup runs in time polynomial in λ.

Accumulate($S = \{x_1, \ldots, x_n\}$) $\rightarrow A$. Accumulate the n items in the set S into an accumulator value A.

Witness(S, x) $\rightarrow w$ or $\bot$. If $x \notin S$, return $\bot$. Otherwise, return a witness w that x was accumulated in Accumulate(S). To be useful, the length of w should be short (constant or logarithmic) in the size of S.

Verify(A, x, w) $\rightarrow \{0, 1\}$. Return "1" if the value-witness pair (x, w) is valid for the accumulator A. Return "0" otherwise.

Camenisch and Lysyanskaya, following Barić and Pfitzmann [4], define an accumulator as *secure*, if for all polynomial-time adversaries $\mathcal{A}$:

$$\Pr[\mathsf{pp} \leftarrow \mathsf{Setup}(\lambda); \ (S, x^*, w^*) \leftarrow \mathcal{A}(\mathsf{pp}); \ x^* \notin S;$$
$$A \leftarrow \mathsf{Accumulate}(S) : \mathsf{Verify}(A, x^*, w^*) = 1] \leq \mathsf{negl}(\lambda).$$

If an accumulator satisfies this definition, then it is infeasible for an adversary to prove that a value x was accumulated in a value A if it was not.

Zero-Knowledge Proof of Knowledge of an Accumulated Value. In many applications, it is useful for a prover to be able to convince a verifier that the prover knows *some* value inside of an accumulator without revealing *which* value the prover knows. Such a proof protocol should satisfy the standard properties of soundness, completeness, and zero-knowledgeness [11, Sec. 2.9]. Camenisch and Lysyanskaya construct one such proof-of-knowledge protocol for the strong-RSA accumulator [8] and we exhibit a protocol for a Merkle-tree-style accumulator in Section 5.3.

5.2   Construction

Given a collision-resistant hash function $H : D \times D \rightarrow D$, which operates on a domain D such that $S \subseteq D$, it is possible to construct a simple accumulator using Merkle trees. For example, given a set $S = \{x_1, x_2, x_3, x_4\}$, the accumulator value A is the value $A \leftarrow H(H(x_1, x_2), H(x_3, x_4))$. A witness w_i that an element x_i is in the accumulator is the set of $O(\log |S|)$ nodes along the Merkle tree needed to verify a path from x_i to the root (labeled A).

The limitation of this accumulator construction is that it no longer admits simple zero-knowledge proofs of knowledge of (x, w) pairs, unless H has a very special form. For instance, if H is a standard cryptographic hash function (e.g., SHA-256), there is no straightforward zero-knowledge protocol for proving knowledge in zero knowledge of a preimage under H. By instantiating H with the function $H(x, y) = x^7 + 3r^7 \bmod N$, as we demonstrate in the following section, it is possible to execute this zero-knowledge proof succinctly.

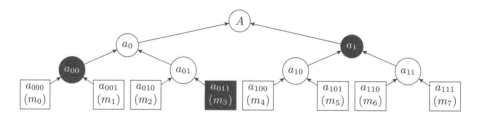

Fig. 1. A perfect Merkle tree with eight leaves rooted at A. The shaded nodes are a witness to the fact that m_2 is accumulated in A. The tree invariant is $a_i = H(a_{i0}, a_{i1})$.

We first recall the standard construction of Merkle trees [21] and then describe the zero-knowledge proof construction. The construction from a general collision-resistant hash function family $\{\mathcal{H}_\lambda\}_{\lambda=1}^\infty$ follows.

Setup$(\lambda) \rightarrow H$. Given a security parameter λ as input, sample a λ-secure collision-resistant hash function H from $\mathcal{H}_\lambda$. Setup runs in time polynomial in λ.

Accumulate$(S = \{x_1, \ldots, x_n\}) \rightarrow A$. If $|S|$ is not a power of two, insert "dummy" elements into S (e.g., by duplicating the first element of S) until $|S|$ is a power of two. Construct a perfect Merkle tree of depth $d = \log_2 |S|$ using the hash function H with the members of S as its leaves and return the root A. Figure 1 depicts an example tree of depth three.

Witness$(S, x) \rightarrow w$ or $\bot$. If $x \notin S$, return $\bot$. Otherwise, let the path from A to the message x be: $P = (A, a_{b_1}, a_{b_1 b_2}, a_{b_1 b_2 b_3}, \ldots, a_{b_1 \ldots b_d})$, where a_{i0} is the left child of node a_i, a_{i1} is the right child of node a_i, and d is the number of edges between the root and leaf labeled x in the tree. The first component of the witness is the list of siblings of the nodes in the path P: $w_\alpha = (a_{\bar{b}_1}, a_{b_1 \bar{b}_2}, a_{b_1 b_2 \bar{b}_3}, \ldots, a_{b_1 \ldots \bar{b}_d})$. The second component of the witness is a bit vector indicating where x is located in the tree: $w_\beta = (b_1, b_2, \ldots, b_{d-1}, b_d)$. The witness is $w = (w_\alpha, w_\beta)$.

Verify$(A, x, w) \rightarrow \{0, 1\}$. Interpret the witness as (w_α, w_β) such that $w_\alpha = (w_1, \ldots, w_d)$ and $w_\beta = (b_1, \ldots, b_d)$. To verify the witness, let $t_d = x$ and recompute the intermediate nodes of the tree from the leaf back to the root. Specifically, compute test nodes t_i for $i = d - 1, \ldots, 0$:

$$t_i = \begin{cases} H(t_{i+1}, w_{i+1}) : \text{if } b_i = 0 \\ H(w_{i+1}, t_{i+1}) : \text{if } b_i = 1 \end{cases}$$

Return "1" if $A = t_0$ and "0" otherwise.

5.3 Proof of Knowledge of an Accumulated Value

When instantiated with a general hash function H, the Merkle-tree accumulator of the prior section does not admit a succinct proof of knowledge of an accumulated value. When instantiated with our new hash function $H(x, y) = x^7 + 3y^7 \bmod N$, however, there *is* a succinct proof of knowledge that the prover knows an opening of a Pedersen commitment C_m such that some leaf of the accumulator Merkle tree has label m. The proof requires a group $\mathbb{G} = \langle g \rangle = \langle h \rangle$ of order N, as in Section 4.3. The proof length is $\log |S|$, for a set S of elements accumulated.

The Setup algorithm outputs an RSA modulus $N \leftarrow \mathsf{RSAgen}(\lambda)$ such that $\gcd(\phi(N), 7) = 1$ and such that no one knows the factorization of N. The hash function H is $H(x, y) = x^7 + 3y^7 \bmod N$ and the accumulator domain D is $\mathbb{Z}_N^*$.

The high-level idea is that, if the prover wants to convince the verifier that a particular value m is accumulated in A, the prover commits to the values of all of the nodes in the Merkle tree along the path from the root to the leaf labeled m. The prover also commits to all of the witness values needed to recreate the path from the leaf labeled m down to the tree root. The prover can then convince the verifier in zero knowledge that these commitments together contain a path to *some* leaf in the tree, without revealing which one.

Assume that the prover has a value-witness pair (x, w) which convinces a verifier that x is accumulated in A. Denote the node values along the path from the root node, with value A, to the leaf node, with value x, in the Merkle tree as: $p = (p_0, p_1, \ldots, p_d)$. Note that $p_0 = A$ and $p_d = x$.

The prover now commits to every value p_i in this path *and* to the values of the left and right children of p_i in the Merkle tree. If the value of the left child is ℓ_i and the right child is r_i, the commitments are, for $i = 0, \ldots, d - 1$:

$$P_i = g^{p_i} h^{s_i} \qquad L_i = g^{\ell_i} h^{s_i'} \qquad R_i = g^{r_i} h^{s_i''}$$

The prover opens P_0 by publishing (p_0, s_0) and the verifier ensures that $p_0 = A$ and that $P_0 = g^{p_0} h^{s_0}$.

The prover now can prove, for $i = 0, \ldots, d - 1$, that each (P_i, L_i, R_i) tuple is well-formed using a standard discrete logarithm proof:

$$\mathsf{PoK}_\alpha \{\ell, r, s, s', s'' : P_i = g^{\ell^7 + 3r^7} h^s \wedge L_i = g^\ell h^{s'} \wedge R_i = g^r h^{s''}\}.$$

The prover then must prove that it knows an opening of the commitment P_{i+1} such that the opening is equal to an opening of either L_i or R_i. For $i = 0, \ldots, d - 1$, the prover proves:

$$\mathsf{PoK}_\beta\{p, s, s_\ell, s_r : P_{i+1} = g^p h^s \wedge (L_i = g^p h^{s_\ell} \vee R_i = g^p h^{s_r})\}.$$

The complete proof is the set of commitment pairs $\{(P_i, L_i, R_i)\}_{i=0}^d$, the $2d$ proofs of knowledge, and the opening (p_0, r_0) of the root commitment P_0. The total length is $O(d) = O(\log |S|)$, since the tree has depth $d = \log |S|$ and each of the elements of the proof has length which is constant in $|S|$.

Security. The completeness and zero-knowledgeness properties follows from the properties of the underlying zero-knowledge proofs used and from the fact that Pedersen commitments are perfectly hiding.

To show soundness, we must demonstrate that if the verifier accepts, it can extract a value-witness pair (x^*, w^*) for the original Merkle tree with non-negligible probability by rewinding the prover. Starting at the root and working towards the leaves of the tree, we will be able to extract the prover's witness for each of the proofs of knowledge with non-negligible probability.

By induction on i, we can show that after d steps, the verifier will be able to extract the value-witness pair (x, w). The base case of the induction is $i = 0$ and the verifier can extract a preimage of A under H. From each of the i PoK_αs, the verifier extracts an element of the witness w_α (the preimage of p_i under H). From each of the i PoK_βs, the verifier extracts an element of the witness w_β (whether the next node in the path is the left or right child of p_i).

6   Claw-Free Functions, Signatures, and Chameleon Hashes

In this section, we describe a few other applications arising from the assumed collision-freeness of the Zagier polynomial.

Claw-Free Functions and Signatures. Assumption 11 immediately gives rise to a family of *trapdoor claw-free functions* [14]. For each RSA modulus N selected as in Section 4.2, we can define a function family:

$$\mathcal{F}_N := \{f_a \,|\, a \in \mathbb{Z}_N^*\} \quad \text{where} \quad f_a(x) = x^7 + 3a^7 \bmod N.$$

Following Damgård [14], a function family $\mathcal{F}_N$ is *claw free* if, given $\mathcal{F}_N$, it is difficult to find a "claw" (x, y, a, b) such that $f_a(x) = f_b(y)$. For all p.p.t. adversaries $\mathcal{A}$, we require that:

$$\Pr[\, N \leftarrow \mathsf{RSAgen}(\lambda), \ (x, y, a, b) \leftarrow \mathcal{A}(N) \ : \ f_a(x) = f_b(y) \,] \leq \mathsf{negl}(\lambda).$$

The claw-freeness of $\mathcal{F}_N$ follows from Assumption 11, since a claw in $\mathcal{F}_N$ implies a collision in $f(x, y) = x^7 + 3y^7 \bmod N$. Additionally, the function family $\mathcal{F}_N$

is trapdoor claw-free, since anyone with knowledge of the factors of N can find claws easily by choosing (x, y, a) arbitrarily and solving for b.

This family $\mathcal{F}_N$ is not quite a family of trapdoor claw-free permutations, since the range of two functions f_a and f_b in $\mathcal{F}_N$ are not necessarily equal (i.e., $f_b^{-1}(f_a(x))$ is sometimes undefined). However, the fraction of choices of (a, b, x) for which this event occurs is negligible, so it is possible to treat $\mathcal{F}_N$ as if it were a family of trapdoor claw-free permutations. In particular, this function family leads to a signature scheme secure against adaptive chosen message attacks in the standard model by way of the Goldwasser-Micali-Rivest signature construction [17].

Chameleon Hash. This commitment scheme immediately gives rise to a new chameleon hash function. A chameleon hash, as defined by Krawczyk and Rabin, is a public hash function $H(m, r)$ with a secret "trapdoor" [20]. A chameleon hash function has three properties:

1. Without the trapdoor, it is difficult to find collisions in H. That is, it is hard to find colliding pairs (m, r) and (m', r') such that $H(m, r) = H(m', r')$.
2. Given the trapdoor, there is an efficient algorithm which takes (m, r, m') as input and outputs a value r' such that $H(m, r) = H(m', r')$.
3. For any pair of messages m and m' in the message space $\mathcal{M}$, the distributions $H(m, r)$ and $H(m', r')$ are statistically close if r and r' are chosen at random.

Chameleon hashes are useful in building secure signature schemes in the standard model [16] and for a number of other applications [20].

To derive a chameleon hash scheme from our commitment scheme, set the public key to the RSA modulus N, and the secret key to the factorization of N. The hash function H is then $H(m, r) = m^7 + 3r^7 \bmod N$. Without the factors of N, it is difficult to find collisions but anyone with knowledge of the factors of N (the "trapdoor") can find collisions.

Chameleon hashes based on Pedersen commitments require two modular exponentiations to evaluate, while ours requires just a few modular multiplications.

7   Conclusion and Future Work

We have used arithmetic properties of bivariate polynomials over $\mathbb{Q}$ to reason about their cryptographic properties in the ring $\mathbb{Z}_N$. Using one particular low-degree polynomial, f_{ZAG}, we build a new statistically hiding commitment scheme, a conceptually simple cryptographic accumulator, and a computationally efficient chameleon hash function. To gain confidence in Conjecture 10 it would be interesting to prove it in the *generic ring* model [1]. We leave that for future work.

Acknowledgments. We are grateful to Bjorn Poonen for information about injective polynomials, to Steven Galbraith and Antoine Joux for comments on our cryptographic assumptions, and to Don Zagier for recounting his rationale for conjecturing the injectivity of $f(x, y) = x^7 + 3y^7$ over $\mathbb{Q}$. We thank Joe Zimmerman for helpful conversations about this work. This work was supported by DARPA, an NSF research grant, and an NSF Graduate Research Fellowship under Grant No. DGE-114747.

References

1. Aggarwal, D., Maurer, U.: Breaking RSA generically is equivalent to factoring. In: Joux, A. (ed.) EUROCRYPT 2009. LNCS, vol. 5479, pp. 36–53. Springer, Heidelberg (2009)
2. Ash, A., Gross, R.: Elliptic Tales: Curves, Counting, and Number Theory. Princeton University Press (2012)
3. Ateniese, G., de Medeiros, B.: Identity-based chameleon hash and applications. In: Juels, A. (ed.) FC 2004. LNCS, vol. 3110, pp. 164–180. Springer, Heidelberg (2004)
4. Barić, N., Pfitzmann, B.: Collision-free accumulators and fail-stop signature schemes without trees. In: Fumy, W. (ed.) EUROCRYPT 1997. LNCS, vol. 1233, pp. 480–494. Springer, Heidelberg (1997)
5. Benaloh, J.C., de Mare, M.: One-way accumulators: A decentralized alternative to digital signatures. In: Helleseth, T. (ed.) EUROCRYPT 1993. LNCS, vol. 765, pp. 274–285. Springer, Heidelberg (1994)
6. Boneh, D., Gentry, C., Hamburg, M.: Space-efficient identity based encryption without pairings. In: FOCS, pp. 647–657 (2007)
7. Browkin, J., Brzeziński, J.: Some remarks on the abc-conjecture. Mathematics of Computation 62(206), 931–939 (1994)
8. Camenisch, J., Lysyanskaya, A.: Dynamic accumulators and application to efficient revocation of anonymous credentials. In: Yung, M. (ed.) CRYPTO 2002. LNCS, vol. 2442, pp. 61–76. Springer, Heidelberg (2002)
9. Camenisch, J., Stadler, M.: Efficient group signature schemes for large groups. In: Kaliski Jr., B.S. (ed.) CRYPTO 1997. LNCS, vol. 1294, pp. 410–424. Springer, Heidelberg (1997)
10. Camenisch, J., Stadler, M.: Proof systems for general statements about discrete logarithms. Tech. Rep. 260, Dept. of Computer Science, ETH Zurich (March 1997)
11. Camenisch, J.: Group Signature Schemes and Payment Systems Based on the Discrete Logarithm Problem. Ph.D. thesis, Swiss Federal Institute of Technology Zürich (ETH Zürich) (1998)
12. Catalano, D., Gennaro, R., Howgrave-Graham, N., Nguyen, P.Q.: Paillier's cryptosystem revisited. In: ACM Conference on Computer and Communications Security, pp. 206–214 (2001)
13. Cornelissen, G.: Stockage diophantien et hypothese abc généralisée. Comptes Rendus de l'Académie des Sciences-Series I-Mathematics 328(1), 3–8 (1999)
14. Damgård, I.B.: The Application of Claw Free Functions in Cryptography. Ph.D. thesis, Aarhus University (May 1988)
15. Damgård, I.B.: A design principle for hash functions. In: Brassard, G. (ed.) CRYPTO 1989. LNCS, vol. 435, pp. 416–427. Springer, Heidelberg (1990)
16. Gennaro, R., Halevi, S., Rabin, T.: Secure hash-and-sign signatures without the random oracle. In: Stern, J. (ed.) EUROCRYPT 1999. LNCS, vol. 1592, pp. 123–139. Springer, Heidelberg (1999)
17. Goldwasser, S., Micali, S., Rivest, R.L.: A digital signature scheme secure against adaptive chosen-message attacks. SIAM Journal on Computing 17(2), 281–308 (1988)

18. Hindry, M., Silverman, J.H.: Diophantine geometry: an introduction, vol. 201. Springer (2000)
19. Kilian, J., Petrank, E.: Identity escrow. In: Krawczyk, H. (ed.) CRYPTO 1998. LNCS, vol. 1462, pp. 169–185. Springer, Heidelberg (1998)
20. Krawczyk, H., Rabin, T.: Chameleon hashing and signatures. In: NDSS, pp. 143–154 (2000)
21. Merkle, R.C.: A digital signature based on a conventional encryption function. In: Pomerance, C. (ed.) CRYPTO 1987. LNCS, vol. 293, pp. 369–378. Springer, Heidelberg (1988)
22. Merkle, R.C.: One way hash functions and DES. In: Brassard, G. (ed.) CRYPTO 1989. LNCS, vol. 435, pp. 428–446. Springer, Heidelberg (1990)
23. Micali, S., Rabin, M., Kilian, J.: Zero-knowledge sets. In: FOCS, pp. 80–91 (2003)
24. Miers, I., Garman, C., Green, M., Rubin, A.D.: Zerocoin: Anonymous distributed e-cash from Bitcoin. IEEE Security and Privacy, 397–411 (2013)
25. Naor, M.: On cryptographic assumptions and challenges. In: Boneh, D. (ed.) CRYPTO 2003. LNCS, vol. 2729, pp. 96–109. Springer, Heidelberg (2003)
26. Ong, H., Schnorr, C.P., Shamir, A.: An efficient signature scheme based on quadratic equations. In: STOC, pp. 208–216 (1984)
27. Pedersen, T.P.: Non-interactive and information-theoretic secure verifiable secret sharing. In: Feigenbaum, J. (ed.) CRYPTO 1991. LNCS, vol. 576, pp. 129–140. Springer, Heidelberg (1992)
28. Pollard, J., Schnorr, C.: An efficient solution of the congruence. IEEE Transactions on Information Theory 33(5), 702–709 (1987)
29. Poonen, B.: Varieties without extra automorphisms III: hypersurfaces. Finite Fields and their Applications 11(2), 230–268 (2005)
30. Poonen, B.: Multivariable polynomial injections on rational numbers. arXiv preprint arXiv:0902.3961v2 (June 2010)
31. Schwenk, J., Eisfeld, J.: Public key encryption and signature schemes based on polynomials over $\mathbb{Z}_n$. In: Maurer, U.M. (ed.) EUROCRYPT 1996. LNCS, vol. 1070, pp. 60–71. Springer, Heidelberg (1996)
32. Shallit, J.: An exposition of Pollard's algorithm for quadratic congruences (October 1984)
33. Shamir, A.: On the generation of multivariate polynomials which are hard to factor. In: STOC, pp. 796–804. ACM (1993)
34. Zagier, D.: Personal communication (June 2014)

A Proof of Statistical Hiding

This appendix presents a proof that the commitment scheme of Section 4.2 is statistically hiding. To demonstrate that the statistical hiding property holds, we show that for *any* message $m \in \mathbb{Z}_N^*$, the distribution of the value of a commitment c to m is statistically close to uniform.

The commitment c is generated by sampling a random value $r \leftarrow_R \mathbb{Z}_N^*$ and letting $c \leftarrow m^7 + 3r^7$. Since $r \in \mathbb{Z}_N^*$, and since $\gcd(7, \phi(N)) = 1$, the RSA function $f(x) = x^7 \bmod N$ defines a permutation on $\mathbb{Z}_N^*$. Thus, there are exactly $|\mathbb{Z}_N^*| = \phi(N)$ possible commitments to m, and each of these values occurs with equal probability.

Let the random variable C take on the value of the commitment to m and let U be a random variable uniformly distributed over $\mathbb{Z}_N$. Then:

$$\Pr[C = c_0] = \frac{1}{\phi(N)}; \qquad \Pr[U = c_0] = \frac{1}{N}$$

The statistical distance between these distributions is:

$$\Delta(C, U) = \frac{1}{2} \sum_{c_0 \in \mathbb{Z}_N} |\Pr[C = c_0] - \Pr[U = c_0]|$$

$$= \frac{1}{2} \sum_{c_0 \in \mathbb{Z}_N} \left| \frac{N - \phi(N)}{N\phi(N)} \right| = \frac{(p + q - 1)}{2\phi(N)} \leq \mathsf{negl}(\lambda).$$

Cryptographic Schemes Based on the **ASASA** Structure: Black-Box, White-Box, and Public-Key (Extended Abstract)

Alex Biryukov, Charles Bouillaguet, and Dmitry Khovratovich

CSC& SnT, University of Luxembourg, Luxembourg, and University of Lille-1, France
{alex.biryukov,dmitry.khovratovich}@uni.lu,
charles.bouillaguet@univ-lille1.fr

Abstract. In this paper we pick up an old challenge to design public key or white-box constructions from symmetric cipher components. We design several encryption schemes based on the ASASA structure ranging from fast and generic symmetric ciphers to compact public key and white-box constructions based on generic affine transformations combined with specially designed low degree non-linear layers. While explaining our design process we show several instructive attacks on the weaker variants of our schemes[1].

Keywords: ASASA, multivariate cryptography, white-box cryptography, cryptanalysis, algebraic, symmetric.

1 Introduction

Since the development of public key cryptography in the late 1970's it has been an open challenge to diversify the set of problems on which such primitives were built as well as to find faster alternatives, since most public key schemes were several orders of magnitude slower than symmetric ones. One of the directions was to design public key schemes from symmetric components. As public key encryption requires trapdoors, they have been hidden in secret affine layers [39], field representations [43], biased S-boxes and round functions [45]; however most of these schemes were broken [35,42]. We recall that a typical symmetric cipher is built from layers of affine transformations (A) and S-boxes (S), a design principle dating back to Shannon. It is thus natural to see what designs can be made from such components. Whereas the classical cipher AES-128 consists of 10 rounds with 19 layers in total, it is striking that a lot of effort has been put into designing public-key schemes with only 3 layers, using the ASA (affine-substitution-affine) structure. This has indeed been the mainstream of what is known as *multivariate cryptography*. However, in this case, the non-linear layer is usually an ad-hoc monolithic function over the full state, as opposed to an array of independent S-boxes.

---

[1] The full version of our paper is available at Eprint [8].

P. Sarkar and T. Iwata (Eds.): ASIACRYPT 2014, PART I, LNCS 8873, pp. 63–84, 2014.

It has been known that the scheme SASAS with two affine and three nonlinear layers is vulnerable to a structural attack if the nonlinear layer consists of several independent S-boxes [10]. The scheme ASA, though secure for a random monolithic S-box, has been shown weak in concrete multivariate proposals. In the seemingly unrelated area of *white-box cryptography* the ASA approach to build obfuscated lookup tables failed multiple times. This suggests exploring the shortest scheme unbroken so far — the ASASA construction with injective S-boxes — in the application to symmetric (black-box), public-key, and white-box cryptography. Let us overview the related areas.

Retrospective of Multivariate Cryptography

The idea of multivariate cryptography dates back to the Shannon's idea that recovering the secrets in any cryptographic scheme could be reduced to solving particular systems of (boolean) equations. Since nearly all forms of cryptology *implicitly* rely on the hardness of solving some kind of equation systems, then it must be possible to design cryptographic schemes that *explicitly* rely on the hardness of this problem. In multivariate public-key schemes, the public-key itself is a system of polynomial equations in several variables. It is well-known that solving such systems is NP-hard, even when the polynomials are quadratic (hence the name of the MQ problem, which stands for Multivariate Quadratic polynomial systems). An additional advantage of the MQ cryptosystems is that they seem invulnerable to quantum algorithms and hence are candidates for Post-Quantum cryptography.

Multivariate polynomials have been used in cryptography in the early 1980's with the purpose of designing RSA variants with faster decryption. At this time, Imai and Matsumoto designed the first public-key scheme explicitly based on the hardness of MQ. It made it to the general crypto community a few years later under the name C* [39].

Several years later, in 1995, Patarin [42] found a devastating attack against C*, allowing to decrypt and to forge signatures very efficiently. Thereafter many multivariate scheme have been proposed (we counted at least 20 of them), including a plethora of bogus and vainly complicated proposal with a short lifespan. A few constructions stood out and received more attention than the others because of their simplicity and their elegance, such as HFE [43] and UOV [34].

However, the practical break of the first HFE challenge, supposed to offer 80 bits of security, in 2003 [29], and the demise of SFLASH in 2007 [24], just after the NESSIE consortium proposed it to be standardized, shattered the hopes and trust of the cryptographic community at large in multivariate cryptography. This brought the multivariate fashion to a stop.

The main problem in multivariate crypto is that the selection of candidates for the nonlinear layer S is scarce (we will discuss this in Section 4). What remains usually has so strong a structure within, that it can be detected and exploited even in the presence of unknown A layers. A very recent example is the promising matrix-based scheme ABC [48]. In the last years, a few researchers started designing public-key schemes based on the hardness of *random instances* of the MQ problem [46], though no drop-in replacement for conventional public-key encryption schemes has been proposed. Still, they are promising because there is a concensus that random instances are hard, and all known algorithms are exponential and impractical on random systems.

This overview clearly indicates the need of a larger structure for multivariate cryptosystems, and suggests truly random polynomials in this context, which we use in our schemes.

Retrospective of White-Box Cryptography

In a parallel development a notion of *white-box cryptography* (WBC) has been introduced in [17]. The initial motivation was to embed symmetric secret keys into the implementation of popular standards like AES or DES in a way that binds the attacker to the specific implementation for DRM purposes. Several proposals have been made [15,16] with the main idea to obfuscate key-dependent parts of the cipher and publish them as lookup tables, so that the entire encryption routine becomes just a sequence of table lookups. The obfuscation constitutes of wrapping the nonlinear transformation (S) with random affine transformations (A) so that the affine layers would cancel each other after composition.

As a result, the lookup tables are just instantiations of the ASA structure. Moreover, since the nonlinear layers of AES and DES consist of independent S-boxes, the resulting ASA structure is very weak and can be attacked by a number of methods [7]. As demonstrated by Biryukov and Shamir [10], even as large structure as SASAS is weak if the S-layers consist of smaller S-boxes. Surprisingly overlooked by the designers of white-box schemes, the generic attack [10] exploits multiset and differential properties of SASAS and applies to all the published white-box proposals so far. It appears that the mainstream ciphers are just poor choice for white-box implementations due to high diffusion properties and the way how the key is injected.

To formalize the problem, two notions have been suggested [47,49]. The **weak white-box implementation** of a cryptographic primitive protects the key and its derivatives i.e. aims to prevent the *key-recovery attack*. This ensures that unauthorized users can not obtain any compact information (e.g. the key or the set of subkeys) to decrypt the protected content.

The **strong white-box implementation** of a primitive protects from the *plaintext-recovery attack*, i.e. does not allow to decrypt given the encryption routine with the embedded key. Such an implementation may replace the public-key cryptosystems in many applications, in particular if it is based on an existing symmetric cipher and is reasonably fast for a legitimate user. The existing white-box implementations of AES and DES [16,17] do not comply with in this notion, since they are easily invertible, which is strikingly different from the black-box implementations of these ciphers. So far the only proposed candidate is the pairing-based obfuscator scheme with poor performance [47].

The ASASA-based designs may not only hide the key for the weak white-box implementation, but also provide non-invertibility aiming for the strong white-box construction.

Our Contributions

We continue to explore the design space of compact schemes built from layers of affine mappings and S-boxes. We first note that there is no known generic attack on the 5-layered ASASA scheme with injective S-boxes in the flavour of [10], which makes the ASASA structure a promising framework for future white-box, black-box, and public-key schemes. Based on this principle, we propose and analyze the following constructions in this paper:

- Two *public-key / strong white-box* variants of the ASASA symmetric scheme: one is based on Daemen's quadratic S-boxes [19] (previously used in various hash functions) and another based on random expanding S-boxes. (Section 2). We explore standard cryptanalytic attacks such as differential, linear and others, the recent decomposition attacks [30], and a new interpolation attack on weakened variants of our schemes (Section 3). We demonstrate that our set of parameters offers a comfortable security margin with respect to the existing attacks.
- A concrete instantiation for a fast symmetric ASASA-based blockcipher with secret S-boxes and affine layers and comparable with AES in it's encryption/decryption speed (Section 4).
- A concept of *memory-hard white-box implementation* for a symmetric block-cipher and a concrete family of ciphers with tunable memory requirements (Section 5). It prevents key recovery and requires the adversary to share the entire set of lookup tables to allow an unauthorized user to decrypt. Therefore, the cipher solves the problem of *weak white-box* implementation.

Due to the space limits, some references and attacks on the weakened variants of our schemes are not present in this paper and are available at [8].

2   Asymmetric ASASA Schemes: Strong White-Box and Public-Key

The first ASASA cryptosystem, designed by Patarin and Goubin, was a public-key scheme with non-bijective S-boxes and was easily broken by Biham, exploiting this property in [5]. Shortly afterwards, Biryukov and Shamir explored multi-layer schemes with bijective S-boxes and demonstrated a generic attack on the structure SASAS with two affine layers [10]. The outer S-boxes are recovered with a variant of the Square attack, whereas the inner affine layers are peeled off with linear algebra methods. It was clearly demonstrated that these properties disappear in larger schemes, and no attack on ASASA or other larger structures has been proposed since.

2.1   Strong White-Box Security

We start with the notion of the strong white-box security that summarizes the discussion in [49].

Definition 1. *Let the pair of algorithms* (E, D) *be a private-key encryption scheme, which takes key K as parameter. Let $\mathcal{O}_{E_K}$ be a function that computes E_K. We say that $\mathcal{O}_{E_K}$ is a secure strong white-box implementation for E_K if it is computationally hard to obtain $\mathcal{D}'$ equivalent to D_K given full access to $\mathcal{O}_{E_K}$.*

In other words, an adversary should be unable to decrypt given the white-box implementation $\mathcal{O}_{E_K}$ of E_K. This notion closely resembles the definition of a trapdoor permutation used to construct a public-key encryption scheme. As we see, our asymmetric proposals are suitable for both notions.

2.2 Outline

We propose several asymmetric instantiations of the ASASA structure, which may serve both in the white-box and public-key setting. We have not found any reasonable use for lookup tables in this framework[2] and hence look for polynomial-based S-boxes. In order to keep the reasonable size of the description, we restrict to polynomials of degree two over some finite field, so that the resulting scheme has degree four. This approach brings us to the area of *multivariate cryptography*, which aims to design cryptographic primitives based on multivariate polynomials over finite field.

Let us introduce the following notations. The public key/white-box implementation is exposed as a set of polynomials **b**, which is constructed out of the following composition:

$$\mathbf{b} = \mathcal{U} \circ \mathbf{a}_2 \circ \mathcal{T} \circ \mathbf{a}_1 \circ \mathcal{S}, \qquad (1)$$

where $\mathbf{a}_1, \mathbf{a}_2$ are nonlinear transformations, and $\mathcal{U}, \mathcal{T}, \mathcal{S}$ are affine transformations.

There have been many proposals for nonlinear layers in the ASA structure, and various attacks exploited these choices. Most attacks are not evidently translated into degree 4, as they compute, e.g., differentials of the public key, which are linear functions the ASA

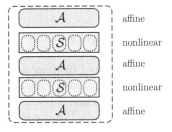

Fig. 1. The ASASA structure: two nonlinear layers surrounded by affine layers

case. The notable exception is the decomposition attack [30,31], that will be discussed in Section 2.3.

We offer two fresh ideas for the nonlinear layers in ASASA. The first candidate is the so called χ-function. It derives from invertible cellular automata and was brought into symmetric cryptography by Daemen. To the best of our knowledge, it has never been used in multivariate cryptography.

The second candidate is a set of random injective S-boxes of degree 2. Since the families of low degree permutations are small and do not absorb much randomness, we propose to use expanding S-boxes, which can be key-dependent.

---

[2] So far all attempts to hide a trapdoor in lookup table-based designs failed. We investigated this problem and conjecture that such scheme just does not exist, at least given the state-of-the-art in the design of preimage-resistant functions.

Having the expansion rate of 2, it is rather easy to obtain injective transformations and still keep them quadratic[3].

Limitations for expanding schemes. Whatever construction is used, an expanding scheme has a clear limitation in the public-key and white-box setting. It implies that only a tiny subset of potential ciphertexts is decryptable, which makes the encryption and decryption process non-interchangeable. As a result, the expanding scheme can be used for encryption only and can not produce signatures. Also in the white-box context, it can not be used for decrypting the content. On the other hand, it can still be used to ensure tamper-resistance of software [41].

2.3 Defeating Decomposition Algorithm with Perturbations

The authors of recently published decomposition algorithms [30, 31] claim to break ASASA schemes with quadratic nonlinear layers with complexity $O(n^9)$, where n is the number of variables. The decomposition problem is formulated as follows: given a set of polynomials $h = (h_1, \ldots, h_u)$ over polynomial ring $\mathbb{K}[x_1, \ldots, x_n]$ ($\mathbb{K}$ denoting an arbitrary field) find any $f = (f_1, \ldots, f_u)$ and $g = (g_1, \ldots, g_n)$ over $\mathbb{K}[x_1, \ldots, x_n]$ whose composition is equal to h:

$$h = (h_1, \ldots, h_u) = (f_1(g_1, \ldots, g_n), \ldots, f_u(g_1, \ldots, g_n)).$$

and their degree being smaller than h.

In the context of the ASASA structure with quadratic S-boxes, the sets f and g, that are produced by a decomposition algorithm, are linearly equivalent to the internal ASA structures. This does not fully constitute a break, since the adversary still needs to invert both ASA constructions. The proposed algorithms also have not been applied to the parameters and fields that we choose. Nevertheless, it is desirable to find some countermeasure.

Our idea is to introduce some perturbation just after the second S layer in the form of several key-dependent secret polynomials of degree 4. A similar approach has been used by Ding in his modification of C* [21] and HFE [22]. In some cases (notably HFE), the "perturbation" would be identified and removed [25], thanks to a differential attack exploiting properties of the non-linear transformations. The use of perturbation polynomials has been also linked to the LWE (Learning with Error) framework in [32], but the full application of LWE to multivariate cryptography is still to be explored in the future.

Denoting the perturbation polynomials as another nonlinear transformation $\mathbf{a}_p$ we obtain the modified public key $\mathbf{b}_p$:

$$\mathbf{b}_p = \mathcal{U} \circ [\mathbf{a}_p + (\mathbf{a}_2 \circ \mathcal{T} \circ \mathbf{a}_1 \circ \mathcal{S})], \tag{2}$$

so

$$\mathbf{b}_p(x) = \mathbf{b}(x) + \mathcal{U}\mathbf{a}_p(x).$$

---

[3] Our experiments show that S-boxes with an even smaller rate of 1.75 can be found.

Hence the perturbation polynomials are mixed by the last affine transformation and spread over the public key. The encryption process remains exactly the same, while for decryption we have to guess the values of these polynomials. Suppose that we work over $\mathbb{F}_2$ so that $\mathbf{a}_p$ is sparse and contains only w polynomials. Let each polynomial be nonzero in $q \cdot 2^n$ points. Then the noise on average consists of qw bit flips, and we guess

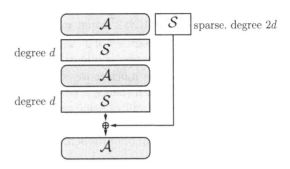

Fig. 2. Small perturbations to defeat decomposition attacks as injection of sparse high-degree polynomials

their positions after about $\binom{w}{qw}$ attempts. For instance, $\mathbf{a}_p$ with 8 non-zero polynomials of weight $\approx 2^{n-1}$ requires 2^6 trial decryptions on average.

We distinguish true plaintexts from false ones either by recomputing the perturbation polynomials or by using expanding S-boxes so that noisy bits prohibit inversion. Padding the plaintexts with zero bits also helps but disallows turning encryption to decryption. The position of noisy bits does not matter much, since it would be concealed by the affine transformation. However, if we filter out noise with expanding S-boxes, it makes sense to spread the noisy bits so that an S-box can still be inverted in the presence of noise.

2.4 χ-Scheme

Our first idea was to build the nonlinear transformation out of a popular quadratic S-box χ [19, Section 6.6.2], which has been used in several hash functions including SHA-3/Keccak [4]. The transformation χ can be defined for every odd length $k = 2t + 1$ and has the following features:

- It has degree 2 in the forward direction, but degree $(t + 1)$ in the backward direction.
- It can be efficiently inverted for every size [8].
- Its differential and linear properties have been widely studied [19].

The S-box χ of length k is defined as follows:

$$\chi(x_0, x_1, x_2, \ldots, x_{k-1}) = (y_0, y_1, y_2, \ldots, y_{k-1}),$$

where

$$y_i = x_i \oplus x_{i+1}x_{i+2} \oplus x_{i+2},$$

and indices are computed modulo k.

Regardless of the χ length (and hence the size of the S-box), we can formulate properties of the whole scheme and its features:

1. For the standard block size of 128 bits, we get approximately (since the S-box size might not divide 128) 2^7 input variables. Thus each output coordinate of **b** is a polynomial of degree 4 of 2^7 variables, has about $\binom{2^7}{4} \approx 2^{24.5}$ terms, so the full scheme description is about $2^{24.5+7-3} = 2^{28.5}$ bytes, or 300 MBytes.

2. The private key size is much more compact, and is dominated by three matrices with 2^{14} bits each (hence 2^{13} bytes in total). If the matrices are deterministically produced out of some secret key (e.g., 128-bit), the description is even smaller.

3. The inverse polynomial of our schemes has degree $(t+1)^2$ for S-boxes of size $2t+1$.

The S-box size (length of χ) has negligible effect on the performance because of the internal structure of χ and its inverse. Hence it only affects the security of the scheme. We choose a single S-box **a** of length 127, so that its inverse has degree 64, and will show later that a system with small S-boxes is insecure.

In order to defeat decomposition algorithms and hide the ASASA structure we suggest using perturbation polynomials. More precisely, we propose 24 random polynomials of degree 4 for the perturbation layer a_p. We pad each plaintext with 8 zero bits, so that for each guess the probability to fit the padding is 2^{-8}. As a result, we get 2^{16} candidate plaintexts, and then check if we correctly computed the noise. This filters out all wrong plaintexts with probability $1 - 2^{-8}$.

Overall security. We have found a number of attacks on the χ-scheme in different variants (Section 3), so it appears that its algebraic structure yields it vulnerable. Nevertheless, the variant with added perturbation remains unbroken, and we offer it as cryptanalytic challenge, but not for the practical use. We expect the perturbation theory to develop in the near future, which would suggest a more secure set of parameters.

2.5   Scheme with Expanding S-Boxes

This variant provides a more compact description of the scheme since we may switch to a larger field. First, we want the nonlinear layer be a degree-2 polynomial over $\mathbb{F}_q, q > 2$, and define the linear (affine) transformations over the same field. Though a few examples of bijective transformations of degree 2 over a field not equal to $\mathbb{F}_2$ exist [36], they appear to be vulnerable to Groebner basis attacks in our own experiments. As a solution, we suggest expanding S-boxes, whose output is twice as big as the input. It is rather easy to design *injective* S-boxes of degree 2 with this property. Indeed, a random function with expansion rate 2 has no collisions with probability around $1/2$, and hence there are enough injective transformations of the desired form.

Here is the summary of the scheme:

- Input length 128 bits (32 variables), output length 512 bits (128 variables);
- All polynomials and affine transformations are defined over $\mathbb{F}_{16}$;

- S-boxes map 16 bits to 32 bits and hence are described by 8 degree-2 polynomials over $\mathbb{F}_{16}$ of four variables. The inverse is computed with lookup tables of size 2^{16}.
- The first nonlinear layer has 8 S-boxes and doubles the state size to 256 bits. The second layer has 16 S-boxes and further doubles to 512 bits. Accordingly, the affine transformations S, T, U operate on 128-, 256-, and 512-bit states, respectively.

The output of the scheme is a set of 2^7 degree-4 polynomials over $\mathbb{F}_{16}$ over 32 input variables (each variable is encoded with 4 bits). There are $\binom{2^5}{4} \approx 2^{16.5}$ possible terms, hence, taking 4-bit constants into account, each polynomial is described by $2^{20.5}$ bits, or $2^{20.5+7-3} = 2^{24.5}$ bytes, which is about 24 MBytes.

The private key is smaller: affine layers contain $2^{7+7+1} + 2^{6+6+1} + 2^{5+5+1} \approx 2^{14.2}$ elements of $\mathbb{F}_{16}$. The 48 S-boxes are described as $2^{5.5+3}$ polynomials of $21 \approx 2^{4.5}$ terms each, hence 2^{13} elements, plus a few noise polynomials. In total, the private key fits into 2^{14} bytes.

We also suggest using perturbation polynomials here. Due to the large expansion rate, we can use rather dense perturbation layer $\mathbf{a}_p$ and still ensure a unique decryption. We use two random polynomials over $\mathbb{F}_{16}$ of degree four at each S-box, hence 32 polynomials in total. While decrypting we face $16^2 = 2^8$ options for each S-box output. As a result, the probability of having non-unique decryption of the last S layer is $2^4 \cdot 2^{16+8-32} = 2^{-4}$, and if this happens the next layer filters out wrong candidates.

As we already mentioned, the expanding character of the scheme allows only public-key and white-box encryption, but not signature generation.

3   Security Analysis of Our White-Box/Public-Key Schemes

In this section we apply various attacks to weakened versions of our schemes, thus demonstrating the design rationale behind them. We demonstrate that the added perturbations are crucial in both schemes, and that they must be secret. We also show that S-boxes in the χ-scheme must be large, that linearity (in contrast to affinity) of A may weaken the scheme, and that the expanding S-boxes should not be biased (these results are presented mainly in [8]). Our attacks are summarized in Table 1.

These attacks allow us to evaluate the security margin of the unbroken variants of our schemes. Since only the perturbation protects the χ-scheme from a number of practical attacks, we conclude that it is rather fragile, but might become a good candidate for a strong white-box implementation when the complexity of generic algorithms applied to the perturbed version is better understood. In contrast, the expanding scheme appears to be more resistant to generic attacks, and we propose it as a ready-to-use public-key encryption scheme and a strong white-box implementation.

Table 1. Summary of our attacks on the weakened versions of our schemes. D stands for the complexity of decomposition attacks.

Weakening	Attack complexity	Attack type	Reference
Expanding scheme			
Public perturbation	$2^{45} + D$	Interpolation	Section 3.2
Biased S-boxes (bias= 1/8)	2^{88}	LPN	Section 3.4
χ-scheme			
Public perturbation	$2^{57} + D$	Interpolation	Section 3.2
No perturbation	$\approx 2^{40}$	Groebner-basis	Section 3.3
Small S-boxes	2^{45}	Algebraic	[8]

3.1   Generic Attacks

Given the public-key of a multivariate scheme, an attacker may directly try to solve the multivariate polynomial equations using a generic algorithm. If the public-key is a vector of m polynomials in n over $\mathbb{F}_q$, then a plaintext can always be found by exhaustive search in time $\mathcal{O}(q^n)$. The other main family of algorithms to solve systems of polynomial equations are Groebner-basis algorithms, such as Buchberger's algorithm and all its derivatives [27, 28].

Without going into details (the interested reader is referred to a standard textbook such as [18]), given a system of polynomial equations $f_1 = \cdots = f_m = 0$ in $x_1, \ldots, x_n$, a *Groebner basis* of the ideal spanned by the f_i's is an equivalent system of equations with nice properties. If the system admits a single solution $(a_1, \ldots, a_n)$, then a Groebner basis is precisely the vector of polynomials: $x_1 - a_1, \ldots, x_n - a_n$. It follows that if a Groebner basis can be computed, then the system of equations can be solved.

Groebner basis algorithms work by performing *polynomial elimination, i.e.,* by trying to eliminate some terms by summing suitable multiples of other polynomials. The complexities of these algorithms are difficult to analyze [2]. They are essentially exponential in the highest degree reached by the polynomials created and manipulated by the algorithms during their execution. On "generic" systems of n equations in n variables, this degree is typically n. However, in some special cases it can be lower. For instance, the first HFE Challenge could be broken because in HFE, for some ranges of parameters, this degree was roughly $\mathcal{O}(\log n)$.

3.2   Interpolation Attack on the **ASASA** Scheme with Public Perturbation Polynomials

We stressed that the perturbation polynomials must be secret. A reader may wonder why this is required, since these polynomials are seemingly mixed by the last affine transformation $\mathcal{U}$.

In this subsection we outline an attack that peels off the perturbation polynomials and recovers the core ASASA scheme in almost practical time. Suppose we work over a field $\mathbb{F}_2$ and the scheme adds perturbation polynomials at r bit positions after the nonlinear transformation $\mathbf{a}_2$ (cf. Eq. (2)), and the total number of variables in the scheme is n. Then we collect N plaintexts x_i such that

$$\mathbf{a}_p(x_i) = 0.$$

Since polynomials of $\mathbf{a}_p$ do not have any structure, finding a common zero is an NP-hard problem, and we expect that 2^r plaintexts must be tried to find a right one. Hence the naive complexity of this step is $N2^r$ evaluations[4] of $\mathbf{a}_p$.

Then we evaluate the right plaintexts on the perturbed scheme $\mathbf{b}_p$. Since $\mathbf{a}_p$ is zero, we have

$$\mathbf{b}_p(x_i) = \mathcal{U} \circ \mathbf{a}_2 \circ \mathcal{T} \circ \mathbf{a}_1 \circ \mathcal{S}(x_i).$$

Therefore, we know the evaluation of the ASASA scheme without perturbations on N plaintexts. Since the scheme has degree 4, the polynomial coefficients can be recovered by the Lagrange interpolation. There are $\sum_{i=0}^{4} \binom{n}{i}$ monomials of degree 4 or smaller, hence N must slightly exceed $\sum_{i=0}^{4} \binom{n}{i}$ to allow for linear dependencies among plaintexts. For the typical value $n = 2^7$ we need about 2^{25} right plaintexts to fully recover the core ASASA polynomials and then launch the decomposition attack. However, the interpolation itself is not a trivial procedure, since we deal with a multivariate function. Only recently an algorithm with complexity quadratic in the number of monomials has been proposed [1]. Equipped with it, we recover a single polynomial in 2^{50} bit operations, and the entire $\mathbf{b}_p$ in 2^{57} operations. In turn, 2^{25} right plaintexts can be obtained for 16 noisy bits in 2^{41} evaluations of $\mathbf{a}_p$, and for 24 noisy bits – in 2^{49} evaluations, which is close to 2^{55} bit operations. Therefore, the total complexity of recovering (x) is about 2^{57} bit operations.

This attack clearly shows that the perturbation polynomials must not be public and should not have any structure that would allow the adversary to find their common zeros. We do not see how the attack can be applied to secret polynomials.

3.3   Algebraic Attack on the Plain χ-Scheme

Although χ has been used successfully in the symmetric world, it turns out to be a complete disaster in a multivariate context. An ASA construction where $S = \chi$, with $n = 127$ variables over $\mathbb{F}_2$ is broken in a few seconds by a direct Groebner basis computation. A two-layer ASASA construction is not more secure, and can be broken in less than two hours using the implementation of the F4 algorithm of the MAGMA computer algebra system [14] (and 100Gbytes of RAM). This happens because a Groebner basis can be computed by manipulating polynomials of small, constant degree (typically 3 or 6).

Let us give some detailed explanation for the insecurity of the ASA construction. We work within the polynomial ring $R = \mathbb{F}_2[x_0, \ldots, x_{n-1}]$, and we consider

---

[4] Finding subsequent solutions might be easier, but this step is not a dominant in our attack complexity.

the ideal of R:

$$\mathcal{I} = \left\langle f_0, \ldots, f_{n-1}, \quad x_0{}^2 - x_0, \ldots, x_{n-1}{}^2 - x_{n-1} \right\rangle$$

where $f_i = x_i + x_{i+2} + x_{i+1}x_{i+2} + a_i$ (all indices are taken modulo n), and where the a_i are constants. Any solution (in the x_i's) making all the polynomials in $\mathcal{I}$ vanish simultaneously, is a solution of $\chi(x_1, \ldots, x_n) = (a_0, \ldots, a_{n-1})$. Such a solution always exists, and is unique.

We will show that there are many linear polynomials in this ideal, and that they can be "easily" discovered (by manipulating small-degree polynomials). Indeed:

$$x_{i+1} \cdot f_i - x_{i+2} \cdot \left(x_{i+1}{}^2 - x_{i+1}\right) + f_{i-1} = (x_{i-1} + x_{i+1}) - (a_{i-1} + a_{i+1})$$

The expression on the left-hand side is a polynomial combination of elements of $\mathcal{I}$, therefore it belongs to $\mathcal{I}$. As a consequence, the linear polynomial on the right-hand side can be found inside $\mathcal{I}$ after performing a few steps of polynomial elimination on polynomials of degree less than 3.

After these n linear relations have been found, another few steps of polynomial elimination allows all the variables but one to disappear. This shows that a Groebner basis of the ideal $\mathcal{I}$ can be computed in polynomial time. Now, performing a (random) linear change of coordinate in $\mathcal{I}$, or replacing the generators of $\mathcal{I}$ by (random) linear combinations thereof does not change this fact. As a conclusion, the ASA construction, where S is the χ-function, falls victim to a direct algebraic attack, by running any Groebner basis algorithm on the equations defining the "white-box".

This reasoning extends to the ASASA construction where both non-linear layers are χ (however, this time the degree is 6). It is an open question of how much the added perturbation slows the Groebner-basis attacks (our implementation does not break the selected noise parameters in reasonable time).

3.4   Attack on the Expanding Scheme with Biased S-Boxes

If S-box output bits are biased, an attack exploiting this bias can be applied. The last affine transformation can be viewed as affine over $\mathbb{F}_2$, so the further analysis without loss of generality applies to any field of characteristic two.

We target a single biased bit b after the second layer of expanding S-boxes: the probability $\mathbb{P}[b = 1]$ of its equality to 1 is equal to $p \neq \frac{1}{2}$. If y is a ciphertext, then following previous notations, the biased bit is the b-th component of $U^{-1} \cdot y$. In other terms, if $\overline{u}$ denotes the b-th line of U^{-1}, then $\langle \overline{u}, \overline{y} \rangle = b$.

Now, assume we collect a large number (say N) of ciphertexts. We stack them vertically into a matrix C, which thus has N rows. Let us also assume that b is biased towards zero. Then we have the "noisy linear system":

$$\overline{u} \cdot C = \overline{e},$$

where $\overline{e}$ is a vector of i.i.d. random variables following the Bernoulli distribution with mean p. Recovering $\overline{u}$ is exactly an instance of the Learning Parity with Noise (LPN) problem.

The best known algorithms to solve LPN are variants of the BKW algorithm [11], whose complexity is of order $\mathcal{O}\left(2^{n/\log n}\right)$. The only actual implementation (along with algorithmic improvements) is described in [37], and some more tweaks are given in [3]. The actual complexities of these algorithms depend on the bias (their efficiency decreases when the bias gets closer to zero).

With $n = 512$ variables, and if $\mathbb{P}[b = 1] = 1/8$, then the implementation of [37] is said to require 2^{80} bits of memory (plus the time needed to sort this much memory 80 times). However, time-memory tradeoffs, plus algorithmic improvements, allow [3] to conclude that the same problem can be solved in 2^{59} bits memory and less than 2^{100} bits operations. If 2^{80} bits of memory are available, then the running-time could be decreased to 2^{88} bit operations.

This beats more naive approaches, such as, for instance, enumerating all the possible sparse possibilities for the first n components of $\bar{e}$, and solving the corresponding linear system for each trial. The above instance would require more than 2^{120} operations to be solved using the naive approach.

Of course, the attack has to be repeated for each row of U^{-1}, and possibly twice for each row (assuming that the targeted bit is biased towards zero, or towards one). Note that the above estimates are extremely pessimistic; in random expanding S-boxes of degree 2, the biases we observed experimentally are much lower than what was used above (we observed $\mathbb{P}[b = 1] \approx 0.49$).

After U is recovered, we can view the output of expanding S-boxes, and are likely to recover them by interpolation due to low degree.

4   Black-Box ASASA Schemes

Given rather low performance and large key size of the public-key ASASA schemes, a reader may wonder if significant performance increase can be achieved with lower security goals. We answer this question twofold. First, we propose a generic black-box symmetric cipher based on the five layerASASA. The cipher is expected to have a very fast software implementation thanks to vector instructions in modern processors. Secondly, we use a small version of this cipher as a building block in achieving weak white-box security (Section 5).

4.1   Design

We propose a symmetric cipher with a classical set of parameters, widely used in AES and other designs. It has a block of $n = 128$ bits with $m = 8$ bit S-boxes and a choice of key-sizes $128 - 256$ bits. Let us outline specific parameters for linear and nonlinear layers.

Affine layers. A key-dependent $n \times n$ affine transformation can be produced out of the master key K by any secure key derivation function H_K (for example a fast stream cipher, or a block cipher in the counter mode (more details in [8])), and checking that the resultant matrix is invertible, this can be done in $O(n^3)$ steps, and we also generate an n-bit constant [5]. The branch number of the matrix [20] determines the minimum number of active S-boxes in a differential trail, and thus the upper bound on the trail probability. Since the matrix is random, we expect the branch number to be close to the maximum possible (the number of S-boxes n/m plus one). Note that for each affine layer a new matrix is generated.

Nonlinear layers. Typically, nonlinear layers of symmetric ciphers consist of several small S-boxes, which have a compact description [12,20]. For the ASASA scheme we propose to use 32 randomly generated 8-bit invertible S-boxes, which are all different and key-dependent. We note that efficiency of generic attacks on the SASAS structure [10] increases if smaller S-boxes are used, and thus it may be interesting in the future to explore full block size non-linear layers, for which such attack would not work.

Large S-box alternatives. The choice for large block algorithmic S-boxes is surprisingly limited. Unless the S-boxes are themselves multi-layer permutations (e.g., fixed-key ciphers), a compact description is typically delivered in the algebraic form as a function over an appropriate finite field. The resulting *permutation polynomials* have become an active research topic in the recent years. The well known example is X^{2^k+1} over $\mathbb{F}_{2^n}$ (scheme C* [39]); the more recent and interesting include $\left(X^{2^k} + X + a\right)^{-l} + X$ over $\mathbb{F}_{2^n}$ by Zeng et al. [50] (derived from Helleseth-Zinoviev polynomials) (more references in [8]). It thus can be an interesting second challenge to break the symmetric ASASA scheme with known block-wide non-linear layers. Note however that fixed S-boxes do not offer implementation advantage and thus we would keep S-boxes secret and randomly generated in the main variant of our scheme. Implementation details

Implementation. The implementation details can be found in the full version of the paper [8].

4.2   Security Analysis

Differential and linear attacks. We expect the secret linear layers to hide all differential [6] and linear [38] properties of the cipher, since it becomes impossible to figure out any high-probability differential or linear trail. It can be argued, however, that the existence of high-probability characteristics may lead to efficient distinguishers. For instance, Dunkelman and Keller showed in [26] that if for every α the differential probabilities $\{\alpha \to \beta\}$ are much higher (or much lower) than for a random permutation, then this can be used as a distinguisher. The authors further suggested the parameter of effective linearity that essentially measures the average probability of the boomerang difference quartet (α, K)

---

[5] Non-anonymous final version of this paper will link to implementations of our schemes and challenges gradually increasing complexity for the interested cryptanalysts.

over all possible α, K and is supposed to take even unknown characteristics into account. In the full version of this paper [8] we show that all these methods do not lead to attacks much faster than exhaustive key search.

Algebraic attacks. We expect the random S-box of width m to have the algebraic degree $m - 1 = 7$. As a result, the entire scheme can be represented by a polynomial of degree 49 over $\mathbb{F}_2$. As observed by Meier [40], the low degree can be detected by applying differentials of the same order to the ciphertext. Therefore, an attacker can distinguish the ASASA construction from random given 2^{49} chosen data and time. However, this does not lead to the disclosure of the plaintext, and it is unclear how this property can be exploited.

Other attacks. The boomerang and impossible differential attack can be also of concern. We have tried basic and improved versions of these attacks, and in all cases the randomness of the affine layers prevented us from mounting an attack. However, it is possible to build boomerang quartets in the known-key setting by activating a single S-box at both sides of the boomerang. Whether such properties can be carried out to the secret-key setting is the object of the future research. Impossible differential attacks typically rely on truncated differentials with probability 1 which exist in some ciphers due to incomplete diffusion. Since in our case the random affine layers provide complete diffusion and since the entrance into and the exit from the scheme are both guarded by these affine layers, chosen plaintext attacks have little chance of predicting truncated values somewhere inside the scheme.

Our scheme should be more secure than a two-round Even-Mansour cipher [13], where the subkeys are simply xored to the internal state (as opposed to applying a full-blown secret affine transformation to the internal state). The recent attack on the 2-round Even-Mansour [23] explicitly requires the access to the internal permutation and thus can not be immediately used in our setting.

The meet-in-the-middle attacks [33] do not apply to our scheme, because the amount of key material used to compute any matching variable is too large (several S-boxes and a large part of the affine transformation). The cube attacks do not apply, since there is no compact polynomial representation of the scheme.

Structural attack. Finally, we investigate the structural attack from [10]. We will see that even though it does not apply to the 128-bit ASASA cipher, it allows to bound the security level of schemes with smaller block, which are used in Section 5. First, we recall the main property preserved by the SASA structure with m-bit S-boxes:

Theorem 1 ([10]). *Let $\{x_1, x_2, \ldots, x_{2^m}\}$ be the inputs to the SASA structure such that the input bits to one S-box take all possible combinations whereas the other bits are constant. Then the XOR of all outputs is the all-zero bit vector:*

$$\bigoplus_i \mathbf{SASA}(x_i) = (0, 0, \ldots, 0).$$

Now consider the input y to some m-bit S-box in the first S layer of the ASASA scheme E. It is an affine function of the scheme input y:

$$y = M \cdot x \oplus c,$$

where M is an $(m \times n)$-matrix and c is a constant. Let L be an $(n \times n)$-matrix such that

$$M \times L = \begin{pmatrix} M' \, 0 \, 0 \cdots 0, \end{pmatrix} \tag{3}$$

where M' is a $(m \times m)$- submatrix. If we apply E to $L \cdot x$, then y depends only on the first m bits of x. Thus we compose inputs $\{x_1, x_2, \ldots, x_{2m}\}$ as in Theorem 1, multiply them by L and apply E. The output bits must sum to 0 bit-wise. This property allows to recover the outer affine layer and eventually all the components of E. Equation (3) holds with probability $2^{-(n-m)m}$, which makes the attack impractical for large n. However, for small n it might be efficient. Therefore, the maximum security level of the ASASA scheme with n variables and m-bit S-boxes does not exceed $(n - m)m$ bits.

For our design, this gives the upper bound of 960 bits, which is far larger than the key length which is used to generate the affine and nonlinear layers. As a result, we claim the 128-bit security level for our design, even though a small factor over exhaustive key search might be saved by biclique attacks or by exploiting the full codebook. This is still higher security than what is offered by 2048 bit RSA.

5   Proposal for Weak White-Box Security: ASASA-Based Block Cipher

5.1   Weak White-Box Security

Definition 2. *Let the pair of algorithms (E, D) be a private-key encryption scheme, which takes key K as a parameter. We call $\mathfrak{F}(K)$ the equivalent key set for key K, if from any element from $\mathfrak{F}(K)$ it is easy to get an algorithm equivalent to E_K, i.e. there is an (efficient) algorithm*

$$\mathcal{A}(\mathcal{K}) \to \mathcal{E}',$$

where $\mathcal{E}'$ equivalent to E_K.

Definition 3. *The function $\mathcal{O}_{E_K}$ is a T-secure weak white-box implementation for E_K if it is computationally hard to obtain $\mathcal{K} \in \mathfrak{F}(K)$ of length less than T given full access to $\mathcal{O}_{E_K}$.*

In other words, an adversary who gets a secure weak white-box implementation is unable to find out any compact (shorter than T) equivalent representation of it. In the practical sense, an adversary who wants to share a protected implementation of the encryption routine, would have to share the entire code. Such ciphers are motivated by DRM applications, which aim to prohibit the users of protected content from sharing the information needed to decrypt it. Clearly, in this context there is little practical difference between sharing the key and sharing, say, the set of subkeys as long as the other cipher operations are independent of the key. Therefore, naive methods of key protection, e.g. transforming it with a preimage-resistant hash function, would not prevent an attack. Ideally, the adversary would have to isolate and extract the entire decryption routine, which might be hard per se.

5.2 Weak White-Box Cipher Proposal

In this section we propose a blockcipher family, which conforms to the weak white-box security notion, so that it is computationally infeasible to derive a key or any other compact secret information from the white-box implementation.

We further say that the white-box implementation is *memory-hard* if it requires a pre-specified and large enough amount of memory in the spirit of memory-hard key-derivation functions [44]. This concept is even stronger than the T-secure weak WB implementations, as an adversary is unable to reduce the implementation size at all and thus would have to publish the entire set of lookup tables. In contrast to earlier white-box designs, we offer a set of ciphers with a wide range of memory requirements.

Our memory-hard cipher consists of a number of smaller components, which are exposed as lookup tables in the white-box implementation. Each component is either a small-block ASASA cipher, adapted from the construction in Section 4, or just a single S-box. The S-boxes are minimum 8-bit wide to avoid equivalence problems [9], but 10-, and 12-bit ones are also used. All S-boxes and affine layers are derived in a deterministic way from the secret key. In

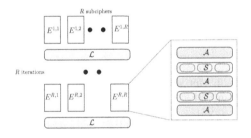

Fig. 3. Blockcipher family for weak white-box security

fact, it is enough to have linear, not affine, layers, since the constant can be kept in the S-boxes. To estimate memory requirements, we assume for simplicity that each table output fits an integer number of bytes (e.g., 2 bytes for 10-bit S-boxes).

We propose the SPN structure for the cipher, i.e. we alternate layers of smaller ciphers (denote their number by R) with a public linear transformation $\mathcal{L}$. Any transformation with good diffusion shall be fine. Recalling that AES can be partitioned into 5 rounds with 4 32-bit Super S-boxes in each, we propose R layers for similar security margin. The cipher's pseudocode is as follows (Figure 3):

– Repeat R times;
  • Apply R parallel ASASA-based distinct blockciphers;
  • Apply the linear transformation $\mathcal{L}$ to the entire state.

We outline specific parameters and memory requirements for 64- and 128-bit blockciphers in Table 2. The 16-bit S layer has two 8-bit S-boxes, the 18-bit – 8-bit and 10-bit S-boxes, the 20-bit – two 10-bit S-boxes, and the 24-bit – three 8-bit S-boxes. We see that whereas the black-box implementation is a few dozen KBytes, the white-box implementation can be made large enough in the range from 2 MBytes to several GBytes.

Table 2. Parameters and memory requirements of white-box and black-box implementations for the 128-bit blockcipher. We assume that n-bit component occupies $\lceil \frac{n}{8} \rceil 2^n$ bytes of memory in the white-box implementation.

Rows	Component	Components in row	Security level (bits)	White-box memory	Black-box memory
		64-bit block			
4	ASASA	$4\times$(16-bit)	64	2MB	16 KB
4	ASASA	$3\times$(18-bit) + 10-bit	64	9 MB	32 KB
		128-bit block			
8	ASASASA	$8\times$(16-bit)	128	8 MB	96 KB
8	ASASA	24-bit + $6\times$(16-bit) +8-bit	64	384 MB	64 KB
5	ASASASA	$4\times$(28-bit) + (16-bit)	128	20 GB	130 KB

Table 3. Our schemes in comparison, along with (presumably) secure parameters for UOV and HFE

Scheme	Field	# vars	# polys	Degree	Private key	PK/white-box	Ref.
Black-box ASASA	$\mathbb{F}_2$	128	-	-	14 KB	196 KB	Sec. 4
χ-scheme*	$\mathbb{F}_2$	127	127	4	8 KB	300 MB	Sec. 2
Expanding scheme	$\mathbb{F}_{16}$	32	128	4	16 KB	24 MB	Sec. 2
Memory-hard cipher	$\mathbb{F}_2$	64	-	-	16-32 KB	2-8 MB	Sec. 5
Memory-hard cipher	$\mathbb{F}_2$	128	-	-	64-130 KB	8 MB – 20 GB	Sec. 5
HFE	$\mathbb{F}_{16}$	64	64	2	48 KB	520 KB	[43]
UOV	$\mathbb{F}_{256}$	78	26	2	71 KB	80 KB	[34]

* — several variants broken in this paper.

Security Analysis. Our ASASA components have very small block and only a few S-boxes in the S layer. Some attacks that are infeasible on the 128-bit block, may have practical complexity on the 16-bit block. The best attack we could find was presented in Section 4.2 and has complexity $2^{(n-m)m}$ for m-bit S-boxes and the n-bit block. As a result, the 16-bit ASASA components with 8-bit S-boxes have maximum security level of 64 bits, the 20-bit components — 100 bits, and 24-bit components with 8-bit S-boxes — 128 bits.

An easy way to increase the security level is to add two more layers, thus producing ASASASA components. This yields a 50% increase of the private key size, but no increase in the white-box implementation size. Since we have not found a way to expand our attack to this structure, we conjecture its security level to 128 bits. In Table 2 we provide both variants so that a protocol designer may choose between them according to his own requirements.

6   Conclusion

We have explored deeply the state of the art in black-box, white-box, and multivariate public key cryptography, and concluded that the ASASA structure is the minimal generic construction which is still unbroken. We constructed cipher candidates for all these settings. We designed two ASASA schemes for public-key cryptography based on multivariate polynomials. We showed how to avoid existing attacks on multivariate schemes, including the recent powerful decomposition algorithms by adding appropriate perturbation functions. In the traditional black-box setting we offered a cryptanalytic challenge of a fast cipher with small random S-boxes and random affine layers.

We proposed several solutions for white-box cryptography, both in weak and strong security notions. In the weak model, we designed a memory-hard cipher, which prohibits key extraction and requires an adversary to spend a large, predefined amount of memory. It is based on small ASASA components. We showed how our multivariate schemes can be used as strong white-box implementations, as they are not invertible without the key and allow fast encryption and decryption for legitimate users. We compare the implementation size of our schemes with other unbroken MQ-systems in Table 3.

Our findings indicate a number of future research directions. First, it would be interesting to explore algorithmic large S-boxes in the black-box ASASA structure, e.g. instantiated with recently found permutation polynomials. Secondly, a theory of perturbation layers as a countermeasure to generic decomposition algorithms needs to be developed, possibly along the concept of LWE (Learning with Error). Thirdly, we suggest investigating the actual security level of small-block (16-,20-, 24-bit) ASASA schemes to figure out which components are suitable for weak white-box implementations. Finally, open question is to develop constructions with smaller descriptions (e.g., within 1 MByte), which are bijective, suitable for digital signatures, and allow strong white-box implementations.

Acknowledgements. We thank Willi Meier for fruitful comments and the discussion about the results of this paper. We also thank anonymous reviewers for comments, which helped to improve the paper.

References

1. Armknecht, F., Carlet, C., Gaborit, P., Künzli, S., Meier, W., Ruatta, O.: Efficient computation of algebraic immunity for algebraic and fast algebraic attacks. In: Vaudenay, S. (cd.) EUROCRYPT 2006. LNCS, vol. 4004, pp. 147–164. Springer, Heidelberg (2006)
2. Bardet, M., Faugère, J.-C., Salvy, B.: On the complexity of Gröbner basis computation of semi-regular overdetermined algebraic equations. In: Proc. International Conference on Polynomial System Solving (ICPSS), pp. 71–75 (2004)
3. Bernstein, D.J., Lange, T.: Never trust a bunny. In: Hoepman, J.-H., Verbauwhede, I. (eds.) RFIDSec 2012. LNCS, vol. 7739, pp. 137–148. Springer, Heidelberg (2013)
4. Bertoni, G., Daemen, J., Peeters, M., Van Assche, G.: The Keccak reference, version 3.0 (2011), http://keccak.noekeon.org/Keccak-reference-3.0.pdf
5. Biham, E.: Cryptanalysis of Patarin's 2-Round Public Key System with S Boxes (2R). In: Preneel, B. (ed.) EUROCRYPT 2000. LNCS, vol. 1807, pp. 408–416. Springer, Heidelberg (2000)
6. Biham, E., Shamir, A.: Differential Cryptanalysis of the Data Encryption Standard. Springer (1993)
7. Billet, O., Gilbert, H., Ech-Chatbi, C.: Cryptanalysis of a white box AES implementation. In: Handschuh, H., Hasan, M.A. (eds.) SAC 2004. LNCS, vol. 3357, pp. 227–240. Springer, Heidelberg (2004)
8. Biryukov, A., Bouillaguet, C., Khovratovich, D.: Cryptographic schemes based on the ASASA structure: Black-box, white-box, and public-key. Cryptology ePrint Archive, Report 2014/474 (2014), http://eprint.iacr.org/
9. Biryukov, A., De Cannière, C., Braeken, A., Preneel, B.: A toolbox for cryptanalysis: Linear and affine equivalence algorithms. In: Biham, E. (ed.) EUROCRYPT 2003. LNCS, vol. 2656, pp. 33–50. Springer, Heidelberg (2003)
10. Biryukov, A., Shamir, A.: Structural cryptanalysis of SASAS. In: Pfitzmann, B. (ed.) EUROCRYPT 2001. LNCS, vol. 2045, pp. 394–405. Springer, Heidelberg (2001)
11. Blum, A., Kalai, A., Wasserman, H.: Noise-tolerant learning, the parity problem, and the statistical query model. J. ACM 50(4), 506–519 (2003)
12. Bogdanov, A., Knudsen, L.R., Leander, G., Paar, C., Poschmann, A., Robshaw, M.J.B., Seurin, Y., Vikkelsoe, C.: PRESENT: An ultra-lightweight block cipher. In: Paillier, P., Verbauwhede, I. (eds.) CHES 2007. LNCS, vol. 4727, pp. 450–466. Springer, Heidelberg (2007)
13. Bogdanov, A., Knudsen, L.R., Leander, G., Standaert, F.-X., Steinberger, J.P., Tischhauser, E.: Key-alternating ciphers in a provable setting: Encryption using a small number of public permutations. In: Pointcheval, D., Johansson, T. (eds.) EUROCRYPT 2012. LNCS, vol. 7237, pp. 45–62. Springer, Heidelberg (2012)
14. Bosma, W., Cannon, J.J., Playoust, C.: The Magma Algebra System I: The User Language. J. Symb. Comput. 24(3/4), 235–265 (1997)
15. Carlier, V., Chabanne, H., Dottax, E.: Grey box implementation of block ciphers preserving the confidentiality of their design. IACR Cryptology ePrint Archive, 2004:188 (2004)
16. Chow, S., Eisen, P.A., Johnson, H., van Oorschot, P.C.: White-box cryptography and an AES implementation. In: Nyberg, K., Heys, H.M. (eds.) SAC 2002. LNCS, vol. 2595, pp. 250–270. Springer, Heidelberg (2003)
17. Chow, S., Eisen, P., Johnson, H., van Oorschot, P.C.: A white-box DES implementation for DRM applications. In: Feigenbaum, J. (ed.) DRM 2002. LNCS, vol. 2696, pp. 1–15. Springer, Heidelberg (2003)

18. Cox, D.A., Little, J., O'Shea, D.: Ideals, Varieties, and Algorithms: An Introduction to Computational Algebraic Geometry and Commutative Algebra. Undergraduate Texts in Mathematics. Springer-Verlag New York, Inc., Secaucus (1991)

19. Daemen, J.: Cipher and Hash Function Design Strategies based on linear and differential cryptanalysis. PhD thesis, Katholieke Universiteit Leuven, Leuven, Belgium (March 1995)

20. Daemen, J., Rijmen, V.: The Design of Rijndael. AES — the Advanced Encryption Standard. Springer (2002)

21. Ding, J.: A new variant of the matsumoto-imai cryptosystem through perturbation. In: Bao, F., Deng, R., Zhou, J. (eds.) PKC 2004. LNCS, vol. 2947, pp. 305–318. Springer, Heidelberg (2004)

22. Ding, J., Schmidt, D.: Cryptanalysis of hFEv and internal perturbation of HFE. In: Vaudenay, S. (ed.) PKC 2005. LNCS, vol. 3386, pp. 288–301. Springer, Heidelberg (2005)

23. Dinur, I., Dunkelman, O., Keller, N., Shamir, A.: Key recovery attacks on 3-round Even-Mansour, 8-step LED-128, and full AES^2. In: Sako, K., Sarkar, P. (eds.) ASIACRYPT 2013, Part I. LNCS, vol. 8269, pp. 337–356. Springer, Heidelberg (2013)

24. Dubois, V., Fouque, P.-A., Shamir, A., Stern, J.: Practical cryptanalysis of SFLASH. In: Menezes, A. (ed.) CRYPTO 2007. LNCS, vol. 4622, pp. 1–12. Springer, Heidelberg (2007)

25. Dubois, V., Granboulan, L., Stern, J.: Cryptanalysis of HFE with internal perturbation. In: Okamoto, T., Wang, X. (eds.) PKC 2007. LNCS, vol. 4450, pp. 249–265. Springer, Heidelberg (2007)

26. Dunkelman, O., Keller, N.: A new criterion for nonlinearity of block ciphers. IEEE Transactions on Information Theory 53(11), 3944–3957 (2007)

27. Faugère, J.-C.: A new efficient algorithm for computing Gröbner bases (F4). Journal of Pure and Applied Algebra 139(1-3), 61–88 (1999)

28. Faugère, J.-C.: A New Efficient Algorithm for Computing Gröbner Bases Without Reduction to Zero (F5). In: Mora, T. (ed.) ISSAC 2002: Proceedings of the 2002 International Symposium on Symbolic and Algebraic Computation, pp. 75–83. ACM Press, New York (2002) ISBN: 1-58113-484-3

29. Faugère, J.-C., Joux, A.: Algebraic Cryptanalysis of Hidden Field Equation (HFE) Cryptosystems Using Gröbner Bases. In: Boneh, D. (ed.) CRYPTO 2003. LNCS, vol. 2729, pp. 44–60. Springer, Heidelberg (2003)

30. Faugère, J.-C., Perret, L.: An efficient algorithm for decomposing multivariate polynomials and its applications to cryptography. J. Symb. Comput. 44(12), 1676–1689 (2009)

31. Faugère, J.-C., von zur Gathen, J., Perret, L.: Decomposition of generic multivariate polynomials. In: ISSAC 2010, pp. 131–137. ACM (2010)

32. Huang, Y.-J., Liu, F.-H., Yang, B.-Y.: Public-key cryptography from new multivariate quadratic assumptions. In: Fischlin, M., Buchmann, J., Manulis, M. (eds.) PKC 2012. LNCS, vol. 7293, pp. 190–205. Springer, Heidelberg (2012)

33. Isobe, T.: A single-key attack on the full GOST block cipher. J. Cryptology 26(1), 172–189 (2013)

34. Kipnis, A., Patarin, J., Goubin, L.: Unbalanced Oil and Vinegar Signature Schemes. In: Stern, J. (ed.) EUROCRYPT 1999. LNCS, vol. 1592, pp. 206–222. Springer, Heidelberg (1999)

35. Kipnis, A., Shamir, A.: Cryptanalysis of the HFE public key cryptosystem by relinearization. In: Wiener, M. (ed.) CRYPTO 1999. LNCS, vol. 1666, pp. 19–30. Springer, Heidelberg (1999)

36. Laigle-Chapuy, Y.: A note on a class of quadratic permutations over f_{2^n}. In: Boztaş, S., Lu, H.-F(F.) (eds.) AAECC 2007. LNCS, vol. 4851, pp. 130–137. Springer, Heidelberg (2007)

37. Levieil, É., Fouque, P.-A.: An improved LPN algorithm. In: De Prisco, R., Yung, M. (eds.) SCN 2006. LNCS, vol. 4116, pp. 348–359. Springer, Heidelberg (2006)

38. Matsui, M.: Linear cryptanalysis method for DES cipher. In: Helleseth, T. (ed.) EUROCRYPT 1993. LNCS, vol. 765, pp. 386–397. Springer, Heidelberg (1994)

39. Matsumoto, T., Imai, H.: Public quadratic polynomial-tuples for efficient signature-verification and message-encryption. In: Günther, C.G. (ed.) EUROCRYPT 1988. LNCS, vol. 330, pp. 419–453. Springer, Heidelberg (1988)

40. Meier, W.: Personal communication (2014)

41. Michiels, W., Gorissen, P.: Mechanism for software tamper resistance: an application of white-box cryptography. In: Digital Rights Management Workshop, pp. 82–89. ACM (2007)

42. Patarin, J.: Cryptanalysis of the matsumoto and imai public key scheme of eurocrypt '88. In: Coppersmith, D. (ed.) CRYPTO 1995. LNCS, vol. 963, pp. 248–261. Springer, Heidelberg (1995)

43. Patarin, J.: Hidden fields equations (HFE) and isomorphisms of polynomials (IP): Two new families of asymmetric algorithms. In: Maurer, U.M. (ed.) EUROCRYPT 1996. LNCS, vol. 1070, pp. 33–48. Springer, Heidelberg (1996)

44. Percival, C.: Stronger key derivation via sequential memory-hard functions (2009) (self-published)

45. Rijmen, V., Preneel, B.: A family of trapdoor ciphers. In: Biham, E. (ed.) FSE 1997. LNCS, vol. 1267, pp. 139–148. Springer, Heidelberg (1997)

46. Sakumoto, K., Shirai, T., Hiwatari, H.: Public-key identification schemes based on multivariate quadratic polynomials. In: Rogaway, P. (ed.) CRYPTO 2011. LNCS, vol. 6841, pp. 706–723. Springer, Heidelberg (2011)

47. Saxena, A., Wyseur, B., Preneel, B.: Towards security notions for white-box cryptography. In: Samarati, P., Yung, M., Martinelli, F., Ardagna, C.A. (eds.) ISC 2009. LNCS, vol. 5735, pp. 49–58. Springer, Heidelberg (2009)

48. Tao, C., Diene, A., Tang, S., Ding, J.: Simple matrix scheme for encryption. In: Gaborit, P. (ed.) PQCrypto 2013. LNCS, vol. 7932, pp. 231–242. Springer, Heidelberg (2013)

49. Wyseur, B.: White-Box Cryptography. PhD thesis, Katholieke Universiteit Leuven, Leuven, Belgium (March 2009)

50. Zeng, X., Zhu, X., Hu, L.: Two new permutation polynomials with the form $(x^{2^k} + x + d)^s + x$ over $\mathbb{F}_2^n$. Appl. Algebra Eng. Commun. Comput. 21(2), 145–150 (2010)

Beyond $2^{c/2}$ Security in Sponge-Based Authenticated Encryption Modes

Philipp Jovanovic[1], Atul Luykx[2], and Bart Mennink[2]

[1] Fakultät für Informatik und Mathematik, Universität Passau, Germany
jovanovic@fim.uni-passau.de
[2] Dept. Electrical Engineering, ESAT/COSIC, KU Leuven, and iMinds, Belgium
{atul.luykx,bart.mennink}@esat.kuleuven.be

Abstract. The Sponge function is known to achieve $2^{c/2}$ security, where c is its capacity. This bound was carried over to keyed variants of the function, such as SpongeWrap, to achieve a $\min\{2^{c/2}, 2^\kappa\}$ security bound, with κ the key length. Similarly, many CAESAR competition submissions are designed to comply with the classical $2^{c/2}$ security bound. We show that Sponge-based constructions for authenticated encryption can achieve the significantly higher bound of $\min\{2^{b/2}, 2^c, 2^\kappa\}$ asymptotically, with $b > c$ the permutation size, by proving that the CAESAR submission NORX achieves this bound. Furthermore, we show how to apply the proof to five other Sponge-based CAESAR submissions: Ascon, CBEAM/STRIBOB, ICEPOLE, Keyak, and two out of the three PRIMATEs. A direct application of the result shows that the parameter choices of these submissions are overly conservative. Simple tweaks render the schemes considerably more efficient without sacrificing security. For instance, NORX64 can increase its rate and decrease its capacity by 128 bits and Ascon-128 can encrypt three times as fast, both without affecting the security level of their underlying modes in the ideal permutation model.

Keywords: Authenticated encryption, CAESAR, Ascon, CBEAM, ICE-POLE, Keyak, NORX, PRIMATEs, STRIBOB.

1 Introduction

Authenticated encryption schemes, cryptographic functions that aim to provide both privacy and integrity of data, have gained renewed attention in light of the recently commenced CAESAR competition [1]. A common approach to building such schemes is to design a mode of operation for a block cipher, as in CCM [2], OCB1-3 [3,4,5], and EAX [6]. Nevertheless a significant fraction of the CAESAR competition submissions use modes of operation for *permutations*.

Most of the permutation-based modes follow the basic design of the Sponge construction [7]: their output is computed from a state value, which in turn is repeatedly updated using key, nonce, associated data, and plaintext by calling a permutation. The state is divided into a rate part of r bits, through which the

P. Sarkar and T. Iwata (Eds.): ASIACRYPT 2014, PART I, LNCS 8873, pp. 85–104, 2014.

user enters plaintext, and a capacity part of c bits, which is out of the user's control.

The security of the Sponge construction as a hash function follows from the fact that the user can only affect the rate, hence an adversary only succeeds with significant probability if it makes on the order of $2^{c/2}$ permutation queries, as this many are needed to produce a collision in the capacity [7]. Keyed versions of the Sponge construction, such as KeyedSponge [8] and SpongeWrap [9], are proven up to a similar bound of 2^{c-a}, assuming a limit of 2^a on online complexity, but are additionally restricted by the key size κ to 2^κ. The permutation-based CAESAR candidates are no exception and recommend parameters based on either the $2^{c/2}$ bound or 2^{c-a} bound, as shown in Table 1.

1.1 Our Results

Contrary to intuition, a wide range of permutation-based authenticated encryption schemes achieve a *significantly* higher mode security level: we prove that the bound is limited by approximately $\min\{2^{(r+c)/2}, 2^c, 2^\kappa\}$ as opposed to $\min\{2^{c/2}, 2^\kappa\}$. The main proof in this work concerns NORX mode [10], but we demonstrate its applicability to the CAESAR submissions Ascon [11], CBEAM[1] [13,14], ICEPOLE [15], Keyak [16], two out of three PRIMATEs [17], and STRIBOB[2] [19,20]. Additionally, we note that it directly applies to SpongeWrap [9] and DuplexWrap [16], upon which Keyak is built.

Our results imply that all of these CAESAR candidates have been overly conservative in choosing their parameters, since a smaller capacity would have lead to the same bound. For instance, Ascon-128 could take $(c, r) = (128, 192)$ instead of $(256, 64)$, NORX64 (the proposed mode with 256-bit security) could increase its rate by 128 bits, and GIBBON-120 and HANUMAN-120 could increase their rate by a factor of 4, all without affecting their mode security levels.

These observations only concern the *mode* security, where characteristics of the underlying permutation are set aside. Specifically, the concrete security of the underlying permutations plays a fundamental role in the choice of parameters. For instance, the authors of Ascon, NORX, and PRIMATEs [11,10,17] acknowledge that non-random properties of some of the underlying primitives exist. Although these properties are harmless, a non-hermetic design approach for the primitives affects the parameter choices.

1.2 Outline

We present our security model in Section 2. A security proof for NORX is derived in Section 3. In Section 4 we show that the proof of NORX generalizes to other CAESAR submissions, as well as to SpongeWrap and DuplexWrap. The work is concluded in Section 5, where we also discuss possible generalizations to Artemia [21] and π-Cipher [22].

---

[1] CBEAM was withdrawn after an attack by Minaud [12], but we focus on modes of operation.

[2] Both CBEAM and STRIBOB use the BLNK Sponge mode [18].

Table 1. Parameters and the achieved mode security levels of seven CAESAR submissions. We remark that ICEPOLE consists of three configurations (two with security level 128 and one with security level 256) and Keyak of four configurations (one with an 800-bit state and three with a 1600-bit state).

	b	c	r	κ	τ	**security**
Ascon [11]	320	192	128	96	96	**96**
	320	256	64	128	128	**128**
CBEAM [?]	256	190	66	128	64	**128**
ICEPOLE [15]	1280	254	1026	128	128	**128**
	1280	318	962	256	128	**256**
Keyak [16]	800	252	548	128..224	128	**128..224**
	1600	252	1348	128..224	128	**128..224**
NORX [10]	512	192	320	128	128	**128**
	1024	384	640	256	256	**256**
GIBBON/ HANUMAN [17]	200	159	41	80	80	**80**
	280	239	41	120	120	**120**
STRIBOB [19]	512	254	258	192	128	**192**

2 Security Model

For $n \in \mathbb{N}$, let $\mathsf{Perm}(n)$ denote the set of all permutations on n bits. When writing $x \xleftarrow{\$} \mathcal{X}$ for some finite set $\mathcal{X}$, we mean that x gets sampled uniformly at random from $\mathcal{X}$. For $x \in \{0,1\}^n$, and $a, b \leq n$, we denote by $[x]^a$ and $[x]_b$ the a leftmost and b rightmost bits of x, respectively. For tuples $(j, k), (j', k')$ we use lexicographical order: $(j, k) > (j', k')$ means that $j > j'$, or $j = j'$ and $k > k'$.

Let Π be an authenticated encryption scheme, which is specified by an encryption function $\mathcal{E}$ and a decryption function $\mathcal{D}$:

$$(C, A) \longleftarrow \mathcal{E}_K(N; H, M, T) \quad \text{and} \quad M/\bot \longleftarrow \mathcal{D}_K(N; H, C, T; A).$$

Here N denotes a nonce value, H a header, M a message, C a ciphertext, T a trailer, and A an authentication tag. The values (H, T) will be referred to as associated data. If verification is correct, then the decryption function $\mathcal{D}_K$ outputs M, and $\bot$ otherwise. The scheme Π is also determined by a set of parameters such as the key size, state size, and block size, but these are left implicit. In addition, we define $\$$ to be an ideal version of $\mathcal{E}_K$, where $\$$ returns $(C, A) \xleftarrow{\$} \{0,1\}^{|M|+\tau}$ on input of a new $(N; H, M, T)$.

We follow the convention in analyzing modes of operation for permutations by modeling the underlying permutations as being drawn uniformly at random from $\mathsf{Perm}(b)$, where b is a parameter determined by the scheme. We note that irregularities in the underlying permutation may invalidate the underlying assumption.

An adversary $\mathcal{A}$ is a probabilistic algorithm that has access to one or more oracles $\mathcal{O}$, denoted $\mathcal{A}^{\mathcal{O}}$. By $\mathcal{A}^{\mathcal{O}} = 1$ we denote the event that $\mathcal{A}$, after interacting with $\mathcal{O}$, outputs 1. We consider adversaries $\mathcal{A}$ that have unbounded computational power and whose complexities are solely measured by the number of queries made to their oracles. These adversaries have query access to the underlying idealized permutations, $\mathcal{E}_K$ or its counterpart \$, and possibly $\mathcal{D}_K$. The key K is randomly drawn from $\{0,1\}^\kappa$ at the beginning of the security experiment. The security definitions below follow [23,24].

Privacy
Let $\mathbf{p}$ denote the list of underlying idealized permutations of Π. We define the advantage of an adversary $\mathcal{A}$ in breaking the privacy of Π as follows:

$$\mathbf{Adv}_\Pi^{\mathrm{priv}}(\mathcal{A}) = \left| \mathbf{Pr}_{\mathbf{p},K}\left(\mathcal{A}^{\mathbf{p}^\pm, \mathcal{E}_K} = 1 \right) - \mathbf{Pr}_{\mathbf{p},\$}\left(\mathcal{A}^{\mathbf{p}^\pm, \$} = 1 \right) \right| ,$$

where the probabilities are taken over the random choices of $\mathbf{p}, \$, K$, and the random choices of $\mathcal{A}$, if any. The fact that the adversary has access to both the forward and inverse permutations in $\mathbf{p}$ is denoted by $\mathbf{p}^\pm$. We assume that adversary $\mathcal{A}$ is nonce-respecting, which means that it never makes two queries to $\mathcal{E}_K$ or \$ with the same nonce. By $\mathbf{Adv}_\Pi^{\mathrm{priv}}(q_p, q_\mathcal{E}, \lambda_\mathcal{E})$ we denote the maximum advantage taken over all adversaries that query $\mathbf{p}^\pm$ at most q_p times, and that make at most $q_\mathcal{E}$ queries of total length at most $\lambda_\mathcal{E}$ blocks to $\mathcal{E}_K$ or \$. We remark that this privacy notion is also known as the CPA security of an (authenticated) encryption scheme.

Integrity
As above, let $\mathbf{p}$ denote the list of underlying idealized permutations of Π. We define the advantage of an adversary $\mathcal{A}$ in breaking the integrity of Π as follows:

$$\mathbf{Adv}_\Pi^{\mathrm{auth}}(\mathcal{A}) = \mathbf{Pr}_{\mathbf{p},K}\left(\mathcal{A}^{\mathbf{p}^\pm, \mathcal{E}_K, \mathcal{D}_K} \text{ forges} \right) ,$$

where the probability is taken over the random choices of $\mathbf{p}, K$, and the random choices of $\mathcal{A}$, if any. Here, we say that "$\mathcal{A}$ forges" if $\mathcal{D}_K$ ever returns a valid message (other than $\perp$) on input of $(N; H, C, T; A)$ where (C, A) has never been output by $\mathcal{E}_K$ on input of a query $(N; H, M, T)$ for some M. We assume that adversary $\mathcal{A}$ is nonce-respecting, which means that it never makes two queries to $\mathcal{E}_K$ with the same nonce. Nevertheless, $\mathcal{A}$ is allowed to repeat nonces in decryption queries. By $\mathbf{Adv}_\Pi^{\mathrm{auth}}(q_p, q_\mathcal{E}, \lambda_\mathcal{E}, q_\mathcal{D}, \lambda_\mathcal{D})$ we denote the maximum advantage taken over all adversaries that query $\mathbf{p}^\pm$ at most q_p times, that make at most $q_\mathcal{E}$ queries of total length at most $\lambda_\mathcal{E}$ blocks to $\mathcal{E}_K$, and at most $q_\mathcal{D}$ queries of total length at most $\lambda_\mathcal{D}$ blocks to $\mathcal{D}_K/\perp$.

3 NORX

We introduce NORX at a level required for the understanding of the security proof, and refer to Aumasson et al. [10] for the formal specification. Let p be a permutation on b bits. All b-bit state values are split into a rate part of r bits and a capacity part of c bits. We denote the key size of NORX by κ bits, the nonce size by ν bits, and the tag size by τ bits. The header, message, and trailer can be of arbitrary length, and are padded using 10*1-padding to a length of a multiple of r bits. Throughout, we denote the r-bit header blocks by $H_1, \ldots, H_u$, message blocks by $M_1, \ldots, M_v$, ciphertext blocks by $C_1, \ldots, C_v$, and trailer blocks by $T_1, \ldots, T_w$.

Unlike other permutation-based schemes, NORX allows for parallelism in the encryption part, which is described using a parameter $D \in \{0, \ldots, 255\}$ corresponding to the number of parallel chains. Specifically, if $D \in \{1, \ldots, 255\}$ NORX has D parallel chains, and if $D = 0$ it has v parallel chains, where v is the block length of M or C.

NORX consists of five proposed parameter configurations: NORXW-R-D for $(W, R, D) \in \{(64, 4, 1), (32, 4, 1), (64, 6, 1), (32, 6, 1), (64, 4, 4)\}$. The parameter R denotes the number of rounds of the underlying permutation p, and W denotes the word size which we use to set $r = 10W$ and $c = 6W$. The default key and tag size are $\kappa = \nu = 4W$. The corresponding parameters for the two different choices of W, 64 and 32, are given in Table 1.

Although NORX starts with an initialization function init which requires the parameters (D, R, τ) as input, as soon as our security experiment starts, we consider (D, R, τ) fixed and constant. Hence we can view init as a function that maps (K, N) to $(K\|N\|0^{b-\kappa-\nu}) \oplus \mathsf{const}$, where const is irrelevant to the mode security analysis of NORX, and will be ignored in the remaining analysis.

After init is called, the header H is compressed into the rate, then the state is branched into D states (if necessary), the message blocks are encrypted in a streaming way, the D states are merged into one state (if necessary), the trailer is compressed, and finally the tag A is computed. All rounds are preceded with a domain separation constant XORed into the capacity: 01 for header compression, 02 for message encryption, 04 for trailer compression, and 08 for tag generation. If $D \neq 1$, domain separators 10 and 20 are used for branching and merging, along with pairwise distinct lane indices id_k for $k = 1, \ldots, D$ (if $D = 1$ we write $id_1 = 0$).

The privacy of NORX is proven in Section 3.1 and the integrity in Section 3.2. In both proofs we consider an adversary that makes q_p permutation queries and $q_{\mathcal{E}}$ encryption queries of total length $\lambda_{\mathcal{E}}$. In the proof of integrity, the adversary can additionally make $q_{\mathcal{D}}$ decryption queries of total length $\lambda_{\mathcal{D}}$. To aid the analysis, we compute the number of permutation calls made via the $q_{\mathcal{E}}$ encryption queries. The exact same computation holds for decryption queries with the parameters defined analogously.

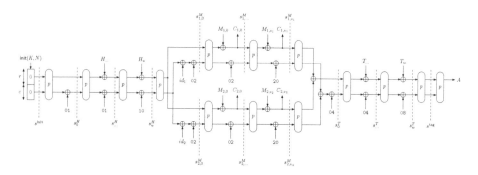

Fig. 1. NORX with $D = 2$

· Consider a query to $\mathcal{E}_K$, consisting of u header blocks, v message blocks, and w trailer blocks. We denote its corresponding state values by

$$
\left(s^{\text{init}}; \; s_0^H, \ldots, s_u^H; \; \begin{bmatrix} s_{1,0}^M, \ldots, s_{1,v_1}^M \\ \vdots \qquad \vdots \\ s_{D,0}^M, \ldots, s_{D,v_D}^M \end{bmatrix}; \; s_0^T, \ldots, s_w^T; \; s^{\text{tag}} \right), \tag{1}
$$

as outlined in Figure 1. Here, $\sum_{k=1}^{D} v_k = v$. If there are no branching and merging phases, i.e. $D = 1$, then the state values corresponding to the branching and merging, $\{s_{1,0}^M, \ldots, s_{D,0}^M\}$ and s_0^T, are left out of the tuple. Note that the length of this tuple equals the number of primitive calls made in this encryption query, as every state value corresponds to the input of exactly one primitive call. A simple calculation shows that if the jth $\mathcal{E}_K$ query is of length $u + v + w$ blocks, it results in $u + v + w + 3$ state values if $D = 1$, in $u + v + w + D + 4$ state values if $D > 1$, and in $u + 2v + w + 4$ state values if $D = 0$.[3] We denote the number of state values by $\sigma_{\mathcal{E},j}$, where the dependence on D is suppressed as D does not change during the security game. In other words, $\sigma_{\mathcal{E},j}$ denotes the number of primitive calls in the jth query to $\mathcal{E}_K$. Furthermore, we define $\sigma_{\mathcal{E}}$ to be the total number of primitive evaluations via the encryption queries, and find that

$$
\sigma_{\mathcal{E}} := \sum_{j=1}^{q_{\mathcal{E}}} \sigma_{\mathcal{E},j} \leq \begin{cases} 2\lambda_{\mathcal{E}} + 4q_{\mathcal{E}}, & \text{if } D = 0, \\ \lambda_{\mathcal{E}} + 3q_{\mathcal{E}}, & \text{if } D = 1, \\ \lambda_{\mathcal{E}} + (D+4)q_{\mathcal{E}}, & \text{if } D > 1. \end{cases} \tag{2}
$$

This bound is rather tight. Particularly, for $D = 0$ an adversary can meet this bound by only making queries without header and trailer. For queries to $\mathcal{D}_K$ we define $\sigma_{\mathcal{D},j}$ and $\sigma_{\mathcal{D}}$ analogously.

---

[3] For $D = 0$, the original specification dictates an additional $10^{b-2}1$-padding for every complete message block. This means that lanes $1, \ldots, v-1$ consist of two rounds. We do not take this padding into account, noting that it is unnecessary for the security analysis.

3.1   Privacy of NORX

Theorem 1. *Let $\Pi = (\mathcal{E}, \mathcal{D})$ be NORX based on an ideal underlying primitive p. Then,*

$$\mathbf{Adv}_{\Pi}^{\mathrm{priv}}(q_p, q_{\mathcal{E}}, \lambda_{\mathcal{E}}) \leq \frac{3(q_p + \sigma_{\mathcal{E}})^2}{2^{b+1}} + \left(\frac{8eq_p\sigma_{\mathcal{E}}}{2^b}\right)^{1/2} + \frac{rq_p}{2^c} + \frac{q_p + \sigma_{\mathcal{E}}}{2^\kappa},$$

where $\sigma_{\mathcal{E}}$ is defined in (2).

Theorem 1 can be interpreted as implying that NORX provides privacy security as long as the total complexity $q_p + \sigma_{\mathcal{E}}$ does not exceed $\min\{2^{b/2}, 2^\kappa\}$ and the total number of primitive queries q_p, also known as the offline complexity, does not exceed $2^c/r$. See Table 1 for the security level of the various parameter choices of NORX.

The proof is based on the observation that NORX is indistinguishable from a random scheme as long as there are no collisions among the (direct and indirect) evaluations of p. Due to uniqueness of the nonce, state values from evaluations of $\mathcal{E}_K$ collide with probability approximately $1/2^b$. Regarding collisions between direct calls to p and calls via $\mathcal{E}_K$: while these may happen with probability about $1/2^c$, they turn out not to significantly influence the bound. The latter is demonstrated in part using the principle of multiplicities [25]: roughly stated, the maximum number of state values with the same rate part. The formal security proof is more detailed. Furthermore, we remark that, at the cost of readability and simplicity of the proof, the bound could be improved by a constant factor.

Proof. We consider any adversary $\mathcal{A}$ that has access to either $(p^\pm, \mathcal{E}_K)$ or $(p^\pm, \$)$ and whose goal is to distinguish these two worlds. For brevity, we write

$$\mathbf{Adv}_{\Pi}^{\mathrm{priv}}(\mathcal{A}) = \Delta_{\mathcal{A}}(p^\pm, \mathcal{E}_K; p^\pm, \$) . \tag{3}$$

We start with replacing $p^\pm$ by a random function, as this simplifies the analysis. This is done with a PRP-PRF switch [26, 27], in which we make a transition from $p^\pm$ to a primitive $f^\pm$ defined as follows. This primitive $f^\pm$ maintains an initially empty list $\mathcal{F}$ of query/response tuples (x, y). For $\mathcal{F}$, we denote its set of domain and range values by $\mathrm{dom}(\mathcal{F})$ and $\mathrm{rng}(\mathcal{F})$, respectively. For a forward query $f(x)$ with $x \in \mathrm{dom}(\mathcal{F})$, the corresponding value $y = \mathcal{F}(x)$ is returned. For a new forward query $f(x)$, the response y is randomly drawn from $\{0,1\}^b$, then if y is in $\mathrm{rng}(\mathcal{F})$ the primitive aborts, otherwise the tuple (x, y) is added to $\mathcal{F}$. The description for f^{-1} is similar. The usage of $\mathcal{F}$ will remain implicit in the remaining usage of $f^\pm$. Now, $p^\pm$ and $f^\pm$ behave identically as long as the latter does not abort. Given that the adversary triggers at most $q_p + \sigma_{\mathcal{E}}$ evaluations of f, such an abort happens with probability at most $\binom{q_p+\sigma_{\mathcal{E}}}{2}/2^b \leq (q_p+\sigma_{\mathcal{E}})^2/2^{b+1}$. This PRP-PRF switch needs to be applied to both the real and ideal world, to get

$$\Delta_{\mathcal{A}}(p^\pm, \mathcal{E}_K; p^\pm, \$) \leq \Delta_{\mathcal{A}}(f^\pm, \mathcal{E}_K; f^\pm, \$) + \frac{(q_p + \sigma_{\mathcal{E}})^2}{2^b} . \tag{4}$$

We restrict our attention to $\mathcal{A}$ with oracle access to $(f^{\pm}, F)$, where $F \in \{\mathcal{E}_K, \$\}$. Without loss of generality, we can assume that the adversary only queries full blocks and that no padding rules are involved. We can do this because the padding rules are injective, allowing the proof to carry over to the case of fractional blocks with 10*1-padding.

We introduce some terminology. Queries to $f^{\pm}$ are denoted (x_i, y_i) for $i = 1, \ldots, q_p$, while queries to F are written as elements $(N_j; H_j, M_j, T_j; C_j, A_j)$ for $j = 1, \ldots, q_{\mathcal{E}}$. If $F = \mathcal{E}_K$, the state values are denoted as in (1), subscripted with a j:

$$\left(s_j^{\text{init}}; \; s_{j,0}^H, \ldots, s_{j,u}^H; \; \begin{bmatrix} s_{j,1,0}^M, \ldots, s_{j,1,v_1}^M \\ \vdots \qquad \vdots \\ s_{j,D,0}^M, \ldots, s_{j,D,v_D}^M \end{bmatrix}; \; s_{j,0}^T, \ldots, s_{j,w}^T; \; s_j^{\text{tag}} \right). \tag{5}$$

If the structure of (5) is irrelevant we refer to the tuple as $(s_{j,1}, \ldots, s_{j,\sigma_{\mathcal{E},j}})$, where we use the convention to list the elements of the matrix column-wise. In this case, we write $\text{parent}(s_{j,k})$ to denote the state value that lead to $s_{j,k}$, with $\text{parent}(s_{j,1}) := \varnothing$ and $\text{parent}(s_{j,0}^T) := (s_{j,1,v_1}^M, \ldots, s_{j,D,v_D}^M)$. We remark that the characteristic structure of NORX, with the D parallel states, only becomes relevant in the two technical lemmas that will be used at the end of the proof. We point out that $s_{j,1}$ corresponds to the initial state value of the evaluation, which requires special attention throughout the remainder of the proof.

We define two collision events, guess and hit. Let $i \in \{1, \ldots, q_p\}$, $j, j' \in \{1, \ldots, q_{\mathcal{E}}\}$, $k \in \{1, \ldots, \sigma_{\mathcal{E},j}\}$, and $k' \in \{1, \ldots, \sigma_{\mathcal{E},j'}\}$:

$$\begin{aligned} \mathsf{guess}(i; j, k) &\equiv x_i = s_{j,k}\,, \\ \mathsf{hit}(j, k; j', k') &\equiv \text{parent}(s_{j,k}) \neq \text{parent}(s_{j',k'}) \wedge s_{j,k} = s_{j',k'}\,. \end{aligned}$$

Event $\mathsf{guess}(i; j, k)$ corresponds to a primitive call in an encryption query hitting a direct primitive query, or vice versa, while $\mathsf{hit}(j, k; j', k')$ corresponds to non-trivial primitive calls colliding in encryption queries. We write $\mathsf{guess} = \vee_{i;j,k} \mathsf{guess}(i; j, k)$, $\mathsf{hit} = \vee_{j,k;j',k'} \mathsf{hit}(j, k; j', k')$, and set $\mathsf{event} = \mathsf{guess} \vee \mathsf{hit}$.

The remainder of the proof is divided as follows. In Lemma 1 we prove that $(f^{\pm}, \mathcal{E}_K)$ and $(f^{\pm}, \$)$ are indistinguishable as long as $\neg\mathsf{event}$ holds. In other words,

$$\Delta_{\mathcal{A}}(f^{\pm}, \mathcal{E}_K; f^{\pm}, \$) \leq \Pr\left(\mathcal{A}^{f^{\pm}, \mathcal{E}_K} \text{ sets event} \right). \tag{6}$$

Then, in Lemma 2 we bound this term by $\dfrac{q_p \sigma_{\mathcal{E}} + \sigma_{\mathcal{E}}^2/2}{2^b} + \left(\dfrac{8 e q_p \sigma_{\mathcal{E}}}{2^b} \right)^{1/2} + \dfrac{r q_p}{2^c} +$ $\dfrac{q_p + \sigma_{\mathcal{E}}}{2^{\kappa}}$. Noting that $\dfrac{q_p \sigma_{\mathcal{E}} + \sigma_{\mathcal{E}}^2/2}{2^b} \leq \dfrac{(q_p + \sigma_{\mathcal{E}})^2}{2^{b+1}}$, this completes the proof via equations (3,4,6). $\square$

Lemma 1. *Given that* event *does not occur,* $(f^{\pm}, \mathcal{E}_K)$ *and* $(f^{\pm}, \$)$ *are indistinguishable.*

Proof. The outputs of $f^{\pm}$ are sampled uniformly at random in both $(f^{\pm}, \mathcal{E}_K)$ and $(f^{\pm}, \$)$, except when such an output collides with a state of an $\mathcal{E}_K$ evaluation in the real world. However, this event is excluded by assuming ¬**guess**, hence it suffices to only consider queries to the big oracle $F \in \{\mathcal{E}_K, \$\}$.

Let N_j be a new nonce used in the F-query $(N_j; H_j, M_j, T_j)$, with corresponding ciphertext and authentication tag (C_j, A_j). Denote the query's state values as in (5). Let u, v, and w denote the number of padded header blocks, padded message blocks, and padded trailer blocks, respectively.

By the definition of $\$$, in the ideal world we have $(C_j, A_j) \xleftarrow{\$} \{0,1\}^{|M_j|+\tau}$. We will prove that (C_j, A_j) is identically distributed in the real world, under the assumption that **guess** $\vee$ **hit** does not occur. Denote the message blocks of M_j by $M_{j,k,\ell}$ for $k = 1, \dots, D$ and $\ell = 1, \dots, v_k$.

We know that $s_{j,u}^H$ is new and that $f(s_{j,u}^H)$ does not collide with any other f-query because otherwise ¬**event** would have been violated. Since $s_{j,k,0}^M = f(s_{j,u}^H) \oplus id_k$ we conclude that $s_{j,k,0}^M$ is new for $k = 1, \dots, D$, as otherwise **event** would be set. Similarly, $s_{j,k,\ell}^M$ is new for all $\ell > 0$. The ciphertext blocks $C_{j,k,\ell}$ are computed as

$$C_{j,k,\ell} = M_{j,k,\ell} \oplus [f(s_{j,k,\ell-1}^M)]^r.$$

As the state value $s_{j,k,\ell-1}^M$ has not been evaluated by f before (neither directly nor indirectly via an encryption query), $f(s_{j,k,\ell-1}^M)$ outputs a uniformly random value from $\{0,1\}^b$, hence $C_{j,k,\ell} \xleftarrow{\$} \{0,1\}^r$. We remark that similar reasoning shows that a ciphertext block corresponding to a truncated message block is uniformly randomly drawn as well, yet from a smaller set. The fact that $A_j \xleftarrow{\$} \{0,1\}^\tau$ follows the same reasoning, using that s_j^{tag} is a new input to f. Thus, $A_j = [f(s_j^{\text{tag}})]^\tau \xleftarrow{\$} \{0,1\}^\tau$. $\square$

Lemma 2. $\Pr\left(\mathcal{A}^{f^{\pm}, \mathcal{E}_K} \text{ sets event}\right) \leq \dfrac{q_p \sigma_\mathcal{E} + \sigma_\mathcal{E}^2/2}{2^b} + \left(\dfrac{8 e q_p \sigma_\mathcal{E}}{2^b}\right)^{1/2} + \dfrac{r q_p}{2^c} + \dfrac{q_p + \sigma_\mathcal{E}}{2^\kappa}.$

Proof. Consider the adversary interacting with $(f^{\pm}, \mathcal{E}_K)$, and let $\Pr(\textbf{guess} \vee \textbf{hit})$ denote the probability we aim to bound. For $i \in \{1, \dots, q_p\}$, define

$$\mathsf{key}(i) \quad \equiv \quad [x_i]^\kappa = K,$$

and $\mathsf{key} = \vee_i \mathsf{key}(i)$. Event $\mathsf{key}(i)$ corresponds to a primitive query hitting the key. Let $j \in \{1, \dots, q_\mathcal{E}\}$ and $k \in \{1, \dots, \sigma_{\mathcal{E},j}\}$, and consider any threshold $\rho \geq 1$, then define

$$\mathsf{multi}(j,k) \quad \equiv \quad$$
$$\left[\max_{\alpha \in \{0,1\}^r} \left|\{j' \leq j, 1 < k' \leq k : \alpha \in \{[s_{j',k'}]^r, [f(s_{j',k'})]^r\}\}\right|\right] > \rho.$$

Event $\mathsf{multi}(j,k)$ is used to bound the number of states that collide in the rate part. Note that state values $s_{j',1}$ are not considered here as they will be covered by key. We define $\mathsf{multi} = \mathsf{multi}(q_{\mathcal{E}}, \sigma_{\mathcal{E},q_{\mathcal{E}}})$, which is a monotone event. By basic probability theory,

$$\mathbf{Pr}\,(\mathsf{guess} \vee \mathsf{hit}) \leq \mathbf{Pr}\,(\mathsf{guess} \vee \mathsf{hit} \mid \neg(\mathsf{key} \vee \mathsf{multi})) + \mathbf{Pr}\,(\mathsf{key} \vee \mathsf{multi})\,. \quad (7)$$

In the remainder of the proof, we bound these probabilities as follows (a formal explanation of the proof technique is given in Appendix A): we consider the ith forward or inverse primitive query (for $i \in \{1, \ldots, q_p\}$) or the kth state of the jth construction query (for $j \in \{1, \ldots, q_{\mathcal{E}}\}$ and $k \in \{1, \ldots, \sigma_{\mathcal{E},j}\}$), and bound the probability that this evaluation makes $\mathsf{guess} \vee \mathsf{hit}$ satisfied, under the assumption that this query does not set $\mathsf{key} \vee \mathsf{multi}$ and also that $\mathsf{guess} \vee \mathsf{hit} \vee \mathsf{key} \vee \mathsf{multi}$ has not been set before. For the analysis of $\mathbf{Pr}\,(\mathsf{key} \vee \mathsf{multi})$ a similar technique is employed.

Event guess. This event can be set in the ith primitive query (for $i = 1, \ldots, q_p$) or in any state evaluation of the jth construction query (for $j = 1, \ldots, q_{\mathcal{E}}$). Denote the state values of the jth construction query as in (5). Consider any evaluation, assume this query does not set $\mathsf{key} \vee \mathsf{multi}$ and assume that $\mathsf{guess} \vee \mathsf{hit} \vee \mathsf{key} \vee \mathsf{multi}$ has not been set before. Firstly, note that $x_i = s_j^{\mathrm{init}}$ for some i, j would imply $\mathsf{key}(i)$ and hence invalidate our assumption. Therefore, we can exclude s_j^{init} from further analysis on guess. For $i = 1, \ldots, q_p$, let $j_i \in \{1, \ldots, q_{\mathcal{E}}\}$ be the number of encryption queries made before the ith primitive query. Similarly, for $j = 1, \ldots, q_{\mathcal{E}}$, denote by $i_j \in \{1, \ldots, q_p\}$ the number of primitive queries made before the jth encryption query.

- Consider a primitive query (x_i, y_i) for $i \in \{1, \ldots, q_p\}$, which may be a forward or an inverse query, and assume it has not been queried to $f^\pm$ before. If it is a forward query x_i, by $\neg\mathsf{multi}$ there are at most ρ state values s with $[x_i]^r = [s]^r$, and thus $x_i = s$ with probability at most $\rho/2^c$. Here, we remark that the capacity part of s is unknown to the adversary and it guesses it with probability at most $1/2^c$. A slightly more complicated reasoning applies for inverse queries. Denote the query by y_i. By $\neg\mathsf{multi}$ there are at most ρ state values s with $[y_i]^r = [f(s)]^r$, hence $y_i = f(s)$ with probability at most $\rho/2^c$. If y_i equals $f(s)$ for any of these states, then $x_i = s$, otherwise $x_i = s$ with probability at most $\sum_{j=1}^{j_i} \sigma_{\mathcal{E},j}/2^b$. Therefore the probability that guess is set via a direct query is at most $\frac{q_p \rho}{2^c} + \sum_{i=1}^{q_p} \sum_{j=1}^{j_i} \frac{\sigma_{\mathcal{E},j}}{2^b}$;
- Next, consider the probability that the jth construction query sets guess, for $j \in \{1, \ldots, q_{\mathcal{E}}\}$. For simplicity, first consider $D = 1$, hence the message is processed in one lane and we can use state labeling $(s_{j,1}, \ldots, s_{j,\sigma_{\mathcal{E},j}})$. We range from $s_{j,2}$ to $s_{j,\sigma_{\mathcal{E},j}}$ (recall that $s_{j,1} = s_j^{\mathrm{init}}$ can be excluded) and consider the probability that this state sets guess assuming it has not been set before. Let $k \in \{2, \ldots, \sigma_{\mathcal{E},j}\}$. The state value $s_{j,k}$ equals $f(s_{j,k-1}) \oplus v$, where v is some value determined by the adversarial input prior to the evaluation of $f(s_{j,k-1})$, including input from (H_j, M_j, T_j) and constants serving as domain separators. By assumption, $\mathsf{guess} \vee \mathsf{hit}$ has not been set before, and $f(s_{j,k-1})$ is thus

randomly drawn from $\{0,1\}^b$. It hits any x_i ($i \in \{1, \ldots, i_j\}$) with probability at most $i_j/2^b$. Next, consider the general case $D > 1$. We return to the labeling of (5). A complication occurs for the branching states $s^M_{j,1,0}, \ldots, s^M_{j,D,0}$ and the merging state $s^T_{j,0}$. Starting with the branching states, these are computed from $s^H_{j,u}$ as

$$
\begin{pmatrix} s^M_{j,1,0} \\ \vdots \\ s^M_{j,D,0} \end{pmatrix} = f(s^H_{j,u}) \oplus \begin{pmatrix} v_1 \\ \vdots \\ v_D \end{pmatrix},
$$

where $v_1, \ldots, v_D$ are some distinct values determined by the adversarial input prior to the evaluation of the jth construction query. These are distinct by the XOR of the lane numbers $id_1, \ldots, id_D$. Any of these nodes equals x_i for $i \in \{1, \ldots, q_p\}$ with probability at most $i_j D/2^b$. Finally, for the merging node $s^T_{j,0}$ we can apply the same analysis, noting that it is derived from a sum of D new f-evaluations. Concluding, the jth construction query sets guess with probability at most $i_j \sigma_{\mathcal{E},j}/2^b$ (we always have in total at most $\sigma_{\mathcal{E},j}$ new state values). Summing over all $q_{\mathcal{E}}$ construction queries, we get $\sum_{j=1}^{q_{\mathcal{E}}} i_j \sigma_{\mathcal{E},j}/2^b$.

Concluding,

$$
\mathbf{Pr}\left(\text{guess} \mid \neg(\text{key} \vee \text{multi})\right) \le \frac{q_p \rho}{2^c} + \sum_{i=1}^{q_p} \sum_{j=1}^{j_i} \frac{\sigma_{\mathcal{E},j}}{2^b} + \sum_{j=1}^{q_{\mathcal{E}}} \frac{i_j \sigma_{\mathcal{E},j}}{2^b} = \frac{q_p \rho}{2^c} + \frac{q_p \sigma_{\mathcal{E}}}{2^b}.
$$

Here we use that $\sum_{i=1}^{q_p} \sum_{j=1}^{j_i} \sigma_{\mathcal{E},j} + \sum_{j=1}^{q_{\mathcal{E}}} \sum_{k=1}^{\sigma_{\mathcal{E},j}} i_j = q_p \sigma_{\mathcal{E}}$, which follows from a simple counting argument.

Event hit. We again employ ideas of guess, and particularly that as long as guess $\vee$ hit is not set, we can consider all new state values (except for the initial states) to be randomly drawn from a set of size 2^b. Particularly, we can refrain from explicitly discussing the branching and merging nodes (the detailed analysis of guess applies) and label the states as $(s_{j,1}, \ldots, s_{j,\sigma_{\mathcal{E},j}})$. Clearly, $s_{j,1} \ne s_{j',1}$ for all j, j' by uniqueness of the nonce. Any state value $s_{j,k}$ for $k > 1$ (at most $\sigma_{\mathcal{E}} - q_{\mathcal{E}}$ in total) hits an initial state value $s_{j',1}$ only if $[s_{j,k}]^\kappa = K$, which happens with probability at most $\sigma_{\mathcal{E}}/2^\kappa$, assuming $s_{j,k}$ is generated randomly. Finally, any two other states $s_{j,k}, s_{j',k'}$ for $k, k' > 1$ collide with probability at most $\binom{\sigma_{\mathcal{E}} - q_{\mathcal{E}}}{2}/2^b$. Concluding, $\mathbf{Pr}\left(\text{hit} \mid \neg(\text{key} \vee \text{multi})\right) \le \binom{\sigma_{\mathcal{E}}}{2}/2^b + \sigma_{\mathcal{E}}/2^\kappa$.

Event key. For $i \in \{1, \ldots, q_p\}$, the query sets key(i) if $[x_i]^\kappa = K$, which happens with probability $1/2^\kappa$ (assuming it did not happen in queries $1, \ldots, i-1$). The adversary makes q_p attempts, and hence $\mathbf{Pr}\left(\text{key}\right) \le q_p/2^\kappa$.

Event multi. We again use the principles from the analysis for guess of construction queries (note that this part does not rely on multi itself). Particularly, consider a new state value $s_{j,k-1}$; then for a fixed state value $x \in \{0,1\}^b$ it satisfies $f(s_{j,k-1}) = x$ or $s_{j,k} = f(s_{j,k-1}) \oplus v = x$ for some predetermined v with probability at most $2/2^b$. Now, let $\alpha \in \{0,1\}^r$. More than ρ state values hit α with probability at most $\binom{\sigma_{\mathcal{E}}}{\rho} (2/2^r)^\rho \leq \left(\frac{2e\sigma_{\mathcal{E}}}{\rho 2^r}\right)^\rho$, using Stirling's approximation ($x! \geq (x/e)^x$ for any x). Considering any possible choice of α, we obtain $\mathbf{Pr}\,(\text{multi}) \leq 2^r \left(\frac{2e\sigma_{\mathcal{E}}}{\rho 2^r}\right)^\rho$.

Addition of the four bounds via (7) gives

$$\mathbf{Pr}\,(\text{guess} \vee \text{hit}) \leq \frac{q_p\sigma_{\mathcal{E}} + \sigma_{\mathcal{E}}^2/2}{2^b} + \frac{q_p\rho}{2^c} + \frac{q_p + \sigma_{\mathcal{E}}}{2^\kappa} + 2^r \left(\frac{e\sigma_{\mathcal{E}}}{\rho 2^r}\right)^\rho.$$

Putting $\rho = \max\left\{r, \left(\frac{2e\sigma_{\mathcal{E}}2^c}{q_p 2^r}\right)^{1/2}\right\}$ gives

$$\mathbf{Pr}\,(\text{guess} \vee \text{hit}) \leq \frac{q_p\sigma_{\mathcal{E}} + \sigma_{\mathcal{E}}^2/2}{2^b} + 2\left(\frac{2eq_p\sigma_{\mathcal{E}}}{2^b}\right)^{1/2} + \frac{rq_p}{2^c} + \frac{q_p + \sigma_{\mathcal{E}}}{2^\kappa},$$

assuming $2eq_p\sigma_{\mathcal{E}}/2^b < 1$ (which we can do, as the bound would otherwise be void anyway). This completes the proof. □

3.2   Authenticity of NORX

Theorem 2. *Let $\Pi = (\mathcal{E}, \mathcal{D})$ be NORX based on an ideal underlying primitive p. Then,*

$$\mathbf{Adv}_\Pi^{\text{auth}}(q_p, q_{\mathcal{E}}, \lambda_{\mathcal{E}}, q_{\mathcal{D}}, \lambda_{\mathcal{D}}) \leq \frac{(q_p + \sigma_{\mathcal{E}} + \sigma_{\mathcal{D}})^2}{2^b} + \left(\frac{8eq_p\sigma_{\mathcal{E}}}{2^b}\right)^{1/2} + \frac{rq_p}{2^c} +$$

$$\frac{q_p + \sigma_{\mathcal{E}} + \sigma_{\mathcal{D}}}{2^\kappa} + \frac{(q_p + \sigma_{\mathcal{E}} + \sigma_{\mathcal{D}})\sigma_{\mathcal{D}}}{2^c} + \frac{q_{\mathcal{D}}}{2^\tau},$$

where $\sigma_{\mathcal{E}}, \sigma_{\mathcal{D}}$ are defined in (2).

The bound is more complex than the one of Theorem 1, but intuitively implies that NORX offers integrity as long as it offers privacy and the number of forgery attempts $\sigma_{\mathcal{D}}$ is limited, where the total complexity $q_p + \sigma_{\mathcal{E}} + \sigma_{\mathcal{D}}$ should not exceed $2^c/\sigma_{\mathcal{D}}$. See Table 1 for the security level for the various parameter choices of NORX. Needless to say, the exact bound is more fine-grained.

Proof. We consider any adversary $\mathcal{A}$ that has access to $(p^\pm, \mathcal{E}_K, \mathcal{D}_K)$ and attempts to make $\mathcal{D}_K$ output a non-⊥ value. As in the proof of Theorem 1, we apply a PRP-PRF switch to find

$$\mathbf{Adv}_\Pi^{\text{auth}}(\mathcal{A}) = \mathbf{Pr}\left(\mathcal{A}^{p^\pm, \mathcal{E}_K, \mathcal{D}_K} \text{ forges}\right)$$

$$\leq \mathbf{Pr}\left(\mathcal{A}^{f^\pm, \mathcal{E}_K, \mathcal{D}_K} \text{ forges}\right) + \frac{(q_p + \sigma_{\mathcal{E}} + \sigma_{\mathcal{D}})^2}{2^{b+1}}. \tag{8}$$

Then we focus on $\mathcal{A}$ having oracle access to $(f^\pm, \mathcal{E}_K, \mathcal{D}_K)$. As before, we assume without loss of generality that the adversary only makes full-block queries.

We inherit terminology from Theorem 1. The state values corresponding to encryption and decryption queries will both be labeled (j, k), where j indicates the query and k the state value within the jth query. If needed we will add another parameter $\delta \in \{\mathcal{D}, \mathcal{E}\}$ to indicate that a state value $s_{\delta,j,k}$ is in the jth query to oracle δ, for $\delta \in \{\mathcal{D}, \mathcal{E}\}$ and $j \in \{1, \ldots, q_\delta\}$. Particularly, this means we will either label the state values as in (5) with a δ appended to the subscript, or simply as $(s_{\delta,j,1}, \ldots, s_{\delta,j,\sigma_{\delta,j}})$.

As before, we employ the collision events guess and hit, but expanded to the new notation with $\delta = \mathcal{E}$. Next, we define two $\mathcal{D}$-related collision events $\mathcal{D}$guess and $\mathcal{D}$hit. Let $i \in \{1, \ldots, q_p\}$, $(\mathcal{D}, j, k)$ be a decryption query index, and (δ', j', k') be an encryption or decryption query index:

$$\mathcal{D}\text{guess}(i; j, k) \equiv x_i = s_{\mathcal{D},j,k},$$
$$\mathcal{D}\text{hit}(j, k; \delta', j', k') \equiv \text{parent}(s_{\mathcal{D},j,k}) \neq \text{parent}(s_{\delta',j',k'}) \wedge s_{\mathcal{D},j,k} = s_{\delta',j',k'},$$

We write $\mathcal{D}\text{guess} = \vee_{i;j,k} \mathcal{D}\text{guess}(i; j, k)$ and $\text{hit} = \vee_{j,k;\delta',j',k'} \mathcal{D}\text{hit}(j, k; \delta', j', k')$, and define $\text{event} = \text{guess} \vee \text{hit} \vee \mathcal{D}\text{guess} \vee \mathcal{D}\text{hit}$.

Observe that from (8) we get

$$\mathbf{Pr}\left(\mathcal{A}^{f^\pm,\mathcal{E}_K,\mathcal{D}_K} \text{ forges}\right) \leq \mathbf{Pr}\left(\mathcal{A}^{f^\pm,\mathcal{E}_K,\mathcal{D}_K} \text{ forges} \mid \neg\text{event}\right) + \qquad (9)$$
$$\mathbf{Pr}\left(\mathcal{A}^{f^\pm,\mathcal{E}_K,\mathcal{D}_K} \text{ sets event}\right).$$

A bound on the probability that $\mathcal{A}$ sets event is derived in Lemma 3.

The remainder of this proof centers on the probability that $\mathcal{A}$ forges given that event does not happen. Such a forgery requires that $[f(s_{\mathcal{D},j}^{\text{tag}})]^\tau = A_j$ for some decryption query j. By $\neg$event, we know that $s_{\mathcal{D},j}^{\text{tag}}$ is a new state value for all $j \in \{1, \ldots, q_\mathcal{D}\}$, hence f's output under $s_{\mathcal{D},j}^{\text{tag}}$ is independent of all other values and uniformly distributed for all j. As a result, we know that the jth forgery attempt is successful with probability at most $1/2^\tau$. Summing over all $q_\mathcal{D}$ queries, we get

$$\mathbf{Pr}\left(\mathcal{A}^{f^\pm,\mathcal{E}_K,\mathcal{D}_K} \text{ forges} \mid \neg\text{event}\right) \leq \frac{q_\mathcal{D}}{2^\tau},$$

and the proof is completed via (8,9) and the bound of Lemma 3, where we again use that $\dfrac{q_p \sigma_\mathcal{E} + \sigma_\mathcal{E}^2/2}{2^b} \leq \dfrac{(q_p + \sigma_\mathcal{E} + \sigma_\mathcal{D})^2}{2^{b+1}}$. □

Lemma 3. $\mathbf{Pr}\left(\mathcal{A}^{f^\pm,\mathcal{E}_K,\mathcal{D}_K} \text{ sets event}\right) \leq \dfrac{q_p \sigma_\mathcal{E} + \sigma_\mathcal{E}^2/2}{2^b} + \left(\dfrac{8 e q_p \sigma_\mathcal{E}}{2^b}\right)^{1/2} + \dfrac{r q_p}{2^c} +$

$\dfrac{q_p + \sigma_\mathcal{E} + \sigma_\mathcal{D}}{2^\kappa} + \dfrac{(q_p + \sigma_\mathcal{E})\sigma_\mathcal{D} + \sigma_\mathcal{D}^2/2}{2^c}.$

The proof of Lemma 3 is given in the full version of this paper [28].

4 Other CAESAR Submissions

In this section we discuss how the mode security proof of NORX generalizes to the CAESAR submissions Ascon, the BLNK mode underlying CBEAM/ STRIBOB, ICEPOLE, Keyak, and two out of the three PRIMATEs. Before doing so, we make a number of observations and note how the proof can accommodate small design differences.

- NORX uses domain separation constants at all rounds, but this is not strictly necessary and other solutions exist. In the privacy and integrity proofs of NORX, and more specifically at the analysis of state collisions caused by a decryption query in Lemma 3, the domain separations are only needed at the transitions between variable-length inputs, such as header to message data or message to trailer data. This means that the proofs would equally hold if there were simpler transitions at these positions, such as in Ascon. Alternatively, the domain separation can be done by using a different primitive, as in GIBBON and HANUMAN, or a slightly more elaborated padding, as in BLNK, ICEPOLE, and Keyak;
- The extra permutation evaluations at the initialization and finalization of NORX are not strictly necessary: in the proof we consider the monotone event that no state collides assuming no earlier state collision occurred. For instance, in the analysis of $\mathcal{D}$hit in the proof of Lemma 3, we necessarily have a new input to p at some point, and *consequently* all next inputs to p are new (except with some probability);
- NORX starts by initializing the state with $\text{init}(K, N) = (K\|N\|0^{b-\kappa-\nu}) \oplus$ const for some constant const and then permuting this value. Placing the key and nonce at different positions of the state does not influence the security analysis. The proof would also work if, for instance, the header is preceded with $K\|N$ or a properly padded version thereof and the starting state is 0^b;
- In a similar fashion, there is no problem in defining the tag to be a different τ bits of the final state; for instance, the rightmost τ bits;
- Key additions into the capacity part *after* the first permutation are harmless for the mode security proof. Particularly, as long as these are done at fixed positions, these have the same effect as XORing a domain separation constant.

These five modifications allow one to generalize the proof of NORX to Ascon, CBEAM and STRIBOB, ICEPOLE, Keyak, and two PRIMATEs, GIBBON and HANUMAN. The only major difference lies in the fact none of these designs accommodates a trailer, hence all are functions of the form

$$(C, A) \longleftarrow \mathcal{E}_K(N; H, M) \qquad \text{and} \qquad M/\bot \longleftarrow \mathcal{D}_K(N; H, C; A),$$

except for one instance of ICEPOLE which accommodates a secret message number. Additionally, these designs have $\sigma_\delta \le \lambda_\delta + q_\delta$ for $\delta \in \{\mathcal{D}, \mathcal{E}\}$ (or $\sigma_\delta \le \lambda_\delta + 2q_\delta$ for CBEAM/STRIBOB). We always write $H = (H_1, \dots, H_u)$ and $M = (M_1, \dots, M_v)$ whenever notation permits. In below sections we elaborate on these designs separately, where we slightly deviate from the alphabetical order to suit the presentation. Diagrams of all modes are given in Figure 2. The parameters and achieved provable security levels of the schemes are given in Table 1.

4.1 Ascon

Ascon is a submission by Dobraunig et al. [11] and is depicted in Figure 2a. It is originally defined based on two permutations p_1, p_2 that differ in the number of underlying rounds. We discard this difference, considering Ascon with one permutation p.

Ascon initializes its state using init that maps (K, N) to $(0^{b-\kappa-\nu} \| K \| N) \oplus$ const, where const is determined by some design-specific parameters set prior to the security experiment. The header and message can be of arbitrary length, and are padded to length a multiple of r bits using 10*-padding. An XOR with 1 separates header processing from message processing. From the above observations, it is clear that the proofs of NORX directly carry over to Ascon.

4.2 ICEPOLE

ICEPOLE is a submission by Morawiecki et al. [15] and is depicted in Figure 2c. It is originally defined based on two permutations, p_1 and p_2, that differ in the number of underlying rounds. We discard this difference, considering ICEPOLE with one permutation p.

ICEPOLE initializes its state as NORX does, be it with a different constant. The header and message can be of arbitrary length, and are padded as follows. Every block is first appended with a frame bit: 0 for header blocks $H_1, \ldots, H_{u-1}$ and message block M_v, and 1 for header block H_u and message blocks $M_1, \ldots, M_{v-1}$. Then, the blocks are padded to length a multiple of r bits using 10*-padding. In other words, every padded block of r bits contains at most $r - 2$ data bits. This form of domain separation using frame bits suffices for the proof to go through. One variant of ICEPOLE also allows for a secret message number M_{secret}, which consists of one block and is encrypted prior to the processing of the header, similar to the message. As this secret message number is of fixed length, no domain separation is required and the proof can easily be adapted. From above observations, it is clear that the proofs of NORX directly carry over to ICEPOLE. Without going into detail, we note that the same analysis can be generalized to the parallelized mode of ICEPOLE [15].

4.3 Keyak

Keyak is a submission by Bertoni et al. [16]. The basic mode for the serial case is depicted in Figure 2d, yet due to its hybrid character it is slightly more general in nature. It is built on top of SpongeWrap [9].

Keyak initializes its state by 0^b, and concatenates K, N, and H using a special padding rule:

$$\mathsf{H}_{\text{pad}}(K, N, H) = \mathsf{keypack}(K, 240) \| \mathsf{enc}_8(1) \| \mathsf{enc}_8(0) \| N \| H,$$

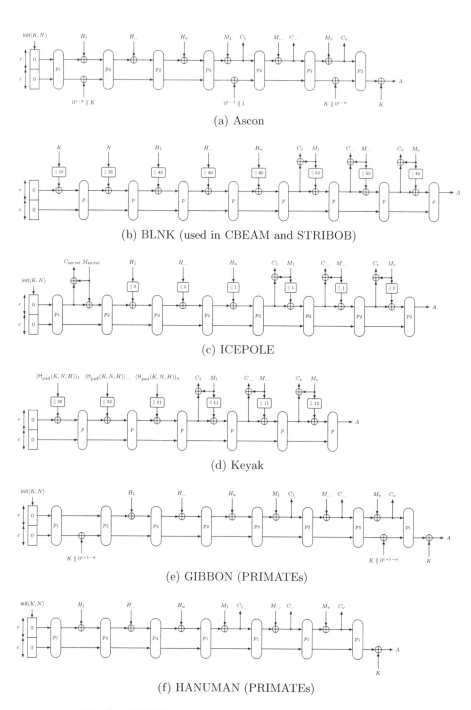

(a) Ascon

(b) BLNK (used in CBEAM and STRIBOB)

(c) ICEPOLE

(d) Keyak

(e) GIBBON (PRIMATEs)

(f) HANUMAN (PRIMATEs)

Fig. 2. CAESAR submission modes discussed in Section 4

where $\mathsf{enc}_8(x)$ is an encoding of x as a byte and $\mathsf{keypack}(K, \ell) = \mathsf{enc}_8(\ell/8)\|K\|$ $10^{-\kappa-1 \bmod (\ell-8)}$. The key-nonce-header combination $\mathsf{H}_{\mathrm{pad}}(K, N, H)$ and message M can be of arbitrary length, and are padded as follows: first, every block is appended with two frame bits, being 00 for header blocks $(\mathsf{H}_{\mathrm{pad}}(K, N, H))_1, \ldots,$ $(\mathsf{H}_{\mathrm{pad}}(K, N, H))_{u-1}$ and 01 for $(\mathsf{H}_{\mathrm{pad}}(K, N, H))_u$, and 11 for message blocks $M_1, \ldots, M_{v-1}$ and 10 for M_v. Then, the blocks are padded to length a multiple of r bits using 10^*1-padding. In other words, every padded block of r bits contains at most $r - 2$ data bits. This form of domain separation using frame bits suffices for the proof to go through. Due to above observations, our proof readily generalizes to SpongeWrap [9] and DuplexWrap [16], and thus to Keyak. Without going into detail, we note that the same analysis can be generalized to the parallelized mode of Keyak [16]. Additionally, Keyak also supports sessions, where the state is re-used for a next evaluation. Our proof generalizes to this case, simply with a more extended description of (1).

4.4   BLNK (CBEAM and STRIBOB)

CBEAM and STRIBOB are submissions by Saarinen [13, 14, 19, 20]. Minaud identified an attack on CBEAM [12], but we focus on the modes of operation. Both modes are based on the BLNK Sponge mode [18], which is depicted in Figure 2b.

The BLNK mode initializes its state by 0^b, compresses K into the state (using one or two permutation calls, depending on κ), and does the same with N. Then, the mode is similar to SpongeWrap [9], though using a slightly more involved domain separation system similar to the one of NORX. Due to above observations, our proof readily generalizes to BLNK [18], and thus to CBEAM and STRIBOB.

4.5   PRIMATEs: GIBBON and HANUMAN

PRIMATEs is a submission by Andreeva et al. [17], and consists of three algorithms: APE, GIBBON, and HANUMAN. The APE mode is the more robust one, and significantly differs from the other two, and from the other CAESAR submissions discussed in this work, in the way that ciphertexts are derived and because the mode is secure against nonce misusing adversaries up to common prefix [27]. We now focus on GIBBON and HANUMAN, which are depicted in Figures 2e and 2f. GIBBON is based on three related permutations $\mathbf{p} = (p_1, p_2, p_3)$, where the difference in p_2, p_3 is used as domain separation of the header compression and message encryption phases (the difference of p_1 from (p_2, p_3) is irrelevant for the mode security analysis). Similarly, HANUMAN uses two related permutations $\mathbf{p} = (p_1, p_2)$ for domain separation.

GIBBON and HANUMAN initialize their state using init that maps (K, N) to $0^{b-\kappa-\nu}\|K\|N$. The header and message can be of arbitrary length, and are padded to length a multiple of r bits using 10^*-padding. In case the true header (or message) happens to be a multiple of r bits long, the 10^*-padding is considered to spill over into the capacity. From above observations, it is clear that the proofs of NORX directly carry over to GIBBON and HANUMAN. A small difference appears due to the usage of two different permutations: we need to make two PRP-PRF switches for each world. Concretely this means that the first term in Theorem 1 becomes $\frac{5(q_p+\sigma_{\mathcal{E}})^2}{2^{b+1}}$ and the first term in Theorem 2 becomes $\frac{3(q_p+\sigma_{\mathcal{E}}+\sigma_D)^2}{2^{b+1}}$.

5   Conclusions

In this work we analyzed one of the Sponge-based authenticated encryption designs in detail, NORX, and proved that it achieves security of approximately $\min\{2^{b/2}, 2^c, 2^\kappa\}$, significantly improving upon the traditional bound of $\min\{2^{c/2}, 2^\kappa\}$. Additionally, we showed that this proof straightforwardly generalizes to five other CAESAR modes, Ascon, BLNK (of CBEAM/STRIBOB), ICEPOLE, Keyak, and PRIMATEs. Our findings indicate an overly conservative parameter choice made by the designers, implying that some designs can improve speed by a factor of 4 at barely any security loss.

It is expected that the security proofs also generalize to the modes of Artemia [21] and π-Cipher [22]. However, they deviate slightly more from the other designs. Artemia is based on the JH hash function [29] and XORs data blocks in both the rate and capacity part. It does not use domain separations, rather it encodes the lengths of the inputs into the padding at the end [30]. Therefore, a generalization of the proof of NORX to Artemia is not entirely straightforward. π-Cipher, on the other hand, is structurally different in the way it maintains state. A so-called "common internal state" is used throughout the evaluation. For the processing of the header (or similarly the message) the state is forked into u chains to process $H_1, \ldots, H_u$ in parallel, resulting in u tag values, which are added into the common internal state. Due to this design property, the deviation of π-Cipher from NORX is too large to simply claim that the proof carries over.

The results in this work are derived in the ideal permutation model, where the underlying primitive is assumed to be ideal. We acknowledge that this model does not perfectly reflect the properties of the primitives. For instance, it is stated by the designers of Ascon, NORX, and PRIMATEs that non-random (but harmless) properties of the underlying permutation exist. Furthermore, it is important to realize that the proofs of security for the modes of operation in the ideal model do not have a direct connection with security analysis performed on the permutations, as is the case with block ciphers modes of operation. Nevertheless, we can use these proofs as heuristics to guide cryptanalysts to focus on the underlying permutations, rather than the modes themselves.

Acknowledgements. The authors would like to thank their co-designers of NORX and PRIMATEs and the designers of Ascon and Keyak for the discussions. In particular, we thank Samuel Neves for his useful comments. This work was supported in part by the Research Fund KU Leuven, OT/13/071, and in part by the Research Council KU Leuven: GOA TENSE (GOA/11/007).

Atul Luykx is supported by a Ph.D. Fellowship from the Institute for the Promotion of Innovation through Science and Technology in Flanders (IWT-Vlaanderen). Bart Mennink is a Postdoctoral Fellow of the Research Foundation – Flanders (FWO).

References

1. CAESAR: Competition for Authenticated Encryption: Security, Applicability, and Robustness (2014), http://competitions.cr.yp.to/caesar.html
2. Whiting, D., Housley, R., Ferguson, N.: AES Encryption and Authentication Using CTR Mode and CBC-MAC. IEEE 802.11-02/001r2 (2002)
3. Rogaway, P., Bellare, M., Black, J., Krovetz, T.: OCB: a block-cipher mode of operation for efficient authenticated encryption. In: Reiter, M.K., Samarati, P. (eds.) ACM Conference on Computer and Communications Security, pp. 196–205. ACM (2001)
4. Rogaway, P.: Efficient instantiations of tweakable blockciphers and refinements to modes OCB and PMAC. In: Lee, P.J. (ed.) ASIACRYPT 2004. LNCS, vol. 3329, pp. 16–31. Springer, Heidelberg (2004)
5. Krovetz, T., Rogaway, P.: The software performance of authenticated-encryption modes. In: Joux, A. (ed.) FSE 2011. LNCS, vol. 6733, pp. 306–327. Springer, Heidelberg (2011)
6. Bellare, M., Rogaway, P., Wagner, D.: The EAX mode of operation. In: Roy, B., Meier, W. (eds.) FSE 2004. LNCS, vol. 3017, pp. 389–407. Springer, Heidelberg (2004)
7. Bertoni, G., Daemen, J., Peeters, M., Van Assche, G.: Sponge functions. In: ECRYPT Hash Function Workshop (2007)
8. Bertoni, G., Daemen, J., Peeters, M., Van Assche, G.: On the security of the keyed sponge construction. In: Symmetric Key Encryption Workshop (SKEW 2011) (2011)
9. Bertoni, G., Daemen, J., Peeters, M., Van Assche, G.: Duplexing the sponge: Single-pass authenticated encryption and other applications. In: Miri, A., Vaudenay, S. (eds.) SAC 2011. LNCS, vol. 7118, pp. 320–337. Springer, Heidelberg (2012)
10. Aumasson, J., Jovanovic, P., Neves, S.: NORX v1 (2014), Submission to CAESAR competition
11. Dobraunig, C., Eichlseder, M., Mendel, F., Schläffer, M.: Ascon v1 (2014), Submission to CAESAR competition
12. Minaud, B.: Re: CBEAM Withdrawn as of today! (2014), CAESAR mailing list
13. Saarinen, M.: CBEAM r1 (2014), Submission to CAESAR competition
14. Saarinen, M.: CBEAM: Efficient authenticated encryption from feebly one-way ϕ functions. In: Benaloh (ed.) [9], pp. 251–269
15. Morawiecki, P., Gaj, K., Homsirikamol, E., Matusiewicz, K., Pieprzyk, J., Rogawski, M., Srebrny, M., Wójcik, M.: ICEPOLE v1 (2014), Submission to CAESAR competition
16. Bertoni, G., Daemen, J., Peeters, M., Van Assche, G., Van Keer, R.: Keyak v1 (2014), Submission to CAESAR competition
17. Andreeva, E., Bilgin, B., Bogdanov, A., Luykx, A., Mendel, F., Mennink, B., Mouha, N., Wang, Q., Yasuda, K.: PRIMATEs v1 (2014), Submission to CAESAR competition

18. Saarinen, M.: Beyond modes: Building a secure record protocol from a crypto-graphic sponge permutation. In: Benaloh (ed.) [9], pp. 270–285
19. Saarinen, M.: STRIBOB r1 (2014), Submission to CAESAR competition
20. Saarinen, M.: Authenticated encryption from GOST R 34.11-2012 LPS permutation. In: CTCrypt 2014 (2014)
21. Alizadeh, J., Aref, M., Bagheri, N.: Artemia v1 (2014), Submission to CAESAR competition
22. Gligoroski, D., Mihajloska, H., Samardjiska, S., Jacobsen, H., El-Hadedy, M., Jensen, R.: π-Cipher v1 (2014), Submission to CAESAR competition
23. Bellare, M., Namprempre, C.: Authenticated encryption: Relations among notions and analysis of the generic composition paradigm. J. Cryptology 21(4), 469–491 (2008)
24. Iwata, T., Ohashi, K., Minematsu, K.: Breaking and repairing GCM security proofs. In: Safavi-Naini, R., Canetti, R. (eds.) CRYPTO 2012. LNCS, vol. 7417, pp. 31–49. Springer, Heidelberg (2012)
25. Bertoni, G., Daemen, J., Peeters, M., Van Assche, G.: Sponge-based pseudo-random number generators. In: Mangard, S., Standaert, F.-X. (eds.) CHES 2010. LNCS, vol. 6225, pp. 33–47. Springer, Heidelberg (2010)
26. Bellare, M., Rogaway, P.: The security of triple encryption and a frame-work for code-based game-playing proofs. In: Vaudenay, S. (ed.) EUROCRYPT 2006. LNCS, vol. 4004, pp. 409–426. Springer, Heidelberg (2006)
27. Andreeva, E., Bilgin, B., Bogdanov, A., Luykx, A., Mennink, B., Mouha, N., Yasuda, K.: APE: Authenticated permutation-based encryption for lightweight cryptography. In: Cid, C., Rechberger, C. (eds.) FSE. LNCS. Springer (2014)
28. Jovanovic, P., Luykx, A., Mennink, B.: Beyond $2^{c/2}$ security in sponge-based authenticated encryption modes. Cryptology ePrint Archive, Report 2014/373 (2014), Full version of this paper
29. Wu, H.: The Hash Function JH (2011) Submission to NIST's SHA-3 competition
30. Bagheri, N.: Padding of Artemia (2014), CAESAR mailing list
31. Benaloh, J. (ed.): CT-RSA 2014. LNCS, vol. 8366. Springer, Heidelberg (2014)

A   Proof Technique Used in Lemma 2

Formally, the proof technique used in Lemma 2 relies on the following paradigm. Note that there is an ordering of the $q_p + \sigma_{\mathcal{E}}$ primitive queries, and we can reformulate guess(ℓ), hit(ℓ), key(ℓ), and multi(ℓ) for $\ell = 1, \ldots, q_p + \sigma_{\mathcal{E}}$ analogously. Defining event(ℓ) = guess(ℓ) $\vee$ hit(ℓ) and help(ℓ) = key(ℓ) $\vee$ multi(ℓ), then

$$\mathbf{Pr}\left(\text{event}\right) \leq \mathbf{Pr}\left(\text{event}(q_p + \sigma_{\mathcal{E}}) \mid \neg\text{event}(1 \cdots q_p + \sigma_{\mathcal{E}} - 1) \wedge \neg\text{help}(1 \cdots q_p + \sigma_{\mathcal{E}})\right) + \mathbf{Pr}\left(\text{event}(1 \cdots q_p + \sigma_{\mathcal{E}} - 1) \vee \text{help}(1 \cdots q_p + \sigma_{\mathcal{E}})\right),$$

and inductively $\mathbf{Pr}\left(\text{event}\right) \leq \sum_{\ell=1}^{q_p + \sigma_{\mathcal{E}}} \mathbf{Pr}\left(\text{event}(\ell) \mid \neg\text{event}(1 \cdots \ell - 1) \wedge \neg\text{help}(1 \cdots \ell)\right) + \mathbf{Pr}\left(\text{help}(\ell) \mid \neg\text{help}(1 \cdots \ell - 1)\right)$. This formulation would however merely reduce the readability of the proof.

How to Securely Release Unverified Plaintext in Authenticated Encryption[*]

Elena Andreeva[1,2], Andrey Bogdanov[3], Atul Luykx[1,2], Bart Mennink[1,2], Nicky Mouha[1,2], and Kan Yasuda[1,4]

[1] Department of Electrical Engineering, ESAT/COSIC, KU Leuven, Belgium
{firstname.lastname}@esat.kuleuven.be
[2] iMinds, Belgium
[3] Department of Mathematics, Technical University of Denmark, Denmark
anbog@dtu.dk
[4] NTT Secure Platform Laboratories, Japan
yasuda.kan@lab.ntt.co.jp

Abstract. Scenarios in which authenticated encryption schemes output decrypted plaintext before successful verification raise many security issues. These situations are sometimes unavoidable in practice, such as when devices have insufficient memory to store an entire plaintext, or when a decrypted plaintext needs early processing due to real-time requirements. We introduce the first formalization of the *releasing unverified plaintext* (RUP) setting. To achieve privacy, we propose using *plaintext awareness* (PA) along with IND-CPA. An authenticated encryption scheme is PA if it has a *plaintext extractor*, which tries to fool adversaries by mimicking the decryption oracle, *without* the secret key. Releasing unverified plaintext to the attacker then becomes harmless as it is infeasible to distinguish the decryption oracle from the plaintext extractor. We introduce two notions of plaintext awareness in the symmetric-key setting, PA1 and PA2, and show that they expose a new layer of security between IND-CPA and IND-CCA. To achieve integrity, INT-CTXT in the RUP setting is required, which we refer to as INT-RUP. These new security notions are compared with conventional definitions, and are used to make a classification of symmetric-key schemes in the RUP setting. Furthermore, we re-analyze existing authenticated encryption schemes, and provide solutions to fix insecure schemes.

1 Introduction

The goal of authenticated encryption (AE) is to simultaneously provide data privacy and integrity. AE decryption conventionally consists of two phases: plaintext computation and verification. As reflected in classical security models, plaintext coming from decryption is output only upon successful verification.

[*] This work was supported in part by the Research Council KU Leuven: GOA TENSE (GOA/11/007) and OT/13/071. Elena Andreeva, Bart Mennink, and Nicky Mouha are Postdoctoral Fellows of the Research Foundation – Flanders (FWO). Atul Luykx is supported by a Ph.D. Fellowship from the Institute for the Promotion of Innovation through Science and Technology in Flanders (IWT-Vlaanderen).

P. Sarkar and T. Iwata (Eds.): ASIACRYPT 2014, PART I, LNCS 8873, pp. 105–125, 2014.

Nevertheless, there are settings where releasing plaintext before verification is desirable. For example, it is necessary if there is not enough memory to store the entire plaintext [21] or because real-time requirements would otherwise not be met [15, 38]. Even beyond these settings, using dedicated schemes secure against the release of unverified plaintext can increase efficiency. For instance, to avoid releasing unverified plaintext into a device with insecure memory [37], the two-pass Encrypt-then-MAC composition can be used: a first pass to verify the MAC, and a second to decrypt the ciphertext. However, a single pass AE scheme suffices if it is secure against the release of unverified plaintext.

If the attacker cannot observe the unverified plaintext directly, it may be possible to determine properties of the plaintext through a side channel. This occurs, for example, in the padding oracle attacks introduced by Vaudenay [39], where an error message or the lack of an acknowledgment indicates whether the unverified plaintext was correctly padded. Canvel et al. [18] showed how to mount a padding oracle attack on the then-current version of OpenSSL by exploiting timing differences in the decryption processing of TLS. As shown by Paterson and AlFardan [1, 30] for TLS and DTLS, it is very difficult to prevent an attacker from learning the cause of decryption failures.

The issue of releasing unverified plaintext has also been acknowledged and explicitly discussed in the upcoming CAESAR competition:[1] *"Beware that security questions are raised by any authenticated cipher that handles a long ciphertext in one pass without using a large buffer: releasing unverified plaintext to applications often means releasing it to attackers and also requires an analysis of how the applications will react."*

For several AE schemes, including OCB [27], AEGIS [41], ALE [15], and FIDES [15], the designers explicitly stress that unverified plaintext cannot be released. Although the issue of *releasing unverified plaintext* (RUP) in AE is frequently discussed in the literature, it has largely remained unaddressed even in recent AE proposals, likely due to a lack of comprehensive study.

We mention explicitly that we do *not* recommend omitting verification, which remains essential to preventing incorrect plaintexts from being accepted. To ensure maximal security, unverified plaintext must be kept hidden from adversaries. However, our scenario assumes that the attacker can see the unverified plaintext, or any information relating to it, before verification is complete. Furthermore, issues related to the behavior of applications which process unverified plaintext are beyond the scope of this paper; careful analysis is necessary in such situations.

1.1 Security Under Release of Unverified Plaintext

AE security is typically examined using indistinguishability under chosen plaintext attack (IND-CPA) for privacy and integrity of ciphertexts (INT-CTXT) for integrity, and a scheme which achieves both is indistinguishable under chosen ciphertext attack (IND-CCA), as shown by Bellare and Namprempre [9] and Katz and Yung [26]. However, in the RUP situation adversaries can also observe

---

[1] http://competitions.cr.yp.to/features.html

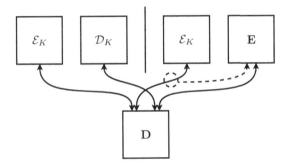

Fig. 1. The two plaintext aware settings (PA1 and PA2) used in the paper, where **D** is an adversary. Not shown in the figure is the type of IV used by the $\mathcal{E}_K$ oracle (cf. Sect. 3.2). **Left:** Real world, with encryption oracle $\mathcal{E}_K$ and decryption oracle $\mathcal{D}_K$. **Right:** Simulated world, with encryption oracle $\mathcal{E}_K$ and plaintext extractor **E**. The plaintext extractor **E** is a stateful algorithm without knowledge of the secret key K, nor access to the encryption oracle $\mathcal{E}_K$. The dotted line indicates that **E** has access to the encryption queries made by adversary **D**, which only holds in the PA1 setting.

unverified plaintext, which the conventional definitions do not take into account. To address this gap we introduce two definitions: integrity under releasing unverified plaintext (INT-RUP) and *plaintext awareness* (PA). For integrity we propose using INT-RUP and for privacy both IND-CPA and PA. In the full version of this paper [3], we discuss how the combination of INT-RUP, IND-CPA, and PA measures the impact of releasing unverified plaintext on security.

INT-RUP. The goal of an adversary under INT-CTXT is to produce new ciphertexts which pass verification, with only access to the encryption oracle. We translate INT-CTXT into the RUP setting, called INT-RUP, by allowing the adversary to observe unverified plaintexts. We formalize this by separating plaintext computation from verification, and giving the adversary access to a plaintext-computing oracle.

Plaintext Awareness (PA). We introduce PA as a new symmetric-key notion to achieve security in the RUP setting. Informally, we define a scheme to be PA if the adversary cannot gain any additional knowledge about the plaintext from decryption queries besides what it can derive from encryption queries.

Our PA notion only involves encryption and decryption, and can thus be defined both for encryption schemes as well as for AE schemes that release unverified plaintext.

At the heart of our new PA notion is the *plaintext extractor*, shown in Fig. 1. We say that an encryption scheme is PA if it has an efficient plaintext extractor, which is a stateful algorithm that mimicks the decryption oracle in order to fool the adversary. It cannot make encryption nor decryption queries, and does not know the secret key. We define two notions of plaintext awareness: PA1 and PA2. The extractor is given access to the history of queries made to the encryption oracle in PA1, but not in PA2. Hence PA1 is used to model RUP scenarios in which the goal of the adversary is to gain knowledge beyond what it knows from

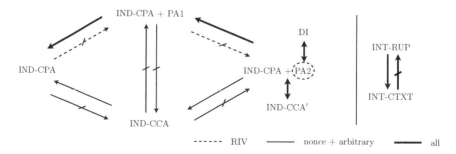

Fig. 2. Implications and separations between the IND-CPA, IND-CPA+PA1, IND-CPA+PA2, IND-CCA, IND-CCA', PA2, and DI security notions (left) and INT-CTXT and INT-RUP (right). Dashed lines refer to relations that hold if the IV is random and thin solid lines in case of nonce or arbitrary IV. We use a thick solid line if the relation holds under all IV cases.

the query history. For situations in which the goal of the adversary is to decrypt one of the ciphertexts in the query history, we require PA2.

Relations among Notions. PA for public-key encryption was introduced by Bellare and Rogaway [11], and later defined without random oracles by Bellare and Palacio [10]. In the symmetric-key setting, our definition of PA is somewhat similar, however there are important technical differences which make the public-key results inapplicable to the symmetric-key setting.

Relations among the PA and conventional security definitions for encryption (see Sect. 3.3) are summarized in Fig. 2. We consider three IV assumptions: random IV, nonce IV (non-repeating value), and arbitrary IV (value that can be reused), as explained in Sect. 3.2. The statements of the theorems and proofs can be found in the full version of this paper [3].

The motivation for having two separate notions, PA1 and PA2, is as follows. As we prove in this work, if the plaintext extractor has access to the query history (PA1), then there are no implications between IND-CPA+PA1 and IND-CCA. However, if we modify plaintext awareness so that the plaintext extractor no longer has access to the query history (PA2), then we can prove that IND-CPA+PA2 implies IND-CCA'. IND-CCA' is a strengthened version of IND-CCA, where we allow the adversary to re-encrypt the outputs of the decryption oracle. Note that such a re-encryption is always allowed in the public-key setting, but not in the symmetric-key setting where the key required for encryption is secret. Furthermore, we also prove that PA2 is equivalent to the notion of *decryption independence* (DI). DI captures the fact that encryption and decryption under the same key are only related to each other as much as encryption and decryption under different keys.

Finally, although INT-RUP clearly implies INT-CTXT, the opposite is not necessarily true.

Motivating Examples. To get an intuition for PA1 (shown in Fig. 1) and how it relates to the RUP setting, we provide two motivating examples with CTR mode. For simplicity, we define the encryption function of CTR mode as

$\mathcal{E}_K(\text{IV}, M) = E_K(\text{IV}) \oplus M$, where the message M and the initialization value IV consist of one block each, and E_K is a block cipher with a secret key K. The corresponding decryption function is $\mathcal{D}_K(\text{IV}, C) = E_K(\text{IV}) \oplus C$. As shown in [13], CTR mode is IND-CPA but not IND-CCA, a result that holds for nonce IVs (unique non-repeating values) as well as for random IVs.

1. *Nonce IV CTR mode is not PA1.* Following Rogaway [31], we assume that an adversary is free to specify the IV for encryption and decryption queries, as long as it does not make two encryption queries with the same nonce IV, N. In the attack, an adversary first makes a decryption query (N, C) with nonce N and one-block ciphertext C to obtain a message M. The correct decryption of M is $E_K(N) \oplus C$ as output by the decryption oracle. The adversary then computes the keystream $\kappa := M \oplus C$. Now in a second query (N, M'), this time to the encryption oracle, the adversary obtains C' where $C' = M' \oplus \kappa$.

   The scheme fails to be plaintext aware as it is infeasible for any plaintext extractor to be consistent with subsequent encryption queries. Specifically, the plaintext extractor cannot compute κ at the time of the first decryption query for the following reasons: it does not know the secret key K, it is not allowed to do encryption queries, and an encryption query with N has not yet been recorded in the query history.

2. *Random IV CTR mode is PA1.* In this setting, the IV used in encryption is chosen randomly by the environment, and therefore out of the attacker's control. However, the adversary can still freely choose the IV for its decryption queries. In this random IV setting, the attack in the nonce IV example does not apply. To see this, consider an adversary which queries the decryption of (IV_1, C) with a one-block ciphertext C. It can compute the keystream associated to IV_1, but does not control when IV_1 is used in encryption. Thus, a plaintext extractor can be defined as outputting a random plaintext M in response to the (IV_1, C) query.

   But what if an adversary makes additional decryption queries with the same IV? Suppose the adversary makes decryption query $(IV_1, C \oplus \Delta)$. Since the plaintext extractor is a stateful algorithm, it can simply output $M \oplus \Delta$ to provide consistency. Furthermore, if an adversary makes encryption queries, these will be seen by the PA1 plaintext extractor. Therefore, the plaintext extractor can calculate the keystream from these queries, and respond to any decryption queries in a consistent way. A proof that random IV CTR mode is PA1 is provided in Prop. 2.

AE schemes such as GCM [28] and CCM [40] reduce to CTR mode in the RUP setting. This is because the adversary does not need to forge a ciphertext in order to obtain information about the corresponding (unverified) plaintext. By requiring that the underlying encryption scheme of an AE scheme is PA1, we ensure that the adversary does not gain any information from decryption queries, meaning no decryption query can be used to find an inconsistency with any past or future queries to the encryption or decryption oracles.

1.2 Analysis of Authenticated Encryption Schemes

Given the formalization of AE in the RUP setting, we categorize existing AE schemes based on the type of IV used by the encryption function: random IV, nonce IV, and arbitrary IV. Then, we re-analyze the security of several recently proposed AE schemes as well as more established AE schemes. In order to do so, we split the decryption algorithms into two parts, plaintext computation and verification, as described in Sect. 3.1.

For integrity, we show that OCB [33] and COPA [4] succumb to attacks by using unverified plaintext to construct forgeries. For privacy an overview of our results can be seen in Table 1, where we also include the encryption-only modes CTR and CBC as random IV examples. We draw a distinction between the schemes that are *online* and the schemes that are not, where an online scheme is one that is able to produce ciphertext blocks as it receives plaintext blocks.

Most of the schemes in Table 1 fail to achieve PA1. As a result, we demonstrate techniques to restore PA1 for nonce IV and arbitrary IV schemes. For the former, we introduce the *nonce decoy* technique, and for the latter the PRF-to-IV method, which converts a random IV PA1 scheme into an arbitrary IV PA1 scheme. For online arbitrary IV schemes, we demonstrate that PA1 security can be achieved only if the ciphertext is substantially longer than the plaintext, or the *decryption* is offline. We show that McOE-G [20] achieves PA1 if the plaintext is padded so that the ciphertext becomes twice as long. We also prove that APE [2], an online deterministic AE scheme with offline decryption, achieves PA1.

Finally we show that the nonce decoy preserves INT-RUP, and the PRF-to-IV method turns any random IV scheme into an INT-RUP arbitrary IV scheme.

1.3 Background and Related Work

The definition of encryption and AE has been extended and generalized in different ways. In 2004, Rogaway [32] introduced nonce IV encryption schemes, in contrast with prior encryption modes that used a random IV, as in the CBC mode standardized by NIST in 1980 [29].

Rogaway and Shrimpton [34] formalized deterministic AE (DAE), where an IV input is optional and can therefore take arbitrary values. Secure DAE differs from secure nonce IV AE schemes in that DAE privacy is possible only up to message repetition, namely an adversary can detect repeated encryptions. Unfortunately, DAE schemes cannot be online. To resolve this issue, Fleischmann et al. [20] explored online DAE schemes, where privacy holds only up to repetitions of messages with identical prefixes or up to the longest common prefix.

Tsang et al. [38] gave syntax and security definitions of AE for streaming data. Bellare and Keelveedhi [6] considered a stronger security model where data may be key-dependent. Boldyreva et al. reformulated AE requirements and properties to handle ciphertext fragmentation in [16], and enhanced the syntax and security definitions so that the verification oracle is allowed to handle multiple failure events in [17]. Our formalization can be interpreted as a special case of the work in [17], yet the emphasis and results differ.

Table 1. PA1 and PA2 security of deterministic and non-deterministic schemes, separated as described in Sect. 3.1. In the columns for PA1 and PA2, ✓ means secure (there exists an extractor), and ✗ means insecure (there exists an attack). Proofs for the security results in this table can be found in Sect. 5.

IV type	Online	Scheme	PA1	PA2	Remark
random	✓	CTR, CBC [29]	✓	✗	
nonce	✓	OCB [33]	✗	✗	
	✓	GCM [28], SpongeWrap [14]	✗	✗	
	✗	CCM [40]	✗	✗	not online [35]
arbitrary	✓	COPA [4]	✗	✗	privacy up to prefix
	✓	McOE-G [20]	✗	✗	//
	✓	APE [2]	✓	✗	//, backwards decryption
	✗	SIV [34], BTM [23], HBS [24]	✓	✗	privacy up to repetition
	✗	Encode-then-Encipher [12]	✓	✓	//, VIL SPRP, padding

2   Preliminaries

Symbols. Given two strings A and B in $\{0,1\}^*$, we use $A\|B$ and AB interchangeably to denote the concatenation of A and B. The symbol $\oplus$ denotes the bitwise XOR operation of two strings. Addition modulo 2^n is denoted by $+$, where n usually is the bit length of a block. For example, in the CTR mode of operation of a block cipher, we increment the IV value by addition $IV+i \pmod{2^n}$, where n is the block size, the n-bit string $IV = IV_{n-1}\cdots IV_1 IV_0 \in \{0,1\}^n$ is converted to an integer $2^{n-1}IV_{n-1}+\cdots+2IV_1+IV_0 \in \{0,1,\ldots,2^n-1\}$, and the result of addition is converted to an n-bit string in the reverse way. By $K \xleftarrow{R} \mathsf{K}$ we mean that K is chosen uniformly at random from the set K. All algorithms and adversaries are considered to be "efficient".

Adversaries and Advantages. An adversary is an oracle Turing machine. Let $\mathbb{D}$ be some class of computationally bounded adversaries; a class $\mathbb{D}$ can consist of a single adversary $\mathbf{D}$, i.e. $\mathbb{D} = \{\mathbf{D}\}$, in which case we simply write $\mathbf{D}$ instead of $\mathbb{D}$. For convenience, we use the notation

$$\underset{\mathbb{D}}{\Delta}(f\,;\,g) := \sup_{\mathbf{D}\in\mathbb{D}} \left|\Pr[\mathbf{D}^f = 1] - \Pr[\mathbf{D}^g = 1]\right|$$

to denote the supremum of the distinguishing advantages over all adversaries distinguishing oracles f and g, where the notation $\mathbf{D}^\mathcal{O}$ indicates the value output by $\mathbf{D}$ after interacting with oracle $\mathcal{O}$. The probabilities are defined over the random coins used in the oracles and the random coins of the adversary, if any. Multiple oracles are separated by a comma, e.g. $\Delta(f_1, f_2\,;\,g_1, g_2)$ denotes distinguishing the combination of f_1 and f_2 from the combination of g_1 and g_2.

If $\mathbf{D}$ is distinguishing $(f_1, f_2, \ldots, f_k)$ from $(g_1, g_2, \ldots, g_k)$, then by $\mathcal{O}_i$ we mean the ith oracle that $\mathbf{D}$ has access to, i.e. either f_i or g_i depending upon which oracles it is interacting with. By $\mathcal{O}_i \hookrightarrow \mathcal{O}_j$ we describe a set of actions that $\mathbf{D}$ can perform: first $\mathbf{D}$ queries $\mathcal{O}_i$, and then at some point in the future $\mathbf{D}$ queries $\mathcal{O}_j$ with the output of $\mathcal{O}_i$, assuming the output of $\mathcal{O}_i$ can be used directly as the input for $\mathcal{O}_j$. If the oracles $\mathcal{O}_i$ and $\mathcal{O}_j$ represent a family of algorithms indexed by inputs, then the indices must match. For example, say that $\mathcal{E}_K^{N,A}$ and $\mathcal{D}_K^{N,A}$ are families indexed by (N, A). Then $\mathcal{E}_K \hookrightarrow \mathcal{D}_K$ describes a set of actions, which includes querying $\mathcal{E}_K^{N,A}(M)$ to receive C, and then at some point in the future querying $\mathcal{D}_K^{N,A}(C)$, where K, N, A, and C are reused.

Our security definitions follow [9] and are given in terms of adversary advantages. A scheme is said to be secure with respect to some definition if it is negligible with respect to all adversaries with time complexity polynomial in the security parameter. As in [9], positive results are given as explicit bounds, whereas negative results, i.e. separations, are given in asymptotic terms, which can easily be converted into concrete bounds.

Online Functions. A function $f : \mathsf{M} \to \mathsf{C}$ is said to be n-*online* if there exist functions $f_i : \{0,1\}^i \to \{0,1\}^{c_i}$ and $f_i' : \{0,1\}^i \to \{0,1\}^{c_i'}$ such that $c_i > 0$, and for all $M \in \mathsf{M}$ we have

$$f(M) = f_n(M_1) \, f_{2n}(M_1 M_2) \cdots f_{jn}(M_1 M_2 \cdots M_j) \, f_{|M|}'(M) \ ,$$

where $j = \lfloor (|M| - 1)/n \rfloor$ and M_i is the ith n-bit block of M. Often we just say f is *online* if the value n is clear from context.

3 AE Schemes: Syntax, Types, and Security

3.1 New AE Syntax

A conventional AE scheme $\Pi = (\mathcal{E}, \mathcal{D})$ consists of an encryption algorithm $\mathcal{E}$ and a decryption algorithm $\mathcal{D}$:

$$(C, T) \leftarrow \mathcal{E}_K^{IV,A}(M) \ ,$$
$$M/\bot \leftarrow \mathcal{D}_K^{IV,A}(C, T) \ ,$$

where $K \in \mathsf{K}$ is a key, $IV \in \mathsf{IV}$ an initialization value, $A \in \mathsf{A}$ associated data, $M \in \mathsf{M}$ a message, $C \in \mathsf{C}$ the ciphertext, $T \in \mathsf{T}$ the tag, and each of these sets is a subset of $\{0,1\}^*$. The *correctness condition* states that for all K, IV, A, and M, $\mathcal{D}_K^{IV,A}(\mathcal{E}_K^{IV,A}(M)) = M$. A secure AE scheme should return $\bot$ when it does not receive a valid (C, T) tuple.

In order to consider what happens when unverified plaintext is released, we disconnect the decryption algorithm from the verification algorithm so that the decryption algorithm always releases plaintext. A separated AE scheme is a triplet $\Pi = (\mathcal{E}, \mathcal{D}, \mathcal{V})$ of keyed algorithms — encryption $\mathcal{E}$, decryption $\mathcal{D}$, and verification $\mathcal{V}$ — such that

$$(C, T) \leftarrow \mathcal{E}_K^{IV,A}(M) \ ,$$
$$M \leftarrow \mathcal{D}_K^{IV,A}(C, T) \ ,$$
$$\top/\bot \leftarrow \mathcal{V}_K^{IV,A}(C, T) \ ,$$

where K, IV, A, M, C, and T are defined as above. Note that in some deterministic schemes IV may be absent, in which case we can expand the interface of such schemes to receive IV input with which it does nothing. Furthermore, for simplicity we might omit A if there is no associated data. The special symbols $\top$ and $\bot$ indicate the success and failure of the verification process, respectively.

As in the conventional setting we impose a *correctness condition*: for all K, IV, A, and M such that $\mathcal{E}_K^{IV,A}(M) = (C,T)$, we require $\mathcal{D}_K^{IV,A}(C,T) = M$ and $\mathcal{V}_K^{IV,A}(C,T) = \top$.

Relation to Conventional Syntax. Given a separated AE scheme $\Pi = (\mathcal{E}, \mathcal{D}, \mathcal{V})$, we can easily convert it into a conventional AE scheme $\overline{\Pi} = (\mathcal{E}, \overline{\mathcal{D}})$. Remember that the conventional decryption oracle $\overline{\mathcal{D}}_K^{IV,A}(C,T)$ outputs M where $M = \mathcal{D}_K^{IV,A}(C,T)$ if $\mathcal{V}_K^{IV,A}(C,T) = \top$, and $\bot$ otherwise.

The conversion in the other direction is not immediate. While the verification algorithm $\mathcal{V}$ can be easily "extracted" from $\overline{\mathcal{D}}$ (i.e., one can easily construct $\mathcal{V}$ using $\overline{\mathcal{D}}$ — just replace M with $\top$), it is not clear if one can always "naturally" extract the decryption algorithm $\mathcal{D}$ from $\overline{\mathcal{D}}$. However, all practical AE schemes that we are aware of can be constructed from a triplet $(\mathcal{E}, \mathcal{D}, \mathcal{V})$ as above, and hence their decryption algorithms $\overline{\mathcal{D}}$ are all separable into $\mathcal{D}$ and $\mathcal{V}$.

3.2 Types of AE Schemes

Classification Based on IVs. In order to achieve semantic security [22], AE schemes must be probabilistic or stateful [5]. Usually the randomness or state is focused into an IV [32]. How the IV is used restricts the scheme's syntax and the types of adversaries considered in the security notions:

1. **Random IV.** The environment chooses a random IV for each encryption, thus an adversary has no control over the choice of IV for each encryption. The generated IV must be sent along with the ciphertext so that the receiver can decrypt.
2. **Nonce IV.** A distinct IV must be used for each encryption, thus an adversary can choose but does not repeat nonce IV values in its encryption queries. How the parties synchronize the nonce is typically left implicit.
3. **Arbitrary IV.** No restrictions on the IV are imposed, thus an adversary may choose any IV for encryption. Often a deterministic AE scheme does not even have an IV input, in which case an IV can be embedded into the associated data A, which gets authenticated along with the plaintext M but does not get encrypted; A is sent in the clear.

In all IV cases the adversary can arbitrarily choose the IV input values to the decryption oracle. In some real-world protocols the decryption algorithm can be stateful [7], but such schemes are out of the scope of this paper, and schemes designed to be secure with deterministic decryption algorithms will be secure in those settings as well.

While random and nonce IV schemes can achieve semantic security, arbitrary IV schemes cannot, and therefore reduce to deterministic security. In the latter case, the most common notions are *"privacy up to repetition"* which is used

Table 2. The type of random oracle needed depending upon the class of AE scheme considered

IV type	type of encryption	
	online	offline
random	random oracle	random oracle
nonce	random oracle	random oracle
arbitrary	random-up-to-prefix oracle	random-up-to-repetition oracle

for DAE [34] and *"privacy up to prefix"* which is used for authenticated online encryption [20]. In any case, we write $ to indicate the ideal oracle from which an adversary tries to distinguish the real encryption oracle $\mathcal{E}_K$, regardless of the IV type. This means that the ideal $ oracle should be either the random oracle, random-up-to-repetition oracle, or random-up-to-prefix oracle, depending upon the IV. Each of the cases with their respective random oracles are listed in Table 2. In order to avoid redundancy in the wording of the definitions, whenever we write $\Delta(\mathcal{E}_K, \ldots ; \$, \ldots)$, it is understood that the $ oracle is the one appropriate for the AE scheme consisting of $\mathcal{E}$.

Online Encryption/Decryption Algorithms. A further distinction is made between online schemes and the others. An AE scheme with online encryption is one in which the ciphertext can be output as the plaintext is received, namely we require that for each (K, IV, A) the resulting encryption function is online as a function of the plaintext M.

Although decryption in AE schemes can never be online due to the fact that the message needs to be verified before it is output, we still consider schemes which can compute the plaintext as the ciphertext is received. In particular, a scheme with online decryption is one in which this plaintext-computing algorithm, viewed as a function of the ciphertext and tag input, is online. Note that in some schemes the tag could be received before the ciphertext, in which case we still consider $\mathcal{D}$ to be online (even though our new syntax implies that the tag is always received after the ciphertext).

3.3   Conventional Security Definitions under the New Syntax

Let $\Pi = (\mathcal{E}, \mathcal{D}, \mathcal{V})$ denote an AE scheme as a family of algorithms indexed by the key, IV, and associated data. With the new separated syntax we reformulate the conventional security definitions, IND-CPA, IND-CCA, and INT-CTXT. As mentioned above, the security notions are defined in terms of an unspecified $, where the exact nature of $ depends on the type of IV allowed (cf. Table 2). In the definitions the only fixed input to the algorithms is the key, indicated by writing $\mathcal{E}_K$ and $\mathcal{D}_K$; all other inputs, such as the IV and associated data, can be entered by the adversary.

Definition 1 (IND-CPA Advantage). *Let **D** be a computationally bounded adversary with access to one oracle $\mathcal{O}$, and let $K \xleftarrow{R} \mathsf{K}$. Then the IND-CPA* *advantage of* $\mathbf{D}$ *relative to* Π *is given by*

$$\mathsf{CPA}_\Pi(\mathbf{D}) := \underset{\mathbf{D}}{\Delta}(\mathcal{E}_K\,;\,\$) \ .$$

Definition 2 (IND-CCA Advantage). *Let* $\mathbf{D}$ *be a computationally bounded adversary with access to two oracles* $\mathcal{O}_1$ *and* $\mathcal{O}_2$, *such that* $\mathbf{D}$ *never queries* $\mathcal{O}_1 \hookrightarrow \mathcal{O}_2$ *nor* $\mathcal{O}_2 \hookrightarrow \mathcal{O}_1$, *and let* $K \xleftarrow{R} \mathsf{K}$. *Then the IND-CCA advantage of* $\mathbf{D}$ *relative to* Π *is given by*

$$\mathsf{CCA}_\Pi(\mathbf{D}) := \underset{\mathbf{D}}{\Delta}(\mathcal{E}_K, \mathcal{D}_K\,;\,\$, \mathcal{D}_K) \ .$$

Note that IND-CCA as defined above does not apply to the random IV setting. When a random IV is used, the adversary is not prohibited from querying $\mathcal{O}_2 \hookrightarrow \mathcal{O}_1$. We introduce a version of IND-CCA below, which can be applied to all IV settings.

Definition 3 (IND-CCA' Advantage). *Let* $\mathbf{D}$ *be an adversary as in Def. 2, except* $\mathbf{D}$ *may now query* $\mathcal{O}_2 \hookrightarrow \mathcal{O}_1$, *and let* $K \xleftarrow{R} \mathsf{K}$. *Then the IND-CCA' advantage of* $\mathbf{D}$ *relative to* Π *is given by*

$$\mathsf{CCA'}_\Pi(\mathbf{D}) := \underset{\mathbf{D}}{\Delta}(\mathcal{E}_K, \mathcal{D}_K\,;\,\$, \mathcal{D}_K) \ .$$

Definition 4 (INT-CTXT Advantage). *Let* $\mathbf{F}$ *be a computationally bounded adversary with access to two oracles* $\mathcal{E}_K$ *and* $\mathcal{V}_K$, *such that* $\mathbf{F}$ *never queries* $\mathcal{E}_K \hookrightarrow \mathcal{V}_K$. *Then the INT-CTXT advantage of* $\mathbf{F}$ *relative to* Π *is given by*

$$\mathsf{CTXT}_\Pi(\mathbf{F}) := \Pr\left[\mathbf{F}^{\mathcal{E}_K, \mathcal{V}_K} \text{ forges}\right] \ ,$$

where the probability is defined over the random key K *and random coins of* $\mathbf{F}$. *Here, "forges" means that* $\mathcal{V}_K$ *returns* $\top$ *to the adversary.*

4   Security under Release of Unverified Plaintext

4.1   Security of Encryption

We introduce the notion of plaintext-aware encryption of symmetric-key encryption schemes. An analysis of existing plaintext-aware schemes can be found in Sect. 5. The formalization is similar to the one in the public-key setting [10]. Let $\Pi = (\mathcal{E}, \mathcal{D})$ denote an encryption scheme.

Definition 5 (PA1 Advantage). *Let* $\mathbf{D}$ *be an adversary with access to two oracles* $\mathcal{O}_1$ *and* $\mathcal{O}_2$. *Let* $\mathbf{E}$ *be an algorithm with access to the history of queries made to* $\mathcal{O}_1$ *by* $\mathbf{D}$, *called a PA1-extractor. We allow* $\mathbf{E}$ *to maintain state across invocations. The PA1 advantage of* $\mathbf{D}$ *relative to* $\mathbf{E}$ *and* Π *is*

$$\mathsf{PA1}_\Pi^{\mathbf{E}}(\mathbf{D}) := \underset{\mathbf{D}}{\Delta}(\mathcal{E}_K, \mathcal{D}_K\,;\,\mathcal{E}_K, \mathbf{E}) \ ,$$

where $K \xleftarrow{R} \mathsf{K}$, *and the probability is defined over the key* K, *the random coins of* $\mathbf{D}$, *and the random coins of* $\mathbf{E}$.

The adversary $\mathbf{D}$ tries to distinguish the case in which its second oracle $\mathcal{O}_2$ is given by $\mathcal{D}_K$ versus the case in which $\mathcal{O}_2$ is given by $\mathbf{E}$. The task of $\mathbf{E}$ is to mimic the outputs of $\mathcal{D}_K$ given only the history of queries made to $\mathcal{E}_K$ by $\mathbf{D}$ (the key is not given to $\mathbf{E}$). Note that $\mathbf{D}$ is allowed to make queries of the form $\mathcal{E}_K \hookrightarrow \mathbf{E}$; these can easily be answered by $\mathbf{E}$ via the query history.

PA2 is a strengthening of PA1 where the extractor no longer has access to the query history of $\mathcal{E}_K$; the extractor becomes a simulator for the decryption algorithm. Note that in order for this to work, we cannot allow the adversaries to make queries of the form $\mathcal{E}_K \hookrightarrow \mathbf{E}$.

Definition 6 (PA2 Advantage). *Let $\mathbf{D}$ be an adversary as in Def. 5, with the added restriction that it may not ask queries of the form $\mathcal{O}_1 \hookrightarrow \mathcal{O}_2$. Let $\mathbf{E}$ be an algorithm, called a PA2-extractor. We allow $\mathbf{E}$ to maintain state across invocations. The PA2 advantage of $\mathbf{D}$ relative to $\mathbf{E}$ and Π is*

$$\mathsf{PA2}_\Pi^\mathbf{E}(\mathbf{D}) := \underset{\mathbf{D}}{\Delta}(\mathcal{E}_K, \mathcal{D}_K \,;\, \mathcal{E}_K, \mathbf{E}) \ ,$$

where $K \xleftarrow{R} \mathsf{K}$, and the probability is defined over the key K, the random coins of $\mathbf{D}$, and the random coins of $\mathbf{E}$.

An equivalent way of describing PA2 is via *decryption independence* (DI), which means that the adversary cannot distinguish between encryption and decryption under the same key and under different keys. The equivalence between PA2 and DI is proven in [3].

Definition 7 (Decryption Independence). *Let $\mathbf{D}$ be a distinguisher accepting two oracles not making queries of the form $\mathcal{O}_1 \hookrightarrow \mathcal{O}_2$, then the DI advantage of $\mathbf{D}$ relative to Π is*

$$\mathsf{DI}_\Pi(\mathbf{D}) := \underset{\mathbf{D}}{\Delta}(\mathcal{E}_K, \mathcal{D}_K \,;\, \mathcal{E}_K, \mathcal{D}_L) \ ,$$

where $K, L \xleftarrow{R} \mathsf{K}$ are independent.

4.2   Security of Verification

Integrity when releasing unverified plaintext is a modification of INT-CTXT (Def. 4) to include the decryption oracle as a means to obtain unverified plaintext. Let $\Pi = (\mathcal{E}, \mathcal{D}, \mathcal{V})$ be an AE scheme with separate decryption and verification.

Definition 8 (INT-RUP Advantage). *Let $\mathbf{F}$ be a computationally bounded adversary with access to three oracles $\mathcal{E}_K$, $\mathcal{D}_K$, and $\mathcal{V}_K$, such that $\mathbf{F}$ never queries $\mathcal{E}_K \hookrightarrow \mathcal{V}_K$. Then the INT-RUP advantage of $\mathbf{F}$ relative to Π is given by*

$$\mathsf{INT\text{-}RUP}_\Pi(\mathbf{F}) := \Pr\left[\mathbf{F}^{\mathcal{E}_K, \mathcal{D}_K, \mathcal{V}_K} \text{ forges}\right] \ ,$$

where the probability is defined over the key K and random coins of $\mathbf{F}$. Here, "forges" means the event of the oracle $\mathcal{V}_K$ returning $\top$ to the adversary.

5   Achieving Plaintext Awareness

5.1   Why Existing Schemes Do Not Achieve PA1

In conventional AE schemes such as OCB, GCM, SpongeWrap, CCM, COPA, and McOE-G, a ciphertext is computed using some bijective function, and then a tag is appended to the ciphertext. The schemes achieve AE because the tag prevents all ciphertexts from being valid. But if the tag is no longer checked, then we cannot achieve PA1, as explained below.

Let $\Pi = (\mathcal{E}_K, \mathcal{D}_K)$ be a nonce or arbitrary IV encryption scheme, then we can describe Π as follows,

$$\mathcal{E}_K^{IV,A}(M) = E_K^{IV,A}(M) \parallel F_K^{IV,A}(M) ,$$

where E_K is length-preserving, i.e. $|E_K^{IV,A}(M)| = |M|$. One can view $F_K^{IV,A}(M)$ as the tag-producing function from a scheme such as GCM. In the following proposition we prove that if Π is IND-CPA and PA1, then E_K cannot be bijective for each (IV, A), assuming either a nonce or arbitrary IV. Note that the proposition only holds if Π is a nonce or arbitrary IV scheme.

Proposition 1. *Say that E_K is bijective for all (IV, A), then there exists an adversary $\mathbf{D}$ such that for all extractors $\mathbf{E}$, there exists an adversary $\mathbf{D}_1$ such that*

$$1 - \mathsf{CPA}_\Pi(\mathbf{D}_1) \leq \mathsf{PA1}_\Pi^{\mathbf{E}}(\mathbf{D}) ,$$

where $\mathbf{D}$ makes one $\mathcal{O}_1$ query, one $\mathcal{O}_2$ query, and $\mathbf{D}_1$ is as efficient as $\mathbf{D}$ plus one query to $\mathbf{E}$.

Proof. See [3]. □

We conclude that in order for a nonce or arbitrary IV scheme to be PA1 and IND-CPA, E_K must either not be bijective, or not be length-preserving.

5.2   PA1 Random IV Schemes

We illustrate Def. 5 and the idea of an extractor by considering the CTR mode with a random IV.

Example 1 (RIV-CTR Extractor). Let $F : \{0,1\}^k \times \{0,1\}^n \to \{0,1\}^n$ be a PRF. For $M_i \in \{0,1\}^n$, $1 \leq i \leq \ell$, define RIV-CTR encryption as

$$\mathcal{E}_K^{C_0}(M_1 \cdots M_\ell) = F_K(C_0 + 1) \oplus M_1 \parallel \cdots \parallel F_K(C_0 + \ell) \oplus M_\ell ,$$

where C_0 is selected uniformly at random from $\{0,1\}^n$ for each encryption, and decryption as

$$\mathcal{D}_K^{C_0}(C_1 \cdots C_\ell) = F_K(C_0 + 1) \oplus C_1 \parallel \cdots \parallel F_K(C_0 + \ell) \oplus C_\ell .$$

We can define an extractor $\mathbf{E}$ for RIV-CTR as follows. Initially, $\mathbf{E}$ generates a random key K' which it will use via $F_{K'}$. Let $(C_0, C_1 \cdots C_\ell)$ denote an input to $\mathbf{E}$. Using C_0, the extractor searches its history for a ciphertext with C_0 as IV.

1. If such a ciphertext exists, we let $(C'_1 \cdots C'_m, M'_1 \cdots M'_m)$ denote the longest corresponding $\mathcal{E}_K$ query-response pair. Define $\kappa_i := C'_i \oplus M'_i$ for $1 \leq i \leq \min\{\ell, m\}$. Notice that κ_i corresponds to the keystream generated by F_K for $1 \leq i \leq \ell$. For $m < i \leq \ell$ we generate κ_i by $F_{K'}(C_0 + i)$.
2. If there is no such ciphertext, then we generate κ_i as $F_{K'}(C_0 + i)$ for $1 \leq i \leq \ell$.

Then we set $\mathbf{E}^{C_0}(C_1 \cdots C_\ell) = (C_1 \oplus \kappa_1, C_2 \oplus \kappa_2 \parallel \cdots \parallel C_\ell \oplus \kappa_\ell)$.

Proposition 2. *Let* $\mathbf{D}$ *be a PA1 adversary for RIV-CTR making queries whose lengths in number of blocks sum up to* σ, *then*

$$\mathsf{PA1}^{\mathbf{E}}_{RIV\text{-}CTR}(\mathbf{D}) \leq \underset{\mathbf{D}_1}{\Delta}\left(F_K, F_K \, ; \, F_K, F_{K'}\right) + \frac{\sigma^2}{2^n} \, ,$$

where $\mathbf{D}_1$ *is an adversary which may not make the same query to both of its oracles, and makes a total of* σ *queries with the same running time as* $\mathbf{D}$.

We refer to a proof of this proposition to the full version of the paper [3]. Here, we also describe and analyze an extractor for the CBC mode.

In the following subsections we discuss ways of achieving PA1 assuming a nonce and arbitrary IV. Our basic building block will be a random IV PA1 scheme.

5.3   PA1 Nonce IV Schemes

Nonce IV schemes are not necessarily PA1 in general. For example, CTR mode with a nonce IV is not PA1 and in [3] we show that IND-CPA is distinct from PA1. Furthermore, coming up with a generic technique which transforms nonce IV schemes into PA1 schemes in an efficient manner is most likely not possible.

If we assume that the nonce IV scheme, when used as a random IV scheme, is PA1, then there is an efficient way of making the nonce IV scheme PA1. Note that we already have an example of a scheme satisfying our assumption: nonce IV CTR mode is not PA1, but RIV-CTR is.

Nonce Decoy. The *nonce decoy* method creates a random-looking IV from the nonce IV and forces the decryption algorithm to use the newly generated IV. Note that we are not only transforming the nonce into a random nonce: the solution depends entirely on the fact that the decryption algorithm does *not* recompute the newly generated IV from the nonce IV.

Let $\Pi = (\mathcal{E}, \mathcal{D}, \mathcal{V})$ be a nonce-IV-based AE scheme. For simplicity assume $\mathsf{IV} := \{0,1\}^n$, so that IVs are of a fixed length n. We prepare a pseudo-random function $G_{K'} : \mathsf{IV} \to \mathsf{IV}$ with an independent key K'. We then construct an AE scheme $\widetilde{\Pi} = (\widetilde{\mathcal{E}}, \widetilde{\mathcal{D}}, \widetilde{\mathcal{V}})$ as follows.

$\widetilde{\mathcal{E}}^{IV,A}_{K,K'}(M)$:	$\widetilde{\mathcal{D}}^{IV,A}_{K,K'}(\widetilde{C}, T)$:	$\widetilde{\mathcal{V}}^{IV,A}_{K,K'}(\widetilde{C}, T)$:
$\widetilde{IV} \leftarrow G_{K'}(IV)$	$\widetilde{IV}\|C \leftarrow \widetilde{C}$	$\widetilde{IV}^* \leftarrow G_{K'}(IV)$
$(C, T) \leftarrow \mathcal{E}^{\widetilde{IV},A}_K(M)$	$M \leftarrow \mathcal{D}^{\widetilde{IV},A}_K(C, T)$	$\widetilde{IV}\|C \leftarrow \widetilde{C}$
$\widetilde{C} \leftarrow \widetilde{IV}\|C$	**return** M	$b \leftarrow \mathcal{V}^{\widetilde{IV},A}_K(C, T)$
return $(\widetilde{C}, T)$		**return** $(\widetilde{IV}^* = \widetilde{IV}$ and $b = \top)?\top : \bot$

Note that the decryption algorithm $\widetilde{\mathcal{D}}$ does not make use of K' or IV. If the decryption algorithm recomputes $\widetilde{IV}$ using K' and IV, then $\widetilde{\Pi}$ will not be PA1. Furthermore, one can combine $\widetilde{\mathcal{D}}$ and $\widetilde{\mathcal{V}}$ in order to create a scheme which rejects ciphertexts when the IV it receives does not come from an encryption query.

The condition that Π with random IVs be PA1 is necessary and sufficient in order for $\widetilde{\Pi}$ to be PA1, assuming G is a PRF; see [3] for the proof of this statement. In Sect. 6.2 we discuss what the nonce decoy does for INT-RUP.

5.4   PA1 Arbitrary IV Schemes

PRF-to-IV. Using a technique similar to MAC-then-Encrypt [9], we can turn a random IV PA1 scheme into an arbitrary IV PA1 scheme.

The idea behind the *PRF-to-IV* method is to evaluate a VIL PRF over the input to the scheme and then to use the resulting output as an IV for the random IV encryption scheme. Let $\Pi = (\mathcal{E}, \mathcal{D}, \mathcal{V})$ be a random IV PA1 scheme taking IVs from $\{0,1\}^n$, and let $G : \{0,1\}^k \times \{0,1\}^* \to \{0,1\}^n$ be a VIL PRF.

$$\widetilde{\mathcal{E}}_{K,K'}^{IV,A}(M):$$
$$\widetilde{IV} \leftarrow G_{K'}(IV\|A\|M)$$
$$(C,T) \leftarrow \mathcal{E}_K^{\widetilde{IV},A}(M)$$
$$\mathbf{return}\ \ (C, \widetilde{IV}\|T)$$

$$\widetilde{\mathcal{D}}_{K,K'}^{IV,A}(C, \widetilde{IV}\|T):$$
$$M \leftarrow \mathcal{D}_K^{\widetilde{IV},A}(C,T)$$
$$\mathbf{return}\ \ M$$

$$\widetilde{\mathcal{V}}_{K,K'}^{IV,A}(C, \widetilde{IV}\|T):$$
$$M \leftarrow \widetilde{\mathcal{D}}_{K,K'}^{IV,A}(C, \widetilde{IV}\|T)$$
$$IV^* \leftarrow G_{K'}(IV\|A\|M)$$
$$b \leftarrow \mathcal{V}_K^{\widetilde{IV},A}(C,T)$$
$$\mathbf{return}\,(\widetilde{IV} = IV^* \text{ and } b = \top)?\top : \bot$$

The PRF-to-IV method is more robust than the nonce decoy since $\widetilde{\mathcal{D}}$ really only can use $\widetilde{IV}$ to decrypt properly.

The condition that Π with random IVs be PA1 is necessary and sufficient in order for $\widetilde{\Pi}$ to be PA1, assuming G is a VIL-PRF; see [3] for the proof of this statement. Note that the PRF-to-IV method is the basic structure behind SIV, BTM, and HBS. We show that the PRF-to-IV method is INT-RUP in Sect.6.2.

Online Encryption. Since the PRF needs to be computed over the entire message before the message is encrypted again, the PRF-to-IV method does not allow for online encryption. Recall that an encryption scheme has online encryption if for all (K, IV, A), the resulting function is online. Examples of such schemes include COPA and McOE-G.

If we want encryption and decryption to both be online in the arbitrary IV setting, then a large amount of ciphertext expansion is necessary, otherwise a distinguisher similar to the one used in the proof of Prop. 1 can be created.

An encryption scheme $\Pi = (\mathcal{E}, \mathcal{D})$ is online if for some n there exist functions f_i and f_i' such that

$$\mathcal{E}_K(M) = f_n(M_1)\,f_{2n}(M_1 M_2) \cdots f_{jn}(M_1 M_2 \cdots M_j)\,f'_{|M|}(M)\ ,$$

where $j = \lfloor (|M| - 1)/n \rfloor$ and M_i is the ith n-bit block of M. If the encryption scheme has online decryption as well, then the decryption algorithm can start decrypting each "block" of ciphertext, or

$$\mathcal{D}_K(f_n(M_1)\, f_{2n}(M_1 M_2) \cdots f_{in}(M_1 M_2 \cdots M_i)) = M_1 M_2 \cdots M_i \ ,$$

for all $i \leq j$.

Proposition 3. *Let $\Pi = (\mathcal{E}, \mathcal{D})$ be an encryption scheme where $\mathcal{E}$ is n-online for all K, IV, and A, and $\mathcal{D}$ is online as well, then there exists a PA1-adversary $\mathbf{D}$ such that for all extractors $\mathbf{E}$ there exists an IND-CPA adversary $\mathbf{D}_1$ such that*

$$1 - \mathsf{CPA}_\Pi(\mathbf{D}_1) \leq \mathsf{PA1}_\Pi^{\mathbf{E}}(\mathbf{D}) \ ,$$

where $\mathbf{D}$ makes one $\mathcal{O}_1$ query, one $\mathcal{O}_2$ query, and $\mathbf{D}_1$ is as efficient as $\mathbf{D}$ plus one query to $\mathbf{E}$.

Proof. See [3]. □

Example 2. In certain scenarios, padding the plaintext is sufficient for PA1. Doing so makes schemes such as McOE-G secure in the sense of PA1, while keeping encryption and decryption online. The cost is a substantial expansion of the ciphertext. For the case of McOE-G, the length of the ciphertext becomes roughly twice the size of its plaintext.

It is important to note that McOE-G is based on an n-bit block cipher, and each n-bit message block is encrypted (after it is XORed with some state values) via the block cipher call. Since the underlying block cipher is assumed to be a strong pseudo-random function (SPRP), we can pad a message $M = M_1 M_2 \cdots M_\ell$ (each M_i is an $n/2$-bit string) as $0^{n/2} M_1 \parallel 0^{n/2} M_2 \parallel \cdots \parallel 0^{n/2} M_\ell$ and then encrypt this padded message using McOE-G. So each block cipher call processes $0^{n/2} M_i$ for some i. This "encode-then-encipher" scheme [12] is PA1 as shown in [3].

Example 3. If we do not require the decryption to be online, then we can achieve PA1 without significant ciphertext expansion. An example of a scheme that falls into this category is the recently-introduced APE mode [2], whose decryption is backward (and hence not online). A proof of this is given in [3].

5.5   PA2 Schemes

Most AE schemes are proven to be IND-CPA and INT-CTXT, which allows one to achieve IND-CCA [9] assuming verification works correctly. In order to be as efficient as possible, the underlying encryption schemes in the AE schemes are designed to only achieve IND-CPA and not IND-CCA, since achieving IND-CCA for encryption usually requires significantly more operations. For example, GCM, SIV, BTM, and HBS all use CTR mode for encryption, yet CTR mode is not IND-CCA. Since IND-CPA+PA2 is equivalent to IND-CCA', none of these schemes achieve PA2.

A scheme such as APE also cannot achieve IND-CCA' because its decryption is online "in reverse". If $(\mathcal{E}_K, \mathcal{D}_K)$ denotes APE, then an adversary can query

$\mathcal{E}_K(M_1 M_2) = C_1 C_2$ and then $\mathcal{D}_K(C'_1 C_2)$, which equals $M'_1 M_2$. But if an adversary interacts with $(\$, \mathcal{D}_K)$ (see Def. 3), then $\mathcal{D}_K(C'_1 C_2)$ will most likely not output $M'_1 M_2$.

Existing designs which do achieve PA2 include those which are designed to be IND-CCA', such as the solutions presented by Bellare and Rogaway [12], Desai [19], and Shrimpton and Terashima [36]. These solutions cannot be online, and they are usually at least "two-pass", meaning the input is processed at least twice.

6 Integrity in the INT-RUP Setting

6.1 INT-RUP Attack

Several AE schemes become insecure if unverified plaintext is released. In Proposition 4, we explain that OCB [33] and COPA [4] are not secure in the RUP setting.

The strategy of our attack is similar to that of Bellare and Micciancio on the XHASH hash function [8]. However, our attack is an improved version that solves a system of linear equations in $GF(2)$ with only half the number of equations and variables.

The attack works by first querying the encryption oracle under nonce N to get a valid ciphertext and tag pair. Then, two decryption queries are made under the same nonce N. Using the resulting plaintexts a system of linear equations is set up, which when solved will give the a forgery with high probability. A formal description of the attack is given in [3].

Proposition 4. *For OCB and COPA, for all $\ell \geq n$ there exists an adversary* **A** *such that*
$$\mathsf{INT\text{-}RUP}_\Pi(\mathbf{A}) \geq 1 - 2^{n-\ell} \ ,$$

where **A** *makes one encryption query and two decryption queries, each consisting of ℓ blocks of n bits. Then, the adversary solves a system of linear equations in $GF(2)$ with n equations and ℓ unknowns.*

6.2 Nonce Decoy and PRF-to-IV

In Sect. 5 we introduced a way of turning a random IV PA1 scheme into a nonce IV PA1 scheme, the nonce decoy, and a way of turning a random IV PA1 scheme into an arbitrary IV PA1 scheme, the PRF-to-IV method. Here we consider what happens to INT-RUP when the two methods are applied.

The nonce decoy adds some integrity to the underlying random IV PA1 scheme. Using the notation from Sect. 5.3, Π needs to be a slightly lighter form of INT-RUP in order for $\widetilde{\Pi}$ to be INT-RUP. Concretely, Π only needs to be INT-RUP against adversaries which use IVs which are the result of an encryption query. Furthermore, this requirement on Π is sufficient to prove that $\widetilde{\Pi}$ is INT-RUP.

Naturally if Π is INT-RUP, then $\widetilde{\Pi}$ is INT-RUP as well. In fact, if Π is INT-RUP against adversaries which use IVs which are the result of an encryption query, then $\widetilde{\Pi}$ is INT-RUP. These statements and their proofs can be found in [3].

The PRF-to-IV method is a much stronger transform than the nonce decoy. Following the notation from Sect. 5.4, we do not need to assume anything about the underlying random IV scheme Π in order to prove that $\widetilde{\Pi}$ is INT-RUP.

7   Conclusions

Many practical applications desire that an AE scheme can securely output plaintext before verification. We formalized security under the release of unverified plaintext (RUP) to adversaries by separating decryption and verification.

Two symmetric-key notions of plaintext awareness (PA1 and PA2) were introduced. In the RUP setting, privacy is achieved as a combination of IND-CPA and PA1 or PA2. For integrity, we introduced the INT-RUP notion as an extension of INT-CTXT, where a forger may abuse unverified plaintext. We connected our notions of privacy and integrity in the RUP setting to existing security notions, and saw that the relations and separations depended on the IV type.

The CTR and CBC modes with a random IV achieve IND-CPA+PA1, but this is non-trivial for nonce-based or deterministic encryption schemes. Our results showed that many AE schemes such as GCM, CCM, COPA, and McOE-G are not secure in the RUP setting. We provided remedies for both nonce-based and deterministic AE schemes. For the former case, we introduced the *nonce decoy* technique, which allowed to transform a nonce to a random-looking IV. The PRF-to-IV method converts random IV PA1 schemes into arbitrary IV PA1 schemes. We showed that deterministic AE schemes cannot be PA1, unless the decryption is offline (as in APE) or there is significant ciphertext expansion.

Future Work. Given that our PRF-to-IV method is rather inefficient, we leave it as an open problem to efficiently modify any encryption-only scheme into an AE scheme that is INT-RUP. A related problem is to fix OCB and COPA to be INT-RUP in an efficient way. The PA1 solutions we provide all start with the assumption that the nonce IV or arbitrary IV scheme is PA1 when a random IV is used instead. An interesting problem is to find alternative solutions to constructing nonce IV and arbitrary IV PA1 schemes. A problem of theoretical interest is to find a non-pathological random IV encryption scheme that is not PA1. In some applications, formalizing security in the RUP setting as IND-CPA+PA1 and INT-RUP may be sufficient. It is interesting to investigate how well this formalization reflects the problems encountered in real-world implementations, to see where PA2 may also be necessary, and how blockwise adaptive adversaries [25] play a role in the RUP setting. Finally, our paper does not address the behavior of applications which use unverified plaintext. A further understanding of the security risks involved in using unverified plaintext in applications is necessary.

Acknowledgments. The authors would like to thank Martijn Stam and the reviewers for their valuable comments.

References

1. AlFardan, N.J., Paterson, K.G.: Lucky Thirteen: Breaking the TLS and DTLS Record Protocols. In: IEEE Symposium on Security and Privacy, pp. 526–540. IEEE Computer Society (2013)
2. Andreeva, E., Bilgin, B., Bogdanov, A., Luykx, A., Mennink, B., Mouha, N., Yasuda, K.: APE: Authenticated Permutation-Based Encryption for Lightweight Cryptography. In: FSE. LNCS, Springer (2014)
3. Andreeva, E., Bogdanov, A., Luykx, A., Mennink, B., Mouha, N., Yasuda, K.: How to Securely Release Unverified Plaintext in Authenticated Encryption. Cryptology ePrint Archive, Report 2014/144 (2014), full version of this paper
4. Andreeva, E., Bogdanov, A., Luykx, A., Mennink, B., Tischhauser, E., Yasuda, K.: Parallelizable and Authenticated Online Ciphers. In: Sako, K., Sarkar, P. (eds.) ASIACRYPT 2013, Part I. LNCS, vol. 8269, pp. 424–443. Springer, Heidelberg (2013)
5. Bellare, M., Desai, A., Jokipii, E., Rogaway, P.: A Concrete Security Treatment of Symmetric Encryption. In: FOCS, pp. 394–403. IEEE Computer Society (1997)
6. Bellare, M., Keelveedhi, S.: Authenticated and Misuse-Resistant Encryption of Key-Dependent Data. In: Rogaway, P. (ed.) CRYPTO 2011. LNCS, vol. 6841, pp. 610–629. Springer, Heidelberg (2011)
7. Bellare, M., Kohno, T., Namprempre, C.: Breaking and provably repairing the SSH authenticated encryption scheme: A case study of the encode-then-encrypt-and-mac paradigm. ACM Tr. Inf. Sys. Sec. 7(2), 206–241 (2004)
8. Bellare, M., Micciancio, D.: A New Paradigm for Collision-Free Hashing: Incrementality at Reduced Cost. In: Fumy, W. (ed.) EUROCRYPT 1997. LNCS, vol. 1233, pp. 163–192. Springer, Heidelberg (1997)
9. Bellare, M., Namprempre, C.: Authenticated Encryption: Relations among Notions and Analysis of the Generic Composition Paradigm. In: Okamoto, T. (ed.) ASIACRYPT 2000. LNCS, vol. 1976, pp. 531–545. Springer, Heidelberg (2000)
10. Bellare, M., Palacio, A.: Towards Plaintext-Aware Public-Key Encryption Without Random Oracles. In: Lee, P.J. (ed.) ASIACRYPT 2004. LNCS, vol. 3329, pp. 48–62. Springer, Heidelberg (2004)
11. Bellare, M., Rogaway, P.: Optimal Asymmetric Encryption. In: De Santis, A. (ed.) EUROCRYPT 1994. LNCS, vol. 950, pp. 92–111. Springer, Heidelberg (1995)
12. Bellare, M., Rogaway, P.: Encode-Then-Encipher Encryption: How to Exploit Nonces or Redundancy in Plaintexts for Efficient Cryptography. In: Okamoto, T. (ed.) ASIACRYPT 2000. LNCS, vol. 1976, pp. 317–330. Springer, Heidelberg (2000)
13. Bellare, M., Rogaway, P.: Introduction to modern cryptography. In: UCSD CSE 207 Course Notes (September 2005)
14. Bertoni, G., Daemen, J., Peeters, M., Van Assche, G.: Duplexing the Sponge: Single-Pass Authenticated Encryption and Other Applications. In: Miri, A., Vaudenay, S. (eds.) SAC 2011. LNCS, vol. 7118, pp. 320–337. Springer, Heidelberg (2012)
15. Bogdanov, A., Mendel, F., Regazzoni, F., Rijmen, V., Tischhauser, E.: ALE: AES-Based Lightweight Authenticated Encryption. In: Moriai, S. (ed.) FSE 2013. LNCS, vol. 8424, pp. 447–466. Springer, Heidelberg (2014)

16. Boldyreva, A., Degabriele, J.P., Paterson, K.G., Stam, M.: Security of Symmetric Encryption in the Presence of Ciphertext Fragmentation. In: Pointcheval, D., Johansson, T. (eds.) EUROCRYPT 2012. LNCS, vol. 7237, pp. 682–699. Springer, Heidelberg (2012)

17. Boldyreva, A., Degabriele, J.P., Paterson, K.G., Stam, M.: On Symmetric Encryption with Distinguishable Decryption Failures. In: Moriai, S. (ed.) FSE 2013. LNCS, vol. 8424, pp. 367–390. Springer, Heidelberg (2014)

18. Canvel, B., Hiltgen, A.P., Vaudenay, S., Vuagnoux, M.: Password Interception in a SSL/TLS Channel. In: Boneh, D. (ed.) CRYPTO 2003. LNCS, vol. 2729, pp. 583–599. Springer, Heidelberg (2003)

19. Desai, A.: New Paradigms for Constructing Symmetric Encryption Schemes Secure against Chosen-Ciphertext Attack. In: Bellare, M. (ed.) CRYPTO 2000. LNCS, vol. 1880, pp. 394–412. Springer, Heidelberg (2000)

20. Fleischmann, E., Forler, C., Lucks, S.: McOE: A Family of Almost Foolproof On-Line Authenticated Encryption Schemes. In: Canteaut, A. (ed.) FSE 2012. LNCS, vol. 7549, pp. 196–215. Springer, Heidelberg (2012)

21. Fouque, P.-A., Joux, A., Martinet, G., Valette, F.: Authenticated On-Line Encryption. In: Matsui, M., Zuccherato, R.J. (eds.) SAC 2003. LNCS, vol. 3006, pp. 145–159. Springer, Heidelberg (2004)

22. Goldwasser, S., Micali, S.: Probabilistic Encryption and How to Play Mental Poker Keeping Secret All Partial Information. In: STOC 1982, pp. 365–377. ACM (1982)

23. Iwata, T., Yasuda, K.: BTM: A Single-Key, Inverse-Cipher-Free Mode for Deterministic Authenticated Encryption. In: Jacobson Jr., M.J., Rijmen, V., Safavi-Naini, R. (eds.) SAC 2009. LNCS, vol. 5867, pp. 313–330. Springer, Heidelberg (2009)

24. Iwata, T., Yasuda, K.: HBS: A Single-Key Mode of Operation for Deterministic Authenticated Encryption. In: Dunkelman, O. (ed.) FSE 2009. LNCS, vol. 5665, pp. 394–415. Springer, Heidelberg (2009)

25. Joux, A., Martinet, G., Valette, F.: Blockwise-Adaptive Attackers: Revisiting the (In)Security of Some Provably Secure Encryption Models: CBC, GEM, IACBC. In: Yung, M. (ed.) CRYPTO 2002. LNCS, vol. 2442, pp. 17–30. Springer, Heidelberg (2002)

26. Katz, J., Yung, M.: Complete characterization of security notions for probabilistic private-key encryption. In: STOC, pp. 245–254. ACM (2000)

27. Krovetz, T., Rogaway, P.: The OCB Authenticated-Encryption Algorithm (June 2013), http://datatracker.ietf.org/doc/draft-irtf-cfrg-ocb

28. McGrew, D.A., Viega, J.: The Security and Performance of the Galois/Counter Mode (GCM) of Operation. In: Canteaut, A., Viswanathan, K. (eds.) INDOCRYPT 2004. LNCS, vol. 3348, pp. 343–355. Springer, Heidelberg (2004)

29. NIST: DES Modes of Operation. FIPS 81 (December 1980)

30. Paterson, K.G., AlFardan, N.J.: Plaintext-Recovery Attacks Against Datagram TLS. In: NDSS. The Internet Society (2012)

31. Rogaway, P.: Authenticated-encryption with associated-data. In: ACM Conference on Computer and Communications Security 2002, pp. 98–107. ACM (2002)

32. Rogaway, P.: Nonce-Based Symmetric Encryption. In: Roy, B., Meier, W. (eds.) FSE 2004. LNCS, vol. 3017, pp. 348–359. Springer, Heidelberg (2004)

33. Rogaway, P., Bellare, M., Black, J.: OCB: A Block-Cipher Mode of Operation for Efficient Authenticated Encryption. ACM Tr. Inf. Sys. Sec. 6(3), 365–403 (2003)

34. Rogaway, P., Shrimpton, T.: A Provable-Security Treatment of the Key-Wrap Problem. In: Vaudenay, S. (ed.) EUROCRYPT 2006. LNCS, vol. 4004, pp. 373–390. Springer, Heidelberg (2006)

35. Rogaway, P., Wagner, D.: A Critique of CCM. Cryptology ePrint Archive, Report 2003/070 (2003)
36. Shrimpton, T., Terashima, R.S.: A modular framework for building variable-input-length tweakable ciphers. In: Sako, K., Sarkar, P. (eds.) ASIACRYPT 2013, Part I. LNCS, vol. 8269, pp. 405–423. Springer, Heidelberg (2013)
37. Tsang, P.P., Smith, S.W.: Secure cryptographic precomputation with insecure memory. In: Chen, L., Mu, Y., Susilo, W. (eds.) ISPEC 2008. LNCS, vol. 4991, pp. 146–160. Springer, Heidelberg (2008)
38. Tsang, P.P., Solomakhin, R.V., Smith, S.W.: Authenticated streamwise on-line encryption. Dartmouth Computer Science Technical Report TR2009-640 (2009)
39. Vaudenay, S.: Security Flaws Induced by CBC Padding - Applications to SSL, IPSEC, WTLS.. In: Knudsen, L.R. (ed.) EUROCRYPT 2002. LNCS, vol. 2332, pp. 534–546. Springer, Heidelberg (2002)
40. Whiting, D., Housley, R., Ferguson, N.: Counter with CBC-MAC (CCM). Request For Comments 3610 (2003)
41. Wu, H., Preneel, B.: AEGIS: A Fast Authenticated Encryption Algorithm. In: Lange, T., Lauter, K., Lisoněk, P. (eds.) SAC 2013. LNCS, vol. 8282, pp. 185–202. Springer, Heidelberg (2013)

Forging Attacks on Two Authenticated Encryption Schemes COBRA and POET

Mridul Nandi

Indian Statistical Institute, Kolkata, India
mridul.nandi@gmail.com

Abstract. In FSE 2014, an authenticated encryption mode COBRA [4], based on pseudorandom permutation (PRP) blockcipher, and POET [3], based on Almost XOR-Universal (AXU) hash and strong pseudorandom permutation (SPRP), were proposed. Few weeks later, COBRA mode and a simple variant of the original proposal of POET (due to a forging attack [13] on the original proposal) with AES as an underlying blockcipher, were submitted to CAESAR, a competition [1] of authenticated encryption (AE). In this paper, we show a forging attack on the mode COBRA based on any n-bit blockcipher. Our attack on COBRA requires about $O(n)$ queries with success probability of about $1/2$. This disproves the claim proved in the FSE 2014 paper. We also show both privacy and forging attack on the parallel version of POET, denoted POET-m. In case of the modes POET and POE (the underlying modes for encryption), we demonstrate a distinguishing attack making only one encryption query when we instantiate the underlying AXU hash function with some other AXU hash function, namely a uniform random involution. Thus, our result violates the designer's main claim (Theorem 8.1 in [1]). However, the attacks can not be extended to the specifications of POET submitted to the CAESAR competition.

Keywords: Authenticated Encryption, COBRA, POET, Distinguishing Attack and Forging Attack.

1 Introduction

The common application of cryptography is to implement a secure channel between two or more users and then to exchange information over that channel. These users can initially set up their one-time shared key. Otherwise, a typical implementation first calls a key-exchange protocol for establishing a shared key or a session key (used only for the current session). Once the users have a shared key, either through the initial key set-up or key-exchange, they use this key to authenticate and encrypt the transmitted information using efficient symmetric-key algorithms such as a *message authentication code* Mac($\cdot$), *pseudorandom function* Prf($\cdot$) and (possibly tweakable symmetric-key) *encryption* Enc($\cdot$) respectively. The encryption Enc provides privacy or confidentiality of the *plaintext M*. The message authentication code Mac and pseudorandom function

P. Sarkar and T. Iwata (Eds.): ASIACRYPT 2014, PART I, LNCS 8873, pp. 126–140, 2014.

Prf provide data-integrity authenticating the transmitted message (M, A), a pair of plaintext M and an *associated data* A. Mac also provides user-authenticity (protecting from impersonation). An **Authenticated Encryption** scheme (or simply **AE**) serves both of the purposes in an integrated manner. An authenticated encryption scheme AE has two functionalities one of which, called tagged encryption, essentially combines message authentication code and encryption, and the other combines verification and decryption algorithms.

1. **Tagged Encryption** $\mathsf{AE.enc}_k$: On an input message M from a message space $\mathcal{M} \subseteq \{0,1\}^*$ and an associated data A from an associated data space $\mathcal{D} \subseteq \{0,1\}^*$, it returns a **tagged ciphertext**[1] $Z \in \{0,1\}^*$.

2. **Verified Decryption** $\mathsf{AE.dec}_k$: On an input tagged ciphertext Z and an associated data A, it returns a plaintext M when $Z = \mathsf{AE.enc}_k(M, A)$, called valid tagged ciphertext. Otherwise, for all invalid tagged ciphertext Z, it returns a special symbol $\perp$.

Note that both algorithms take the shared key k from a keys-space $\mathcal{K} = \{0,1\}^{L_{\mathrm{key}}}$ where L_{key} denotes the key-size. The key includes keys for an underlying block-cipher, masking keys etc. Some constructions derive more keys by invoking the blockcipher with different constant inputs.

PRIVACY AND AUTHENTICITY ADVANTAGE. Informally speaking, an AE scheme is said to have *privacy* if the tagged ciphertext behave like a uniform random string for any adaptively chosen plaintext. More formally, let A be an oracle adversary which can make queries to AE.enc adaptively. Let \$ be a random oracle which returns a uniform random string for every new query. We define the *privacy advantage* of A against AE to be

$$\mathbf{Adv}_{\mathsf{AE}}^{\mathrm{priv}}(A) := \left| \Pr[A^{\mathsf{AE.enc}_K} = 1] - \Pr[A^{\$} = 1] \right|$$

where the two probabilities are taken under \$, K (usually chosen uniformly from the key-space $\mathcal{K}$) and the random coins of A. Similarly, we define the *authenticity advantage* of A as

$$\mathbf{Adv}_{\mathsf{AE}}^{\mathrm{auth}}(A) := \Pr[A^{\mathsf{AE.enc}_K} = Z, \ \mathsf{AE.dec}_K(Z) \neq \perp \text{ and } Z \text{ is fresh}]$$

where the probability is taken under K and the random coins of A. By fresh, we mean that Z is not a response of an encryption query of A.

1.1 Two AE Schemes **COBRA** and **POET** Submitted to **CAESAR**

CAESAR [1] is an ongoing competition for authenticated encryption schemes. The final goal of the competition is to identify a portfolio of different authenticated encryption schemes depending on different applications and environments. Fifty seven schemes have been submitted. AES-COBRA and POET are two such

---

[1] A tagged ciphertext usually consists of a ciphertext and tag. However, there may not exist a clear separation between ciphertext and tag.

submissions. Variants of these two schemes have been published before in FSE 2014. In [13] Guo et al. demonstrated a forging attack against POET making only one encryption query. So the designers of POET modified it accordingly to resist this forging attack and submitted the revised version to CAESAR.

1.2 Our Contribution

In this paper, we investigate the resistance of the two authenticated encryption schemes COBRA and POET against forging and privacy attacks. The paper is essentially divided in two sections: Section 4 describes the forging attack for COBRA and Section 5 describes forging and privacy analysis on the POET-mode and its parallel variant, called POET-m.

1. **Attack on COBRA.** In this paper, we show a forging attack on the submitted version of AES-COBRA. In fact, the attack works for the mode COBRA based on any blockcipher. Thus it disproves the claim stated in [4]. The authenticity advantage of our proposed algorithm is about $1/2$ and it makes about $2n$ encryption queries where n is the plaintext size of the underlying blockcipher.

2. **Analysis of POET and POET-m.** The designers of POET have recommended a parallel version, denoted POET-m. We provide distinguishing and forging attacks on it. Moreover, the designers claimed security of POET for an arbitrary AXU or almost XOR universal hash function (the formal definition of AXU hash function is given in Section 2). Here we disprove their claim by showing a distinguishing attack on a special choice of AXU, namely a uniform random involution. Thus, the security proof of the claims have flaws. We also extend this to a forging attack. All these attack algorithms make very few encryption queries and succeed with probability close to one.

We would like to note that while the COBRA is affected by our attack, the instantiation of the POET candidate which uses specific AXU hash functions is not affected.

2 Basics of Almost XOR Universal (AXU) Hash

2.1 Notation and Basics

In this paper, we fix a positive integer n which denotes the block size of the underlying blockcipher. We mostly use AES (advanced encryption standard) [11] with 128 bit key size as the underlying blockcipher and in this case $n = 128$.

BINARY FIELD. We identify the set $\{0,1\}^n$ as the binary field of size 2^n. An n bit string $\alpha = \alpha_0\alpha_1...\alpha_{n-1}$, $\alpha_i \in \{0,1\}$ can be equivalently viewed as a polynomial $\alpha(x) = \alpha_0 + \alpha_1 x + \cdots + \alpha_{n-1}x^{n-1}$. For notational simplicity, we write the concatenation of two binary strings α and β as $\alpha\beta$. The field addition between two n bit strings is bit-wise addition $\oplus$ (we also use "+"). Let us fix a primitive polynomial $p(x)$ of degree n. Field multiplication between two n-bit strings α and β can be defined as the binary string corresponding to the polynomial $\alpha(x)\beta(x)$ mod $p(x)$. We denote the multiplication of α and β as $\alpha \cdot \beta$. Thus, the zero polynomial $\mathbf{0}$ and the constant $\mathbf{1}$ polynomial are the additive and multiplicative identity respectively. Moreover, x is a primitive element since the polynomial $p(x)$ is primitive.

2.2 Almost XOR Universal (AXU) Hash

Universal hash functions and its close variants *strongly universal, AXU-hash* [9, 12, 18, 20–23] are information theoretic notions which are used as building blocks of several cryptographic constructions, e.g., *message authentication code* [9, 24], domain extension of pseudorandom function [5, 8], extractor [17], quasi-randomness and other combinatorial objects [12, 21].

Definition 1 (AXU Hash Function). *A function family $F_L : \mathcal{M} \rightarrow \{0,1\}^n$ indexed by $L \in \mathcal{L}$ is called ϵ-AXU [18] if for all $x \neq x' \in \mathcal{M}$ and $\delta \in \{0,1\}^n$, $\Pr_L[F_L(x) \oplus F_L(x') = \delta] \leq \epsilon$ where L is chosen uniformly from $\mathcal{L}$.*

Examples. FIELD MULTIPLIER. Let $L \in \{0,1\}^n$ be chosen uniformly then $F_L(x) = L \cdot x$ (field multiplication on $\{0,1\}^n$) is 2^{-n}-AXU.

POLYNOMIAL HASH. Polynomial hash [20] is one of the popular universal hash which can be computed efficiently by Horner's rule [14] (same as computation of CBC message authenticated code [2, 6]).

Definition 2. *[20] We define the polynomial-hash indexed by $L \in \{0,1\}^n$ over the domain $(\{0,1\}^n)^+ := \cup_{i=1}^{\infty} \{0,1\}^{ni}$ as*

$$poly_L(a_d, a_{d-1}, \ldots, a_0) = a_0 + a_1 \cdot L \cdots + a_{d-1} \cdot L^{d-1} + a_d \cdot L^d$$

where $a_0, a_1, \ldots, a_d \in \{0,1\}^n$ and L^i denotes $L \cdot L \cdots L$ (i times).

It is easy to see that the function mapping $(a_1, \ldots, a_d)$ to $a_1 \cdot L + \cdots + a_d \cdot L^d$ is $\frac{d}{2^n}$-AXU hash function over the domain $(\{0,1\}^n)^d$. To see this, let $(a_1, \ldots, a_d) \neq (b_1, \ldots, b_d)$ and $c \in \{0,1\}^n$. So,

$$c + (a_1 - b_1) \cdot L + \cdots + (a_d - b_d) \cdot L^d$$

is a non-zero polynomial and hence it has at most d distinct roots of L.

FOUR-ROUND AES. The AES (for 128 bit keys) has ten rounds. However, it has been shown that four-round AES has good differential properties. More formally, Daemen et al. in [10] showed that four-round AES is a family of 2^{-113}-AXU under the simplified assumption that all four round keys are uniform and independent.

UNIFORM RANDOM INVOLUTION. The uniform random function from $\{0,1\}^n$ to itself is an 2^{-n}-AXU hash function. A function $f : \{0,1\}^n \rightarrow \{0,1\}^n$ is called an **involution** if f is inverse of itself (so it must be permutation). Let I_n denote a random involution whose responses are defined according to the following procedure: After responding to every query, it updates two sets: the set of all queries D and the set of all responses R. On a query $x \notin D \cup R$, it returns an element chosen uniformly from the set $\{0, 1\}^n \setminus (D \cup R)$. If $x \in D$ then it returns the previous response corresponding to x. Similarly, if $x \in R$ then it returns the previous query $y \in D$ for which the response was x.

Lemma 1. *The uniform random involution I_n (as defined above) is an $\frac{2}{2^n - 2}$-AXU hash function.*

Proof. Let $x \neq x' \in \{0, 1\}^n$ and $\delta \in \{0, 1\}^n$. Let us assume that $x \oplus x' \neq \delta$. By conditioning $I_n(x) = y$, we must have $I_n(x') = y \oplus \delta$ which happens with probability at most $1/(2^n - 2)$. Note that if $y = x'$ or $y = x \oplus \delta$ then the probability is zero. So $\Pr[I_n(x) \oplus I_n(x') = \delta] \leq \frac{1}{2^n - 2}$. Now assume $x \oplus x' = \delta$. So $\Pr[I_n(x) = x'] \leq \frac{1}{2^n - 2}$. When $I_n(x) \neq x'$, by similar argument as before we also have differential probability bounded above by $\frac{1}{2^n - 2}$. This proves the Lemma. □

2.3   Combination of AXU Hash Functions

Compositions of AXU Hash Functions. Now we show that property of being an AXU-hash function does not preserve under composition with same key. In other words, there exists an ϵ-AXU F_L for a "small" ϵ such that $F_L \circ F_L$ is not even δ-AXU for any $\delta < 1$. In particular, if we choose F_L to be a uniform random involution then F_L is $\frac{1}{2^n - 2}$-AXU whereas the composition $F_L \circ F_L$ is the identity function. Trivially, a similar result holds if we apply the CBC mode for a uniform random involution I_n. The CBC mode applied to a function f is defined as follows:

$$CBC^f(x_1, \ldots, x_d) = y_d, \text{ where } y_i = f(y_{i-1} \oplus x_i), 1 \leq i \leq d$$

and $y_0 = 0^n$. So when $d = 2$ and $x_2 = 0$, $CBC^{I_n}(x_1, 0) = I_n(I_n(x_1)) = x_1$ and so CBC^{I_n} is not δ-AXU for any $\delta < 1$. However, it is true for some specific choices of F_L, e.g. when F_L is field multiplier. In this case, CBC^{F_L} is nothing but the poly-hash which has been shown to be $d/2^n$-AXU (see the paragraph immediately after Definition 2).

Sum of AXU Hash Functions. Now we consider another method of domain extension of AXU hash function. Given an ϵ-AXU F_L, we define the sum hash

$$F_L^{\text{sum}}(x_1, \ldots, x_\ell) = F_L(x_1) \oplus \cdots \oplus F_L(x_\ell).$$

Note that if F_L is linear (which is true for the field multiplier) then the sum hash can be simplified as $F_L^{\text{sum}}(x_1, \ldots, x_\ell) = F_L(x_1 \oplus \ldots \oplus x_\ell)$ for which a collision can be found easily. So it can not be δ-AXU for any $\delta < 1$. However, this does not work when we consider a uniform random function or involution and we concatenate a counter to message blocks.

3   Description of **COBRA**

COBRA is an authenticated encryption mode based on blockcipher. It was originally published in FSE 2014 [4]. Later the same mode with AES as the underlying blockcipher, called AES-COBRA, was submitted to CAESAR [1]. The mode can be viewed as hash then ECB (or electronic code-book) type encryption where hash function is poly-hash and ECB is applied on a double block, i.e., $2n$ bit plaintext. The double block encryption is defined by two-round Feistel structure [15].[2] As it uses Feistel structure, it is inverse-free. In other words, even though it is based on AES blockcipher, the decryption of COBRA does not require AES decryption.

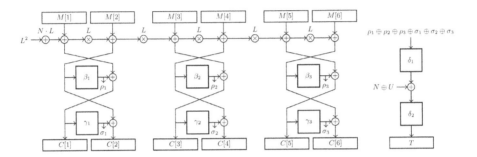

Fig. 3.1. COBRA Modes for ciphertext and tag generation for three double blocks message. U is obtained from associated data, N is nonce and L is the hash key.

3.1   Encryption Mode for **COBRA**

COBRA is defined for any messages of size at least n bits. Now we briefly describe how the encryption algorithm of COBRA works for all inputs $M \in (\{0,1\}^{2n})^+$. In addition to a message M, it also takes a nonce $N \in \{0,1\}^n$ and an associated data A, and outputs a tagged ciphertext (C, T) where $|C| = |M|$ and $T \in \{0,1\}^n$. Readers are referred to [1, 4] for complete description of the algorithm (i.e., how it behaves for other input sizes). We write $M = M_1 \| \cdots \| M_d$ for some positive integer d where $M_1, \ldots, M_d \in \{0,1\}^{2n}$. We also write $M_i = (M_i[1], M_i[2])$ where $M_i[1], M_i[2] \in \{0,1\}^n$ are also called blocks and M_i's are called **double blocks**. Let $\beta_i{}'$ and γ_i's be independent uniform random (or pseudorandom) permutations over $\{0,1\}^n$ for all $i \geq 1$. We describe the COBRA-mode based on these permutations.[3] It uses the two-round Feistel structure which is defined as follows:

$$\mathsf{LR}_i(X[1], X[2]) = (Y[1], Y[2]), \quad X[1], X[2] \in \{0,1\}^n$$

---

[2] The 3 and 4 rounds security analysis is given in [15] (see [16] for characterization of Luby-Rackoff constructions).

[3] These are actually derived from a single blockcipher using the standard masking algorithm (i.e., XEX construction [19]).

where

1. $Y[1] = X[1] \oplus \beta_i(X[2])$ and
2. $Y[2] = X[2] \oplus \gamma_i(X[1])$.

It is easy to see that it is invertible and the inverse function $2\mathsf{LR}_i^{-1}(Y[1], Y[2]) = (X[1], X[2])$ where $X[2] = \gamma_i(Y[1]) \oplus Y[2]$ and $X[1] = \beta_i(X[2]) \oplus Y[1]$.

Algorithm: COBRA Encryption
Input: $(M_1[1], M_1[2], \ldots, M_d[1], M_d[2]) \in (\{0,1\}^n)^{2d}$, $N \in \{0,1\}^n$
Output: $(C_1, C_2, ..., C_d) \in (\{0,1\}^{2n})^d$

---

1   **for** $i = 1$ to d

2        $P_i[1] = \mathsf{poly}_L(1, N, M_1[1], M_1[2], \ldots, M_i[1]);$

3        $P_i[2] = \mathsf{poly}_L(1, N, M_1[1], M_1[2], \ldots, M_i[1], M_i[2]);$

4        $C_i = \mathsf{LR}_i(P_i[1], P_i[2]);$

5   **end for loop**

6   **Return** $(C_1, C_2, ..., C_d)$

Algorithm 1. COBRA encryption algorithm for a nonce $N \in \{0,1\}^n$, and a messages M of sizes multiple of $2n$. Note that the associated data has no influence on the ciphertext. It is used for computing the tag.

3.2   Tag Generation and Verified Decryption Algorithm

The final tag T is computed from nonce N and U (depends only on the associated data A) and

$$S := \bigoplus_{i=1}^{d}(P_i[1] \oplus P_i[2] \oplus C_i[1] \oplus C_i[2]).$$

We simply denote the tag by $T(N, U, S)$. One can find the details of the construction of T in [1, 4]. The verified decryption algorithm takes a tagged ciphertext $(C_1, \ldots, C_d, T)$ where C_i's are double blocks and $T \in \{0,1\}^n$. It works as follows:

1. It first computes $P_i = \mathsf{LR}_i^{-1}(C_i)$, $1 \leq i \leq d$.
2. It returns $\bot$ if $T \neq T(N, U, S)$.
3. Else it returns $(M_1, \ldots, M_d)$ where $M_i[2] = L \cdot P_i[1] \oplus P_i[2]$ and $M_i[1] = L \cdot P_{i-1}[2] \oplus P_i[1]$, $1 \leq i \leq d$.

4   Forging Attack on **COBRA**

We first state the following fact which plays key role in our forging attack.

Fact 1.[7]   *Let $h \in \{0,1\}^n$ be a fixed element and $h_0^1, h_1^1, \ldots, h_0^s, h_1^s$ be chosen uniformly from $\{0,1\}^n$. Then, the probability that there exists $b_1, \ldots, b_s \in \{0,1\}$ such that $\bigoplus_i h_{b_i}^i = h$ is at least $1 - 2^{n-s}$. Furthermore, the sequence $b_1, \ldots, b_s$ can be efficiently computed.*

Key Idea of the Forging Attack. Now we describe the main idea of our forging attack. Our attack fixes nonce and associated data and so we simply denote the tag $T(N, U, S)$ by $T(S)$. Suppose M^0 is an encryption query with $T^0 := T(S^0)$ as a tag where S^0 denotes the S-value for the message M^0. Suppose C_i^0 and C_i^1 are the two i^{th} double-block ciphertexts for two different queries, $1 \le i \le s$. By Fact 1, we can find $b_1, \ldots, b_s$ such that $\oplus_{i=1}^{s}(C_i^{b_i}[0] \oplus C_i^{b_i}[1]) = S^0$. So if we can choose messages such that $\oplus_{i=1}^{s}(P_i^{b_i}[0] \oplus P_i^{b_i}[1]) = 0^n$ happens with high probability then $(C_1^{b_1}, \ldots, C_s^{b_s}, T^0)$ is a valid tagged ciphertext. As polyhash is linear, we can ensure that $\oplus_{i=1}^{s}(P_i^{b_i}[0] \oplus P_i^{b_i}[0]) = 0^n$ holds with high probability for suitably chosen queries.

Forging Algorithm $\mathcal{F}_0$. Now let us fix a positive integer ℓ whose exact value will be determined later. We define the following messages

$$M^i := ((0,0)^{i-1}, (0,1), (0,0)^{\ell-i}), \quad 1 \le i \le \ell.$$

Let M^0 be the all zero block message. Our forging algorithm makes $\ell + 1$ many queries, namely M^i's.

Forging Algorithm $\mathcal{F}_0$ for COBRA.

1. It makes encryption queries M^i and obtains responses (C^i, T^i), $0 \le i \le \ell$.
2. Let $C^0 = (C_1^0[1], C_1^0[2], \cdots, C_\ell^0[1], C_\ell^0[2])$ and $h_0^i = C_i^0[1] \oplus C_i^0[2]$.
3. For $i = 1$ to ℓ
       let $C^i = (C_1^i[1], C_1^i[2], \cdots, C_\ell^i[1], C_\ell^i[2])$ and $h_1^i = C_i^i[1] \oplus C_i^i[2]$.
4. Let $h = h_0^\ell \oplus (\bigoplus_{i=1}^{\ell-1} h_0^i)$ (the sum of the ciphertext blocks for M^0).
5. Based on Fact 1, it finds a sequence $b_1, \ldots, b_{\ell-1} \in \{0,1\}$, $\bigoplus_{i=1}^{\ell-1} h_{b_i}^i = h_1^\ell \oplus h$.

6. If there is no such sequence then it aborts else it proceeds.
7. If $b_1 \oplus \cdots \oplus b_{\ell-1} \ne 1$ then it aborts.
8. Else it makes the forgery $(C^* := (C_1^*, \ldots, C_\ell^*), T^0)$ where for all $1 \le i \le \ell-1$

$$C_i^* = \begin{cases} C_i^i[1] \| C_i^i[2] & \text{if } b_i = 1, \\ C_i^0[1] \| C_i^0[2] & \text{if } b_i = 0. \end{cases}$$

and $C^*[\ell] = C_\ell^\ell[1] \| C_\ell^\ell[2]$.

Now we compute the success probability of the forging attack. The forging algorithm aborts in two cases. We show that the abort probabilities are small. Moreover, given that it does not abort, we also show that the forging attack works perfectly.

Theorem 1. *The forgery algorithm $\mathcal{F}_0$ has success probability at least $\frac{1}{2} \times (1 - 2^{-n})$ when we set $\ell = 2n$.*

Proof. In the ideal case, h_0^i and h_1^i are independently and (almost) uniformly drawn from $\{0,1\}^n$ as these are xor of two blocks of the ith double-block ciphertext for fresh queries M^i and M^0 respectively. Note that M^i and M^0 have different double block values in the ith position. By Fact 1, with probability at least $1/2$, we can efficiently find $b_1, \ldots, b_{\ell-1}$ such that $\oplus_{j=1}^{\ell-1} h_{b_j}^j = h \oplus h_1^\ell$.

Claim. Let us assume that we have found such $b_1, \ldots, b_{\ell-1} \in \{0,1\}$ which happens with probability at least $1 - 2^{n-\ell}$. Then,

$$b_1 \oplus \cdots \oplus b_{\ell-1} = 1 \Rightarrow (C^*, T) \text{ is a valid ciphertext tag pair.}$$

We first note that (C^*, T) is a fresh tagged ciphertext as the last double block of ciphertext is different from those of all other tagged ciphertexts. To prove that tagged ciphertext is valid, we first compute S^* and S^0 for the given forged ciphertext and M^0 respectively where S^0 denotes the S values for the message M^0.

Computation of S^0. Computation of S^0 is straightforward from its definition.

$$S^0 := (\oplus_{j=1}^\ell (P_j^0[1] \oplus P_j^0[2])) \oplus (\oplus_{i=1}^\ell (C_i^0[1] \oplus C_i^0[2]).$$

Now note that $P_i^0[1] = \mathsf{poly}_L(1, N, 0^{2i-2}, 0)$ and $P_0^i[2] = \mathsf{poly}_L(1, N, 0^{2i-1}, 0)$. Let $\Sigma = \bigoplus_{i=1}^\ell (\mathsf{poly}_L(1, N, 0^{2i-1}, 0) \oplus \mathsf{poly}_L(1, N, 0^{2i-2}, 0))$. So

$$S^0 = h \oplus (\bigoplus_{i=1}^\ell (\mathsf{poly}_L(1, N, 0^{2i-1}, 0) \oplus \mathsf{poly}_L(1, N, 0^{2i-2}, 0)) = h \oplus \Sigma.$$

Computation of S^*. Now we compute S^* under the assumption that the first abort does not hold, i.e., we have found $b_1, \ldots, b_{\ell-1}$ such that $\oplus_{j=1}^{\ell-1} h_{b_j}^j = h \oplus h_1^\ell$. Note that the xor of ciphertext blocks which is equal to $\oplus_{j=1}^{\ell-1} h_{b_j}^j \oplus h_1^\ell = h$.

Now we decrypt the forged ciphertext double blocks by applying $2\mathsf{LR}^{-1}$. Let $P_i^* := (P_i^*[1], P_i^*[2])$ be the ith double block of forged ciphertext after we apply Luby-Rackoff two round decryption. Similarly, we denote P_i values for M^j query as P_i^j. As all ciphertext double blocks C_i^* have appeared in responses of queries at the same position, the P_i^* values, $1 \leq i \leq \ell$ are given as below.

$$P_i^* = \begin{cases} P_i^i[1] \| P_i^i[2] & \text{if } b_i = 1, \\ P_i^0[1] \| P_i^0[2] & \text{if } b_i = 0. \end{cases}$$

and $P_\ell^* = P_\ell^\ell[1] \| P_\ell^\ell[2]$. Note that

1. $P_i^i[1] = \mathsf{poly}_L(1, N, 0^{2i-1})$ and $P_i^i[2] = \mathsf{poly}_L(1, N, 0^{2i-1}1)$,
2. $P_i^0[1] = \mathsf{poly}_L(1, N, 0^{2i-1})$ and $P_i^0[2] = \mathsf{poly}_L(1, N, 0^{2i})$.

By linearity of poly_L, we can simply write for $1 \leq i \leq \ell - 1$,

1. $P_i^*[1] \oplus P_i^*[2] = \mathsf{poly}_L(1, N, 0^{2i-1}) \oplus \mathsf{poly}_L(1, N, 0^{2i}) \oplus b_i$ and
2. $P_\ell^*[1] \oplus P_\ell^*[2] = \mathsf{poly}_L(1, N, 0^{2\ell-1}) \oplus \mathsf{poly}_L(1, N, 0^{2\ell}) \oplus 1$.

So $\bigoplus_{j=1}^{\ell}(P_j^*[1] \oplus P_j^*[2]) = \Sigma \oplus 1 \oplus (\oplus_{j=1}^{\ell-1} b_j)$ and

$$S^* = \oplus(\Sigma \oplus 1 \oplus (\oplus_{j=1}^{\ell-1} b_j)).$$

Now if the second abort does not hold then (i.e., $\oplus_j b_j = 1$) we have $S^* = h \oplus \Sigma = S^0$. This proves the claim.

PROBABILITIES OF THE ABORT EVENTS. Now we informally argue that the second abort probability is $\Pr[\oplus_{j=1}^{\ell-1} b_j = 1] = 1/2$. Note that $C_i^i[1]$, $C_i^i[2]$, $C_i^0[1]$, $C_i^0[2]$'s are independent and so are h_0^i, h_1^i for all $1 \leq i \leq \ell$. Thus by conditioning h_0^ℓ, h_1^ℓ, choices of b_i's are independent and uniform. So the probability is $1/2$. By Fact 1, the first abort does not hold with probability $1 - 2^{n-\ell}$ and now we claim the second abort does not hold with probability $1/2$. Hence success probability of forging is at least $\frac{1}{2}(1 - 2^{n-\ell})$ which is almost $1/2$ if we set $\ell = 2n$.   □

Remark 1. Note that in the above attack, we can verify whether the forged ciphertext tag pair is valid without querying it. So we can repeat this process n times (we choose bit 0 or 1 in different position instead of the last bit as described above) to succeed with probability of about $1 - 2^{-n}$.

Remark 2. In the above analysis we make several probabilistic assumptions to make the analysis clean and simple. Here we list these.

1. We assume that h_j^i's are independent and uniform. However, for a fixed i, h_1^i and h_0^i are not completely independent as these are generated from ideal online random permutation. However, for these $4n$ outputs $C_i^i[1], C_i^i[2]$, $C_i^0[1], C_i^0[2]$ these are statistically very close to the uniform distribution with distance about $\binom{4n}{2}/2^n$.
2. True distributions of b_i's may not be uniform and independent. It actually depends on how we define b_i's as there could be more than one choice of b_i's. However, all of these choices would lead to abort with probability of about $1/2$ or less.

5 Security Analysis of **POET** and **POET**-m

In this section we analyze POET and its parallel variant POET-m for some positive integer m.

Algorithm: POET-m Encryption
Input: $(M_1, M_2, ..., M_\ell) \in (\{0,1\}^n)^\ell$
Output: $(C_1, C_2, ..., C_\ell, T) \in (\{0,1\}^n)^{\ell+1}$

---

1   **for** $i = 1$ to $\ell - 1$
2      $X_i = \tau \oplus F_{L^{top}}(M_1 \oplus L_1) \oplus F_{L^{top}}(M_2 \oplus L_2) \oplus \cdots \oplus F_{L^{top}}(M_i \oplus L_i)$.
3      $Y_i = E_K(X_i)$;
4      $C_i = F_{L^{bot}}(Y_{i-1} \oplus Y_i) \oplus L_i$;
5   **end for loop**
6    $X_\ell = F_{L^{top}}(X_{\ell-1}) \oplus M_\ell$.
7    $Y_\ell = E_K(X_\ell)$;
8    $C_\ell = F_{L^{bot}}(Y_{\ell-1} \oplus Y_\ell)$;
9    $X_{\ell+1} = F_{L^{top}}(X_\ell) \oplus S \oplus \tau$.
10   $Y_{\ell+1} = E_K(X_{\ell+1})$;
11   $T = F_{L^{bot}}(Y_\ell) \oplus Y_{\ell+1} \oplus S$;
12   **Return** $(C_1, C_2, ..., C_\ell, T)$

Algorithm 2. POET-m encryption algorithm for a messages M of sizes ℓn with $\ell < m$. Let τ be an n-bit element which is derived from associated data. The elements $L_1, \ldots, L_{m-1}$ are derived keys and S is a key derived from length of the message.

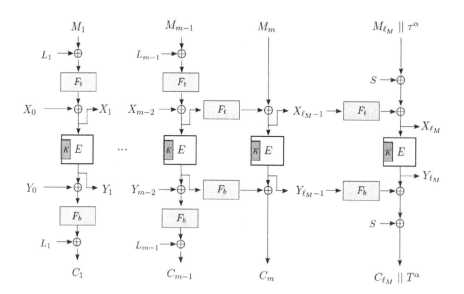

Fig. 5.1. POET-m Mode for ciphertext and tag generation. $X_0 = Y_0 = \tau$ is obtained from the associated data. We denote $F_{L^{top}}$ and $F_{L^{bot}}$ simply by F_t and F_b respectively.

5.1   POET-m and Its Security Analysis

POET-m. We first describe ciphertext generation algorithm of parallel version POET-m. We consider F_L to be the field multiplier hash in which message block is multiplied by the key L. We describe how POET-m works for all messages $(M_1, \ldots, M_\ell)$ with $\ell < m$. Let τ be an n-bit element which is derived from associated data. The elements $L_1, \ldots, L_{m-1}$ are keys derived by invoking pseudorandom permutation on different constants (see [1, 3] for details). Note that the input of the blockcipher X_i is a sum hash. When we instantiate the AXU by field multiplier we can simplify the sum hash (due to linearity). We have

$$X_i = \tau \oplus L^{top} \cdot (M_1 \oplus \cdots \oplus M_i) \oplus L'$$

where L' is the remaining part depending only on keys. We use this expression to mount the attack.

Privacy Attack on POET-m. We first demonstrate a distinguishing attack on POET-m distinguishing it from uniform random online cipher when $m > 4$. We make two queries

1. $M = (M_1, M_2, M_3, M_4)$ and
2. $M' = (M'_1, M'_2, M'_3 := M_3, M'_4)$ such that $M_1 \neq M'_1$ and $M_1 \oplus M_2 = M'_1 \oplus M'_2$.

We denote the corresponding internal variables by X, C's and X', C''s. It is easy to see that $X_2 = X'_2$ and $X_3 = X'_3$ and hence $C_3 = C'_3$ with probability one. This equality of third ciphertext block happens with probability 2^{-n} for uniform random online cipher. So we have a distinguisher which succeeds with probability almost one. The presence of fourth block makes sure that X_i's are defined as above (as the final block is processed differently). We can keep all other inputs, for example nonce, associated data etc., the same.

Forging Attack on POET-m. Now we see how we can exploit the above weakness in sum of AXU hash to mount a forging attack on the construction. We can forge when the number of message blocks is less than m and the last block is complete (as described in Algorithm 2). We first simply describe how the decryption algorithm works. Assume $m > 3$ and let C_1, C_2, C_3, T be an input for decryption where $C_i, T \in \{0, 1\}^n$. We note the following observations:

1. Y_i depends on $C_1 \oplus \cdots \oplus C_i$ for $i \leq 3$.
2. Verification algorithm depends on X_3, Y_3, T and some fixed values depending on associated data and key.

We make one query $M = (M_1, M_2, M_3)$ and obtain the response (C, T) where $C = (C_1, C_2, C_2)$. Let $C' = (C'_1, C'_2, C'_3 := C_3) \neq C$ such that $C_1 \oplus C_2 = C'_1 \oplus C'_2$. We denote the corresponding internal variable by X, C's and X', C''s. It is easy to see that by choices of C' and the first observation $Y_3 = Y'_3$ and so $X_3 = X'_3$. Again by the second observation, we see that verification algorithm depends on X_3, Y_3, T (or X'_3, Y'_3, T') and some fixed information based on key and associated data. So, whenever verification algorithm passes for X_3, Y_3, it must pass for X_3', Y_3'. Thus, (C_1', C_2', C_3, T) is a valid forge.

Note that the above attack is a single query forging attack and hence it is also applicable to situations where nonce can not be reused.

5.2   Security Analysis of POET mode

POET: We now describe ciphertext generation algorithm of POET, i.e. POE the underlying encryption algorithm. In [1], the following theorem (restated) was claimed:

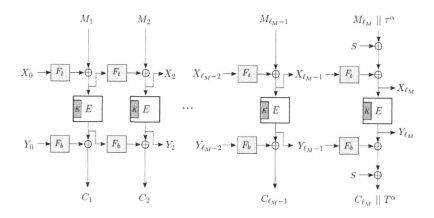

Fig. 5.2. POET Mode for ciphertext and tag generation. $X_0 = Y_0 = \tau$ is obtained from the associated data. In this figure, let F_t and F_b any independent ϵ-AXU hash functions.

Theorem 8.1 of POET Submission in [1]. Let E be a uniform random permutation, F_t and F_b be independent ϵ-AXU hash functions. Then, for any privacy adversary A making at most q queries of a total length of at most σ blocks, we have

$$\mathbf{Adv}^{\mathrm{priv}}_{\mathsf{POET}^E}(A) \leq \epsilon\sigma^2 + \frac{\sigma^2}{2^n - \sigma}.$$

Here we consider F_t and F_b to be any arbitrary AXU functions as mentioned above in Theorem 8.1 of the submission POET in [1]. Given messages $(M_1, \ldots, M_\ell)$, we compute for $1 \leq i \leq \ell - 1$ as follows:

$$C_i = F_b(Y_{i-1}) \oplus Y_i, \ Y_i = E_K(X_i), \ X_i = F_t(X_{i-1}) \oplus M_i$$

where $X_0 = Y_0 = \tau$. The last ciphertext block is computed differently and we do not need its description for our distinguishing attack. Note that X_i is computed by CBC^{F_t}. If F_t is a uniform random involution then as we have seen before, after applying CBC mode to F it does not remain δ-AXU for all $\delta < 1$. We use this property to obtain a distinguisher.

Privacy Attack on POET with Uniform Random Involution F_t. Now we demonstrate a privacy attack on POET distinguishing it from uniform random cipher when F_L is instantiated with uniform random involution. In this attack we only make a single query and so it is also nonce-respecting. This would violate Theorem 8.1 of the submission POET in [1] (online permutation security of POE). We believe that the theorem remains valid when F_L is instantiated with field multiplier (however, proof needs to be revised). The attack is described below.

Claim. $\Pr[C_2 = C_4] = 1$ where $(C_1, C_2, \ldots)$ is the response of $(M_1, 0, 0, 0, \cdots)$ to POET with involution F_t.

We prove the claim by using the involution property of F. We can easily see that

1. $X_3 = F(F(X_1)) = X_1$ and
2. similarly, $X_4 = X_2 = F(X_1)$.

So $Y_1 = Y_3$ and $Y_2 = Y_4$ and hence $C_2 = C_4$. Note, we can choose any arbitrary nonce and associated data. This proves the claim.

In an ideal case, we observe $C_2 = C_4$ with probability 2^{-n}. So the distinguisher of POET has advantage at least $1 - 2^{-n}$.

6   Conclusion

In this paper, we demonstrate forging attack on COBRA with practical complexity. Hence the theorem proved in [4] is wrong. We also demonstrate forging and distinguishing attack on POET-m for one particular recommended choice of AXU hash function. We also disprove the security claim for POET by presenting a distinguishing attack on a different choice of AXU hash function (not in the recommended list). However, these attacks on POET do not carry over to the versions submitted to CAESAR.

Acknowledgement. This work is supported by Centre of Excellence in Cryptology at Indian Statistical Institute, Kolkata. Author would also like to thank all anonymous reviewers who provided us very useful comments to improve the quality of the paper.

References

1. CAESAR submissions (2014), `http://competitions.cr.yp.to/caesar-submissions.html`
2. ISO/IEC 9797. Data cryptographic techniques-Data integrity mechanism using a cryptographic check function employing a blockcipher algorithm (1989)
3. Abed, F., Fluhrer, S., Forler, C., List, E., Lucks, S., McGrew, D., Wenzel, J.: Pipelineable on-line encryption. In: Fast Software Encryption. LNCS. Springer 3:320–337 (to appear)
4. Andreeva, E., Luykx, A., Mennink, B., Yasuda, K.: Cobra: A parallelizable authenticated online cipher without block cipher inverse. In: Fast Software Encryption. LNCS. Springer (to appear, 2014)

5. Bellare, M.: New proofs for nmac and hmac: Security without collision-resistance. IACR Cryptology ePrint Archive, 2006:43 (2006)
6. Bellare, M., Kilian, J., Rogaway, P.: The security of cipher block chaining. In: Desmedt, Y.G. (ed.) CRYPTO 1994. LNCS, vol. 839, pp. 341–358. Springer, Heidelberg (1994)
7. Bellare, M., Micciancio, D.: A new paradigm for collision-free hashing: Incrementality at reduced cost. In: Fumy, W. (ed.) EUROCRYPT 1997. LNCS, vol. 1233, pp. 163–192. Springer, Heidelberg (1997)
8. Black, J., Halevi, S., Krawczyk, H., Krovetz, T., Rogaway, P.: UMAC: Fast and Secure Message Authentication. In: Wiener, M. (ed.) CRYPTO 1999. LNCS, vol. 1666, pp. 216–233. Springer, Heidelberg (1999)
9. Carter, L., Wegman, M.N.: Universal Classes of Hash Functions. J. Comput. Syst. Sci. 18(2), 143–154 (1979)
10. Daemen, J., Lamberger, M., Pramstaller, N., Rijmen, V., Vercauteren, F.: Computational aspects of the expected differential probability of 4-round aes and aes-like ciphers. Computing 85(1-2), 85–104 (2009)
11. Daemen, J., Rijmen, V.: The Design of Rijndael: AES - The Advanced Encryption Standard (2002), http://csrc.nist.gov/CryptoToolkit/aes/rijndael/Rijndael-ammended.pdf
12. Gilbert, E.N., MacWilliams, F.J., Sloane, N.J.: Codes which detect deception. Bell System Technical Journal 53(3), 405–424 (1974)
13. Guo, J., Jean, J., Peyrin, T., Wang, L.: Breaking poet authentication with a single query. Technical report, Cryptology ePrint Archive, Report 2014/197 (2014), http://eprint.iacr.org
14. Horner, W.G.: Philosophical Transactions. Royal Society of London 109, 308–335 (1819)
15. Luby, M., Rackoff, C.: How to construct pseudo-random permutations from pseudorandom functions. In: Williams, H.C. (ed.) CRYPTO 1985. LNCS, vol. 218, pp. 447–447. Springer, Heidelberg (1986)
16. Nandi, M.: The characterization of luby-rackoff and its optimum single-key variants. In: Gong, G., Gupta, K.C. (eds.) INDOCRYPT 2010. LNCS, vol. 6498, pp. 82–97. Springer, Heidelberg (2010)
17. Nisan, N., Zuckerman, D.: Randomness is linear in space. J. Comput. Syst. Sci. 52(1), 43–52 (1996)
18. Rogaway, P.: Bucket hashing and its application to fast message authentication. In: Coppersmith, D. (ed.) CRYPTO 1995. LNCS, vol. 963, pp. 29–42. Springer, Heidelberg (1995)
19. Rogaway, P.: Efficient instantiations of tweakable blockciphers and refinements to modes ocb and pmac. In: Lee, P.J. (ed.) ASIACRYPT 2004. LNCS, vol. 3329, pp. 16–31. Springer, Heidelberg (2004)
20. Shoup, V.: On fast and provably secure message authentication based on universal hashing. In: Koblitz, N. (ed.) CRYPTO 1996. LNCS, vol. 1109, pp. 313–328. Springer, Heidelberg (1996)
21. Stinson, D.R.: On the connections between universal hashing, combinatorial designs and error-correcting codes. Congressus Numerantium 114, 7–27 (1996)
22. Stinson, D.R.: Universal hashing and authentication codes. In: Feigenbaum, J. (ed.) CRYPTO 1991. LNCS, vol. 576, pp. 74–85. Springer, Heidelberg (1992)
23. Stinson, D.R.: Universal hashing and authentication codes. Des. Codes Cryptography 4(4), 369–380 (1994)
24. Wegman, M.N., Carter, L.: New hash functions and their use in authentication and set equality. J. Comput. Syst. Sci. 22(3), 265–279 (1981)

Low Probability Differentials and the Cryptanalysis of Full-Round CLEFIA-128[*]

Sareh Emami[1], San Ling[2], Ivica Nikolić[2], Josef Pieprzyk[3]
and Huaxiong Wang[2]

[1] Macquarie University, Australia
[2] Nanyang Technological University, Singapore
[3] Queensland University of Technology, Australia

Abstract. So far, low probability differentials for the key schedule of block ciphers have been used as a straightforward proof of security against related-key differential analysis. To achieve resistance, it is believed that for cipher with k-bit key it suffices the upper bound on the probability to be 2^{-k}. Surprisingly, we show that this reasonable assumption is incorrect, and the probability should be (much) lower than 2^{-k}. Our counter example is a related-key differential analysis of the well established block cipher CLEFIA-128. We show that although the key schedule of CLEFIA-128 prevents differentials with a probability higher than 2^{-128}, the linear part of the key schedule that produces the round keys, and the Feistel structure of the cipher, allow to exploit particularly chosen differentials with a probability as low as 2^{-128}. CLEFIA-128 has 2^{14} such differentials, which translate to 2^{14} pairs of weak keys. The probability of each differential is too low, but the weak keys have a special structure which allows with a divide-and-conquer approach to gain an advantage of 2^7 over generic analysis. We exploit the advantage and give a membership test for the weak-key class and provide analysis of the hashing modes. The proposed analysis has been tested with computer experiments on small-scale variants of CLEFIA-128. Our results do not threaten the practical use of CLEFIA.

Keywords: CLEFIA, cryptanalysis, weak keys, CRYPTREC, differentials.

1 Introduction

CLEFIA [13] is a block cipher designed by Sony. It is advertised as a fast encryption algorithm in both software and hardware and it is claimed to be highly secure. The efficiency comes from the generalized Feistel structure and the byte orientation of the algorithm. The security is based on the novel technique called Diffusion Switching Mechanism, which increases resistance against linear and differential attacks, in both single and related-key models. These and several other attractive features of CLEFIA-128 have been widely recognized, and the cipher

---

[*] The researcher is supported by the Singapore National Research Foundation Fellowship 2012 NRF-NRFF2012-06.

P. Sarkar and T. Iwata (Eds.): ASIACRYPT 2014, PART I, LNCS 8873, pp. 141–157, 2014.

has been submitted for standardization (and already standardized) by several bodies: CLEFIA was submitted to IETF (Internet Engineering Task Force) [1], it is on the Candidate Recommended Ciphers List[1] of CRYPTREC (Japanese government standardization body), and it is one of the only two[2] lightweight block ciphers recommended by the ISO/IEC standard [8].

A significant body of analysis papers has been published on the round-reduced versions of CLEFIA [18,19,14,17,15,10,16,9,6], all for the single-key model, but the analysis based on related keys is missing. Often this type of analysis can cover a higher number of rounds but requires the cipher to have a relatively simple and almost linear key schedule. CLEFIA, however, has a highly non-linear key schedule, equivalent roughly to 2/3 of the state transformation and designed with an intention to make the cipher resistant against analysis based on related-key differentials. Using a widely accepted approach, the designers have proved that no such analysis could exist as the key schedule has only low probability ($\leq 2^{-128}$ for CLEFIA with 128-bit keys) differential characteristics. Note, we will not try to exploit the fact that some characteristics can be grouped into a differential that has a much higher probability than the individual characteristics. Our results go a step further and we show that key schedule differentials with a probability as low as 2^{-128}, can still be used in analysis. This happens when they have a special structure, namely, the input/output differences of the differentials are not completely random, but belong to a set that, as in the case of CLEFIA-128, is described with a linear relation.

We exploit the special form of the key schedule: a large number of non-linear transformations at the beginning of the key schedule is followed by light linear transformations that are used to produce the round keys. In the submission paper of CLEFIA-128, the proof of related-key security is based only on the non-linear part as this part guarantees that the probability of any output difference is 2^{-128}. In contrast, our analysis exploits the linear part and we show that there are 2^{14} of the above low probability differences which, when supplied to the linear part, produce a special type of iterative round key differences. CLEFIA-128 is a Feistel cipher and, as shown in [5], iterative round key differences lead to an iterative differential characteristic in the state that holds with probability 1. Therefore we obtain related-key differentials with probability 1 in the state and 2^{-128} in the key schedule. The low probability (2^{-128}) of each of the 2^{14} iterative round key differences means that for each of them there is only one pair of keys that produces such differences, or in total 2^{14} pairs for all of them – these pairs form the weak-key class of the cipher. When we target each pair independently, we cannot exploit the differentials. However, the whole set of 2^{14} pairs has a special structure and we can target independently two smaller sets of sizes 2^7 and thus obtain the advantage of 2^7 over generic analysis. As we will see in the paper, the special structure of the weak key class is due to the linear part of the key schedule, therefore we exploit the weakness of this part twice (the first time for producing iterative round key differences).

---

[1] This is the final stage of evaluation, before becoming CRYPTREC standard.

[2] The second one is PRESENT [7].

We further analyze the impact of the 2^{14} pairs of keys and the advantage of 2^7 that we gain over generic analysis. First we show that CLEFIA-128 instantiated with any pair of weak keys can be analyzed, namely we present a membership test for the weak class. Next, for the hashing mode of CLEFIA-128, i.e. when the cipher is used in single-block-length hash constructions, we show that differential multi-collisions [4] can be produced with a complexity lower than for an ideal cipher.

The paper is organized as follows. We start with a description of CLEFIA-128 given in Section 2. We present the main results related to the analysis of the key schedule and the production of the class of 2^{14} pairs of weak-keys in Section 3. The differential membership test is given in Section 4. We present the analysis of the hashing mode of the cipher in Section 5 and in Section 6 we conclude the paper.

2   Description of CLEFIA-128

CLEFIA is a 128-bit cipher that supports 128, 192, and 256-bit keys. We analyze CLEFIA with 128-bit keys that is referred as CLEFIA-128. Before we define the cipher, we would like to make an important note. To simplify the presentation, we consider CLEFIA-128 without whitening keys [3]. Our analysis applies to the original CLEFIA-128 as shown in Appendix B. We proceed now with a brief description of CLEFIA-128. It is an 18-round four-branch Feistel (see Fig. 3 of Appendix A) that updates two words per round. A definition of the state update function is irrelevant to our analysis (see [13] for a full description) and further we focus on the key schedule only.

A 128-bit master key K is input to a 12-round Feistel $GFN_{4,12}$(with the same round function as the one in the state, refer to Fig. 3 of Appendix A) resulting in a 128-bit intermediate key L. All the 36 round keys[4] $RK_i, i = 0, \ldots, 35$ are produced by applying a linear transformation to the master key K and the intermediate key L as shown below ($\oplus$ stands for the XOR operation and $||$ is concatenation):

$$
\begin{aligned}
RK_0||RK_1||RK_2||RK_3 &\leftarrow L &\oplus S_1,\\
RK_4||RK_5||RK_6||RK_7 &\leftarrow \Sigma(L) \oplus K &\oplus S_2,\\
RK_8||RK_9||RK_{10}||RK_{11} &\leftarrow \Sigma^2(L) &\oplus S_3,\\
RK_{12}||RK_{13}||RK_{14}||RK_{15} &\leftarrow \Sigma^3(L) \oplus K &\oplus S_4,\\
RK_{16}||RK_{17}||RK_{18}||RK_{19} &\leftarrow \Sigma^4(L) &\oplus S_5,\\
RK_{20}||RK_{21}||RK_{22}||RK_{23} &\leftarrow \Sigma^5(L) \oplus K &\oplus S_6,\\
RK_{24}||RK_{25}||RK_{26}||RK_{27} &\leftarrow \Sigma^6(L) &\oplus S_7,\\
RK_{28}||RK_{29}||RK_{30}||RK_{31} &\leftarrow \Sigma^7(L) \oplus K &\oplus S_8,\\
RK_{32}||RK_{33}||RK_{34}||RK_{35} &\leftarrow \Sigma^8(L) &\oplus S_9,
\end{aligned}
$$

---

[3] There are four whitening keys: two are added to the plaintext, and two to the ciphertext.

[4] Two round keys are used in every round, thus there are $2 \cdot 18 = 36$ keys in total.

where S_i are predefined 128-bit constants, and Σ is a linear function defined further. In short, each four consecutive round keys RK_{4i}, RK_{4i+1}, RK_{4i+2}, RK_{4i+3} are obtained by XOR of multiple applications of Σ to L, possibly the master key K, and the constant S_i. The resulting 128-bit sequence is divided into four 32-bit words and each is assigned to one of the round key words. The linear function Σ (illustrated in Fig. 1) is a simple 128-bit permutation used for diffusion. The function $\Sigma : \{0,1\}^{128} \to \{0,1\}^{128}$ is defined as follows:

$$X_{128} \to Y_{128}$$
$$Y = X[120 - 64]X[6 - 0]X[127 - 121]X[63 - 7],$$

where $X[a - b]$ is a bit sequence from the a-th bit to the b-th bit of X.

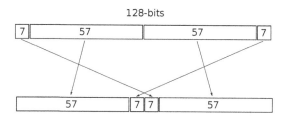

Fig. 1. The function Σ. The numbers denote the size of the bit sequence.

We would like to make a note about the notations of XOR differences used throughout the paper. To emphasize that a difference is in the word X, we use ΔX, otherwise, if it irrelevant or clear from the context we use simply Δ.

3   Weak Keys for CLEFIA-128

In the related-key model, the security of a cipher is analyzed by comparing two encryption functions obtained by two unknown but related keys. Given a specific relation[5] between keys, if the pair of encryption functions differs from a pair of random permutations, then the cipher has a weakness and can be subject to related-key analysis. Sometimes the analysis is applicable only when the pairs of related keys belong a relatively small subset of all possible pairs of keys. The subset is called the *weak-key class* of the cipher and the number of pairs of keys is the size of the class.

We will show that a weak-key class in CLEFIA-128 consists of pairs of keys $(K, \tilde{K} = K \oplus \mathcal{L}_1(D))$, where D can take approximately 2^{14} different 128-bit values, such that for *any* plaintext P, the following relation holds:

$$E_K(P) \oplus E_{\tilde{K}}(P \oplus \mathcal{L}_2(D)) = \mathcal{L}_3(D), \tag{1}$$

---

[5] Some relations are prohibited as they lead to trivial attacks, see [3] for details.

where $\mathcal{L}_1, \mathcal{L}_2, \mathcal{L}_3$ are linear functions defined below. The property can be seen as a related-key differential, with the difference $\mathcal{L}_1(D)$ for the master key, $\mathcal{L}_2(D)$ for the plaintext and $\mathcal{L}_3(D)$ for the ciphertext. From Equation (1), it follows that once D is defined, the probability of the differential is precisely one.

In the state of CLEFIA-128, the probability of a differential characteristic is one if in each Feistel round, there is no incoming difference to the non-linear round function. This happens when the differences in the state and in the round key cancel each other. Consequently, the input difference to the round function becomes zero[6]. An illustration of the technique for four rounds of CLEFIA-128 is given in Fig. 2. Notice that the input state difference at the beginning of the first round $(\Delta_1, \Delta_2, \Delta_3, \Delta_4)$ is the same as the output difference after the fourth round, i.e. it is iterative with the period of 4 rounds. Therefore, we will obtain a differential characteristic with probability 1 (in the state) for the full-round CLEFIA-128 if *we can produce 4-round iterative round key differences*.

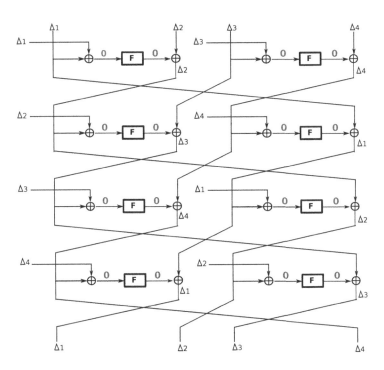

Fig. 2. Iterative related-key differential characteristic for 4 rounds of the CLEFIA-128 that is true with probability 1. The symbols $\Delta_1, \Delta_2, \Delta_3, \Delta_4$ denote word differences.

Each round of the state uses two round keys, thus the above 4-round iterative characteristic requires the round key differences to have a period of 8, i.e. $\Delta RK_i = \Delta RK_{i+8}$. Moreover, an additional condition has to

---
[6] A similar idea is given in [5].

hold. Note that in Fig. 2, the differences in the consecutive round keys are $(\Delta_1, \Delta_3, \Delta_2, \Delta_4, \Delta_3, \Delta_1, \Delta_4, \Delta_2)$, that is among the 8 round key differences, the first four are different, while the remaining four are only permutations of the first. These two conditions can be summarized as follows:

Condition 1 - For all i, it should hold $\Delta RK_i = \Delta RK_{i+8}$.

Condition 2 - For all i divisible by 8, it should hold $\Delta RK_i = \Delta RK_{i+5}$, $\Delta RK_{i+1} = \Delta RK_{i+4}$, $\Delta RK_{i+2} = \Delta RK_{i+7}$, $\Delta RK_{i+3} = \Delta RK_{i+6}$. This can be rewritten as $(\Delta RK_{i+4}, \Delta RK_{i+5}, \Delta RK_{i+6}, \Delta RK_{i+7}) = \pi(\Delta RK_i, \Delta RK_{i+1}, \Delta RK_{i+2}, \Delta RK_{i+3})$, where π is 4-word permutation $(0, 1, 2, 3) \rightarrow (1, 0, 3, 2)$.

Further we show how to find the set of differences for which the two conditions hold.

Condition 1. From the definition of the key schedule

$$RK_{8i+0}\|RK_{8i+1}\|RK_{8i+2}\|RK_{8i+3} \quad \leftarrow \Sigma^{2i}(L) \qquad \oplus S_{2i+1}$$
$$RK_{8i+8}\|RK_{8i+9}\|RK_{8i+10}\|RK_{8i+11} \quad \leftarrow \Sigma^{2i+2}(L) \qquad \oplus S_{2i+3},$$

it follows that Condition 1 for the first 4 (out of 8) round key differences in an octet of round keys can be expressed as

$$\Delta L = \Sigma^2(\Delta L). \tag{2}$$

We will obtain the same equation if we consider the remaining 4 round key differences. To satisfy Condition 1, we have to find possible values for ΔL such that Equation (2) holds. This can be achieved easily as (2) is a system of 128 linear equations with 128 unknowns (refer to the definition of Σ), and has solutions of the form (expressed as concatenation of bit sequences):

$$\Delta L = a_1 a_2 t b_2 b_1 b_2 b_1 b_2 b_1 b_2 a_2 a_1 a_2 a_1 a_2 a_1 a_2 t b_1 b_2, \tag{3}$$

where a_1, a_2 are any 7-bit values, t is the most significant bit of a_1 and the 7-bit values b_1, b_2 are defined as $tb_2b_1 = a_1 a_2 t$. Thus there are $2^7 \cdot 2^7 = 2^{14}$ solutions.

Condition 2. From the definition of the key schedule

$$RK_{8i+0}\|RK_{8i+1}\|RK_{8i+2}\|RK_{8i+3} \quad \leftarrow \Sigma^{2i}(L) \qquad \oplus S_{2i+1},$$
$$RK_{8i+4}\|RK_{8i+5}\|RK_{8i+6}\|RK_{8i+7} \quad \leftarrow \Sigma^{2i+1}(L) \oplus K \qquad \oplus S_{2i+2},$$

we see that Condition 2 can be expressed as

$$\pi(\Delta L) = \Sigma(\Delta L) \oplus \Delta K,$$

where π is 4-word permutation $(0, 1, 2, 3) \rightarrow (1, 0, 3, 2)$. Thus when ΔL is fixed (to one of the values from (3)), the difference in the master key ΔK can be determined as

$$\Delta K = \pi(\Delta L) \oplus \Sigma(\Delta L). \tag{4}$$

Summary. We have shown above that Conditions 1 and 2 can be achieved simultaneously as there are 2^{14} values for ΔL_i (see Equation (3)) with corresponding values of ΔK_i (see Equation (4)). It means that given the difference in the master key ΔK_i and the difference of the intermediate key ΔL_i (i.e. the differential in the 12-round Feistel $GFN_{4,12}$ of the key schedule is $\Delta K_i \rightarrow \Delta L_i$), the differences in the round keys are going to be of the requested form as shown below:

$$\Delta RK_0||\Delta RK_1||\Delta RK_2||\Delta RK_3 = \Delta_1||\Delta_3||\Delta_2||\Delta_4,$$
$$\Delta RK_4||\Delta RK_5||\Delta RK_6||\Delta RK_7 = \Delta_3||\Delta_1||\Delta_4||\Delta_2,$$

$$\cdots$$

$$\Delta RK_{28}||\Delta RK_{29}||\Delta RK_{30}||\Delta RK_{31} = \Delta_3||\Delta_1||\Delta_4||\Delta_2,$$
$$\Delta RK_{32}||\Delta RK_{33}||\Delta RK_{34}||\Delta RK_{35} = \Delta_1||\Delta_3||\Delta_2||\Delta_4,$$

where $\Delta_1||\Delta_3||\Delta_2||\Delta_4 = \Delta L_i$. As a result, we have obtained the necessary differences in the round keys and we can use the 4-round iterative characteristic from Fig. 2.

Now we can easily specify the description of the weak-key class given by Equation (1). The value of D coincides with the values of ΔL from Equation (3). Therefore the first linear function $\mathcal{L}_1$ is defined as $\mathcal{L}_1(D) = \pi(D) \oplus \Sigma(D)$. The input difference in the plaintext is the same as the input difference in the first four round keys (which is again ΔL), but the order of the words is slightly different – instead of $(\Delta_1, \Delta_3, \Delta_2, \Delta_4)$ it is $(\Delta_1, \Delta_2, \Delta_3, \Delta_4)$, see Fig. 2. Hence, we introduce the 4-word permutation $\pi_2 : (0, 1, 2, 3) \rightarrow (0, 2, 1, 3)$ that corrects the order. With this notation, the second linear function $\mathcal{L}_2$ is defined as $\mathcal{L}_2(D) = \pi_2(D)$. Finally, $\mathcal{L}_3$ is defined similarly. `CLEFIA-128` has 18 rounds, thus the last 4-round iterative characteristic (for the rounds 17,18) will be terminated after the second round, with an output difference $(\Delta_2, \Delta_3, \Delta_4, \Delta_1)$. It differs from ΔL only in the order of the four words, hence we introduce $\pi_3 : (0, 1, 2, 3) \rightarrow (3, 1, 0, 2)$ and conclude that $\mathcal{L}_3(D) = \pi_3(D)$.

In the weak-key class the pairs of keys are defined as $(K, K \oplus \pi(D) \oplus \Sigma(D))$ and for any plaintext P, it holds

$$E_K(P) \oplus E_{K \oplus \pi(D) \oplus \Sigma(D)}(P \oplus \pi_2(D)) = \pi_3(D). \tag{5}$$

A pair of keys belongs to this class if for any of the 2^{14} values $D = \Delta L$ defined by Equation (3), the 12-round Feistel $GFN_{4,12}$ in the key schedule, on input difference $\Delta K = \pi(\Delta L) \oplus \Sigma(\Delta L)$ gives the output difference ΔL, i.e. $GFN_{4,12}(K \oplus \pi(\Delta L) \oplus \Sigma(\Delta L)) \oplus GFN_{4,12}(K) = \Delta L$. Therefore not all of the keys K have a related key and form a pair in the weak-key class, but only those for which the differential in the Feistel permutation holds.

We deal with a 12-round Feistel permutation and thus the probability of the differential $\pi(\Delta L) \oplus \Sigma(\Delta L) \rightarrow \Delta L$ is low. We assume it is 2^{-128} (as proven by the designers), which is the probability of getting fixed output difference from a fixed input difference in a random permutation. However, even when we model the Feistel permutation by a random one, *there still exist 2^{14} key schedule* *differentials that have a probability of 2^{-128} and that result in iterative round key differences.*

In CLEFIA-128, there are 2^{128} possible keys K, and therefore for a specific value of D, the number of related keys $(K, K \oplus \pi(D) \oplus \Sigma(D))$ is the same. The probability of the differential in the Feistel permutation is 2^{-128}, thus among all of the pairs, only one will pass the differential. However, there are 2^{14} possible values for D, hence the size of the weak-key class is 2^{14}.

4   Membership Test for the Weak-Key Class

An analysis technique that succeeds when the related keys belong to the weak-key class is called a membership test. For the weak-key class of CLEFIA-128, the membership test will be a differential distinguisher that succeeds always and whose data, time and memory complexities are equal to 2^8. That is to say that we can decide with probability 1 whether the underlying cipher is CLEFIA-128 with weak keys or other (possibly ideal) cipher.

Given a pair of weak keys $(K, K \oplus \pi(D) \oplus \Sigma(D))$, it is easy to distinguish CLEFIA-128 (see Equation (5)) with only a single pair of related plaintexts $(P, P \oplus \pi_2(D))$ but D has to be known. If it is unknown, we will have to try all 2^{14} possible values of D (as D coincides with one of ΔL_i). Consequently, we are going to end up with a brute force attack on the space of weak keys. To address this problem, we have to be able to detect the correct value of ΔL efficiently.

Finding the correct ΔL_i can be performed much faster if we take into account the additional properties of the difference in the intermediate key. All 2^{14} values of ΔL_i (see Equation (3)) can be defined as XOR of two elements from two different sets each of cardinality 2^7 as shown below

$$\Delta L_i = \Delta L_i(a_1, a_2) = a_1 a_2 t b_2 b_1 b_2 b_1 b_2 b_1 b_2 a_2 a_1 a_2 a_1 a_2 a_1 a_2 t b_1 b_2 =$$
$$= G^1(a_1) \oplus G^2(a_2),$$
$$a_1 = 0, \ldots, 2^7 - 1, a_2 = 0, \ldots, 2^7 - 1,$$

where $G^1(a_1)$ is a 128-bit word that is the same as ΔL on the bits that depend on a_1 and has 0's for the bits that depend on a_2 while $G^2(a_2)$ is the opposite, i.e. coincides with ΔL on bits for a_2 and has 0's for bits that depend on a_1[7].

Using the representation helps to detect the correct ΔL by finding collisions on two specific sets. Assume the pair $(K, \tilde{K} = K \oplus \pi(\Delta L) \oplus \Sigma(\Delta L))$ belongs to the weak-key class. For a randomly chosen plaintext P, let us define two pools, each with 2^7 chosen plaintexts:

$$P_i^1 = \pi_2(P \oplus G^1(a_1^i)), a_1^i = 0, 1, \ldots, 2^7 - 1,$$
$$P_i^2 = \pi_2(P \oplus G^2(a_2^i)), a_2^i = 0, 1, \ldots, 2^7 - 1.$$

---

[7] Recall that each bit of b_1, b_2, t is equal to a single bit of either a_1 or a_2.

Next, we obtain two pools of ciphertexts with $(K, \tilde{K})$ as encryption keys, i.e. $C_i^1 = E_K(P_i^1), C_i^2 = E_{\tilde{K}}(P_i^2)$. Finally, we compute two sets V^1, V^2:

$$V^1 = \{V_i^1 | V_i^1 = \pi_2^{-1}(P_i^1) \oplus \pi_3^{-1}(C_i^1)\},$$
$$V^2 = \{V_i^2 | V_i^2 = \pi_2^{-1}(P_i^2) \oplus \pi_3^{-1}(C_i^2)\}.$$

The crucial observation is that the sets V^1 and V^2 will always collide, i.e. there exist V_i^1 and V_j^2 such that $V_i^1 = V_j^2$. This comes from the following sequence:

$$
\begin{aligned}
V_i^1 \oplus V_j^2 &= \\
&= \pi_2^{-1}(P_i^1) \oplus \pi_3^{-1}(C_i^1) \oplus \pi_2^{-1}(P_j^2) \oplus \pi_3^{-1}(C_j^2) = \\
&= \pi_2^{-1}(P_i^1 \oplus P_j^2) \oplus \pi_3^{-1}(E_K(P_i^1) \oplus E_{\tilde{K}}(P_j^2)) = \\
&= \pi_2^{-1}(\pi_2(G^1(a_1^i) \oplus G^2(a_2^j))) \oplus \\
&\quad \oplus \pi_3^{-1}(E_K(P_i^1) \oplus E_{\tilde{K}}(P_i^1 \oplus \pi_2(G^1(a_1^i) \oplus G^2(a_2^j)))) = \\
&= \Delta L' \oplus \pi_3^{-1}(E_K(P_i^1) \oplus E_{\tilde{K}}(P_i^1 \oplus \pi_2(\Delta L'))),
\end{aligned}
$$

where $\Delta L' = G^1(a_1^i) \oplus G^2(a_2^i)$. Note that $\Delta L'$ can take all possible 2^{14} values (as a_1^i, a_2^j take all 2^7 values), and therefore for some particular i, j, it must coincide with ΔL. In such case, the difference in the plaintext is $\pi_2(\Delta L)$, and thus for the ciphertext we obtain

$$E_K(P_i^1) \oplus E_{\tilde{K}}(P_i^1 \oplus \pi_2(\Delta L)) = \pi_3(\Delta L)$$

Then $V_i^1 \oplus V_j^2 = \Delta L \oplus \pi_3^{-1}(\pi_3(\Delta L)) = 0$.

The possibility to create the sets independently and then to find a collision between them is the main idea of the membership test on CLEFIA-128. It works according to the following steps.

1. Choose at random a plaintext P.
2. Create a pool of 2^7 plaintexts $P_i^1 = \pi_2(P \oplus G^1(a_1^i))$ and ask for the corresponding ciphertext C_i^1 obtained with encryption under the first key, i.e. $C_i^1 = E_K(P_i^1)$. Compute the set V^1 composed of elements $V_i^1 = \pi_2^{-1}(P_i^1) \oplus \pi_3^{-1}(C_i^1)$.
3. Create a pool of 2^7 plaintexts $P_i^2 = \pi_2(P \oplus G^2(a_2^i))$ and ask for the corresponding ciphertext C_i^2 obtained with encryption under the second key, i.e. $C_i^2 = E_{\tilde{K}}(P_i^2)$. Compute the set V^2 composed of elements $V_i^2 = \pi_2^{-1}(P_i^2) \oplus \pi_3^{-1}(C_i^2)$.
4. Check for collisions between V^1 and V^2. If such a collision exists, then output that the examined cipher is CLEFIA-128. Otherwise, it is an ideal cipher.

The total data complexity of the membership test is $2^7 + 2^7 = 2^8$ plaintexts. The time complexity of each of the steps 2,3 is 2^7 encryptions, while the collision at step 4 can be found with 2^7 operations and 2^7 memory that is used to store one of the sets V^1 or V^2. Therefore, given a pair of keys from the weak-key class, we can distinguish CLEFIA-128 in 2^8 data, time and memory.

To confirm the correctness of the membership test, we implemented it for a small-scale variant of CLEFIA-128. Each word was shrunk to 8-bit value, thus the whole state became 32 bits. The Sbox from AES was taken as the round function F, and random 8-bit values were chosen as constants. The chunks in the linear function Σ were taken of size $5, 11$ (compared to the $7, 57$ in the original version). The expected size of the weak-key class in this toy version is 2^{10} (because $X = \Sigma^2(X)$ has 2^{10} solutions), while in practice we obtained $960 = 2^{9.9}$ solutions. For a random key pair chosen from this class, we were able to distinguish the cipher after 2^6 encryptions which confirms our findings to a large extent.

5   Analysis of the Hashing Modes of CLEFIA-128

In this section we analyze the impact of the weak-key class on hashing modes of CLEFIA-128. We show that compression functions built upon single-block-length modes instantiated with CLEFIA-128 exhibit non-random properties that come in a form of differential multicollisions. The analysis of hashing modes of a cipher is usually reduced to finding open-key distinguishers for the cipher. Note, open-key distinguishers come in a form of known-key (the adversary has the knowledge of the key, but cannot control it) and chosen-key (the adversary can choose the value of the key). Our analysis applies to the second case, i.e. we show non-randomness of the hashing modes of CLEFIA-128 when the adversary can control the key.

First, let us find a pair of keys (K_1, K_2) that belong to the weak-key class – we stress that the task is to find the pair explicitly, i.e. to produce the two values that compose a weak-key pair. From the previous analysis we have seen that a pair is a weak-key pair if for one of the 2^{14} values of ΔL defined previously: 1) the difference $\Delta K = K_1 \oplus K_2$ satisfies $\Delta K = \pi(\Delta L) \oplus \Sigma(\Delta L)$, and 2) the 12-round Feistel in the key schedule $GFN_{4,12}$ produces output difference ΔL, i.e. $GFN_{4,12}(K_1) \oplus GFN_{4,12}(K_2) = \Delta L$. The two conditions can be generalized as search for a pair that satisfies the differential $\pi(\Delta L) \oplus \Sigma(\Delta L) \to \Delta L$ through the 12-round Feistel in the key schedule.

Recall that the difference ΔL is an XOR of two elements (defined as $G^1(a_1)$ and $G^2(a_2)$) from sets of size 2^7, i.e. $\Delta L = G^1(a_1) \oplus G^2(a_2)$. Therefore we get that:

$$\Delta K = \pi(\Delta L) \oplus \Sigma(\Delta L) = \pi(G^1(a_1) \oplus G^2(a_2)) \oplus \Sigma(G^1(a_1) \oplus G^2(a_2)) =$$
$$= [\pi(G^1(a_1)) \oplus \Sigma(G^1(a_1))] \oplus [\pi(G^2(a_2)) \oplus \Sigma(G^2(a_2))] =$$
$$= T^1(a_1) \oplus T^2(a_2),$$

where $T^1(a_1) = \pi(G^1(a_1)) \oplus \Sigma(G^1(a_1)), T^2(a_2) = \pi(G^2(a_2)) \oplus \Sigma(G^2(a_2))$ are two linear functions (as π, Σ, G^1, G^2 are linear), and therefore the difference in the keys of a weak-key pair is an XOR of two sets as well. Using this fact, we can find a weak-key pair as follows:

1. Create a set $\Delta\mathcal{K}$ of 2^{14} values $T^1(a_1) \oplus T^2(a_2), a_1 = 0, \ldots, 2^7 - 1, a_2 = 0, \ldots, 2^7 - 1$.
2. Randomly choose a key K.
3. Create a set V_1 of 2^7 pairs

$$(K_1, K_1 \oplus \pi(GFN_{4,12}(K_1)) \oplus \Sigma(GFN_{4,12}(K_1))),$$

   where $K_1 = K \oplus T^1(a_1), a_1 = 0, \ldots, 2^7 - 1$. Index the set V_1 by the second elements.
4. Create a set V_2 of 2^7 pairs

$$(K_2, K_2 \oplus \pi(GFN_{4,12}(K_2)) \oplus \Sigma(GFN_{4,12}(K_2))),$$

   where $K_2 = K \oplus T^2(a_2), a_2 = 0, \ldots, 2^7 - 1$. Index V_2 as well by the second elements.
5. Check for collisions between V^1 and V^2 on the second (and indexed) elements. If such a collision exists, then confirm the key pair is weak by checking if the xor difference of the first elements belongs to $\Delta\mathcal{K}$. If so, then output that found pair (K_1, K_2) and exit. Otherwise, go to step 2.

The above algorithm will output a correct weak-key pair after repeating around 2^{114} times the steps 2-5. For each randomly chosen key K, there are 2^{14} pairs of keys (K_1, K_2) with difference $K_1 \oplus K_2 = K \oplus T^1(a_1) \oplus K \oplus T^2(a_2) = T^1(a_1) \oplus T^2(a_2) = \pi(\Delta L_i) \oplus \Sigma(\Delta L_i)$. If the output difference of 12-round Feistel is precisely the same ΔL_i (an event that happens with probability 2^{-128}), i.e. if $GFN_{4,12}(K_1) \oplus GFN_{4,12}(K_2) = \Delta L_i$, then

$$\pi(GFN_{4,12}(K_1) \oplus GFN_{4,12}(K_2)) \oplus \Sigma(GFN_{4,12}(K_1) \oplus GFN_{4,12}(K_2)) = \pi(\Delta L_i) \oplus \Sigma(\Delta L_i),$$

and therefore

$$K_1 \oplus K_2 = \pi(GFN_{4,12}(K_1) \oplus GFN_{4,12}(K_2)) \oplus \Sigma(GFN_{4,12}(K_1) \oplus GFN_{4,12}(K_2)),$$

which is equivalent to

$$K_1 \oplus \pi(GFN_{4,12}(K_1)) \oplus \Sigma(GFN_{4,12}(K_1)) = K_2 \oplus \pi(GFN_{4,12}(K_2)) \oplus \Sigma(GFN_{4,12}(K_2)).$$

Therefore a collision between V_1 and V_2 suggests a possible weak-key pair. The suggested pair is weak-key only if the input and the output differences satisfy the differential, thus with probability 2^{-128}. As we take 2^{114} random keys K, and for each there are 2^{14} pairs, with overwhelming probability, one will be a weak-key pair. To avoid false positives, we add step 1 and the additional checking at step 5, i.e. we make sure that the difference between the keys is $\pi(\Delta L_i) \oplus \Sigma(\Delta L_i)$ for some of the 2^{14} good values of ΔL_i. Hence, the algorithm will produce a weak-key pair in $2^{14} + 2^{114} \times 2 \times 2^7 \approx 2^{122}$ time and 2^{14} memory.

We can use the found pair to show weakness of CLEFIA-128 when used for cryptographic hashing. More precisely, we consider hashing based on single-block-length[8] modes, where a compression function is built from a block cipher. If the compression function uses CLEFIA-128 then we can find a pair of weak keys in 2^{122} time using the described algorithm. Once such pair (K_1, K_2) is found, we can produce any number of differential multicollisions [4] for any of the 12 modes investigated by Preneel et al. [12], including the popular Davies-Meyer, Matyas-Meyer-Oseas modes. For instance, for the Davies-Meyer mode, i.e. when the compression function $C(H, M)$ is defined as $C(H, M) = E_M(H) \oplus H$, the differential multicollisions have the form

$$C(H_i, K_1) \oplus C(H_i \oplus \pi_2(\Delta L), K_1 \oplus \pi(\Delta L) \oplus \Sigma(\Delta L)) =$$
$$= E_{K_1}(H_i) \oplus H_i \oplus E_{K_1 \oplus \pi(\Delta L) \oplus \Sigma(\Delta L)}(H_i \oplus \pi_2(\Delta L))) \oplus H_i \oplus \pi_2(\Delta L) =$$
$$= E_{K_1}(H_i) \oplus E_{K_1 \oplus \pi(\Delta L) \oplus \Sigma(\Delta L)}(H_i \oplus \pi_2(\Delta L)) \oplus \pi_2(\Delta L) =$$
$$= \pi_3(\Delta L) \oplus \pi_2(\Delta L),$$

for $i = 0, 1, \ldots$. Note that we do not need to call the compression functions as $C(H_i, K_1) \oplus C(H_i \oplus \pi_2(\Delta L), K_1 \oplus \pi(\Delta L) \oplus \Sigma(\Delta L)) = \pi_3(\Delta L) \oplus \pi_2(\Delta L)$ as long as $(K_1, K_1 \oplus \pi(\Delta L) \oplus \Sigma(\Delta L))$ form a weak-key pair. Consequently, we can produce an arbitrary number of differential multicollisions with the complexity 2^{122}. On the other hand, the proven lower bound (see [4]) in the case of ideal cipher is 2^{128}. A distinguisher for the hashing based on CLEFIA-128 has already been presented by Aoki at ISITA'12 [2]. It works in the framework of middletext distinguishers [11] (open-key version of the integral attack), where the adversary starts with a set of particularly chosen states in the middle of the cipher, then from them (and the knowledge of the key) produces the set of plaintexts and the set of ciphertexts, and finally shows that these two sets have some property that cannot be easily reproduced if the cipher was ideal. For CLEFIA-128, Aoki showed how to choose 2^{112} starting middle states that result in 17-round middletext distinguisher, and then added one more round where he used subkey guesses, to obtain the 18-round distinguisher. We want to point out that there is a substantial difference, between our result and that of Aoki. We do not fix the values neither of the plaintexts nor of the ciphertexts, and our analysis is applicable as long as the pair of chaining values has the required difference – the values can be arbitrary and even unknown.

6   Conclusion

The analysis of CLEFIA-128 presented in this paper shows existence of a weak-key class that consists of 2^{14} pairs of keys. We have shown how to exploit the pairs in two different scenarios: hashing mode of CLEFIA-128 and membership test for the weak-key class. In the hashing mode (or open-key mode in general)

---

[8] The state and key sizes in CLEFIA-128 coincide, thus we can construct only single-block-length compression functions.

we have shown that a weak-key pair can be found in around 2^{122} time, and such pair can be used to produce differential multicollisions faster than the generic 2^{128}. Furthermore, we have shown a membership test for the weak-key class that has 2^8 time and data complexity, compared to the generic 2^{14}. The main ideas of the analysis have been verified with computer experiments on small-scale variants of CLEFIA-128.

The analysis is invariant of three important security features that presumably increase the strength of a cipher. First, the non-linear part of the key schedule can be any random permutation (not necessarily a 12-round Feistel). Our analysis would still work as we do not need high probability differentials for this permutation. Second, the state update functions (in CLEFIA-128 F_0, F_1 are one round substitution-permutation networks) can be arbitrary functions or permutations, including several layers of SP – the difference never goes into them, hence, the probability of the characteristic in the state would stay 1. Finally, the number of rounds in CLEFIA-128 plays absolutely no role in our analysis – even if CLEFIA-128 had 1000 rounds, the complexity of the analysis would stay the same.

To prevent future analysis as ours, we have to clearly understand what are the main drawbacks of the design. The weak-key class and the three analysis invariances are results of these drawbacks (not their cause) and provide clues on what the actual cause might be. The invariance of the state update function is due to the Feistel structure of the cipher – this construction can lead to probability 1 characteristics as it can cancel round key and state differences. To maintain the cancellation through arbitrary number of rounds (invariance of the number of rounds), the round key differences have to be iterative. The key schedule prevents high probability iterative (or any fixed value) differences as they have to be produced from a difference in the key that goes initially through a 12-round Feistel modeled as random permutation. The Feistel, however, produces low probability (2^{-128}) differences (invariance of the random permutation), and 2^{14} of them become iterative round key differences due to the linear function used after the Feistel. That is, because of the linear function, with 2^{-128} we can have a special type of differences in 36 rounds keys (1152 bits !). Therefore, the analysis of CLEFIA-128 holds due to the Feistel structure of the cipher and the weak linear function that is used to produce the round keys.

To conclude, our work shows that *low probability differentials (around 2^{-k} for a cipher with k-bit key and n-bit state) for the key schedule of Feistel ciphers, cannot be used as a sole proof of resistance against related-key differential analysis.*. A safe upper bound on the probability of such differentials, which proves and provides security against related-key analysis, is not 2^{-k} but 2^{-2k-n} – this comes from the fact that there can be as many as 2^{2k} pairs of weak keys, and their combined probability should be below 2^{-n}.

References

1. RFC 6114. CLEFIA, http://www.rfc-editor.org/rfc/rfc6114.txt
2. Aoki, K.: A middletext distinguisher for full CLEFIA-128. In: ISITA, pp. 521–525. IEEE (2012)
3. Bellare, M., Kohno, T.: A theoretical treatment of related-key attacks: RKA-PRPs, RKA-PRFs, and applications. In: Biham, E. (ed.) EUROCRYPT 2003. LNCS, vol. 2656, pp. 491–506. Springer, Heidelberg (2003)
4. Biryukov, A., Khovratovich, D., Nikolić, I.: Distinguisher and related-key attack on the full AES-256. In: Halevi, S. (ed.) CRYPTO 2009. LNCS, vol. 5677, pp. 231–249. Springer, Heidelberg (2009)
5. Biryukov, A., Nikolić, I.: Complementing feistel ciphers. In: Moriai, S. (ed.) FSE 2013. LNCS, vol. 8424, pp. 3–18. Springer, Heidelberg (2014)
6. Bogdanov, A., Geng, H., Wang, M., Wen, L., Collard, B.: Zero-correlation linear cryptanalysis with FFT and improved attacks on ISO standards camellia and CLE-FIA. In: Lange, T., Lauter, K., Lisoněk, P. (eds.) SAC 2013. LNCS, vol. 8282, pp. 306–323. Springer, Heidelberg (2013)
7. Bogdanov, A., Knudsen, L.R., Leander, G., Paar, C., Poschmann, A., Robshaw, M.J.B., Seurin, Y., Vikkelsoe, C.: PRESENT: An ultra-lightweight block cipher. In: Paillier, P., Verbauwhede, I. (eds.) CHES 2007. LNCS, vol. 4727, pp. 450–466. Springer, Heidelberg (2007)
8. ISO/IEC 29192-2. Information technology - Security techniques - Lightweight cryptography - Part 2: Block ciphers, http://www.iso.org/iso/iso_catalogue/catalogue_tc/catalogue_detail.htm?csnumber=56552
9. Li, Y., Wu, W., Zhang, L.: Improved integral attacks on reduced-round CLEFIA block cipher. In: Jung, S., Yung, M. (eds.) WISA 2011. LNCS, vol. 7115, pp. 28–39. Springer, Heidelberg (2012)
10. Mala, H., Dakhilalian, M., Shakiba, M.: Impossible differential attacks on 13-round CLEFIA-128. J. Comput. Sci. Technol. 26(4), 744–750 (2011)
11. Minier, M., Phan, R.C.-W., Pousse, B.: Distinguishers for ciphers and known key attack against rijndael with large blocks. In: Preneel, B. (ed.) AFRICACRYPT 2009. LNCS, vol. 5580, pp. 60–76. Springer, Heidelberg (2009)
12. Preneel, B., Govaerts, R., Vandewalle, J.: Hash functions based on block ciphers: A synthetic approach. In: Stinson, D.R. (ed.) CRYPTO 1993. LNCS, vol. 773, pp. 368–378. Springer, Heidelberg (1994)
13. Shirai, T., Shibutani, K., Akishita, T., Moriai, S., Iwata, T.: The 128-bit block-cipher CLEFIA (Extended abstract). In: Biryukov, A. (ed.) FSE 2007. LNCS, vol. 4593, pp. 181–195. Springer, Heidelberg (2007)
14. Sun, B., Li, R., Wang, M., Li, P., Li, C.: Impossible differential cryptanalysis of CLEFIA. IACR Cryptology ePrint Archive, 2008:151 (2008)
15. Tang, X., Sun, B., Li, R., Li, C.: Impossible differential cryptanalysis of 13-round CLEFIA-128. Journal of Systems and Software 84(7), 1191–1196 (2011)
16. Tezcan, C.: The improbable differential attack: Cryptanalysis of reduced round CLEFIA. In: Gong, G., Gupta, K.C. (eds.) INDOCRYPT 2010. LNCS, vol. 6498, pp. 197–209. Springer, Heidelberg (2010)
17. Tsunoo, Y., Tsujihara, E., Shigeri, M., Saito, T., Suzaki, T., Kubo, H.: Impossible differential cryptanalysis of CLEFIA. In: Nyberg, K. (ed.) FSE 2008. LNCS, vol. 5086, pp. 398–411. Springer, Heidelberg (2008)

18. Wang, W., Wang, X.: Improved impossible differential cryptanalysis of CLEFIA. IACR Cryptology ePrint Archive, 2007:466 (2007)

19. Zhang, W., Han, J.: Impossible differential analysis of reduced round CLEFIA. In: Yung, M., Liu, P., Lin, D. (eds.) Inscrypt 2008. LNCS, vol. 5487, pp. 181–191. Springer, Heidelberg (2009)

A   Specification on `CLEFIA-128`

B   Analysis of `CLEFIA-128` with Whitening Keys

The whitening keys are the four words $WK_i, i = 0, 1, 2, 3$, defined as $WK_0 \| WK_1 \| WK_2 \| WK_3 = K$, i.e. they are the words of the master key K. The first two are XOR-ed to the second and the fourth plaintext words, and the remaining two to the second and the fourth ciphertext words (see Fig 3).

To index the whitening words, we define two linear functions on 128-bit words (or four 32-bit words). Assume X is 128-bit word, such that $X = a|b|c|d$, where a, b, c, d are 32-bit words. Then $l(X) : \{0, 1\}^{128} \to \{0, 1\}^{128}$ is defined as $l(X) = l(a|b|c|d) = 0|a|0|b$. Similarly $r(X) : \{0, 1\}^{128} \to \{0, 1\}^{128}$ is defined as $r(X) = r(a|b|c|d) = 0|c|0|d$.

Now we can easily specify the weak-key class:

- the key difference remains the same,
- the plaintext difference, instead of $\pi_2(\Delta L)$, should be $\pi_2(\Delta L) \oplus l(\Delta K)$,
- the ciphertext difference, instead of $\pi_3(\Delta L)$, should be $\pi_3(\Delta L) \oplus r(\Delta K)$.

As $\Delta K = \pi(\Delta L) \oplus \Sigma(\Delta L)$, it follows that the weak-key class for the original `CLEFIA-128` is defined as 2^{14} pairs of keys $(K, K \oplus \pi(\Delta L) \oplus \Sigma(\Delta L))$ such that for any plaintext P holds:

$$E_K(P) \oplus E_{K \oplus \pi(\Delta L) \oplus \Sigma(\Delta L)}(P \oplus \pi_2(\Delta L) \oplus l(\pi(\Delta L) \oplus \Sigma(\Delta L))) =$$
$$\pi_3(\Delta L) \oplus r(\pi(\Delta L) \oplus \Sigma(\Delta L)).$$

Let us focus on the membership test. We define the plaintexts pools as:

$$P_i^1 = P \oplus \pi_2(G^1(a_1^i)) \oplus l(T^1(a_1^i)), a_1^i = 0, 1, \ldots, 2^7 - 1,$$
$$P_i^2 = P \oplus \pi_2(G^2(a_2^i)) \oplus l(T^2(a_2^i)), a_2^i = 0, 1, \ldots, 2^7 - 1.$$

This way, the difference between each two plaintext from two different pools is $\pi_2(\Delta L') \oplus l(\Delta K)$, i.e. it is as required by the class.

To define the sets V^1, V^2 that lead to a collision, first we have to understand how a collision can occur. In the previous membership test (on `CLEFIA-128` without whitening keys), we used the trick that the difference in both the plaintext and the ciphertext is ΔL, but with permuted words (that is why we applied π_2^{-1}, π_3^{-1}). Here it is not the same: in the plaintext the difference is ΔL and two more words of ΔK, while in the ciphertext it is ΔL and the remaining two words of ΔK. Hence, XOR of these values does not trivially produce zero as the two words from l and the two from r are different.

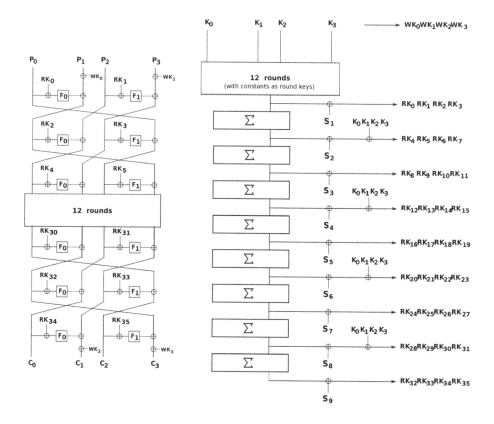

Fig. 3. The encryption function of CLEFIA-128 at the left, and the key schedule at the right. P_0, P_1, P_2, P_3 are 32-bit plaintext words, C_0, C_1, C_2, C_3 are the ciphertext words, K_0, K_1, K_2, K_3 are the key words, RK_i, WK_j are the round and whitening keys, respectively, and S_i are 128-bit constants. Finally, F_0, F_1 are the two state update functions, while Σ is a linear function (permutation).

Nevertheless, we can achieve collisions. Assume $\Delta L = a|b|c|d$. Then the difference Δ_P in the plaintext is

$$\Delta_P = \pi_2(a|b|c|d) \oplus l(\pi(a|b|c|d) \oplus \Sigma(a|b|c|d)) =$$
$$a|c|b|d \oplus l(b|a|d|c) \oplus l(\Sigma(a|b|c|d)) =$$
$$a|c + b|b|d + a \oplus l(\Sigma(a|b|c|d)).$$

Note, $l(\Sigma(a|b|c|d)$ has zeros at the first and at the third words.
Similarly, the difference Δ_C in the ciphertext is

$$\Delta_C = \pi_3(a|b|c|d) \oplus r(\pi(a|b|c|d) \oplus \Sigma(a|b|c|d)) =$$
$$c|b|d|a \oplus r(b|a|d|c) \oplus r(\Sigma(a|b|c|d)) =$$
$$c|b + d|d|a + c \oplus r(\Sigma(a|b|c|d)).$$

Again, in the sum r influences only the second and the fourth word.

Let us introduce a function f, that acts on the four 32-bit words of a 128-bit state and it XORs the first word to the fourth word, and the third word to the second word, i.e. $f(x|y|z|t) = (x|y+z|z|t+x)$. Then

$$f(\Delta_P) = a|c|b|d \oplus l(\Sigma(a|b|c|d)),$$
$$f(\Delta_C) = c|b|d|a \oplus r(\Sigma(a|b|c|d)).$$

The function Σ is linear and therefore $\Sigma(a|b|c|d) = \Sigma(a|0|0|0) + \Sigma(0|b|0|0) + \Sigma(0|0|c|0) + \Sigma(0|0|0|d)$. Let us denote these four values with Σ_a, Σ_b, Σ_c, and Σ_d. Furthermore, with superscripts we denote the four 32-bit words of Σ_x, e.g. Σ_a^2 is the second (most significant) word of Σ_a. This allows us to remove the functions l, r from the terms, and as a result we obtain

$$f(\Delta_P) = a|c + \Sigma_a^1 + \Sigma_b^1 + \Sigma_c^1 + \Sigma_d^1|b|d + \Sigma_a^2 + \Sigma_b^2 + \Sigma_c^2 + \Sigma_d^2,$$
$$f(\Delta_C) = c|b + \Sigma_a^3 + \Sigma_b^3 + \Sigma_c^3 + \Sigma_d^3|d|a + \Sigma_a^4 + \Sigma_b^4 + \Sigma_c^4 + \Sigma_d^4.$$

Next, we define a function $g(x|y|z|t)$ that from x, z computes $\Sigma_x^1, \ldots, \Sigma_x^4$, $\Sigma_z^1, \ldots, \Sigma_z^4$ and it adds Σ_x^4, Σ_z^4 to the first word, Σ_x^1, Σ_z^1 to the second, Σ_x^3, Σ_z^3 to the third, and Σ_x^2, Σ_z^2 to the fourth. Similarly, for Δ_C we define $h(x|y|z|t)$ that from x, z computes $\Sigma_x^1, \ldots, \Sigma_x^4$ and it adds Σ_x^1, Σ_z^1 to the first word, Σ_x^3, Σ_z^3 to the second, Σ_x^2, Σ_z^2 to the third, and Σ_x^4, Σ_z^4 to the fourth. Thus we get

$$g(f(\Delta_P)) = a + \Sigma_a^4 + \Sigma_b^4|c + \Sigma_c^1 + \Sigma_d^1|b + \Sigma_a^3 + \Sigma_b^3|d + \Sigma_c^2 + \Sigma_d^2,$$
$$h(f(\Delta_C)) = c + \Sigma_c^1 + \Sigma_d^1|b + \Sigma_a^3 + \Sigma_b^3|d + \Sigma_c^2 + \Sigma_d^2|a + \Sigma_a^4 + \Sigma_b^4.$$

Obviously $h(f(\Delta_C)) = \pi_4(g(f(\Delta_P)))$, where $\pi_4(0, 1, 2, 3) \rightarrow (3, 0, 1, 2)$. Therefore the sets V_1, V_2 are defined as:

$$V^1 = \{V_i^1 | V_i^1 = \pi_4(g(f(P_i^1))) \oplus h(f(C_i^1))\},$$
$$V^2 = \{V_i^2 | V_i^2 = \pi_4(g(f(P_i^2))) \oplus g(f(C_i^2))\},$$

and a collision between this two sets suggests that $\Delta L'$ coincides with ΔL. Thus the membership test for CLEFIA-128 with whitening keys has the same complexity as before (without whitening).

Automatic Security Evaluation and (Related-key) Differential Characteristic Search: Application to SIMON, PRESENT, LBlock, DES(L) and Other Bit-Oriented Block Ciphers[*]

Siwei Sun[1,2], Lei Hu[1,2], Peng Wang[1,2], Kexin Qiao[1,2], Xiaoshuang Ma[1,2], and Ling Song[1,2]

[1]State Key Laboratory of Information Security, Institute of Information Engineering, Chinese Academy of Sciences, Beijing 100093, China
[2]Data Assurance and Communication Security Research Center, Chinese Academy of Sciences, Beijing 100093, China
{sunsiwei,hulei,wpeng,qiaokexin,maxiaoshuang,songling}@iie.ac.cn

Abstract. We propose two systematic methods to describe the differential property of an S-box with linear inequalities based on logical condition modelling and computational geometry respectively. In one method, inequalities are generated according to some conditional differential properties of the S-box; in the other method, inequalities are extracted from the H-representation of the convex hull of all possible differential patterns of the S-box. For the second method, we develop a greedy algorithm for selecting a given number of inequalities from the convex hull. Using these inequalities combined with Mixed-integer Linear Programming (MILP) technique, we propose an automatic method for evaluating the security of bit-oriented block ciphers against the (related-key) differential attack with several techniques for obtaining tighter security bounds, and a new tool for finding (related-key) differential characteristics automatically for bit-oriented block ciphers.

Keywords: Automatic cryptanalysis, Related-key differential attack, Mixed-integer Linear Programming, Convex hull.

1 Introduction

Differential cryptanalysis [7] is one of the most well-known attacks on modern block ciphers, based on which many cryptanalytic techniques have been developed, such as truncated differential attack [34], impossible differential attack [9], and boomerang attack [51]. Providing a security evaluation with respect to the differential attack has become a basic requirement for a newly designed practical block cipher to be accepted by the cryptographic community.

---

[*] An extended version of this paper containing more applications and the source code is available at http://eprint.iacr.org/2013/676.

P. Sarkar and T. Iwata (Eds.): ASIACRYPT 2014, PART I, LNCS 8873, pp. 158–178, 2014.

Contrary to the single-key model, where methodologies for constructing block ciphers provably resistant to differential attacks are readily available, the understanding of the security of block ciphers with regard to related-key differential attacks is relatively limited. This limited understanding of the security concerning related-key differential attacks has been greatly improved in recent years for AES-like byte- or word-oriented SPN block ciphers. Along this line of research, two representative papers [10,25] were published in Eurocrypt 2010 and Crypto 2013. In the former paper [10], an efficient search tool for finding differential characteristics both in the state and in the key was presented, and the best differential characteristics were obtained for some byte-oriented block ciphers such as AES, byte-Camellia, and Khazad. In the latter paper [25], Pierre-Alain Fouque *et al.* showed that the full-round AES-128 can not be proven secure against differential attacks in the related-key model unless the exact coefficients of the MDS matrix and the S-Box differential properties are taken into account. Moreover, a variant of Dijkstra's shortest path algorithm for finding the most efficient related-key attacks on SPN ciphers was developed in [25]. In [27], Ivica Nikolic presented a tweak for the key schedule of AES and the new cipher called xAES is resistant against the related-key differential attacks found in AES.

For bit-oriented block ciphers such as PRESENT-80 and DES, Sareh Emami *et al.* proved that no related-key differential characteristic exists with probability higher than 2^{-64} for the *full-round* PRESENT-80, and therefore argue that PRESENT-80 is secure against basic related-key differential attacks [22]. In [48], Sun *et al.* obtained tighter security bounds for PRESENT-80 with respect to the related-key differential attacks using the Mixed-integer Linear Programming (MILP) technique. Alex Biryukov and Ivica Nikolić proposed two methods [11] based on Matsui's tool [42] for finding related-key differential characteristics for DES-like ciphers. For their methods, they stated that "... our approaches can be used as well to search for high probability related-key differential characteristics in any bit-oriented ciphers *with linear key schedule.*"

Sareh Emami *et al.* [22] and Sun *et al.*'s method [48] can not be used to search for actual (related-key) differential characteristics, and Alex Biryukov *et al.*'s method [11] is only applicable to ciphers with linear key schedule.

In this paper, we provide a method based on MILP which can not only evaluate the security (obtain security bound) of a block cipher with respect to the (related-key) differential attacks, but is also able to search for actual (related-key) differential characteristics even if the key schedule algorithm of the block cipher is nonlinear.

The problem of MILP is a class of optimization problems derived from Linear Programming in which the aim is to optimize an objective function under certain constraints. Despite its intimate relationship with discrete optimization problems, such as the set covering problem, 0-1 knapsack problem, and traveling salesman problem, it is only in recent years that MILP has been explicitly applied in cryptographic research [1,17,18,36,46,52,57].

In this paper, we are mainly concerned with the application of MILP method in the (related-key) differential cryptanalysis. A practical approach to evaluate the security of a cipher against differential attack is to determine the lower bound of the number of active S-boxes throughout the cipher. This strategy has been employed in many designs [4,8,15,16,19]. MILP was applied in automatically determining the lower bounds of the numbers of active S-boxes for some word-oriented symmetric-key ciphers, and therefore used to prove their security against differential cryptanalysis [14,44,54] . Laura Winnen [53] and Sun *et al.* [48] extended this method by making it applicable to ciphers involving bit-oriented operations. We notice that such MILP tools [14,44,48,54] for counting the minimum number of active S-boxes are also applied or mentioned in the design and analysis of some authenticated encryption schemes [8,20,21,29,30,31,55,58].

Our Contributions. We find that the constraints presented in [48] are too coarse to accurately describe the differential properties of a specific cipher, since there are a large number of invalid differential patterns of the cipher satisfying all these constraints, which yields a feasible region of the MILP problem much larger than the set of all valid differential characteristics.

In this paper, we propose two methods to tighten the feasible region by cutting off some impossible differential patterns of a specific S-box with linear inequalities: one method is based on logical condition modeling, and the other is a more general approach based on convex hull computation — a fundamental algorithmic problem in computational geometry.

However, the second approach produces too many inequalities so that adding all of them to an MILP problem will make the solving process impractical. Therefore, we develop a greedy algorithm for selecting a given number of linear inequalities from the convex hull.

By adding all or a part of the constraints generated by these methods, we provide MILP based methods for evaluating the security of a block cipher with respect to the (related-key) differential attack, and searching for actual (related-key) differential characteristics. Using these methods, we obtain the following results.

1. The probability of the best related-key differential characteristic of the 24-round PRESENT-80 is upper bounded by 2^{-64}, which is the tightest security bound obtained so far for PRESENT-80.
2. The probability of the best related-key differential characteristic for the full-round LBlock is at most 2^{-60}.
3. We obtain a single-key *differential characteristic* and a single-key *differential* for the 15-round SIMON48 (a lightweight block cipher designed by the U.S. National Security Agency) with probability 2^{-46} and $2^{-41.96}$ respectively, which are the best results published so far for SIMON48.
4. We obtain a 14-round related-key differential characteristic of LBlock with probability 2^{-49} *in no more than 4 hours on a PC*. Note that the probabilities of the best previously published related-key characteristics covering the 13- and 14-round LBlock are 2^{-53} and 2^{-65} [56], respectively.
5. We obtain an 8-round related-key differential characteristic of DESL with probability $2^{-34.78}$ *in 10 minutes on a PC*. To the best of our knowledge, no

related-key differential characteristic covering more than 7 rounds of DESL has been published before.

6. We obtain a 7-round related-key characteristic for PRESENT-128 with probability 2^{-11} and 0 active S-box in its key schedule algorithm, based on which an improved related-key boomerang distinguisher for the 14-round PRESENT-128 and a key-recovery attack on the 17-round PRESENT-128 can be constructed by using exactly the same method presented in [47].

The method presented in this paper is generic, automatic, and applicable to other lightweight ciphers with bit-oriented operations. Due to the page limit, the concrete results concerning the related-key or single-key differential characteristics for LBlock, PRESENT-128, and DES(L) are put into an extended version of this paper available at http://eprint.iacr.org/2013/676.

Organization of the Paper. In Sect. 2, we introduce Mouha *et al.*'s framework and its extension for counting the number of active S-boxes of bit-oriented ciphers automatically with the MILP technique. In Sect. 3, we introduce the concept of valid cutting-off inequalities for tightening the feasible region of an MILP problem, and explore how to generate and select valid cutting-off inequalities. We present the methods for automatic security evaluation with respect to the (related-key) differential attack, and searching for (related-key) differential characteristics in Sect. 4 and Sect. 5. In Sect. 6 we conclude the paper and propose some research directions for bit-oriented ciphers and the application of the MILP technique in cryptography. The application of the methods presented in this paper to PRESENT, LBlock, and SIMON is given in Appendices.

2   Mouha *et al.*'s Framework and Its Extension

2.1   Mouha *et al.*'s Framework for Word-Oriented Block Ciphers

Assume a cipher is composed of the following three word-oriented operations, where ω is the word size:

- XOR, $\oplus : \mathbb{F}_2^\omega \times \mathbb{F}_2^\omega \to \mathbb{F}_2^\omega$;
- Linear transformation $L : \mathbb{F}_{2^\omega}^m \to \mathbb{F}_{2^\omega}^m$ with branch number $\mathcal{B}_L$;
- S-box, $\mathcal{S} : \mathbb{F}_2^\omega \to \mathbb{F}_2^\omega$.

Mouha *et al.*'s framework uses 0-1 variables, which are subjected to certain constraints imposed by the above operations, to denote the word level differences propagating through the cipher (1 for nonzero difference and 0 otherwise).

Firstly, we should include the constraints imposed by the operations of the cipher.

Constraints Imposed by XOR Operations. Suppose $a \oplus b = c$, where a, b, $c \in \mathbb{F}_2^\omega$ are the input and output differences of the XOR operation, the following constraints will make sure that when a, b, and c are not all zero, then there are at least two of them are nonzero:

$$\begin{cases} a + b + c \geq 2d_\oplus \\ d_\oplus \geq a, \ d_\oplus \geq b, \ d_\oplus \geq c \end{cases} \tag{1}$$

where $d_{\oplus}$ is a dummy variable taking values from $\{0, 1\}$. If each one of a, b, and c represents one bit, we should also add the inequality $a + b + c \leq 2$.

Constraints Imposed by Linear Transformation. Let x_{i_k} and $y_{j_k}, k \in \{0, 1, \ldots, m - 1\}$, be 0-1 variables denoting the word-level input and output differences of the linear transformation L respectively. Since for nonzero input differences, there are totally at least $\mathcal{B}_L$ nonzero ω-bit words in the input and output differences, we include the following constraints:

$$\begin{cases} \sum_{k=0}^{m-1} (x_{i_k} + y_{j_k}) \geq \mathcal{B}_L d_L \\ d_L \geq x_{i_k}, d_L \geq y_{j_k}, \quad k \in \{0, \ldots, m - 1\} \end{cases} \tag{2}$$

where d_L is a dummy variable taking values in $\{0, 1\}$ and $\mathcal{B}_L$ is the branch number of the linear transformation.

Then, we set up the **objective function** to be the sum of all variables representing the input words of the S-boxes.

2.2 Extension of Mouha *et al.*'s Framework for Bit-Oriented Ciphers

For bit-oriented ciphers, bit-level representations and additional constraints are needed [48]. For every input and output bit-level difference, a new 0-1 variable x_i is introduced such that $x_i = 1$ if and only if the difference at this bit is nonzero.

For every S-box in the schematic diagram, including the encryption process and the key schedule algorithm, we introduce a new 0-1 variable A_j such that $A_j = 1$ if the input word of the Sbox is nonzero and $A_j = 0$ otherwise.

At this point, it is natural to choose the objective function f, which will be minimized, as $\sum A_j$ for the goal of determining a lower bound of the number of active S-boxes.

For bit-oriented ciphers, we need to include two sets of constraints. The first one is the set of constraints imposed by XOR operations, and the other is due to the S-box operation. After changing the representations to bit-level, the set of constraints imposed by XOR operations for bit-oriented ciphers are the same as that presented in (1). The S-box operation is more tricky.

Constraints Describing the S-box Operation. Suppose $(x_{i_0}, \ldots, x_{i_{\omega-1}})$ and $(y_{j_0}, \ldots, y_{j_{\nu-1}})$ are the input and output bit-level differences of an $\omega \times \nu$ S-box marked by A_t. Firstly, to ensure that $A_t = 1$ holds if and only if $x_{i_0}, \ldots, x_{i_{\omega-1}}$ are not all zero, we require that:

$$\begin{cases} A_t - x_{i_k} \geq 0, \quad k \in \{0, \ldots, \omega - 1\} \\ x_{i_0} + x_{i_1} + \cdots + x_{i_{\omega-1}} - A_t \geq 0 \end{cases} \tag{3}$$

For bijective S-boxes, nonzero input difference must result in nonzero output difference and vice versa:

$$\begin{cases} \omega y_{j_0} + \omega y_{j_1} + \cdots + \omega y_{j_{\nu-1}} - (x_{i_0} + x_{i_1} + \cdots + x_{i_{\omega-1}}) \geq 0 \\ \nu x_{i_0} + \nu x_{i_1} + \cdots + \nu x_{i_{\omega-1}} - (y_{j_0} + y_{j_1} + \cdots + y_{j_{\nu-1}}) \geq 0 \end{cases} \tag{4}$$

Note that the above constraints should not be used for non-bijective S-box such as the S-box of DES(L) [37].

Finally, the Hamming weight of the $(\omega+\nu)$-bit word $x_{i_0}\cdots x_{i_{\omega-1}}y_{j_0}\cdots y_{j_{\nu-1}}$ is lower bounded by the branch number $\mathcal{B}_{\mathcal{S}}$ of the S-box for nonzero input difference $x_{i_0}\cdots x_{i_{\omega-1}}$, where $d_{\mathcal{S}}$ is a dummy variable:

$$\begin{cases} \sum_{k=0}^{\omega-1} x_{i_k} + \sum_{k=0}^{\nu-1} y_{j_k} \geq \mathcal{B}_{\mathcal{S}} d_{\mathcal{S}} \\ d_{\mathcal{S}} \geq x_{i_k}, d_{\mathcal{S}} \geq y_{j_t}, \quad k \in \{0,\ldots,\omega-1\}, \quad t \in \{0,\ldots,\nu-1\} \end{cases} \tag{5}$$

where the branch number $\mathcal{B}_{\mathcal{S}}$ of an S-box $\mathcal{S}$, is defined as $\mathcal{B}_{\mathcal{S}} = \min_{a \neq b}\{\mathrm{wt}((a \oplus b)\|(\mathcal{S}(a) \oplus \mathcal{S}(b)) : a, b \in \mathbb{F}_2^{\omega}\}$, and $\mathrm{wt}(\cdot)$ is the standard Hamming weight of an $(\omega+\nu)$-bit word. We point out that constraint (5) is redundant for an invertible S-box with branch number $\mathcal{B}_{\mathcal{S}} = 2$, since in this particular case, all differential patterns not satisfying (5) violate (4).

0-1 Variables. The MILP model proposed above is indeed a Pure Integer Programming Problem since all variables appearing are 0-1 variables. However, in practice we only need to explicitly restrict a part of all variables to be 0-1, while all other variables can be allowed to be any real numbers, which leads to an MILP problem. Following this approach, the MILP solving process may be accelerated as suggested in [17].

3   Tighten the Feasible Region with Valid Cutting-off Inequalities

The feasible region of an MILP problem is defined as the set of all variable assignments satisfying all constraints in the MILP problem. The modelling process presented in the previous sections indicates that every differential path corresponds to a solution in the feasible region of the MILP problem. However, a feasible solution of the MILP model is not guaranteed to be a valid differential path, since our constraints are far from perfect to rule out all invalid differential patterns. For instance, assume x_i and y_i ($0 \leq i \leq 3$) are the bit-level input and output differences of the PRESENT-80 S-box. According to Sect. 2.2, x_i, y_i are subjected to the constraints of (3), (4) and (5). Obviously, $(x_0 \cdots, x_3, y_0, \cdots, y_3) = (1, 0, 0, 1, 1, 0, 1, 1)$ satisfies the above constraints, whereas $0x9 = 1001 \to 0xB = 1011$ is not a valid difference propagation pattern for the PRESENT S-box, which can be seen from the differential distribution table of the PRESENT S-box. Hence, we are actually trying to minimize the number of the active S-boxes over a larger region, and the optimum value obtained in this setting must be smaller than or equal to the actual minimum number of active S-boxes. Although the above fact will not invalidate the lower bound we obtained from our MILP model, this prevents the designers or analysts from obtaining tighter security bounds and valid (related-key) differential characteristics from the feasible region.

The situation would be even worse when modelling an invertible S-box with branch number $\mathcal{B}_S = 2$, which is the minimal value of the branch number for an invertible S-box. In the case of invertible S-box with $\mathcal{B}_S = 2$, the constraints of (3), (4) are enough, and (5) is redundant.

Therefore, we are motivated to look for linear inequalities which can cut off some part of the feasible region of the MILP model while leaving the region of valid differential characteristics intact. For the convenience of discussion, we give the following definition.

Definition 1. *A valid cutting-off inequality is a linear inequality which is satisfied by all possible valid differential patterns, but is violated by at least one feasible solution corresponding to an impossible differential pattern in the feasible region of the original MILP problem.*

3.1 Methods for Generating Valid Cutting-Off Inequalities

In this section, we present two methods for generating valid cutting-off inequalities by analyzing the differential behavior of the underlying S-box.

Modelling Conditional Differential Behaviour. In building integer programming models in practice, sometimes it is possible to model certain logical constraints as linear inequalities. For example, assume x is a continuous variable such that $0 \le x \le M$, where M is a fixed integer, and we know that δ is a 0-1 variable taking value 1 when $x > 0$, that is $x > 0 \;\Rightarrow\; \delta = 1$. It is easy to verify that the above logical condition can be achieved by imposing the constraint $x - M\delta \le 0$.

In fact, there is a surprisingly large number of different types of logical conditions can be imposed in a similar way. We now give a theorem which will be used in the following.

Theorem 1. *If we assume that all variables are 0-1 variables, then the logical condition that $(x_0, \ldots, x_{m-1}) = (\delta_0, \ldots, \delta_{m-1}) \in \{0,1\}^m \subseteq \mathbb{Z}^m$ implies $y = \delta \in \{0,1\} \subseteq \mathbb{Z}$ can be described by the following linear inequality*

$$\sum_{i=0}^{m-1} (-1)^{\delta_i} x_i + (-1)^{\delta+1} y - \delta + \sum_{i=0}^{m-1} \delta_i \ge 0, \tag{6}$$

where δ_i, δ are fixed constants and $\mathbb{Z}$ is the set of all integers.

Proof. We only prove the Theorem for the case $\delta = 0$. For $\delta = 1$, it can be proved in a similar way. We assume

$$(\delta_0, \ldots, \delta_{m-1}) = (\delta_0, \ldots, \delta_{s_1-1}; \delta_{s_1}, \ldots, \delta_{m-1}) = (1, 1, \ldots, 1; 0, 0, \ldots, 0) = \Delta^*.$$

For other 0-1 patterns, it can be permuted into such a form and this will not affect our proof.

Firstly, $(\Delta^*, 0)$ is satisfied by (6), which can be verified directly.

Secondly, we prove that all vectors $(x_0, \ldots, x_{m-1}, y) \in \{0,1\}^{m+1}$ such that $(x_0, \ldots, x_{m-1}) \neq \Delta^*$ are satisfied by (6). In such cases, we have

$$\sum_{i=0}^{m-1} (-1)^{\delta_i} x_i + (-1)^{\delta+1} y - \delta + \sum_{i=0}^{m-1} \delta_i = -\sum_{i=0}^{s_1-1} x_i + \sum_{i=s_1}^{m} x_i - y - 0 + s_1 \geq 0,$$

for $y = 0$ or $y = 1$.

Finally we prove that the vector $(x_0, \ldots, x_{m-1}, y) = (\Delta^*, 1)$ is not satisfied by the linear inequality. In such case, we have

$$\sum_{i=0}^{m-1} (-1)^{\delta_i} x_i + (-1)^{\delta+1} y - \delta + \sum_{i=0}^{m-1} \delta_i = -\sum_{i=0}^{s_1-1} x_i + \sum_{i=s_1}^{m} x_i - 1 - 0 + s_1 < 0.$$

The proof is completed.

For example, the PRESENT S-box has the following conditional differential [26,32,38,33,18] properties, which are referred to as undisturbed bits in [50].

Fact 1. *The S-box of PRESENT-80 has the following properties:*

(i) 1001→***0: If the input difference of the S-box is $0x9 = 1001$, then the least significant bit of the output difference must be 0;

(ii) 0001→***1 and 1000→***1: If the input difference of the S-box is $0x1 = 0001$ or $0x8 = 1000$, then the least significant bit of the output difference must be 1;

(iii) ***1→0001 and ***1→0100: If the output difference of the S-box is $0x1 = 0001$ or $0x4 = 0100$, then the least significant bit of the input difference must be 1; and

(iv) ***0→0101: If the output difference of the S-box is $0x5 = 0101$, then the least significant bit of the input difference must be 0.

From Theorem 1, we have the following fact.

Fact 2. *Let 0-1 variables (x_0, x_1, x_2, x_3) and (y_0, y_1, y_2, y_3) represent the input and output bit-level differences of the S-box respectively, where x_3 and y_3 are the least significant bits. Then the logical conditions in Theorem 1 can be described by the following linear inequalities:*

$$-x_0 + x_1 + x_2 - x_3 - y_3 + 2 \geq 0 \tag{7}$$

$$\begin{cases} x_0 + x_1 + x_2 - x_3 + y_3 \geq 0 \\ -x_0 + x_1 + x_2 + x_3 + y_3 \geq 0 \end{cases} \tag{8}$$

$$\begin{cases} x_3 + y_0 + y_1 + y_2 - y_3 \geq 0 \\ x_3 + y_0 - y_1 + y_2 + y_3 \geq 0 \end{cases} \tag{9}$$

$$-x_3 + y_0 - y_1 + y_2 - y_3 + 2 \geq 0 \tag{10}$$

For example, the linear inequality (7) removes all differential patterns of the form $(x_0, \ldots, x_3, y_0, \ldots, y_3) = (1, 0, 0, 1, *, *, *, 1)$, where $(x_0, \ldots, x_3)$ and $(y_0, \ldots, y_3)$ are the input and output differences of the PRESENT S-box respectively. We call this group of constraints presented in (7), (8), (9), and (10) the constraints of conditional differential propagation (CDP constraints for short). The CDP constraints obtained from Fact 1 and the differential patterns removed by these CDP constraints are given in Table 1.

Table 1. Impossible differential patterns removed by the CDP constraints generated according to the differential properties of the PRESENT S-box. Here, a vector $(\lambda_0, \ldots, \lambda_3, \gamma_0, \ldots, \gamma_3, \theta)$ in the left column denotes a linear inequality $\lambda_0 x_0 + \cdots + \lambda_3 x_3 + \gamma_0 y_0 + \cdots + \gamma_3 y_3 + \theta \geq 0$.

Constraints obtained by logical condition modelling	Impossible differential patterns removed
$(-1, 1, 1, -1, 0, 0, 0, -1, 2)$	(1, 0, 0, 1, 0, 0, 0, 1), (1, 0, 0, 1, 0, 0, 1, 1), (1, 0, 0, 1, 0, 1, 0, 1), (1, 0, 0, 1, 0, 1, 1, 1),
	(1, 0, 0, 1, 1, 0, 0, 1), (1, 0, 0, 1, 1, 0, 1, 1), (1, 0, 0, 1, 1, 1, 0, 1), (1, 0, 0, 1, 1, 1, 1, 1)
$(1, 1, 1, -1, 0, 0, 0, 1, 0)$	(0, 0, 0, 1, 0, 0, 0, 0), (0, 0, 0, 1, 0, 0, 1, 0), (0, 0, 0, 1, 0, 1, 0, 0), (0, 0, 0, 1, 0, 1, 1, 0),
	(0, 0, 0, 1, 1, 0, 0, 0), (0, 0, 0, 1, 1, 0, 1, 0), (0, 0, 0, 1, 1, 1, 0, 0), (0, 0, 0, 1, 1, 1, 1, 0)
$(-1, 1, 1, 1, 0, 0, 0, 1, 0)$	(1, 0, 0, 0, 0, 0, 0, 0), (1, 0, 0, 0, 0, 0, 1, 0), (1, 0, 0, 0, 0, 1, 0, 0), (1, 0, 0, 0, 0, 1, 1, 0),
	(1, 0, 0, 0, 1, 0, 0, 0), (1, 0, 0, 0, 1, 0, 1, 0), (1, 0, 0, 0, 1, 1, 0, 0), (1, 0, 0, 0, 1, 1, 1, 0)
$(0, 0, 0, 1, 1, 1, 1, -1, 0)$	(0, 0, 0, 0, 0, 0, 0, 1), (0, 0, 1, 0, 0, 0, 0, 1), (0, 1, 0, 0, 0, 0, 0, 1), (0, 1, 1, 0, 0, 0, 0, 1),
	(1, 0, 0, 0, 0, 0, 0, 1), (1, 0, 1, 0, 0, 0, 0, 1), (1, 1, 0, 0, 0, 0, 0, 1), (1, 1, 1, 0, 0, 0, 0, 1)
$(0, 0, 0, 1, 1, -1, 1, 1, 0)$	(0, 0, 0, 0, 0, 1, 0, 0), (0, 0, 1, 0, 0, 1, 0, 0), (0, 1, 0, 0, 0, 1, 0, 0), (0, 1, 1, 0, 0, 1, 0, 0),
	(1, 0, 0, 0, 0, 1, 0, 0), (1, 0, 1, 0, 0, 1, 0, 0), (1, 1, 0, 0, 0, 1, 0, 0), (1, 1, 1, 0, 0, 1, 0, 0)
$(0, 0, 0, -1, 1, 1, -1, 1, -1, 2)$	(0, 0, 0, 1, 0, 1, 0, 1), (0, 0, 1, 1, 0, 1, 0, 1), (0, 1, 0, 1, 0, 1, 0, 1), (0, 1, 1, 1, 0, 1, 0, 1),
	(1, 0, 0, 1, 0, 1, 0, 1), (1, 0, 1, 1, 0, 1, 0, 1), (1, 1, 0, 1, 0, 1, 0, 1), (1, 1, 1, 1, 0, 1, 0, 1)

However, there are cases where no such conditional differential property exists. For example, two out of the eight S-boxes of Serpent [6] exhibit no such property. Even when the S-box under consideration can be described with this logical condition modelling technique, the inequalities generated may be not enough to produce a satisfied result. In the following, a more general approach for generating valid cutting-off inequalities is proposed.

Convex Hull of All Possible Differentials for an S-box. The convex hull of a set Q of discrete points in $\mathbb{R}^n$ is the smallest convex set that contains Q. A convex hull in $\mathbb{R}^n$ can be described as the common solutions of a set of finitely many linear (in)equalities as follows:

$$\begin{cases} \lambda_{0,0} x_0 + \cdots + \lambda_{0,n-1} x_{n-1} + \lambda_{0,n} \geq 0 \\ \qquad \cdots \\ \gamma_{0,0} x_0 + \cdots + \gamma_{0,n-1} x_{n-1} + \gamma_{0,n} = 0 \\ \qquad \cdots \end{cases} \tag{11}$$

This is called the H-Representation of a convex hull. Computing the H-representation of the convex hull of a set of finitely many points is a fundamental algorithm in computation geometry with many applications.

If we treat a differential of an $\omega \times \nu$ S-box as a point in $\mathbb{R}^{\omega+\nu}$, then we can get a set of finitely many discrete points which includes all possible differential patterns of this S-box . For example, one possible differential pattern of PRESENT S-box is $0x9 = 1001 \to 0xE = 1110$ which is identified with $(1, 0, 0, 1, 1, 1, 1, 0)$. The set of all possible differential patterns for the S-boxes are essentially sets of finitely many discrete points in high dimensional space, hence we can compute their convex hulls by standard method in computational geometry.

We now define the convex hull of a specific $\omega \times \nu$ S-box to be the set of all linear (in)equalities in the H-Representation of the convex hull $\mathcal{V}_S \subseteq \mathbb{R}^{\omega+\nu}$ of all possible differential patterns of the S-box. The convex hull of a specific S-box can be obtained by using the inequality_generator() function in the sage.geometry.polyhedron class of the SAGE computer algebra system [49]. The convex hull of the PRESENT S-box contains 327 linear inequalities. Any one of these inequalities can be taken as a valid cutting-off inequality.

3.2   Selecting Valid Cutting-off Inequalities from the Convex Hull: A Greedy Approach

The number of (in)equalities in the H-Representation of a convex hull computed from a set of discrete points in n dimensional space is very large in general. For instance, the convex hull $\mathcal{V}_S \subseteq \mathbb{R}^8$ of a 4×4 S-box typically involves several hundreds of linear inequalities. Adding all of them to an MILP problem will make the MILP problem insolvable in practical time. Hence, it is necessary to select a small number, say n, of "best" inequalities from the convex hull. Here by "best" we mean that, among all possible selections of n inequalities, the selected ones maximize the number of removed impossible differentials. Obviously, this is a hard combinatorial optimization problem. Therefore, we design a greedy algorithm, listed in Algorithm 1, to approximate the optimum selection.

This algorithm builds up a set of valid cutting-off inequalities by selecting at each step an inequality from the convex hull which maximizes the number of removed impossible differential patterns from the current feasible region. For instance, We select 6 valid cutting-off inequalities from the convex hull of the PRESENT S-box using Algorithm 1. Compared with the 6 valid cutting-off inequalities obtained by Theorem 1 (see Table 1), they cut off 24 more impossible differential patterns, which leads to a relatively tighter feasible region.

Algorithm 1. Selecting n inequalities from the convex hull $\mathcal{H}$ of an S-box

---

    Input: $\mathcal{H}$: the set of all inequalities in the H-representation of the
                convex hull of an S-box; $\mathcal{X}$: the set of all impossible differential
                patterns of an S-box; n: a positive integer.
    Output: $\mathcal{O}$: a set of n inequalities selected from $\mathcal{H}$

1 $l^* :=$ None; $\mathcal{X}^* := \mathcal{X}$; $\mathcal{H}^* := \mathcal{H}$; $\mathcal{O} := \emptyset$;
2 **for** $i \in \{0, \ldots, n-1\}$ **do**
3        $l^* :=$ The inequality in $\mathcal{H}^*$ which maximizes the number of removed
         impossible differential patterns from $\mathcal{X}^*$;
4        $\mathcal{X}^* := \mathcal{X}^* - \{$removed impossible differential patterns by $l^*\}$;
5        $\mathcal{H}^* := \mathcal{H}^* - \{l^*\}$; $\mathcal{O} := \mathcal{O} \cup \{l^*\}$;
6 **end**
7 return $\mathcal{O}$

---

4 Automatic Security Evaluation

To obtain the security bound of a block cipher with respect to related-key differential attack, we can build an MILP model according to Sect. 2 with the constraints introduced in Sect. 3.1 and Sect. 3.2 included. Then we solve the MILP model using any MILP optimizer, and the optimized solution, say N, is the minimum number of the active S-boxes from which we can deduce that the probability of the best differential characteristic is upper bounded by ϵ^N, where ϵ is the maximum differential probability (MDP) a single S-box.

However, it is computationally infeasible to solve an MILP model generated by an r-round block cipher with large r. In such case, we can turn to the so called simple split approach. We split the r-round block cipher into two parts with consecutive r_1 rounds and r_2 rounds such that $r_1 + r_2 = r$. Then we apply our method to these two parts. Assuming that there are at least N_{r_1} and N_{r_2} active S-boxes in the first and second part respectively, we can deduce that the probability of the best differential characteristic for this r-round cipher is upper bounded by $\epsilon^{(N_{r_1}+N_{r_2})}$. If r_1 and r_2 are still too large, they can be divided into smaller parts further. Note that our method is applicable to both the single-key and related-key models.

4.1 Techniques for Getting Tighter Security Bounds

Technique 1. In the above analysis, we pessimistically (in the sense that we want to prove the security of a cipher) assume that all the active S-boxes take the MDP ϵ. However, this is unlikely to happen in practice, especially in the case that the number of active S-boxes is minimized. Therefore, we have the following strategy for obtaining tighter security bound for a t-round characteristic.

Firstly, compute the set $\mathcal{E}$ of all the differential patterns of an S-box with probabilities greater than or equal to the S-box's MDP ϵ.

Secondly, compute the H-representation $H_{\mathcal{E}}$ of the convex hull of $\mathcal{E}$, and then use the inequalities selected from $H_{\mathcal{E}}$ by Algorithm 1 to generate a t-round model according to Sect. 2 and Sect. 3. Note that the feasible region of this model is smaller than that of a t-round model generated in standard way, since the differential patterns allowed to take in this model is more restrictive. Hence, we hope to get a larger objective value than N_t, which is the result obtained by using the standard t-round model.

Finally, solve the model using a software optimizer. If the objective value is greater than N_t, we know that there is no differential characteristic with only N_t active S-boxes such that all these S-boxes take differential patterns with probability ϵ. And hence, we can conclude that there is at least one active S-box taking a differential pattern with probability less than ϵ in a t-round characteristic with only N_t active S-boxes.

Technique 2. Yet another technique for obtaining tighter security bound is inspired by Alex Biryukov *et al.* and Sareh Emami *et al.*'s (extended) split approach [11,22]. In Sun *et al.*'s work [48], the strategy for proving the security of an n-round iterative cipher against the related-key differential attacks is to use the simple split approach. By employing the MILP technique, compute the minimum number N_t of differentially active S-boxes for any consecutive t-round $(1 \leq t \leq n)$ related-key differential characteristic. Then the lower bound of the number of active S-boxes for the full cipher (n-round) can be obtained by computing $\sum_{j \in I \subseteq \{1,2,\dots\}} N_{t_j}$, where $\sum_{j \in I} t_j = n$. Note that the computational cost is too high to compute N_n directly.

We point out that this simple "split strategy" can be improved to obtain tighter security bound by exploiting more information of a differential characteristic. The main idea is that the characteristic covering round 1 to round m and the characteristic covering round $m+1$ to round $2m$ should not be treated equal although they have the same number of rounds, since the starting difference of a characteristic of round $m+1$ to $2m$ is not as free as that of a characteristic of round 1 to round m. Therefore, we have the following strategy.

Firstly, split an r-round into two parts: round 1 to round r_1, and round $r_1 + 1$ to round $r = r_1 + r_2$.

Secondly, construct an MILP model covering round 1 to round r. Change the objective function to be the sum of all S-boxes covering round $r_1 + 1$ to round r. Add some additional constraints on the number of active S-boxes covering round 1 to round r_1 (One way to obtain such constraints is to solve the model covering round 1 to round r_1).

Finally, solve the model using any software optimizer, and the result is the lower bound of the number of active S-boxes of round $r_1 + 1$ to round r (r_2 rounds in total) for any characteristic covering round 1 to round r.

We have applied the methods presented in this section to PRESENT-80 and LBlock, and the results are given in Appendix A.

5   A Heuristic Method for Finding (Related-key) Differential Characteristics Automatically

To find a (related-key) differential characteristic with relatively high probability covering r rounds of a cipher is the most important step in (related-key) differential cryptanalysis. Most of the tools for searching differential characteristics are essentially based on Matsui's algorithm [42]. In this section, we propose an MILP based heuristic method for finding (related-key) differential characteristics. Compared to other methods, our method is easier to implement, and more flexible.

Thanks to the valid cutting-off inequalities which can describe the property of an S-box according to its differential distribution table, our method can output a good (related-key) differential characteristic directly by employing the MILP technique. The procedure of our method is outlined as follows.

Step 1. For every S-box $\mathcal{S}$, select n inequalities from the convex hull of the set of all possible differential patterns of $\mathcal{S}$ using Algorithm 1, and generate an r-round MILP model in which we require that *all variables involved are 0-1*.

Step 2. Extract a feasible solution of the MILP model by using the Gurobi [45] optimizer.

Step 3. Check whether the feasible solution is a valid (related-key) differential characteristic. If it is a valid characteristic, the procedure terminates. Otherwise, go to step 1, increase the number of selected inequalities from the convex hulls, and repeat the whole process.

We have developed a software by employing the python interface provided by the Gurobi optimizer, which automates the whole process of the above method.

To demonstrate the practicability of our method, we have applied the methods presented in this section to SIMON and the results are given in Appendix B.

On the Quality of the Characteristics. The characteristics found by this method are not guaranteed to be the best. However, if you would like to wait until the optimizer outputs optimum solution, the characteristic found by this method is guaranteed to have the minimum number of active S-boxes. Experimental results show that we get reasonably good solutions.

On the Flexibility of the Searching Algorithm. By adding a small number of additional constraints, our method can be used to search characteristics with specific properties. For example, by setting some given variables marking the activity of some S-boxes to 1, we can search for characteristics with active S-boxes of predefined positions, which may be used in leaked-state forgery attacks [55]; by requiring the output and input variables to be the same, we can search for iterative characteristics; by setting all the variables marking the activity of all the S-boxes in the key schedule algorithm to be 0, we can search for characteristics with 0 active S-boxes in its key schedule algorithm, which may be preferred in the related-key differential attack.

6   Conclusion and Directions for Future Work

In this paper, we bring new constraints into the MILP model to describe the differential properties of a specific S-box, and obtain a more accurate MILP model for the differential behavior of a block cipher. Based on these constraints, we propose an automatic method for evaluating the security of bit-oriented block ciphers with respect to (related-key) differential attack. We also present a new tool for finding (related-key) characteristics automatically.

At this point, several open problems emerge. Firstly, we observe that the MILP instances derived from such cryptographic problems are very hard to solve compared with general MILP problems with the same scale with respect to the numbers of variables and constraints. Hence, it is interesting to develop specific methods to accelerate the solving process of such problems and therefore increase the number of rounds of the cipher under consideration that can be dealt with. Secondly, the method presented in this paper is very general. Is it possible to develop a compiler which can convert a standard description, say a description using hardware description language, of a cipher into an MILP instance to automate the entire security evaluation cycle with respect to (related-key) differential attack?

Finally, the methodology presented in this paper has some limitations which we would like to make clear, and trying to overcome these limitations is a topic deserving further investigation. Firstly, this methodology is only suitable to evaluate the security of constructions with S-boxes, XOR operations and bit permutations, and can not be applied to block cipher like SPECK [5], which involve modulo addition and no S-boxes at all. For tools which can be applied to ARX constructions, we refer the reader to [12,39,40,41,43]. Secondly, in this paper we do not consider the *differential effect* and we assume that the expected differential probability (EDP) π of a characteristic over all keys is (almost) the same as the fixed-key differential probability (DP) π_K for almost all keys (the common hypothesis of *stochastic equivalence* [35]), and that if the lower bound of the EDP for any characteristic of a block cipher is less than 2^{-s}, where s is bigger than the block size or key size, then the block cipher is secure against the (related-key) differential attack. For more in-depth discussion of the essential gap between EDP π and DP π_K, we refer the reader to [13] for more information.

Acknowledgements. The authors would like to thank the anonymous reviewers for their helpful comments and suggestions. The work of this paper is supported by the National Key Basic Research Program of China (2013CB834203, 2014CB340603), the National Natural Science Foundation of China (Grants 61402469, 61472417, 61472415 and 61272477), the Strategic Priority Research Program of Chinese Academy of Sciences under Grant XDA06010702, and the State Key Laboratory of Information Security, Chinese Academy of Sciences.

References

1. Albrecht, M., Cid, C.: Cold boot key recovery by solving polynomial systems with noise. In: Lopez, J., Tsudik, G. (eds.) ACNS 2011. LNCS, vol. 6715, pp. 57–72. Springer, Heidelberg (2011)
2. Biryukov, A., Roy, A., Velichkov, V.: Differential analysis of block ciphers SIMON and SPECK. In: Fast Software Encryption, FSE 2014 (2014)
3. Alkhzaimi, H.A., Lauridsen, M.M.: Cryptanalysis of the SIMON family of block ciphers. Cryptology ePrint Archive, Report 2013/543 (2013), http://eprint.iacr.org/2013/543
4. Aoki, K., Ichikawa, T., Kanda, M., Matsui, M., Moriai, S., Nakajima, J., Tokita, T.: *Camellia*: A 128-bit block cipher suitable for multiple platforms - design and analysis. In: Stinson, D.R., Tavares, S. (eds.) SAC 2000. LNCS, vol. 2012, pp. 39–56. Springer, Heidelberg (2001)
5. Beaulieu, R., Shors, D., Smith, J., Treatman-Clark, S., Weeks, B., Wingers, L.: The SIMON and SPECK families of lightweight block ciphers. Cryptology ePrint Archive, Report 2013/404 (2013), http://eprint.iacr.org/2013/404
6. Biham, E., Anderson, R., Knudsen, L.: Serpent: A new block cipher proposal. In: Vaudenay, S. (ed.) FSE 1998. LNCS, vol. 1372, pp. 222–238. Springer, Heidelberg (1998)
7. Biham, E., Shamir, A.: Differential cryptanalysis of DES-like cryptosystems. Journal of Cryptology 4(1), 3–72 (1991)
8. Bilgin, B., Bogdanov, A., Knežević, M., Mendel, F., Wang, Q.: FIDES: Lightweight authenticated cipher with side-channel resistance for constrained hardware. In: Bertoni, G., Coron, J.-S. (eds.) CHES 2013. LNCS, vol. 8086, pp. 142–158. Springer, Heidelberg (2013)
9. Biryukov, A.: Impossible differential attack. In: Encyclopedia of Cryptography and Security, pp. 597–597. Springer (2011)
10. Biryukov, A., Nikolić, I.: Automatic search for related-key differential characteristics in byte-oriented block ciphers: Application to AES, Camellia, Khazad and others. In: Gilbert, H. (ed.) EUROCRYPT 2010. LNCS, vol. 6110, pp. 322–344. Springer, Heidelberg (2010)
11. Biryukov, A., Nikolić, I.: Search for related-key differential characteristics in DES-like ciphers. In: Joux, A. (ed.) FSE 2011. LNCS, vol. 6733, pp. 18–34. Springer, Heidelberg (2011)
12. Biryukov, A., Velichkov, V.: Automatic search for differential trails in ARX ciphers. In: Benaloh, J. (ed.) CT-RSA 2014. LNCS, vol. 8366, pp. 227–250. Springer, Heidelberg (2014)
13. Blondeau, C., Bogdanov, A., Leander, G.: Bounds in shallows and in miseries. In: Canetti, R., Garay, J.A. (eds.) CRYPTO 2013, Part I. LNCS, vol. 8042, pp. 204–221. Springer, Heidelberg (2013)
14. Bogdanov, A.: On unbalanced feistel networks with contracting MDS diffusion. Designs, Codes and Cryptography 59(1-3), 35–58 (2011)
15. Bogdanov, A.A., Knudsen, L.R., Leander, G., Paar, C., Poschmann, A., Robshaw, M., Seurin, Y., Vikkelsoe, C.: PRESENT: An ultra-lightweight block cipher. In: Paillier, P., Verbauwhede, I. (eds.) CHES 2007. LNCS, vol. 4727, pp. 450–466. Springer, Heidelberg (2007)
16. Borghoff, J., Canteaut, A., Güneysu, T., Kavun, E.B., Knezevic, M., Knudsen, L.R., Leander, G., Nikov, V., Paar, C., Rechberger, C., Rombouts, P., Thomsen, S.S., Yalçın, T.: PRINCE – A low-latency block cipher for pervasive computing

applications. In: Wang, X., Sako, K. (eds.) ASIACRYPT 2012. LNCS, vol. 7658, pp. 208–225. Springer, Heidelberg (2012)

17. Borghoff, J., Knudsen, L.R., Stolpe, M.: Bivium as a mixed-integer linear programming problem. In: Parker, M.G. (ed.) Cryptography and Coding 2009. LNCS, vol. 5921, pp. 133–152. Springer, Heidelberg (2009)

18. Bulygin, S., Walter, M.: Study of the invariant coset attack on PRINTcipher: more weak keys with practical key recovery. Tech. rep., Cryptology ePrint Archive, Report 2012/85 (2012), http://eprint.iacr.org/2012/085.pdf

19. Daemen, J., Rijmen, V., Proposal, A.: Rijndael. In: Proceedings from the First Advanced Encryption Standard Candidate Conference, National Institute of Standards and Technology (NIST) (1998)

20. Andreeva, E., Bilgin, B., Bogdanov, A., Luykx, A., Mendel, F., Mennink, B., Mouha, N., Wang, Q., Yasuda, K.: PRIMATEs v1. CAESAR submission (2014), http://competitions.cr.yp.to/round1/primatesv1.pdf

21. Kavun, E.B., Lauridsen, M.M., Leander, G., Rechberger, C., Schwabe, P., Yalcin, T.: PrØst v1. CAESAR submission (2014), http://competitions.cr.yp.to/round1/proestv1.pdf

22. Emami, S., Ling, S., Nikolic, I., Pieprzyk, J., Wang, H.: The resistance of PRESENT-80 against related-key differential attacks. Cryptology ePrint Archive, Report 2013/522 (2013), http://eprint.iacr.org/

23. Abed, F., List, E., Wenzel, J., Lucks, S.: Differential cryptanalysis of round-reduced SIMON and SPECK. In: Fast Software Encryption, FSE 2014 (2014)

24. Abed, F., List, E., Lucks, S., Wenzel, J.: Differential and linear cryptanalysis of reduced-round SIMON. Cryptology ePrint Archive, Report 2013/526 (2013), http://eprint.iacr.org/526/

25. Fouque, P.A., Jean, J., Peyrin, T.: Structural evaluation of AES and chosen-key distinguisher of 9-round AES-128. In: Canetti, R., Garay, J.A. (eds.) CRYPTO 2013, Part I. LNCS, vol. 8042, pp. 183–203. Springer, Heidelberg (2013)

26. Fuhr, T.: Finding second preimages of short messages for Hamsi-256. In: Abe, M. (ed.) ASIACRYPT 2010. LNCS, vol. 6477, pp. 20–37. Springer, Heidelberg (2010)

27. Nikolić, I.: Tweaking AES. In: Biryukov, A., Gong, G., Stinson, D.R. (eds.) SAC 2010. LNCS, vol. 6544, pp. 198–210. Springer, Heidelberg (2011)

28. Alizadeh, J., Bagheri, N., Gauravaram, P., Kumar, A., Sanadhya, S.K.: Linear cryptanalysis of round reduced SIMON. Cryptology ePrint Archive, Report 2013/663 (2013), http://eprint.iacr.org/2013/663

29. Jean, J., Nikolić, I., Peyrin, T.: Deoxys v1. CAESAR submission (2014), http://competitions.cr.yp.to/round1/deoxysv1.pdf

30. Jean, J., Nikolić, I., Peyrin, T.: Joltik v1. CAESAR submission (2014), http://competitions.cr.yp.to/round1/joltikv1.pdf

31. Jean, J., Nikolić, I., Peyrin, T.: Kiasu v1. CAESAR submission (2014), http://competitions.cr.yp.to/round1/kiasuv1.pdf

32. Knellwolf, S., Meier, W., Naya-Plasencia, M.: Conditional differential cryptanalysis of NLFSR-based cryptosystems. In: Abe, M. (ed.) ASIACRYPT 2010. LNCS, vol. 6477, pp. 130–145. Springer, Heidelberg (2010)

33. Knellwolf, S., Meier, W., Naya-Plasencia, M.: Conditional differential cryptanalysis of trivium and KATAN. In: Miri, A., Vaudenay, S. (eds.) SAC 2011. LNCS, vol. 7118, pp. 200–212. Springer, Heidelberg (2012)

34. Knudsen, L.R.: Truncated and higher order differentials. In: Preneel, B. (ed.) FSE 1994. LNCS, vol. 1008, pp. 196–211. Springer, Heidelberg (1995)

35. Lai, X., Massey, J.L.: Markov ciphers and differential cryptanalysis. In: Davies, D.W. (ed.) EUROCRYPT 1991. LNCS, vol. 547, pp. 17–38. Springer, Heidelberg (1991)
36. Lamberger, M., Nad, T., Rijmen, V.: Numerical solvers and cryptanalysis. Journal of Mathematical Cryptology 3(3), 249–263 (2009)
37. Leander, G., Paar, C., Poschmann, A., Schramm, K.: New lightweight DES variants. In: Biryukov, A. (ed.) FSE 2007. LNCS, vol. 4593, pp. 196–210. Springer, Heidelberg (2007)
38. Lehmann, M., Meier, W.: Conditional differential cryptanalysis of grain-128a. In: Pieprzyk, J., Sadeghi, A.-R., Manulis, M. (eds.) CANS 2012. LNCS, vol. 7712, pp. 1–11. Springer, Heidelberg (2012)
39. Leurent, G.: Construction of differential characteristics in ARX designs application to skein. In: Canetti, R., Garay, J.A. (eds.) CRYPTO 2013, Part I. LNCS, vol. 8042, pp. 241–258. Springer, Heidelberg (2013)
40. Lipmaa, H., Moriai, S.: Efficient algorithms for computing differential properties of addition. In: Matsui, M. (ed.) FSE 2001. LNCS, vol. 2355, pp. 336–350. Springer, Heidelberg (2002)
41. Lipmaa, H., Wallén, J., Dumas, P.: On the additive differential probability of exclusive-or. In: Roy, B., Meier, W. (eds.) FSE 2004. LNCS, vol. 3017, pp. 317–331. Springer, Heidelberg (2004)
42. Matsui, M.: On correlation between the order of S-boxes and the strength of DES. In: De Santis, A. (ed.) EUROCRYPT 1994. LNCS, vol. 950, pp. 366–375. Springer, Heidelberg (1995)
43. Mouha, N., Preneel, B.: Towards finding optimal differential characteristics for ARX: Application to Salsa20. Cryptology ePrint Archive, Report 2013/328 (2013), http://eprint.iacr.org/2013/328
44. Mouha, N., Wang, Q., Gu, D., Preneel, B.: Differential and linear cryptanalysis using mixed-integer linear programming. In: Wu, C.-K., Yung, M., Lin, D. (eds.) Inscrypt 2011. LNCS, vol. 7537, pp. 57–76. Springer, Heidelberg (2012)
45. Optimization, Gurobi: Gurobi optimizer reference manual (2013), http://www.gurobi.com
46. Oren, Y., Kirschbaum, M., Popp, T., Wool, A.: Algebraic side-channel analysis in the presence of errors. In: Mangard, S., Standaert, F.-X. (eds.) CHES 2010. LNCS, vol. 6225, pp. 428–442. Springer, Heidelberg (2010)
47. Özen, O., Varıcı, K., Tezcan, C., Kocair, Ç.: Lightweight block ciphers revisited: Cryptanalysis of reduced round PRESENT and HIGHT. In: Boyd, C., González Nieto, J. (eds.) ACISP 2009. LNCS, vol. 5594, pp. 90–107. Springer, Heidelberg (2009)
48. Sun, S., Hu, L., Song, L., Xie, Y., Wang, P.: Automatic security evaluation of block ciphers with s-bp structures against related-key differential attacks. In: International Conference on Information Security and Cryptology – Inscrypt 2013 (2013)
49. Stein, W., et al.: Sage: Open source mathematical software (2008)
50. Tezcan, C.: Improbable differential attacks on PRESENT using undisturbed bits. Journal of Computional and Applied Mathematics 259, 503–511 (2014)
51. Wagner, D.: The boomerang attack. In: Knudsen, L.R. (ed.) FSE 1999. LNCS, vol. 1636, pp. 156–170. Springer, Heidelberg (1999)
52. Walter, M., Bulygin, S., Buchmann, J.: Optimizing guessing strategies for algebraic cryptanalysis with applications to EPCBC. In: Kutyłowski, M., Yung, M. (eds.) Inscrypt 2012. LNCS, vol. 7763, pp. 175–197. Springer, Heidelberg (2013)

53. Winnen, L.: Sage S-box MILP toolkit,
    http://www.ecrypt.eu.org/tools/sage-s-box-milp-toolkit
54. Wu, S., Wang, M.: Security evaluation against differential cryptanalysis for block
    cipher structures. Tech. rep., Cryptology ePrint Archive, Report 2011/551 (2011),
    http://eprint.iacr.org/2011/551.pdf
55. Wu, S., Wu, H., Huang, T., Wang, M., Wu, W.: Leaked-state-forgery attack against
    the authenticated encryption algorithm ALE. In: Sako, K., Sarkar, P. (eds.) ASI-
    ACRYPT 2013, Part I. LNCS, vol. 8269, pp. 377–404. Springer, Heidelberg (2013)
56. Wu, W., Zhang, L.: LBlock: A lightweight block cipher. In: Lopez, J., Tsudik, G.
    (eds.) ACNS 2011. LNCS, vol. 6715, pp. 327–344. Springer, Heidelberg (2011)
57. Yap, H., Khoo, K., Poschmann, A., Henricksen, M.: EPCBC - A block cipher
    suitable for electronic product code encryption. In: Lin, D., Tsudik, G., Wang, X.
    (eds.) CANS 2011. LNCS, vol. 7092, pp. 76–97. Springer, Heidelberg (2011)
58. Sasaki, Y., Todo, Y., Aoki, K., Naito, Y., Sugawara, T., Murakami,
    Y., Matsui, M., Hirose, S.: Minalpher v1. CAESAR submission (2014),
    http://competitions.cr.yp.to/round1/minalpherv1.pdf

A   On the Security of PRESENT-80, and LBlock with Respect to the Related-Key Differential Attack

A.1   Results on PRESENT-80

We apply the logical condition modelling method presented in Sect. 3.1 to the block cipher PRESENT-80 to determine its security bound with respect to the related-key differential attack. In each of these MILP models, we include one more constraint to ensure that the difference of the initial key register is nonzero, since the case where the difference of the initial key register is zero can be analyzed in the single-key model. Then we employ the Gurobi 5.5 optimizer [45] to solve the MILP instances.

By default the computations are performed on a PC using 4 threads with Intel(R) Core(TM) Quad CPU (2.83GHz, 3.25GB RAM, Windows XP), and a star "*" is appended on a timing data to mark that the corresponding computation is taken on a workstation equipped with two Intel(R) Xeon(R) E5620 CPU(2.4GHz, 8GB RAM, 8 cores).

We compute the number of active S-boxes for PRESENT-80 in the related-key model up to 14 rounds, and the results and a comparison with previous results without using CDP constraints are summarized in Table 2. For example, according to the 6th row of Table 2, the Gurobi optimizer finds that the minimum number of active S-boxes for 6-round PRESENT-80 is at least 5 in no more than 16 seconds by solving the MILP model with CDP constraints

These results clearly demonstrate that the MILP models with CDP constraints lead to tighter security bounds. In particular, we have proved that there are at least 16 active S-boxes in the best related-key differential characteristic for any consecutive 12-rounds of PRESENT-80. Therefore, the probability of the best related-key differential characteristic of 24-round PRESENT-80 is $(2^{-2})^{16} \times (2^{-2})^{16} = 2^{-64}$, leading to the result that the 24-round PRESENT-80 is resistant to basic related-key differential attack based on related-key differential characteristic (rather than differential).

Table 2. Results obtained from MILP models for PRESENT-80

Rounds	With CDP Constraints		Without CDP Constraints	
	# Active S-boxes	# Time(in seconds)	# Active S-boxes	# Time(in seconds)
1	0	1	0	1
2	0	1	0	1
3	1	1	1	1
4	2	1	2	1
5	3	5	3	3
6	5	16	4	10
7	7	107	6	26
8	9	254	8	111
9	10	522	9	171
10	13	4158	12	1540
11	15	18124	13	8136
12	16	50017	15	18102
13	18	137160*	17	49537*
14	20	1316808*	18	685372*
15	–	> 20days	–	> 20days

A.2 Results on LBlock

Up to now, there is no concrete result concerning the security of full-round LBlock [56] against differential attack in the related-key model due to a lack of proper tools for bit-oriented designs.

Since the encryption process of LBlock is nibble-oriented, the security of LBlock against single-key differential attack can be evaluated by those word-oriented techniques. However, the "$\lll$ 29" operations in the key schedule algorithm of LBlock destroy its overall nibble-oriented structure. In this subsection, we apply the method proposed in this paper to LBlock, and some results concerning its security against related-key differential attacks are obtained. Note that the type of constraints given in (5) are removed in our MILP models for LBlock according to the explanations presented in previous sections.

From Table 3, we can deduce that the probability of the best differential characteristic for full LBlock (totally $32 = 11 + 11 + 10$ rounds) is upper bounded by $(2^{-2})^{10} \times (2^{-2})^{10} \times (2^{-2})^{8} = 2^{-56}$, where 2^{-2} is the MDP for a single S-box of LBlock.

In fact, here we have an implicit trade-off between the number of constraints we use and the number of rounds we analyze. For example, we can use less constraints for every S-box and try to analyze more rounds, or we can use more constraints and focus on less rounds (but stronger bounds). However, it is not a simple task to find the best trade-off due to our limited computational power. We do try to analyze more rounds by using only one inequality selected from the convex hull for every S-box. The largest number of rounds we are able to analyze is 13, and we have prove that *there are at least 13 active S-boxes in any related-key characteristic for 13-round LBlock* on a PC in roughly 49 days.

Table 3. Results for related-key differential analysis on LBlock (The #Variables column records the sum of the number of the 0-1 variables and continuous variables in the MILP model).

Rounds	#Variables	#Constraints	#Active S-boxes	Time (in seconds)
1	218+104 = 322	660	0	1
2	292+208 = 500	1319	0	1
3	366+312 = 678	1978	0	1
4	440+416 = 856	2637	0	1
5	514+520 = 1034	3296	1	2
6	588+624 = 1212	3955	2	12
7	662+728 = 1390	4614	3	38
8	736+832 = 1568	5273	5	128
9	810+936 = 1746	5932	6	386
10	884+1040 = 1924	6591	8	19932
11	958+1144 = 2102	7250	10	43793

Then, we try to improve the above result with the two techniques presented in Sect. 4.1. By using the first technique, we can show that *there are at least 13 active S-boxes in a 13-round related-key differential characteristic of LBlock, and there is at least one active S-box taking a differential pattern with probability 2^{-3} in any 13-round related-key differential characteristic of LBlock with only 13 active S-boxes. Therefore, the probability of a 13-round related-key differential characteristic of LBlock is upper bounded by* $(2^{-2})^{12} \times (2^{-3}) = 2^{-27}$.

We now turn to the second technique presented in Sect. 4.1. By adding the constraint that the number of active S-boxes of any characteristic covering round 22 to round 26 (5 rounds in total) has at least 1 active S-box (see Table 3), and at most 12 active S-boxes to a 11-round (round 22 to round 32) MILP model (If this is not the case, it will enable us to get better bounds than the result presented here), we can show that there are at least 3 active S-boxes in a characteristic covering round 27 to round 32 . Combined with Fact 3, we have that the probability of the best related-key differential characteristic for full LBlock is upper bounded by $2^{-27} \times 2^{-27} \times (2^{-2})^3 = 2^{-60}$.

B   Search for Related-Key Characteristics of SIMON48

SIMON [5] is a family of lightweight block ciphers designed by the U.S National Security Agency (NSA). For a detailed description of SIMON and existing attacks, we refer the reader to [2,3,23,24,28].

By treating the AND ($\mathbb{F}_2 \times \mathbb{F}_2 \to \mathbb{F}_2$) operation as a 2×1 S-box, we apply our method to SIMON in the single-key model. For SIMON48 we obtain a 15-round differential characteristic with probability 2^{-46} (see Table 4), which is the best 15-round differential characteristic for 15-round SIMON48 published so far. If we fix the input and output differences to be the differences suggested by the characteristic we found, we can compute the probability of this *differential* by searching all characteristics with probability greater than 2^{-54} in this differential, and the result is $2^{-41.96}$ which is also the best result published so far.

We would like to emphasize that in our MILP models we treat the input bits of the AND operation as *independent input bits*, and the dependencies of the input bits to the AND operation are not considered. Therefore, the characteristic obtained by our method is not guaranteed to be valid. Hence, every time after the Gurobi optimizer outputs a good solution (characteristic), we check its validity and compute its probability by the method presented in [2].

Table 4. Single-key differential characteristic of 15-round SIMON48

Rounds	Left	Right
0	000000001000000000000000	000000100010001000000000
1	000000000010001000000000	000000001000000000000000
2	000000000000100000000000	000000000010001000000000
3	000000000000001000000000	000000000000100000000000
4	000000000000000000000000	000000000000001000000000
5	000000000000001000000000	000000000000000000000000
6	000000100001000000000000	000000000000001000000000
7	000000000010001000000010	000000100001000000000000
8	001000001000001000001000	000000000010001000000010
9	000000000010001000000010	001000001000001000001000
10	000000100001000000000000	000000000010001000000010
11	000000000000001000000000	000000100001000000000000
12	000000000000000000000000	000000000000001000000000
13	000000000000001000000000	000000000000000000000000
14	000000000000100000000000	000000000000001000000000
15	000000000010001000000000	000000000000100000000000

Scrutinizing and Improving Impossible Differential Attacks: Applications to CLEFIA, Camellia, LBlock and SIMON[*]

Christina Boura[1], María Naya-Plasencia[2] and Valentin Suder[2]

[1] Versailles Saint-Quentin-en-Yvelines University, France
christina.boura@prism.uvsq.fr
[2] Inria, France
{Maria.Naya_Plasencia,Valentin.Suder}@inria.fr

Abstract. Impossible differential cryptanalysis has shown to be a very powerful form of cryptanalysis against block ciphers. These attacks, even if extensively used, remain not fully understood because of their high technicality. Indeed, numerous are the applications where mistakes have been discovered or where the attacks lack optimality. This paper aims in a first step at formalizing and improving this type of attacks and in a second step at applying our work to block ciphers based on the Feistel construction. In this context, we derive generic complexity analysis formulas for mounting such attacks and develop new ideas for optimizing impossible differential cryptanalysis. These ideas include for example the testing of parts of the internal state for reducing the number of involved key bits. We also develop in a more general way the concept of using multiple differential paths, an idea introduced before in a more restrained context. These advances lead to the improvement of previous attacks against well known ciphers such as CLEFIA-128 and Camellia, while also to new attacks against 23-round LBlock and all members of the SIMON family.

Keywords: block ciphers, impossible differential attacks, CLEFIA, Camellia, LBlock, SIMON.

1 Introduction

Impossible differential attacks were independently introduced by Knudsen [21] and Biham et al. [5]. Unlike differential attacks [6] that exploit differential paths of high probability, the aim of impossible differential cryptanalysis is to use differentials that have a probability of zero to occur in order to eliminate the key candidates leading to such impossible differentials.

The first step in an impossible differential attack is to find an impossible differential covering the maximum number of rounds. This is a procedure that

---

[*] Partially supported by the French Agence Nationale de la Recherche through the BLOC project under Contract ANR-11-INS-011.

P. Sarkar and T. Iwata (Eds.): ASIACRYPT 2014, PART I, LNCS 8873, pp. 179–199, 2014.

has been extensively studied and there exist algorithms for finding such impossible differentials efficiently [20,19,9]. Once such a maximum-length impossible differential has been found and placed, one extends it by some rounds to both directions. After this, if a candidate key partially encrypts/decrypts a given pair to the impossible differential, then this key certainly cannot be the right one and is thus rejected. This technique provides a sieving of the key space and the remaining candidates can be tested by exhaustive search.

Despite the fact that impossible differential cryptanalysis has been extensively employed, the key sieving step of the attack does not seem yet fully understood. Indeed, this part of the procedure is highly technical and many parameters have to be taken into consideration. Questions that naturally arise concern the way to choose the plaintext/ciphertext pairs, the way to calculate the necessary data to mount the attack, the time complexity of the overall procedure as well as which are the parameters that optimize the attack. However, no simple and generalized way for answering these questions has been provided until now and the generality of most of the published attacks is lost within the tedious details of each application. The problems that arise from this approach is that mistakes become very common and attacks become difficult to verify. Errors in the analysis are often discovered and as we demonstrate in the next paragraph, many papers in the literature present flaws. These flaws include errors in the computation of the time or the data complexity, in the analysis of the memory requirements or of the complexity of some intermediate steps of the attacks. We can cite many such cases for different algorithms, as shown in Table 1. Note however, that the list of flaws presented in this table is not exhaustive.

Table 1. Summary of flaws in previous impossible differential attacks on CLEFIA-128, Camellia, LBlock and SIMON. Symbol ✗ means that the attack does not work, while ✓ says that the corrected attacks work. Error type (1) is when the data complexity is higher than the codebook, error type (2) shows a big computation flaw, error type (3) stands for small complexity flaws, while error type (4) is if the attack cannot be verified without implementation.

Algorithm	# rounds	Ref.	Type of error	Repaira- bility	Where discovered
CLEFIA-128 without without whit. layers	14	[36]	(1)	✗	[13]
CLEFIA-128	13	[30]	(4)	-	[7]
Camellia without FL/FL^{-1} layers	12	[34]	(2)	✗	this paper, similar problem as [33]
Camellia-128	12	[33]	(2)	✗	[25]
Camellia-128/192/256 without FL/FL^{-1} layers	11/13/14	[23]	(3)	✓	[34]
LBlock	22	[26]	(3)	✓	[27]
SIMON (all versions)	14/15/15/16/16/ 19/19/22/22/22	[3]	(1)	✗	Table 1 of [3]
SIMON (all versions)	13/15/17/20/25	[1,2]	(2)	✗	this paper

Instances of such flaws can for example be found in analyses of the cipher CLEFIA. CLEFIA is a lightweight 128-bit block cipher developed by SONY in 2007 [28] and adopted as an international ISO/IEC 29192 standard in lightweight cryptography. This cipher has attracted the attention of many researchers and numerous attacks have been published so far on reduced round versions [31,32,30,24,29,8]. Most of these attacks rely on impossible differential cryptanalysis. However, as pointed out by the designers of CLEFIA [14], some of these attacks seem to have flaws, especially in the key filtering phase. We can cite here a recent paper by Blondeau [7] that challenges the validity of the results in [30], or a claimed attack on 14 rounds of CLEFIA-128 [36], for which the designers of CLEFIA showed that the necessary data exceeds the whole codebook [13]. Another extensively analyzed cipher is the ISO/IEC 18033 standard Camellia, designed by Mitsubishi and NTT [4]. Among the numerous attacks presented against this cipher, some of the more successful ones rely on impossible differential cryptanalysis [34,33,22,25,23]. In the same way as for CLEFIA, some of these attacks were detected to have flaws. For instance, the attack from [33] was shown in [25] to be invalid. We discovered a similar error in the computation that invalidated the attack of [34]. Also, [34] reveals small flaws in [23]. Errors in impossible differential attacks were also detected for other ciphers. For example, in a cryptanalysis against the lightweight block cipher LBlock [26], the time complexity revealed to be incorrectly computed [27]. Another problem can be found in [3], where the data complexity is higher than the amount of data available in the block cipher SIMON, or in [1,2], where some parameters are not correctly computed. During our analysis, we equally discovered problems in some attacks that do not seem to have been pointed out before. In addition to all this, the more the procedure becomes complicated, the more the approach lacks optimality. To illustrate this lack of optimality presented in many attacks we can mention a cryptanalysis against 22-round LBlock [18], that could easily be extended to 23 rounds if a more optimal approach had been used to evaluate the data and time complexities, as well as an analysis of Camellia [22] which we improve in Section 4.

The above examples clearly show that impossible differential attacks suffer from the lack of a unified and optimized approach. For this reason, the first aim of our paper is to provide a general framework for dealing with impossible differential attacks. In this direction, we provide new generic formulas for computing the data, time and memory complexities. These formulas take into account the different parameters that intervene into the attacks and provide a highly optimized way for mounting them. Furthermore, we present some new techniques that can be applied in order to reduce the data needed or to reduce the number of key bits that need to be guessed. In particular we present a new method that helps reducing the number of key bits to be guessed by testing instead some bits of the internal state during the sieving phase. This technique has some similarities with the methods introduced in [15,17], however important differences exist as both techniques are applied in a completely different context. In addition to this, we apply and develop the idea of multiple impossible differentials, introduced in [32], to obtain more data for mounting our attacks. To illustrate the strength of our new approach we consider Feistel constructions and we apply the above ideas to a number of block ciphers, namely CLEFIA, Camellia, LBlock and SIMON.

More precisely, we present an attack as well as different time/data trade-offs on 13-round CLEFIA-128 that improve the time and data complexity of the previous best known attack [25] and improvements in the complexity of the best known attacks against all versions of Camellia [22]. In addition, in order to demonstrate the generality of our method, we provide the results of our attacks against 23-round LBlock and all versions of the SIMON block cipher. The attack on LBlock is the best attack so far in the single-key setting [1], while our attacks on SIMON are the best known impossible differential attacks for this family of ciphers and the best attacks in general for the three smaller versions of SIMON.

Summary of Our Attacks. We present here a summary of our results on the block ciphers CLEFIA-128, Camellia, LBlock and SIMON and compare them to the best impossible differential attacks known for the four analyzed algorithms. This summary is given in Table 2, where we point out with a '*' if the mentioned attack is the best cryptanalysis result on the target cipher or not, i.e. by the best known attack we consider any attack reaching the highest number of rounds, and with the best complexities among them.

The rest of the paper is organized as follows. In Section 2 we present a generic methodology for mounting impossible differential attacks, provide our complexity formulas and show new techniques and improvements for attacking a Feistel-like block cipher using impossible differential cryptanalysis. Section 3 is dedicated to the details of our attacks on CLEFIA and Section 4 presents our applications to all versions of Camellia. Due to lack of space, our applications on LBlock and the SIMON family of ciphers are given in the full version of this paper [11].

2 Complexity Analysis

We provide in this section a complexity analysis of impossible differential attacks against block ciphers as well as some new ideas that help improving the time and data complexities. We derive in this direction new generic formulas for the complexity evaluation of such attacks. The role of these formulas is twofold; on the one hand we aim at clarifying the attack procedure by rendering it as general as possible and on the other hand help at optimizing the time and data requirements. Establishing generic formulas should help mounting as well as verifying such attacks by avoiding the use of complicated procedures often leading to mistakes.

An impossible differential attack consists mainly of two general steps. The first one deals with the discovery of a maximum-length impossible differential, that is an input difference Δ_X and an output difference Δ_Y such that the probability

---

[1] In [12], an independent and simultaneous result on 23-round LBlock with worse time complexity was proposed.

Table 2. Summary of the best impossible differential attacks on CLEFIA-128, Camellia, LBlock and SIMON and presentation of our results. The presence of a '*' mentions if the current attack is the best known attack against the target cipher. Note here that we provide only the best of our results with respect to the time complexity. Other trade-offs can be found in the following sections. [†] see Section 4 for details.

Algorithm	Rounds	Time	Data (CP)	Memory (Blocks)	Reference
CLEFIA-128	13	$2^{121.2}$	$2^{117.8}$	$2^{86.8}$	[24]
using state-test technique	13	$2^{116.90}$	$2^{116.33}$	$2^{83.33}$	Section 3
using multiple impossible differentials	13	$\mathbf{2^{122.26}}$	$\mathbf{2^{111.02}}$	$\mathbf{2^{82.60}}$	Section 3*
combining with state-test technique	13	$\mathbf{2^{116.16}}$	$2^{114.58}$	$2^{83.16}$	[11]*
Camellia-128	11	2^{122}	2^{122}	2^{98}	[22]
	11	$\mathbf{2^{118.43}}$	$\mathbf{2^{118.4}}$	$\mathbf{2^{92.4}}$	Section 4*
Camellia-192	12	$2^{187.2}$	2^{123}	$2^{155.41}$	[22]
	12	$\mathbf{2^{161.06}}$	$\mathbf{2^{119.7}}$	$\mathbf{2^{150.7}}$	Section 4*
Camellia-256	13	$2^{251.1}$	2^{123}	2^{203}	[22]
	13	$\mathbf{2^{225.06}}$	$\mathbf{2^{119.71}}$	$\mathbf{2^{198.71}}$	Section 4*
Camellia-256[†]	14	$2^{250.5}$	2^{120}	2^{120}	[22]
	14	$\mathbf{2^{220}}$	$\mathbf{2^{118}}$	2^{173}	Section 4
LBlock	22	$2^{79.28}$	2^{58}	$2^{72.67}$	[18]
	22	$\mathbf{2^{71.53}}$	2^{60}	$\mathbf{2^{59}}$	[11,10]
	23	$2^{74.06}$	$2^{59.6}$	$2^{74.6}$	[11,10]*
SIMON32/64	19	$2^{62.56}$	2^{32}	2^{44}	[11]*
SIMON48/72	20	$2^{70.69}$	2^{48}	2^{58}	[11]*
SIMON48/96	21	$2^{94.73}$	2^{48}	2^{70}	[11]*
SIMON64/96	21	$2^{94.56}$	2^{64}	2^{60}	[11]
SIMON64/128	22	$2^{126.56}$	2^{64}	2^{75}	[11]
SIMON96/96	24	$2^{94.62}$	2^{94}	2^{61}	[11]
SIMON96/144	25	$2^{190.56}$	2^{128}	2^{77}	[11]
SIMON128/128	27	$2^{126.6}$	2^{94}	2^{61}	[11]
SIMON128/192	28	$2^{190.56}$	2^{128}	2^{77}	[11]
SIMON128/256	30	$2^{254.68}$	2^{128}	2^{111}	[11]

that Δ_X propagates after a certain number of rounds, r_Δ, to Δ_Y is zero. The second step, called the key sieving phase, consists in the addition of some rounds to potentially both directions. These extra added rounds serve to verify which key candidates partially encrypt (resp. decrypt) data to the impossible differential. As this differential is of probability zero, keys showing such behavior are clearly not the right encryption key and are thus removed from the candidate keys space.

We start by introducing the notation used in the rest of the paper. As in this work we are principally interested in the key sieving phase, we start our attack after a maximum impossible differential has been found for the target cipher.

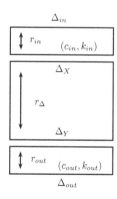

- Δ_X, Δ_Y: input (resp. output) differences of the impossible differential.
- r_Δ: number of rounds of the impossible differential.
- Δ_{in}, Δ_{out}: set of all possible input (resp. output) differences of the cipher.
- r_{in}: number of rounds of the differential path(Δ_X, Δ_{in}).
- r_{out}: number of rounds of the differential path(Δ_Y, Δ_{out}).

The differential $(\Delta_X \rightarrow \Delta_{in})$ (resp. $(\Delta_Y \rightarrow \Delta_{out})$) occurs with probability 1 while the differential $(\Delta_X \leftarrow \Delta_{in})$ (resp. $(\Delta_Y \leftarrow \Delta_{out})$) is verified with probability $\frac{1}{2^{c_{in}}}$ (resp. $\frac{1}{2^{c_{out}}}$), where c_{in} (resp. c_{out}) is the number of bit-conditions that have to be verified to obtain Δ_X from Δ_{in} (resp. Δ_Y from Δ_{out}).

It is important to correctly determine the number of key bits intervening during an attack. We call this quantity *information key bits*. In an impossible differential attack, one starts by determining all the subkey bits that are involved in the attack. We denote by k_{in} the subset of subkey bits involved in the attack during the first r_{in} rounds, and k_{out} during the last r_{out} ones. However, some of these subkey bits can be related between them. For example, two different subkey bits can actually be the same bit of the master key. Alternatively, a bit in the set can be some combination, or can be easily determined by some other bits of the set. The way that the different key bits in the target set are related is determined by the key schedule. The actual parameter that we need to determine for computing the complexity of the attacks is the information key bits intervening in total, that is from an information theoretical point of view, the log of the entropy of the involved key bits, that we denote by $|k_{in} \cup k_{out}|$.

We continue now by describing our attack scenario on $(r_{in} + r_\Delta + r_{out})$ rounds of a given cipher.

2.1 Attack Scenario

Suppose that we are dealing with a block cipher of block size n parametrized by a key K of size $|K|$. Let the impossible differential be placed between the rounds $(r_{in} + 1)$ and $(r_{in} + r_\Delta)$. As already said, the impossible differential implies that it is not feasible that an input difference Δ_X at round $(r_{in} + 1)$ propagates to an output difference Δ_Y at the end of round $(r_{in} + r_\Delta)$. Thus, the goal is, for each given pair of inputs (and their corresponding outputs), to discard the keys that generate a difference Δ_X at the beginning of round $(r_{in} + 1)$ and at the same time, a difference Δ_Y at the output of round $(r_{in} + r_\Delta)$. We need then enough pairs so that the number of non-discarded keys is significantly lower than the a priori total number of key candidates.

Suppose that the first r_{in} rounds have an input truncated difference in Δ_{in} and an output difference Δ_X, which is the input of the impossible differential. Suppose that there are c_{in} bit-conditions that need to be verified so that Δ_{in} propagates to Δ_X and $|k_{in}|$ information key bits involved.

In a similar way, suppose that the last r_{out} rounds have a truncated output difference in Δ_{out} and an input difference Δ_Y, which is the output of the impossible differential. Suppose that there are c_{out} bit-conditions that need to be verified so that Δ_{out} propagates to Δ_Y in the backward direction and $|k_{out}|$ information key bits involved.

We show next how to determine the amount of data needed for an attack.

2.2   Data Complexity

The probability that for a given key, a pair of inputs already satisfying the differences Δ_{in} and Δ_{out} verifies all the $(c_{in} + c_{out})$ bit-conditions is $2^{-(c_{in}+c_{out})}$. In other words, this is the probability that for a pair of inputs having a difference in Δ_{in} and an output difference in Δ_{out}, a key from the possible key set is discarded. Therefore, by repeating the procedure with N different input (or output) pairs, the probability that a trial key is kept in the candidate keys set is $P = (1 - 2^{-(c_{in}+c_{out})})^N$.

There is not a unique strategy for choosing the amount of input (or output) pairs N. This choice principally depends on the overall time complexity, which is influenced by N, and the induced data complexity. Different trade-offs are therefore possible. A popular strategy, generally used by default is to choose N such that only the right key is left after sieving. This amounts to choose P as

$$P = (1 - 2^{-(c_{in}+c_{out})})^N < \frac{1}{2^{|k_{in} \cup k_{out}|}}.$$

In this paper we adopt a different approach that can help reducing the number of pairs needed for the attack and offers better trade-offs between the data and time complexity. More precisely, we permit smaller values of N. By proceeding like this, we will be probably left with more than one key in our candidate keys set and we will need to proceed to an exhaustive search among the remaining candidates, but the total time complexity of the attack will probably be much lower. In practice, we will start considering values of N such that P is slightly smaller than $\frac{1}{2}$ so to reduce the exhaustive search by at least one bit. The smallest value of N, denoted by $N_{\min}$, verifying

$$P = (1 - 2^{-(c_{in}+c_{out})})^{N_{\min}} \simeq e^{-N_{\min} \times 2^{-(c_{in}+c_{out})}} < \frac{1}{2}$$

is approximately $N_{\min} = 2^{c_{in}+c_{out}}$. Then we have to choose $N \geq N_{\min}$.

We provide then a solution for determining the cost of obtaining N pairs such that their input difference belongs to Δ_{in} and their output difference belongs to Δ_{out}. To the best of our knowledge, this is the first generic solution to this problem. We evaluated this cost as

$$C_N = \max \left\{ \min_{\Delta \in \{\Delta_{in}, \Delta_{out}\}} \left\{ \sqrt{N2^{n+1-|\Delta|}} \right\}, N2^{n+1-|\Delta_{in}|-|\Delta_{out}|} \right\}. \quad (1)$$

A detailed explanation on how this formula is derived can be found in the full version of the paper [11]. The cost C_N represents also the amount of needed data. Obviously, as the size of the state is n, the following inequality, should hold:

$$C_N \le 2^n.$$

This inequality simply states that the total amount of data used for the attack cannot exceed the codebook. These conditions are not verified in several cases from [3], as well as in the corrected version of [36] which invalidates the corresponding attacks.

2.3   Time and Memory Complexity

We are going to detail now the computation of the time complexity of the attack. Note that the formulas that we are presenting in this section are the first generic formulas given for estimating the complexity of impossible differential attacks.

By following the early abort technique [23], the attack consists in storing the N pairs and testing out step by step the key candidates, by reducing at each time the size of the remaining possible pairs. The time complexity is then determined by three quantities. The first term is the cost C_N, that is the amount of needed data (see Formula (1)) for obtaining the N pairs, where N is such that $P < 1/2$. The second term corresponds to the number of candidate keys $2^{|k_{in} \cup k_{out}|}$, multiplied by the average cost of testing the remaining pairs. For all the applications that we have studied, this cost can be very closely approximated by $\left(N + 2^{|k_{in} \cup k_{out}|} \frac{N}{2^{c_{in}+c_{out}}}\right) C_E'$, where C_E' is the ratio of the cost of partial encryption to the full encryption. Finally, the third term is the cost of the exhaustive search for the key candidates still in the candidate keys set after the sieving. By taking into account the cost of one encryption C_E, we conclude that the time complexity of the attack is

$$T_{comp} = \left(C_N + \left(N + 2^{|k_{in} \cup k_{out}|} \frac{N}{2^{c_{in}+c_{out}}} \right) C_E' + 2^{|K|} P \right) C_E, \quad (2)$$

where $C_N = \max \left\{ \min_{\Delta \in \{\Delta_{in}, \Delta_{out}\}} \left\{ \sqrt{N2^{n+1-|\Delta|}} \right\}, N2^{n+1-|\Delta_{in}|-|\Delta_{out}|} \right\}$, with N such that $P = (1 - 1/(2^{c_{in}+c_{out}}))^N < 1/2$ and where the last term corresponds to $2^{|K|-|k_{in} \cup k_{out}|} P 2^{|k_{in} \cup k_{out}|}$. Obviously, as we want the attack complexity to be smaller than the exhaustive search complexity, the above quantity should be smaller than $2^{|K|} C_E$.

It must be noted here that this is a minimum estimation of the complexity, that, in practice, and thanks to the idea of Section 2.4, it approximates really well the actual time complexity, as it can be seen in the applications, and in particular, in the tight correspondence shown between the LBlock estimation that we detail in [11] and the exact calculation from [10]. The precise evaluation of C'_E (that is always smaller than 1) can only be done once the attack parameters are known. However, C'_E can be estimated quite by calculating the ratio between the active SBoxes during a partial encryption and the total number of SBoxes (thought it is not always the best approximation, it is a common practice).

Memory complexity. By using the early abort technique [23], the only elements that need to be stored are the N pairs. Therefore, the memory complexity of the attack [2] is determined by N.

2.4   Choosing $\Delta_{in}, \Delta_{out}, c_{in}$ and c_{out}

We explain now, the two possible ways for choosing $\Delta_{in}, \Delta_{out}, c_{in}$ and c_{out}. For this, we introduce the following example that can be visualized in Figure 1 and where we consider an Sbox-based cipher. In this example, we will only talk about Δ_{in} and c_{in}, however the approach for Δ_{out} and c_{out} is identical.

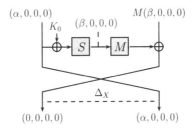

Fig. 1. Choosing Δ_{in} and c_{in}

Suppose that the state is composed of two branches of four nibbles each. The round function is composed of a non-linear layer S, seen as a concatenation of four Sboxes S_0, S_1, S_2 and S_3, followed by a linear layer M. There exist two different ways for choosing $|\Delta_{in}|$ and c_{in}:

1. The most intuitive way is to consider $|\Delta_{in}| = 4 + 4$ and $c_{in} = 4$, as the size of α and of β is 4 bits, and in the first round we want 4 bits to collide. In this case, for a certain key, the average probability that a pair taken out of the $2^{4+4}2^{4+4-1}$ pairs belonging to Δ_{in} leads to Δ_X is 2^{-4}.
2. In general, the difference α can take $2^4 - 1$ different values. However, each value can be associated by the differential distribution table of the Sbox S_0

---

[2] If $N > 2^{|k_{in} \cup k_{out}|}$ we could store the discarded key candidates instead, this is rarely the case. Thus, we can consider a memory complexity of $\min\{N, 2^{|k_{in} \cup k_{out}|}\}$.

to 2^3 output differences on average[3], so the possibilities for the difference β are limited to 2^3. Therefore, we can consider that $|\Delta_{in}| \approx 4 + 3$. But, in this case $c_{in} = 3$, as for each input pair belonging to the $2^{4+3}2^{4+3-1}$ possible ones, there exist on average 2 values that make the differential transition $\alpha \to \beta$ possible (instead of 1 in the previous case).

We can see, by using the generic formulas of Section 2.3, that both cases induce practically the same time complexity, as the difference in N compensates with the difference in $c_{in} + c_{out}$. However, the memory complexity, given by N, is slightly better in case 2. Furthermore, case 2, in which a preliminary pairs filtering is done, allows to reduce the average cost of using the early abort technique [23].

In several papers, for example in [33] and [23], the second case is followed. However, its application is partial (either for the input or the output part) and this with no apparent reason. Note however, that in these papers, the associated c_{out} was not always correctly computed and sometimes, 8-bit conditions were considered when 7-bit conditions should have been accounted for. For reasons of simplicity, we will consider case 1 in our applications and check afterwards the actual memory needed.

2.5   Using Multiple Impossible Differentials to Reduce the Data Complexity

We explain in this section a method to reduce the data complexity of an attack. This method is inspired by the notion of multiple impossible differentials that was introduced by Tsunoo et al. [32] and applied to 12-round CLEFIA-128. The idea in this technique is to consider at once several impossible differentials, instead of just one. We assume, as done in [16], that the differences in Δ_{in} (and in Δout) lie in a closed set. There are two ways in which this can be a priori done:

1. Take rotated versions of a certain impossible differential. We call n_{in} the number of different input pattern differences generated by the rotated versions of the chosen impossible differential.
2. When the middle conditions have several impossible combinations, we can consider the same first half of the differential path together with a rotated version of the second one, in a way to get a different impossible differential. We call n_{out} the number of different output pattern differences generated by the rotated versions of the second part of the path that we will consider. For the sake of simplicity and without loss of generality we will only consider the case of rotating the second half of the path.

It is important to point out that for our analysis to be valid, in both cases the number of conditions associated to the impossible differential attack should stay the same. Both cases can be translated into a higher amount of available

---
[3] This quantity depends on the Sbox. In this example, we consider that all four Sboxes have good cryptographic properties.

data by redefining two quantities, $|\Delta'_{in}|$ and $|\Delta'_{out}|$, that will take the previous roles of $|\Delta_{in}|$ and $|\Delta_{out}|$,

$$|\Delta'_{in}| = |\Delta_{in}| + \log_2(n_{in}) \text{ and } |\Delta'_{out}| = |\Delta_{out}| + \log_2(n_{out}).$$

$|\Delta'_{in}|$ is the log of the total size of the set of possible input differences, and $|\Delta'_{out}|$ is the log of the total size of the set of possible output differences.

In this case, the data complexity C_N is computed with the corrected values for the input sizes and is, as can be easily seen, smaller than if only one path had been used. The time complexity remains the same, except for the C_N term. Indeed, the middle term of Formula (2) remains the same, as for a given pair, the number of key bits involved stays $2^{|k_{in} \cup k_{out}|}$. Equally, as the number of involved possible partial keys is $n_{in} n_{out} 2^{|k_{in} \cup k_{out}|}$, the last term of Formula (2) is now

$$\frac{2^{|K|}}{n_{in} \cdot n_{out} 2^{|k_{in} \cup k_{out}|}} (P \cdot n_{in} \cdot n_{out} \cdot 2^{|k_{in} \cup k_{out}|}) = 2^{|K|} P$$

and so also stays the same.

In Section 3 we present our attacks on CLEFIA. In part of these attacks, we use multiple impossible differentials to reduce the data complexity. Besides, this technique shows particularly useful for mounting attacks on some versions of the SIMON family for which there is not enough available data to mount a valid attack with the traditional method.

2.6   Introducing the *State-Test* Technique

We introduce now a new method that consists in making a test for some part of the internal state instead of guessing the necessary key bits for computing it. This somewhat reminds the techniques presented in [15,17] in the context of meet-in-the-middle attacks. However, the technique that we present in this section, and that we call the *state-test* technique is different since it consists in checking the values of the internal state to verify if we can discard all the involved candidates.

Very often during the key filtering phase of impossible differential attacks, the size of the internal state that needs to be known is smaller than the number of key bits on which it depends. As we will see, focusing on the values that a part of the state can take permits to eliminate some key candidates without considering all the values for the involved key bits. The state-test technique works by fixing s bits of the plaintexts, which allows us to reduce the number of information key bits by s. We will explain how this method works by a small example.

Consider a 32-bit Feistel construction, where each branch can be seen as a concatenation of four nibbles (see Figure 2). Suppose that the round function is composed of a non-linear layer S, seen as a concatenation of four 4-bit invertible Sboxes (S_0, S_1, S_2, S_3) and of a linear layer M on $\mathbb{F}_{2^4}$. We suppose for this example that the branch number of M, that is the minimal number of active Sboxes in any two consecutive rounds, is less than 5. Let $\Delta_X = (\alpha, 0, 0, 0)|(0, 0, 0, 0)$ be the input difference of the impossible differential, placed at the end of the second round and let $\Delta_{in} = (*, *, *, 0)|(*, *, *, *)$ be the difference at the input of the block cipher. Note however that in reality, the leftmost side of Δ_{in} only depends on a 4-bit non-zero difference δ, i.e. $\Delta_{in} = M(\delta, 0, 0, 0)|(*, *, *, *)$.

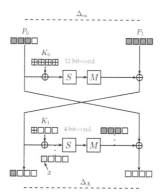

Fig. 2. Grey color stands for nibbles with non-zero difference. Hatched key nibbles correspond to the part of the subkeys that have to be guessed. The nibble x is the part of the state on which we apply the state-test technique.

As can be seen in Figure 2, there are in total 4 active Sboxes and thus there are $c_{in} = 16$ conditions that have to be verified in order to have a transition from Δ_{in} to Δ_X. Therefore, the first step is to collect N pairs such that $P = (1 - 2^{-(c_{in}+c_{out})})^N = (1 - 2^{-c_{in}})^N = (1 - 2^{-16})^N < \frac{1}{2}$. The exact value of N will be chosen in a way to obtain the best trade-off for the complexities. Before describing the new method, we start by explaining how this attack would have worked in the classical way. As we can see in Figure 2, there are 3×4 bits that have to be guessed ($K_{0,0}$, $K_{0,1}$ and $K_{0,2}$) in order to verify the conditions on the first round and there are 2×4 bits that have to be guessed ($K_{0,3}$ and $K_{1,0}$) in order to verify the conditions on the second round.

Therefore, for all N pairs, one starts by testing all the 2^4 possible values for the first nibble of K_0. After this first guess, $N \times 2^{-4}$ pairs remain in average, as there are 4-bit conditions that need to be verified by the guess through the first round. Then one continues by testing the second and the third nibble of K_0 and finally the last nibble of K_0 and the first nibble of K_1. At each step, the amount of data remaining is divided by 2^4. To summarize, we have $|k_{in} \cup k_{out}| = |k_{in}| = 20$ and $2^{c_{in}+c_{out}} = 2^{c_{in}} = 2^4 2^4 2^4 2^4$. Then Formula (2) can be used to evaluate the time complexity of the attack as

$$\left(C_N + \left(N + 2^{20} \frac{N}{2^{16}} \right) C'_E + 2^{20} P 2^{|K|-20} \right) C_E. \tag{3}$$

We will see now how the state-test technique applies to this example and how it permits to decrease the time complexity. Consider the first nibble of the left part of the state after the addition of the subkey K_1. We denote this nibble by x. Note that mathematically, x can be expressed as

$$x = K_{1,0} \oplus P_{1,0} \oplus M(S(K_0 \oplus P_0))_0$$
$$x \oplus P_{1,0} = K_{1,0} \oplus m_0 S_0(K_{0,0} \oplus P_{0,0}) \oplus m_1 S_1(K_{0,1} \oplus P_{0,1})$$
$$\oplus\, m_2 S_2(K_{0,2} \oplus P_{0,2}) \oplus m_3 S_3(K_{0,3} \oplus P_{0,3}), \tag{4}$$

where the m_i's are coefficients in $\mathbb{F}_2^4$.

Suppose now that for all pairs, we fix the last $s = 4$ bits of P_0 to the same constant value. One can verify that this is a reasonable assumption, as by fixing this part of the inputs we still have enough data to mount the attack. Then one starts as before, by guessing the first three nibbles of K_0. After this 12-bit guess, approximately $N \times 2^{-12}$ pairs remain. We know for each pair the input and output differences of the Sbox of the second round as the needed part of K_0 has been guessed. Therefore, by a simple lookup at the differential distribution table of the involved Sbox, we obtain one value for x that verifies the second round conditions in average per pair (about half of the time the transition is not possible, whereas for the other half we find two values). Equation (4) becomes

$$x \oplus P_{1,0} \oplus m_0 S_0(K_{0,0} \oplus P_{0,0}) \oplus m_1 S_1(K_{0,1} \oplus P_{0,1}) \oplus m_2 S_2(K_{0,2} \oplus P_{0,2})$$
$$= K_{1,0} \oplus m_3 S_3(K_{0,3} \oplus P_{0,3}), \tag{5}$$

where the left side of Equation (5), that we denote by x', is known for each pair.

Thus, for each guess of $(K_{0,0}, K_{0,1}, K_{0,2})$, we construct a table of size $N \times 2^{-12}$, where we store these values of x'. The last and more important step consists now in looking if all 2^4 possible values of x' appear in the table. Note here, that as $N \geq 2^{16}$, the size of the table is necessarily greater than or equal to 2^4.

Since $P_{0,3}$ is fixed, the only unknown values in Equation (5) are $K_{1,0}$ and $K_{0,3}$. If all values for x' are in the table and since S_3 is a permutation, for any choice of $K_{1,0}$ and any choice of $K_{0,3}$, there will always exist (at least) one pair such that $K_{1,0} \oplus m_3 S_3(K_{0,3} \oplus P_{0,3})$ is in the table, leading thus to the impossible differential.

As a conclusion, we know that if x' takes all the possible values in the table, we can remove the keys composed by the guessed value $(K_{0,0}, K_{0,1}, K_{0,2})$ from the candidate keys set, as for all the values of $(K_{1,0}, K_{0,3})$, they would imply the impossible differential. If instead, x' does not take all the possible values for a certain value of $(K_{0,0}, K_{0,1}, K_{0,2})$, we can test this partial key combined to all the possibilities of the remaining key bits that verify Equation (5) for the missing x', as they belong to the remaining key candidates.

The main gain of the state-test technique is that it decreases the number of information key bits and therefore the time complexity. For instance, in this example, the variable x' can be seen as 4 information key bits [4] instead of 2×4 key bits we had to guess in the classic approach (the bits of $K_{0,3}$ and of $K_{1,0}$). We have $s = 4$ less bits to guess thanks to the $s = 4$ bits of the plaintext that we have fixed. Thus the time complexity in this case becomes

$$\left(C_N + \left(N + 2^{20-4} \frac{N}{2^{16}} \right) C'_E + 2^{20-4} P 2^{|K|-(20-4)} \right) C_E. \tag{6}$$

One can see now by comparing Equations (6) and (3) that the time complexity is lower with the state-test technique, than with the trivial method. Indeed, the first and the third term of the Equations (6) and (3) remain the same, while the second term is lower in Equation (6). Finally, note that the probability P for a key to be still in the candidate keys set remains the same as before. Indeed, during the attack we detect all and the same candidate keys for which none of the N pairs implies the impossible differential, which are the same candidate keys that we would have detected in a classic attack.

We would like to note here that we have implemented the state-test technique on a toy cipher, having a structure similar to the one that we introduced in this section, and we have verified its correctness.

Application of the state-test technique in parallel for decreasing the probability P. An issue that could appear with this technique is that as we have to fix a part of the plaintexts, s bits, the amount of data available for computing the N pairs is reduced. The probability P associated to an attack is the probability for a key to remain in the candidate keys set. When the amount of available data is small, the number of pairs N that we can construct is equally small and thus the probability P is high. In such a situation, the dominant term of the time complexity (Formula (2)), is in general the third one, i.e. $2^{|K|}P$.

More precisely, we need the sum of $\log_2(C_N)$ and s, the number of plaintext bits that we fix, to be less than or equal to the block size. This limits the size of N that we can consider, leading to higher probabilities P, and could lead, sometimes, to higher time complexities. To avoid this, one can repeat the attack in parallel for several different values, say Y, of the fixed part of the plaintext. In this case, the data and memory needed are multiplied by Y. On the other hand, repeating the attack in parallel permits to detect more efficiently if a guessed key could be the right one. Indeed, for a guessed key, only if none of the tables constructed as described above contains all the values for x', one can test if this guessed key is the correct one.

To summarize, by repeating the state-test technique in parallel, we multiply the available data by Y, as well as the available pairs, and since the attack is done Y times in parallel, the probability P becomes P^Y. The probability decreases

---

[4] Note that we could, equivalently, consider all possible values of x' in the last step, and consider the associated remaining pairs table, that would have a size of $N2^{-16}$ (empty if the key is a good candidate, not empty otherwise), obtaining the same key candidates of 16 bits, 12 from $(K_{0,0}, K_{0,1}, K_{0,2})$ and 4 information key bits from x', with the same complexity as in the previously described method.

much faster than the data or the other terms of the time complexity increase. Therefore, the Formula (2) becomes in this case:

$$\left(C_N \times Y + \left(N \times Y + 2^{|k_{in} \cup k_{out}| - s} \frac{N \times Y}{2^{c_{in} + c_{out}}}\right) C_E' + 2^{|K|} P^Y\right) C_E. \quad (7)$$

In Section 3, we are going to see an application of this technique to 13-round CLEFIA-128, and at the end of Section 4 we show an application on Camellia-256.

3 Application to CLEFIA

CLEFIA is a lightweight 128-bit block cipher designed by Shirai et al. in 2007 [28] and based on a 4-branch generalized Feistel network. It supports keys of size 128, 192 or 256 bits and the total number of iterations, say R, depends on the key size. More precisely, $R = 18$ for the 128-bit version, while $R = 22$ and $R = 26$ for the two following variants. A key-scheduling algorithm is used to generate $2R$ round keys $RK_0, \ldots, RK_{2R-1}$ and 4 whitening keys $WK_0, \ldots, WK_3$. The whitening keys are XORed to the right branches of the first and the last round. CLEFIA's round function design can be visualized in Figure 3. For a more complete description of the specifications one can refer to [28].

We describe now several attacks against 13-round CLEFIA-128.

3.1 Impossible Differential Cryptanalysis of 13-round CLEFIA-128

The authors of [31] noticed that a difference on the internal state of CLEFIA of the form $P^i = 0_{32}|0_{32}|0_{32}|A$ cannot lead to a difference $P^{i+9} = 0_{32}|0_{32}|B|0_{32}$ after 9 rounds, where A and B are 4-byte vectors for which only one byte in a different position is active (e.g. $A = (\alpha, 0_8, 0_8, 0_8)$ and $B = (0_8, \beta, 0_8, 0_8)$). We use this same 9-round impossible differential and place it between rounds 3 and 11. Therefore, for our attack, $r_{in} = r_{out} = 2$ and $r_\Delta = 9$, as in [24].

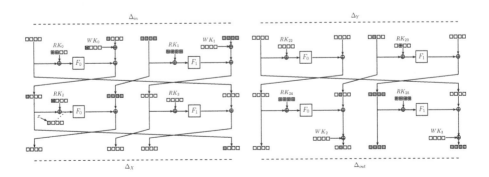

Fig. 3. The attack on CLEFIA-128. Grey color stands for bytes with a non-zero difference, while hatched bytes are the subkey bytes that have to be guessed.

The differential placed on the top and at the bottom of the impossible differential are depicted in Figure 3. We describe now the parameters for our cryptanalysis of 13-round CLEFIA-128. As can be seen in Figure 3 there are $c_{in} + c_{out} = 40 + 40$ bit-conditions that need to be verified so that the difference in the plaintexts $\Delta_{in} = 0_{32}|(*_8, 0_8, 0_8, 0_8)|M_0(*_8, 0_8, 0_8, 0_8)|*_{32}$ propagates to $\Delta_X = 0_{32}|0_{32}|0_{32}|(\alpha, 0_8, 0_8, 0_8)$ and the difference in the ciphertexts $\Delta_{out} = 0_{32}|(0_8, *_8, 0_8, 0_8)|M_1(0_8, *_8, 0_8, 0_8)|*_{32}$ propagates to $\Delta_Y = 0_{32}|0_{32}|(0_8, \beta, 0_8, 0_8)|0_{32}$. In this way, $|\Delta_{in}| = |\Delta_{out}| = 48$.

Following the complexity analysis of Section 2, we need to construct at least $N_{\min} = 2^{80}$ pairs. The cost to construct these pairs is

$$C_{N_{\min}} = \max\left\{ \sqrt{2^{80}2^{129-48}}, 2^{80}2^{129-48-48} \right\} = 2^{113}.$$

Using the state-test technique. We use now the state-test technique, described in Section 2.6 to test the 8 bits of the internal state denoted by x in Figure 3, instead of guessing the whole subkey RK_0 and the XOR of the leftmost byte of RK_2 and WK_0. For doing this, we need to fix part of the 32 leftmost bits of the plaintexts. As the number of needed data is $C_{N_{\min}} = 2^{113}$, we can fix at most $128 - 113 = 15$ bits. However, as each Sbox is applied to 8 bits, we will only fix one byte of this part of the plaintexts. We will guess then 24 bits of the subkey RK_0 which are situated on the other bytes.

During a classical attack procedure, we would need to guess 32 bits of RK_1, 32 bits of RK_0 and 8 bits of $RK_2 \oplus WK_0$, thus $k_{in} = 72$. We would also need to guess 8 bits of $RK_{23} \oplus WK_2$, 32 bits of RK_{24} and 32 bits of RK_{25}, therefore $k_{out} = 72$. However, the subkeys RK_1 and RK_{24} share 22 bits in common. As a consequence, the number of information key bits would be $|k_{in} \cup k_{out}| = 72 + 72 - 22 = 122$. As we will fix 8 bits of the plaintexts, according to Section 2.6, it is the same to say that there will be $|k_{in} \cup k_{out}| - 8 = 122 - 8 = 114$ bits to test. The time complexity of our attack, computed using Formula (2) is then

$$\left(C_N + \left(N + 2^{114} \frac{N}{2^{80}} \right) \frac{18}{104} + 2^{128} P \right) C_E,$$

where the fraction $18/104$ is the ratio of the cost of partial encryption to the full encryption. Since our attack needs at least 2^{113} plaintexts and since we fixed 8 bits out of them, we have $128 - 113 - 8 = 7$ bits of freedom for building structures.

Among all possible trade-offs with respect to the amount of data, the best time complexity is $2^{116.90}C_E$ with $2^{83.33}$ pairs built from $2^{116.33}$ plaintexts.

Using multiple impossible differentials. The authors of [31] noticed that there exist several different 9-round impossible differentials, see [31, Table 1]. In [32], multiple impossible differentials were used to attack 12 rounds of CLEFIA-128. Here, we will apply our formalized approach of this idea presented in Section 2.5, to reduce the data complexity of the attack on 13 rounds of CLEFIA-128.

We use the $n_{in} = 2 \times 4$ different inputs to the impossible differentials, that is $P^i = 0_{32}|A|0_{32}|0_{32}$ and $P^i = 0_{32}|0_{32}|0_{32}|A$, where A can take a difference on only one of the four possible bytes. For each one of them, there are $n_{out} = 3$ different output impossible differences $P^{i+9} = 0_{32}|0_{32}|B|0_{32}$ after 9 rounds, where B has only one byte active in a different position than the active byte in A. We have now $|\Delta'_{in}| = |\Delta_{in}| + \log_2(8) = 48 + 3$ and $|\Delta'_{out}| = |\Delta_{out}| + \log_2(3) = 48 + 1.58$. Since the bit-conditions remain unchanged, $c_{in} + c_{out} = 80$, the minimal number of pairs needed for the attack to work is $N_{min} = 2^{80}$. For this number of pairs, we need $C_{N_{min}} = 2^{113-4.58} = 2^{108.42}$ plaintexts. The number of information key bits is $|k_{in} \cup k_{out}| = 122$. We have then $\left(C_N + \left(N + 2^{122}\frac{N}{2^{80}}\right)\frac{18}{104} + P2^{128}\right)C_E$. Among all the possible trade-offs with respect to the amount of data, the best time complexity we obtained is $2^{122.26}C_E$ with $2^{82.6}$ pairs built from $2^{111.02}$ plaintexts. Recall here that the aim of this approach was to reduce data complexity. Thus, in this attack the gain on the data complexity is the important part[5].

In the full version of this paper [11] we show how to combine the state-test-technique together with multiple differentials in order to reduce at the same time the time and the data complexity for the attacks on CLEFIA-128.

4   Applications to Camellia

Camellia is a 128-bit block cipher designed by Aoki et. al. in 2000 [4]. It is a Feistel-like construction where two key-dependent layers FL and FL^{-1} are applied every 6 rounds to each branch. Whitening keys are equally applied to the first and the last round of the cipher. There exist three different versions of the cipher, that we note Camellia-128, Camellia-192 and Camellia-256, depending on the key size used. The number of iterations is 18 for the 128-bit version and 24 for the other two versions. A detailed description of Camellia's structure can be found in the full version of the paper. For further details, one can refer to [4].

Previous Cryptanalysis. Camellia is since 2005 an international ISO/IEC standard and has therefore attracted a lot of attention from the cryptographic community. Since Camellia has a particular design, involving the so-called FL/FL^{-1} layers, its cryptanalysis can be classified in several categories. Some attacks consider the FL/FL^{-1} functions, while others do not take them into consideration. Equally, some attacks take into account the whitening keys, whereas others don't and finally all attacks do not start from the same round. The best attacks on Camellia in terms of the number of rounds and the complexities are those presented in [22, Section 4.2].

Here we start by presenting improvements of the best attacks that include the FL/FL^{-1} layers and the whitening keys. Next we build an attack using the state-test technique on 14-round Camellia-256 starting from the first round but without the FL/FL^{-1} layers and the whitening keys.

Improvements. We improve here the complexities of the previous attacks that take into account the FL/FL^{-1} layers and the whitening keys on all three

---

[5] In [24], the authors used a loose approximation for C'_E, as $C'_E = 1/104$.

versions of Camellia. By using the complexity analysis introduced in Section 2, we can optimize the complexities of the corresponding attacks from [22]. Note that we use for this the same parameters as in [22]. The parameters of our attacks on 11-round Camellia-128, 12-round Camellia-192 and 13-round Camellia-256 are depicted in Table 3. As can be seen in Table 2, the time complexity of our improved attack on Camellia-128 is $2^{118.43}C_E$, with data complexity $2^{118.4}$ and memory complexity $2^{92.4}$. For Camellia-192, the time, data and memory complexities are $2^{161.06}C_E$, $2^{119.7}$ and $2^{150.7}$ respectively, while for Camellia-256 the corresponding complexities are $2^{225.06}C_E$, $2^{119.71}$ and $2^{198.71}$.

Table 3. Attack parameters against all versions of Camellia

Algorithm	$|\Delta_{in}|$	$|\Delta_{out}|$	r_{in}	r_{out}	r_Δ	c_{in}	c_{out}	$|k_{in} \cup k_{out}|$
Camellia-128	23	80	1	2	8	32	57	96
Camellia-192	80	80	2	2	8	73	73	160
Camellia-256	80	128	2	3	8	73	121	224

Using the State-Test Technique on Camellia-256. We provide here an impossible differential attack on Camellia-256 without FL/FL^{-1} layers and whitening keys by using the state-test technique. Note here, that unlike all previous attacks not starting from the first round in order to take advantage of the key schedule asymmetry, our attack starts from the first round of the cipher. It covers 14 rounds of Camellia-256 which is, to the best of our knowledge, the highest number of rounds attacked for this version. In [22] another attack on 14-round Camellia-256 with FL/FL^{-1} and whitening keys is presented, however, as said before, it does not start from the first round and it uses a specific property of the key schedule at the rounds where it is applied.

In this attack, we consider the same 8-round impossible differential as in [25] and we add 4+2 rounds such that $r_{in} = 4$, $r_{out} = 2$ and $r_\Delta = 8$. We have $|\Delta_{in}| = 128$, $|\Delta_{out}| = 56$, $c_{in} = 120$ and $c_{out} = 48$. Then we need at least $N_{min} = 2^{168}$ plaintexts pairs. The amount of data needed to construct these pairs is $C_{N_{min}} = \max\left\{\sqrt{2^{168}2^{129-128}}, 2^{168}2^{129-184}\right\} = 2^{113}$. There remain then $128 - 113 = 15$ bits of freedom. Thus, we can fix $s = 8$ bits on the ciphertexts to apply the state-test technique on the 8 bits of the internal state at the penultimate round. The number of information key bits is $|k_{in} \cup k_{out}| = 227 - 8 = 219$ since there are 45 bits shared between the subkeys with respect to the key schedule. The best attack is obtained with $N = 2^{118}$ pairs. In this case, the time complexity is $2^{220}C_E$, the data complexity is 2^{118} plaintexts and the memory is 2^{118}.

5   Conclusion

To start with, we have proposed in this paper a generic vision of impossible differential attacks with the aim of simplifying and helping the construction and verification of this type of cryptanalysis. Until now, these attacks were very tedious to mount and even more to verify, and so, very often flaws appeared in the computations. We believe that our objective has been successfully reached, as it can be seen by the high amount of new improved attacks that we have been able to propose, as well as by all the different possible trade-offs for each one of them, something that would be near to unthinkable prior to our work.

Next, the generic and clear vision of impossible differential attacks has allowed us to discover and propose new ideas for improving these attacks. In particular, we have proposed the *state-test* technique, that allows to reduce the number of key bits involved in the attack, and so to reduce the time complexity. We have also formalized and adapted to our generic scenario the notion introduced in [32] of multiple impossible differentials. This option allows reducing the data complexity. Finally, we have proposed several applications for different variants of the Feistel ciphers CLEFIA, Camellia, LBlock and SIMON, providing in most of the cases, the best known attack on reduced-round versions of these ciphers.

We hope that these results will simplify and improve future impossible attacks on Feistel ciphers, as well as their possible combination with other attacks. For instance, in [35] a combination of impossible differential with linear attacks is proposed. We haven't verified these results, but this direction could be promising.

References

1. Abed, F., List, E., Lucks, S., Wenzel, J.: Differential and linear cryptanalysis of reduced-round SIMON. Cryptology ePrint Archive, Report 2013/526 (2013)
2. Abed, F., List, E., Wenzel, J., Lucks, S.: Differential Cryptanalysis of round-reduced Simon and Speck. In: FSE 2014. LNCS. Springer (to appear, 2014)
3. Alkhzaimi, H.A., Lauridsen, M.M.: Cryptanalysis of the SIMON Family of Block Ciphers. Cryptology ePrint Archive, Report 2013/543 (2013)
4. Aoki, K., Ichikawa, T., Kanda, M., Matsui, M., Moriai, S., Nakajima, J., Tokita, T.: *Camellia*: A 128-Bit Block Cipher Suitable for Multiple Platforms - Design and Analysis. In: Stinson, D.R., Tavares, S. (eds.) SAC 2000. LNCS, vol. 2012, pp. 39–56. Springer, Heidelberg (2001)
5. Biham, E., Biryukov, A., Shamir, A.: Cryptanalysis of Skipjack Reduced to 31 Rounds Using Impossible Differentials. In: Stern, J. (ed.) EUROCRYPT 1999. LNCS, vol. 1592, pp. 12–23. Springer, Heidelberg (1999)
6. Biham, E., Shamir, A.: Differential Cryptanalysis of DES-like Cryptosystems. In: Menezes, A., Vanstone, S.A. (eds.) CRYPTO 1990. LNCS, vol. 537, pp. 2–21. Springer, Heidelberg (1991)
7. Blondeau, C.: Improbable Differential from Impossible Differential: On the Validity of the Model. In: Paul, G., Vaudenay, S. (eds.) INDOCRYPT 2013. LNCS, vol. 8250, pp. 149–160. Springer, Heidelberg (2013)
8. Bogdanov, A., Geng, H., Wang, M., Wen, L., Collard, B.: Zero-Correlation Linear Cryptanalysis with FFT and Improved Attacks on ISO Standards Camellia and CLEFIA. In: Lange, T., Lauter, K., Lisoněk, P. (eds.) SAC 2013. LNCS, vol. 8282, pp. 306–323. Springer, Heidelberg (2013)
9. Bouillaguet, C., Dunkelman, O., Fouque, P.-A., Leurent, G.: New Insights on Impossible Differential Cryptanalysis. In: Miri, A., Vaudenay, S. (eds.) SAC 2011. LNCS, vol. 7118, pp. 243–259. Springer, Heidelberg (2012)

10. Boura, C., Minier, M., Naya-Plasencia, M., Suder, V.: Improved Impossible Differential Attacks against Round-Reduced LBlock. Cryptology ePrint Archive, Report 2014/279 (2014)
11. Boura, C., Naya-Plasencia, M., Suder, V.: Scrutinizing and Improving Impossible Differential Attacks: Applications to CLEFIA, Camellia, LBlock and Simon (Full Version). Cryptology ePrint Archive, Report 2014/699 (2014)
12. Chen, J., Futa, Y., Miyaji, A., Su, C.: Impossible differential cryptanalysis of LBlock with concrete investigation of key scheduling algorithm. Cryptology ePrint Archive, Report 2014/272 (2014)
13. CLEFIA Design Team. Comments on the impossible differential analysis of reduced round CLEFIA presented at Inscrypt 2008 (January 8, 2009)
14. CLEFIA Design Team. Private communication (May 2014)
15. Dunkelman, O., Sekar, G., Preneel, B.: Improved Meet-in-the-Middle Attacks on Reduced-Round DES. In: Srinathan, K., Rangan, C.P., Yung, M. (eds.) INDOCRYPT 2007. LNCS, vol. 4859, pp. 86–100. Springer, Heidelberg (2007)
16. Gilbert, H., Peyrin, T.: Super-Sbox Cryptanalysis: Improved Attacks for AES-Like Permutations. In: Hong, S., Iwata, T. (eds.) FSE 2010. LNCS, vol. 6147, pp. 365–383. Springer, Heidelberg (2010)
17. Isobe, T., Shibutani, K.: Generic Key Recovery Attack on Feistel Scheme. In: Sako, K., Sarkar, P. (eds.) ASIACRYPT 2013, Part I. LNCS, vol. 8269, pp. 464–485. Springer, Heidelberg (2013)
18. Karakoç, F., Demirci, H., Harmancı, A.E.: Impossible Differential Cryptanalysis of Reduced-Round LBlock. In: Askoxylakis, I., Pöhls, H.C., Posegga, J. (eds.) WISTP 2012. LNCS, vol. 7322, pp. 179–188. Springer, Heidelberg (2012)
19. Kim, J., Hong, S., Lim, J.: Impossible differential cryptanalysis using matrix method. Discrete Mathematics 310(5), 988–1002 (2010)
20. Kim, J., Hong, S., Sung, J., Lee, C., Lee, S.: Impossible Differential Cryptanalysis for Block Cipher Structures. In: Johansson, T., Maitra, S. (eds.) INDOCRYPT 2003. LNCS, vol. 2904, pp. 82–96. Springer, Heidelberg (2003)
21. Knudsen, L.R.: DEAL – A 128-bit cipher. Technical Report, Department of Informatics, University of Bergen, Norway (1998)
22. Liu, Y., Li, L., Gu, D., Wang, X., Liu, Z., Chen, J., Li, W.: New Observations on Impossible Differential Cryptanalysis of Reduced-Round Camellia. In: Canteaut, A. (ed.) FSE 2012. LNCS, vol. 7549, pp. 90–109. Springer, Heidelberg (2012)
23. Lu, J., Kim, J., Keller, N., Dunkelman, O.: Improving the Efficiency of Impossible Differential Cryptanalysis of Reduced Camellia and MISTY1. In: Malkin, T. (ed.) CT-RSA 2008. LNCS, vol. 4964, pp. 370–386. Springer, Heidelberg (2008)
24. Mala, H., Dakhilalian, M., Shakiba, M.: Impossible Differential Attacks on 13-Round CLEFIA-128. J. Comput. Sci. Technol. 26(4), 744–750 (2011)
25. Mala, H., Shakiba, M., Dakhilalian, M., Bagherikaram, G.: New Results on Impossible Differential Cryptanalysis of Reduced–Round Camellia–128. In: Jacobson Jr., M.J., Rijmen, V., Safavi-Naini, R. (eds.) SAC 2009. LNCS, vol. 5867, pp. 281–294. Springer, Heidelberg (2009)
26. Minier, M., Naya-Plasencia, M.: A Related Key Impossible Differential Attack Against 22 Rounds of the Lightweight Block Cipher LBlock. Inf. Process. Lett. 112(16), 624–629 (2012)
27. Minier, M., Naya-Plasencia, M.: Private communication (May 2013)
28. Shirai, T., Shibutani, K., Akishita, T., Moriai, S., Iwata, T.: The 128-Bit Blockcipher CLEFIA (Extended Abstract). In: Biryukov, A. (ed.) FSE 2007. LNCS, vol. 4593, pp. 181–195. Springer, Heidelberg (2007)

29. Tang, X., Sun, B., Li, R., Li, C.: Impossible differential cryptanalysis of 13-round CLEFIA-128. Journal of Systems and Software 84(7), 1191–1196 (2011)
30. Tezcan, C.: The Improbable Differential Attack: Cryptanalysis of Reduced Round CLEFIA. In: Gong, G., Gupta, K.C. (eds.) INDOCRYPT 2010. LNCS, vol. 6498, pp. 197–209. Springer, Heidelberg (2010)
31. Tsunoo, Y., Tsujihara, E., Shigeri, M., Saito, T., Suzaki, T., Kubo, H.: Impossible Differential Cryptanalysis of CLEFIA. In: Nyberg, K. (ed.) FSE 2008. LNCS, vol. 5086, pp. 398–411. Springer, Heidelberg (2008)
32. Tsunoo, Y., Tsujihara, E., Shigeri, M., Suzaki, T., Kawabata, T.: Cryptanalysis of CLEFIA using multiple impossible differentials. In: Information Theory and Its Applications, ISITA 2008, pp. 1–6 (2008)
33. Wu, W., Zhang, L., Zhang, W.: Improved Impossible Differential Cryptanalysis of Reduced-Round Camellia. In: Avanzi, R.M., Keliher, L., Sica, F. (eds.) SAC 2008. LNCS, vol. 5381, pp. 442–456. Springer, Heidelberg (2009)
34. Wu, W., Zhang, W., Feng, D.: Impossible Differential Cryptanalysis of Reduced-Round ARIA and Camellia. J. Comput. Sci. Technol. 22(3), 449–456 (2007)
35. Yuan, Z., Li, X., Liu, H.: Impossible Differential-Linear Cryptanalysis of Reduced-Round CLEFIA-128. Cryptology ePrint Archive, Report 2013/301 (2013)
36. Zhang, W., Han, J.: Impossible Differential Analysis of Reduced Round CLEFIA. In: Yung, M., Liu, P., Lin, D. (eds.) Inscrypt 2008. LNCS, vol. 5487, pp. 181–191. Springer, Heidelberg (2009)

A Simplified Representation of AES

Henri Gilbert[*]

ANSSI, France
henri.gilbert@ssi.gouv.fr

Abstract. We show that the so-called *super S-box representation* of AES – that provides a simplified view of two consecutive AES rounds – can be further simplified. In the *untwisted representation* of AES presented here, two consecutive AES rounds are viewed as the composition of a non-linear transformation S and an affine transformation R that respectively operate on the four 32-bit columns and on the four 32-bit rows of their 128-bit input. To illustrate that this representation can be helpful for analysing the resistance of AES-like ciphers or AES-based hash functions against some structural attacks, we present some improvements of the known-key distinguisher for the 7-round variant of AES presented by Knudsen and Rijmen at ASIACRYPT 2007. We first introduce a known-key distinguisher for the 8-round variant of AES which constructs a 2^{64}-tuple of (input,output) pairs satisfying a simple integral property. While this new 8-round known-key distinguisher is outperformed for 8 AES rounds by known-key differential distinguishers of time complexity 2^{48} and 2^{44} presented by Gilbert and Peyrin at FSE 2010 and Jean, Naya-Plasencia, and Peyrin at SAC 2013, we show that one can take advantage of its specific features to mount a known-key distinguisher for the 10-round AES with independent subkeys and the full AES-128. The obtained 10-round distinguisher has the same time complexity 2^{64} as the 8-round distinguisher it is derived from, but the highlighted input-output correlation property is more intricate and therefore its impact on the security of the 10-round AES when used as a known key primitive, e.g. in a hash function construction, is questionable. The new known-key distinguishers do not affect at all the security of AES when used as a keyed primitive, for instance for encryption or message authentication purposes.

1 Introduction

In this paper we present an alternative representation of AES. More precisely we show that AES can be viewed as the composition of other elementary transformations than those originally used for the specification of its round function. While one might wonder whether selecting any of the equivalent descriptions of a cipher is more than an arbitrary convention, numerous examples illustrate that the choice of an appropriate description may be very useful for highlighting some

---

[*] This work was partially supported by the French National Research Agency through the BLOC project (contract ANR-11-INS-011).

P. Sarkar and T. Iwata (Eds.): ASIACRYPT 2014, PART I, LNCS 8873, pp. 200–222, 2014.

of its structural features and serve as a starting point for its cryptanalysis or for optimised implementions. To take a simple example, it is well known that while the so-called ladder representation of the Feistel scheme is strictly equivalent to its more traditional twisted representation for any even number of rounds, it is helpful for understanding some attacks against DES and DES-like ciphers, for instance the Davies-Murphy attack [8].

In the case of AES, several alternative representations have been proposed [9,20] to highlight some aspects of its algebraic structure. These representations respectively allow to relate the ciphertext to the plaintext using continued fractions, resp. algebraic equations over $GF(2^8)$. In [2] it was shown that numerous dual ciphers of AES - *i.e.* equivalent descriptions of AES up to fixed, easy to compute and to invert bijective mappings on the plaintexts, the ciphertexts, and the keys - can be obtained by applying appropriately chosen modifications to the irreducible polynomial used to represent $GF(2^8)$, the affine transformation in the S-box, the coefficients of MixColumns, etc. This observation was further extended in [3]. While these dual ciphers can be considered as equivalent representations of AES, these representations essentially preserve the structure of the round function of the AES up to small variations on the exact parameter of each elementary transformation. They are therefore closer to the original AES than the equivalent representations we consider in this paper.

The starting point for the AES representation introduced here is the so-called super S-box (or super-box) representation of two AES rounds which allows to describe two consecutive AES rounds as the composition of one single non-linear operation, namely a range of four parallel 32-bit to 32-bit key-dependent S-boxes and several affine transformations. This representation was introduced in [7] by the designers of AES as a useful notion for the analysis of AES differentials over two rounds. It was subsequently reused in [11,12] and [18] in order to extend so-called rebound attacks on AES-like permutations by at least one round: this improved rebound technique, sometimes referred to as super S-box cryptanalysis, was shown to be applicable in two related contexts, the cryptanalysis of AES-like hash functions and the investigation of so-called known-key distinguishers for AES-like block ciphers. Many recent improved distinguishers for reduced-round versions of AES-like hash functions such as the SHA-3 candidates Grøstl and ECHO are using super S-boxes, e.g. [19,16,15].

We introduce a novel representation of two consecutive AES rounds that results from an extra simplification of the super S-box representation. The simplification relates to the description of the affine transformations that surround the 32-bit super S-boxes. We show that all these transformations can be replaced by one simple 32-bit oriented affine transformation that operates on the rows of the 4×4 matrix of bytes representing the current state. We propose to name the resulting view of two or more generally r AES rounds the *untwisted* representation since it avoids viewing the affine transformations that surround the super S-boxes as column-oriented operations "twisted" by the action the ShiftRows transformation. The untwisted representation thus provides an equivalent description of two consecutive AES rounds as the composition of:

- *a non-linear transformation* denoted by S (a shorthand for "super S-boxes") that consists of the parallel application of four non-linear bijective mappings which operate on the four 32-bit columns of the AES state. These four mappings are essentially super S-boxes up to permutations of the four input bytes and the four output bytes of each column;
- *an affine transformation* denoted by R (a shorthand for "MixRows") that consists of the parallel application of four affine mappings which operate on the four 32-bit rows of the AES state.

S   R

Fig. 1. Equivalent representation of two AES rounds as the composition $R \circ S$ of four parallel non-linear bijections of the columns and four parallel affine bijections of the rows of the input state

As shown in Figure 1, two consecutive AES rounds can thus be viewed as one "super-round" that is the composition $R \circ S$ of S and R. As will be shown more in detail in the sequel, the small price to pay for this simplified view is that in the resulting equivalent representation of $2r$ AES rounds as the composition of r super-rounds, the first (resp. last) super-round is preceded (resp. followed) by a simple affine permutation.

While an alternative representation of a cipher can obviously be regarded in itself neither as a design nor as a cryptanalysis result, we believe that the simplicity of the new representation can play a significant heuristic role in the investigation of structural attacks on reduced-round versions of AES. Indeed, the new representation pushes the advantage of the super S-box representation of highlighting the 32-bit structure underlying the AES transformation one step further.

To illustrate this alternative representation, we present extensions of the known cryptanalytic results on reduced-round versions of AES in the so-called known-key model. The known-key model was first introduced by Knudsen and Rijmen in [17]. Attacks in this model are most often named known-key distinguishers and we will use this terminology in the sequel.[1] An integral known-key distinguisher for the 7-round AES was introduced by Knudsen and Rijmen in [17]. We first present an improvement of this distinguisher whose idea was inspired by the use of the untwisted representation of AES. This provides a

---

[1] This terminology may seem a bit confusing since known-key distinguishers have little to do with the notion of distinguisher one considers in more traditional security models, namely a testing algorithm with an oracle access capability. But on the other hand the wording known-key distinguisher conveys probably less risks of misinterpretation than the wording known-key attack.

known-key distinguisher against the 8-round AES. While this distinguisher is outperformed by the differential known-key distinguishers for the 8-round AES of [12] and [14], whose respective complexities are 2^{48} and 2^{44}, we show that one can take advantage of its specific features, that reflect integral properties of the 8-round AES, to extend it by one outer round at both sides. We thus obtain the first known-key distinguisher for the full 10-round AES. This known-key distinguisher has the same time complexity 2^{64} (now measured as an equivalent number of 10-round AES encryptions) as the one of the 8-round distinguisher it is derived from, but the highlighted input-output correlation property is more intricate. We nevertheless provide some evidence that unlike some generic known distinguishers that are known to exist for block ciphers if the key size is sufficiently small, the obtained distinguisher can reasonably be considered meaningful. While in this paper we will only investigate the security of AES in the known-key model, it is worth mentioning a recent result on the security of AES in a related but even stronger security model, namely the chosen-key distinguisher on the 9-round AES-128 of [10].

The rest of this paper is organized as follows. In Section 2, we introduce the novel representation of two consecutive AES rounds and of $2r$ AES rounds. In Section 3, we propose a definition of the known-key model, *i.e.* we define the adversaries considered in this model and we remind known impossibility results on the resistance of block ciphers to all known-key distinguishers. In Section 4, we show how to use the untwisted representation of AES to mount known-key distinguishers for the 8-round AES and its extension to the full 10-round AES and why the latter distinguisher can be considered meaningful.

2   A New Representation of AES

Notational Conventions and Usual Representation of AES. Throughout this paper we most often denote the composition of two mappings F and G multiplicatively by $F \cdot G$ instead of using the more classical notation $G \circ F$. The advantage of this notation in the context considered here is that when read from left to right it describes the successive transformations that are applied to the input value.

Let us briefly recall the AES features that will be useful for the sequel and the associated notation. Each AES block is represented by a four times four matrix of bytes. While there are three standard versions of AES, of respective key lengths $k = 128$, 192, and 256 bits and respective number of rounds 10, 12, and 14 rounds, for the purpose of this paper we restrict ourselves for the sake of simplicity to the full 10-round AES-128 and reduced-round versions of this cipher.[2] For $r \leq 10$, the r-round version of the AES-128 encryption function is denoted by AES_r and is parametrized by $(r + 1)$ 128-bit subkeys denoted by K_0 to K_r. These subkeys

---

[2] However, since the AES properties we are investigating do not relate to the key schedule but to the data encryption part of the block cipher that is the same for all AES versions, all the presented results are also applicable to reduced-round versions of AES-192 and AES-256.

are derived from a k-bit key K by the key schedule; since the exact features of the AES-128 key schedule are not relevant for the analysis presented here, we do not detail them and refer to the full specification of AES for their description. Each round of the encryption function AES_r is the composition $SB \cdot SR \cdot MC \cdot AK$ of four transformations named SubBytes or SB, ShiftRows or SR, MixColumns or MC, and AddRoundKey or AK. SubBytes applies a fixed 8-bit to 8-bit bijective S-box to each input byte, ShiftRows circularly shifts each of the four 4-byte rows of the input state by 0, 1, 2, and 3 bytes to the left, MixColumns applies to each of the four-byte columns of the input state, viewed as a 4-coordinate vector with $GF(2^8)$ coefficients, a left multiplication by a fixed 4×4 matrix M with $GF(2^8)$ coefficients, and at round $i \in [1; r]$, AddRoundKey or AK consists of a bytewise exclusive or of the input block with subkey K_i.[3] The first round of AES_r is preceded by a key addition with the subkey K_0 and the MixColumns operation is omitted in the last round. In the sequel we will sometimes also have to refer to the variant of AES_r where the MixColumns transformation is kept in the last round: we will denote this variant by AES_{r+}. At the end of Section 4, we will also have to refer to the r-round variant of AES parametrized by $r + 1$ independent subkeys. Depending whether the MixColumns transformation is omitted or kept in the last round, we will denote this variant by AES_r^* or AES_{r+}^*.

Super S-box Representation of 2 Consecutive AES Rounds. The super S-box representation allows to view two consecutive AES rounds as the parallel invocation of four 32-bit to 32 bit mappings named super S-boxes - which are applied to the four columns of the AES state - surrounded by affine applications. More in detail, since the transformations SB and SR commute and the composition of transformations is associative, the composition of two consecutive rounds:

$$SB \cdot SR \cdot MC \cdot AK \cdot SB \cdot SR \cdot MC \cdot AK$$

can be rewritten as:

$$SR \cdot (SB \cdot MC \cdot AK \cdot SB) \cdot SR \cdot MC \cdot AK.$$

We can notice that the middle term in brackets, *i.e.* $SuperSB = (SB \cdot MC \cdot AK \cdot SB)$, where $SuperSB$ stands for "Super S-boxes", is the composition of transformations that all preserve the column-wise structure of the AES state. Thus $SuperSB$ splits up into 4 parallel key-dependent bijective transformations of one column of the input state. It is surrounded by the left, resp right affine transformations SR, resp $SR \cdot MC \cdot AK$. Each super S-box applies its 4-byte input column the composition of 4 parallel S-box invocations, a left multiplication by the MixColumn matrix M, a xor with a 32-bit subkey column, and 4 final parallel S-box invocations.

---

[3] Since AddRoundKey is parametrized by a subkey the use of the notation AK, that suggests a fixed transformation, is a slight abuse of notation, but this notation is convenient in the context of this paper: in the sequel AK just stands for a xor with some constant — whose value does not affect the properties we consider.

Moving to the Untwisted Representation of 2 Consecutive AES Rounds. We now show how to move from the super S-box representation of two consecutive rounds to their untwisted representation as the composition $S \cdot R$ of four parallel column-wise non-linear transformations and four parallel row-wise affine transformations. We first observe that the periodic repetition, in r iterations, of the 2-round pattern associated with the super S-box representation:

$$SR \cdot SuperSB \cdot SR \cdot MC \cdot AK$$

can be equivalently viewed as the periodic repetition in r iterations of the cyclically shifted periodic 2-round pattern:

$$SuperSB \cdot SR \cdot MC \cdot AK \cdot SR$$

up to a minor correction, namely the left composition of the first iteration with SR and the right composition of the last iteration with SR^{-1}. Now in order to move to the aimed 2-round representation the conducting idea is to left and right-compose the $SuperSB$ and $SR \cdot MC \cdot AK \cdot SR$ transformations using well chosen byte permutations P and Q and their inverses P^{-1} and Q^{-1}. Due to the cancellation effect produced by the alternate use of these permutations and their inverse, r iterations of the obtained 2-round description:

$$(Q^{-1} \cdot SuperSB \cdot P^{-1}) \cdot (P \cdot SR \cdot MC \cdot AK \cdot SR \cdot Q)$$

gives, for any choice of the two byte permutations, exactly the same product as r iterations of the 2-round transformation it is derived from, up to a left composition of the first iteration by Q^{-1} and a right composition of the last iteration by Q. In order for the byte permutations P and Q to provide the desired untwisted representation, they must satisfy the two following extra requirements:

- (i) the non-linear transformation $S = Q^{-1} \cdot SuperSB \cdot P^{-1}$ must operate on columns;
- (ii) the affine transformation $R = P \cdot SR \cdot MC \cdot AK \cdot SR \cdot Q$ must operate on rows.

In order to describe the byte permutations satisfying the above requirements that we found, we introduce the following auxiliary byte permutations:

- we denote by T the matrix transposition that operates on 4×4 matrices of bytes as follows:

$$T : \begin{pmatrix} a_0 & a_4 & a_8 & a_{12} \\ a_1 & a_5 & a_9 & a_{13} \\ a_2 & a_6 & a_{10} & a_{14} \\ a_3 & a_7 & a_{11} & a_{15} \end{pmatrix} \mapsto \begin{pmatrix} a_0 & a_1 & a_2 & a_3 \\ a_4 & a_5 & a_6 & a_7 \\ a_8 & a_9 & a_{10} & a_{11} \\ a_{12} & a_{13} & a_{14} & a_{15} \end{pmatrix}$$

- we denote by SC (or SwapColumns) the swapping of the second and fourth columns of the input state:

$$SC : \begin{pmatrix} a_0 & a_4 & a_8 & a_{12} \\ a_1 & a_5 & a_9 & a_{13} \\ a_2 & a_6 & a_{10} & a_{14} \\ a_3 & a_7 & a_{11} & a_{15} \end{pmatrix} \mapsto \begin{pmatrix} a_0 & a_{12} & a_8 & a_4 \\ a_1 & a_{13} & a_9 & a_5 \\ a_2 & a_{14} & a_{10} & a_6 \\ a_3 & a_{15} & a_{11} & a_7 \end{pmatrix}$$

Proposition 1. *The byte permutations*

$$P = SR \cdot T \cdot SR^{-1} \quad : \quad \begin{pmatrix} a_0 & a_4 & a_8 & a_{12} \\ a_1 & a_5 & a_9 & a_{13} \\ a_2 & a_6 & a_{10} & a_{14} \\ a_3 & a_7 & a_{11} & a_{15} \end{pmatrix} \mapsto \begin{pmatrix} a_0 & a_5 & a_{10} & a_{15} \\ a_3 & a_4 & a_9 & a_{14} \\ a_2 & a_7 & a_8 & a_{13} \\ a_1 & a_6 & a_{11} & a_{12} \end{pmatrix}$$

and

$$Q = SR^{-1} \cdot T \cdot SR \cdot SC \quad : \quad \begin{pmatrix} a_0 & a_4 & a_8 & a_{12} \\ a_1 & a_5 & a_9 & a_{13} \\ a_2 & a_6 & a_{10} & a_{14} \\ a_3 & a_7 & a_{11} & a_{15} \end{pmatrix} \mapsto \begin{pmatrix} a_0 & a_7 & a_{10} & a_{13} \\ a_1 & a_4 & a_{11} & a_{14} \\ a_2 & a_5 & a_8 & a_{15} \\ a_3 & a_6 & a_9 & a_{12} \end{pmatrix}$$

satisfy the requirements (i) and (ii) and thus result in the desired untwisted representation.

Proof sketch.
(i): It is easy to see that P, Q, and their inverses operate on columns. Therefore $S = Q^{-1} \cdot SuperSB \cdot P^{-1}$ also operates on columns.
(ii): We can simplify the expression of R:

$$R = P \cdot SR \cdot MC \cdot AK \cdot SR \cdot Q$$
$$= SR \cdot T \cdot SR^{-1} \cdot SR \cdot MC \cdot AK \cdot SR \cdot SR^{-1} \cdot T \cdot SR \cdot SC$$
$$= SR \cdot T \cdot MC \cdot AK \cdot T \cdot SR \cdot SC$$

Since $T \cdot MC \cdot T$ and therefore $T \cdot MC \cdot AK \cdot T$ operates on rows and SR and SC also operate on rows, R operates on rows. □

The linear part of the row-wise affine transformation R determined by P and Q is described by the four following circulant matrices R_i, $i = 0$ to 3. Each matrix R_i operates on a 4-byte row vector x_i that represents row i of the input block of R and produces the 4-byte row vector $y_i = x_i \cdot R_i$ that represents row i of the linear part of the image of the input block by R. The coefficients of the R_i are those of the MixColumns matrix M (in a different order).

$$R_0 = R_2 = \begin{pmatrix} 2 & 3 & 1 & 1 \\ 3 & 1 & 1 & 2 \\ 1 & 1 & 2 & 3 \\ 1 & 2 & 3 & 1 \end{pmatrix} \quad R_1 = R_3 = \begin{pmatrix} 1 & 1 & 2 & 3 \\ 1 & 2 & 3 & 1 \\ 2 & 3 & 1 & 1 \\ 3 & 1 & 1 & 2 \end{pmatrix} \quad M = \begin{pmatrix} 2 & 3 & 1 & 1 \\ 1 & 2 & 3 & 1 \\ 1 & 1 & 2 & 3 \\ 3 & 1 & 1 & 2 \end{pmatrix}$$

Remark. (P, Q) is not the unique pair of byte permutations that satisfy requirements (i) and (ii). Given any permutations σ and τ of the set $\{0, 1, 2, 3\}$, let us denote by C_σ, resp. D_τ the associated column and row permutations, that on input a 4-tuple (x_0, x_1, x_2, x_3) of columns, resp. of rows produces the permuted 4-tuple (x'_0, x'_1, x'_2, x'_3) of columns, resp. of rows given by $x'_{\sigma(i)} = x_i$, resp. $x'_\tau(i) = x_i$, i.e. $x'_i = x_{\sigma^{-1}(i)}$ resp. $x'_i = x_\tau^{-1}(i)$, $i = 0$ to 3. It is easy to see that all the pairs of byte permutations $(P_{\sigma,\tau}, Q_{\sigma,\tau}) = (C_\sigma \cdot D_\tau \cdot P, Q \cdot D_{\tau^{-1}} \cdot C_{\sigma^{-1}})$ also satisfy requirements (i) and (ii). We will however only use (P, Q) in the sequel.

Resulting Untwisted Representation of AES$_{2r+}$ and AES$_{2r}$. The former 2-round untwisted representation of two consecutive AES rounds immediately results in the following equivalent untwisted description of the 2r-round version AES$_{2r+}$ of the encryption function of AES (in which the MixColumns transformation is kept in the last round).

$$AES_{2r+} = AK \cdot IP \cdot (S \cdot R)^r \cdot FP,$$

where the initial and final permutations IP and FP are the byte permutations given by:

$$IP = SR \cdot Q = T \cdot SR \cdot SC;$$
$$FP = Q^{-1} \cdot SR^{-1} = IP^{-1}.$$

This representation AES$_{2r+}$ is illustrated on Figure 2. To confirm the equivalence of the above representation of AES$_{2r+}$ with its usual representation using SB, SR, MC, and AK, implementations based on both representations were checked to provide equal output values on a few input values.

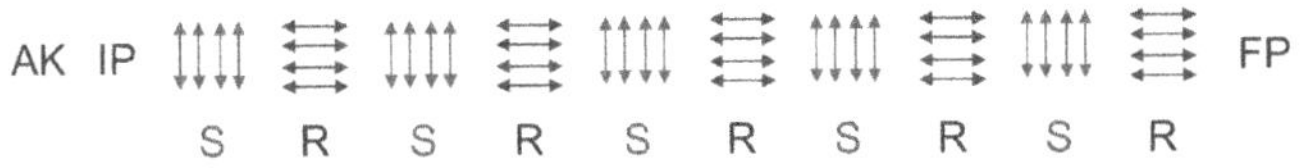

Fig. 2. Equivalent representation of AES$_{10+}$. IP and FP are permutations of the byte positions.

The former representation of AES$_{2r+}$ can be used to derive a first representation of AES$_{2r}$, that will be used in the sequel to mount a known-key distinguisher for AES$_8$. The right composition of AES$_{2r}$ with an appropriate conjugate of MC^{-1} is required in order to cancel out the MixColumns operation in the last round. If one "develops" the last occurrence of R and simplifies the obtained expression, one obtains the equality:

$$AES_{2r} = AK \cdot IP \cdot (S \cdot R)^{r-1} \cdot S \cdot P \cdot SR \cdot AK.$$

We also introduce a second equivalent representation of AES$_{2r}$ that will be used in the sequel to mount a known-key distinguisher for AES$_{10}$: we start from an equivalent representation of the $2(r-1)$-round version AES$_{2(r-1)+}$ of AES, apply a left composition with a full round and a right composition with a last round without MixColumns, and simplify the obtained expression using the equality $R = P \cdot SR \cdot MC \cdot AK \cdot SR \cdot Q$.

$$
\begin{aligned}
AES_{2r} &= (AK \cdot SB \cdot SR \cdot MC) \cdot AES_{2(r-1)+} \cdot (SB \cdot SR \cdot AK) \\
&= AK \cdot SB \cdot SR \cdot MC \cdot AK \cdot SR \cdot Q \cdot (S \cdot R)^{r-1} \cdot Q^{-1} \cdot SR^{-1} \cdot SB \cdot SR \cdot AK \\
&= AK \cdot SB \cdot P^{-1} \cdot R \cdot (S \cdot R)^{r-1} \cdot Q^{-1} \cdot SB \cdot AK \\
&= AK \cdot P^{-1} \cdot SB \cdot R \cdot (S \cdot R)^{r-1} \cdot SB \cdot Q^{-1} \cdot AK
\end{aligned}
$$

Thus AES$_{2r}$ can be equivalently viewed as a middle transformation $R \cdot (S \cdot R)^{r-1}$ preceded and followed by simplified initial and final "external rounds", namely $AK \cdot P^{-1} \cdot SB$ and $SB \cdot Q^{-1} \cdot AK$.

3   The Known-Key Model

We believe that the untwisted AES representation introduced above can potentially help analysing known *structural attacks* of reduced-round versions of AES, AES-like ciphers, or AES-based hash functions.[4] In the next section we will present two "attacks" that substantiate this belief. They both happen to belong to a quite specific class of structural attacks, the so-called known-key distinguishers, and respectively relate to a reduced-round version of AES and the full 10-round AES-128. In this section we introduce the underlying security model, that is named the *known-key* model. This model was inspired from the cryptanalysis of hash functions and first introduced by Knudsen and Rijmen in [17]. The difference between the known-key model and the usual security model considered for block ciphers can be outlined as follows.

- In the usual model, the adversary is given a *black box* (oracle) access to an instance of the encryption function associated with a random *secret* key and its inverse and must find the key or more generally efficiently distinguish the encryption function from a perfect random permutation;
- In the known-key model, the adversary is given a *white box* (*i.e.* full) access to an instance of the encryption function associated with a *known* random key and its inverse and her purpose is to simultaneously control the inputs and the outputs of the primitive, *i.e.* to achieve input-output correlations she could not efficiently achieve with the inputs and outputs of a perfect random permutation to which she would have an oracle access.

We now propose a more detailed definition of the known-key model – *i.e.* of the adversaries considered in this model, that are named *known-key distinguishers*. In order to capture the idea that the goal of such adversaries is to derive an N-tuple of input blocks of the considered block cipher E that is "abnormally correlated" with the corresponding N-tuple of output blocks, we first introduce the notion of T-intractable relation on N-tuples of E blocks. This notion (that is independent of E up to the fact that for the sake of simplicity we are using the time complexity of E as the unit for quantifying time complexities) is closely related to the notion of correlation intractable relation proposed in [6]. It essentially expresses that it is difficult to derive from oracle queries to a random permutation and its inverse an N-tuple of input/output pairs satisfying the relation.

Definition 1 (T-Intractable Relation). *Let $E : (K, X) \in \{0,1\}^k \times \{0,1\}^n \mapsto E_K(X) \in \{0,1\}^n$ denote a block cipher of block size n bits. Let $N \geq 1$ and $\mathcal{R}$ denote an integer and any relation[5] over the set S of N-tuples of n-bit blocks. $\mathcal{R}$*

---

[4] By structural attacks we mean here attacks that unlike statistical attacks, e.g. differential and linear cryptanalysis, do not consider the detail of the algorithm's elementary ingredients such as the S-boxes, but put more emphasis on their overall construction, their use of transformations that preserve the byte structure or the 32-bit structure of the data, etc.

[5] Let us remind that for any set S, a relation $\mathcal{R}$ over S can be defined as a subset of the cartesian product $S \times S$ and that for any pair (a, b) of $S \times S$, $a\mathcal{R}b$ means that (a, b) belongs to this subset.

is said to be T-intractable relatively to E if, given any algorithm $\mathcal{A}'$ that is given an oracle access to a perfect random permutation Π of $\{0,1\}^n$ and its inverse, it is impossible for $\mathcal{A}'$ to construct in time $T' \leq T$ two N-tuples $\mathcal{X}' = (X_i')$ and $\mathcal{Y}' = (Y_i')$ such that $Y_i' = \Pi(X_i')$, $i = 1 \cdots N$ and $\mathcal{X}' \mathcal{R} \mathcal{Y}'$ with a success probability $p' \geq \frac{1}{2}$ over Π and the random choices of $\mathcal{A}'$. The computing time T' of $\mathcal{A}'$ is measured as an equivalent number of computations of E, with the convention that the time needed for one oracle query to Π or Π^{-1} is equal to 1. Thus if q' denotes the number of queries of $\mathcal{A}'$ to Π or Π^{-1}, $q' \leq T'$.

Definition 2 (Known-Key Distinguisher). *Let $E : (K, X) \in \{0,1\}^k \times \{0,1\}^n \mapsto E_K(X) \in \{0,1\}^n$ denote a block cipher of block size n bits. A known-key distinguisher $(\mathcal{R}, \mathcal{A})$ of order $N \geq 1$ consists of (1) a relation $\mathcal{R}$ over the N-tuples of n-bit blocks (2) an algorithm $\mathcal{A}$ that on input a k-bit key K produces in time $T_\mathcal{A}$, i.e. in time equivalent with $T_\mathcal{A}$ computations of E, an N-tuple $\mathcal{X} = (X_i)_{i=1\cdots N}$ of plaintext blocks and an N-tuple $\mathcal{Y} = (Y_i)_{i=1\cdots N}$ of ciphertext blocks related by $Y_i = E_K(X_i)$, The two following conditions must be met:*
(i)   The relation $\mathcal{R}$ must be $T_\mathcal{A}$-intractable relatively to E.
(ii) The validity of $\mathcal{R}$ must be efficiently checkable: we formalize this requirement by incorporating the time for checking whether two N-tuples are related by $\mathcal{R}$ in the computing time $T_\mathcal{A}$ of algorithm $\mathcal{A}$.[6]

It is important for the sequel to notice that in the former definition, while the algorithm $\mathcal{A}$ takes a random key K as input, the relation $\mathcal{R}$ satisfied by the N-tuples of input and output blocks constructed by $\mathcal{A}$ is the same for all values of K and must be efficiently checkable without knowing K.

Example 1. The following example of a known-key distinguisher of order $N = 2$ illustrates the link between the use of block ciphers for hashing purposes and their security in the known-key model. Let E denote a block cipher of key length k bits and block length n bits and (X_1, X_2) and (Y_1, Y_2) denote two pairs of n-bit blocks. We define the relation $(X_1, X_2) \mathcal{R} (Y_1, Y_2)$ by the conditions $X_1 \neq X_2$ and $X_1 \oplus Y_1 = X_2 \oplus Y_2$. The definition of relation $\mathcal{R}$ obviously implies that if E is vulnerable to a known-key distinguisher $(\mathcal{R}, \mathcal{A})$ of complexity $T \ll 2^{\frac{n}{2}}$, then the compression function $h : \{0,1\}^k \times \{0,1\}^n \to \{0,1\}^n : (K, X) \mapsto X \oplus E_K(X)$ derived from E using the Matyas-Meyer-Oseas construction is vulnerable to a collision attack of complexity T that is more powerful than any generic collision attack against h.[7]

In the next example and throughout the rest of this paper, we are using the following notation to describe integral properties of partial AES encryptions and decryptions.

Notation. *Let $F : \{0,1\}^n \to \{0,1\}^n$ denote any mapping over the block space and let us consider the transformation by F of a structure $\mathcal{X}$ of $N = 2^{8m}$ blocks,*

---

[6] This avoids specifying an explicit upper bound on the time complexity for checking whether two N-tuples are related by $\mathcal{R}$. In practice one typically expects the time complexity for checking $\mathcal{R}$ to be at most the one of N computations of E.

[7] It could be shown that if $T \ll 2^{\frac{n}{2}}$, $\mathcal{R}$ is T-intractable.

$m \leq 16$. *An input or output byte* b_i, $i \in \{0, \cdots, 15\}$ *of* F *is said to be constant and marked* C *if it takes one constant value. It is said to be uniform and marked* U *if it takes each of the* 2^8 *possible values exactly* $2^{8(m-1)}$ *times. A* s-*tuple* $(b_{i_1}, \cdots, b_{i_s})$, *where* $s \leq m$ *and* $i_1, \cdots i_s \in \{0, \cdots, 15\}$, *of input or output bytes of* F *is said to be uniform and marked* $U_1, \cdots U_s$ *if* $(b_{i_1}, \cdots, b_{i_s})$ *takes each of the* 2^{8s} *possible* s-*tuple values exactly* $2^{8(m-s)}$ *times.*

Example 2. The known-key distinguisher for AES$_7$ introduced in [17] uses a relation $\mathcal{R}$ of order $N = 2^{56}$ that exploits integral properties of partial AES encryptions and decryptions. The following integral properties are used:

$$\begin{pmatrix} U_1 & C & C & C \\ C & U_2 & C & C \\ C & C & U_3 & C \\ C & C & C & U_4 \end{pmatrix} \overset{+4r}{\to} \begin{pmatrix} U & U & U & U \\ U & U & U & U \\ U & U & U & U \\ U & U & U & U \end{pmatrix} \text{ and } \begin{pmatrix} U_1 & C & C & C \\ U_2 & C & C & C \\ U_3 & C & C & C \\ U_4 & C & C & C \end{pmatrix} \overset{-3r}{\to} \begin{pmatrix} U & U & U & U \\ U & U & U & U \\ U & U & U & U \\ U & U & U & U \end{pmatrix}$$

where $4r$ denotes 4 consecutive AES encryption rounds without MixColumns in the last round and $-3r$ denotes 3 full AES decryption rounds. These properties imply that if a middle structure $\mathcal{Z}$ of $N = 2^{56}$ blocks is chosen as to satisfy the properties of the intermediate block of the scheme below, then by applying 4 forward encryption rounds and 3 backward decryption rounds to this structure one obtains a N-tuple of (plaintext, ciphertext) pairs that satisfy the relation $\mathcal{R}$ that (1) the N input blocks are pairwise distinct and (2) each of the 16 input bytes and each of the 16 output bytes is uniformly distributed.

$$\begin{pmatrix} U & U & U & U \\ U & U & U & U \\ U & U & U & U \\ U & U & U & U \end{pmatrix} \overset{-3r}{\leftarrow} \overset{\mathcal{Z}}{\begin{pmatrix} U_1 & C & C & C \\ U_5 & U_2 & C & C \\ U_6 & C & U_3 & C \\ U_7 & C & C & U_4 \end{pmatrix}} \overset{+4r}{\to} \begin{pmatrix} U & U & U & U \\ U & U & U & U \\ U & U & U & U \\ U & U & U & U \end{pmatrix}$$

While $\mathcal{R}$ could be shown to be N-intractable by the same kind of arguments as those used in the next section, we do not give a detailed proof here. The authors of [17] do not use exactly the same notion of T-intractable relation, but conjecture the related – somewhat stronger – property that "for a randomly chosen 128-bit permutation, finding a collection of 2^{56} texts in similar time, using similar (little) memory and with similar properties as in the case of 7-round AES has a probability of succeeding which is very close to zero".

Example 3. In [12] a known-key distinguisher of order $N = 2$ for AES$_8$ of time complexity $T = 2^{48}$, memory about 2^{32}, and success probability close to 1 is described. The associated relation $\mathcal{R}$ is differential in nature. It is defined as follows: $(X_1, X_2)\mathcal{R}(Y_1, Y_2)$ if and only if $X_1 \neq X_2$, the single non-zero bytes of the input difference $X_1 \oplus X_2$ are the diagonal bytes, *i.e.* the bytes numbered 0, 5, 10, and 15, and the single non-zero bytes of the output difference $Y_1 \oplus Y_2$ are the four bytes numbered 0, 7, 10, and 13. It was shown in [13] that given a perfect random permutation Π, the best method to get an input pair (X_1, X_2) and an output pair $(Y_1, Y_2) = (\Pi(X_1), \Pi(X_2))$ satisfying $(X_1, X_2)\mathcal{R}(Y_1, Y_2)$ is the so-called limited birthday technique, that requires about 2^{65} oracle queries for a target success probability of about $\frac{1}{2}$. With only $T = 2^{48}$ oracle queries, the success probability of this best method would decrease to about 2^{-17}.

Example 4. When applied to block ciphers, so-called zero-sum distinguishers [1,4,5], that thanks to higher order differential properties produce structures $(X_i, Y_i)_{i=1 \cdots N}$ of N (input, output) pairs such that $\bigoplus_{i=1}^{N} X_i = \bigoplus_{i=1}^{N} Y_i = 0$ also represent examples of known-key distinguishers.

Impossiblity Results on the Resistance of Block Ciphers to all Known-Key Distinguishers. Specifying the requirements on the resistance of a block cipher E against known-key distinguishers is a notoriously difficult issue because of an impossibility result that was first pointed out by Canetti, Goldreich, and Halevi in [6]. While the notion of correlation intractability was originally used to state this result, the related notion of resistance against known-key distinguishers can be used to reformulate it as follows:

Proposition 2. *Every block cipher of key length k bits and block length n bits such that $k \leq n$ is vulnerable to a known-key distinguisher of order 1 and complexity about one computation of E.*

Proof sketch. In order to give the intuition of the proof, let us restrict ourselves to the situation where $k = n$. It suffices to use the whole specification of E in the definition of R to get the claimed result. Let us define $X \mathrel{\mathcal{R}} Y$, where X and Y are any n-bit blocks, by the condition $Y = E_X(X)$. Given any known k-bit key K, the easy to compute values $X = K$ and $Y = E_K(K)$ are related by E_K and satisfy $X \mathrel{\mathcal{R}} Y$. However, for any adversary $\mathcal{A}'$ that makes $q << 2^n$ queries to a perfect random permutation Π of the block space, finding X such that $X \mathrel{\mathcal{R}} \Pi(X)$, i.e. $\Pi(X) = E_X(X)$ is very unlikely to succeed: by separately considering the cases where $\mathcal{A}'$ outputs a value X that belongs or does not belong to a queried pair it can indeed be shown that the success probability of $\mathcal{A}'$ is upper bounded by $\frac{q}{2^n} + \frac{1}{2^n - q}$, and is therefore negligible if $q << 2^n$. $\square$

The former proposition can be easily extended as follows.

Proposition 3. *Every block cipher of block length n bits and key length $k = Nn$ bits is vulnerable to a known-key distinguisher of order N and complexity about N computations of E.*[8]

Proof sketch. We just need to replace the relation $\mathcal{R}$ used in the former proof by the following relation $\mathcal{R}_N$ over the N-tuples of blocks: if $\mathcal{X} = (X_i)_{i=1 \cdots N}$ and $\mathcal{Y} = (Y_i)_{i=1 \cdots N}$, $\mathcal{X} \mathcal{R}_N \mathcal{Y}$ iff $\forall i \in [1; N] E_{\mathcal{X}}(X_i) = Y_i$, where $E_{\mathcal{X}}$ denotes the block cipher E parametrized by the Nn-bit key $X_1 || X_2 || \cdots || X_N$. $\square$

---

[8] One can generalize the former result a bit further by noticing that if $k \leq Nn$, then given any easy to compute and easy to invert function $f : \{0, 1\}^{Nn} \to \{0, 1\}^k$, a simple variant of the known-key distinguisher of Proposition 3 can be obtained by replacing $\mathcal{R}_N$ by the relation $\mathcal{R}'_N$ defined by $\mathcal{X} \mathcal{R}'_N \mathcal{Y}$ iff $\forall i \in [1; N] \ E_{f(\mathcal{X})}(X_i) = Y_i$.

To summarize the above impossibility results, for a block cipher E of block and key lengths n and k, generic known-key distinguishers of order N are known to exist iff $k \leq Nn$.

Discussion. If $k > Nn$, any known-key distinguisher of order at most N can be reasonably conjectured to be *meaningful*, *i.e.* to reflect, unlike the *artificial* generic known-key distinguishers of Propositions 2 and 3, a meaningful correlation property of E. Now in the frequently encountered case where $k \leq Nn$, that is met for instance for the known-key distinguisher of [17] where $k = 128$ and $Nn = 2^{56} \times 128$, characterizing which known-key distinguishers of order N should be considered *meaningful* and which ones should be considered *artificial* is a very complex issue. Finding a complete characterization remains an open problem that even lacks a rigorous statement and we will not attempt to solve it here. We will limit ourselves to propose informal criteria allowing to identify two classes of known key distinguishers that have little to do with artificial distinguishers identified so far and can be both reasonably considered meaningful.
– *Informal criterion 1.* One heuristic argument in favour of the view that the known-key distinguisher of Example 2 [17] for AES$_7$ is *meaningful* is the observation that while the description of the generic relations $\mathcal{R}$ and $\mathcal{R}_N$ used in Propositions 2 and 3 involve the specification of E itself, the relation $\mathcal{R}$ used in [17] has no obvious connection with the specification of E. More generally, if a known-key distinguisher uses an intractable relation $\mathcal{R}$ whose specification does not extensively reuse operations of E, this provides some heuristic evidence that it can be considered meaningful.[9]
– *Informal criterion 2.* While the informal criterion 1 sounds like a reasonable sufficient condition, we think it should not be considered as a necessary condition. In other words, known-key distinguishers that do not satisfy it, *i.e.* whose relation $\mathcal{R}$ re-uses some operations of E, should not be systematically ruled out as if they were all *artificial*. We informally state an alternative criterion for highlighting that independently of whether their relation $\mathcal{R}$ reuses operations of E or not, some known-key distinguishers have little to do with existing *artificial* distinguishers. One can observe that in the *artificial* distinguishers $(\mathcal{A}, \mathcal{R})$ of Propositions 2 and 3 and of the generalisation of Proposition 3 in the remark above, algorithm $\mathcal{A}$ produces an N-tuple $\mathcal{X}$ of input blocks from which the value of the whole key can be easily derived: in other words, one exploits the fact that $\mathcal{X}$ "encodes" the value of the entire key. If for a given known-key distinguisher $(\mathcal{A}, \mathcal{R})$ the entire key can neither be derived from the N-tuples of input values $\mathcal{X}$ nor from the N-tuples of output values $\mathcal{Y}$ produced by $\mathcal{A}$ one is brought back to a situation somewhat similar to the case where $k > Nn$ (a condition that obviously prevents $\mathcal{X}$ and $\mathcal{Y}$ from encoding the entire key) and this provides some evidence that $(\mathcal{A}, \mathcal{R})$ has little to do with the *artificial* distinguishers identified so far. We will use this informal criterion at the end of the next section.

---

[9] Giving a rigorous definition of the former informal criterion seems difficult. One might perhaps express that the verification of $\mathcal{R}$ is not substantially sped up by oracle accesses to E.

4   Application: Improved Known-Key Distinguishers for AES$_8$ and AES$_{10}$

4.1   A Known-Key Distinguisher for AES$_8$

Let us now show how to use the first untwisted representation of AES$_{2r}$ introduced in Section 2 in order to mount a known-key distinguisher of order $N = 2^{64}$ for AES$_8$. The distinguisher starts from a suitably chosen middle N-block structure and exploits the forward and backward properties of the final rounds, resp. the initial rounds of the AES$_8$, that are illustrated on Figure 3. These properties result from the fact that the initial and final rounds essentially consist of the composition $S \cdot R \cdot S$, up to simple initial and final transformations.

Property 1. *For any structure* $\mathcal{X}_{(a,b,c,d)} = \{(x \oplus a, b, c, d), x \in \{0,1\}^{32}\}$ *of* 2^{32} *input blocks — where* (a, b, c, d) *denotes an AES block of columns a, b, c, and d — each of the four 4-byte columns of the image of* $\mathcal{X}_{(a,b,c,d)}$ *by* $S \cdot R \cdot S$ *is uniformly distributed.*

This can be easily seen by following the column-wise transitions through transformations S, R, and S on the top of Figure 3 and by observing (1) that S transforms each column bijectively and (2) that if one fixes the second, third, and fourth input columns of R, each of the four output columns of R is a bijective affine function of the first input column. Since moreover $P \cdot SR \cdot AK$ is just a permutation of the byte positions followed by a key addition, each of the 16 bytes of the image of $\mathcal{X}_{(a,b,c,d)}$ by $S \cdot R \cdot S \cdot P \cdot SR \cdot AK$ is uniformly distributed and can be marked "U".

Property 2. *For any structure* $\mathcal{Y}_{(e,f,g,h)} = \{(y \oplus e, f, g, h), y \in \{0,1\}^{32}\}$ *of* 2^{32} *blocks, each four-byte column of the preimage of* $\mathcal{Y}_{(e,f,g,h)}$ *by* $S \cdot R \cdot S$ *is uniformly distributed.*

This can be easily seen by following the column-wise transitions through transformations S^{-1}, R^{-1}, and S^{-1} on the bottom of Figure 3 and observing (1) that S^{-1} transforms each column bijectively and that (2) if one fixes the second, third, and fourth input columns of R^{-1}, each of the four output columns of R^{-1} is a bijective affine function of the first input column. Since moreover IP^{-1} is a permutation of the byte positions, each of the 16 bytes of the preimage of $\mathcal{Y}_{(e,f,g,h)}$ by $AK \cdot IP \cdot S \cdot R \cdot S$ is uniformly distributed and can be marked "U". We are using these properties to mount the known-key distinguisher of order $N = 2^{64}$ for AES$_8$ illustrated on Figure 4, *i.e.* an algorithm $\mathcal{A}$ allowing to efficiently derive from any known key a N-tuple $\left((X_i, Y_i)\right)_{i=1\cdots N}$ of AES$_8$ (input, output) pairs that satisfy the relation $\mathcal{R}$ defined as follows.

$$\mathcal{X}_{(a,b,c,d)}$$

$$\begin{pmatrix} U_1 & C & C & C \\ U_2 & C & C & C \\ U_3 & C & C & C \\ U_4 & C & C & C \end{pmatrix} \xrightarrow{S} \begin{pmatrix} U_1 & C & C & C \\ U_2 & C & C & C \\ U_3 & C & C & C \\ U_4 & C & C & C \end{pmatrix} \xrightarrow{R} \begin{pmatrix} U_1^1 & U_1^2 & U_1^3 & U_1^4 \\ U_2^1 & U_2^2 & U_2^3 & U_2^4 \\ U_3^1 & U_3^2 & U_3^3 & U_3^4 \\ U_4^1 & U_4^2 & U_4^3 & U_4^4 \end{pmatrix} \xrightarrow{S} \begin{pmatrix} U_1^1 & U_1^2 & U_1^3 & U_1^4 \\ U_2^1 & U_2^2 & U_2^3 & U_2^4 \\ U_3^1 & U_3^2 & U_3^3 & U_3^4 \\ U_4^1 & U_4^2 & U_4^3 & U_4^4 \end{pmatrix} \xrightarrow{P \cdot SR \cdot AK} \begin{pmatrix} U & U & U & U \\ U & U & U & U \\ U & U & U & U \\ U & U & U & U \end{pmatrix}$$

$$\mathcal{Y}_{(e,f,g,h)}$$

$$\begin{pmatrix} U_1 & C & C & C \\ U_2 & C & C & C \\ U_3 & C & C & C \\ U_4 & C & C & C \end{pmatrix} \xrightarrow{S^{-1}} \begin{pmatrix} U_1 & C & C & C \\ U_2 & C & C & C \\ U_3 & C & C & C \\ U_4 & C & C & C \end{pmatrix} \xrightarrow{R^{-1}} \begin{pmatrix} U_1^1 & U_1^2 & U_1^3 & U_1^4 \\ U_2^1 & U_2^2 & U_2^3 & U_2^4 \\ U_3^1 & U_3^2 & U_3^3 & U_3^4 \\ U_4^1 & U_4^2 & U_4^3 & U_4^4 \end{pmatrix} \xrightarrow{S^{-1}} \begin{pmatrix} U_1^1 & U_1^2 & U_1^3 & U_1^4 \\ U_2^1 & U_2^2 & U_2^3 & U_2^4 \\ U_3^1 & U_3^2 & U_3^3 & U_3^4 \\ U_4^1 & U_4^2 & U_4^3 & U_4^4 \end{pmatrix} \xrightarrow{(AK \cdot IP)^{-1}} \begin{pmatrix} U & U & U & U \\ U & U & U & U \\ U & U & U & U \\ U & U & U & U \end{pmatrix}$$

Fig. 3. Forward and backward properties of $S \cdot R \cdot S$

Relation $\mathcal{R}$: $(X_i)_{i=1\cdots N}\,\mathcal{R}\,(Y_i)_{i=1\cdots N}$ iff the N blocks X_i are pairwise distinct and for each byte position $j \in \{0, \cdots, 15\}$, the j-th byte of the X_i and the j-th byte of the Y_i are uniformly distributed.

Algorithm $\mathcal{A}$: The conducting idea is that in the untwisted representation of AES_8 in Figure 4, the initial and final rounds of Figure 3 are linked together by the transformation R, that is affine. This allows to construct a structure that simultaneously achieves the requirements on the intput and the output of R^{-1} in order to apply Properties 1 and 2. More in detail, we are using the 2^{64} chosen middle blocks structure $\mathcal{Z} = \mathcal{X}_0 \oplus R\mathcal{Y}_0$, where $\mathcal{X}_0$ and $\mathcal{Y}_0$ are shorthands for $\mathcal{X}_{(0,0,0,0)}$ and $\mathcal{Y}_{(0,0,0,0)}$ and $\mathcal{X}_0 \oplus R\mathcal{Y}_0$ denotes the set $\{X \oplus R(Y), X \in \mathcal{X}_0, Y \in \mathcal{Y}_0\}$. It directly results from the definition of $\mathcal{Z}$ that it can be partitioned into 2^{32} structures $\mathcal{X}_0 \oplus R(y,0,0,0) = \mathcal{X}_{R(y,0,0,0)}$ of 2^{32} blocks each, one for each value $y \in \{0,1\}^{32}$. In other words, $\mathcal{Z}$ can be partitioned into 2^{32} structures of the form $\mathcal{X}_{(a,b,c,d)}$. Therefore, due to Property 1, each byte of the image of $\mathcal{Z}$ by $S \cdot R \cdot S \cdot P \cdot SR \cdot AK$ satisfies property U. Let us denote by L and C the linear and constant parts of the affine mapping R, i.e. the linear mapping and the constant such that $\forall X \in \{0,1\}^{128} R(X) = L(X) \oplus C$. Since the linear mapping and the constant associated with R^{-1} are $L' = L^{-1}$ and $C' = L^{-1}(C)$, the preimage of $\mathcal{Z}$ by R is $R^{-1}(\mathcal{Z}) = L^{-1}(\mathcal{X}_0 \oplus L(\mathcal{Y}_0) \oplus C) \oplus C' = L^{-1}(\mathcal{X}_0) \oplus \mathcal{Y}_0$. Therefore $R^{-1}(\mathcal{Z})$ can be partitioned into 2^{32} structures $\mathcal{Y}_0 \oplus L^{-1}(x,0,0,0) = \mathcal{Y}_{L^{-1}(x,0,0,0)}$ of 2^{32} blocks each[10] – one for each value $x \in \{0,1\}^{32}$. In other words, $R^{-1}(\mathcal{Z})$ can be partitioned into 2^{32} structures of the form $\mathcal{Y}_{(e,f,g,h)}$ and the application of Property 2 to $R^{-1}(\mathcal{Z})$ shows that each byte of the preimage of $R^{-1}(\mathcal{Z})$ by $AK \cdot IP \cdot S \cdot R \cdot S$, i.e. each byte of the preimage of $\mathcal{Z}$ by $AK \cdot IP \cdot S \cdot R \cdot S \cdot R$, satisfies property U.

In summary, we derived from the middle structure $\mathcal{Z}$ a N-tuple $\big((X_i, Y_i)\big)_{i=1\cdots N}$ of AES_8 (input, output) pairs that satisfy relation $\mathcal{R}$. The time complexity of the derivation of such an N-tuple is $T = N = 2^{64}$ AES_8 computations. To complete the proof that we have mounted a known-key distinguisher for AES_8, we just have to show that property $\mathcal{R}$ is T-intractable, i.e. that the success probability of any

---

[10] One can notice that the above partitions of $\mathcal{Z}$ and $R^{-1}(\mathcal{Z})$ do not map into each other through R.

$$R^{-1}(Z)=Y_0 \oplus L^{-1}X_0 \qquad\qquad Z = X_0 \oplus RY_0$$
$$\equiv 2^{32}\times \qquad\qquad\qquad \equiv 2^{32}\times$$

$$\begin{pmatrix} U & U & U & U \\ U & U & U & U \\ U & U & U & U \\ U & U & U & U \end{pmatrix} \xleftarrow{(AK\cdot IP)^{-1}} \begin{pmatrix} U_1^1 & U_1^2 & U_1^3 & U_1^4 \\ U_2^1 & U_2^2 & U_2^3 & U_2^4 \\ U_3^1 & U_3^2 & U_3^3 & U_3^4 \\ U_4^1 & U_4^2 & U_4^3 & U_4^4 \end{pmatrix} \xleftarrow{(SRS)^{-1}} \begin{pmatrix} U_1 & C & C & C \\ U_2 & C & C & C \\ U_3 & C & C & C \\ U_4 & C & C & C \end{pmatrix} \xleftarrow{R^{-1}} \begin{pmatrix} U_1 & C & C & C \\ U_2 & C & C & C \\ U_3 & C & C & C \\ U_4 & C & C & C \end{pmatrix} \cdots$$

$$\cdots \xrightarrow{SRS} \begin{pmatrix} U_1^1 & U_1^2 & U_1^3 & U_1^4 \\ U_2^1 & U_2^2 & U_2^3 & U_2^4 \\ U_3^1 & U_3^2 & U_3^3 & U_3^4 \\ U_4^1 & U_4^2 & U_4^3 & U_4^4 \end{pmatrix} \xrightarrow{P\cdot SR\cdot AK} \begin{pmatrix} U & U & U & U \\ U & U & U & U \\ U & U & U & U \\ U & U & U & U \end{pmatrix}$$

Fig. 4. A known-key distinguisher for AES_8

oracle algorithm $\mathcal{A}^{(\Pi,\Pi^{-1})}$ of overall time complexity upper bounded by N (and therefore of number q of queries also upper bounded by N) is negligible.

Proposition 4. *For any oracle algorithm $\mathcal{A}$ that makes $q \leq N = 2^{64}$ oracle queries to a perfect random permutation Π of $\{0,1\}^n$ (where $n = 128$) or its inverse, the probability that $\mathcal{A}$ successfully outputs a N-tuple $\big((X_i, Y_i)\big)_{i=1\cdots N}$ of (input, output) pairs of Π that satisfy $\mathcal{R}$ is upper bounded by $\frac{1}{2^n-(N-1)}$ and hence by $\frac{1}{2^{n-1}}$.*

Proof. If at least one of the N pairs (X_i, Y_i) output by $\mathcal{A}$ does not result from the query X_i to Π or the query Y_i to Π^{-1}, then the probability that for this pair $Y_i = \Pi(X_i)$ and thus the success probability of $\mathcal{A}$ is upper bounded by $\frac{1}{2^n-(N-1)}$. In the opposite case, i.e. if $q = N$ and all the (X_i, Y_i) result from queries to Π or Π^{-1}, we can assume w.l.o.g. that (X_N, Y_N) results from the N-th query X_N or Y_N of $\mathcal{A}$ to Π or Π^{-1}. But given any pairs $(X_i, Y_i)_{i=1\cdots N-1}$ at most one value of the block Y_N, resp. X_N is such that each of the 16 bytes of $(Y_i)_{i=1\cdots N}$, resp. $(X_i)_{i=1\cdots N}$ be uniformly distributed.[11] However the oracle answer Y_N, resp. X_N is uniformly drawn from $\{0,1\}^n \setminus \{Y_1,\cdots Y_{N-1}\}$, resp. $\{0,1\}^n \setminus \{X_1,\cdots X_{N-1}\}$. Therefore the probability that the answer to the N-th query allows the output of $\mathcal{A}$ to satisfy property $\mathcal{R}$ is also upper bounded by $\frac{1}{2^n-(N-1)}$ in this case. $\square$

Discussion. The known-key distinguisher of order $N = 2^{64}$ for AES_8 presented above has a time complexity of about 2^{64}. It is obviously applicable without modification to the AES_8 variant parametrized by independent subkeys AES_8^*. In both cases, the fact that informal criterion of Section 3 is met, i.e. that the relation $\mathcal{R}$ used by the distinguisher has no obvious connection with the AES specification suggests that the obtained known-key distinguisher can be considered meaningful. While the presented 8-round known-key distinguisher is outperformed by the differential known-key distinguishers for AES_8 of complexities 2^{48} and 2^{44} of [12,14], the strong property expressed by relation $\mathcal{R}$ that

---

[11] This can for instance be deduced from the fact that the X_i and the Y_i must satisfy
$$\bigoplus_{i=1}^N X_i = \bigoplus_{i=1}^N Y_i = 0.$$

each input and output byte is not only balanced as in zero-sum distinguishers, but uniformly distributed turns out to be convenient for further extending the known key distinguisher by two rounds in a provable manner, as will be shown in the rest of this section.

Strengthening Proposition 4 Under a Heuristic Assumption. Let us give some partial evidence that $\mathcal{R}$ is actually T-intractable in a stronger sense than in Proposition 4 above, namely that the success probability of any adversary $\mathcal{A}$ who makes $M > N$ oracle queries to Π or Π^{-1} remains negligible if $M - N$ is not too large. While a rigorous proof requiring no unproven assumptions could be easily derived along the same lines as Proposition 4 for values of M marginally larger than N, e.g. $N + 3$, for larger values of M we make the heuristic assumption that querying both Π and Π^{-1} does not improve the performance of $\mathcal{A}$ over an adversary who only queries one of these oracles. Therefore, we consider an adversary $\mathcal{A}$ who only makes queries to an oracle permutation Π not its inverse, and aims at finding an N-tuple of (input, output) pairs that satisfy the relation $\mathcal{R}$ of Section 4.1. To upper bound the success probability of such an adversary, we observe that given any N-tuple of distinct input blocks X_i and any output byte position $j \in [0; 15]$, the 256-tuple $(N_0, \cdots, N_{255})$ of numbers of occurrences of the values $0, 1, \cdots 255$ for byte j of the blocks $Y_i = \Pi(X_i)$ is nearly governed by a multinomial law. For any 256-tuple $(N_0, \cdots, N_{255})$ such that $\sum_{i=0}^{255} N_i = N$, we denote the multinomial coefficient $\frac{N!}{N_0! N_1! \cdots N_{255}!}$ by $\binom{N}{N_0, \cdots N_{255}}$.

Proposition 5. *For any N-tuple $(X_i)_{i=1 \cdots N}$ of distinct inputs to Π an upper bound on the probability p that for byte positions $j = 0$ to 15, the 256-tuple of numbers of occurrence of the values of byte j of $\Pi(X_i)$ be $(N_0^j, \cdots, N_{255}^j)$ — where for $j=0$ to 15 the 256-tuple $(N_0^j, \cdots, N_{255}^j)$ satisfies $\sum_0^{255} N_i^j = N$ — is given by:*

$$p \leq \prod_{j=0}^{15} \binom{N}{N_0^j, \cdots N_{255}^j} \times (\frac{1}{2^{128} - N + 1})^N.$$

An upper bound on the success probability p_A of an adversary $\mathcal{A}$ who makes $M > N$ queries to Π and no query to Π^{-1} is given by:

$$p_A \leq \binom{M}{N} \times \binom{N}{\frac{N}{256}, \frac{N}{256} \cdots \frac{N}{256}}^{16} \times (\frac{1}{2^{128} - N + 1})^N.$$

Since $N = 2^{64}$, Proposition 5 provides very small upper bounds $p_A \ll \frac{1}{2}$ for values of M of up to $M \approx N + 2^{11}$. But it provides no bound $p_A < \frac{1}{2}$ for slightly larger values, e.g. $M \approx N + 2^{12}$. We do not know whether the bounds of Proposition 5, that relate to the probability that the (input, output) pairs provided by M queries contain one N-tuple, can be significantly improved. Since even in a situation where such N-tuples exist it can be computationally difficult to find one in time T, a potential approach might consist in establishing upper bounds that hold for higer values of M under computational assumptions.

4.2   A Known-Key Distinguisher for the 10-Round AES

In this section we show that the former known-key distinguisher for AES_8 can be extended by two rounds without significant complexity increase. The price to pay for this extension is that the relation $\mathcal{R}$ of the new distinguisher is much less simple and that its description involves operations of the first and last rounds. This raises the question whether the new known-key distinguisher reflects a *meaningful* correlation property of the cipher. Since we can provide more simple arguments supporting this view for AES_{10}^* (*i.e.* the 10-round AES parametrized by 11 independent subkeys), we first describe the application of the new known-key distinguisher to AES_{10}^* and then discuss how this transposes to AES-128.

As shown at the end of Section 2, AES_{10}^* can be equivalently represented by the sequence of transformations

$$AK \cdot P^{-1} \cdot SB \cdot R \cdot (S \cdot R)^4 \cdot SB \cdot Q^{-1} \cdot AK$$

The properties we are using to build a known-key distinguisher on AES_{10}^* are illustrated on Figure 5.

$$(X_i) \xleftarrow{(AK \cdot P^{-1} \cdot SB \cdot R)^{-1}} \begin{pmatrix} U\ U\ U\ U \\ U\ U\ U\ U \\ U\ U\ U\ U \\ U\ U\ U\ U \end{pmatrix} \xleftarrow{(SRS)^{-1}} \begin{pmatrix} U_1\ C\ C\ C \\ U_2\ C\ C\ C \\ U_3\ C\ C\ C \\ U_4\ C\ C\ C \end{pmatrix} \xleftarrow{R^{-1}} \begin{pmatrix} U_1\ C\ C\ C \\ U_2\ C\ C\ C \\ U_3\ C\ C\ C \\ U_4\ C\ C\ C \end{pmatrix} \xrightarrow{SRS} \begin{pmatrix} U\ U\ U\ U \\ U\ U\ U\ U \\ U\ U\ U\ U \\ U\ U\ U\ U \end{pmatrix} \xrightarrow{R \cdot SB \cdot Q^{-1} \cdot AK} (Y_i)$$

$$\mathcal{U} \qquad R^{-1}(\mathcal{Z}) = y_0 \oplus L^{-1} x_0 \qquad \mathcal{Z} = x_0 \oplus R y_0 \qquad \mathcal{V}$$

Fig. 5. Derivation of the N AES_{10} (input,output) pairs used in our known-key distinguisher

Algorithm $\mathcal{A}$: We reuse the same structure $\mathcal{Z}$ of $N = 2^{64}$ intermediate blocks as for the known-key distinguisher on AES_8 presented above, but extend the forward computation and backward computations $S \cdot R \cdot S$ and $(S \cdot R \cdot S \cdot R)^{-1}$, by two outer transformations whose structures are symmetric of each other, namely $(AK \cdot P^{-1} \cdot SB \cdot R)^{-1}$ (backward) and $R \cdot SB \cdot Q^{-1} \cdot AK$ (forward) to get an N-tuple of related AES_{10}^* inputs and outputs. As shown in the former subsection the inputs to the forward and backward outer transformations each consist of four columns that are uniformly distributed and therefore each of the 16 bytes of each of these two states $\mathcal{U}$ and $\mathcal{V}$ is uniformly distributed and can be marked U. However, these states are related to the AES_{10}^* inputs X_i and to the AES_{10}^* outputs Y_i by the outer transformations.

This implies that if we denote by α and β the 128-bit states $P^{-1}(K_0)$ and $Q(K_{10})$ the N-tuple $\mathcal{X} = (X_i)_{i=1\cdots N}$ and $\mathcal{Y} = (Y_i)_{i=1\cdots N}$ are related by the key-dependent relation $\mathcal{R}_{\alpha,\beta}$ defined as follows: $\mathcal{X}\mathcal{R}_{\alpha,\beta}\mathcal{Y}$ if and only if each byte of $R \circ SB(P^{-1}(X_i) \oplus \alpha)$ and each byte of $R^{-1} \circ SB^{-1}(Q(Y_i) \oplus \beta)$ is uniformly distributed. We can now define the following relation $\mathcal{R}$ over the N-tuples of blocks:

Relation $\mathcal{R}$: *Given two N-tuples $\mathcal{X}' = (X_i')_{i=1\cdots N}$ and $\mathcal{Y}' = (Y_i')_{i=1\cdots N}$ $\mathcal{X}'\mathcal{R}\mathcal{Y}'$ if and only if all the X_i', $i = 1 \cdots N$ are pairwise distinct and there exists a pair α', β' of 128-bit states such that $\mathcal{X}'\mathcal{R}_{\alpha',\beta'}\mathcal{Y}'$.*

It is important to understand that though relation $\mathcal{R}$ reflects the existence of values α' and β' that can be conveniently interpreted as subkeys, ckecking $\mathcal{R}$ does not take as input any key or subkey: given two N-tuples $\mathcal{X}'$ and $\mathcal{Y}'$ that can be possibly derived from a random key value K by algorithm $\mathcal{A}$, whether $\mathcal{X}'\mathcal{R}\mathcal{Y}'$ must be efficiently checkable without providing the verifyer with K or any other side information about suitable values of α' and β'.

It immediately results from the definition of $\mathcal{R}$ that the N-tuples $\mathcal{X}$ and $\mathcal{Y}$ derived as described in Figure 5 satisfy property $\mathcal{R}$ and the complexity of the derivation algorithm $\mathcal{A}$ is $T = N = 2^{64}$. To complete the proof that $(\mathcal{R}, \mathcal{A})$ is a known-key distinguisher for AES_{10}^*, we just have to show that $\mathcal{R}$ is efficiently checkable and T-intractable.

$\mathcal{R}$ **is Efficiently Checkable.** Though the involvement in $\mathcal{R}$ of 128-bit constants α' and β' might suggest that checking $\mathcal{R}$ has a huge complexity, this is not the case because the existence of 128-bit states α', β' such that $\mathcal{X}'\mathcal{R}_{\alpha',\beta'}\mathcal{Y}'$ can be split into independent conditions. Let us denote by $sb : \{0,1\}^{32} \rightarrow \{0,1\}^{32}$ a parallel application of four AES S-boxes that from a four-byte row produces a four-byte output row. For $j = 0$ to 3 let us denote by row_j the mapping that from a 128-bit state outputs the row numbered j of this state, and by R_j the linear transformation of row j introduced in Section 2. It is easy to see that the existence of α' and β' is equivalent to the existence of eight 32-bit constants α'_j, $j = 0\cdots3$ and $\beta'_j, j = 0\cdots3$ (representing the rows of α' and β') such that for $j = 0\cdots3$ each of the four bytes of $R_j \circ sb \circ row_j(P^{-1}(X_i) \oplus \alpha'_j)$ and $R_j^{-1} \circ sb^{-1} \circ row_j(Q(Y_i) \oplus \alpha'_j)$ is uniformly distributed. This can be easily done by first computing in a first step the number of occurrences of each of the 2^{32} possible values of the 32-bit words $row_j(P^{-1}(X_i))$ and $row_j(Q(Y_i))$, $j = 0\cdots3$, and then using the obtained distributions of frequencies in a second step for computing, for $j = 0$ to 3 and each of the 2^{32} possible values of α'_j, resp. β'_j the resulting distribution of frequencies of $R_j \circ sb \circ row_j(P^{-1}(X_i) \oplus \alpha'_j)$, resp $R_j^{-1} \circ sb^{-1} \circ row_j(Q(Y_i) \oplus \alpha'_j)$ and checking that at least one of them induces a balanced distribution for each byte position. Since the first step requires 2^{64} very simple operations that are much less complex that one operation of AES_{10}^* and the second step again requires 8 times 2^{64} very simple operations, the overall complexity of checking $\mathcal{R}$ is strictly smaller than $N = 2^{64}$ AES_{10}^* operations.

Remark. The reader might wonder whether the technique we used to derive a known-key distinsguisher for the 10-round AES from a known-key distinguisher for the 8-round AES, by expressing that the 10-round inputs and outputs are related (by one outer round at each side) to intermediate blocks that satisfy the relation used by the 8-round distinguisher does not allow to extend this 8-round known distinguisher by an arbitrary number of rounds. If this was the case, this would of course render this technique highly suspicious. It is easy however to see that the argument showing that 10-round relation $\mathcal{R}$ is efficiently checkable does not transpose for showing that the relations over $r > 10$ rounds one could derive from the 8-round relation by expressing that the r-round inputs and outputs are related by $r - 8 > 2$ outer rounds to intermediate blocks that satisfy the 8-round relation are efficiently checkable. To complete this remark, we explain at the end of this section why the 2-round extension technique we used is not generically applicable to extend any r-round known-key distinguisher to a $r + 2$-round distinguisher.

$\mathcal{R}$ **is T-Intractable.** In order to show that relation $\mathcal{R}$ is T-intractable, we now have to prove that the success probability of any oracle algorithm of overall time complexity upper bounded by $N = 2^{64}$ (and therefore of number q of queries also upper bounded by N) is negligible.

Proposition 6. *For any oracle algorithm $\mathcal{A}$ that makes $q \leq N = 2^{64}$ oracle queries to a perfect random permutation Π of $\{0, 1\}^{128}$ or Π^{-1}, the probability that $\mathcal{A}$ outputs a N-tuple $(X_i, Y_i)_{i=1\cdots N}$ of Π that satisfies and $\forall i \in [1; N]\ Y_i = \Pi(X_i)$ and also satisfies $\mathcal{R}$ is upper bounded by $2^{256} \times (\frac{5^{16}}{2^{128} - (N-5)})^3 \approx 2^{-16.5}$.*

Proof. If at least one of the N pairs (X_i, Y_i) output by $\mathcal{A}$ does not result from a query X_i to Π or a query Y_i to Π^{-1}, then the probability that for this pair $Y_i = \Pi(X_i)$ and consequently the success probability of $\mathcal{A}$ is upper bounded by $\frac{1}{2^n - (N-1)}$. So from now on we only consider the opposite case, *i.e.* $q = N$ and all the (X_i, Y_i) result from queries to Π or Π^{-1}. Given any two 128-bit words α and β, let us upper bound the probability that $\mathcal{A}$ outputs an N-tuple (X_i, Y_i) that satisfies $\forall i \in [1; N]\ Y_i = \Pi(X_i)$ and the relation $\mathcal{R}_{\alpha,\beta}$. The conducting idea is that the constraints on the very last queries to the oracle (Π, Π^{-1}) in order for $\mathcal{R}_{\alpha,\beta}$ to hold are so strong this is extremely unlikely to happen. For the sake of simplicity of this proof, we consider the consider the last 5 queries of $\mathcal{A}$ to the oracle (Π, Π^{-1}): indeed, while considering the d last queries, $d > 5$, might have lead to a tighter upper bound, the chosen value of 5 is sufficient for establishing a suitable upper bound. Since the 5 last queries contain at least 3 queries to either Π or Π^{-1} we can assume *w.l.o.g.* that they contain at least 3 queries X, X', and X'' to Π and we denote the corresponding responses by Y, Y', and Y''. In order for the property $\mathcal{R}_{\alpha,\beta}$ to be satisfied, for each byte position $j \in [0; 15]$, the set of byte values $B_j = \{b \in [0; 255] \mid \sharp\{i \in [1; N-5] \mid R^{-1} \circ SB^{-1}(Q(Y_i) \oplus \beta)[j] = b\} \neq \frac{N}{256}\}$ must contain at most 5 elements (since the last 5 queries can affect the number of occurrences of at most 5 of the 256 byte values and all the unaffected numbers of occurrences must already be $\frac{N}{256}$). Furthermore, in order for property $\mathcal{R}_{\alpha,\beta}$ to be satisfied, one must have $\forall i \in [N - 4; N]\ R^{-1} \circ SB^{-1}(Q(Y_i) \oplus \beta)[j] \in B_j$, *i.e.* $\forall i \in [N - 4; N]\ Y_i \in \mathcal{S} = Q^{-1} \circ SB \circ R(\prod_{j=0}^{15} B_j) \oplus \beta$. Since Q, SB, R, and the xor with β are bijective, the set $\mathcal{S}$ defined above contains $\sharp\mathcal{S} = \sharp\prod_{j=0}^{15} B_j$ elements (where $\prod_{j=0}^{15} B_j$ denotes the Cartesian product of the B_j). Since for j=0 to 15 $\sharp B_j \leq 5$, $\sharp\prod_{j=0}^{15} B'_j \leq 5^{16}$ and hence $\sharp\mathcal{S} \leq 5^{16}$. Therefore the probability that the three blocks Y, Y', and Y'' all belong to $\mathcal{S}$ is upper bounded by $(\frac{5^{16}}{2^{128} - (N-5)})^3$. By summing the obtained upper bound over all the 2^{256} possible values of α, β, one gets the claimed upper bound $2^{256} \times (\frac{5^{16}}{2^{128} - (N-5)})^3 \approx 2^{-16.5}$ on the probability that $\mathcal{R}$ be satisfied. $\square$

In order to give partial evidence that $\mathcal{R}$ is not only N-intractable as shown in Proposition 6 above, but remains M-intractable for $M > N$ if $M - N$ is not too large, we can make the heuristic assumption that the success probabilities of adversaries who are allowed to make oracle queries to both Π and Π^{-1} and adversaries who are allowed to make oracle queries to Π only have the same upper limit. Proposition 5 can be transposed to the 10-round relation $\mathcal{R}$, up to a multiplication of the upper bounds obtained for p and p_A by 2^{256}. This multiplicative factor does not strongly affect the values of $M-N$ one can reach and one still gets very small upper bounds $p_A \ll \frac{1}{2}$ for values of M of up to $M \approx N+2^{11}$.

The Former 2-Round Extension Technique is Not Generic. The reader might wonder why the two-round extension technique introduced above does not allow to extend any r-round known-key distinguisher to an $r+2$-round known-key distinguisher. There are two reasons that can make such an extension fail: firstly, unlike the r-round relation it is derived from, the $r + 2$-round relation may not be efficiently ckeckable; secondly, unlike the r-round relation it is derived from, the $r + 2$-round relation may be insufficiently intractable to mount a $r + 2$-round distinguisher. This second situation occurs in the case of the 8-round differential relation $\mathcal{R}_8$ of order 2 used in [12]. In the full version of this paper we show that unlike $\mathcal{R}_8$, that is T-intractable for $T = 2^{48}$, the 10-round relation $\mathcal{R}_{10}$ derived from $\mathcal{R}_8$ is not intractable at all for $T = 2^{48}$, but simple to achieve with a probabillity about 0.97 with only two queries to a perfect random permutation Π and no extra operation. In other words, the transposition of our technique to the 8-round distinguisher of [12] does not allow to derive a valid 10-round distinguisher.

In the full version of this paper, we also show that while we do not preclude that the use of the stronger property (reflected by a higher-order relation than $\mathcal{R}_8$) that several pairs satisfying the differential relation of [12] can be derived might potentially result in a 10-round distinguisher that outperforms the 10-round distinguisher presented above, giving a rigorous proof (as was done in Proposition 6) seems technically difficult. We leave the investigation of improved 10-round known-key distinguishers and associated proofs – or even plausible heuristic arguments if rigorous proofs turn out to be too difficult to obtain – as an open issue.

Discussion. The known-key distinguisher $(\mathcal{R}, \mathcal{A})$ of order $N = 2^{64}$ for AES$^*_{10}$ presented above has a time complexity of about 2^{64}. Unlike in the former 8-round known-key distinguishers the relation $\mathcal{R}$ involves operations of the AES. However, it is easy to show that the alternative criterion at the end of Section 3 for differentiating certain known-key distinguishers from the artificial known-key distinguishers that result from generic impossibility results is applicable. Indeed, the derivation by $\mathcal{A}$ of the input N-tuple $(X_i)_{i=1 \cdots N}$ from the intermediate structure $\mathcal{Z}$ only involves the 6 first subkeys K_0 to K_5 and the derivation $\mathcal{A}$ of the output N-tuple $(Y_i)_{i=1 \cdots N}$ from the same structure only involves the 5 last subkeys K_6 to K_{11}. Consequently the 5 last subkeys cannot be derived from $(X_i)_{i=1\cdots N}$ and thus the input N-tuples do not "encode" the entire key. Similarly, the 6 first subkeys cannot be derived from $(Y_i)_{i=1\cdots N}$ and thus the output N-tuples do not "encode" the entire key. This suggests that the obtained known-key distinguisher for AES_{10}^* can reasonably be considered meaningful.

While the former known-key distinguisher is obviously applicable without any modification to AES_{10}, i.e. the full AES-128, the former argument vanishes in this case because all subkeys are related by the key schedule: the first subkey, resp. the last subkey can actually be derived from the input, resp. the output N-tuple and because of the key schedule relations this determines the entire key. This does not mean that when applied to AES_{10} the former distinguisher becomes artificial. Actually, the fact that the very same distinguisher is applicable to AES_{10}^* gives a hint that it can still be considered meaningful.[12]

5 Conclusion

As said before, the untwisted representation of AES introduced in this paper is not exclusively intended for the analysis of the security of AES in the known-key model. We think however that the fact that this represention was used to find the two known-key distinguishers presented in Section 4 provides some evidence that this representation is well suited for analysing the resistance of (a reduced-round version of) AES against some structural attacks.

Whether there exists a more simple 10-round known-key or even chosen-key distinguisher for AES than the 10-round known key distinguisher presented in this paper – allowing to highlight a less tenuous deviation from the behaviour of a perfect random permutation, resp. of an ideal cipher remains an interesting open question.

Acknowledgements. We would like to thank Yannick Seurin for helpful discussions and insights.

References

1. Aumasson, J.-P., Meier, W.: Zero-sum distinguishers for reduced Keccak-f and for the core functions of Luffa and Hamsi, Comment on the NIST SHA-3 Hash Competition (2009)
2. Barkan, E., Biham, E.: In How Many Ways Can You Write Rijndael? In: Zheng, Y. (ed.) ASIACRYPT 2002. LNCS, vol. 2501, pp. 160–175. Springer, Heidelberg (2002)
3. Biryukov, A., De Cannière, C., Braeken, A., Preneel, B.: A Toolbox for Cryptanalysis: Linear and Affine Equivalence Algorithms. In: Biham, E. (ed.) EUROCRYPT 2003. LNCS, vol. 2656, pp. 33–50. Springer, Heidelberg (2003)

---

[12] Since the input N-tuple now encodes the entire key, there might exist *artificial* variants of the former known-key distinguisher that produce the same input N-tuples (or the same output N-tuples) but can be extended to AES_r for any value of r. We conjecture however that unlike the known-key distinguisher presented here, such variants would not be applicable to AES_r^*.

4. Boura, C., Canteaut, A.: Zero-Sum Distinguishers for Iterated Permutations and Application to KECCAK-ƒ and Hamsi-256. In: Biryukov, A., Gong, G., Stinson, D.R. (eds.) SAC 2010. LNCS, vol. 6544, pp. 1–17. Springer, Heidelberg (2011)

5. Boura, C., Canteaut, A., De Cannière, C.: Higher-Order Differential Properties of KECCAK and *Luffa*. In: Joux, A. (ed.) FSE 2011. LNCS, vol. 6733, pp. 252–269. Springer, Heidelberg (2011)

6. Canetti, R., Goldreich, O., Halevi, S.: The random oracle methodology, revisited. J. ACM 51(4), 557–594 (2004)

7. Daemen, J., Rijmen, V.: Understanding Two-Round Differentials in AES. In: De Prisco, R., Yung, M. (eds.) SCN 2006. LNCS, vol. 4116, pp. 78–94. Springer, Heidelberg (2006)

8. Davies, D.W., Murphy, S.: Pairs and Triplets of DES S-Boxes. Journal of Cryptology 8(1), 1–25 (1995)

9. Ferguson, N., Schroeppel, R., Whiting, D.L.: A Simple Algebraic Representation of Rijndael. In: Vaudenay, S., Youssef, A.M. (eds.) SAC 2001. LNCS, vol. 2259, pp. 103–111. Springer, Heidelberg (2001)

10. Fouque, P.-A., Jean, J., Peyrin, T.: Structural Evaluation of AES and Chosen-Key Distinguisher of 9-Round AES-128. In: Canetti, R., Garay, J.A. (eds.) CRYPTO 2013, Part I. LNCS, vol. 8042, pp. 183–203. Springer, Heidelberg (2013)

11. Gilbert, H., Peyrin, T.: Super-Sbox Cryptanalysis: Improved Attacks for AES-like permutations. IACR Cryptology ePrint Archive, 2009:531 (2009)

12. Gilbert, H., Peyrin, T.: Super-Sbox Cryptanalysis: Improved Attacks for AES-Like Permutations. In: Hong, S., Iwata, T. (eds.) FSE 2010. LNCS, vol. 6147, pp. 365–383. Springer, Heidelberg (2010)

13. Iwamoto, M., Peyrin, T., Sasaki, Y.: Limited-Birthday Distinguishers for Hash Functions. In: Sako, K., Sarkar, P. (eds.) ASIACRYPT 2013, Part II. LNCS, vol. 8270, pp. 504–523. Springer, Heidelberg (2013)

14. Jean, J., Naya-Plasencia, M., Peyrin, T.: Multiple Limited-Birthday Distinguishers and Applications

15. Jean, J., Naya-Plasencia, M., Peyrin, T.: Improved Rebound Attack on the Finalist Grøstl. In: Canteaut, A. (ed.) FSE 2012. LNCS, vol. 7549, pp. 110–126. Springer, Heidelberg (2012)

16. Jean, J., Naya-Plasencia, M., Schläffer, M.: Improved Analysis of ECHO-256. In: Miri, A., Vaudenay, S. (eds.) SAC 2011. LNCS, vol. 7118, pp. 19–36. Springer, Heidelberg (2012)

17. Knudsen, L.R., Rijmen, V.: Known-Key Distinguishers for Some Block Ciphers. In: Kurosawa, K. (ed.) ASIACRYPT 2007. LNCS, vol. 4833, pp. 315–324. Springer, Heidelberg (2007)

18. Lamberger, M., Mendel, F., Rechberger, C., Rijmen, V., Schläffer, M.: Rebound Distinguishers: Results on the Full Whirlpool Compression Function. In: Matsui, M. (ed.) ASIACRYPT 2009. LNCS, vol. 5912, pp. 126–143. Springer, Heidelberg (2009)

19. Mendel, F., Rechberger, C., Schläffer, M., Thomsen, S.S.: Rebound Attacks on the Reduced Grøstl Hash Function. In: Pieprzyk, J. (ed.) CT-RSA 2010. LNCS, vol. 5985, pp. 350–365. Springer, Heidelberg (2010)

20. Murphy, S., Robshaw, M.: Essential Algebraic Structure within the AES. In: Yung, M. (ed.) CRYPTO 2002. LNCS, vol. 2442, pp. 1–16. Springer, Heidelberg (2002)

Simulatable Leakage: Analysis, Pitfalls, and New Constructions

Jake Longo[1], Daniel P. Martin[1], Elisabeth Oswald[1], Daniel Page[1],
Martijn Stam[1], and Michael J. Tunstall[2]

[1] Department of Computer Science, University of Bristol,
Merchant Venturers Building, Woodland Road,
Bristol, BS8 1UB, United Kingdom
{jake.longo,dan.martin,elisabeth.oswald,daniel.page,
martijn.stam}@bris.ac.uk
[2] Cryptography Research Inc.
425 Market Street, 11th Floor
San Francisco, CA 94105, United States
michael.tunstall@cryptography.com

Abstract. In 2013, Standaert *et al.* proposed the notion of simulatable leakage to connect theoretical leakage resilience with the practice of side channel attacks. Their use of simulators, based on physical devices, to support proofs of leakage resilience allows verification of underlying assumptions: the indistinguishability game, involving real vs. simulated leakage, can be 'played' by an evaluator. Using a concrete, block cipher based leakage resilient PRG and high-level simulator definition (based on concatenating two partial leakage traces), they included detailed reasoning why said simulator (for AES-128) resists state-of-the-art side channel attacks.

In this paper, we demonstrate a distinguisher against their simulator and thereby falsify their hypothesis. Our distinguishing technique, which is evaluated using concrete implementations of the Standaert *et al.* simulator on several platforms, is based on 'tracking' consistency (resp. identifying simulator *in*consistencies) in leakage traces by means of cross-correlation. In attempt to rescue the approach, we propose several alternative simulator definitions based on splitting traces at points of low intrinsic cross-correlation. Unfortunately, these come with significant caveats, and we conclude that the most natural way of producing simulated leakage is by using the underlying construction 'as is' (but with a random key).

Keywords: leakage resilience, side channel attack, simulatable leakage, cross-correlation.

1 Introduction

At Crypto'13, Standaert *et al.* [19] proposed a new notion for leakage resilience involving simulators. The intuition behind their proposal is that if an adversary

P. Sarkar and T. Iwata (Eds.): ASIACRYPT 2014, PART I, LNCS 8873, pp. 223–242, 2014.

cannot tell the difference between real leakage and simulated leakage (from a simulator that does not know the secret key), then clearly the leakage does not reveal any information about the secret key. This offers a middle ground which connects theorists (who desire provably secure scheme) and practitioners (who require empirically verifiable constructions). In this paper we show that while this is a step in the right direction in terms of modelling leakage, the specific simulator given for their construction is in fact distinguishable. We explain why this is the case, and show how to resolve the problem so that their theoretical proof still holds.

1.1   What is Leakage?

A fundamental discrepancy between theory and practice lies in the different understanding of what constitutes 'leakage'. When examining the vast literature on side channels and leakage resilience, there seem to be three different understandings of what constitutes a leakage function. The first two of these ideas are typically found in works written by practitioners (such as [8,11]) and centre around a mathematical description of the physical nature of real-world leakage traces. The third, in contrast, seeks to define leakage functions in more general and powerful terms, and is often found in theoretical contributions.

Leakage understood as the modelling of the physical nature of the observed leakage points. The first understanding is that a leakage function is the mathematical function describing the shape and form of points in leakage traces. Such a function then models the manner in which the operations and data act on the physical environment, alongside other electrical components including the measurement apparatus, and the environment conditions. This understanding of leakage implies that the leakage function fundamentally depends on how the leakage is acquired (because it includes the measurement apparatus). It also implies that the leakage function, whilst being key dependent, is in principle unbounded: every new measurement gives more information.

Leakage understood as modelling of the exploitable information about the key. The second understanding is that of a mathematical function that again describes the leakage traces; however, the conceptual emphasis is that the term 'leakage' refers to leakage about the key. A key must have a finite length and therefore the leakage function can never reveal more than that amount of information. It is hence a function that could (depending on the number of queries to the function), under ideal circumstances, reveal the entire key information but no more than that.

Leakage understood as a mathematical concept largely separated from any physical interpretation. The third understanding is that leakage is a function that has certain restrictions (for example, see [4,5]) such that defining cryptography is still possible, but otherwise is meant to be as general and powerful as possible. Consequently a direct physical interpretation is not intended as such; rather, the idea is pursued that any 'realistic' leakage function is included given the general nature of the definition. The discrepancy that arises from this perspective is that for practitioners, any overhead that is incurred because of 'proof relics' or protection against 'magic' such as future computation attacks [15] is an unnecessary expense.

1.2   Simulatable Leakage

The recent contribution by Standaert *et al.* [19] hence comes as a welcome addition to the current approaches for dealing with the concept of leakage as part of provable security. In a nutshell, the authors suggest that a sensible notion for leakage resilience is that if real leakage cannot be distinguished from simulated leakage (from a simulator that does not have access to the secret key k), it cannot contain any information about the key. This approach removes the problem of having to mathematically define a leakage function: instead one gets a concrete instance of it in the form of an *actual* simulator. Rather than struggling with meaningful definitions for what is leakage, and how to practically derive bounds for it for a concrete device, the new challenge is therefore to define and build (practical) simulators.

The challenge of building concrete leakage simulators for invertible functions was also taken up in [19], where the authors suggest an efficient solution: given some public input x and output c of an invertible scheme f, they explain how a trace can be constructed with a random key k^* that is consistent with the public inputs x and c. This can be done by choosing a random key k^* and computing $c\prime = f_{k^*}(x)$, $x\prime = f_{k^*}^{-1}(c)$, and then $c = f_{k^*}(x\prime)$ to generate leakage traces $L(x) = l^l||\alpha$, $L(x\prime) = \beta||l^r$ that can be 'split up' (as indicated) and concatenated to a new simulated trace $(l^l||l^r)$. The q-sim game, that can be played in practice, consists of the attacker trying to distinguish real traces from simulated traces given q real (or simulated) traces by using whatever state of the art attacks that are available. The rest of [19] consists of two major contributions; first they discuss why state of the art side channel attacks cannot win this game effectively, and, second, they use the game restricted to $q = 2$ to prove a PRG (using AES as the underlying PRF) construction leakage resilient in the standard model.

1.3   Our Contribution

We show there exists a side channel distinguisher (against Standaert *et al.*'s simulator when instantiated with AES) which can effectively distinguish simulated traces: it does so by detecting the fact such traces are constructed via split and concatenation in the inner encryption rounds. We do not require knowledge of input or output, or access to the auxiliary leakage oracles for building templates. Our attack is based on using cross-correlation to check the consistency of the data flow across encryption rounds and in order to pinpoint inconsistencies from splitting and concatenating traces, and works across different real world platforms. Our distinguisher has the property that playing a game with q traces for a single key is equivalent to playing a game with one trace for q keys, this implies we can win the game in an asymptotic setting.

We analyse the properties of (cross-)correlation in this specific application to identify the factors that impact on the ability to win the game efficiently. The factors are the (intrinsic) cross-correlation between points in leakage traces, and the ratio between signal and noise. Whilst changing the ratio between signal and noise is well understood and practically achievable, it is not sufficient with regards to decreasing an adversary's chance to win the q-sim game. Our attempts to work with points that have low intrinsic cross-correlation were only successful in theory: for the concrete instantiations in our paper, which are based on AES, there (in theory) should exist such points because of the nature of Sub-Bytes. Theoretically, the input and output of the SubBytes operation are highly uncorrelated. However, we explain why in practice it remains a challenge to exploit this. Finally we suggest a method that indeed withstands the powerful cross-correlation distinguisher, which is based on instantiating the PRG with a double block cipher construction. Our proposed simulator then uses a meet-in-the-middle technique to determine keys that map x to c without introducing inconsistencies in the data flow. This simulator is somewhat theoretical because of the implied computation cost. However it allows the proof given in [19] to hold once more (remember that this proof crucially depends on the existence of a simulator).

Finally we note in our work that even the computationally expensive simulator still requires noisy traces: it would seem then that the most natural way to produce simulated leakage is to just use the construction 'as is' and run it with a random key. For sufficiently noisy traces even profiling prior to the game will not help an attacker: noisy leakage implies that an adversary must have sufficiently many traces to distinguish real from simulated leakage. By limiting q in the game stage this will be infeasible. With this somewhat simple fact in mind, we can argue that leakage can be simulated for any cryptographic primitive or construction. It however requires implementations with high noise.

2   Simulatable Leakage: Standaert *et al.*'s Model

Before we discuss the model introduced by Standaert *et al.* in [19], we will introduce the required notation. The probabilistic leakage of a block cipher will be given as $\mathbf{BC}_k(x) \rightsquigarrow l \overset{\text{def}}{=} L(k, x)$ where L is the leakage function, x is the plaintext and k the secret key. The leakage function can be described as a vector $l = (l_1, l_2, ...)$. For a block cipher, which typically consists of several encryption rounds, we group those trace points corresponding to a round and indicate this by placing the round number as a superscript l^i. For AES-128 we can represent a leakage as $l = [l^1, \ldots, l^{10}]$. We will later require to split and concatenate leakage vectors. For this purpose we use the short-hand $l^{i,j} = [l^i, \ldots, l^j]$ to denote that we take the parts of the leakage vector that correspond to rounds i up to j. To highlight where we 'split and concatenate' within leakage vectors we use $\|$ to explicitly mark concatenations. We often need to work with sets of leakages and so denote such a set with bold typesetting, i.e. $\mathbf{l}$ now is a set of leakages, $\mathbf{l}^{i,j} = [\mathbf{l}^i, \ldots, \mathbf{l}^j]$ now means that we refer to rounds i until j in all leakages in the set. Finally, if we need to differentiate between points within multiple leakages we will use a subscript, i.e., $\mathbf{l}_u = (l_{1,u}, l_{2,u}, \ldots)$ means we index the u-th point in each leakage vector in $\mathbf{l}$. Now that the notation has been defined, we are ready to discuss the model.

2.1   Model and q-sim Game

Figure 1 describes the q-simulatable leakage game (q-sim) from [19] for completeness. Recall that the intuition captured in the q-sim game is that if an adversary cannot distinguish real (i.e. depending on the secret key) from simulated (i.e. depending on a random key) traces, the real traces cannot contain any information about the secret key. In the game, the adversary can make q queries to the Enc oracle and receives back the encryption of x under key k and either the real leakage $L(k, x)$ or the simulated leakage $S^L(k^*, x, c)$. The adversary can also make s_A queries to the leakage oracle with a chosen key and message, which represents profiling a device (i.e. the adversary can attempt to derive (compute, or represent) the otherwise unknown leakage function). We note that this, in particular, allows an adversary to query the leakage function on the inputs from the game, so templates specifically for the inputs used in the game can be derived. The last oracle which can be called once is Gen which delivers simulated leakage for a chosen message/key pair where either the real or random key in the game is output as the ciphertext. This is to represent the fact that often encryption keys themselves are the result of block cipher invocation, i.e. in practice (and in the constructions discussed in [19]) the encryption key used in round r is generated as the output of the block cipher in the previous round $r - 1$. The adversary's advantage is calculated as $\mathbf{Adv}_{L,S^L,\mathbf{BC}}^{q\text{-sim}}(A) = |\Pr[q\text{-sim}(A, \mathbf{BC}, L, S^L, 1) = 1] - \Pr[q\text{-sim}(A, \mathbf{BC}, L, S^L, 0) = 1]|$.

2.2   Construction

The model given in [19] is used to prove the security of a leakage-resilient PRG which is instantiated using block ciphers (we will continue the pattern started in [19] and will instantiate the block cipher with AES). This construction show on the left in Fig. 2 and the underlying 2PRG is shown on the right in Fig. 2. When considering this within the q-sim game we note that each key is only used twice (once to create the PRG output and one to create the new key) and thus the value of $q = 2$ is the one of interest.

2.3   Simulator

For completeness we also recall the simulator in Fig. 3(a). The simulator takes in a random key k^* as well as a plaintext/ciphertext pair (x, c); note that this pair was created using a key different to k^*. The simulator first encrypts x under

Experiment q-sim$(A, \mathbf{BC}, L, S^L, b)$:

$k, k^* \stackrel{\$}{\leftarrow} \{0,1\}^n$

$(i, j, l) \leftarrow (0, 0, 0)$

$b' \leftarrow A^{Enc,(\cdot), Gen(\cdot, \cdot), Leak(\cdot, \cdot)}()$

Return $b\prime$

proc $Enc(x)$:

$i \leftarrow i + 1$

if $i > q$ then

    Return $\perp$

end if

$c \leftarrow \mathbf{BC}_k(x)$

if $b = 0$ then

    $\Lambda \leftarrow L(k, x)$

else

    $\Lambda \leftarrow S^L(k^*, x, c)$

end if

Return (c, Λ)

proc $Gen(z, x)$:

if $l = 1$ then

    Return $\perp$

end if

$l \leftarrow 1$

if $b = 0$ then

    $\Lambda \leftarrow S^L(z, x, k)$

else

    $\Lambda \leftarrow S^L(z, x, k^*)$

end if

Return Λ

proc $Leak(z, x)$:

$j \leftarrow j + 1$

if $j > s_A$ then

    Return $\perp$

end if

$\Lambda \leftarrow L(z, x)$

Return Λ

Fig. 1. q-simulatable leakage from [19]

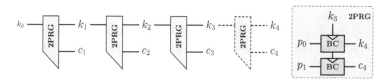

Fig. 2. Left: Leakage resilient PRG. Right: 2PRG construction

k^* and records the leakage. The next step is to decrypt c under k^* to get a new plaintext x'. The final stage is to encrypt x' under k^* (note that this will encrypt to c) and record the leakage. The leakage is the split (in half) and concatenated such that the first part of the new trace corresponds to the leakage on x while the second half corresponds to the leakage on c. This simulator is referred to as split and concatenate ($S_{s\&c}$) simulator and we will refer to traces from this simulator as $s\&c$-traces.

3   The Security Game for the Practical Use of the 2PRG: the p-q-sim Game

It is clear that in the q-sim game, the adversary can get only q queries on a key: this represents only a single round in the PRG. However, based on the construction for the 2PRG, we argue that the appropriate security game, which we denote the p-q-sim game, should take into account that potentially many, say p calls to the 2PRG are made. That is, in the p-q-sim game the adversary

simulator $S_{s\&c}^L(k^*, x, c)$:
$c' \leftarrow \mathbf{AES}_{k^*}(x) \rightsquigarrow l^{1,5}||\alpha$
$x' \leftarrow \mathbf{AES}_{k^*}^{-1}(c)$
$c \leftarrow \mathbf{AES}_{k^*}(x') \rightsquigarrow \beta||l^{6,10}$
Return $l^{1,5}||l^{6,10}$

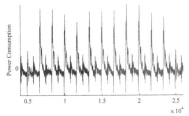

(a) Description of simulator $S_{s\&c}^L$ (b) Simulated $S_{s\&c}^L$ SASEBO-R trace

Fig. 3. Definition for the $S_{s\&c}^L$ simulator and an exemplary trace

gets to make q queries against either p real or p simulated instantiations (with p different keys) and then he must work out if the leakage he is seeing real or simulated leakage for all of the p instantiations.

We argue further that the p-q-sim game is in general more appropriate: in an evaluation context, which is what [19] really consider, the q-sim game would be played for real, and it seems unlikely that an evaluator would only ever play the game once and take q traces (especially if q is small). More likely, one would play this game several times to get a sense of the success rate for q traces.

It would be tempting to believe that an adversary cannot exploit the information leakage across different games because they are based on different keys. Traditional side channel distinguishers require after all to make key-dependent hypotheses: hence whenever a new key is introduced the attacker is presented with a new, fresh challenge. Whilst [19] discusses why the original q-sim game does not hybridise with respect to q, they do not consider the possibility that the game does hybridise in the number of keys p. If the game were to hybridise, then we would have that $\mathbf{Adv}_{L,S^L,\mathbf{BC}}^{p\text{-}q\text{-sim}}(A) \leq p \cdot \mathbf{Adv}_{L,S^L,\mathbf{BC}}^{q\text{-sim}}(A)$; the game gets easier to win if it is played more often.

In the next section we will show a distinguisher that can work across different keys and thus can take advantage of the full (and more realistic) p-q-sim game.

4   Breaking the Split-and-Concatenate Simulator

Whilst the proposed simulator can be instantiated with any invertible function, we continue to follow Standaert *et al.*'s exposition and use implementations of AES-128 as running examples. To explain why cross-correlation is an efficient distinguisher we briefly recall how a typical side channel trace relates to the information flow in an AES implementation.

4.1   Properties of Real World Leakage

As argued in previous work [8,3], any implementation of AES will happen as a sequence of steps that correspond to the processing of intermediate values. For instance, in a serial implementation, the state bytes will be accessed sequentially

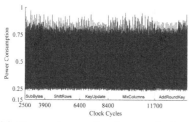
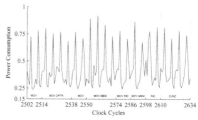

(a) Power trace showing a single AES (b) Power trace showing individual as-
round.   sembly instructions

Fig. 4. Power traces for an 8051 8-bit microcontroller

and updated according to the AES round function. As all high level functions
(SubBytes, MixColumns, etc.) are processed by some gates at the lowest level,
a clock signal is involved that governs (to some extent) when data flows between gates. All changes in gates, in each clock cycle, produce some form of
leakage (time for signals to travel, power consumed, radiation emitted) that
becomes available through observing the device. Figure 4 illustrates the power
consumption measurements from an AES implementation on a simple 8-bit microcontroller (i.e. all intermediate values are represented as bytes). Evidently,
we can see patterns in Fig. 4(a), which represents a single AES round.

In case of this very simple processor, we can even identify the effect (with
regard to shape and height) of individual instructions in the power consumption
by zooming into the trace further, see Fig. 4(b). It shows now only a small
part of the first round that corresponds to performing the SubBytes operation.
The instructions used for this purpose are register transfers, (MOV and MOVC), a
register increment (INC), and conditional branches which correspond to a loop
that runs through the 16 state bytes.

In a parallel implementation, each operation will touch multiple state bytes
but the sequence of round functions must remain sequential. SubBytes is sometimes implemented using combinational logic in dedicated hardware [22]. A combinatorial logic circuit is not governed by a clock but the output from such a
circuit will typically be connected to a synchronous storage element. Referring
to Fig. 5(c) we can see that there are 10 visible peaks relating to the AES-128
rounds.

4.2   Cross-Correlation as a Distinguisher

The term correlation is often used to refer to a broad class of statistical dependencies in data. There are different metrics to measure correlation. Most
commonly used (at least in side channel attacks) is the Pearson correlation coefficient, and we use this metric when we refer to correlation in this article. In
the context of side channel analysis, correlation has been applied in DPA before
[2] and its properties as a good distinguisher are well understood [12].

Cross-correlation is a term coined in signal processing and is commonly used to identify (and measure) similarities of wave forms. It has been used in the context of side channel analysis before, e.g. [13,18,21].

We make the simple observation that the cross-correlation of signals of length one (i.e. points) equates to computing the correlation between trace points (l_u, l_v) (as opposed to the correlation with key-dependent predictions in the DPA context). Equation (1) recalls how (Pearson) correlation is defined and estimated.

$$\rho(l_u, l_v) = \frac{Cov(l_u, l_v)}{\sqrt{Var(l_u) \cdot Var(l_v)}} \tag{1}$$

$$= \frac{\sum_i (l_{i,u} - \bar{l}_u) \cdot (l_{i,v} - \bar{l}_v)}{\sqrt{\sum_i (l_{i,u} - \bar{l}_u)^2 \cdot \sum_i (l_{i,v} - \bar{l}_v)^2}}$$

Cross-Correlation Traces. We recall that cryptographic algorithms are implemented as step-wise processes (with varying degrees of parallelism). Although AES mixes keying material and input efficiently, and so the correlation between key, input, and output is small, we can expect a high correlation between the subsequent states, e.g. we expect a high correlation between the input and output of ShiftRows, as well as states that operate on the same data (even though they might be separated in time). This in turn implies that we can expect a high correlation between subsequent points within leakage traces (see [11, Ch. 4]), as well as points that are related to the same intermediate values. Further to that we can also expect high correlations between data that is related to the program state but independent of the states, e.g. the value of the program counter, pointers to memory locations, etc. Hence any implementation will lead to a specific cross-correlation trace depending on the data flow.

By producing a cross-correlation trace that shows the cross-correlation of *all* pairs of points, i.e. $\{\rho(l_u, l_v), \forall(u, v)\}$, we can consequently track the consistency of data flow. Such a trace however would be very long (the length would be the square of the original traces' length). We hence opted to reduce the cross-correlation data by selecting the highest cross-correlation value for each u over all 'distant' pairs (l_u, l_v), i.e. for each u we took $\tilde{\rho}_u = max\{\rho(l_u, l_v) \forall v\}$ where v is not within a small window around u. Hence our cross-correlation traces $\tilde{\rho} = \{\tilde{\rho}_u, \forall u\}$ have the same length as the actual leakage traces, and they have uninformative trivial cross-correlation removed because neighbouring points are not considered. Consequently the cross-correlation traces show the effect of data consistency, and any 'dip' implies some form of discontinuity in the data flow.

We provide in Fig. 5 some cross-correlation traces for illustration. We chose two devices with contrasting architectures to demonstrate that cross-correlation works irrespective of the underlying device. The first device features a highly serial implementation of AES where each step only touches at most one byte of the state. It is representative of AES implementations in the low cost market. The second device features a highly parallel implementation of AES. Such an

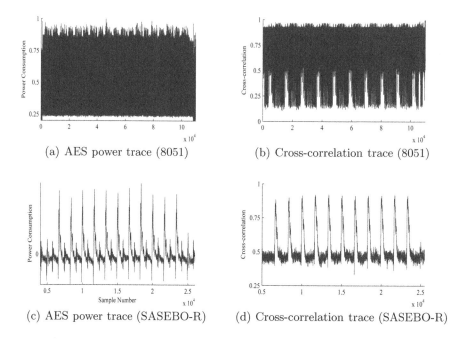

(a) AES power trace (8051)   (b) Cross-correlation trace (8051)

(c) AES power trace (SASEBO-R)   (d) Cross-correlation trace (SASEBO-R)

Fig. 5. AES power traces and cross-correlation plots

implementation is more likely to be found in high end products where implementation speed and security is considered an important factor.

4.3   Detecting *s&c*-traces

In the proof given in [19], an adversary plays the q-sim game and so has access to the oracles *Enc*, *Leak*, and *Gen* (an adversary may also query them in the p-q-sim game). Our distinguisher does not require any access to the oracles *Leak* and *Gen*.

To detect the *s&c*-traces we apply the cross-correlation method to a set of q leakage traces. The detection is then based on the absence of cross-correlation that would otherwise be present in traces that have not been simulated, i.e. a set of points which leak on the same data no longer show a significant correlation with respect to each other.

The q-sim (and p-q-sim) games define that there exists a simulator that is secure for all adversaries (with set computational limits). We may hence assume the adversary's knowledge includes all implementation details of the PRF as well as the principle of the simulator, i.e. where the traces are split and concatenated. Recall that a cross-correlation trace shows 'patterns' which are based on the relationships between following and related intermediate values. Consequently, even a cross-correlation trace from the simulated leakages will show such patterns, see right hand side of Fig. 5. Only at the round where the simulated traces have been split and concatenated a different pattern will occur. Consequently an attacker gets a 'fingerprint' for how the cross-correlation trace should look (in the case of the s&c-simulator) by examining the beginning and end of the cross-correlation trace. Any significant deviation from the fingerprint (i.e. any discrepancy between the two) will hence identify the s&c-traces.

4.4   Experiments for Real Devices

Practical attacks are often played down because they are device specific and hence it is often not possible to draw general conclusions from a single attack. However, the heart of the proposal in [19] is to be able to implement a secure simulator on some real world devices. We consequently did not want to resort to 'pure simulations' and opted to fully implement simulators for several real world devices. By choosing different devices we can refute the argument that our analysis outcomes are not valid in general: the devices we choose have different architectures that lead to different AES implementations, different leakage models and different noise characteristics.

AES can be implemented in different ways. Serial implementations are often found on small processors (8-bit or 32-bit). These are software-only implementations and we have used two different widely-used processors to instantiate such a software implementation for our attacks. Parallel implementations can be found as dedicated hardware implementations. In practice 32-bit implementations would be considered as suitable for constrained devices such as smart cards, whereas highly parallel 128-bit architectures would be used when throughput is a practical concern. We opted to use a highly parallel 128-bit architecture to provide contrasting results to the software implementations. In the following, we give a brief overview of the results obtained from each architecture. Details regarding the acquisition setup and target devices can be found in Appendix A.

Software Implementations. We used a general purpose microcontroller which features an 8051 instruction set as the first device for our attacks. The cross-correlation plots for this architecture reveal detailed information about the data flow during the execution of an algorithm. In our running example (AES-128), we are able to detect the order in which state bytes and functions are accessed, the operations performed and for a masked implementation, when and where each masked is applied. In Fig. 6(a) and 6(b) we show a portion of the cross-correlation for real and simulated leakage traces respectively. There is a clear break in the cross-correlation as a result of the concatenation between traces with inconsistent states. We repeated the s&c experiment with an ARM[1] based 32-bit architecture and implementation. Like for the 8051, we can track the AES data flow and so the simulated leakage trace, once again, leads to a drop in the cross-correlation.

---

[1] The cross-correlation plots for this device are available in the full paper [10].

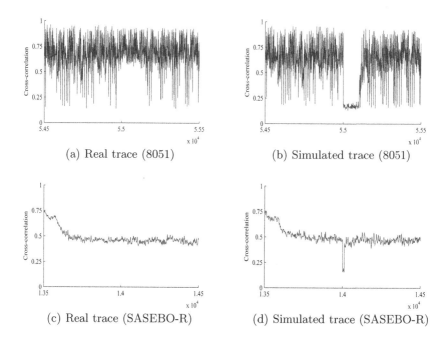

(a) Real trace (8051)   (b) Simulated trace (8051)

(c) Real trace (SASEBO-R)   (d) Simulated trace (SASEBO-R)

Fig. 6. Cross-correlation distinguisher plots

Hardware Implementation. The SASEBO-R ASIC boasts a large number of cryptographic functions implemented as dedicated logic. Unlike the two previous devices, the information leaked is no longer dependant on processor operations but on combinatorial switching. Figures 5(c) and 3(b) show the power consumption over an execution of AES and a simulated trace generated by the $s\&c$ simulator respectively. The cross-correlation distinguisher now plays on the coherency of the combinatorial switching rather than the state leakage at each clock cycle. As with the software implementations, we can easily identify the simulated leakage traces, see Figs. 6(c) and 6(d). One key difference over the software implementation is cross-correlation no longer reveals any information about the data flow or what operations are being performed but simply that there exists some data dependency in the power consumption of the device.

4.5 Measures to Secure the Split-and-Concatenate Simulator

Recall that we estimate the cross-correlation between trace points, which implies varying inputs and/or keys. The number of different inputs per key is q, whereas the number of different keys is p. We may hence decide to increase p and keep q small, which implies that we can break the PRG construction of [19] because it has multiple rounds. As our distinguisher hybridises, the construction looses security every time a round is leaked, on even though fresh keys may be used!

We now discuss how traditional engineering-style countermeasures impact on the success of winning the game. We begin by showing how noise on leakage traces will help to make winning the game harder. We then explain why masking and hiding approaches are unlikely to defeat the cross-correlation distinguisher.

Increasing the Noise. Just as we can write down a formula for the impact of the signal-to-noise ratio (SNR) on a correlation-based DPA attack (see, e.g. [10, Ch. 4.3.1]), we can express the impact of the SNR on our proposed cross-correlation attack.

For this we write leakage points as a direct sum of an (unknown) signal ($\mathbf{S}$) plus independent (Gaussian) noise ($\mathbf{N}$), i.e. $\mathbf{l}_u = \mathbf{S}_u + \mathbf{N}_u$, with $\mathbf{N}_u \sim \mathcal{N}(0, \sigma_u)$, and $\mathbf{l}_v = \mathbf{S}_v + \mathbf{N}_v$, with $\mathbf{N}_v \sim \mathcal{N}(0, \sigma_v)$ [3,11]. The signal to noise ratio (SNR) is defined as $SNR = \frac{Var(\mathbf{S})}{Var(\mathbf{N})}$. The respective SNRs at the points u and v are then $SNR_u = Var(\mathbf{S}_u)/Var(\mathbf{N}_u)$ and $SNR_v = Var(\mathbf{S}_v)/Var(\mathbf{N}_v)$. Making the simplifying assumption that $SNR_u \approx SNR_v = SNR$, it turns out (using the same technique as [10, Ch. 4.3.1]) that

$$\rho(\mathbf{l}_u, \mathbf{l}_v) = \rho(\mathbf{S}_u, \mathbf{S}_v) \cdot \frac{1}{1 + \frac{1}{SNR}}$$

In comparison to a DPA attack (we refer to [10, Ch. 6.3]), the impact of the SNR is potentially stronger on this distinguisher. However, because this distinguisher hybridises over different keys, it is also easier to gather more leakage traces to compensate for this. In particular, it follows that asymptotically over an increasing number of keys p we can still win the game for any small q.

More Complex/Parallel Architectures. Another important observation at this point is that in contrast to DPA attacks, the 'relative size' of the intermediate value (in terms of its bit length in contrast to the overall device state) itself is less important. At first glance this goes against the intuition from DPA style attacks where practice has shown that they become harder for architectures that employ a larger data-path (and so the intermediate values that are attacked contribute only a small amount to the overall leakage). For instance, DPA style attacks on 32-bit processors often only predict 8 bits of the 32-bit state, so the remaining 24 bits are noise. In addition, in DPA style attacks using correlation one requires to 'model' the leakage behaviour, and especially for dedicated hardware, standard models such as the Hamming weight or Hamming distance are less than ideal approximations.

The cross-correlation distinguisher however does not require to model the device leakage and, importantly, we are effectively using the entire state information because we are working with points and do not have to make any predictions. This explains the contrast to typical DPA style attacks, and so the highly effective nature of the distinguisher.

Masking and Hiding Approaches. It would be tempting to assume that any countermeasure against correlation-based DPA attacks would automatically work against the cross-correlation distinguisher because both distinguishers share the same statistical method.

simulator $S_{2c}^L(k^*, x, c)$:
$c', ST_5 \leftarrow \mathbf{AES}_{k^*}(x) \rightsquigarrow l^{1,4}||\alpha$
$x' \leftarrow \mathbf{AES}_{k^*}^{-1}(c)$
$c, ST_6 \leftarrow \mathbf{AES}_{k^*}(x') \rightsquigarrow \beta||l^{6,10}$
Construct $k^\#$ using ST_5, ST_6
$ST_6 \leftarrow \mathbf{AES1}_{k^\#}(ST_5) \rightsquigarrow l^5$
Return $l^{1,4}||l^5||l^{6,10}$

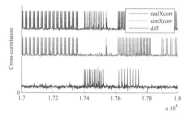

Fig. 7. Simulator description and cross-correlation comparison for S_{2c}

The two main engineering approaches to distinguish classes of countermeasures are hiding and masking [11]. Hiding countermeasures typically change the SNR and so increase the number of leakage traces that are needed for a successful attack.

Masking (i.e. secret sharing) countermeasures aim to make exploiting the information impossible by distributing it over different intermediate values (and hence leakage points), such that it becomes increasingly infeasible to 'recombine' that information. In practice however it is not possible to implement secret sharing with many shares. Typically only two, or at most three, shares are used and the masks (i.e. randomness) are not refreshed in between rounds or in between invocations of an intermediate value [7,6]. Consequently, practical masking schemes maintain the consistency between the subsequent transformations on the state and so the cross-correlation distinguisher remains applicable.

5   The Challenge of Making Secure Simulators

Given that traditional countermeasures are not suitable to rescue the split-and-concatenate simulator, we have to come up with new ideas. Two approaches are (intuitively) worth pursuing. Firstly, this is to try and maintain state consistency across the concatenated traces. Secondly, this is to split where there is a 'natural' discontinuity in the data flow.

5.1   Maintaining State Consistency

Over the execution of any algorithm there exists a degree of consistency between the intermediate values, which is disrupted by splitting traces. Hence, we attempted to design a 'state aware' simulator by generating an 'intermediate round' that ensures such consistency.

This simulator shown in Fig. 7 and the extra notation can be understood as follows; ST_i is the state of AES at the start of the ith round and $\mathbf{AES1}_k$ runs a single round of AES on round key k.

The simulator S_{2c} operates similarly to $S_{s\&c}$ by first performing an encryption of x under the key k^*. The leakage captured from this corresponds to the first four rounds $l^{1,4}$. We also store the state ST_5. Next, the encryption of x' under

k^* is performed and the leakage of the last 5 rounds is captured $l^{6,10}$ along with ST_6. To connect the two otherwise disconnected states we generate an extra trace $\textbf{AES1}_{k\#}(ST_5) \rightsquigarrow l^5$. Note that finding the key $k^\#$ which maps ST_5 to ST_6 is simple considering only a single round of AES.

We proceeded to implement the S_{2c} simulator for the 8051 and the resulting cross-correlation was once again able to detect the simulated traces. This time it detected the discontinuity in the round key schedule. This shows how hard it is to achieve state consistency because we have to take into account the AES state as well as the AES key schedule (in the p-q-sim game), and the fact that different instructions can leak subtly differently.

5.2 Leveraging an Algorithm-Dependent Data-Flow Discontinuity

Given our running example is AES, the natural candidate intermediate value is SubBytes, because the input and output of SubBytes are (almost) statistically uncorrelated. We can hence expect that there is also only a low correlation between the corresponding trace points, which implies that the data flow across SubBytes is somewhat discontinued. We tested this idea first on an 8051 software emulator which produces noise-free leakage on the data processed at each instruction. The simulated traces were indeed indistinguishable from real traces (taken from the emulator) produced. Consequently we attempted to implement this on real devices.

On real devices, one has two implementation choices for SubBytes. Either a table is stored and hence a SubBytes computation corresponds to a table lookup operation. This is suitable for somewhat 'serial' implementations, as typically only a single instance of the table is held in logic. This is the option used in our AES software implementations. Alternatively, one implements it as combinational circuit, which is hence suitable for dedicated hardware platforms. This option is used in the hardware AES on the SASEBO-R.

Our new simulator S_{sbc} 'tweaks' the $S_{s\&c}^{L}$ simulator construction to perform the split-and-concatenate at the S-box lookup rather than points of 'no activity'. However, we note that for a sequential lookup, the simulator is required to perform splice at each S-box rather than a simple split-and-concatenate as illustrated by Fig. 8(a) (hence we effectively 'chop up' a trace).

A Practical Attempt for an 8051 Processor. Implementing this for a real 8051 device reveals the challenges of real world (imperfect) leakage. Although we could pinpoint the exact location of the SubBytes operation, it so happens that each low-level operation is performed over multiple clock cycles and hence leaks multiple times for each operand. To be precise: consider the lookup $r_0 = M[A]$, where a register r_0 is loaded with the contents of memory address A. The resulting leakage trace resembles the form $[\mathcal{L}(A)\|\mathcal{L}(M[A])\|\mathcal{L}(A)\|\mathcal{L}(M[A])]$ for some leakage function $\mathcal{L}$ in our 8051 processor.

As a result, the simulator for the 8051 needs to splice multiple times within each S-box lookup. This behaviour is clearly very architecture specific. Figure 8(b) shows the cross-correlation resulting from splicing at each S-box; the

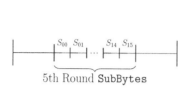
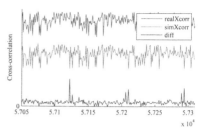

(a) A visual illustration for S_{sbc} (b) Cross-correlation comparison for S_{sbc}

Fig. 8. Illustration for a serial S_{sbc} simulator and a cross-correlation comparison of S_{sbc} for the 8051 microcontroller

top plot (printed in blue) shows the cross-correlation trace as derived from real traces. The middle plot (printed in red) shows the cross-correlation trace as derived from simulated (S_{sbc}) traces. Whilst a visual detection seems difficult at first, an adversary with information about the time points (or a fingerprint, which we explained previously can be constructed even from simulated traces) can spot the difference. This is made clear by the lowest plot (printed in black) that shows that there is a distinct difference between real and simulated cross-correlation.

A Practical Attempt for the SASEBO-R. We now consider the implications of a parallel combinatorial SubBytes function as performed by the SASEBO-R ASIC. Pinpointing the SubBytes operation is no longer such a trivial task as each round function is evaluated as a combinatorial circuit rather than being governed by an external clock. Attempting to model the leakage in an ideal setting would also require significant knowledge of both the design and layout of the ASIC. We hence resorted to a exhaustive search over a whole encryption round in order to determine whether or not a point existed that would allow us to build the S_{sbc} simulator for such a device.

Perhaps unsurprisingly, we were unable to identify points that did not produce a significant drop in the cross-correlation. This is primarily due to the relation between evaluation stages of a combinatorial circuit. Without further insight on the design of the ASIC, it is impossible to determine what processes were taking place that made building a viable S_{sbc} simulator impractical.

6   A Sound Simulator

The previous section has shown that the intuitions for building secure simulators failed to translate to the practical devices that we considered. Furthermore, *how* they failed leaves us with little confidence that using other platforms, e.g. embedded platforms (with other processors) or a different combinational circuit

simulator $S^L(x, c)$:

Perform a meet–in–the–middle attack to learn a valid $(k_i^*, k_i^{*'})$

$\mathbf{BC}_{k_i^*}(\mathbf{BC}_{k_i^{*'}}(x)) \rightsquigarrow \Lambda$

Return Λ

Fig. 9. A generic simulator S secure against the cross-correlation distinguisher

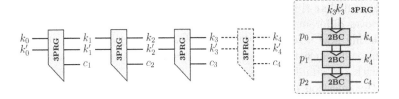

Fig. 10. The adjusted PRG construction

for SubBytes, would be any more successful. The fundamental hurdle is simply that real world leakage is complex, and cryptographic algorithms are necessarily implemented via step-wise processes. Hence there is a specific data flow that will be somehow disrupted when using a split-and-concatenate approach. Without substantial alterations to the devices' designs this cannot be easily changed.

Splitting *within* one instantiation of a block cipher seems hence futile: but how about considering constructions that are based on two somewhat independent block cipher calls?

6.1   Doubling the Cipher

Now we discuss an approach based on using a double block cipher (i.e. a block cipher **2BC** that consists of two sequential computations of a block cipher **BC** with independent keys k_i and k_i'): $c = \mathbf{BC}_{k_i}(\mathbf{BC}_{k_i'}(x))$. In this construction there is a natural discontinuity between the first and the second encryption with regards to the key state and so the data flow across the 'boundary' between the first and the second encryption. This makes this boundary an ideal place to 'split' traces, and a generic simulator S that uses a meet–in–the–middle technique [9] to find a suitable pair of keys follows immediately (see Fig. 9).

For completeness we show that this simulator can be plugged into the PRG construction from [19] maintaining the correctness of the proof. We only switch out the underlying PRF from AES to double AES, and subsequently we need to switch the 2PRG for a 3PRG for the extra rekeying material. The proof given in [19] can be expanded for any constant number of calls to the PRF and thus the construction will not need to be reproved secure. The resulting construction can be seen in Fig. 10.

6.2   Some Final Considerations

In the case of AES, this simulator requires approximately 2^{65} AES encryptions (because of the meet–in–the–middle technique) per valid trace. This is computationally too expensive to be practical; yet the simulator is secure against the cross-correlation distinguisher per design so a practical implementation is not necessary in that regard.

However, its security against standard DPA attacks still needs to be considered. Recall that this already is an advantage because DPA attacks do not hybridise over different keys!

Considering now a standard DPA on S^L one would notice (in the process of performing such an attack) that no key hypothesis ever achieves a good correlation with the simulated traces. Hence for a DPA distinguisher we need to consider the question of how many traces are necessary to decide with some certainty that all key candidates are equally likely. We leave this as an option question but note that the usual arguments for DPA success (or lack thereof) will apply. These are that for a (reasonably) low SNR, and a small number of leakage traces q, an attack does not succeed. In particular for the purposes of the PRG construction from [19], which limits q to be two, practical instantiation of S on an ASIC such as the SASEBO-R should be feasible.

Acknowledgements. Daniel Martin and Elisabeth Oswald have been supported in part by EPSRC via grant EP/I005226/1, and Daniel Page by EP/H001689. Jake Longo Galea has been supported in part by a studentship under the EPSRC Doctoral Training Partnership (DTP) scheme.

References

1. Atmel. AT89S8253 Datasheet, http://www.atmel.com/Images/doc3286.pdf
2. Brier, E., Clavier, C., Olivier, F.: Correlation power analysis with a leakage model. In: Joye, M., Quisquater, J.-J. (eds.) CHES 2004. LNCS, vol. 3156, pp. 16–29. Springer, Heidelberg (2004)
3. Chari, S., Jutla, C.S., Rao, J.R., Rohatgi, P.: Towards sound approaches to counteract power-analysis attacks. In: Wiener, M. (ed.) CRYPTO 1999. LNCS, vol. 1666, pp. 398–412. Springer, Heidelberg (1999)
4. Dziembowski, S., Pietrzak, K.: Leakage-resilient cryptography. In: FOCS, pp. 293–302 (2008)
5. Faust, S., Rabin, T., Reyzin, L., Tromer, E., Vaikuntanathan, V.: Protecting circuits from leakage: the computationally-bounded and noisy cases. In: Gilbert, H. (ed.) EUROCRYPT 2010. LNCS, vol. 6110, pp. 135–156. Springer, Heidelberg (2010)
6. Fumaroli, G., Martinelli, A., Prouff, E., Rivain, M.: Affine masking against higher-order side channel analysis. In: Biryukov, A., Gong, G., Stinson, D.R. (eds.) SAC 2010. LNCS, vol. 6544, pp. 262–280. Springer, Heidelberg (2011)
7. Herbst, C., Oswald, E., Mangard, S.: An AES smart card implementation resistant to power analysis attacks. In: Zhou, J., Yung, M., Bao, F. (eds.) ACNS 2006. LNCS, vol. 3989, pp. 239–252. Springer, Heidelberg (2006)

8. Kocher, P.C., Jaffe, J., Jun, B.: Differential power analysis. In: Wiener, M. (ed.) CRYPTO 1999. LNCS, vol. 1666, pp. 388–397. Springer, Heidelberg (1999)
9. Lai, X., Massey, J.L.: Hash functions based on block ciphers. In: Rueppel, R.A. (ed.) EUROCRYPT 1992. LNCS, vol. 658, pp. 55–70. Springer, Heidelberg (1993)
10. Longo Galea, J., Martin, D., Oswald, E., Page, D., Stam, M., Tunstall, M.: Simulatable leakage: analysis, pitfalls, and new construction. Cryptology ePrint Archive, Report 2014/357, https://eprint.iacr.org/2014/357
11. Mangard, S., Oswald, E., Popp, T.: Power Analysis Attacks: Revealing the Secrets of Smart Cards. Springer (2008)
12. Mangard, S., Oswald, E., Standaert, F.-X.: One for all - all for one: unifying standard differential power analysis attacks. IET Information Security 5(2), 100–110 (2011)
13. Messerges, T.S., Dabbish, E., Sloan, R.H.: Power analysis attacks of modular exponentiation in smartcards. In: Koç, Ç.K., Paar, C. (eds.) CHES 1999. LNCS, vol. 1717, pp. 144–157. Springer, Heidelberg (1999)
14. NXP. LPC2124 Datasheet, http://www.keil.com/dd/docs/datashts/philips/lpc2114_2124.pdf
15. Pietrzak, K.: A leakage-resilient mode of operation. In: Joux, A. (ed.) EUROCRYPT 2009. LNCS, vol. 5479, pp. 462–482. Springer, Heidelberg (2009)
16. SASEBO. SASEBO Crypto LSI Specification, http://www.rcis.aist.go.jp/files/special/SASEBO/CryptoLSI-ja/CryptoLSI2_Spec_Ver1.0_English.pdf
17. SASEBO. SASEBO-R Specification, http://www.rcis.aist.go.jp/files/special/SASEBO/SASEBO-R-ja/SASEBO-R_Spec_Ver1.0_English.pdf
18. Sauvage, L., Guilley, S., Flament, F., Danger, J.-L., Mathieu, Y.: Blind cartography for side channel attacks: Cross-correlation cartography. Int. J. Reconfig. Comp. 2012(15), 1–9 (2012)
19. Standaert, F.-X., Pereira, O., Yu, Y.: Leakage-resilient symmetric cryptography under empirically verifiable assumptions. In: Canetti, R., Garay, J.A. (eds.) CRYPTO 2013, Part I. LNCS, vol. 8042, pp. 335–352. Springer, Heidelberg (2013)
20. IAIK TU. DPA Demo Board, https://www.iaik.tugraz.at/content/research/implementation_attacks/impa_lab_infrastructure/
21. Witteman, M.F., van Woudenberg, J.G.J., Menarini, F.: Defeating RSA multiply-always and message blinding countermeasures. In: Kiayias, A. (ed.) CT-RSA 2011. LNCS, vol. 6558, pp. 77–88. Springer, Heidelberg (2011)
22. Wolkerstorfer, J., Oswald, E., Lamberger, M.: An ASIC implementation of the AES sBoxes. In: Preneel, B. (ed.) CT-RSA 2002. LNCS, vol. 2271, pp. 67–78. Springer, Heidelberg (2002)

A   Acquisition Setup and Target Devices

In this appendix we outline the equipment used to gather the side channel data provide a brief note about each of the target devices used throughout the paper.

A.1   Acquisition Setup

The hardware used throughout the experiments follows a typical acquisition setup commonly found in the literature [8] [11, Ch. 3]. The measurement apparatus used for each of the experiments is as follows:

- Tektronix DPO7104 1Ghz Digital Oscilloscope.
- Tektronix P7330 High-performance differential probe.
- TTI EX354 Stable bench power supply.
- Agilent 33220 Signal generator.

The power consumption for each device was captured by measuring the drop across a resistor placed in the ground return path for each of the devices.

A.2   The AT89S8253 8051 Microcontroller

The AT89S8253 [1] is an 8-bit microcontroller which represents the lower end market for hardware. The device can be found in smartcards and is well documented in the side channel community for it's Hamming weight leakage model. This was used in conjunction with the DPA Demo board from IAIK-TU[20].

The AES implementation was limited to an 8-bit serial implementation due to the architectural constraints. Each of the S-box operations were executed as a table lookup. The device was clocked at 12Mhz throughout all experiments and the oscilloscope set to capture the power signal at 200Ms/s.

A.3   LPC2124 ARM7TDMI NXP Microcontroller

The LPC2124 [14] microcontroller is a 32-bit RISC microcontroller with a 4 stage pipeline. This device serves to represent the mid-range market of microcontrollers with 32-bit architectures. A custom board was designed and used to facilitate power measurement.

The AES implementation consisted primarily of 32-bit operations to build each of the round functions (AddRoundKey, ShiftRows etc.) with the exception of SubBytes which was performed as an 8-bit lookup table. The device was clocked at 14Mhz throughout all experiments and the oscilloscope set to capture the power signal at 250Ms/s.

A.4   SASEBO-R Cryptographic LSI

The SASEBO (Side-channel Attack Standard Evaluation Board) project aimed to provide development kits to facilitate side channel research. The SASEBO-R [17] board is specifically designed to fit a cryptographic LSIs [16]. The AES core used throughout this paper is the AES2 instantiation. Both the clock and power regulation for the target ASIC is managed on the SASEBO-R board. The oscilloscope was set to capture the power signal at 2Gs/s.

Multi-target DPA Attacks: Pushing DPA Beyond the Limits of a Desktop Computer

Luke Mather, Elisabeth Oswald, and Carolyn Whitnall

Department of Computer Science, University of Bristol,
Merchant Venturers Building, Woodland Road,
Bristol, BS8 1UB, United Kingdom
{luke.mather,elisabeth.oswald,carolyn.whitnall}@bris.ac.uk

Abstract. Following the pioneering CRYPTO '99 paper by Kocher et al., differential power analysis (DPA) was initially geared around low-cost computations performed using standard desktop equipment with minimal reliance on device-specific assumptions. In subsequent years, the scope was broadened by, e.g., making explicit use of (approximate) power models. An important practical incentive of so-doing is to reduce the data complexity of attacks, usually at the cost of increased computational complexity. It is this trade-off which we seek to explore in this paper. We draw together emerging ideas from several strands of the literature—high performance computing, post-side-channel global key enumeration, and effective combination of separate information sources—by way of advancing (non-profiled) 'standard DPA' towards a more realistic threat model in which trace acquisitions are scarce but adversaries are well resourced. Using our specially designed computing platform (including our parallel and scalable DPA implementation, which allows us to work efficiently with as many as 2^{32} key hypotheses), we demonstrate some dramatic improvements that are possible for 'standard DPA' when combining DPA outcomes for several intermediate targets. Unlike most previous 'information combining' attempts, we are able to evidence the fact that the improvements apply even when the exact trace locations of the relevant information (i.e. the 'interesting points') are not known *a priori* but must be searched simultaneously with the correct subkey.

1 Introduction

Differential power analysis (DPA) was initially conceived as a computationally 'cheap' way to recover secret information from side-channel leakages, under the assumption that trace measurements could be easily acquired [14]. Over time, the emphasis has changed and several directions have been pursued in the literature, e.g. attacks using power models [6] and attacks using several trace points [7] ([15] surveys the many variations of DPA style attacks). Across all these directions, one 'measure' of attack success has emerged and now dominates the scientific discourse with regards to attack efficiency. This measure is the number of power traces needed to identify the correct (sub)key[1].

---

[1] The overall key recovery works according to a divide-and-conquer strategy; each (for example) byte of the key is attacked and recovered individually.

P. Sarkar and T. Iwata (Eds.): ASIACRYPT 2014, PART I, LNCS 8873, pp. 243–261, 2014.
© International Association for Cryptologic Research 2014

What is the purpose of considering (sub)key recovery attacks? From a practical perspective any strategy is considered successful if it reveals 'enough' information about the (global) key to enable a brute force search. It is crucial that side-channel resistance, like other aspects of security, be considered with respect to realistic threat models. Real world adversaries are then (arguably) mostly interested in exploring trade-offs between the number of leakage traces available and the computational resources dedicated to extracting as much information as possible from those traces. Recent work by Veyrat-Charvillon et al. [25,26] presents an algorithm for searching the candidate space containing the key and a means to estimate its size if the enumeration capabilities of the analyst are below those of a better-resourced adversary.

Resources, from the point of view of a contemporary DPA adversary, include not only sophisticated measurement equipment but crucially also processing capabilities that directly map to the *time* necessary to mount and complete attacks [8]. Moradi et al.'s recent work [18] demonstrates how the use of a handful of modern graphics cards allows for dramatic increase in processing capabilities, enabling an attack on 32-bit key hypotheses in a known point scenario (the leakage point corresponding to the attacked operation was determined *a priori* via a known key attack).

In this submission we explore the possibilities for sophisticated use of modern processing capabilities (such as those associated with high performance computing (HPC), albeit restricted to the setting of a few machines or a 'small' cluster) to facilitate 'multi-target' DPA attacks. Multi-target DPA consists in amalgamating outcomes from multiple single-target attacks with the aim of reducing global key entropy more quickly than an individual single-target attack. For example, against a sequential AES implementation, multi-target DPA could amalgamate the outcomes of standard attacks on the AddRoundKey, SubBytes, and MixColumns operations. We will show later that we can do this meaningfully, and also efficiently, for correlation-based DPA attacks—even in realistic scenarios where the exact leakage points for those target functions are not known and must each be searched within windows of the trace. Most importantly, we show that such attacks can dramatically out-perform single-target attacks and are by far the best strategy to minimise the number of leakage traces required.

1.1   Our Contribution

An adversary who is capable of attacking large numbers of key hypotheses has a greater choice of intermediate target functions to attack. For instance, possible AES targets include the output of AES MixColumns (involving four bytes of the secret key) as well as the (implementation-dependent) intermediate computations for MixColumns (involving two or three key bytes at once). Given the potential plethora of intermediate value combinations for a sequential AES implementation (as typically found on micro-processors) we investigate the effectiveness of some of the possible combinations with respect to the reduction on key guessing entropy. We also touch on the possibility of combining different distinguisher outputs and explain when this is (or is not) going to be helpful.

We also take inspiration from the suggestion of Veyrat-Charvillon et al. [25,26] (originally for the purposes of a key enumeration algorithm) that *probability distributions* on the subkeys can be derived from the outcome of a DPA attack. We propose an alternative (more conservative) heuristic for assigning 'probability' scores to subkeys, and show how these can be used to simply and usefully combine information from multiple standard univariate DPA attacks in a strategy inspired by Bayesian updating.

This research is rooted in our developed capability to efficiently process large numbers of key hypotheses over many repeat experiments; our architecture (which we sketch out) is influenced by the design of modern HPC platforms.

We structure our contribution as follows. We briefly provide the relevant preliminaries and then discuss prior literature (Section 2). We then introduce our specialised attack framework and explain our attack strategy, including our method of assigning and updating 'probability' scores, in Section 3. Section 4 reports the results of our experiments with simulated leakage data, exploring what can be achieved by combining the outcomes of attacks against different target functions, as well as investigating the potential to combine different DPA *strategies*. In Section 5 we report the outcomes of some practical attacks against traces measured from an ARM 7 microcontroller, including scenarios in which the precise locations of the intermediate targets in the traces are unknown.

1.2   Preliminaries: Differential Power Analysis

We consider a 'standard DPA attack' scenario as defined in [16], and briefly explain the underlying idea as well as introduce the necessary terminology here. We assume that the power consumption P of a cryptographic device depends on some internal value (or state) $F_{k^*}(X)$ which we call the *target*: a function $F_{k^*} : \mathcal{X} \to \mathcal{Z}$ of some part of the known plaintext—a random variable $X \overset{R}{\in} \mathcal{X}$— which is dependent on some part of the secret key $k^* \in \mathcal{K}$. Consequently, we have that $P = L \circ F_{k^*}(X) + \varepsilon$, where $L : \mathcal{Z} \to \mathbb{R}$ describes the data-dependent component and ε comprises the remaining power consumption which can be modeled as independent random noise (this simplifying assumption is common in the literature—see, again, [16]). The attacker has N power measurements corresponding to encryptions of N known plaintexts $x_i \in \mathcal{X}$, $i = 1, \ldots, N$ and wishes to recover the secret key k^*. The attacker can accurately compute the internal values as they would be under each key hypothesis $\{F_k(x_i)\}_{i=1}^{N}$, $k \in \mathcal{K}$ and uses whatever information he possesses about the true leakage function L to construct a prediction model $M : \mathcal{Z} \to \mathcal{M}$.

DPA is motivated by the intuition that the model predictions under the correct key hypothesis should give more information about the true trace measurements than the model predictions under an incorrect key hypothesis. A distinguisher D is some function which can be applied to the measurements and the hypothesis-dependent predictions in order to quantify the correspondence between them. For a given such comparison statistic, D, the *estimated* vector from a practical instantiation of the attack is $\hat{\mathbf{D}}_N = \{\hat{D}_N(L \circ F_{k^*}(\mathbf{x}) + \mathbf{e}, M \circ F_k(\mathbf{x}))\}_{k \in \mathcal{K}}$ (where

$\mathbf{x} = \{x_i\}_{i=1}^N$ are the known inputs and $\mathbf{e} = \{e_i\}_{i=1}^N$ is the observed noise). Then the attack is *o-th order successful* if $\#\{k \in \mathcal{K} : \hat{\mathbf{D}}_N[k^*] \leq \hat{\mathbf{D}}_N[k]\} \leq o$.

The *success rate* of a DPA attack is the probability that the correct key is ranked first by the distinguisher (the *o-th order success rate* is the probability it is ranked among the *o* first candidates); the *guessing entropy* is the expected number of candidates to test before reaching the correct one [24]. These metrics are often associated with the *subkeys* targeted in the 'divide-and-conquer' paradigm rather than with the global key when the partial outcomes are finally combined; we use the terms accordingly, unless explicitly stated.

Unless stated otherwise, we use the (estimate of the) Pearson correlation coefficient as distinguisher, in combination with a Hamming weight power model.

2   Related Literature

Our work unites and advances three broad areas of the literature: resource-intensive side-channel strategies, post-SCA global key enumeration, and optimal combination of multiple sources of exploitable information.

Resource-intensive strategies. Such strategies have for a long time been considered mainly relevant in single-trace settings (e.g. SPA attacks using algebraic methods [19,20]); this has only lately begun to change, with a few recent studies making use of modern graphics cards to speed up DPA attacks [3,18]. These articles essentially use GPUs within a single machine to speed up the processing of standard correlation DPA attacks. Our more ambitious approach is to distribute all the different components of a DPA attack (*including* workloads related to combination functionality) across several cards and several machines.

Post-SCA global key enumeration. Recent work by Veyrat-Charvillon et al. [25,26] focuses on the opportunity for a well-resourced adversary to view side-channel analysis as an auxiliary phase in an enhanced global key search, rather than a stand-alone 'win-or-lose' attack. They present an algorithm for searching, based on probability distributions for each of the subkeys (derived from DPA outcomes) [25]. In the case of profiling DPA with Gaussian templates, the true leakage distributions conditioned on each subkey hypothesis are known, and the probabilities are naturally produced in the Bayesian template matching. In the case of non-profiling DPA, these conditional leakage distributions are *not* known; an attack does *not* produce a probability distribution on the subkey candidates but a set of distinguisher scores (for example, correlations) associated with each candidate. Deriving probabilities from these scores is tricky; the method suggested in [25] is to use the hypothesis-dependent fitted leakage models after a non-profiled linear regression ('stochastic') attack as estimates on the 'true' conditional distributions. However, non-profiled linear regression-based DPA specifically relies on the fact that the models built under incorrect key hypotheses are *invalid*. Consequently, the hypothesised functions do *not* describe the true data-dependent deterministic behaviour of the trace measurements, and so they are useless for (statistical) inference. For this reason, we opt for a different ('safer') heuristic for assigning 'probability' scores, as explained in Section 3.1.

Combining multiple sources of information. Whilst profiling attacks with multi-variate Gaussian templates [7] naturally exploit multiple trace points, notions of 'multivariate' non-profiled DPA are varied in nature and intention. In particular, techniques designed to defeat protected implementations are best considered separately from attempts to enhance trace efficiency, and we now focus on the latter. Already in an unprotected implementation, information on a given sub-key generally leaks via more than one target function (AddRoundKey, SubBytes and MixColumns, for example, in the case of AES) and moreover each of those target functions can be seen to leak at more than one trace point. In some cases, an adversary may even have opportunity to observe multiple side-channels simultaneously (timing, power consumption, electromagnetic radiation...).

In the realms of both profiled and non-profiled DPA, several efforts have been made to combine information from multiple trace points in such a way as to optimise the (trace) efficiency of an attack. Dimensionality reduction techniques such as principal component analysis or linear discriminant analysis can be used to transform the (often collinear) trace measurements into a reduced number of linearly uncorrelated variables, together accounting for the important variation in the original data [1,4,22]. In this way it is even possible to combine information from different side-channels, such as power and electro-magnetic radiation [22]. Such methods can be very effective if the leakage associated with a particular intermediate value is concentrated into a single component giving rise to a stronger attack outcome than the 'best' of any individual point in the raw dataset. A recent work by Hajra et al. [12] achieves a similar end via signal processing techniques. They show how to maximise the signal-to-noise ratio (SNR) (and consequently demonstrate the success rate of a univariate correlation DPA) by finding the linear Finite Impulse Response (FIR) filter coefficients for the leakage signal. Hutter et al. [13] also seek to enhance DPA efficiency by incorporating multiple sources of information, but take an entirely different approach in which the combination is instead made at the trace acquisition stage. They measure the *difference* in consumption between two identical devices operating on different data, which they reason has a higher data-dependent signal because all environmental and operation-dependent noise is cancelled out.

Other suggestions involve performing separate attacks (against different targets, power models or using different distinguishers) and then attempting to combine the *distinguishing vectors* themselves in a meaningful way. Doget et al. [9] present options for combining difference-of-means (DoM) style outcomes in order to avoid the 'suboptimality' associated with attacks exploiting only one or a few of the bits at a time. Whitnall et al. [28] try applying a multivariate extension of the mutual information to the AddRoundKey and an S-box jointly, but find that it is *less* efficient than the corresponding attack against the S-box alone, and moreover would not scale easily beyond a two-target scenario due to the complex nature of the statistic. Souissi et al. [21] suggest to combine different *distinguishers* (namely, Pearson's and Spearman's correlation) applied against the same or different leakage points by taking either the sum or the maximum of the two, and show that the former works better, and is most effective if the trace points contain non-equivalent information. Most directly related to our study is a paper by Elaabid et al. [10] which suggests to (pointwise) *multiply* correlation distinguishing vectors together in order to enhance distinguishing outcomes. They do this for four *known* leakage points, all relating to the same target function and power model, and find that it substantially improves over the outcomes achieved for any one of those leakage points taken individually. Our own combining approach is different: we first convert distinguishing vectors to 'probability' scores and view the multiplication as a Bayesian updating-like procedure. Moreover, we focus on combinations between *different target values* (rather than different leakage points for the same value) with potentially different-sized subkey hypotheses.

3   Methodology

3.1   Assigning Probabilities

The attempt of [25] to estimate 'genuine' probabilities on the subkey hypotheses in the non-profiled setting (see Section 2), by using the recovered models derived from a linear regression based attacks, is expensive as well as unsuitable for our purpose. Ignoring the fact that the incorrect key hypotheses (using their approach) recover invalid models, the method of [25,26] may be viewed as one possible heuristic to assign probabilities to key guesses. It preserves the ranking of the keys as they appear in the distinguishing vector produced by a non-profiled linear regression-based DPA. However, because of the nature of the formula used it dramatically exaggerates the apparent distance between the high- and low-ranked key candidates. If the implied key is the right one it reinforces this 'correct' result. But if it is not the right one it reinforces the misleading result. In their application (i.e. key enumeration) this may cause a less efficient key search. However, we are aiming to combine distinguisher results, and hence key rankings, and mixing in a grossly exaggerated incorrect key ranking may destroy the effectiveness of the method.

Embracing the heuristic nature of the task of obtaining (from distinguishing vectors) scores which may be handled as though probabilities, we suggest the conversion be kept simple and conservative. Our approach firstly transforms the distinguishing vector to be positive-valued with a baseline of zero (in a manner appropriate to the statistic—e.g. the absolute value for correlation, subtraction of the minimum for the mutual information) before secondly normalising the scores to sum to one. We draw analogy between this idea and the notion of *subjective probability* basic to a Bayesian view of statistics: both involve human-allocated scores derived from one's current best knowledge about reality.

3.2   Combining Probabilities

A Bayesian interpretation views probabilities as measures of uncertainty on hypotheses. Each time new information becomes available, the current state of knowledge can be *updated* via Bayes' theorem:

$$\mathbb{P}(H|B) = \frac{\mathbb{P}(B|H)\mathbb{P}(H)}{\mathbb{P}(B)},$$

where H is some hypothesis (for example, a guess on the key, "$K = k$"), and B is some data (for example, a set of trace measurements $\mathbf{l} = L \circ F_{k^*}(\mathbf{x}) + \mathbf{e}$).

Suppose that we have probabilities for $(K = k)$ conditioned on two sources of data $\mathbf{l}_1, \mathbf{l}_2$, which are conditionally independent given $K = k$ so that $\mathbb{P}(\mathbf{l}_1, \mathbf{l}_2|K = k) = \mathbb{P}(\mathbf{l}_1|K = k)\mathbb{P}(\mathbf{l}_2|K = k)$. This is a natural assumption for the leakages of two target intermediate values: they are related via their shared dependence on the underlying key, but as long as they are separated in the trace, we would not expect any dependency in the residual variances after the key is taken into account. In this case, the task of combining the conditional probabilities is straightforward (see [2]):

$$\begin{aligned}
\mathbb{P}(K = k|\mathbf{l}_1, \mathbf{l}_2) &= \frac{\mathbb{P}(\mathbf{l}_1, \mathbf{l}_2|K = k)\mathbb{P}(K = k)}{\mathbb{P}(\mathbf{l}_1, \mathbf{l}_2)} \\
&= \frac{\mathbb{P}(\mathbf{l}_1|K = k)\mathbb{P}(\mathbf{l}_2|K = k)\mathbb{P}(K = k)}{\mathbb{P}(\mathbf{l}_1, \mathbf{l}_2)} \\
&= \frac{\mathbb{P}(\mathbf{l}_1)\mathbb{P}(\mathbf{l}_2)}{\mathbb{P}(\mathbf{l}_1, \mathbf{l}_2)} \times \frac{\mathbb{P}(K = k|\mathbf{l}_1)\mathbb{P}(K = k|\mathbf{l}_2)}{\mathbb{P}(K = k)},
\end{aligned}$$

(via Bayes' theorem again, since $\mathbb{P}(\mathbf{l}_i|K = k) = \mathbb{P}(K = k|\mathbf{l}_i)\mathbb{P}(\mathbf{l}_i)/\mathbb{P}(K = k)$). Since $a = \mathbb{P}(\mathbf{l}_1)\mathbb{P}(\mathbf{l}_2)/\mathbb{P}(\mathbf{l}_1, \mathbf{l}_2)$ does not depend on the key hypothesis we can treat it as a normalisation constant which just needs to be computed so as to satisfy $\sum_{k \in \mathcal{K}} \mathbb{P}(K = k|\mathbf{l}_1, \mathbf{l}_2) = 1$. In the typical case that all keys are *a priori* equally likely, the denominator in the second product term is $\frac{1}{|\mathcal{K}|}$ (constant for all key hypotheses) and simply gets absorbed into the normalising constant. Thus, conditional probabilities on the key candidates can be updated with the introduction of any new, independent information via a simple multiplication-and-normalisation step.

3.3   Parallelised Attack Architecture

Combining multiple distinguishing vectors and attacking target functions involving 24 or more bits of the key are both computationally demanding tasks, and necessitate the use of parallelised computation. We elected to use the OpenCL language and a set of graphics cards to parallelise the computation needed to attack up to 32-bits of a key, the combination and normalisation of distinguishing vectors, and finally the statistics necessary for evaluating the effectiveness of each combined attack.

We took inspiration from modern HPC facilities, in which a significant amount of the computing power is delivered by GPUs. Hence our experimental setup consists of several (up to 6) workstations, each containing two discrete GPUs

Fig. 1. Average keys per second recorded during DPA attacks on 32-bits of the input to the MixColumns operation for a variety of different sample sizes. Implementations are a 'naive' single-threaded CPU implementation, a parallelised OpenCL CPU-based implementation, and the two fastest OpenCL GPU implementations.

(the cost per machine is approximately 2000 GBP). These were various pairs of high-end AMD and Nvidia cards, installed in our own workstations or within the Bluecrystal Phase 3 supercomputing facility[2]. In total, including all the functionality used to fully produce and analyse our experimental results, we were able to complete at least 2^{50} operations on combined distinguishing vectors, in very roughly a couple of weeks of computation time.

The most computationally demanding function was performing a 32-bit DPA attack on the MixColumns operation. Here we decided to share the cost over multiple GPUs, with each work group inside a single card computing a partial piece of the distinguishing vector using a portion of the traces and a subset of the key hypotheses, followed by a global reduction to compute the final vector. Fig. 1 shows the performance of our OpenCL attack implementation for a variety of devices, in terms of the number of key hypotheses tested per second.

We note that these benchmark timings are not likely to be optimal. We did not try to improve the memory coalescence of our kernels, nor did we try to perform any other non-trivial optimisation beyond maximising kernel occupancy, and so there may be *considerable* headroom in key-search throughput still to be gained. It is clear from the extremely cheap price for a dual GPU setup, coupled with the considerable performance increases observed with the introduction of new GPU architectures, that an adversary can acquire very large side-channel key-search capabilities at minimal financial cost.

Bartkewitz et al. [3] use Nvidia's CUDA technology and a Tesla C2070 to parallelise 8-bit CPA attacks on the SubBytes operation, and focus on maximising trace data throughput in an 8-bit setting. Our more ambitious goal is to optimise for large key-search problems as well as for trace data throughput. In this context Moradi et al. [18] utilise 4 Nvidia Tesla GPUs to attack 32-bits of key using 60,000 traces, and are able to attack a single time-point every 33 minutes. A direct comparison is not possible as we are using slightly more modern hardware and the exact computational costs included in the benchmarking are not clear—however we might expect to be able to perform a similar attack in approximately 20 minutes.

---

[2] Bluecrystal is managed by the Advanced Computing Research Centre at the University of Bristol—see http://www.bris.ac.uk/acrc/

4   Experiments with Simulated Data

The goal of our combining strategy is to reduce (relative to 'standard univariate DPA') the guessing entropy on the subkeys (and consequently on the global key). Many types of combination are possible. We study the effect of combining outcomes from different targets as well as, secondarily, the effect of combining outcomes from different distinguishers applied to the same target. We do this initially for simulated trace measurements so that we can take into account different noise levels (i.e. by varying the SNR) as well as the impact of using an imperfect power model. Both aspects have practical relevance.

4.1   Combining Outcomes from Different Targets

We simulated leakages of AES AddRoundKey, SubBytes, and three 8-bit interim values in the computation of MixColumns: one involving two key bytes (namely $GFm2(state_i \oplus state_{i+1})$ where $state_i$ is the ith state byte after the SubBytes operation, and $GFm2$ denotes doubling in Rijndael's finite field), one involving three key bytes (namely $GFm2(state_i \oplus state_{i+1}) \oplus state_{i+1} \oplus state_{i+2}$), and one involving four key bytes (namely $GFm2(state_i \oplus state_{i+1}) \oplus state_{i+1} \oplus state_{i+2} \oplus state_{i+3}$) [3].

In the case of the 16-bit multi-target attack we necessarily hypothesise over two key bytes (in order to incorporate the MixColumns leakage). The experiments each involve two AddRoundKey correlation-based DPA attacks (which are then combined into probabilities on the full 16-bit subkeys via multiplication), two S-box attacks (combined likewise), and the one MixColumns attack, before multiplying each possible target function pair together, as well as multiplying all three together. Similarly, for the 24-bit multi-target attack we hypothesise over three key bytes. The experiments in the 24-bit attack then involve three AddRoundKey attacks, three S-box attacks, and the one attack on an interim MixColumns value. We amalgamate probabilities by multiplication as in the 16-bit case. The 32-bit multi-target attack proceeds in the same fashion: we combine four AddRoundkey attack results and four SubBytes results into the MixColumns attack result. The graphs in Fig. 2(a) show these different scenarios for a single column of the AES state.

In the following paragraphs we analyse these graphs with respect to three questions that are relevant for practice. Firstly, what is the impact of a (low) SNR with regards to our multi-target strategy? As we base our DPA attacks on correlation distinguishers, we would hope that, similarly to single-target attacks, multi-target attacks will 'scale' alongside the SNR. Secondly, we are interested in how the size of the key hypotheses impacts on the guessing entropy, and lastly, in how multi-target attacks behave when the attacker's power model is imprecise.

---

[3] This targets a single intermediate byte. The relative effectiveness of combining all four attacks on all the possible intermediate bytes would also be interesting to investigate, but generating results requires time and so is left as future work.

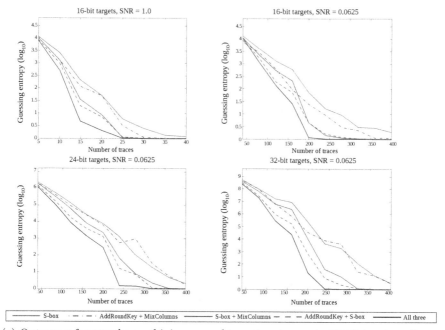

(a) Outcomes for attacks combining several targets using up to 32-bit key hypotheses

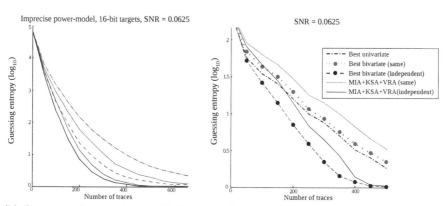

(b) Outcomes for attacks combining several distinguishers for the same target (the S-box output)

Fig. 2. Simulation results

Impact of SNR. The top two graphs in Fig. 2(a) show the subkey guessing entropies (for a 16-bit key guess) as the number of traces increases, for the attacks against simulated Hamming weight leakages with two SNR levels. Aside from the fact that all attacks require increased numbers of traces as the SNR decreases (as we would expect) the scenarios exhibit similar outcomes. The attacks on S-boxes are effective at reducing uncertainty on the key (the results for these are printed in red), but are clearly outperformed by all three 'bivariate' combinations—even the one between the MixColumns sub-computation and AddRoundKey. The combination between all three further reduces the enumeration work required.

Impact of larger distinguishing vectors. The top right and the bottom graphs in Fig. 2(a) show the subkey guessing entropies for increasing subkey sizes (16-bit in the top right, 24-bit in the bottom left, and 32-bit in the bottom right). In all three experiments the multi-target attacks outperform the single target attacks. Note that the guessing entropy range naturally increases with the size of the key hypothesis and is in no way an indicator of attack degradation. For the 16-bit attack the guessing entropy is out of 2^{16} and eight such guesses need to be combined to get a global key with guessing entropy between 1 and 2^{128}. For the 32-bit attack the guessing entropy is out of 2^{32} but only four such guesses need to be combined. It is the global guessing entropy which ultimately matters and the subkeys always need to be combined at some point – incorporating information at (e.g.) the 32-bit level simply increases the scope of intermediate targets exploitable by the attacker. For both hypothesis sizes, the outcomes suggest that we are able to succeed with roughly half the number of leakage traces when using the best multi-target attack (for a fixed subkey guessing entropy, the best multi-target attacks require roughly half of the traces required by the best single-target attack). It is possible to estimate global key guessing entropies based on these results by assuming that the attacks on the other 'chunks' of the key would behave identically. For instance, in the 16-bit case, if all eight 16-bit attacks give identical outcomes, we could estimate global key entropies by raising the results of a single 16-bit attack to the power eight. However, this does not necessarily translate into practice, so we will instead show actual global key guessing entropies when we come to discuss attacks on real data.

Impact of imperfect power model. The left picture of Fig. 2(b) shows the outcomes (against a 16-bit subkey target: the legend from Fig. 2(a) applies) in the case where the Hamming weight is *not* a perfect match to the leakage, because of the presence of a constant reference state (representing an address, for example) of Hamming weight 1. The most striking impact of this distortion occurs for attacks that include AddRoundKey as a target, which are no longer able to identify the correct key as a likely candidate. This is because the Hamming distance of the AddRoundKey from the reference state when the correct key is guessed is the same as the Hamming weight of the AddRoundKey when the key guess is the correct key XORed with the reference state. In effect, an incorrect key is masquerading as the correct one, and the correlation DPA against AddRound-Key will naturally preference this. (The same cannot happen for the S-box, for example, because the key XOR is *inside* the highly nonlinear transformation, with the Hamming distance being taken *afterwards*).

Nonetheless, in this case where the reference state is itself of low weight, incorporating AddRoundKey information still produces marginal reductions on the guessing entropies after S-box and MixColumns (separately, and combined).

Greater imprecision of the power model will more strongly impact on AddRound-Key attacks; it may be advisable to exclude it as a target in such cases.

4.2   Combining Outputs from the Same Target

One might ask whether or not the outcomes of different *distinguishing statistics* or *power models* can likewise be combined to some advantage.

Using different distinguishing statistics. Suppose we run three different attacks against the leakage of an AES S-box, e.g.: mutual information [11], Kolmogorov–Smirnov [27], and the variance ratio [23], all using a Hamming weight power model. The distinguishing vectors are transformed to have a baseline of zero and to sum to one, for use as heuristic 'probability' scores. We would then like to know whether the combined outcomes improve upon the individual ones.

The right picture in Fig. 2(b) shows what happens when we attempt this in the example scenario of Hamming weight leakage with SNR 0.0625. When the same measurements are used for all of the attacks, combining the outcomes actually *increases* the guessing entropy. By contrast, when independent measurements are used in each case (i.e., each distinguisher has been applied against a different point in the trace leaking the same information but with independent noise), there is some scope to refine the information on the key by combining outcomes—although all three outcomes together on average produce worse results than the best combination of two. We found that it was generally the addition of mutual information which degraded the outcome, as it required substantially more data to estimate to an equivalent degree of precision.

This is very much in line with what we might expect, and acts as a noteworthy warning: it is the addition of *new information* which improves attack outcomes—exploiting the same measurements using the same power models but with different distinguishers does *not* contribute anything further. In the context of our heuristic 'probability' distributions such a practice could be particularly dangerous, as it still serves to exaggerate the magnitude of the peaks, thus giving a false sense of increased certainty. Note that the multiplication step implicitly assumes *independence* of the separate score vectors, which is clearly violated in the case that they are all based on the same leakage information.

Using different power models. In the light of the ineffectiveness of combining information about the same target, we briefly revisit previous work by Bevan and Knudsen [5]. They suggest to combine eight difference-of-means attacks, each targeting a distinct bit of the intermediate value, by 'summing over the distinguisher results' (in our approach we convert them into 'probability' distributions on the set of 2^8 subkeys, as per Section 3.1). Since each attack exploits a *separated portion* of the overall leaked value we may expect that each new bit attacked helps to further reduce the candidate search space—and, indeed, our experiments confirm this (see Appendix A). Such a technique is hence very useful in leakage scenarios which are unfamiliar to an attacker, which is often the case when attacking dedicated hardware.

5   Practical Attacks

We tested our strategy in practice using a dataset of 10,000 traces from an ARM7 microcontroller running an unprotected implementation of AES. The 10,000 traces were divided up in 200 sets of 50 traces each to conduct sufficient repeat experiments to report reasonably precise estimates for the guessing entropies in the same vein as our simulated attacks. Multi-target attacks, similar to multivariate attacks, are greatly helped by knowledge about where the attacked intermediate values leak in the traces. Consider for instance a (multivariate) template attack: it is much harder for an adversary to conduct such an attack when in the profiling phase a similar device is available but not the exact implementation (of, say, AES). In such a case an adversary could still build templates for microprocessor instructions during profiling, but in the attack phase the adversary would need to find the specific trace points at which to apply the templates. Similarly, knowing precisely where the single-target leakages occur is helpful for a multi-target attack. We consequently focus initially on a 'known point' scenario and then make a first attempt at relaxing this assumption.

5.1   Practical Attacks against Known Interesting Points

We applied two multi-target attacks (one involving 16-bit, one 32-bit key hypotheses) under the assumption that interesting trace points are known, running 200 repeat experiments for increasing samples of up to size 50. For each 16-bit subkey, correlation DPA attacks were performed against the two corresponding AddRoundKey operations, the S-boxes and the MixColumns sub-computation $GFm2(state_i \oplus state_{i+1})$, where $state_i$ is the state byte corresponding to the ith key byte after the S-box substitutions and (in this implementation) the ShiftRows operation. For each 32-bit subkey, correlation DPA attacks are performed against four AddRoundKey operations, four S-boxes, and the 32-bit MixColumns computation $GFm2(state_i \oplus state_{i+1}) \oplus state_{i+1} \oplus state_{i+2} \oplus state_{i+3}$.

The first two graphs in Fig. 3(a) show the guessing entropies on the first key-byte pair and the final global guessing entropies, estimated by multiplying the eight subkey guessing entropies together (the outcomes for the other seven subkey guesses can be found in the Appendix of the full version of our paper, see [17]).[4] They largely, but not perfectly, match up with our observations for simulated traces. This is an important point: theory and practice rarely perfectly align, even in the case of a relatively 'simple' platform like the ARM7. In the practical experiments, AddRoundKey and the MixColumns sub-value are consistently unable to identify the correct key alone (at least, not within 50 traces). However, the two together produce guessing entropies to rival the effectiveness of the S-box attack, and both produce improvements in combination with the S-box. All three together produce the best guessing entropies for many of the

---

[4] The more refined rank estimation methodology of [26] indicates that this simple method of approximating global guessing entropies underestimates the rank by 20 to 40 binary orders of magnitude.

16-bit subkeys, although they are sometimes outperformed by the two-target S-box+AddRoundKey attacks, which achieve a marginal advantage overall.

The second two graphs in Fig. 3(a) show the guessing entropies for the first 32-bit subkey and the final global guessing entropies. The global entropies were estimated by multiplying the four subkey entropies together. We observed varying behaviour for our combined attacks on different subkeys; our targeted Mix-Columns computation does not leak nearly as much information in the middle 8 bytes of the state as it does in the first and final 4 bytes. Consequently, despite (as suggested by our simulated attacks) observing strong performance of the combined three-target attacks in the latter two cases, in the global setting this advantage is diminished, and the 'trivariate' attack produces similar performance to the combined four-byte S-box+AddRoundKey attacks. It is noteworthy that even in the presence of this variable leakage, most combined attacks outperform the S-box attack. Graphs and data for each of the four separate subkey attacks can be found in the Appendix of the full version of our paper [17].

5.2 Practical Attacks where Interesting Points Are *A Priori* Unknown

The natural next question to ask is whether we can relax the assumption that the leakage points are precisely known. We made some preliminary inroads using 'desktop-level' resources (whilst our GPU machines were occupied with other experiments), focusing, for computational feasibility, on 8-bit key hypotheses. The three targets we selected to combine were AddRoundKey, the S-box outputs, and the interim MixColumns value $GFm2(state_i \oplus state_{i+1})$ with the assumption that the second involved key byte of the two is known.

We relaxed the 'known point' assumption by visually inspecting the AES traces in order to identify the intervals in which each of the three target functions are contained. The first round takes about 1,400 clock cycles in total and the (non-overlapping) windows we selected for experimentation were of widths 240, 230, and 180 for AddRoundKey, SubBytes and MixColumns respectively. Within these windows we took an 'exhaustive search' approach. First, we subjected each point to a standard DPA attack against the associated target function, and computed the 'probability' scores. We then pairwise combined them in each of the three possible configurations, and finally we combined all three. We tried two strategies: in the first, we took (for each configuration) the combined vector with the largest peak as the one most likely to correspond to the correct key and pair/triple of leakage points, and in the second we took the N_t combined vectors with the largest peaks and multiplied these together (for different values of N_t), so achieving a sort of 'majority vote'.

The left side of Fig. 3(b) shows the average guessing entropy for each of the attacks using the first 'maximum peak' strategy. The AddRoundKey attack in an unknown point scenario performs very badly. Further analysis of the trace window reveals that there are other points exhibiting strong correlations with AddRoundKey $\oplus R$, for R some other (possibly address?) value in $\{0, 255\}$ (see

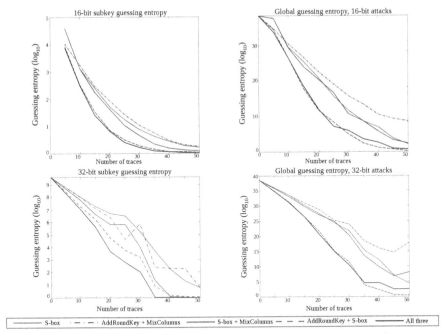

(a) Outcomes for multi-target attacks in a known points scenario

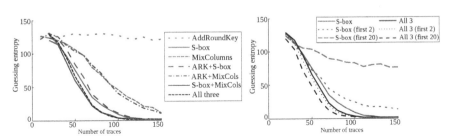

(b) Outcomes for multi-target attacks in a known interval scenario

Fig. 3. Practical results

Fig. 5 in Appendix B).[5] Moreover, at these points the *correct* key correlations are *low*, so that the contribution to the combined leakage is highly distorting (as opposed to when an 'imperfect but close' leakage prediction is made, in which case the combination can still improve distinguishability). In the presence of such misleading leakage information, it is reassuring that the attack outcomes are *robust* to the combining step.

---

[5] Note that the leakage of the S-box is less vulnerable to such distortions: a non-zero reference state will not masquerade as an alternative key hypothesis, as the key addition happens *inside* the S-box.

The combined MixColumns and S-box attack exhibits lower guessing entropies than either of the two taken individually. The trivariate attack (as expected from the above) does not really add much to this, but again we reflect that the inclusion of AddRoundKey at least does not seem to harm the outcome.

The right side of Fig. 3(b) shows the advantage gained by multiplying the top-ranked few 'probability' vectors for the trivariate attack, as well as (for comparison) for the S-box attack on its own. Interestingly, even the addition of the second ranked vector degrades the S-box attack, whereas the product combining for the top-ranked triples reduces the guessing entropy at least up to $N_t = 20$. The subsequent total improvement over the S-box outcome on its own indicates this as a potentially worthwhile strategy for key recovery in an unknown point scenario.

From a practical perspective, a useful forward approach for multi-target attacks would be to 'try out' (for a concrete device and implementation) different combinations of targets, and different point selection strategies, to see which give the best results. We want to caution against drawing too many conclusions from these last experiments: they clearly represent a first step only!

6   Conclusion

We have shown how to amalgamate single-target 'standard' DPA attacks (using a correlation distinguisher and a Hamming weight power model) into multi-target attacks capable of increasing information on the correct key by combining DPA outcomes that are treated as heuristic probabilities. Leveraging our modern HPC-inspired computing platform, we are able to efficiently handle key hypotheses of up to 32 bits using a small cluster of simple workstations containing consumer graphics cards. Such a capability allows us to combine many intermediate targets; in this work we made the first serious attempt to explore the characteristics of successful combinations. Our results indicate that combining S-box+AddRoundKey or additionally including an intermediate MixColumns computation typically produces the strongest results. Multi-target attacks scale predictably with noise and are robust with regards to imprecise power models. Our primary investigative effort is mainly on 'known' (leakage) point attacks, in line with assumptions generally made for multivariate attacks. When leakage points are not known, an exhaustive search in suitable visually-identified trace windows, together with a 'majority vote'-style approach to decide on 'peaks', leads to improved practical attacks even in this challenging scenario.

Our definition of multi-target attacks and intuitive and efficient combination technique opens up many interesting new research questions: e.g. is there any single best combination of intermediate values for a given cipher? How effectively can we combine power and EM attack results in this way? Could we even move further on and include results from the second encryption round? What other strategies for combining in unknown point scenarios exist? How could we use this against implementations when masking and hiding are used? For better or worse, these are "interesting times"—to call to mind the fabled Chinese curse.

Acknowledgements. This work has been supported in part by EPSRC via grant EP/I005226/1. This work was carried out using the computational facilities of the Advanced Computing Research Centre, University of Bristol— http://www.bris.ac.uk/acrc/.

References

1. Archambeau, C., Peeters, E., Standaert, F.-X., Quisquater, J.-J.: Template Attacks in Principal Subspaces. In: Goubin, L., Matsui, M. (eds.) CHES 2006. LNCS, vol. 4249, pp. 1–14. Springer, Heidelberg (2006)
2. Bailer-Jones, C., Smith, K.: Combining probabilities. Technical Report GAIA-C8-TN-MPIA-CBJ-053, Max Planck Institute for Astronomy, Heidelberg (January 2010)
3. Bartkewitz, T., Lemke-Rust, K.: A high-performance implementation of differential power analysis on graphics cards. In: Prouff, E. (ed.) CARDIS 2011. LNCS, vol. 7079, pp. 252–265. Springer, Heidelberg (2011)
4. Batina, L., Hogenboom, J., van Woudenberg, J.: Getting More from PCA: First Results of Using Principal Component Analysis for Extensive Power Analysis. In: Dunkelman, O. (ed.) CT-RSA 2012. LNCS, vol. 7178, pp. 383–397. Springer, Heidelberg (2012)
5. Bévan, R., Knudsen, E.: Ways to Enhance Differential Power Analysis. In: Lee, P.J., Lim, C.H. (eds.) ICISC 2002. LNCS, vol. 2587, pp. 327–342. Springer, Heidelberg (2003)
6. Brier, E., Clavier, C., Olivier, F.: Correlation Power Analysis with a Leakage Model. In: Joye, M., Quisquater, J.-J. (eds.) CHES 2004. LNCS, vol. 3156, pp. 16–29. Springer, Heidelberg (2004)
7. Chari, S., Rao, J., Rohatgi, P.: Template Attacks. In: Kaliski Jr., B.S., Koç, Ç.K., Paar, C. (eds.) CHES 2002. LNCS, vol. 2523, pp. 51–62. Springer, Heidelberg (2003)
8. Common Criteria, Technical editor: BSI. Application of Attack Potential to Smart Cards (2009),
   http://www.commoncriteriaportal.org/files/supdocs/CCDB-2009-03-001.pdf
9. Doget, J., Prouff, E., Rivain, M., Standaert, F.-X.: Univariate side channel attacks and leakage modeling. J. Cryptographic Engineering 1(2), 123–144 (2011)
10. Elaabid, M., Meynard, O., Guilley, S., Danger, J.-L.: Combined Side-Channel Attacks. In: Chung, Y., Yung, M. (eds.) WISA 2010. LNCS, vol. 6513, pp. 175–190. Springer, Heidelberg (2011)
11. Gierlichs, B., Batina, L., Tuyls, P., Preneel, B.: Mutual Information Analysis. In: Oswald, E., Rohatgi, P. (eds.) CHES 2008. LNCS, vol. 5154, pp. 426–442. Springer, Heidelberg (2008)
12. Hajra, S., Mukhopadhyay, D.: SNR to Success Rate: Reaching the Limit of Non-Profiling DPA. Cryptology ePrint Archive, Report 2013/865 (2013), http://eprint.iacr.org/
13. Hutter, M., Kirschbaum, M., Plos, T., Schmidt, J.-M., Mangard, S.: Exploiting the Difference of Side-Channel Leakages. In: Schindler, W., Huss, S.A. (eds.) COSADE 2012. LNCS, vol. 7275, pp. 1–16. Springer, Heidelberg (2012)
14. Kocher, P.C., Jaffe, J., Jun, B.: Differential Power Analysis. In: Wiener, M. (ed.) CRYPTO 1999. LNCS, vol. 1666, pp. 388–397. Springer, Heidelberg (1999)

15. Mangard, S., Oswald, E., Popp, T.: Power Analysis Attacks: Revealing the Secrets of Smart Cards
16. Mangard, S., Oswald, E., Standaert, F.-X.: One for All – All for One: Unifying Standard DPA Attacks. IET Information Security 5(2), 100–110 (2011)
17. Mather, L., Oswald, E., Whitnall, C.: Multi-target DPA attacks: Pushing DPA beyond the limits of a desktop computer. Cryptology ePrint Archive, Report 2014/365 (2014), http://eprint.iacr.org/
18. Moradi, A., Kasper, M., Paar, C.: Black-Box Side-Channel Attacks Highlight the Importance of Countermeasures. In: Dunkelman, O. (ed.) CT-RSA 2012. LNCS, vol. 7178, pp. 1–18. Springer, Heidelberg (2012)
19. Renauld, M., Standaert, F.-X.: Combining Algebraic and Side-Channel Cryptanalysis against Block Ciphers. In: 30th Symposium on Information Theory in the Benelux (2009)
20. Renauld, M., Standaert, F.-X., Veyrat-Charvillon, N.: Algebraic Side-Channel Attacks on the AES: Why Time also Matters in DPA. In: Clavier, C., Gaj, K. (eds.) CHES 2009. LNCS, vol. 5747, pp. 97–111. Springer, Heidelberg (2009)
21. Souissi, Y., Bhasin, S., Guilley, S., Nassar, M., Danger, J.-L.: Towards Different Flavors of Combined Side Channel Attacks. In: Dunkelman, O. (ed.) CT-RSA 2012. LNCS, vol. 7178, pp. 245–259. Springer, Heidelberg (2012)
22. Standaert, F.-X., Archambeau, C.: Using Subspace-Based Template Attacks to Compare and Combine Power and Electromagnetic Information Leakages. In: Oswald, E., Rohatgi, P. (eds.) CHES 2008. LNCS, vol. 5154, pp. 411–425. Springer, Heidelberg (2008)
23. Standaert, F.-X., Gierlichs, B., Verbauwhede, I.: Partition vs. Comparison Side-Channel Distinguishers: An Empirical Evaluation of Statistical Tests for Univariate Side-Channel Attacks against Two Unprotected CMOS Devices. In: Lee, P.J., Cheon, J.H. (eds.) ICISC 2008. LNCS, vol. 5461, pp. 253–267. Springer, Heidelberg (2009)
24. Standaert, F.-X., Malkin, T.G., Yung, M.: A Unified Framework for the Analysis of Side-Channel Key Recovery Attacks. In: Joux, A. (ed.) EUROCRYPT 2009. LNCS, vol. 5479, pp. 443–461. Springer, Heidelberg (2009)
25. Veyrat-Charvillon, N., Gérard, B., Renauld, M., Standaert, F.-X.: An Optimal Key Enumeration Algorithm and Its Application to Side-Channel Attacks. In: Knudsen, L.R., Wu, H. (eds.) SAC 2012. LNCS, vol. 7707, pp. 390–406. Springer, Heidelberg (2013)
26. Veyrat-Charvillon, N., Gérard, B., Standaert, F.-X.: Security Evaluations beyond Computing Power. In: Johansson, T., Nguyen, P.Q. (eds.) EUROCRYPT 2013. LNCS, vol. 7881, pp. 126–141. Springer, Heidelberg (2013)
27. Veyrat-Charvillon, N., Standaert, F.-X.: Mutual Information Analysis: How, When and Why? In: Clavier, C., Gaj, K. (eds.) CHES 2009. LNCS, vol. 5747, pp. 429–443. Springer, Heidelberg (2009)
28. Whitnall, C., Oswald, E.: A Comprehensive Evaluation of Mutual Information Analysis Using a Fair Evaluation Framework. In: Rogaway, P. (ed.) CRYPTO 2011. LNCS, vol. 6841, pp. 316–334. Springer, Heidelberg (2011)

A   Combining Difference-of-Means Outcomes

Fig. 4 shows the reduction in subkey guessing entropy as an increasing number of difference-of-means (against different individual bits) are combined via our strategy.

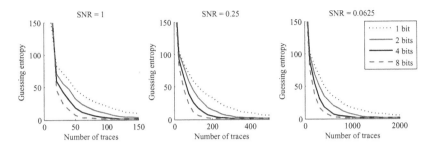

Fig. 4. Combining the outcomes of up to eight difference-of-means attacks against Hamming weight leakage of the AES S-box

B   Unknown Point Attacks: Problem of Rival Peaks

Fig. 5 illustrates the difficulty of separating the true key from strong rival candidates when the relevant 'interesting points' in the trace are not known. As described in Section 5, this introduces distorting information into the point search, which reduces the ability to increase an attack's effectiveness by the addition of AddRoundKey outcomes.

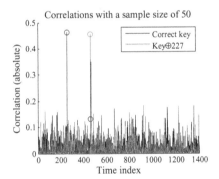
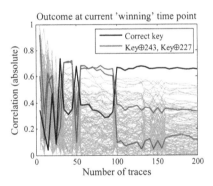

Fig. 5. Left: Example of a fixed XOR offset from the key producing a rival peak in the AddRoundKey correlation attack against the ARM7 traces. Right: The evolution of an AddRoundKey correlation attack against the ARM7 traces, showing the confounding effect of strong rival candidates.

GLV/GLS Decomposition, Power Analysis, and Attacks on ECDSA Signatures with Single-Bit Nonce Bias

Diego F. Aranha[1], Pierre-Alain Fouque[2], Benoît Gérard[3,4],
Jean-Gabriel Kammerer[3,5], Mehdi Tibouchi[6], and Jean-Christophe Zapalowicz[7]

[1] Institute of Computing, University of Campinas
dfaranha@ic.unicamp.br
[2] Université de Rennes 1 and Institut Universitaire de France
fouque@irisa.fr
[3] DGA–MI, Rennes
[4] IRISA, benoit.gerard@irisa.fr
[5] IRMAR, Université de Rennes 1
jean-gabriel.kammerer@m4x.org
[6] NTT Secure Platform Laboratories
tibouchi.mehdi@lab.ntt.co.jp
[7] Inria
jean-christophe.zapalowicz@inria.fr

Abstract. The fastest implementations of elliptic curve cryptography in recent years have been achieved on curves endowed with nontrivial efficient endomorphisms, using techniques due to Gallant–Lambert–Vanstone (GLV) and Galbraith–Lin–Scott (GLS). In such implementations, a scalar multiplication $[k]P$ is computed as a double multiplication $[k_1]P + [k_2]\psi(P)$, for ψ an efficient endomorphism and k_1, k_2 appropriate half-size scalars. To compute a random scalar multiplication, one can either select the scalars k_1, k_2 at random, hoping that the resulting $k = k_1 + k_2\lambda$ is close to uniform, or pick a uniform k instead and decompose it as $k_1 + k_2\lambda$ afterwards. The main goal of this paper is to discuss security issues that may arise using either approach.

When k_1 and k_2 are chosen uniformly at random in $[0, \sqrt{n})$, $n = \text{ord}(P)$, we provide a security proofs under mild assumptions. However, if they are chosen as random integers of $\lfloor \frac{1}{2} \log_2 n \rfloor$ bits, the resulting k is slightly skewed, and hence not suitable for use in schemes like ECDSA. Indeed, for GLS curves, we show that this results in a bias of up to 1 bit on a suitable multiple of $k \bmod n$, and that this bias is practically exploitable: while lattice-based attacks cannot exploit a single bit of bias, we demonstrate that an earlier attack strategy by Bleichenbacher makes it possible. In doing so, we set a record by carrying out the *first ECDSA full key recovery using a single bit of bias*.

On the other hand, computing k_1 and k_2 by decomposing a uniformly random $k \in [0, n)$ avoids any statistical bias, but the decomposition algorithm may leak side-channel information. Early proposed algorithms relied on lattice reduction and exhibited a significant amount of timing channel leakage. More recently, constant-time approaches have also been

P. Sarkar and T. Iwata (Eds.): ASIACRYPT 2014, PART I, LNCS 8873, pp. 262–281, 2014.

proposed, but we show that they are amenable to power analysis: we describe a template attack that can be combined with classical lattice-based attacks on ECDSA to achieve full key recovery on physiscal devices.

Keywords: Elliptic Curve Cryptography, GLV/GLS Method, Bleichenbacher's ECDSA Attacks, Side-Channel Analysis.

1   Introduction

The GLV/GLS Techniques. Many record implementations of elliptic curve cryptography in software, including, most recently, works such as [27,5,10], rely on elliptic curves endowed with fast endomorphisms, as constructed by the methods due to Gallant–Lambert–Vanstone (GLV) [15], Galbraith–Lin–Scott (GLS) [13], and generalizations thereof. In such implementations, the fast endomorphism ψ on the elliptic curve $E/\mathbb{F}_q$ is used to speed up full size scalar multiplications $[k]P$ by computing them as multi-exponentiation $[k_1]P + [k_2]\psi(P)$, where k_1 and k_2 are roughly half of the size of k. Indeed, on a prime order subgroup of $E(\mathbb{F}_q)$, ψ acts by multiplication by some constant λ, and thus, for a generator P of that subgroup, we have $[k_1]P + [k_2]\psi(P) = [k_1 + k_2\lambda]P$.

In order to compute random scalar multiplications with those techniques, two types of approaches have been considered, as far back as in the earliest presentations of the GLV method (such as Gallant's talk at ECC'99 [14]).

On the one hand, k_1 and k_2 can simply be chosen uniformly at random in a suitable half-length interval. This approach, which we call the *recomposition technique* (since k is "recomposed" as $k = k_1 + k_2\lambda$), results in a very simple implementation, and has been used in several implementation records including [27], but Gallant expressed concerns about possible biases in the resulting scalar k. Such concerns have been partially vindicated by some numerical evidence provided by Brumley and Nyberg [7], who also described a relatively general way to choose intervals for k_1 and k_2 so that the resulting choice of k is in fact secure (in the sense that it has high entropy). However, the Brumley–Nyberg method is a bit cumbersome, and no attack so far has been demonstrated against arbitrary half-length uniform choices of k_1 and k_2, so that the security picture is somewhat unclear.

On the other hand, one can also pick k at random and subsequently deduce half-length values k_1 and k_2, which eliminates concerns regarding possible biases in the distribution of k. This *decomposition technique* usually relies on lattice reduction in dimension 2 (or equivalently, continued fractions, a generalized Euclidean algorithm, etc.), as originally described in the GLV paper [15], and is significantly more computationally demanding than recomposition. Simplifications of this method have later been proposed (particularly in [28]), as well as higher-dimensional generalizations [25] to tackle decompositions involving several endomorphisms (as recently used in [31,16] for instance).

ECDSA Attacks. The success of GLV/GLS method in implementations lately makes it desirable to reconsider these decomposition and recomposition techniques from a security viewpoint. We do so in this paper in the context of ECDSA signatures, one of the most widely deployed elliptic curve cryptographic schemes, and an interesting target for the cryptanalyst (like other Schnorr-like signature schemes) due to its sensitivity to biases in the distribution of nonce values, as demonstrated by the powerful attack due to Howgrave-Graham and Smart [17] based on lattice reduction techniques, which breaks (EC)DSA when some of the most significant bits of the nonces are known. This attack was analyzed in further details by Nguyen and Shparlinski [23,24] and carried out in practice in many contexts, including against physical devices (see e.g. [22,6] for some examples). The basic idea is to express the key recovery problem as an instance of the Hidden Number problem (HNP), which reduces to the closest vector problem (CVP) in a suitable lattice. Since CVP is tractable in low-dimensional lattices, many practical instances of ECDSA can be broken depending on key size and the number of leaked nonce bits. The largest problem instance broken so far is the case of 2-bit nonce leaks on 160-bit curves, tackled by Liu and Nguyen [19] using the most advanced known techniques for lattice reduction (BKZ 2.0 [9]). Breaking 2-bit leaks on 256-bit curves, or 4-bit leaks on 384-bit curves seems currently out of reach (see the discussions in [9,21]).

In any case, there is a hard limit to what can be achieved using lattice reduction: due to the underlying structure of the HNP lattice, it is impossible to attack (EC)DSA using a single-bit nonce leak with lattice reduction. In that case, the "hidden lattice point" corresponding to the HNP solution will not be the closest vector even under the Gaussian heuristic (see [26]), so that lattice techniques cannot work. To break this "lattice barrier", the only known alternate attack is an algorithm due to Bleichenbacher [3] which predates the attack of Howgrave-Graham and Smart, but was generally considered of mostly theoretical interest until it was recently revisited by De Mulder et al. [21] to attack 384-bit curves. Bleichenbacher devised his attack to demonstrate a vulnerability in DSS at the time, in which DSA nonces were generated by picking a random value of ℓ_n bits, where ℓ_n is the bit length of the group order n, and then to reduce it modulo n. Bleichenbacher showed that the resulting bias could be exploited in a very interesting way, obtaining a key recovery using about 2^{41} signatures and about 2^{47} time and 2^{41} memory complexities. At that time, it was not possible to mount this attack and only simulations on reduced numbers were possible and the paper was never published.

In the first stage, Bleichenbacher's algorithm reduces the signatures from 160 bits to say 40 bits using linear combinations of the original signatures and then, during a second phase, a Discrete Fourier Transform is used to recover the most significant bits of the secret key. The bias of the reduced signatures is higher than the bias of the original signatures, that's the reason why Fourier technique is needed to extract this information. This algorithm is very similar to Blum, Kalai and Wasserman algorithms [4,18] for solving LPN and LWE problems. For 384-bit order, the first stage of Bleichenbacher original attack is not sufficiently efficient to reduce the signatures and more advanced techniques based on LLL and BKZ are needed if the number of leaked bits is high enough [21].

The modification of the first stage is not possible if less than one bit of nonces is available and we turn back to Bleichenbacher's original attack which requires a high number of signatures.

Our Contributions. Our first contribution is the first implementation of Bleichenbacher's attack against ECDSA with a single-bit on nonce bias. We carry out this attack on the standardized SECG P160 R1 elliptic curve. On this 160-bit curve, we use 2^{33} ECDSA signatures, and achieve a full key recovery in a few hours of wall-clock time on a 64-core workstation. The most time-consuming part of the attack is the first phase, in which a sorting algorithm is executed several times. This is the first key recovery from a single bit of bias, which paves the way to new applications. We stress again that this record cannot be achieved using lattice reduction techniques based on HNP problem, since even if the HNP lattice satisfies the Gaussian heuristic, a condition for finding the hidden lattice point is that the number of known bits of the nonce must be greater than $\log_2(\sqrt{\pi e/2}) \approx 1.0471$ (hence at least 2) [26], irrespective of the underlying lattice reduction algorithm.

As a second contribution, we show a security proof for the recomposition method on curves obtained by the quadratic GLS method once the values k_1 and k_2 are uniformly distributed in the interval $[0, \sqrt{n})$, where n is the prime group order. We prove that the statistical distance between this distribution and the uniform distribution in $[0, n)$ is negligible. Furthermore, if k_1 and k_2 are taken at random in a small interval of the form $[0, 2^m)$, where $m = \lfloor \frac{1}{2} \log_2 n \rfloor$, the bias on the distribution on k used in Bleichenbacher's attack is negligible. However, we show that the bias of the distribution on tk where t is the trace is sufficiently large and a Bleichenbacher's attack allows to recover the secret key. We also implement this attack and the complexities are similar to the previous part.

Finally, we study the decomposition technique proposed in GLV with the implementation described by Park *et al.* in [28]. To this end, we propose a very efficient side-channel attack that uses the leakage on the multiplication in order to recover some of the least significant bits of the nonces. Consequently, we can thus use lattice techniques to recover the secret key.

2   Preliminaries

2.1   Bias Definition and Properties

The measurement of the bias of random variables represents a significant part of our analyses. We thus recall the definition of the bias which was proposed by Bleichenbacher in [3].

Definition 1. *Let X be a random variable over $\mathbb{Z}/n\mathbb{Z}$. The bias $B_n(X)$ is defined as*

$$B_n(X) = E(e^{2\pi i X/n}) = B_n(X \bmod n),$$

where $E(X)$ represents the mean.

Similarly, the sampled bias of a set of points $V = (v_1, \cdots, v_L)$ in $\mathbb{Z}/n\mathbb{Z}$ is defined by

$$B_n(V) = \frac{1}{L} \sum_{j=0}^{L-1} e^{2\pi i v_j / n}.$$

The bias as defined above presents some useful properties we recall in Lemma 1.

Lemma 1. *Let $0 < T \leq n$ be a bound and X, Y random variables uniformly distributed on the interval $[0, T - 1]$.*

(a) If X is uniformly distributed on the interval $[0, n - 1]$, then $B_n(X) = 0$.

(b) If X and Y are independent, then $B_n(X + Y) = B_n(X)B_n(Y)$.

(c) $B_n(-X) = \overline{B_n(X)}$ where $\overline{a}$ denotes the conjugate of a.

(d) $B_n(X) = \frac{1}{T} \left| \frac{\sin(\pi T/n)}{\sin(\pi/n)} \right|$ and $B_n(X)$ is real-valued with $0 \leq B_n(X) \leq 1$.

(e) Let a be an integer with $|a|T \leq n$ and $Y = aX$, then $B_n(Y) = \frac{1}{T} \frac{\sin(\pi a T/n)}{\sin(\pi a/n)}$

2.2 ECDSA Signature Generation

ECDSA is a NIST standard and we describe the signature generation in Algorithm 1.

Algorithm 1. ECDSA signature. P is a base point of order n and $H : \{0,1\}^* \to [0, n-1]$ is a cryptographic hash function. The private key is an element $x \in \mathbb{Z}/n\mathbb{Z}$ and the public key is denoted by (p, n, H, P, Q) with $Q = [x]P$.

1: **function** $\text{SIGN}_{\text{ECDSA}}(m)$
2:    $k \xleftarrow{\$} [0, n-1]$
3:    $(u, v) \leftarrow [k]P$
4:    $r \leftarrow u \bmod n$; **if** $r = 0$ **then goto** step 2;
5:    $s \leftarrow k^{-1}(H(m) + rx) \bmod n$; **if** $s = 0$ **then goto** step 2;
6:    **return** (r, s)
7: **end function**

---

3   Bleichenbacher's Attack on Single Bit Bias

In this part, we present our results on an ECDSA signature generation scheme where the nonce k is 1-bit biased. We demonstrate that an attack proposed some years ago by Bleichenbacher can succeed in retrieving the secret key in about 2^{37} time and 2^{33} memory complexities given 2^{33} signatures, for 160-bit order. This attack was initially focusing on the DSA signature generation scheme but can be applied without any modification to ECDSA we consider in this paper.

The main idea consists in using the fact that the nonces k_j are chosen from a biased random variable $\mathbf{K}$, *i.e.* k are not randomly and uniformly generated on $[0, n - 1]$. Because the values k_j are biased and linked with the secret key x by the equations which are used for the signature computations, these signatures, correctly manipulated, also present a bias which will only be significant for the correct value of x. In other words the bias plays the role of the distinguisher in this attack.

Obviously, for cryptographic sizes, evaluating the bias for all values in $[0, n-1]$ is impractical. However, Bleichenbacher observed that it is possible to "broaden the peak" of the bias in such a way that, with a value close the correct value of x, the bias will remain significant. Thus the bias computations can be performed on a more sparse set of candidates thanks to the Fast Fourier Transform. In return, it requires a non-negligible work on the signatures which reduces the bias, and the attack returns an approximation of the secret key, *i.e.* its most significant bits. The attack can be iterated to retrieve more bits of the secrets and as soon as sufficiently many bits of x are known, Pollard's lambda method [29] can be used to derive the remaining bits. Algorithm 2 presents the main steps of the attack.

Algorithm 2. Bleichenbacher's attack given S ECDSA signatures. The parameters S, ℓ and ι have to be chosen accordingly to the bias.

Require: S biased ECDSA signatures (r_j, s_j) computed using a single secret key x.
Ensure: The ℓ most significant bits of x.

1: **Preprocessing**
2: **for** $j = 0$ **to** $S - 1$ **do**
3:     $h_j \leftarrow H(m_j) \cdot s^{-1} \bmod n$
4:     $c_j \leftarrow r_j \cdot s_j^{-1} \bmod n$
5: **end for**

6: **Reduction of the c_j values (Sort-and-Difference Algorithm)**
7: $A \leftarrow [(c_j, h_j)]_{0 \leq j \leq S-1}$
8: **for** $i = 1$ **to** ι **do**
9:     Sort A by the c_j values $\triangleright c_j \leq c_{j+1}$
10:     **for** $j = 0$ **to** $S - \iota$ **do**
11:         $A[j] \leftarrow A[j+1] - A[j]$ $\triangleright A[j] = (c_{j+1} - c_j, h_{j+1} - h_j)$
12:     **end for**
13: **end for**
14: Only keep the pairs (c_j, h_j) such that $c_j < 2^\ell$
15: Denote by L the number of such pairs

16: **Bias computation using the inverse FFT**
17: $Z \leftarrow (0, \cdots, 0)$ a vector of size 2^ℓ
18: **for** $j = 0$ **to** $L - 1$ **do**
19:     $Z_{c_j} \leftarrow Z_{c_j} + e^{2\pi i h_j / n}$
20: **end for**
21: $W \leftarrow \text{iFFT}(Z)$ $\triangleright$ Inverse FFT computation. The output is also a vector of
        complex numbers.
22: Find the value m such that $|Z_m|$ is maximal
23: **return** $msb_\ell(mn/2^\ell)$

3.1   Attack Analysis

We first explain why the bias can serve as a distinguisher and while doing so explain the goal of the preprocessing phase, as it was done in [21], for the sake of completeness. For that purpose, consider S ECDSA signatures (r_j, s_j) with biased nonces k_j. We have the following relation due to step 5 of Algorithm 1:

$$k_j = H(m_j)s_j^{-1} + r_j s_j^{-1} x \bmod n \quad \text{for} \quad 0 \le j \le S - 1.$$

Now let $h_j = H(m_j)s_j^{-1} \bmod n$ and $c_j = r_j s_j^{-1} \bmod n$. Then the set $\{h_j + c_j x\}_{j=0}^{S-1} = \{k_j\}_{j=0}^{S-1}$ will show a significant nonzero sampled bias. Moreover, for any $w \ne x$, the sampled bias from $V_w = \{h_j + c_j w\}_{j=0}^{S-1}$ will be relatively small. Since h_j and c_j are publicly computable, we thus have a way to determine the correct value of x by testing all the value $w \in [0, n - 1]$.

To have a practical test, we have to broaden the peak of the bias such that values of w close to the correct value x will also show a significant bias. The peak will be broad if the c_j are relatively small. More precisely, by denoting 2^ℓ a bound such that $0 \le c_j < 2^\ell$, then we can find an approximation of x by evaluating the sampled bias of 2^ℓ evenly-spaced values of w between 0 and $n - 1$.

The reduction of the c_j, second phase in Algorithm 2, can be done using a sort-and-difference algorithm. From S pairs (c_j, h_j), we first sort them according to their first element. Then we subtract each c_j from the next largest one and we take the differences of the corresponding h_j as well. We thus obtain a list of $S - 1$ pairs (c'_j, h'_j) whose values c'_j are on average $\log(S)$ bits smaller. More details about the analysis of this reduction are given later. This reduction algorithm can be repeated in order to achieve the bound 2^ℓ: once the MSB of x are known, one can rewrite the system and attack the next top bits, by integrating the learnt MSB into the c_j as was done in [21].

Now let $w_m = mn/2^\ell$, with $m \in [0, 2^\ell - 1]$, be 2^ℓ evenly-spaced values between 0 and $n - 1$. For sake of clarity, we keep the notation (c_j, h_j) for the reduced pairs with $c_j < 2^\ell$ and we consider having L such pairs. Then

$$B_n(V_{w_m}) = \frac{1}{L} \sum_{j=0}^{L-1} e^{2\pi i (h_j + c_j mn/2^\ell)/n} = \sum_{t=0}^{2^\ell - 1} \left(\frac{1}{L} \sum_{\{j \mid c_j = t\}} e^{2\pi i h_j/n} \right) e^{2\pi i t m/2^\ell}$$

$$= \sum_{t=0}^{2^\ell - 1} Z_t e^{2\pi i t m/2^\ell}$$

with $Z_t = \frac{1}{L} \sum_{\{j \mid c_j = t\}} e^{2\pi i h_j/n}$. $B_n(V_{w_m})$ can be viewed as the inverse Fast Fourier Transform of the vector $Z = (Z_0, \cdots, Z_{2^\ell - 1})$. Thus the multiple bias computations can be performed very efficiently using the FFT. From Step 17 to 20 in Algorithm 2, we compute this vector Z. Step 21 outputs a vector of the ·sampled bias for the 2^ℓ candidates, i.e. iFFT$(Z) = (B_n(V_{w_0}), B_n(V_{w_1}), \cdots, B_n(V_{w_{2^\ell - 1}}))$. Finally, the value of $w_m = mn/2^\ell$ with the largest sampled bias should share its ℓ most significant bits with the secret key x.

Choosing the Parameters. We first give some properties which will help to define the parameters for the attack. We can estimate the sampled bias for a wrong candidate w_m, *i.e.* a value w_m which do not share some most significant bits with the secret key x. More precisely, it can be shown that for w_m either significantly larger or significantly smaller than x, we have $B_n(V_{w_m}) \approx \frac{1}{\sqrt{L}}$, which corresponds to the average distance from the origin for a random walk on the complex plane.

The second property concerns the c_j reduction phase and gives a relation between the number of signatures S and the number of reduced pairs L.

Proposition 1. *Consider S ECDSA signatures of the form (c_j, h_j) and $\gamma \in \mathbb{Z}$. The percentage of signatures (c'_j, h'_j) after the first application of the sort-and-difference algorithm such that $c'_j < 2^{\log q - \log S + \gamma}$ can be approximated by $1 - e^{-2^\gamma}$.*

Lemma 2. *Let $X_1, \ldots, X_N$ be N independent uniformly distributed random variables over $[0, 1]$, and for all i, denote by $X_{(i)}$ the i-th order statistic of the X_j's (namely, $X_{(i)}$ is the i-th smallest among the X_j's). Then, the random variables $Y_i = X_{(i+1)} - X_{(i)}$ for $i = 1, \ldots, N - 1$ are identically distributed, and all follow the beta distribution $B(1, N)$, of probability density function (hereafter pdf) $f(t) = N \cdot (1 - t)^{N-1}$. As a result, for any constant $\alpha > 0$, we have $\Pr[Y_i \leq \alpha/N] = 1 - e^{-\alpha} + O(1/N)$.*

Proof. Indeed, a standard formula [11, 2.2.1] expresses the joint pdf of $X_{(i)}$ and $X_{(i+1)}$ as:

$$f_{i,i+1}(u, v) = \begin{cases} \frac{N!}{(i-1)!(N-i-1)!} u^{i-1}(1-v)^{N-i-1} & \text{for } 0 \leq u \leq v \leq 1, \\ 0 & \text{otherwise.} \end{cases}$$

Hence, the pdf f_i of Y_i is given by:

$$f_i(t) = \int_0^{1-t} f_{i,i+1}(u, u + t)dt \quad \text{for } t \in [0, 1].$$

The change of variable $u = (1 - t)w$ gives:

$$\begin{aligned} f_i(t) &= (1-t) \int_0^1 f_{i,i+1}\big((1-t)w, (1-t)w + t\big)dw \\ &= c(1-t) \int_0^1 (1-t)^{i-1}w^{i-1}(1 - w - t + wt)^{N-i-1}dw \\ &= c(1-t)^i \int_0^1 w^{i-1}(1-t)^{N-i-1}(1-w)^{N-i-1}dw \\ &= c(1-t)^{N-1} \int_0^1 w^{i-1}(1-w)^{N-i-1}dw, \end{aligned}$$

where $c = \frac{N!}{(i-1)!(N-i-1)!}$. In particular, we have $f_i(t) = c'(1-t)^{N-1}$ for some constant c' and all $t \in [0,1]$, and since $\int_0^1 f_i = 1$, we must have $f_i(t) = N(1-t)^{N-1} = f(t)$ as required. As a result, we obtain:

$$\Pr\left[Y_i \leq \frac{\alpha}{N}\right] = \int_0^{\alpha/N} N(1-t)^{N-1}dt = 1 - \left(1 - \frac{\alpha}{N}\right)^N$$
$$= 1 - \exp\left(N \cdot (-\alpha/N + O(1/N^2))\right) = 1 - e^{-\alpha} + O(1/N).$$

This concludes the proof.   □

As an example consider a modulus n of size 160. Starting from 2^{40} ECDSA signatures, after one iteration of the sort-and-difference algorithm, about 86.5% of them will have a value $c'_j < 2^{121}$. The percentage drops to 22.1% if we consider only those ones with a value $c'_j < 2^{118}$. Note that this proposition is only true for the first iteration of the algorithm where we really can consider variables as uniformly random and independently distributed. Clearly they are not after this: if after the first round variables were uniformly distributed, the ratio between $\gamma = -2$ and $\gamma = 1$ would be $0.125 = 1/2^3$ where it is ≈ 0.255. Sadly, it appears that the ratio progress in our disfavor when we want to iterate, *i.e.* the ratio after ι iterations is less than $(1 - e^{-2^\gamma})^\iota$. We thus do not have a lower bound. However the ratio can be experimentally determined and Table 1 gives an overview for different values of γ up to 6 iterations.

Table 1. Experimental ratio between the ECDSA signatures of the form (c'_j, h'_j) such that $c'_j < 2^{\log n - \iota \cdot (\log S + \gamma)}$, and the S initial signatures, after ι iterations of the sort-and-difference algorithm

γ	-2	-1	0	1	2
1st iteration	0.22	0.39	0.63	0.86	0.98
2nd iteration	0.031	0.12	0.36	0.75	0.94
3rd iteration	$3.2 \ 10^{-3}$	0.025	0.17	0.64	0.89
4th iteration	$3.0 \ 10^{-4}$	$4.6 \ 10^{-3}$	0.069	0.53	0.84
5th iteration	$2.0 \ 10^{-5}$	$6.7 \ 10^{-4}$	0.022	0.40	0.79
6th iteration	$2.8 \ 10^{-6}$	$9.5 \ 10^{-5}$	$6.5 \ 10^{-3}$	0.28	0.73

Given S signatures, we have to choose a pair (γ, ι) such that $\log n - \iota \cdot (\log S + \gamma) = \ell$ is sufficiently small to perform a FFT in $2^\ell \log \ell$ time and 2^ℓ memory complexities. The algorithm complexity is $O(S \log(S) + \ell \log(\ell))$. Now a verification is necessary to be sure that this set of parameters will give a successful attack. Indeed denote by $B_n(\mathbf{K})$ the initial bias which is fully determined by the number of most (or least) significant bits of the k_i which are known or set to zero (see Table 2 for some values). From properties (b) and (c) of the Lemma 1, each iteration of the sort-and-difference algorithm reduces the bias by raising it to the square of its norm (assuming that the variables are independant): indeed, let X, Y be uniformly distributed and independent random variables on

$[0, n-1]$, then $B_n(X) = B_n(Y)$ and $B_n(X-Y) = B_n(X)\overline{B_n(Y)} = |B_n(X)|^2$. The final bias is then approximated by $|B_n(\mathbf{K})|^{2^\iota}$. Thus the following inequality holds since $B_n(V_{w_m}) \approx 1/\sqrt{L}$:

$$|B_n(\mathbf{K})|^{2^\iota} \gg 1/\sqrt{L},$$

where L represents as before the number of reduced pairs (c_j, h_j) with $c_j < 2^\ell$. Using Table 1 which gives the ratio L/S for different choices of pairs (γ, ι), we obtain a relation between S, ι, ℓ and n.

Note that contrary to previous reports in the literature [21,3], we do not need to center the k_j around 0. Indeed sort-and-difference algorithm performs only subtractions and does not mix subtractions and additions as is common with lattice reduction or generalized birthday algorithms.

Table 2. Some values of bias for large n, when b most (or least) significant bits of k are known, using Property (d) of Lemma 1

b	1	2	3	4	5
$B_n(\mathbf{K})$	0.6366198	0.9003163	0.9744954	0.9935869	0.9983944

3.2 Implementation

We successfully implemented the attack. As our target, we chose the SECG P160 R1 curve, published in 2000 by the SECG consortium [8] and still considered secure. We fixed the most significant bit in the nonces and checked (with the help of the secret) that we indeed got the expected bias: ≈ 0.63662. Our C++ implementation was based on the RELIC toolkit (using its provided plain C integer arithmetic) [1] and FFTW [12]. We parallelized it in a straightforward manner (including (quick)sorting phases) and tested it on a multicore machine.

We generated 2^{33} signatures and performed 4 sort-and-difference reduction phases. 450 millions (which is 52.5%) of our initial 2^{33} signatures had their c_j reduced down to 32 bits, as was expected from table 1. The bias after 4 reduction steps was 0.000743558 which is slightly greater than the expected $0.63662^{2^4} \approx 0.00072792$. We then computed a FFT on 32 bits (we selected the reduced c_j smaller than 2^{32}). The best candidate had a score approximately 35% greater than the second. Both corresponding MSB of the secret differed only by the 31st and 32nd most significant bits. The 3rd and 4th candidates were also very close to the two first ones, with score approximately 1/3 of the best candidate. Then, there was a number of random values with maximal score approximately 1/6 of the best one. We repeated the experiment several (5) times and got similar results, always finding at least the 30 MSB of the secret with the best candidate. We couldn't repeat it more because of the high computational resources involved.

The total memory used by the signatures and FFT tables was slightly more than 1 terabyte. To recover 32 bits of the private key, the attack took approximately 1150 CPU-hours, most of it being data exchange, which we can decompose as follows:

- 70%: parallelised quicksort (the most memory-intensive phase)
- 18%: signature generation (approximately 250 to 430 kilocycles per signature depending on the CPU, excluding hash computations)
- 10%: candidate selection and FFT table preparation
- $\approx 1\%$: the FFT itself.

We did not use more parallelizable sorts like Batcher odd-even mergesort [2] but this would clearly be the next thing to do from a performance perspective.

Next steps of the attack to recover the following bits of the secret were done as in [21]. Basically, it amounts to a replay of algorithm 2 on the initial signatures, putting the previously found MSB of the secret into h_j. Write the private key x as $x_0 2^m + x_1$ where x_0 is the recovered m MSB at the first round. Then $(h_j) + (c_j)x = (h_j + c_j x_0 2^m) + (c_j x_1)$ and we want to recover the MSB of x_1. We proceed as in the first round, except that we now keep the c_j that are smaller than $2^{\ell+m}$ instead of 2^ℓ (thus when $\ell = m$ we just have to stop the reduction one iteration earlier). Then we build the FFT table as $Z[c_j/2^m] = Z[c_j/2^m] + e^{2\pi i h'_j/n}$. The FFT recovers the next most significant chunk of bits of the secret key. The computation restart makes it necessary to go back from the initial signatures, but there's no need to keep them in memory during the reduction. In practice we had barely enough memory to keep them, but in order to reduce memory usage they should either be stored on disk and retrieved to iterate the secret recovery, or tracked down through the reduction and rebuilt afterwards.

In practice, it is advisable to take a small security margin and reinject only 30 bits of the computed MSB of the secret to account for small variations of sampling around the peak. In any case, if we recurse with a wrong secret, the FFT will not detect any peak. Experiments indeed showed no peak in this case, with the highest score not being statistically different from the other ones. This paves the way for a time/memory tradeoff: suppose the hardware is limited in memory and can only work on (say) 2^{31} signatures and 2^{30} FFT size instead of the 2^{33} needed for attacking 160 bits with 4 iterations with the previous algorithm. We first reduce the c_j from 160 to 40 bits with 4 reductions as usual. We then simply guess the 10 MSB of the secret and build 2^{30}-sized FFT tables accordingly. The guess will be correct on the only one FFT among the 2^{10} which shows a significant peak. Since FFTs are particularly efficient, much more than sorting, this is of practical importance. Alternatively, if it's possible to compute 2^{41} signatures, we can select only the expected $1/2^{10}$ fraction of signatures whose corresponding c_j have their 10 MSB already zeroes, that is to say that have 150 bits instead of 160 and can be reduced to 30 with 4 iterations. Finally, since the FFT table takes less memory than the signatures (a complex number occupies 16 bytes whereas a signature requires at least 40), we could improve the attack further by either carrying out several FFTs in parallel when guessing some bits of the secret, or by increasing the size of the FFT table slightly (with a corresponding increase of the selection bound on c_j). This would have two advantadges. Firstly, it would improve the sampling around the peak and reduce the uncertainty. Secondly, the bound increase implies that some signatures would be selected after the third round of reduction instead of the fourth, thus having a much better bias and hopefully revealing more precise information about the secret.

Our experiments targeted a 160-bit curve, but it should be pointed out that larger curves are susceptible to this attack as well. Roughly speaking, one can carry out the key recovery attack with 1-bit nonce bias on an N-bit curve in time $\approx 2^{N/5} \log_2(N/5)$ and memory $\approx 2^{N/5}$. For example, a 256-bit curve can be attacked in time $\approx 2^{58}$ and memory $\approx 2^{52}$: generate 2^{52} signatures, perform 4 reduction steps (removing $4 \cdot 51 = 204$ bits on approximately 53% of the data), keep signatures with c_j less than 2^{52} and carry out the FFT on a table of size 2^{52}. One signature is $64 = 2^6$ bytes, so that the total memory needed for the attack is 2^{18} terabytes of storage, which corresponds to 65536 of today's 4 TB disks. This does not appear to be out of reach of well-funded adversaries.

4   Security Analysis of the Recomposition Technique

The results presented so far had no direct connection with GLV/GLS curves. We now turn to such curves, and first discuss in this section the security of what we called the "recomposition technique" for GLV/GLS coefficients (namely, choose k_1 and k_2 uniformly at random in some interval $[0, K)$ to obtain $k = k_1 + k_2\lambda \bmod n$), whereas the next section will focus on the "decomposition technique".

To fix ideas, we consider an elliptic curve E obtained by the quadratic GLS method over a prime field [13, §2.1]. In other words, there is an elliptic curve E_0 over the prime field $\mathbb{F}_p$ such that E is the quadratic twist of E_0 over $\mathbb{F}_{p^2}$. If we denote by $p + 1 - t$ the order of $E_0(\mathbb{F}_p)$ (where t is bounded as $|t| \le 2\sqrt{p}$ by the Hasse–Weil theorem), the order n of $E(\mathbb{F}_{p^2})$ satisfies:

$$n = (p-1)^2 + t^2. \tag{1}$$

We assume that this order n is prime, which is the main case of interest. Then, E is endowed with an efficient endomorphism ψ (obtained by conjugating the Frobenius map with the twisting isomorphism) which acts on the cyclic group $E(\mathbb{F}_{p^2})$ by multiplication by

$$\lambda \equiv t^{-1}(p-1) \pmod{n}. \tag{2}$$

In particular, $\lambda^2 \equiv (p-1)^2/t^2 \equiv -t^2/t^2 \equiv -1 \pmod{n}$.

In this setting, we first prove in §4.1 that if k_1 and k_2 are chosen uniformly at random in $[0, \sqrt{n})$, then $k = k_1 + k_2\lambda$ is statistically close to uniform in $\mathbb{Z}/n\mathbb{Z}$, so that such a choice of (k_1, k_2) can be used securely in any cryptographic protocol (and in particular ECDSA). On the other hand, we show in §4.2 that if k_1 and k_2 are chosen in $[0, 2^m)$ where $m = \lfloor \frac{1}{2} \log_2 n \rfloor$ instead, then $k = k_1 + k_2\lambda$ may not be close to uniform anymore, and we show that a variant of Bleichenbacher's attack can apply. In §4.3, we describe an implementation of that attack on a 160-bit GLS curve, similar to the attack of §3.

4.1   A Secure Choice of (k_1, k_2)

Let E be a curve of prime order n over $\mathbb{F}_{p^2}$ obtained by the quadratic GLS method as above. In view of (1), we have:

$$(p-1)^2 \leq n \leq (p-1)^2 + (2\sqrt{p})^2 = (p+1)^2,$$

and the inequalities are in fact strict, since n is prime. Thus, we have $p - 1 < \sqrt{n} < p + 1$, and it follows that the distribution of $k = k_1 + k_2\lambda$ for (k_1, k_2) uniform in $[0, \sqrt{n})^2$ is statistically close to the distribution of the same k for (k_1, k_2) uniform in $[0, p-1)^2$. We will thus concentrate on the latter, and show that it is close to uniform in $\mathbb{Z}/n\mathbb{Z}$ using the following lemma.

Lemma 3. *The following map is injective.*

$$F \colon [0, p-1)^2 \longrightarrow \mathbb{Z}/n\mathbb{Z}$$
$$(k_1, k_2) \longmapsto k_1 + k_2\lambda.$$

Proof. Consider two distinct pairs $(k_1, k_2) \neq (k_1', k_2')$ such that $F(k_1, k_2) = F(k_1', k_2')$. We have:

$$(x - x') + (y - y')\lambda \equiv 0 \pmod{n}$$
$$(x - x')^2 \equiv \lambda^2(y - y')^2 \pmod{n}$$
$$(x - x')^2 + (y - y')^2 \equiv 0 \pmod{n},$$

since $\lambda^2 \equiv -1 \pmod{n}$. Thus, the positive integer $(x-x')^2+(y-y')^2$ is divisible by n, and it is also smaller than $2(p-1)^2 < 2n$, so we must have $(x - x')^2 + (y - y')^2 = n$. In other words, $(x - x')^2 + (y - y')^2$ is a decomposition of n as a sum of two squares. Now it is well-known that, as a prime number, n has at most one decomposition as a sum of two squares up to order and sign (see e.g. [20, §3.6]), and $(p - 1)^2 + t^2$ is one such representation. As a result, we must have either $x - x' = \pm(p - 1)$ or $y - y' = \pm(p - 1)$, and neither is possible since those difference are bounded by $p - 2$ in absolute value. Hence, F is injective as required.   □

Theorem 1. *The distribution of the values $k = k_1 + k_2\lambda$ for (k_1, k_2) uniform in $[0, p-1)^2$ is statistically close to the uniform distribution on $\mathbb{Z}/n\mathbb{Z}$. More precisely, the statistical distance:*

$$\Delta_1 = \sum_{k \in \mathbb{Z}/n\mathbb{Z}} \left| \Pr\left[k = k_1 + k_2\lambda \; ; \; (k_1, k_2) \overset{\$}{\leftarrow} [0, p-1)^2\right] - \frac{1}{n} \right|$$

is given by $\Delta_1 = 2t^2/n$, which is negligible.

Proof. Indeed, since the function F above is injective by Lemma 3, the probability $\Pr\left[k = k_1 + k_2\lambda \; ; \; (k_1, k_2) \overset{\$}{\leftarrow} [0, p-1)^2\right]$ is equal to $1/(p-1)^2$ for each of the $(p-1)^2$ points in the image of F, and 0 for each of the $n - (p-1)^2 = t^2$ points outside of that image. Therefore:

$$\Delta_1 = (p-1)^2 \cdot \left| \frac{1}{(p-1)^2} - \frac{1}{n} \right| + t^2 \cdot \left| 0 - \frac{1}{n} \right| = 1 - \frac{(p-1)^2}{n} + \frac{t^2}{n} = \frac{2t^2}{n}$$

as required. This is bounded above by $8p/(p-1)^2$, which is indeed negligible.   □

Remark 1. Theorem 1 means that it is secure, in any ECC protocol instantiated over the GLS curve E, to sample random scalars k by picking k_1 and k_2 uniformly in $[0, p-1)$, or equivalently $[0, \sqrt{n})$.

As we can see, the proof relies on the particular arithmetic properties of the quadratic GLS method (mainly the fact that $\lambda = \sqrt{-1}$ in $\mathbb{Z}/n\mathbb{Z}$), so that the result does not readily extend to different settings, like the GLV method on a curve of CM discriminant -3. And indeed, in that case, Brumley and Nyberg have provided evidence that choosing (k_1, k_2) uniformly in $[0, \sqrt{n})$ may not yield a close to uniform distribution for k [7, Example 3]. They suggest an alternate approach to select intervals to choose k_1 and k_2 from and still achieve high entropy in a more general setting, but since the quadratic GLS method is one of the most used variants of GLV/GLS, we believe Theorem 1 is of significant practical interest.

4.2 Breaking Insecure Choices of (k_1, k_2) with Bleichenbacher's Attack

In the quadratic GLS setting, we have just seen that choosing (k_1, k_2) uniformly in $[0, \sqrt{n})^2$ yields a close-to-uniform distribution of $k = k_1 + k_2\lambda$. However, we can reasonably suspect that if we choose k_1 and k_2 uniformly in $[0, 2^m)$, $m = \lfloor \frac{1}{2} \log_2 n \rfloor$ (i.e. uniform bitstrings of length just under half of the size of n), the distribution of k will no longer be uniform. This is not immediately visible on the bias, however.

Indeed, if we let $T = 2^m$ and define K_1, K_2 as independent uniform random variables over $[0, T)$ and K as the random variable in $\mathbb{Z}/n\mathbb{Z}$ given by $K = K_1 + K_2\lambda$, we have, by Lemma 1:

$$B_n(K) = B_n(K_1) \cdot B_n(\lambda K_2) = \frac{1}{T} \left| \frac{\sin(\pi T/n)}{\sin(\pi/n)} \right| \cdot \frac{1}{T} \left| \frac{\sin(\pi \lambda T/n)}{\sin(\pi \lambda/n)} \right|.$$

The first factor is very close to 1, but the second factor is usually negligible. For example, on the 160-bit GLS curve (3) below, we have $T = 2^{79}$ and $B_n(\lambda K_2) \approx 1.52/T$. As a result, Bleichenbacher's attack does not apply directly to this setting in general.

However, since $\lambda \equiv t^{-1}(p-1) \pmod{n}$, we claim that there is a significant bias on the values $t \cdot k$. Indeed, we have:

$$B_n(tK) = B_n(tK_1) \cdot B_n\big((p-1)K_2\big)$$
$$= \frac{1}{T}\left|\frac{\sin(\pi t T/n)}{\sin(\pi t/n)}\right| \cdot \frac{1}{T}\left|\frac{\sin(\pi(p-1)T/n)}{\sin(\pi(p-1)/n)}\right|$$
$$= \frac{1}{T}\frac{\pi t T/n + O((tT/n)^3)}{\pi t/n + O((t/n)^3)} \cdot \frac{1}{T}\frac{|\sin(\pi(p-1)T/n)|}{\pi(p-1)/n + O(((p-1)/n)^3)}$$
$$= \Big(1 + O\big((tT/n)^2 + (p/n)^2\big)\Big) \cdot \left|\frac{\sin(\pi(p-1)T/n)}{\pi(p-1)T/n}\right|.$$

The big-O in the first factor is negligible since $tT/n = \Theta(p^{1/2} \cdot p/p^2) = \Theta(p^{-1/2})$ and $p/n = \Theta(p^{-1})$. On the other hand, $(p-1)T/n \approx T/\sqrt{n}$ is roughly between 0.5 and 1 depending on how close n is to a power of two. Thus, the bias is significant in general, and is maximal when $(p-1)T/n$ is smallest (close to $1/2$), which happens when n is just under a power of two. The bias $B_n(tK)$ is then close to $1/(\pi \cdot 1/2) = 2/\pi \approx 0.637$.

It is then straightforward to adapt Bleichenbacher's attack to this setting by targetting the values $t \cdot k$ instead of k. We can then break ECDSA signatures that use nonces of the form $k = k_1 + k_2\lambda$ above using that variant. An implementation of that attack is discussed in the next subsection.

4.3   Implementation of Bleichenbacher's Attack in the GLS Setting

We carry out the attack described above on the 160-bit GLS curve E defined as follows. Over the 80-bit prime field[1] $\mathbb{F}_p$, $p = 255 \cdot 2^{72} + 1$, we define $E_0 \colon y^2 = x^3 - 3x/23 + 104$. Then, the elliptic curve E is the quadratic twist of E_0 over $\mathbb{F}_{p^2} = \mathbb{F}_p(\sqrt{23})$, namely:

$$E \colon y^2 = x^3 - 3x + 104 \cdot \sqrt{23}^3 \text{ over } \mathbb{F}_{p^2}. \tag{3}$$

The order of $E_0(\mathbb{F}_p)$ is $p + 1 - t$ for $t = 776009485427$, and $E(\mathbb{F}_{p^2})$ is of prime order $n = (p-1)^2 + t^2$. The theoretical value of the bias $B_n(tK)$, computed using the exact formula above, is then ≈ 0.634.

We performed the recovery of 32 MSB of a private key as in section 3.2. We computed 2^{33} signatures and unrolled the attack on $(tc_j \bmod n, th_j \bmod n)$ instead of (c_j, h_j). We checked the bias and obtained ≈ 0.634116 which is close to the theory. In practice the attack took about 2000 CPU-hours, with 56% for the signature generation, 37% for the four sort-and-difference reduction steps, 5% for the candidate selection and FFT table preparation and less than 0.5% for the FFT itself. In wall-clock time terms, except for the signature generation which took (much) longer, other phases were identical as 3.2. We attribute this unexpected increase in signing time to threshold effects: for example, representing elements on a prime field with $\approx 2^{160}$ elements needs only 3 64-bit words, whereas a on $\mathbb{F}_{p^2}$ we needed $4 \times 2 = 8$ words.

---

[1] This is an example of "optimal prime field" (OPF). See e.g. [32].

5   Security Analysis of the Decomposition Technique

In this section, we analyze the security of algorithms for computing the decomposition of the nonces used in the GLV method from a side-channel analytic perspective. Many techniques have been proposed, including [15,28]. The original GLV method [15] based on LLL reduction of a lattice that depends on the nonce k, and variants thereof, have an execution time that depends on k, and are therefore vulnerable to timing attacks.

Therefore, we examine the security of a potentially more secure approach, the Park *et al.* [28] decomposition technique, using more involved power analysis technique.

5.1   Decomposition Algorithm

Park *et al.* provide an alternative decomposition to the GLV paper [15] which reduces the theoretical bound for the decomposition using the theory of μ-Euclidian algorithm and is a little bit faster. The algorithm requires two short and independent vectors v_1 and v_2 of the two-dimensional lattice $L = \{(x,y) : x + y\lambda = 0 \bmod n\}$. We can find these vectors during a precomputation time using the Gauss reduction. The algorithm consists in finding a vector in the lattice $L = \mathbb{Z}v_1 + \mathbb{Z}v_2$ that is close to $(k,0)$ using linear algebra. Then, (k_1, k_2) is determined by the equation:

$$(k_1, k_2) = (k, 0) - (\lfloor b_1 \rceil v_1 + \lfloor b_2 \rceil v_2),$$

where $(k, 0) = b_1 v_1 + b_2 v_2$ is an element of $\mathbb{Q} \times \mathbb{Q}$.

Algorithm 3. Decomposition technique of Park *et al.* in [28]

Require: $k \approx n$, the shortest vectors $v_1 = (x_1, y_1), v_2 = (x_2, y_2)$
Ensure: (k_1, k_2) such that $k = k_1 + k_2\lambda \pmod{n}$

1: $D = x_1 y_2 - x_2 y_1, a_1 = y_2 k, a_2 = -y_1 k$
2: $z_i = \lfloor a_i/D \rceil$ for $i = 1, 2$
3: $k_1 = k - (z_1 x_1 + z_2 x_2), k_2 = z_1 y_1 + z_2 y_2$ **return** (k_1, k_2)

The decomposition technique depicted in Algorithm 3 makes many computations involving the sensitive nonce k. Particularly, the computation of a_1 (*resp.* a_2) is based on a multiplication of the nonce k by y_2 (*resp.* y_1) which is assumed to be known since it is a precomputed value obtained from public parameters using a deterministic algorithm.

Suppose now that we obtain the knowledge of the least significant byte of ℓ nonces $k_1, \cdots, k_\ell$. The best strategy for finding the secret key x consists in performing classical lattice attacks as proposed in [17,23,24]. For a 160-bit modulus, the lattice attack works consistently for $\ell \gtrsim 27$. However the side-channel attack may sometimes fail, *i.e.* the returned byte of some k_j can be a wrong value.

Thus, by denoting $0 < c < 1$ the confidence rate, the side-channel attack has to be performed on $m > \lceil 27/c \rceil$ signatures. Then:

- Select 27 signatures at random among them.
- Perform the attack using these signatures.
- If the attack fails, goto the first step.

The probability of success at each iteration of the lattice attack is $\binom{m \cdot c}{27} / \binom{m}{27}$. As an example, suppose we obtain $m = 200$ signatures, and can guess the least significant byte with 90% accuracy ($c = 0.9$). Then the probability of success of the lattice attack is about 4.7% and 21 lattice reductions have to be performed on average. Since LLL reductions are cheap, much lower success probabilities are tractable as well.

In the following, we discuss the side-channel attack that aims at recovering the first byte of the nonce targeting the two aforementioned multiplications. We present the attack in the particular case of a 8-bit implementation (that corresponds to the device we used in experiments). Note that this attack may also work for 16-bit implementation but in this case the computational cost will be larger and the success rate smaller.

5.2   Side-Channel Attack on this Implementation

The details of the attack highly rely on the way the multiplication is implemented. Depending of the underlying algorithm, the attack may be more or less difficult. We present here the attack corresponding to the implementation we target but we will discuss adaptations to different algorithms. The multiplication we target is a schoolbook multiplication with the nonce being scanned in the outer loop. Algorithm 4 outlines the implementation of such multiplication for ℓ_n-bit nonces and $\ell_{n/2}$-bit b.

Algorithm 4. Multiplication $v = kb$ of $k = \sum_{i=0}^{\ell_n/8} k_i 2^{8i}$ times $b = \sum_{i=0}^{\ell_{n/2}/8} b_i 2^{8i}$

Require: ℓ_n-bit k and $\ell_{n/2}$-bit b two integers, $v = 0$
Ensure: $v = k \times b$

1: $v \leftarrow 0$
2: **for** $i = 0$ to $i < \ell_n/8$ **do**
3:     $c_0 \leftarrow 0$
4:     **for** $j = 0$ to $j < \ell_{n/2}/8$ **do**
5:         $v_{i+j} = (k_i \times b_j + c_j)$ & 0xFF
6:         $c_{j+1} = (k_i \times b_j + c_j) \gg 8$
7:     **end for**
8: **end for**   **return** v

The idea is to take profit of all operations involving the first nonce-byte in the inner loop to recover its value. This can be done by propagating a probability distibution from an operation to another and updating it with the corresponding leakages. Since the nonce bits have to be recovered using a single trace (the nonce is randomly generated for each signature) we place ourselves in the context of a profiled attack. The application of such an attack in a non-profiled setting is left as an open question.

Template Attack on One Step. One step of the inner loop consists in a multiplication of the first byte of the nonce k_0, a byte of the auxiliary input b_i and the carry c_i. This results in a value v_i and a new carry c_{i+1}. We may obtain leakages for each of these variables. We denote by capital letters the output distributions of template exploitation corresponding to small letter variables. For instance, after processing the leakage corresponding to c_i, the attacker gets a distribution

$$C_i = \big(\Pr(c_i = 0), \Pr(c_i = 1), \ldots, \Pr(c_i = 255)\big).$$

Since these variables may be manipulated more than once during the computation, different leakage points may be combined by multiplicating probabilities then normalizing the resulting distribution. More precisely, let $l_1, l_2, \ldots, l_l$ be leakages corresponding to variable k_0, then the distribution K_0 obtained from these leakages is computed as

$$\Pr(k_0 = x) = \frac{1}{Z} \prod_{j=1}^{l} \Pr(k_0 = x|l_j),$$

where the normalizing coefficient Z is given by $\sum_x \prod_{j=1}^{l} \Pr(k_0 = x|l_j)$.

Propagating and Updating Distribution. Let us now discuss how to take profit of all the leakages of the inner loop to gain information on the byte k_0. The main idea is to gather all the information from all variables of a given step i into distribution K_0 and C_{i+1} then do the same at step $i + 1$ using the newly updated distributions. From a probabilistic point of view we should compute the joint distribution of variables of step i then compute marginalized distributions K_0 and C_{i+1}. The following algorithm updates the distributions K_0 and C_{i+1} according to the distributions of variables b_i, v_i and c_i.

Algorithm 5. Information propagation for one step of the multiplication inner loop

---

Require: distributions K_0, B_i, V_i, C_i and C_{i+1}
Ensure: K_0' and C_{i+1}' updated distribution

1: $K_0' = (0, 0, \ldots, 0)$
2: **for** $0 \leq k, b, c < 256$ **do**
3: $\quad 2^8 \cdot u + v \leftarrow k \times b + c$
4: $\quad K_0'(k) \leftarrow K_0'(k) + K_0(k) \cdot B_i(b) \cdot C_{i-1}(c) \cdot V_i(v) \cdot C_i(u)$
5: $\quad C_{i+1}'(u) \leftarrow C_{i+1}'(u) + K_0(k) \cdot B_i(b) \cdot C_i(c) \cdot V_i(v) \cdot C_{i+1}(u)$
6: **end for**
7: **return** $K_0'/\sum_k K_0'(k)$ and $C_{i+1}'/\sum_u C_{i+1}'(u)$

The attacker starts with using Algorithm 5 for the first step. Then she uses the newly updated distributions K_0 and C_1 and the initial distributions B_1, V_1 and C_2 as inputs of Algorithm 5 and so on ... At the end, the attacker gets the final distribution K_0 from which she can derive the most likely value of the least significant bit (or more).

References

1. Aranha, D.F., Gouvêa, C.P.L.: RELIC is an Efficient LIbrary for Cryptography, http://code.google.com/p/relic-toolkit/
2. Batcher, K.E.: Sorting networks and their applications. In: Proceedings of the Spring Joint Computer Conference, AFIPS 1968 (Spring), pp. 307–314. ACM, New York (1968)
3. Bleichenbacher, D.: On the generation of one-time keys in DL signature schemes. Presentation at IEEE P1363 Working Group Meeting (2000)
4. Blum, A., Kalai, A., Wasserman, H.: Noise-tolerant learning, the parity problem, and the statistical query model. J. ACM 50(4), 506–519 (2003)
5. Bos, J.W., Costello, C., Hisil, H., Lauter, K.: High-Performance Scalar Multiplication Using 8-Dimensional GLV/GLS Decomposition. In: Bertoni, G., Coron, J.-S. (eds.) CHES 2013. LNCS, vol. 8086, pp. 331–348. Springer, Heidelberg (2013)
6. Brumley, B.B., Hakala, R.M.: Cache-Timing Template Attacks. In: Matsui, M. (ed.) ASIACRYPT 2009. LNCS, vol. 5912, pp. 667–684. Springer, Heidelberg (2009)
7. Brumley, B.B., Nyberg, K.: On Modular Decomposition of Integers. In: Preneel, B. (ed.) AFRICACRYPT 2009. LNCS, vol. 5580, pp. 386–402. Springer, Heidelberg (2009)
8. Certicom Research. Standards for efficient cryptography, SEC 1: Elliptic curve cryptography, Version 1.0 (September 2000)
9. Chen, Y., Nguyen, P.Q.: BKZ 2.0: Better Lattice Security Estimates. In: Lee, D.H., Wang, X. (eds.) ASIACRYPT 2011. LNCS, vol. 7073, pp. 1–20. Springer, Heidelberg (2011)
10. Costello, C., Hisil, H., Smith, B.: Faster Compact Diffie–Hellman: Endomorphisms on the x-line. In: Nguyen, P.Q., Oswald, E. (eds.) EUROCRYPT 2014. LNCS, vol. 8441, pp. 183–200. Springer, Heidelberg (2014)
11. David, H.A., Nagaraja, H.N.: Order Statistics. Wiley (2003)
12. Frigo, M., Johnson, S.G.: The design and implementation of FFTW3. Proceedings of the IEEE 93(2), 216–231 (2005); Special issue on Program Generation, Optimization, and Platform Adaptation
13. Galbraith, S.D., Lin, X., Scott, M.: Endomorphisms for faster elliptic curve cryptography on a large class of curves. J. Cryptology 24(3), 446–469 (2011)
14. Gallant, R.: Efficient multiplication on curves having an endomorphism of norm 1. In: Workshop on Elliptic Curve Cryptography (1999)
15. Gallant, R.P., Lambert, R.J., Vanstone, S.A.: Faster Point Multiplication on Elliptic Curves with Efficient Endomorphisms. In: Kilian, J. (ed.) CRYPTO 2001. LNCS, vol. 2139, pp. 190–200. Springer, Heidelberg (2001)
16. Guillevic, A., Ionica, S.: Four-Dimensional GLV via the Weil Restriction. In: Sako, Sarkar (eds.) [30], pp. 79–96
17. Howgrave-Graham, N., Smart, N.P.: Lattice Attacks on Digital Signature Schemes. Des. Codes Cryptography 23(3), 283–290 (2001)

18. Levieil, É., Fouque, P.-A.: An Improved LPN Algorithm. In: De Prisco, R., Yung, M. (eds.) SCN 2006. LNCS, vol. 4116, pp. 348–359. Springer, Heidelberg (2006)

19. Liu, M., Nguyen, P.Q.: Solving BDD by enumeration: An update. In: Dawson, E. (ed.) CT-RSA 2013. LNCS, vol. 7779, pp. 293–309. Springer, Heidelberg (2013)

20. McKean, H., Moll, V.: Elliptic curves: function theory, geometry, arithmetic. Cambridge University Press (1999)

21. Mulder, E.D., Hutter, M., Marson, M.E., Pearson, P.: Using Bleichenbacher's solution to the hidden number problem to attack nonce leaks in 384-bit ECDSA: extended version. J. Cryptographic Engineering 4(1), 33–45 (2014)

22. Naccache, D., Nguyên, P.Q., Tunstall, M., Whelan, C.: Experimenting with Faults, Lattices and the DSA. In: Vaudenay, S. (ed.) PKC 2005. LNCS, vol. 3386, pp. 16–28. Springer, Heidelberg (2005)

23. Nguyen, P.Q., Shparlinski, I.: The Insecurity of the Digital Signature Algorithm with Partially Known Nonces. J. Cryptology 15(3), 151–176 (2002)

24. Nguyen, P.Q., Shparlinski, I.: The Insecurity of the Elliptic Curve Digital Signature Algorithm with Partially Known Nonces. Des. Codes Cryptography 30(2), 201–217 (2003)

25. Nguyên, P.Q., Stehlé, D.: Low-Dimensional Lattice Basis Reduction Revisited. In: Buell, D.A. (ed.) ANTS 2004. LNCS, vol. 3076, pp. 338–357. Springer, Heidelberg (2004)

26. Nguyen, P.Q., Tibouchi, M.: Lattice-Based Fault Attacks on Signatures. In: Fault Analysis in Cryptography. Information Security and Cryptography, pp. 201–220. Springer (2012)

27. Oliveira, T., López, J., Aranha, D.F., Rodríguez-Henríquez, F.: Two is the fastest prime: lambda coordinates for binary elliptic curves. J. Cryptographic Engineering 4(1), 3–17 (2014)

28. Park, Y.-H., Jeong, S., Kim, C.H., Lim, J.: An Alternate Decomposition of an Integer for Faster Point Multiplication on Certain Elliptic Curves. In: Naccache, D., Paillier, P. (eds.) PKC 2002. LNCS, vol. 2274, pp. 323–334. Springer, Heidelberg (2002)

29. Pollard, J.M.: Kangaroos, monopoly and discrete logarithms. J. Cryptology 13(4), 437–447 (2000)

30. Sako, K., Sarkar, P. (eds.): ASIACRYPT 2013, Part I. LNCS, vol. 8269. Springer, Heidelberg (2013)

31. Smith, B.: Families of Fast Elliptic Curves from $\mathbb{Q}$-curves. In: Sako, Sarkar (eds.) [30], pp. 61–78

32. Wenger, E., Großschädl, J.: An 8-bit AVR-based elliptic curve cryptographic RISC processor for the Internet of Things. In: MICRO Workshops, pp. 39–46. IEEE Computer Society (2012)

Soft Analytical Side-Channel Attacks

Nicolas Veyrat-Charvillon[1], Benoît Gérard[2], and François-Xavier Standaert[3]

[1] IRISA-CAIRN, Campus ENSSAT, 22305 Lannion, France
[2] DGA Maîtrise de l'Information, 35998 Rennes, France
[3] ICTEAM/ELEN/Crypto Group, Université catholique de Louvain, Belgium

Abstract. In this paper, we introduce a new approach to side-channel key recovery, that combines the low time/memory complexity and noise tolerance of standard (divide and conquer) differential power analysis with the optimal data complexity of algebraic side-channel attacks. Our fundamental contribution for this purpose is to change the way of expressing the problem, from the system of equations used in algebraic attacks to a code, essentially inspired by low density parity check codes. We then show that such codes can be efficiently decoded, taking advantage of the sparsity of the information corresponding to intermediate variables in actual leakage traces. The resulting soft analytical side-channel attacks work under the same profiling assumptions as template attacks, and directly exploit the vectors of probabilities produced by these attacks. As a result, we bridge the gap between popular side-channel distinguishers based on simple statistical tests and previous approaches to analytical side-channel attacks that could only exploit hard information so far.

1 Introduction

The great majority of side-channel attacks published in the literature follow a divide and conquer strategy (DC). That is, they first attack independent parts of the key separately (divide), and then combine these pieces of information (conquer). Information on individual parts of the key is obtained by studying correlations between key-dependent leakage predictions and the actual side-channel measurements. The information can then be combined either by simply concatenating the most probable values of each key part together, or by using an enumeration algorithm [27,28]. Examples of distinguishers exploiting such a strategy include Kocher et al.'s Differential Power Analysis (DPA) [13], Brier et al.'s Correlation Power Analysis (CPA) [2], Gierlichs et al.'s Mutual Information Analysis (MIA) [9], Chari et al.'s Template Attacks (TA) [4] and Schindler et al.'s Stochastic Approach (SA) [23]. The popularity of these tools is due to their simplicity and versatility: they can be adapted to essentially any implementation, have low time complexity and work in a gray box manner. That is, they do not require a precise understanding of the underlying hardware, but their data complexity is highly dependent on the quality of the adversary's leakage predictions. Therefore, the knowledge of implementation details and some engineering intuition can usually be exploited to improve their time and data complexity.

P. Sarkar and T. Iwata (Eds.): ASIACRYPT 2014, PART I, LNCS 8873, pp. 282–296, 2014.
© International Association for Cryptologic Research 2014

In this context, one fundamental question regarding DC distinguishers is whether they are sufficient for security evaluations. That is, are the security levels estimated with such tools close enough to the worst-case? In view of the previously listed qualities (and in particular, their excellent time complexity), the most likely drawback candidate for DC strategies is a suboptimal data complexity. As a result, a number of research works have investigated whether the application of analytical strategies (i.e. targeting the full key at once) could provide improved results. To the best of our knowledge, one of the first attempts in this direction was Mangard's Simple Power Analysis (SPA) against the AES key expansion algorithm [15]. The Side-Channel Collision Attacks (SCCA) in [1,24,25] were next interesting steps, in which the key is recovered by solving a set of (mostly) linear equations corresponding to the first cipher round(s). Following, Algebraic Side-Channel Attacks (ASCA) were introduced in [21,22] and probably constitute the most representative example of analytical strategy to date. Under certain conditions, they are able to extract the key of an AES implementation from a single leakage trace, in an unknown plaintext/ciphertext scenario.

So to some extent, ASCA could be viewed as an extreme opposite to DC attacks, with a minimum data complexity coming at the cost of a (much) more complex and sensitive solving phase – hence raising questions regarding their practical relevance. For example in the first papers from Renauld et al., the adversary represents the target block cipher and its leakages as an instance of satisfiability problem that she sends to a generic SAT solver (other types of solvers, e.g. based on Gröbner bases, have also been analyzed [3]). The main issue with this approach is a very weak resistance to noise, since the solver essentially needs to be fed with correct hard information. For this purpose, the usual strategy was to group certain leakage values according to a model with lower cardinality, e.g. the well-known Hamming weight one, in order to trade robustness for informativeness. Improved heuristics are presented in [17,29]. More recently, Oren et al. proposed to replace the use of a solver by that of an optimizer, leading to Tolerant ASCA (TASCA) able to exploit more general models [18,19]. Yet, even these last attempts were quite inefficient in exploiting soft information, mainly because of the difficulty to translate a vector of probabilities (e.g. as provided by classical TA) into an optimizer-friendly format. In fact TASCA essentially encode these vectors as exhaustive hard information, hence limiting the number of leaking operations that could be included in the optimizer to a couple of rounds (compared to the full cipher in ASCA), because of memory issues. Eventually, the results in [10] provide yet another powerful approach to analytical side-channel attacks, based smart enumeration and specialized to the AES, but so far they also remain limited to the exploitation of hard information.

This state-of-the-art seems to suggest that the probabilistic information provided by side-channel leakages can be easily exploited with DC attacks, while analytical strategies require a preprocessing step to translate this soft information into hard one. In this paper, we argue that this intuition is flawed, and in fact relates to the way of formulating the problem rather than to its nature. That is, while previous analytical attacks were expressing the target block ciphers and their leakages as equations, we propose to describe the same problem as a code. As a result, and for the first time, we detail a Soft Analytical Side-Channel Attack (SASCA) that combines the best of two worlds, namely the noise robustness and low time complexity of DC strategies with the low data complexity of analytical ones. In this respect, our first contribution is to exhibit a natural way to encode a side-channel cryptanalysis problem. Next, we show that we can efficiently decode such problems thanks to the Belief Propagation algorithm (BP). Using these new tools, we are able (i) for low noise levels: to attack the AES FURIOUS implementation that was targeted in previous works on ASCA/TASCA with a single leakage trace, with significantly reduced time and memory complexities, (ii) for large noise levels: to attack the same implementation with multiple plaintexts, but with 2^3 to 2^4 less traces than a standard TA. Summarizing, the proposed technique bridges the gap between DPA and ASCA.

Related Works. While the motivation for SASCA quite directly derives from previous works in ASCA/TASCA, its mathematical modeling fundamentally differs from them and is in fact much closer to some results exploiting techniques from coding theory. In particular, the application of Hidden Markov Models in the context of time-randomized implementations [5,12] or side-channel disassemblers [6], and the decoding of Low Density Parity Check (LDPC) codes in the context of SCCA [8] were sources of inspiration for the following work.

Cautionary Note. In order to show the applicability of SASCA at different noise levels, our empirical results are based on simulated experiments. Yet, we insist that SASCA is (in general) just as realistic as any TA, since it relies on the same assumptions for the profiling phase (i.e. the knowledge of a single key – see Appendix A). Furthermore, we paid attention to exploit exhaustive templates (i.e. used 256 profiles per intermediate value attacked) which can be generalized to any leakage function and corresponds to the worst-case time complexity.

2   Soft Analytical Side-Channel Attacks

We first emphasize the differences between previous solver- or optimizer-based approaches to analytical side-channel attacks and our decoder-based solution. We then describe the BP algorithm and discuss its connection to the exploitation of side-channel leakage. We finally detail how to describe an AES implementation as a factor graph, that can be efficiently decoded by BP. The following descriptions assume a profiled attack scenario, as usual in worst-case evaluations [26].

2.1   Solving (or Optimizing) vs. Decoding

In the course of a profiled side-channel attack, the adversary extracts information from leakage traces. This information comes from the processing of intermediate values throughout the cryptographic computations. By comparing these leakages with previously estimated templates, she obtains for each target value X_i a conditional posterior distribution $\Pr[X_i|L]$. Provided the device is not perfectly side-channel resistant, most of the posterior distributions will have an entropy lower than that of a uniform distribution. In this context, the most interesting pieces of information relate to the encryption key. For this purpose, not only the leakages that directly correspond to key bytes – informally denoted as SPA leakages – are exploited, but also those of intermediate variables that depend on both the key and the (usually known) plaintext – informally denoted as DPA leakages – such as the SBOX outputs in the first AES round, typically. For example, starting from the posterior probability of the output value S_{out} given the leakage L_{out}, one can deduce its image before the substitution layer:

$$\Pr[S_{in} = v | L_{out}] = \Pr[S_{out} = \mathrm{SBOX}(v) | L_{out}].$$

For a known plaintext value P, one can compute a posterior distribution on a key byte K by unrolling the computation one step further:

$$\Pr[K = k | P = p, L_{out}] = \Pr[S_{out} = \mathrm{SBOX}(k \oplus p) | L_{out}].$$

These simple equations show that it is possible to derive information about the key using intermediate variables. Furthermore, one can easily combine the leakage obtained from multiple plaintexts, by marginalizing $\Pr(K = k)$ over the corresponding traces: this is in fact what DC attacks do. Next, and since multiple key-dependent variables can usually be found within cryptographic implementations, a natural problem is to find ways to exploit them efficiently. But this is exactly where the DC strategy faces limitations. Namely, combining the leakage of these intermediate variables is trivial as long as they *only* depend on a single key byte, e.g. the SBOX inputs and outputs in a first block cipher round. One just deals with the additional variable as with an additional plaintext in this case. Taking the example of AES, this can even be extended to the first MIXCOLUMNS operation, if 32-bit key hypotheses are performed by the adversary. But the DC approach is inherently limited to the exploitation of predictable parts of the key. So as soon as the diffusion is complete (which very rapidly occurs in modern ciphers and therefore corresponds to most of their intermediate computations), the leakages are left unexploited by such strategies. This limitation directly leads to the main problem we tackle in this paper, namely: *How to efficiently exploit the leakage of any intermediate variable in a side-channel attack?*

Previous ASCA were a first attempt to answer this question, by trying to solve a system of equations describing the target cryptographic algorithm, complemented with the information extracted. These attacks typically begin by sieving intermediate values, keeping only the most probable ones. A usual approach is to coalesce the leakages by Hamming weights for this purpose. The set of remaining values is then verified one against another (e.g. using heuristic SAT-solvers). Unfortunately, this algebraic approach cannot easily deal with the probability distributions output by TA, which are thus discretized and sieved. Whenever the measurement noise is not negligible, this introduces "errors" that are fatal to algebraic solvers. As mentioned in introduction, optimizers allow mitigating this problem, but are still limited in the exhaustive way they encode the probabilities (which is too expensive for describing more than a couple of AES rounds).

Our method works differently, by operating directly on the posterior distributions of the intermediate values extracted from leakage traces, and propagating the information throughout the computation steps of the algorithm. When attacking a cryptographic implementation, we first build a large graphical model containing the intermediate variables, which are linked by constraints corresponding to the atomic operations executed. For instance, the exclusive-OR and SBOX functions are usually found in software implementations of the AES. Next, the goal is to find the marginal distribution of the key, given the distributions of all the intermediate variables. While this is generally a hard problem, we observe that an important feature of cryptographic algorithms is that intermediate values tend to appear only in a few places. A similar behavior is present in Gallager codes [7], also called Low-Density Parity Check codes (LDPC). In such a code, codeword bits are linked together by a small number of parity constraints (i.e. linear in the codeword size). Decoding such a construction is generally performed via application of the BP algorithm, also known as sum-product algorithm. Our application in the following sections is a (conceptually) simple extension, where values are not limited to bits, and parity constraints go beyond exclusive-ORs.

2.2   The Belief-Propagation Algorithm

Our description of the BP algorithm is largely based on the (excellent) description provided in [14, chapter 26]. Let us consider a set of N variables $\mathbf{x} \equiv \{x_n\}_{n=1}^{N}$, and define a function P^* of $\mathbf{x}$ which is a product of M factors:

$$P^*(\mathbf{x}) = \prod_{m=1}^{M} f_m(\mathbf{x}_m),$$

where each factor $f_m(\mathbf{x}_m)$ is a function of a subset $\mathbf{x}_m$ of $\mathbf{x}$. The P^* function is typically depicted using a *factor graph*, in which circles correspond to variables x_i and squares to functions f_m. An edge is drawn between x_i and f_m if $x_i \in \mathbf{x}_m$, meaning that the m-th factor depends on the i-th variable. For example, the parity functions and factor graph of a simple 3-repetition code are shown below:

$$f_1(x_1) \quad = \Pr(x_1 = 1)$$
$$f_2(x_2) \quad = \Pr(x_2 = 1)$$
$$f_3(x_3) \quad = \Pr(x_3 = 1)$$
$$f_4(x_1, x_2) = \begin{cases} 1 & \text{if } x_1 \oplus x_2 = 0 \\ 0 & \text{otherwise} \end{cases}$$
$$f_5(x_2, x_3) = \begin{cases} 1 & \text{if } x_2 \oplus x_3 = 0 \\ 0 & \text{otherwise} \end{cases}$$

The task we are interested in is that of *marginalization*. That is, we aim to be able to compute the following function:

$$Z_n(x_n) = \sum_{\mathbf{x}, x_n = x_n} P^*(\mathbf{x}),$$

and more importantly it normalized version $P_n(x_n) = Z_n(x_n)/Z$, where:

$$Z = \sum_{\mathbf{x}} \prod_{m=1}^{M} f_m(\mathbf{x}_m).$$

These tasks are intractable in general. Even when the factor functions are limited to three variables, the cost of computing the exact marginal is believed to grow exponentially with the number of variables N. The BP algorithm can circumvent this problem and compute marginals efficiently as long as the factor graph is tree-like. We will denote by $\mathcal{N}(m)$ the set of variables involved in factor f_m, by $\mathcal{M}(n)$ the set of factors where variable x_n appears, and shorthand the set of variables in $\mathbf{x}_m$ with x_n excluded as: $\mathbf{x}_m \setminus n \equiv \{x_{n'} : n' \in \mathcal{N}(m) \setminus n\}$. The algorithm works by passing two types of messages along the edges of the factor graph, from variables to factors ($q_{n \to m}$) and from factors to variables ($r_{m \to n}$). The sets of messages are updated using two rules:

$$q_{n \to m}(x_n) = \prod_{m' \in \mathcal{M}(n) \setminus m} r_{m' \to n}(x_n).$$

$$r_{m \to n}(x_n) = \sum_{\mathbf{x}_m \setminus n} \left(f_m(\mathbf{x}_m) \prod_{n' \in \mathcal{N}(m) \setminus n} q_{n' \to m}(x'_n) \right).$$

Convergence should occur after a finite number of iterations, at most equal to the longest path. Once the network has converged, the marginal function (also called *belief*) of a variable x_n can be recovered by multiplying together all incoming messages at the corresponding node:

$$Z_n(x_n) = \prod_{m \in \mathcal{M}(n)} r_{m \to n}(x_n).$$

The normalized value $P_n(x_n) = Z_n(x_n)/Z$ is easily obtained by summing together the marginal functions $Z = \sum_{x_n} Z_n(x_n)$. As already mentioned, the BP algorithm returns the exact marginals as long as the factor graph is a tree-shaped graphical model. Yet, in many useful cases such as decoding, the graph contains cycles. Fortunately, BP can be applied directly on general factor graphs as well, raising the so-called "loopy" BP. While this version does not guarantee to return the correct marginals, and may even not converge to a fixed point in some cases, it usually gives results that are good enough for most applications.

2.3   Efficient Representation of an AES Implementation

Our method for SASCA consists in an application of the BP algorithm to the decoding of keys using plaintexts, ciphertexts and side-channel traces. In this section, we illustrate it in the context of an implementation of the AES in an 8-bit device. For this purpose, the x_i variables defined in the description of the BP algorithm will represent the intermediate values handled by the cryptographic algorithm, and the parity functions will be separated into two sets:

- The first set corresponds to the *a priori* knowledge on the variables acquired through side-channel leakages, denoted as $f_i(x_i) = \Pr[x_i = v|L]$.
- The second set corresponds to the operations executed by the implementation. In the case of a binary operation $\text{OP}(x_{i_1}, x_{i_2})$, the function is defined by:

$$f_i(x_{i_1}, x_{i_2}, x_{i_3}) = \begin{cases} 1 & \text{if } \text{OP}(x_{i_1}, x_{i_2}) = x_{i_3}, \\ 0 & \text{otherwise.} \end{cases}$$

Based on these notations, an adversary first has to encode the AES computations in a form that is compatible with the BP algorithm. For illustration, and because it is publically available, we will describe how to build a factor graph for the AES FURIOUS implementation (http://point-at-infinity.org/avraes).

Concretely, our program takes in a description which is very similar to the assembly code of AES FURIOUS, with the memory related operations left out, but where any assignment requires a newly named variable. Namely, variable nodes are denoted by names starting with a capital letter, such as K[2,4]_0 for the intermediate key in row 2 and column 4 of key scheduling round 0, which also happens to be the second master key byte (noted $K_{2,4}^0$ in the factor graph), or SB[2,1]_0 for the SBOX output in row 2 and column 1 of round 0 ($SB_{2,1}^0$ in the factor graph). These variable nodes correspond to intermediate values computed during encryption, such as the state (ST), key addition or MIXCOLUMNS intermediate results (AK and MC), outputs of XTIME operations (XT), ... Besides, factor node names start with an underscore such as _Xor (exclusive OR) and _Xtime (polynomial multiplication by x). They correspond to instructions executed during encryption. For example, Table 1 gives samples of the correspondence between the assembly code, input description and factor graph. Note that the factor nodes for the prior probabilities of the variables are not drawn.

Table 1. Factor graph representation of an AES encryption

Assembly code	Graph description	Factor graph

ld H1, Y+
eor ST11, H1
mov ZL, ST11
lpm ST11, Z
	*
	_Xor AK[1,1]_0 ST[1,1]_0 K[1,1]_0
	*
	_Sbox SB[1,1]_0 AK[1,1]_0
		$K_{1,1}^0$ $ST_{1,1}^0$ $AK_{1,1}^0$ $SB_{1,1}^0$
		XOR SBOX

mov H3, ST11
eor H3, ST21
mov ZL, H3
lpm H3, Z
	*
	_Xor MC[3,1]_0 SB[1,1]_0 SB[2,1]_0
	*
	_Xtime XT[1,1]_0 MC[3,1]_0
		$SB_{1,1}^0$ $SB_{2,1}^0$ $MC_{3,1}^0$ $XT_{1,1}^0$
		XOR XTIME

mov ZL, ST24 lpm H3, Z eor ST11, H3 eor ST11, H1	* _Sbox SK[1,1]_1 K[2,4]_0 _Xor XK[1,1]_1 SK[1,1]_1 K[1,1]_0 _XorCst K[1,1]_1 XK[1,1]_1 0x1	$K_{2,4}^0$ $SK_{1,1}^1$ $K_{1,1}^0$ $XK_{1,1}^1$ $K_{1,1}^1$ SBOX XOR XORCST

There are two notable differences between SASCA and the classical decoding of LDPC codes. First, variable nodes are not binary digits, but rather elements of

GF(2^8). Second, factor nodes are not limited to exclusive OR's, but may include any of the variety of functions used in cryptographic implementations (e.g. XOR, SBOX, XTIME). However, these factor nodes are not much more complex than for classical decoding, as illustrated with our three previous examples:

$$\text{XOR}(A, B, C) = \begin{cases} 1 & \text{if } A \oplus B = C, \\ 0 & \text{otherwise.} \end{cases}$$

$$\text{SBOX}(A, B) = \begin{cases} 1 & \text{if } A = S(B), \\ 0 & \text{otherwise.} \end{cases}$$

$$\text{XTIME}(A, B) = \begin{cases} 1 & \text{if } A = Xt(B), \\ 0 & \text{otherwise.} \end{cases}$$

This natural representation of operations is very efficient, as opposed to the contrived way AES encryptions are translated to SAT instances (roughly, it corresponds to 1,200 equations and variables in **GF**(2^8) compared to 18,000 equations in 10,000 variables in SAT-based ASCA). Taking advantage of it, the SASCA adversary then tries to compute the key marginal probability for $P_n(K)$ given the leakages. For this purpose, one simply has to incorporate the implicit factor nodes corresponding to prior knowledge on variable nodes, as given by the templates of the side-channel attack. For instance, the factor for the output of the first SBOX in the first round $f_m(SB_{1,1}^0)$ is the posterior distribution $Pr[SB_{1,1}^0|L]$. In addition, any known value (for instance the plaintext bytes) has a prior knowledge with entropy zero, and any value that does not leak (either because it is protected or precomputed) has a uniform prior. Eventually, the loopy BP algorithm propagates information throughout the factor graph: if successful (i.e. in case of convergence), it should return the approximate marginal probabilities of the key bytes $P_n(K_{1,1}^0)$ to $P_n(K_{4,4}^0)$, i.e. the answer we are looking for.

2.4   Attacking with Several Traces

The ability to efficiently exploit (i.e. combine the information of) several leakage traces is one of the reasons that have made DPA attacks so popular – since it typically leads to the noise vs. data complexity tradeoff that is at the core of most side-channel attacks. It also remains one of the main practical issue for ASCA and follow-up works. So far, the only way several traces can be useful is when they are repetitions of the same encryption (without randomizations), so that the noise can be averaged out. By contrast, adding traces corresponding to multiple plaintexts could only be managed with the construction of larger systems, that are too memory consuming for TASCA, and increasing the probability that one piece of hard information in such systems is incorrect for ASCA.

Interestingly, SASCA are able to improve the key recovery success rate with each additional trace observed. Practically, the factor graph used for decoding is first replicated for each trace. Yet, since the master key stays the same during the course of the attack, the part of our factor graph corresponding the key

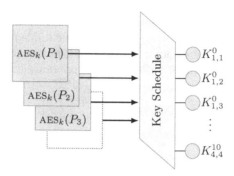

Fig. 1. Factor graph connections for several traces

scheduling also remains constant: it forms a kind of "backbone" where all the encryption rounds connect, as depicted in Figure 1. As a result, whenever several messages are used, the probability distributions are propagated from each replicated graph towards the key schedule. The impact of such propagation is in fact very similar to the one resulting from using several traces in a classical TA, where probabilities are multiplied together and the success rate increases.

3   Experimental Results

We now validate the method described in the previous section with illustrative simulated attacks against the AES FURIOUS implementation. For this purpose, we assume a setup that is essentially similar to the one used to demonstrate the applicability of ASCA to the AES in [22]. The only difference is that we will consider implementations with and without the key scheduling leakages. As previously explained (and illustrated in Table 1), all the operations found in the assembly code are translated into factor nodes, excluding memory related operations. For illustration, we considered Hamming weight leakages affected by a noise of variance σ_n^2, but the attack is independent of this choice: any function could be incorporated without performance penalty. The only important parameter in our case is the informativeness of the leakages which, in the first-order setting we investigate, can be measured with a Signal-to-Noise Ratio (SNR) [16]. Since the signal (i.e. variance) of a Hamming weight leakage function for 8-bit intermediate values equals 2, one can simply derive the SNR as $2/\sigma_n^2$. For illustration, we compared our results with the ones of two standard TA. Namely, one univariate exploiting only the first-round S-box output leakages, and one bivariate exploiting the first-round S-box input and output leakages.

The results of our experiments are shown in figure 2. The x-axis corresponds to the number of messages used for the attack (in log scale), and the y-axis is a stack of success rate curves for decreasing SNRs (i.e. increasing noise levels). An alternative view is provided in Figure 3, which sums up these simulation results by showing the data complexity gains of SASCA over TA. It appears from both figures that these gains are significant and consistently observed for

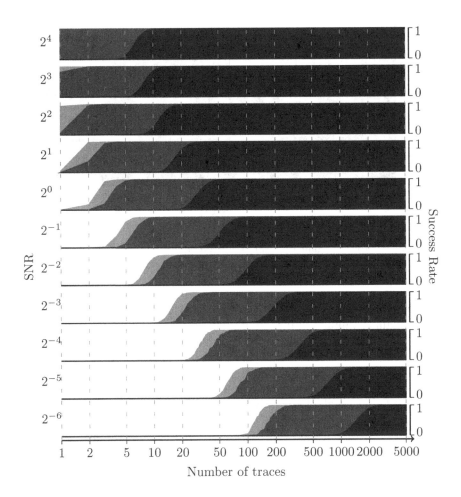

Fig. 2. Attacks results for our simulated FURIOUS implementation. Each graph gives the success rate (SR, ranging from 0 to 1) for a given signal-to-noise ratio (SNR, ranging from 2^4 down to 2^{-6}) as a function of the number of traces (in logarithmic scale, ranging from 1 to 5000). The attacks are:

- univariate TA targetting the SBOX output (in dark gray ■),
- bivariate TA targetting the SBOX input and output (in blue ■),
- SASCA attack ignoring the key schedule leakages (in violet ■),
- SASCA attack exploiting all the intermediate values (in orange ■).

any noise level. Eventually, the unknown inputs and outputs scenario is detailed for SASCA in Figure 4. We see that its impact is limited if the key scheduling leaks (confirming the results from [15]) and more significant otherwise.

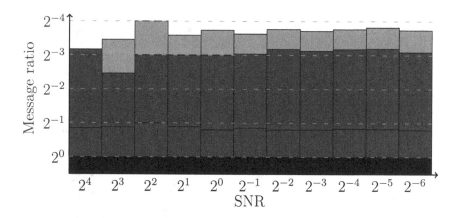

Fig. 3. Data complexity gain of SASCA compared to TA given as the fraction of measurements needed to reach a success rate of 0.9 (same colors as in Figure 2)

Discussion. Compared to previous results in ASCA/TASCA, our new tools bring two main advantages. First, from the SNR point of view, these works were typically limited to scenarios where a single leakage trace was enough to recover the master key (i.e. to SNRs $> 2^2$). We can deal with any SNR. Second, the time and memory complexity of the BP decoding is much improved compared to SAT-solver based ASCA and optimizer-based TASCA. Our implementation deals with a factor graph of size proportional to the number of messages, with a relatively high (yet easily tractable in practice) constant of approximately 16M per message. Its computation time is proportional to both the diameter of the graph (constant after the second message) which sets the number of decoding iterations, and the number of measurements which sets the amount of messages exchanged at each iteration. This makes the evolution of the time and memory complexity of SASCA quite comparable to the one in divide and conquer TA (i.e. linear in the number of messages). Yet, decoding the AES encryption factor graph with the BP algorithm implies a larger computation time of approximately one second per message in our prototype implementation, running on an Intel i7-2720QM. This (constant) overhead is the main penalty to enjoy the substantially smaller data complexities of SASCA (i.e. similar to ASCA/TASCA) which is, as expected, the main advantage of analytical strategies over DC ones.

As detailed in Appendix A, the practical relevance of such attacks is quite similar to TA, since it requires the same profiling assumptions (i.e. the knowledge of a single key). Admittedly, the profiling effort is significantly more expensive for SASCA, since it requires characterizing all the target intermediate values. But since all these target values can be profiled independently, building their templates can be done quite efficiently (with essentially the same amount of measurements as needed to characterize the first-round operations exploited in TA), and is easily automated with standard side-channel attack techniques.

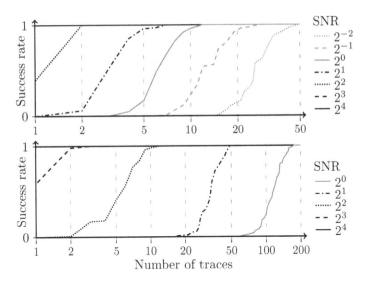

Fig. 4. SASCA with unknown input and output for different SNRs.The x-axis is the number of traces used for the attack (in log scale), and the y-axis gives the probability of key recovery. The top graph corresponds to a leaky key schedule, and the lower graph gives the results for a leak-free key schedule.

4   Conclusions

By modeling the side-channel analysis problem adequately, SASCA bring the missing link between standard DC distinguishers and analytical strategies for key recoveries. As a result and for the first time, we are able to efficiently exploit the probabilistic information of all the leaking operations in a software implementation. Our resulting attacks are optimal in data complexity and efficient in time and memory. Yet, we note that the tools exploited in this first instantiation of SASCA can certainly be improved. For example, the BP algorithm performs too many computations for our needs. Indeed, it propagates every distribution throughout the factor graph whereas in practice, we are mostly interested in the key. Hence, further works could exploit the propagation of messages only towards the schedule (i.e. perform Bayesian inference). This would additionally allow the attack to be performed one message at a time, by accumulating information retrieved from each trace onto the nodes of the key schedule, hence reducing the memory requirements to that of a single trace (i.e. 16M).

In view of the improved noise robustness of SASCA, an important open problem is to determine whether the strong results obtained with this new type of analytical strategy also apply to implementations protected with countermeasures. Masking, shuffling and leakage-resilient cryptography appear as the most interesting targets in this respect. Besides, the experiments in this work considered a worst-case scenario where the adversary could take advantage of all the leaking operations of an AES implementation (i.e. assuming the knowledge of the source code, essentially). But the investigation of an intermediate scenario where the adversary would exploit less leaking observations (e.g. the ones he could guess without knowing the source code) and its resulting time and data complexity is another interesting scope for additional investigations.

Acknowledgements. François-Xavier Standaert is a research associate of the Belgian Fund for Scientific Research (FNRS-F.R.S.). This work has been funded in parts by the European Commission through the ERC project 280141 (acronym CRASH) and by the PAVOIS project (ANR 12 BS02 002 01).

References

1. Bogdanov, A., Kizhvatov, I., Pyshkin, A.: Algebraic methods in side-channel collision attacks and practical collision detection. In: Chowdhury, D.R., Rijmen, V., Das, A. (eds.) INDOCRYPT 2008. LNCS, vol. 5365, pp. 251–265. Springer, Heidelberg (2008)
2. Brier, E., Clavier, C., Olivier, F.: Correlation power analysis with a leakage model. In: Joye, Quisquater (eds.) [11], pp. 16–29
3. Carlet, C., Faugère, J.-C., Goyet, C., Renault, G.: Analysis of the algebraic side channel attack. Journal of Cryptographic Engineering 2(1), 45–62 (2012)
4. Chari, S., Rao, J.R., Rohatgi, P.: Template attacks. In: Kaliski Jr., B.S., Koç, Ç.K., Paar, C. (eds.) CHES 2002. LNCS, vol. 2523, pp. 13–28. Springer, Heidelberg (2003)
5. Durvaux, F., Renauld, M., Standaert, F.-X., van Oldeneel tot Oldenzeel, L., Veyrat-Charvillon, N.: Efficient removal of random delays from embedded software implementations using hidden markov models. In: Mangard, S. (ed.) CARDIS 2012. LNCS, vol. 7771, pp. 123–140. Springer, Heidelberg (2013)
6. Eisenbarth, T., Paar, C., Weghenkel, B.: Building a side channel based disassembler. Transactions on Computational Science 10, 78–99 (2010)
7. Gallager, R.G.: Low-density parity-check codes. IRE Transactions on Information Theory 8(1), 21–28 (1962)
8. Gérard, B., Standaert, F.-X.: Unified and optimized linear collision attacks and their application in a non-profiled setting. In: Prouff, Schaumont (eds.) [20], pp. 175–192
9. Gierlichs, B., Batina, L., Tuyls, P., Preneel, B.: Mutual information analysis. In: Oswald, E., Rohatgi, P. (eds.) CHES 2008. LNCS, vol. 5154, pp. 426–442. Springer, Heidelberg (2008)
10. Guo, S., Zhao, X., Zhang, F., Wang, T., Shi, Z.J., Standaert, F.-X., Ma, C.: Exploiting the incomplete diffusion feature: A specialized analytical side-channel attack against the AES and its application to microcontroller implementations. IEEE Transactions on Information Forensics and Security 9(6), 999–1014 (2014)
11. Joye, M., Quisquater, J.-J. (eds.): CHES 2004. LNCS, vol. 3156. Springer, Heidelberg (2004)
12. Karlof, C., Wagner, D.: Hidden markov model cryptanalysis. In: Walter, C.D., Koç, Ç.K., Paar, C. (eds.) CHES 2003. LNCS, vol. 2779, pp. 17–34. Springer, Heidelberg (2003)
13. Kocher, P., Jaffe, J., Jun, B.: Differential power analysis. In: Wiener, M. (ed.) CRYPTO 1999. LNCS, vol. 1666, pp. 388–397. Springer, Heidelberg (1999)

14. MacKay, D.J.C.: Information theory, inference, and learning algorithms. Cambridge University Press (2003)
15. Mangard, S.: A simple power-analysis (SPA) attackon implementations of the AES key expansion. In: Lee, P.J., Lim, C.H. (eds.) ICISC 2002. LNCS, vol. 2587, pp. 343–358. Springer, Heidelberg (2003)
16. Mangard, S., Oswald, E., Standaert, F.-X.: One for all - all for one: unifying standard differential power analysis attacks. IET Information Security 5(2), 100–110 (2011)
17. Mohamed, M.S.E., Bulygin, S., Zohner, M., Heuser, A., Walter, M., Buchmann, J.: Improved algebraic side-channel attack on AES. Journal of Cryptographic Engineering 3(3), 139–156 (2013)
18. Oren, Y., Kirschbaum, M., Popp, T., Wool, A.: Algebraic side-channel analysis in the presence of errors. In: Mangard, S., Standaert, F.-X. (eds.) CHES 2010. LNCS, vol. 6225, pp. 428–442. Springer, Heidelberg (2010)
19. Oren, Y., Renauld, M., Standaert, F.-X., Wool, A.: Algebraic side-channel attacks beyond the hamming weight leakage model. In: Prouff, Schaumont (eds.) [20], pp. 140–154
20. Prouff, E., Schaumont, P. (eds.): CHES 2012. LNCS, vol. 7428. Springer, Heidelberg (2012)
21. Renauld, M., Standaert, F.-X.: Algebraic side-channel attacks. In: Bao, F., Yung, M., Lin, D., Jing, J. (eds.) Inscrypt 2009. LNCS, vol. 6151, pp. 393–410. Springer, Heidelberg (2010)
22. Renauld, M., Standaert, F.-X., Veyrat-Charvillon, N.: Algebraic side-channel attacks on the AES: Why time also matters in DPA. In: Clavier, C., Gaj, K. (eds.) CHES 2009. LNCS, vol. 5747, pp. 97–111. Springer, Heidelberg (2009)
23. Schindler, W., Lemke, K., Paar, C.: A stochastic model for differential side channel cryptanalysis. In: Rao, J.R., Sunar, B. (eds.) CHES 2005. LNCS, vol. 3659, pp. 30–46. Springer, Heidelberg (2005)
24. Schramm, K., Leander, G., Felke, P., Paar, C.: A collision-attack on AES: Combining side channel- and differential-attack. In: Joye, Quisquater (eds.) [11], pp. 163–175
25. Schramm, K., Wollinger, T., Paar, C.: A new class of collision attacks and its application to DES. In: Johansson, T. (ed.) FSE 2003. LNCS, vol. 2887, pp. 206–222. Springer, Heidelberg (2003)
26. Standaert, F.-X., Malkin, T., Yung, M.: A unified framework for the analysis of side-channel key recovery attacks. In: Joux, A. (ed.) EUROCRYPT 2009. LNCS, vol. 5479, pp. 443–461. Springer, Heidelberg (2009)
27. Veyrat-Charvillon, N., Gérard, B., Renauld, M., Standaert, F.-X.: An optimal key enumeration algorithm and its application to side-channel attacks. In: Knudsen, L.R., Wu, H. (eds.) SAC 2012. LNCS, vol. 7707, pp. 390–406. Springer, Heidelberg (2013)
28. Veyrat-Charvillon, N., Gérard, B., Standaert, F.-X.: Security evaluations beyond computing power. In: Johansson, T., Nguyen, P.Q. (eds.) EUROCRYPT 2013. LNCS, vol. 7881, pp. 126–141. Springer, Heidelberg (2013)
29. Zhang, F., Zhao, X., Guo, S., Wang, T., Shi, Z.: Improved algebraic fault analysis: A case study on piccolo and applications to other lightweight block ciphers. In: Prouff, E. (ed.) COSADE 2013. LNCS, vol. 7864, pp. 62–79. Springer, Heidelberg (2013)

A Attack Requirements

In this section, we provide a brief discussion of the profiling step that precedes the application of SASCA. In particular, we argue that the profiling overhead and required knowledge for this purpose are similar to those of standard TA.

Profiling Overhead. Similarly to classical TA, SASCA require profiling the leakage corresponding to their target intermediate values. In this respect, the only difference is that they can take advantage of many such values, whereas DC strategies only exploit the first round(s) leakages. In general, one can assume that all target intermediate values leak a similar amount of information. And if it is not the case, it is usually the first round(s) leakages that have lower SNRs. As a result, and given that the set of profiling traces corresponds to random inputs, one can essentially build all the SASCA templates with the same traces as for a TA, by simply re-organizing these traces according to the target intermediate values. This process can be automated based on the implementation knowledge, and its computational cost grows linearly with the number of targets. Concretely, this cost should be small for most concrete implementations, and if needed can be speeded up by assuming sets of intermediate values to leak according to the same model (possibly at the cost of some information loss).

Required Knowledge. Since templates are built by grouping the leakage traces according to some target intermediate values, it requires being able to predict these values. Both for TA and SASCA, this is usually achieved thanks to some key knowledge (or a profiling device). So both attacks can be based on the same assumptions. In fact, their main difference is that any intimate knowledge of the target implementations can – *but does not have to* – be exploited by SASCA (while, e.g. the middle round leakages are useless for DC attacks). The experiments in this paper consider a worst-case scenario where the adversary knows the implementation source code. Another extreme scenario would be to consider only "standard" attack points that can be guessed from the algorithms specifications (e.g. S-boxes inputs/outputs), which would reduce the gain of SASCA compared to TA. Any intermediate situation could be investigated, corresponding to various tradeoffs between implementation details and attack efficiency.

On the Enumeration of Double-Base Chains with Applications to Elliptic Curve Cryptography

Christophe Doche

Department of Computing
Macquarie University, Australia
christophe.doche@mq.edu.au

Abstract. The Double-Base Number System (DBNS) uses two bases, 2 and 3, in order to represent any integer n. A Double-Base Chain (DBC) is a special case of a DBNS expansion. DBCs have been introduced to speed up the scalar multiplication $[n]P$ on certain families of elliptic curves used in cryptography. In this context, our contributions are twofold. First, given integers n, a, and b, we outline a recursive algorithm to compute the number of different DBCs with a leading factor dividing $2^a 3^b$ and representing n. A simple modification of the algorithm allows to determine the number of DBCs with a specified length as well as the actual expansions. In turn, this gives rise to a method to compute an optimal DBC representing n, i.e. an expansion with minimal length. Our implementation is able to return an optimal expansion for most integers up to 2^{60} bits in a few minutes. Second, we introduce an original and potentially more efficient approach to compute a random scalar multiplication $[n]P$, based on the concept of controlled DBC. Instead of generating a random integer n and then trying to find an optimal, or at least a short DBC to represent it, we propose to directly generate n as a random DBC with a chosen leading factor $2^a 3^b$ and length ℓ. To inform the selection of those parameters, in particular ℓ, which drives the trade-off between the efficiency and the security of the underlying cryptosystem, we enumerate the total number of DBCs having a given leading factor $2^a 3^b$ and a certain length ℓ. The comparison between this total number of DBCs and the total number of integers that we wish to represent a priori provides some guidance regarding the selection of suitable parameters. Experiments indicate that our new Near Optimal Controlled DBC approach provides a speedup of at least 10% with respect to the NAF for sizes from 192 to 512 bits. Computations involve elliptic curves defined over $\mathbb{F}_p$, using the Inverted Edwards coordinate system and state of the art scalar multiplication techniques.

Keywords: Double-base number system, elliptic curve cryptography.

1   Introduction

1.1   Elliptic Curve Cryptography

An *elliptic curve* E defined over a field K is a nonsingular projective plane cubic together with a point with coordinates in K. For cryptographic applications, the

P. Sarkar and T. Iwata (Eds.): ASIACRYPT 2014, PART I, LNCS 8873, pp. 297–316, 2014.

field K is always finite. In practice, it is a large prime field $\mathbb{F}_p$ or a binary field $\mathbb{F}_{2^d}$. We refer to [23] for a mathematical presentation of elliptic curves and to [1, 16] for a discussion focused on cryptographic applications.

There are different ways to represent the curve E, in particular with a Weierstraß equation or in Edwards form [13, 3]. Irrespective of the representation, the set of points lying on the curve E can be endowed with an abelian group structure. This property has been exploited for about twenty five years to implement public-key cryptographic primitives.

The core operation in elliptic curve cryptography is the *scalar multiplication*, which consists in computing $[n]P$ given a point P on the curve E and some integer n. Several methods exist relying on different representations of n. One of the simplest approach relies on the *non-adjacent form* (NAF) [20, 19], which allows to compute $[n]P$ with t doublings and $t/3$ additions on average, where t is the binary length of n. The approach discussed next is more sophisticated and has recently received increasing attention.

1.2   Double-Base Number System

The *Double-Base Number System* (DBNS) was introduced by Dimitrov and Cooklev [5] and later used in the context of elliptic curve cryptography [6]. With this system, an integer n is represented as

$$n = \sum_{i=1}^{\ell} c_i 2^{a_i} 3^{b_i}, \text{ with } c_i \in \{-1, 1\}. \tag{1}$$

This representation is highly redundant and an expansion can easily be found with a greedy-type approach. The principle is to find at each step the best approximation of a given integer in terms of a $\{2,3\}$-*integer, i.e.* an integer of the form $2^a 3^b$. Then compute the difference and reapply the process until we reach zero.

Example 1. *Applying this approach to* $n = 542788$, *we find that*

$$542788 = 2^8 3^7 - 2^3 3^7 + 2^4 3^3 - 2.3^2 - 2.$$

In [7], Dimitrov *et al.* show that for any integer n, this greedy approach returns a DBNS expansion of n having at most $O\left(\frac{\log n}{\log \log n}\right)$ terms. However, in general this system is not well suited for scalar multiplications. For instance, in order to compute $[542788]P$ from the DBNS expansion given in Example 1, it seems that we need more than 8 doublings and 7 triplings unless we can use extra storage to keep certain intermediate results. But, if we are lucky enough that the terms in the expansion can be ordered in such a way that their powers of 2 and 3 are both decreasing, then it becomes trivial to obtain $[n]P$.

1.3   Double-Base Chain

The concept of *Double-Base Chain* (DBC), introduced in [6], corresponds to an expansion of the form

$$\sum_{i=1}^{\ell} c_i 2^{a_i} 3^{b_i}, \quad \text{with } c_i \in \{-1, 1\} \tag{2}$$

$$\text{such that } a_1 \geqslant a_2 \geqslant \cdots \geqslant a_\ell \text{ and } b_1 \geqslant b_2 \geqslant \cdots \geqslant b_\ell. \tag{3}$$

Equivalently, (3) means that $2^{a_\ell} 3^{b_\ell} \mid \cdots \mid 2^{a_2} 3^{b_2} \mid 2^{a_1} 3^{b_1}$. It guarantees that exactly a_ℓ doublings, b_ℓ triplings, $\ell - 1$ additions, and at most two variables are sufficient to compute $[n]P$. It is straightforward to adapt the greedy algorithm to return a DBC.

Example 2. *A modified greedy algorithm returns the following DBC*

$$542788 = 2^{14}3^3 + 2^{12}3^3 - 2^{10}3^2 - 2^{10} + 2^6 + 2^2.$$

The DBC expansion returned by the greedy approach is always at least as long than its DBNS counterpart. Furthermore, it has been shown in [18] that for any size t, there exists a t-bit integer n such that any DBC representing n needs at least $\Omega(t)$ terms. But the DBC has the advantage to offer a much more direct and easy way to compute a scalar multiplication. The most natural approach is probably to proceed from right-to-left. With this method, each term $2^{a_i} 3^{b_i}$ is computed individually and all the terms are added together. This can be implemented using two variables. The left-to-right method, which can be seen as a Horner-like scheme, needs only one variable. Simply initialize it with $[c_1 2^{a_1-a_2} 3^{b_1-b_2}]P$, then add $c_2 P$ and apply $[2^{a_2-a_3} 3^{b_2-b_3}]$ to the result. Repeating this process eventually gives $[n]P$, as illustrated with the chain of Example 2

$$[542788]P = [2^2]\big([2^4]\big([2^4]\big([3^2]\big([2^2 3]([2^2]P + P) - P\big) - P\big) + P\big) + P\big).$$

Note that there are other methods to compute a DBC, see for instance a tree-based algorithm developed in [9]. There exist also several variants and generalizations of the DBC. For instance, the extended DBC [10] relies on nontrivial coefficients and precomputed points in order to obtain shorter chains. There is also a notion of joint DBC [11, 12] for double scalar multiplications of the form $[n]P + [m]Q$. Next we are interested to find the best possible chains for a given integer. To this end, we introduce the following.

Definition 1. *We call the largest {2,3}-integer of a DBC chain in absolute value, i.e. $2^{a_1} 3^{b_1}$ in (2), the* leading factor *of the chain. It encapsulates the total number of doublings and of triplings necessary to compute $[n]P$.*

Among all the different DBCs with a leading factor dividing $2^a 3^b$ and representing n, the DBCs with minimal length play a special role as they minimize the number of additions required to compute $[n]P$. This observation gives rise to the following definition.

Definition 2. *Given integers a, b, and n, a DBC with a leading factor dividing $2^a 3^b$ and representing n is said to be* optimal *for n, if its length ℓ is minimal across all the DBCs with leading factor dividing $2^a 3^b$ and representing n.*

Remark 1. *For the purpose of this study, we slightly modify the definitions of a Double-Base expansion and of a DBC so that we can precisely and meaningfully enumerate them. Concretely, we require that each term $2^{a_i} 3^{b_i}$ appears at most once in any expansion or chain. In practice, expansions always fulfill this property. Also, this requirement is not a real constraint since $2^{a_i} 3^{b_i} + 2^{a_i} 3^{b_i} = 2^{a_i+1} 3^{b_i}$. From now on, when we use the terms double-base expansion or DBC, this restriction is implied.*

Definition 3. *An* unsigned Double-Base Chain *is a DBC of the form (2) such that all the coefficients c_i's are equal to 1 and satisfying (3).*

Some properties of the set containing all the unsigned DBCs of a given integer n, in particular its structure and cardinality, are studied in [17]. Next, we investigate the number of signed DBCs representing a given integer.

2   Enumerating DBCs Representing a Given Integer

2.1   Partition Problem

Given an integer n, the number $p(n)$ of partitions of n of the form

$$n = d_k + \cdots + d_2 + d_1 \quad \text{with} \quad d_1 \mid d_2 \mid \cdots \mid d_k$$

is studied by Erdős and Loxton in [14]. The authors also introduce $p_1(n)$ as the number of partitions of n of the form $n = d_k + \cdots + d_2 + 1$ with $d_1 \mid d_2 \mid \cdots \mid d_k$. They observe that $p(n) = p_1(n) + p_1(n+1)$ and that

$$p_1(n) = \sum_{d \mid n-1, d>1} p_1 \left(\frac{n-1}{d} \right).$$

2.2   Enumerating DBCs

Mimicking their approach, we introduce $q(a, b, n)$, the number of signed partitions of n of the form

$$n = d_k \pm d_{k-1} \pm \cdots \pm d_2 \pm d_1 \quad \text{with} \quad d_1 \mid d_2 \mid \cdots \mid d_k \mid 2^a 3^b.$$

Clearly, $q(a, b, n)$ corresponds to the number of DBCs with a leading factor dividing $2^a 3^b$ and representing n. Note that in the signed version, it is necessary to take into account a and b, the largest powers of 2 and 3. Indeed, we observe that $1 = 2^k - \sum_{i=0}^{k-1} 2^i$ for any $k > 0$. This shows that the number of signed representations of any integer is infinite. Obviously, the problem disappears when we bound the leading factor of an expansion by $2^a 3^b$. Similarly, we introduce $q_1(a, b, n)$ as the number of partitions of n of the form

$$n = d_k \pm d_{k-1} \pm \cdots \pm d_2 + 1 \quad \text{with} \quad d_2 \mid \cdots \mid d_k \mid 2^a 3^b$$

and $q_{\bar{1}}(a, b, n)$ as the number of partitions of n of the form

$$n = d_k \pm d_{k-1} \pm \cdots \pm d_2 - 1 \quad \text{with} \quad d_2 \mid \cdots \mid d_k \mid 2^a 3^b.$$

In the following, we denote the valuation of u at 2 and 3 by $\mathrm{val}_2(u)$ and $\mathrm{val}_3(u)$, respectively.

Proposition 1. *We have*

1. $q(a, b, n) = q_1(a, b, n) + q_{\bar{1}}(a, b, n) + q_{\bar{1}}(a, b, n - 1).$
2.

$$q_1(a, b, n) = \sum_{\substack{d \mid \gcd(n-1, 2^a 3^b) \\ d > 1}} q_1\left(a - \mathrm{val}_2(d), b - \mathrm{val}_3(d), \frac{n-1}{d}\right)$$

$$+ \sum_{\substack{d \mid \gcd(n-1, 2^a 3^b) \\ d > 1}} q_{\bar{1}}\left(a - \mathrm{val}_2(d), b - \mathrm{val}_3(d), \frac{n-1}{d}\right)$$

3.

$$q_{\bar{1}}(a, b, n) = \sum_{\substack{d \mid \gcd(n+1, 2^a 3^b) \\ d > 1}} q_1\left(a - \mathrm{val}_2(d), b - \mathrm{val}_3(d), \frac{n+1}{d}\right)$$

$$+ \sum_{\substack{d \mid \gcd(n+1, 2^a 3^b) \\ d > 1}} q_{\bar{1}}\left(a - \mathrm{val}_2(d), b - \mathrm{val}_3(d), \frac{n+1}{d}\right).$$

4. $q_1(a, b, 1) = 1$, *if* $a \geqslant 0$ *and* $b \geqslant 0$, *and* $q_1(a, b, 1) = 0$ *otherwise.*
5. $q_{\bar{1}}(a, b, 1) = a$, *if* $a \geqslant 0$ *and* $b \geqslant 0$, *and* $q_{\bar{1}}(a, b, 1) = 0$ *otherwise.*

Proof.

1. We observe that any DBC representing n must end by 1, -1, or a term that is a nontrivial divisor of the leading factor. These three sets form a partition of all the DBCs representing n. By definition, the cardinality of the first two sets is $q_1(a, b, n)$ and $q_{\bar{1}}(a, b, n)$. There exists a bijection between this last set and the set of DBCs representing $n - 1$ ending with -1. Note that we could also compute $q(a, b, n)$ as $q_1(a, b, n) + q_{\bar{1}}(a, b, n) + q_1(a, b, n + 1)$.
2. Let us consider a DBC with a leading factor dividing $2^a 3^b$, ending with 1, and representing n. Then this DBC can be written $\sum_i c_i 2^{a_i} 3^{b_i} \pm d + 1$ where $d > 1$ and $d \mid 2^{a_i} 3^{b_i}$ for all i. If we denote $\alpha = \mathrm{val}_2(d)$ and $\beta = \mathrm{val}_3(d)$, we see that the chain $\sum_i c_i 2^{a_i - \alpha} 3^{b_i - \beta} \pm 1$ represents $(n - 1)/d$. We note that its leading

factor must divide $2^{a-\alpha}3^{b-\beta}$ and it ends by 1 or -1. Also, by construction, the factor d is a divisor of $n-1$ and of 2^a3^b. Reciprocally, take $d = 2^\alpha3^\beta$ a common divisor of $n-1$ and 2^a3^b. Then for any DBC with a leading factor dividing $2^{a-\alpha}3^{b-\beta}$ and representing $(n-1)/d$, it corresponds a unique DBC with a leading factor dividing 2^a3^b, finishing with 1 and representing n.

3. The proof is similar to 2., except that we need to consider DBCs of the form $\sum_i c_i 2^{a_i} 3^{b_i} \pm d - 1$.

4. We assume that each term $2^{a_i}3^{b_i}$ appears at most once, cf Remark 1. With this constraint in mind, it is easy to check that there is a unique DBC ending with 1 and representing 1, namely the chain 1.

5. Regarding the DBCs representing 1 and ending with -1, we note that for any $k > 0$, we have $2^k - \sum_{i=0}^{k-1} 2^i = 1$. In particular, the previous formula for $k = 1$ up to a gives rise to a total number of a different DBCs with a leading factor dividing 2^a3^b, ending with -1, and representing 1. It is easy to see that there is no other solution. This shows that $q_{\bar{1}}(a,b,1) = a$, when $a \geqslant 0$ and $b \geqslant 0$. □

Using Proposition 1, it is possible to compute $q(a,b,n)$ recursively, for any tuple (a,b,n).

Example 3. *We have $q(14,5,542788) = 2092690$. In other words, there are 2092690 different DBCs with a leading factor dividing $2^{14}3^{10}$ and representing 542788.*

Remark 2. *The approach is highly recursive but precomputing small values can greatly speed up computations. For instance, precomputing $q_1(a,b,n)$ and $q_{\bar{1}}(a,b,n)$ for all $(a,b,n) \in [0,30] \times [0,20] \times [1,1000]$ allows to deal with numbers of size up to 30 bits in a few seconds.*

2.3   Enumerating DBCs of Bounded Length

A simple modification of the algorithm outlined above allows to determine the total number of different DBCs of length less or equal to ℓ with a leading factor dividing 2^a3^b and representing an integer n. Namely, we introduce a new parameter ℓ to keep track of the length of the DBC. It is straightforward to check that

$$q(a,b,\ell,n) = q_1(a,b,\ell,n) + q_{\bar{1}}(a,b,\ell,n) + q_{\bar{1}}(a,b,\ell+1,n-1).$$

Additionally, $q_1(a,b,\ell,n)$ and $q_{\bar{1}}(a,b,\ell,n)$ satisfy relations similar to the ones expressed in Proposition 1. For instance,

$$q_1(a,b,\ell,n) = \sum_{\substack{d|\gcd(n-1,2^a3^b)\\d>1}} q_1\left(a - \mathrm{val}_2(d), b - \mathrm{val}_3(d), \ell-1, \frac{n-1}{d}\right)$$

$$+ \sum_{\substack{d|\gcd(n-1,2^a3^b)\\d>1}} q_{\bar{1}}\left(a - \mathrm{val}_2(d), b - \mathrm{val}_3(d), \ell-1, \frac{n-1}{d}\right).$$

Finally, it is easy to see that $q_1(a, b, \ell, 1) = \min(1, \max(0, \ell))$ and $q_{\bar{1}}(a, b, \ell, 1) = \min(a, \max(0, \ell - 1))$. This gives rise to Algorithms 1 and 2.

Algorithm 1. $q_1(a, b, \ell, n)$

INPUT: An integer n and parameters a, b, and ℓ.
OUTPUT: The number of DBCs representing n, ending with 1, having a leading factor dividing $2^a 3^b$, and a length less than or equal to ℓ.

1.	**if** $n \leqslant 0$ or $a < 0$ or $b < 0$ or $\ell \leqslant 0$ **then return** 0
2.	**else if** $n = 1$ **then**
3.	**if** $a \geqslant 0$ and $b \geqslant 0$ **then return** $\min(1, \max(0, \ell))$
4.	**else return** 0
5.	**else if** $n > 1$ **then**
6.	$D \leftarrow \gcd(n - 1, 2^a 3^b)$
7.	$s \leftarrow 0$
8.	**for** each divisor $d > 1$ of D **do**
9.	$s \leftarrow s + q_1\left(a - \mathrm{val}_2(d), b - \mathrm{val}_3(d), \ell - 1, \frac{n-1}{d}\right)$
10.	$s \leftarrow s + q_{\bar{1}}\left(a - \mathrm{val}_2(d), b - \mathrm{val}_3(d), \ell - 1, \frac{n-1}{d}\right)$
11.	**return** s

Algorithm 2. $q_{\bar{1}}(a, b, \ell, n)$

INPUT: An integer n and parameters a, b, and ℓ.
OUTPUT: The number of DBCs representing n, ending with -1, having a leading factor dividing $2^a 3^b$, and a length less than or equal to ℓ.

1.	**if** $n \leqslant 0$ or $a < 0$ or $b < 0$ or $\ell \leqslant 0$ **then return** 0
2.	**else if** $n = 1$ **then**
3.	**if** $a \geqslant 0$ and $b \geqslant 0$ **then return** $\min\left(a, \max(0, \ell - 1)\right)$
4.	**else return** 0
5.	**else if** $n > 1$ **then**
6.	$D \leftarrow \gcd(n + 1, 2^a 3^b)$
7.	$s \leftarrow 0$
8.	**for** each divisor $d > 1$ of D **do**
9.	$s \leftarrow s + q_1\left(a - \mathrm{val}_2(d, 2), b - \mathrm{val}_3(d, 3), \ell - 1, \frac{n+1}{d}\right)$
10.	$s \leftarrow s + q_{\bar{1}}\left(a - \mathrm{val}_2(d, 2), b - \mathrm{val}_3(d, 3), \ell - 1, \frac{n+1}{d}\right)$
11.	**return** s

Example 4. *Using Algorithms 1 and 2, we see that among the* 2092690 *different DBCs with a leading factor dividing* $2^{14}3^{10}$ *and representing* 542788, *there are three optimal chains of length* 5, 81 *chains of length* 6, 843 *of length* 7, 5005 *of length* 8, 19715 *of length* 9, 56148 *of length* 10, *and so on. The total number is bounded as for instance, there cannot be a DBC of length greater or equal to* 26 *since the leading factor is at most* $2^{14}3^{10}$.

2.4   Optimal DBCs

Using the algorithms described in the previous part, it is simple to determine the optimal length of a DBC representing an integer n with a leading factor dividing 2^a3^b. Simply compute $q(a, b, \ell, n)$ for increasing values of $\ell \geqslant 1$ until a positive cardinality is returned. Also, along with the total number of DBCs, it is possible to return the list of all the actual DBCs representing an integer, by introducing a few simple modifications in the Algorithms 1 and 2. We note that we can further modify Algorithms 1 and 2 so that we compute only the DBCs having a specified length. Also, in case we are only interested in finding an optimal chain for a given integer n, we can implement a simple early abort technique to terminate the search once a DBC of a certain given size has been found. This is possible because these algorithms perform a depth-first search.

Example 5. *Among the three optimal DBCs of length* 5 *with leading factor dividing* $2^{14}3^{10}$ *and representing* 542788, *one is*

$$2^83^7 - 2^63^5 - 2^63^3 + 2^63 + 2^2.$$

The running time of this approach is largely driven by the length of the optimal chain that is returned. Typically, it takes a few seconds for chains of length 12 up to a few hours for length 15. In general, it is practical to determine an optimal DBC for integers of size around 60 to 70 bits. See Section 5.1 and Table 1 for details including actual experiments and timings of our C++ implementation that is available from our homepage, see [8].

So it is clear that computing an optimal DBC for a scalar of size around 200 bits, i.e. the kind of size typically used in elliptic curve cryptography, is completely out of reach with this approach. Instead, we consider another approach to efficiently perform a random scalar multiplication $[n]P$.

3   Enumerating DBCs with Given Parameters

Instead of computing the number of DBCs representing a given integer n, this time we want to count the number of different DBCs with a given leading factor 2^a3^b and a given length ℓ.

Remark 3. *The same problem is straightforward for DBNS expansions. Indeed, we see from (1) that there are* $2^\ell \binom{(a+1)(b+1)}{\ell}$ *different expansions of length* ℓ *and such that* $\max a_i = a$ *and* $\max b_i = b$. *Note that all the expansions are different in this count, but the integers they represent are not necessarily all different.*

It is more involved to determine the number of unsigned DBCs (see Definition 3) and of DBCs with a given leading factor 2^a3^b and a given length ℓ.

3.1   First Properties

Definition 4. *Let $S_\ell(a, b)$ denote the number of unsigned DBCs of length ℓ with a leading factor equal to $2^a 3^b$. Let $T_\ell(a, b)$ denote the number of unsigned DBCs of length ℓ with a leading factor dividing $2^a 3^b$.*

Proposition 2. *Let $\ell \geqslant 1$. We have:*

1. $S_{\ell+1}(a, b) = T_\ell(a, b) - S_\ell(a, b)$.
2.

$$T_{\ell+1}(a, b) = \sum_{i=0}^{a} \sum_{j=0}^{b} \left[(a - i + 1)(b - j + 1) - 1 \right] S_\ell(i, j).$$

3. *$S_\ell(a, b)$ and $T_\ell(a, b)$ are both symmetrical polynomials.*
4. *The leading terms of $S_\ell(a, b)$ and of $T_\ell(a, b)$ are respectively $\frac{(ab)^{\ell-1}}{((\ell-1)!)^2}$ and $\frac{(ab)^\ell}{\ell!^2}$.*

Proof. The first three relations are a simple consequence of the definitions of $S_\ell(a, b)$ and $T_\ell(a, b)$. To prove 4. we first note that $S_\ell(a, b)$ is of degree $2\ell - 2$ and $T_\ell(a, b)$ is of degree 2ℓ. This can be shown by induction based on $S_1(a, b) = 1$, $T_1(a, b) = (a+1)(b+1)$, and using 1. and 2. We can now prove 4. by induction. The property is true for $S_1(a, b)$ and $T_1(a, b)$. Also, by 1. and given that $S_{\ell+1}(a, b)$ and $T_\ell(a, b)$ are of the same degree, it is clear that their leading terms are equal. So by the induction hypothesis, it is clear that the property holds for $S_\ell(a, b)$, for all $\ell \geqslant 1$. Now assuming it holds for $T_{\ell-1}(a, b)$, let us show that it holds for $T_\ell(a, b)$. Using the induction hypothesis, we observe that the leading term of

$$T_\ell(a, b) = \sum_{i=0}^{a} \sum_{j=0}^{b} \left[(a - i + 1)(b - j + 1) - 1 \right] S_{\ell-1}(i, j)$$

is equal to the leading term of

$$\frac{1}{((\ell - 2)!)^2} \sum_{i=0}^{a} \sum_{j=0}^{b} (a - i)(b - j)(ij)^{\ell-2}.$$

Next, we note that the leading term of $\sum_{i=0}^{a} i^k$ is $\frac{1}{(k+1)} a^{k+1}$. We deduce that the leading term of

$$\sum_{i=0}^{a} \sum_{j=0}^{b} (a - i)(b - j)(ij)^{\ell-2}$$

is

$$(ab)^\ell \left(\frac{1}{\ell^2} + \frac{1}{(\ell - 1)^2} - \frac{2}{\ell(\ell - 1)} \right) = \frac{(ab)^\ell}{((\ell - 1)\ell)^2}.$$

It follows that the leading term of $T_\ell(a, b)$ is $\frac{(ab)^\ell}{\ell!^2}$, as expected. $\qquad\square$

Remark 4. *The number of signed DBCs of length ℓ with a leading factor equal to $2^a 3^b$ and dividing $2^a 3^b$ can be easily deduced from $S_\ell(a, b)$ and $T_\ell(a, b)$, respectively. Namely, it is only necessary to multiply by a factor 2^ℓ. Note that all those DBCs represent positive and negative integers. But it is easy to see that the sign of the integer represented by a chain corresponds to the sign of the largest term of the chain. See Lemma 1 in Section 4. So if we are only interested in DBCs representing positive values, the multiplication factor between unsigned and signed DBCs should be $2^{\ell-1}$.*

3.2 Explicit Computations

Recall that $S_1(a, b) = 1$ and $T_1(a, b) = (a + 1)(b + 1)$. Proposition 2 can then be used to explicitly determine the polynomials S_ℓ and T_ℓ of rank $\ell \geqslant 2$ recursively. For instance, we have $S_2(a, b) = ab + a + b$ from 1. and $T_2(a, b) = \frac{1}{4}(ab + 2a + 2b)T_1(a, b)$ using 2. We can then compute $S_3(a, b)$, then $T_3(a, b)$, and so on.

In practice, however, the complexity of those polynomials rapidly grows with ℓ and it becomes quickly impossible to compute them formally. Fortunately, we are only interested by the value of these polynomials at a specific pair (a_0, b_0). This can be done very efficiently using some precomputations and Lagrange interpolation. Since S_ℓ is a polynomial of degree $\ell - 1$ in a and $\ell - 1$ in b, it is enough to know the value of S_ℓ at ℓ^2 pairs (a_i, b_j), for $(i, j) \in [1, \ell]^2$ in order to compute $S_\ell(a_0, b_0)$. First, for each $i \in [1, \ell]$, we interpolate with respect to the second coordinate based on the values $S_\ell(a_i, b_j)$, for $j \in [1, \ell]$. We obtain ℓ polynomials in variable b. Specializing those polynomials at b_0, we obtain ℓ values and a second Lagrange interpolation, followed by a specialization at a_0 gives $S_\ell(a_0, b_0)$. Note that in order to find the Lagrange polynomial $P(x)$ interpolating the points $(x_k, f(x_k))$, it is faster, in our case, to use the following formulas

$$P(x) = w(x) \sum_{k=1}^{\ell} \frac{f(x_k)}{w'(x_k)(x - x_k)} \quad \text{with} \quad w(x) = \prod_{j=1}^{\ell} (x - x_j)$$

rather than a more classical approach such as Aitken method. For each length ℓ, the ℓ^2 precomputed values can be obtained with Proposition 2. There is a similar approach for evaluating T_ℓ at (a_0, b_0).

Our PARI/GP implementation allows to deal efficiently with length ℓ up to 150. For most pairs (a, b), it takes less than 50ms to evaluate $S_\ell(a, b)$ or $T_\ell(a, b)$. In any case, at most a few seconds are necessary. The corresponding precomputations require about 45 MB. Only 10 MB are necessary to handle lengths ℓ up to 100. See [8] to access the actual implementation.

3.3 Generalization to Multi-Base Chains

It is easy to generalize the previous results to Multi-Base Chains. Let $p_1, \ldots, p_k$ be k pairwise coprime bases. A *Multi-Base Chain* (MBC) allows to represent a positive integer n as

$$n = \sum_{i=1}^{\ell} c_i p_1^{a_{1,i}} \ldots p_k^{a_{k,i}}, \quad \text{with} \quad c_1 = 1 \quad \text{and} \quad c_i = \pm 1, \quad \text{for} \quad i > 1$$

and $a_{j,1} \geqslant a_{j,2} \geqslant \cdots \geqslant a_{j,\ell}$, for all $j \in [1,k]$. An *unsigned Multi-Base Chain* is similar to a Multi-Base Chain except that all the c_i's are equal to 1. In any case, we assume that the term $p_1^{a_{1,i}} \ldots p_k^{a_{k,i}}$ appears at most once in any expansion.

Definition 5. *Let $\underline{a}$ denote the vector $(a_1, \ldots, a_k)$ and let $S_\ell(\underline{a})$ be the number of unsigned Multi-Base Chains of length ℓ satisfying $a_{j,1} = a_j$, for all j. Also, let $T_\ell(\underline{a})$ be the number of unsigned Multi-Base Chains of length ℓ satisfying $a_{j,1} \leqslant a_j$, for all j.*

The following Proposition is a simple generalization of Proposition 2. The proof is also similar.

Proposition 3. *Let $\ell \geqslant 1$. We have*

1. $S_{\ell+1}(\underline{a}) = T_\ell(\underline{a}) - S_\ell(\underline{a})$.
2.

$$T_{\ell+1}(\underline{a}) = \sum_{i_1=0}^{a_1} \cdots \sum_{i_k=0}^{a_k} \left[\prod_{j=1}^{k} (a_j - i_j + 1) - 1 \right] S_\ell(\underline{a}).$$

3. *$S_\ell(\underline{a})$ and $T_\ell(\underline{a})$ are both symmetrical polynomials.*
4. *The leading terms of $S_\ell(\underline{a})$ and of $T_\ell(\underline{a})$ are respectively $\frac{(a_1 \ldots a_k)^{\ell-1}}{(\ell-1)!^k}$ and $\frac{(a_1 \ldots a_k)^\ell}{\ell!^k}$.*

Remark 5. *Again the number of MBCs of length ℓ with a leading factor equal or dividing $p_1^{a_1,1} \ldots p_k^{a_k,1}$ can be easily deduced from $S_\ell(\underline{a})$ or $T_\ell(\underline{a})$. Namely, it is only necessary to multiply by a factor $2^{\ell-1}$.*

Example 6. *For $k = 3$, we have*

$$S_1(\underline{a}) = (a_1 + 1)(a_2 + 1)(a_3 + 1),$$
$$T_1(\underline{a}) = 1,$$
$$S_2(\underline{a}) = \frac{1}{8}(a_1 a_2 a_3 + 2a_1 a_2 + 2a_1 a_3 + 2a_2 a_3 + 4a_1 + 4a_2 + 4a_3)S_1(\underline{a}),$$
$$T_2(\underline{a}) = a_1 a_2 a_3 + a_1 a_2 + a_1 a_3 + a_2 a_3 + a_1 + a_2 + a_3.$$

4 Controlled DBC for Scalar Multiplication

For cryptographic applications, we propose a new way to perform a random scalar multiplication based on the concept of *controlled DBC*. The idea is to directly generate a random DBC expansion instead of choosing a random integer n and then finding a corresponding DBC to represent it.

Definition 6. *Given a leading factor $2^a 3^b$ and a given length ℓ, the* controlled
DBC approach *refers to the generation of a DBC expansion*

$$\sum_{i=1}^{\ell} c_i 2^{a_i} 3^{b_i}, \ \text{with } c_i \in \{-1, 1\}$$

*such that $c_1 = 1$, $a_1 = a$, $b_1 = b$, and whose $\ell - 1$ remaining terms $c_i 2^{a_i} 3^{b_i}$ are
selected to satisfy $a_1 \geqslant a_2 \geqslant \cdots \geqslant a_\ell$ and $b_1 \geqslant b_2 \geqslant \cdots \geqslant b_\ell$.*

This has two main advantages. Although very efficient, the greedy approach
still requires some time to return a DBC. No conversion is necessary with this
approach. Furthermore, there is no guarantee that the DBC expansion returned
by the greedy approach is optimal. In fact, we have evidence that the greedy
method returns a DBC that is far from optimal in general, especially for large
integers. See Section 5.2 and Figure 2. By choosing the DBC expansion first, in
particular its leading factor as well as its length, we can get closer to the average
optimal length. As a result, we can perform a scalar multiplication faster than
with the DBC obtained with the greedy approach by saving many additions. This
approach raises a few questions, in particular, regarding a suitable selection of
the length. For a given size and a given leading factor, it is possible to estimate
the length which corresponds heuristically to the average optimal length of a
DBC representing integers of that size with that leading factor. See Definition 7
for the notion of Near Optimal Length.

First, let us address the range of the integers that can be represented a priori
with a DBC having a leading term equal to $2^a 3^b$.

4.1 Integer Range

The following result provides an answer.

Lemma 1. *Any DBC with leading factor $2^a 3^b$ belongs to the interval*

$$\left[\frac{3^b + 1}{2}, \ 2^{a+1} 3^b - \frac{3^b + 1}{2} \right].$$

*It follows that the sign of the integer represented by a DBC with leading factor
equal to $2^a 3^b$ is driven by the sign of the coefficient of the leading factor in the
DBC.*

Proof. It is not difficult to see that the largest integer represented with a DBC
having a leading factor equal to $2^a 3^b$ can be constructed with a greedy-type
approach. In other words, it is enough to pick the largest available term at each
step to end up with the largest possible integer. Starting from $2^a 3^b$, the next
term in the DBC is of the form $2^i 3^j$ with $i \leqslant a$, $j \leqslant b$, and $(i, j) \neq (a, b)$.
Assuming that $a \geqslant 1$, clearly, $2^{a-1} 3^b$ is the largest possible integer we can pick.

If $a = 0$, then there is no choice but to pick 3^{b-1}. Repeating this argument, we deduce that the largest integer that can be represented is

$$2^a 3^b + \sum_{j=0}^{a-1} 2^j 3^b + \sum_{j=0}^{b-1} 3^j = 2^{a+1} 3^b + \frac{3^b - 1}{2}.$$

Similarly, the smallest integer corresponds to $\frac{3^b + 1}{2}$. Finally, it is obvious that if a DBC starts with $-2^a 3^b$, then the integers that can be represented with this DBC belong to the interval

$$\left[-2^{a+1} 3^b + \frac{3^b + 1}{2}, \ -\frac{3^b + 1}{2} \right].$$

So integers represented by a DBC starting with $2^a 3^b$ are always positive and those represented by a DBC starting with $-2^a 3^b$ are always negative.   □

The work in Section 3 gives the exact cardinality of the set containing all the DBCs with selected parameters. It is then tempting to select a length ℓ giving rise to as many DBCs as there are integers in the interval given in Lemma 1. However, this is ignoring that in general an integer has many different DBCs representations.

4.2 Redundancy and Near Optimal Length

In the controlled DBC approach, we need to be careful in selecting the length ℓ, as generating DBCs that are not long enough could compromise the security of the cryptosystem by severely restricting the number of scalars that can be represented with those chains. What length is then long enough? See Definition 7 for the notion Near Optimal Length addressing this question.

For various leading factors up to $2^{30} 3^{10}$ and length between 1 and 12, we have computed the number of different optimal representations of integers having an optimal DBC with this particular leading factor and length. For every selection of parameters, we consider between $10,000$ and 100 such integers. We then compute the average number of optimal DBCs for each length between 1 and 12, taking into account all the possible leading factors. This search was carried out with the algorithms developed in Section 2. The data fit an exponential regression of the form $y = \exp(0.4717x - 1.1683)$ with $R^2 = 0.9975$, see Figure 1.

To double-check the relevance of this estimate, we investigate DBCs having a leading factor of the form $2^{3\ell}$. We know that in this case the optimal length is ℓ, which corresponds to the NAF. We then compute the number of DBCs with a leading factor equal to 3ℓ and a length equal to ℓ using what we have done in Section 3. Dividing this quantity by $2^{3\ell+1}$, which corresponds approximately to the number of integers that can be represented a priori, we should obtain an estimate of the average number of optimal DBCs representing an integer, i.e. something close to $\exp(0.4717\ell - 1.1683)$. For all $\ell \in [10, 100]$, the ratio between

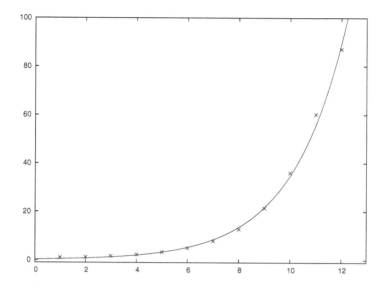

Fig. 1. Curve $\exp(0.4717x - 1.1683)$ fitting the experimental data

these two quantities lies in the interval $[0.0974, 3.384]$. This tends to confirm the relevance of our estimate, at least for relatively small values of b (0 in this case).

Definition 7. *For a leading factor equal to $2^a 3^b \simeq 2^t$, the* Near Optimal Length *corresponds to the integer value ℓ minimizing*

$$\left| 2^{\ell-1} S_\ell(a, b) - 2^t \lceil \exp(0.4717\ell - 1.1683) \rceil \right|.$$

Indeed, we expect that the average number of different DBC expansions of length ℓ representing the same integer is close to $\lceil \exp(0.4717\ell - 1.1683) \rceil$. Heuristically, we also expect that this redundancy factor multiplied by 2^t is equal to $2^{\ell-1} S_\ell(a, b)$ for the average optimal length ℓ.

4.3   Applications to Elliptic Curve Cryptography

For a chosen coordinate system representing a point on an elliptic curve and the corresponding complexities of a doubling, a tripling, and a mixed addition, it is possible to determine the optimal parameters, i.e. leading factor $2^a 3^b$ and length ℓ, which minimize the overall cost of a scalar multiplication with that particular coordinate system, without compromising the security of the system.

Definition 8. *For a given coordinate system and a bit size t, the* Near Optimal Controlled (NOC) DBC *method refers to the generation of a Controlled DBC with Near Optimal Length, which minimizes the costs of a scalar multiplication.*

In practice, we first select the bit size t, then consider all the possible pairs (a, b) such that $2^a 3^b \simeq 2^t$. For each pair (a, b), we work out the corresponding Near Optimal Length ℓ. Then we can compute the overall complexity to perform a scalar multiplication based on a controlled DBC with leading factor $2^a 3^b$ and length ℓ. It is then a matter of selecting the pair (a, b) corresponding to the lowest complexity overall. See Figure 2 and Tables 2 and 3.

5   Experiments

We have implemented the work described in Section 3 in C++ using NTL 6.0.0 [21] built on top of GMP 5.1.2 [15]. The approach described in Section 4 is implemented in PARI/GP 2.7.1 [22]. See [8] to access the actual C++ and PARI/GP implementations. All the programs are executed on a quad core i7-2620 at 2.70Ghz.

5.1   Optimal DBC Search

Given an integer n, the running time of Algorithms 1 and 2 to find the optimal length of a DBC representing n with a leading factor dividing $2^a 3^b$ is largely driven by the length ℓ of this optimal expansion. It usually takes several minutes for DBCs of length 14. See Table 1.

Table 1. Average running times to find an optimal DBC of length ℓ

Length ℓ	9	10	11	12	13	14
Time in s	1.08	5.21	28.52	66.38	214.80	757.91

Considering integers related to π, the longest optimal DBC that we have been able to compute corresponds to the 69-bit integer 314159265358979323846 with a leading factor equal to $2^{38} 3^{19}$ and length 18. It takes about 22 hours to show that there is no expansion of length less than or equal to 17 and it takes a bit less than six hours to return an optimal expansion of length 18 with the early abort technique mentioned in Section 2.4. Interestingly, the greedy approach returns a DBC of length 18 so that we can obtain an optimal DBC in no time, in that particular case.

5.2   Comparison between Greedy and Near Optimal Length

We have run some tests for sizes 192, 256, 320, 384, 448, and 512 bits. For each size t, we have considered various leading factors of the form $2^a 3^b \simeq 2^t$. More precisely, we fix a between $t/2$ and t, compute the corresponding b, and then compute the average length of the DBCs returned by the greedy method for 5,000 random integers. We also compute the Near Optimal Length of a

DBC with leading factor equal to $2^a 3^b$, see Definition 7 in Section 4.2. Our computations indicate that considering controlled DBCs that are 20 to 30% shorter than those returned by the greedy algorithm should not significantly reduce the set of integers that can be represented. See Figure 2, which shows a comparison for size $t = 320$. The x-coordinate axis corresponds to a between 160 and 315. The y-coordinate axis corresponds to the average length of the DBCs.

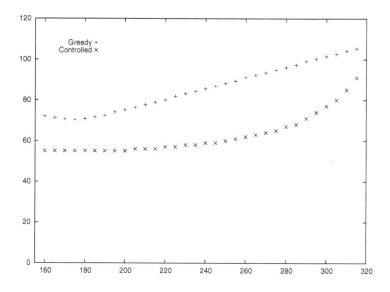

Fig. 2. Comparison between the average length of the DBCs returned by the greedy method and the Near Optimal Length for size 320 bits

5.3   Scalar Multiplication

In this part, we are interested in the potential savings introduced by our new scalar multiplication framework described in Section 4, in particular using the notion of Near Optimal Controlled DBC, see Definition 8.

In the following, we select the Inverted Edwards coordinate system [4] for a curve defined over a large prime field $\mathbb{F}_p$. This system offers a very fast doubling and a reasonably cheap mixed addition and tripling [2]. More precisely, the respective costs of a doubling, mixed addition, and tripling are $3M + 4S$, $8M + S$, and $9M + 4S$, where M and S stand respectively for a multiplication and a squaring in $\mathbb{F}_p$. To allow easy comparisons and as customary, we assume that $S = 0.8M$.

Until now, computing $[n]P$ for a random n, in Inverted Edwards coordinates with a DBC was not really worth it. Indeed, only the greedy method was fast enough to return a DBC in a reasonable time and the overall savings obtained were marginal with respect to the NAF, whose recoding can be achieved much faster. With the NAF, we perform t doublings and approximately $t/3$ mixed additions in order to compute $[n]P$ where n is of size t bits.

In Table 2, we display the parameters, costs, and speedups corresponding to different methods, for various sizes between 192 and 512. First, we consider the Near Optimal Controlled DBC approach, then the greedy method, and finally the NAF. LF stands for the leading factor and ℓ is the length of the corresponding expansion. The costs are expressed in terms of the number of multiplications needed to compute $[n]P$ but do not take into account the effort to produce each expansion. Regarding the NOC DBC, we determine for each size the optimal leading factor $2^a 3^b$ and corresponding Near Optimal Length ℓ minimizing the costs of the scalar multiplication, as explained in Section 5.2. Similarly, for the greedy approach we rely on the computations of Section 5.2.

Table 2. Theoretical comparison between NOC, greedy, and NAF methods

	NOC			Greedy			NAF			Speedups	
Size	LF	ℓ	Cost	LF_1	ℓ_1	$Cost_1$	LF_2	ℓ_2	$Cost_2$	S_1	S_2
192	$2^{151}3^{26}$	37	1570.20	$2^{116}3^{48}$	44.63	1688.74	2^{192}	64.00	1744.80	7.02%	10.01%
256	$2^{198}3^{37}$	48	2092.60	$2^{153}3^{65}$	58.73	2249.62	2^{256}	85.33	2329.33	6.98%	10.16%
320	$2^{260}3^{38}$	62	2612.40	$2^{180}3^{89}$	70.80	2816.04	2^{320}	106.67	2913.87	7.23%	10.35%
384	$2^{297}3^{55}$	71	3128.40	$2^{217}3^{106}$	84.74	3375.51	2^{384}	128.00	3498.40	7.32%	10.58%
448	$2^{369}3^{50}$	86	3645.80	$2^{254}3^{123}$	98.73	3935.42	2^{448}	149.33	4082.93	7.36%	10.71%
512	$2^{406}3^{67}$	95	4161.80	$2^{286}3^{143}$	112.07	4495.22	2^{512}	170.67	4667.47	7.42%	10.83%

To validate these theoretical results, we have developed an implementation in C++ using NTL 6.0.0 [21] built on top of GMP 5.1.2 [15]. The program is compiled and executed on a quad core i7-2620 at 2.70Ghz. For $t = 192$, 256, 320, 384, 448, and 512, we generate a random prime number p_t having bit size t. For each p_t, we then create a total of 100 curves of the form

$$E : x^2 + y^2 = c^2(1 + dx^2 y^2)$$

defined over $\mathbb{F}_{p_t}$, where c and d are small random values. For each curve E, we determine a random point P on E. Next, we select 100 random scalars in the interval $[0, p_t - 1]$. The corresponding NAF and greedy DBC expansions with a leading factor as in Table 2 are then computed for each scalar. For each t, we also directly create 100 random DBC expansions of length ℓ returned by the controlled DBC approach. Since we only want to assess the efficiency of the scalar multiplication, our only constraint is to generate a DBC with the specified length ℓ and leading factor as in Table 2. In practice, the method used to generate the expansions should be thoroughly designed and analyzed to ensure that the integers that are produced are uniformly distributed. This will be the object of some future work.

The experiments confirm the theoretical complexity analysis provided in Table 2, especially regarding S_2. The discrepancy between the theoretical and the experimental values of S_1 can be explained by a ratio M/S that is closer to 0.95 in NTL rather than 0.8 as initially assumed.

See Table 3 for actual timings. Note that the respective times necessary to compute the expansions for each method are not counted.

Table 3. Comparison of running times of NOC, greedy, and NAF methods

Size	NOC Time in ms	Greedy Time in ms	NAF Time in ms	Speedups S_1	S_2
192	0.822	0.861	0.939	4.58%	12.49%
256	1.444	1.531	1.642	5.73%	12.08%
320	2.446	2.584	2.766	5.35%	11.58%
384	3.511	3.703	3.960	5.17%	11.33%
448	5.088	5.392	5.729	5.65%	11.20%
512	6.569	6.982	7.408	5.91%	11.32%

6   Conclusion and Future Work

In this article, we have introduced new techniques to compute an optimal DBC representing a given integer. The algorithms that we have developed allow to tackle sizes of around 60 to 70 bits in a reasonable time.

We have also developed a new way to produce DBCs, namely the controlled DBC approach, which allows to directly create a DBC expansion instead of selecting an integer and converting it to DBC format. This idea raises a few issues regarding the choice of parameters, in particular the length of the expansion.

We use heuristics to estimate the average length of an optimal DBC expansion representing an integer of a certain bit size with a given leading factor. This estimate is based on the enumeration of the DBCs with given parameters and the expected number of different optimal DBCs representing the same integer.

For a given size and coordinate system, these heuristics allow to determine the optimal parameters, i.e. leading factor and length, which minimize the overall costs of a scalar multiplication of that size. This gives rise to the concept of Near Optimal Controlled DBC. Our experiments show speedups for this approach in excess of 10% over the NAF and of about 5% over the greedy method. Those computations do not take into account the time necessary to produce the expansions. So the interest of this new method is even greater as the expansions do not have to be computed unlike for the greedy and NAF methods.

In future, we aim at studying the redundancy of DBCs more accurately in order to find an upper bound on the number of DBCs of a certain length, representing an integer of a certain size.

Also, given a leading factor, once we have an estimate of the length of the expansion, the problem remains to actually create random controlled DBC expansions, such that the corresponding integers are uniformly distributed.

This question is not addressed in the present paper and will be the object of some future work.

Acknowledgments. The author would like to thank the GAATI group and the Department of Mathematics of the University of French Polynesia for hosting him while this research was carried out.

References

Numbers in curly brackets at the end specify the pages where the citations occur.

1. Avanzi, R.M., Cohen, H., Doche, C., Frey, G., Lange, T., Nguyen, K., Vercauteren, F.: Handbook of Elliptic and Hyperelliptic Curve Cryptography. Discrete Mathematics and its Applications. Chapman & Hall/CRC, Boca Raton (2005) {298}
2. Bernstein, D.J., Lange, T.: Explicit-formulas database, http://www.hyperelliptic.org/EFD/ {312}
3. Bernstein, D.J., Lange, T.: Faster addition and doubling on elliptic curves. In: Kurosawa, K. (ed.) ASIACRYPT 2007. LNCS, vol. 4833, pp. 29–50. Springer, Heidelberg (2007) {298}
4. Bernstein, D.J., Lange, T.: Inverted Edwards Coordinates. In: Boztaş, S., Lu, H.-F(F.) (eds.) AAECC 2007. LNCS, vol. 4851, pp. 20–27. Springer, Heidelberg (2007) {312}
5. Dimitrov, V.S., Cooklev, T.: Hybrid Algorithm for the Computation of the Matrix Polynomial $I + A + \cdots + A^{N-1}$. IEEE Trans. on Circuits and Systems 42(7), 377–380 (1995) {298}
6. Dimitrov, V.S., Imbert, L., Mishra, P.K.: Efficient and Secure Elliptic Curve Point Multiplication Using Double-Base Chains. In: Roy, B. (ed.) ASIACRYPT 2005. LNCS, vol. 3788, pp. 59–78. Springer, Heidelberg (2005) {298, 299}
7. Dimitrov, V.S., Jullien, G.A., Miller, W.C.: An Algorithm for Modular Exponentiation. Information Processing Letters 66(3), 155–159 (1998) {298}
8. Doche, C.: C++ and PARI/GP implementations to compute optimal and enumerate Double-Base Chains, http://www.comp.mq.edu.au/~doche {304, 306, 311}
9. Doche, C., Habsieger, L.: A Tree-Based Approach for Computing Double-Base Chains. In: Mu, Y., Susilo, W., Seberry, J. (eds.) ACISP 2008. LNCS, vol. 5107, pp. 433–446. Springer, Heidelberg (2008) {299}
10. Doche, C., Imbert, L.: Extended Double-Base Number System with applications to Elliptic Curve Cryptography. In: Barua, R., Lange, T. (eds.) INDOCRYPT 2006. LNCS, vol. 4329, pp. 335–348. Springer, Heidelberg (2006) {299}
11. Doche, C., Kohel, D.R., Sica, F.: Double-Base Number System for Multi-scalar Multiplications. In: Joux, A. (ed.) EUROCRYPT 2009. LNCS, vol. 5479, pp. 502–517. Springer, Heidelberg (2009) {299}
12. Doche, C., Sutantyo, D.: New and Improved Methods to Analyze and Compute Double-Scalar Multiplications. IEEE Trans. Comput. 63(1), 230–242 (2014) {299}
13. Edwards, H.M.: A normal form for elliptic curves. Bull. Amer. Math. Soc (N.S.) 44(3), 393–422 (2007) (electronic) {298}

14. Erdős, P., Loxton, J.H.: Some problems in partitio numerorum. J. Austral. Math. Soc. Ser. A 27(3), 319–331 (1979) {300}
15. Free Software Foundation. GNU Multiple Precision Library {311, 313}
16. Hankerson, D., Menezes, A.J., Vanstone, S.A.: Guide to Elliptic Curve Cryptography. Springer, Berlin (2003) {298}
17. Imbert, L., Philippe, F.: Strictly chained (p, q)-ary partitions. Contrib. Discrete Math. 5(2), 119–136 (2010) {300}
18. Lou, T., Sun, X., Tartary, C.: Bounds and Trade-offs for Double-Base Number Systems. Information Processing Letters 111(10), 488–493 (2011) {299}
19. Morain, F., Olivos, J.: Speeding up the Computations on an Elliptic Curve using Addition-Subtraction Chains. Inform. Theor. Appl. 24, 531–543 (1990) {298}
20. Reitwiesner, G.: Binary arithmetic. Adv. Comput. 1, 231–308 (1962) {298}
21. Shoup, V.: NTL: A Library for doing Number Theory {311, 313}
22. The PARI Group, Bordeaux. PARI/GP, version 2.7.1 (2014) {311}
23. Washington, L.C.: Elliptic Curves. Discrete Mathematics and its Applications. Number theory and cryptography. Chapman & Hall/CRC, Boca Raton (2003) {298}

Kummer Strikes Back: New DH Speed Records

Daniel J. Bernstein[1,2], Chitchanok Chuengsatiansup[2], Tanja Lange[2],
and Peter Schwabe[3]

[1] Department of Computer Science, University of Illinois at Chicago
Chicago, IL 60607–7045, USA
djb@cr.yp.to
[2] Department of Mathematics and Computer Science
Technische Universiteit Eindhoven
P.O. Box 513, 5600 MB Eindhoven, The Netherlands
c.chuengsatiansup@tue.nl, tanja@hyperelliptic.org
[3] Radboud University Nijmegen, Digital Security Group
P.O. Box 9010, 6500 GL Nijmegen, The Netherlands
peter@cryptojedi.org

Abstract. This paper sets new speed records for high-security constant-time variable-base-point Diffie–Hellman software: 305395 Cortex-A8-slow cycles; 273349 Cortex-A8-fast cycles; 88916 Sandy Bridge cycles; 88448 Ivy Bridge cycles; 54389 Haswell cycles. There are no higher speeds in the literature for any of these platforms.

The new speeds rely on a synergy between (1) state-of-the-art formulas for genus-2 hyperelliptic curves and (2) a modern trend towards vectorization in CPUs. The paper introduces several new techniques for efficient vectorization of Kummer-surface computations.

Keywords: performance, Diffie–Hellman, hyperelliptic curves, Kummer surfaces, vectorization.

1 Introduction

The Eurocrypt 2013 paper "Fast cryptography in genus 2" by Bos, Costello, Hisil, and Lauter [17] reported 117000 cycles on Intel's Ivy Bridge microarchitecture for high-security constant-time scalar multiplication on a genus-2 Kummer surface. The eBACS site for publicly verifiable benchmarks [13] confirms 119032 "cycles to compute a shared secret" (quartiles: 118904 and 119232) for the kumfp127g software from [17] measured on a single core of h9ivy, a 2012 Intel Core i5-3210M running at 2.5GHz. The software is not much slower on Intel's previous microarchitecture, Sandy Bridge: eBACS reports 122716 cycles (quartiles: 122576 and 122836) for kumfp127g on h6sandy, a 2011 Intel Core i3-2310M running at 2.1GHz. (The quartiles demonstrate that rounding to a

This work was supported by the National Science Foundation under grant 1018836 and by the Netherlands Organisation for Scientific Research (NWO) under grants 639.073.005, 613.001.011, and through the Veni 2013 project 13114. Permanent ID of this document: 1c5c0ead2524267af6b4f6d9114f10f0. Date: 2014.09.25.

P. Sarkar and T. Iwata (Eds.): ASIACRYPT 2014, PART I, LNCS 8873, pp. 317–337, 2014.

multiple of 1000 cycles, as in [17], loses statistically significant information; we follow eBACS in reporting medians of exact cycle counts.)

The paper reported that this was a "new software speed record" ("breaking the 120k cycle barrier") compared to "all previous genus 1 and genus 2 implementations" of high-security constant-time scalar multiplication. Obviously the genus-2 cycle counts shown above are better than the (unverified) claim of 137000 Sandy Bridge cycles by Longa and Sica in [40] (Asiacrypt 2012) for constant-time elliptic-curve scalar multiplication; the (unverified) claim of 153000 Sandy Bridge cycles by Hamburg in [34] for constant-time elliptic-curve scalar multiplication; the 182708 cycles reported by eBACS on h9ivy for curve25519, a constant-time implementation by Bernstein, Duif, Lange, Schwabe, and Yang [11] (CHES 2011) of Bernstein's Curve25519 elliptic curve [9]; and the 194036 cycles reported by eBACS on h6sandy for curve25519.

One might conclude from these figures that genus-2 hyperelliptic-curve cryptography (HECC) solidly outperforms elliptic-curve cryptography (ECC). However, two newer papers claim better speeds for ECC, and a closer look reveals a strong argument that HECC should have trouble competing with ECC.

The first paper, [44] by Oliveira, López, Aranha, and Rodríguez-Henríquez (CHES 2013 best-paper award), is the new speed leader in eBACS for *non-constant-time* scalar multiplication; the paper reports a new Sandy Bridge speed record of 69500 cycles. Much more interesting for us is that the paper claims 114800 Sandy Bridge cycles for *constant-time* scalar multiplication, beating [17]. eBACS reports 119904 cycles, but this is still faster than [17].

The second paper, [24] by Faz-Hernández, Longa, and Sánchez, claims 92000 Ivy Bridge cycles or 96000 Sandy Bridge cycles for constant-time scalar multiplication; a July 2014 update of the paper claims 89000 Ivy Bridge cycles or 92000 Sandy Bridge cycles. These claims are not publicly verifiable, but if they are even close to correct then they are faster than [17].

Both of these new papers, like [40], rely heavily on curve endomorphisms to eliminate many doublings, as proposed by Gallant, Lambert, and Vanstone [27] (Crypto 2001), patented by the same authors, and expanded by Galbraith, Lin, and Scott [26] (Eurocrypt 2009). Specifically, [44] uses a GLS curve over a binary field to eliminate 50% of the doublings, while also taking advantage of Intel's new pclmulqdq instruction to multiply binary polynomials; [24] uses a GLV+GLS curve over a prime field to eliminate 75% of the doublings.

One can also use the GLV and GLS ideas in genus 2, as explored by Bos, Costello, Hisil, and Lauter starting in [17] and continuing in [18] (CHES 2013). However, the best GLV/GLS speed reported in [18], 92000 Ivy Bridge cycles, provides only 2^{105} security and is not constant time. This is less impressive than the 119032 cycles from [17] for constant-time DH at a 2^{125} security level, and less impressive than the reports in [44] and [24].

The underlying problem for HECC is easy to explain. All known HECC addition formulas are considerably slower than the state-of-the-art ECC addition formulas at the same security level. Almost all of the HECC options explored in

[17] are bottlenecked by additions, so they were doomed from the outset, clearly incapable of beating ECC.

The one exception is that HECC provides an extremely fast *ladder* (see Section 2), built from extremely fast *differential* additions and doublings, considerably faster than the Montgomery ladder frequently used for ECC. This is why [17] was able to set DH speed records.

Unfortunately, differential additions do not allow arbitrary addition chains. Differential additions are incompatible with standard techniques for removing most or all doublings from fixed-base-point single-scalar multiplication, and with standard techniques for removing many doublings from multi-scalar multiplication. As a consequence, differential additions are incompatible with the GLV+GLS approach mentioned above for removing many doublings from single-scalar multiplication. This is why the DH speeds from [17] were quickly superseded by DH speeds using GLV+GLS. A recent paper [22] (Eurocrypt 2014) by Costello, Hisil, and Smith shows feasibility of combining differential additions and use of endomorphisms but reports 145000 Ivy Bridge cycles for constant-time software, much slower than the papers mentioned above.

1.1. Contributions of This Paper. We show that HECC has an important compensating advantage, and we exploit this advantage to achieve new DH speed records. The advantage is that we are able to heavily *vectorize* the HECC ladder.

CPUs are evolving towards larger and larger vector units. A low-cost low-power ARM Cortex-A8 CPU core contains a 128-bit vector unit that every two cycles can compute two vector additions, each producing four sums of 32-bit integers, or one vector multiply-add, producing two results of the form $ab + c$ where a, b are 32-bit integers and c is a 64-bit integer. Every cycle a Sandy Bridge CPU core can compute a 256-bit vector floating-point addition, producing four double-precision sums, and at the same time a 256-bit vector floating-point multiplication, producing four double-precision products. A new Intel Haswell CPU core can carry out two 256-bit vector multiply-add instructions every cycle. Intel has announced future support for 512-bit vectors ("AVX-512").

Vectorization has an obvious attraction for a chip manufacturer: the costs of decoding an instruction are amortized across many arithmetic operations. The challenge for the algorithm designer is to efficiently vectorize higher-level computations so that the available circuitry is performing useful work during these computations rather than sitting idle. What we show here is how to fit HECC with surprisingly small overhead into commonly available vector units. This poses several algorithmic challenges, notably to minimize the permutations required for the Hadamard transform (see Section 4). We claim broad applicability of our techniques to modern CPUs, and to illustrate this we analyze all three of the microarchitectures mentioned in the previous paragraph.

Beware that different microarchitectures often have quite different performance. A paper that advertises a "better" algorithmic idea by reporting new record cycle counts on a new microarchitecture, not considered in the previous literature, might actually be reporting an idea that *loses* performance on *all* microarchitectures. We instead emphasize HECC performance on the widely deployed Sandy Bridge microarchitecture, since Sandy Bridge was shared as a target by the recent ECC speed-record papers listed above. We have now set a new Sandy Bridge DH speed record, demonstrating the value of vectorized HECC. We have also set DH speed records for Ivy Bridge, Haswell, and Cortex-A8.

1.2. Constant Time: Importance and Difficulty. See full version of this paper online at https://eprint.iacr.org/2014/134.

1.3. Performance Results. eBACS shows that on a single core of h6sandy our DH software ("kummer") uses just 88916 Sandy Bridge cycles (quartiles: 88868 and 89184). On a single core of h9ivy our software uses 88448 cycles (quartiles: 88424 and 88476). On a single core of titan0, an Intel Xeon E3-1275 V3 (Haswell), our software uses 54389 cycles (quartiles: 54341 and 54454). On h7beagle, a TI Sitara AM3359 (Cortex-A8-slow), our software uses 305395 cycles (quartiles: 305380 and 305413). On h4mx515e, a Freescale i.MX515 (Cortex-A8-fast), our software uses 273349 cycles (quartiles: 273337 and 273387).

1.4. Cycle-Count Comparison. Table 1.5 summarizes reported high-security DH speeds for Cortex-A8, Sandy Bridge, Ivy Bridge, and Haswell.

This table is limited to software that *claims* to be constant time, and that claims a security level close to 2^{128}. This is the reason that the table does not include, e.g., the 767000 Cortex-A8 cycles and 108000 Ivy Bridge cycles claimed in [18] for constant-time scalar multiplication on a Kummer surface; the authors claim only 103 bits of security for that surface. This is also the reason that the table does not include, e.g., the 69500 Sandy Bridge cycles claimed in [44] for non-constant-time scalar multiplication.

The table does not attempt to report whether the listed cycle counts are from software that actually meets the above security requirements. In some cases inspection of the software has shown that the security requirements are violated; see Section 1.2. "Open" means that the software is reported to be open source, allowing third-party inspection.

Our speeds, on the same platform targeted in [17], solidly beat the HECC speeds from [17]. Our speeds also solidly beat the Cortex-A8, Sandy Bridge, and Ivy Bridge speeds from all available ECC software, including [11], [15], [22], and [44]; solidly beat the speeds claimed in [34] and [40]; and are even faster than the July 2014 Sandy Bridge/Ivy Bridge DH record claimed in [24], namely 92000/89000 cycles using unpublished software for GLV+GLS ECC. For Haswell, despite Haswell's exceptionally fast binary-field multiplier, our speeds beat the 55595 cycles from [44] for a GLS curve over a binary field. We set our new speed records using an HECC ladder that is conceptually much simpler than GLV and GLS, avoiding all the complications of scalar-dependent precomputations, lattice size issues, multi-scalar addition chains, endomorphism-rho security analysis, Weil-descent security analysis, and patents.

Table 1.5. Reported high-security DH speeds for Cortex-A8, Sandy Bridge, Ivy Bridge, and Haswell. Cycle counts from eBACS are for `curve25519`, `kumfp127g`, `gls254prot`, and our `kummer` on `h7beagle` (Cortex-A8-slow), `h4mx515e` (Cortex-A8-fast), `h6sandy` (Sandy Bridge), `h9ivy` (Ivy Bridge), and `titan0` (Haswell). Cycle counts not from SUPERCOP are marked "?". ECC has $g = 1$; genus-2 HECC has $g = 2$. See text for security requirements.

arch	cycles	ladder	open	g	field	source of software	
A8-slow	497389	yes	yes	1	$2^{255} - 19$	[15]	CHES 2012
A8-slow	305395	yes	yes	2	$2^{127} - 1$	new (this paper)	
A8-fast	460200	yes	yes	1	$2^{255} - 19$	[15]	CHES 2012
A8-fast	273349	yes	yes	2	$2^{127} - 1$	new (this paper)	
Sandy	194036	yes	yes	1	$2^{255} - 19$	[11]	CHES 2011
Sandy	153000?	yes	no	1	$2^{252} - 2^{232} - 1$	[34]	
Sandy	137000?	no	no	1	$(2^{127} - 5997)^2$	[40]	Asiacrypt 2012
Sandy	122716	yes	yes	2	$2^{127} - 1$	[17]	Eurocrypt 2013
Sandy	119904	no	yes	1	2^{254}	[44]	CHES 2013
Sandy	96000?	no	no	1	$(2^{127} - 5997)^2$	[24]	CT-RSA 2014
Sandy	92000?	no	no	1	$(2^{127} - 5997)^2$	[24]	July 2014
Sandy	88916	yes	yes	2	$2^{127} - 1$	new (this paper)	
Ivy	182708	yes	yes	1	$2^{255} - 19$	[11]	CHES 2011
Ivy	145000?	yes	yes	1	$(2^{127} - 1)^2$	[22]	Eurocrypt 2014
Ivy	119032	yes	yes	2	$2^{127} - 1$	[17]	Eurocrypt 2013
Ivy	114036	no	yes	1	2^{254}	[44]	CHES 2013
Ivy	92000?	no	no	1	$(2^{127} - 5997)^2$	[24]	CT-RSA 2014
Ivy	89000?	no	no	1	$(2^{127} - 5997)^2$	[24]	July 2014
Ivy	88448	yes	yes	2	$2^{127} - 1$	new (this paper)	
Haswell	145907	yes	yes	1	$2^{255} - 19$	[11]	CHES 2011
Haswell	100895	yes	yes	2	$2^{127} - 1$	[17]	Eurocrypt 2013
Haswell	55595	no	yes	1	2^{254}	[44]	CHES 2013
Haswell	54389	yes	yes	2	$2^{127} - 1$	new (this paper)	

2   Fast Scalar Multiplication on the Kummer Surface

This section reviews the smallest number of field operations known for genus-2 scalar multiplication. Sections 3 and 4 optimize the performance of those field operations using 4-way vector instructions.

Vectorization changes the interface between this section and subsequent sections. What we actually optimize is not individual field operations, but rather pairs of operations, pairs of pairs, etc., depending on the amount of vectorization available from the CPU. Our optimization also takes advantage of sequences of operations such as the output of a squaring being multiplied by a small constant. What matters in this section is therefore not merely the *number* of field multiplications, squarings, etc., but also the *pattern* of those operations.

2.1. Only 25 Multiplications. Almost thirty years ago Chudnovsky and Chudnovsky wrote a classic paper [21] optimizing scalar multiplication inside the elliptic-curve method of integer factorization. At the end of the paper they

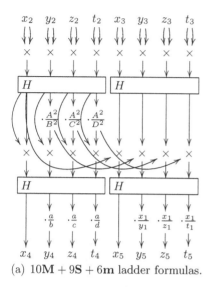
(a) $10\mathbf{M} + 9\mathbf{S} + 6\mathbf{m}$ ladder formulas.

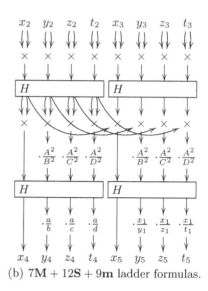
(b) $7\mathbf{M} + 12\mathbf{S} + 9\mathbf{m}$ ladder formulas.

Fig. 2.2. Ladder formulas for the Kummer surface. Inputs are $X(Q - P) = (x_1 : y_1 : z_1 : t_1)$, $X(P) = (x_2 : y_2 : z_2 : t_2)$, and $X(Q) = (x_3 : y_3 : z_3 : t_3)$; outputs are $X(2P) = (x_4 : y_4 : z_4 : t_4)$ and $X(P + Q) = (x_5 : y_5 : z_5 : t_5)$. Formulas in (a) are from Gaudry [30]; diagrams are copied from Bernstein [10].

also considered the performance of scalar multiplication on Jacobian varieties of genus-2 hyperelliptic curves. After mentioning various options they gave some details of one option, namely scalar multiplication on a Kummer surface.

A Kummer surface is related to the Jacobian of a genus-2 hyperelliptic curve in the same way that x-coordinates are related to a Weierstrass elliptic curve. There is a standard rational map X from the Jacobian to the Kummer surface; this map satisfies $X(P) = X(-P)$ for points P on the Jacobian and is almost everywhere exactly 2-to-1. Addition on the Jacobian does not induce an operation on the Kummer surface (unless the number of points on the surface is extremely small), but scalar multiplication $P \mapsto nP$ on the Jacobian induces scalar multiplication $X(P) \mapsto X(nP)$ on the Kummer surface. Not every genus-2 hyperelliptic curve can have its Jacobian mapped to the standard type of Kummer surface over the base field, but a noticeable fraction of curves can; see [31].

Chudnovsky and Chudnovsky reported $14\mathbf{M}$ for doubling a Kummer-surface point, where $\mathbf{M}$ is the cost of field multiplication; and $23\mathbf{M}$ for "general addition", presumably differential addition, computing $X(Q+P)$ given $X(P), X(Q)$, $X(Q-P)$. They presented their formulas for doubling, commenting on a "pretty symmetry" in the formulas and on the number of multiplications that were actually squarings. They did not present their formulas for differential addition.

Two decades later, in [30], Gaudry reduced the total cost of differential addition and doubling, computing $X(2P), X(Q + P)$ given $X(P), X(Q), X(Q - P)$, to $25\mathbf{M}$, more precisely $16\mathbf{M} + 9\mathbf{S}$, more precisely $10\mathbf{M} + 9\mathbf{S} + 6\mathbf{m}$, where $\mathbf{S}$ is

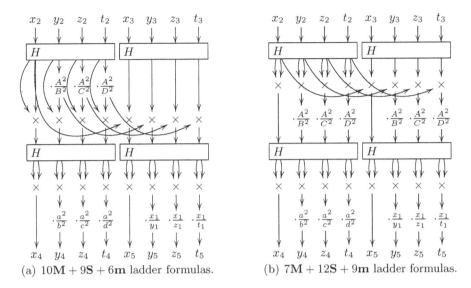

(a) $10\mathbf{M} + 9\mathbf{S} + 6\mathbf{m}$ ladder formulas.　　(b) $7\mathbf{M} + 12\mathbf{S} + 9\mathbf{m}$ ladder formulas.

Fig. 2.4. Ladder formulas for the squared Kummer surface. Compare to Figure 2.2.

the cost of field squaring and $\mathbf{m}$ is the cost of multiplication by a curve constant. An ℓ-bit scalar-multiplication ladder therefore costs just $10\ell\mathbf{M} + 9\ell\mathbf{S} + 6\ell\mathbf{m}$.

Gaudry's formulas are shown in Figure 2.2(a). Each point on the Kummer surface is expressed projectively as four field elements $(x : y : z : t)$; one is free to replace $(x : y : z : t)$ with $(rx : ry : rz : rt)$ for any nonzero r. The "H" boxes are Hadamard transforms, each using 4 additions and 4 subtractions; see Section 4. The Kummer surface is parametrized by various constants $(a : b : c : d)$ and related constants $(A^2 : B^2 : C^2 : D^2) = H(a^2 : b^2 : c^2 : d^2)$. The doubling part of the diagram, from $(x_2 : y_2 : z_2 : t_2)$ down to $(x_4 : y_4 : z_4 : t_4)$, uses $3\mathbf{M} + 5\mathbf{S} + 6\mathbf{m}$, matching the $14\mathbf{M}$ reported by Chudnovsky and Chudnovsky; but the rest of the picture uses just $7\mathbf{M} + 4\mathbf{S}$ extra, making remarkable reuse of the intermediate results of doubling. Figure 2.2(b) replaces $10\mathbf{M} + 9\mathbf{S} + 6\mathbf{m}$ with $7\mathbf{M} + 12\mathbf{S} + 9\mathbf{m}$, as suggested by Bernstein in [10]; this saves time if $\mathbf{m}$ is smaller than the difference $\mathbf{M} - \mathbf{S}$.

2.3. The Original Kummer Surface vs. The Squared Kummer Surface. Chudnovsky and Chudnovsky had actually used slightly different formulas for a slightly different surface, which we call the "squared Kummer surface". Each point $(x : y : z : t)$ on the original Kummer surface corresponds to a point $(x^2 : y^2 : z^2 : t^2)$ on the squared Kummer surface. Figure 2.4 presents the equivalent of Gaudry's formulas for the squared Kummer surface, relabeling $(x^2 : y^2 : z^2 : t^2)$ as $(x : y : z : t)$; the squarings at the top of Figure 2.2 have moved close to the bottom of Figure 2.4.

The number of field operations is the same either way, as stated in [10] with credit to André Augustyniak. However, the squared Kummer surface has a computational advantage over the original Kummer surface, as pointed out by

Bernstein in [10]: constructing surfaces in which all of $a^2, b^2, c^2, d^2, A^2, B^2, C^2, D^2$ are small, producing fast multiplications by constants in Figure 2.4, is easier than constructing surfaces in which all of $a, b, c, d, A^2, B^2, C^2, D^2$ are small, producing fast multiplications by constants in Figure 2.2.

2.5. Preliminary Comparison to ECC. A Montgomery ladder step for ECC costs $5\mathbf{M}+4\mathbf{S}+1\mathbf{m}$, while a ladder step on the Kummer surface costs $10\mathbf{M}+9\mathbf{S}+6\mathbf{m}$ or $7\mathbf{M}+12\mathbf{S}+9\mathbf{m}$. Evidently ECC uses only about half as many operations. However, for security ECC needs primes around 256 bits (such as the convenient prime $2^{255} - 19$), while the Kummer surface can use primes around 128 bits (such as the even more convenient prime $2^{127} - 1$), and presumably this saves more than a factor of 2.

Several years ago, in [10], Bernstein introduced 32-bit Intel Pentium M software for generic Kummer surfaces (i.e., $\mathbf{m} = \mathbf{M}$) taking about 10% fewer cycles than his Curve25519 software, which at the time was the speed leader for ECC. Gaudry, Houtmann, and Thomé, as reported in [32, comparison table], introduced 64-bit software for Curve25519 and for a Kummer surface; the second option was slightly faster on AMD Opteron K8 but the first option was slightly faster on Intel Core 2. It is not at all clear that one can reasonably extrapolate to today's CPUs.

Bernstein's cost analysis concluded that HECC could be as much as 1.5× faster than ECC on a Pentium M (cost 1355 vs. cost 1998 in [10, page 31]), depending on the exact size of the constants $a^2, b^2, c^2, d^2, A^2, B^2, C^2, D^2$. This motivated a systematic search through small constants to find a Kummer surface providing high security and high twist security. But this was more easily said than done: genus-2 point counting was much more expensive than elliptic-curve point counting.

2.6. The Gaudry–Schost Kummer Surface. Years later, after a 1000000-CPU-hour computation relying on various algorithmic improvements to genus-2 point counting, Gaudry and Schost announced in [33] that they had found a secure Kummer surface $(a^2 : b^2 : c^2 : d^2) = (11 : -22 : -19 : -3)$ over $\mathbf{F}_p$ with $p = 2^{127} - 1$. This is exactly the surface that was used for the HECC speed records in [17]. We obtain even better speeds for the same surface.

Note that, as mentioned by Bos, Costello, Hisil, and Lauter in [17], the constants $(1 : a^2/b^2 : a^2/c^2 : a^2/d^2) = (1 : -1/2 : -11/19 : -11/3)$ in Figure 2.4 are projectively the same as $(-114 : 57 : 66 : 418)$. The common factor 11 between $a^2 = 11$ and $b^2 = -22$ helps keep these integers small. The constants $(1 : A^2/B^2 : A^2/C^2 : A^2/D^2) = (1 : -3 : -33/17 : -33/49)$ are projectively the same as $(-833 : 2499 : 1617 : 561)$.

3 Decomposing Field Multiplication

The only operations in Figures 2.2 and 2.4 are the H boxes, which we analyze in Section 4, and field multiplications, which we analyze in this section. Our goal here is to obtain the smallest possible number of CPU cycles for $\mathbf{M}$, $\mathbf{S}$, etc. modulo $p = 2^{127} - 1$.

This prime has been considered before, for example in [8] and [10]. What is new here is fitting arithmetic modulo this prime, for the pattern of operations shown in Figure 2.4, into the vector abilities of modern CPUs. There are four obvious dimensions of vectorizability:

- Vectorizing across the "limbs" that represent a field element such as x_2. The most obvious problem with this approach is that, when f is multiplied by g, each limb of f needs to communicate with each limb of g and each limb of output. A less obvious problem is that the optimal number of limbs is CPU-dependent and is usually nonzero modulo the vector length. Each of these problems poses a challenge in organizing and reshuffling data inside multiplications.
- Vectorizing across the four field elements that represent a point. All of the multiplications in Figure 2.4 are visually organized into 4-way vectors, except that in some cases the vectors have been scaled to create a multiplication by 1. Even without vectorization, most of this scaling is undesirable for any surface with small a^2, b^2, c^2, d^2: e.g., for the Gaudry–Schost surface we replace $(1 : a^2/b^2 : a^2/c^2 : a^2/d^2)$ with $(-114 : 57 : 66 : 418)$. The only remaining exception is the multiplication by 1 in $(1 : x_1/y_1 : x_1/z_1 : x_1/t_1)$ where $X(Q - P) = (x_1 : y_1 : z_1 : t_1)$. Vectorizing across the four field elements means that this multiplication costs $1\mathbf{M}$, increasing the cost of a ladder step from $7\mathbf{M} + 12\mathbf{S} + 12\mathbf{m}$ to $8\mathbf{M} + 12\mathbf{S} + 12\mathbf{m}$.
- Vectorizing between doubling and differential addition. For example, in Figure 2.4(b), squarings are imperfectly paired with multiplications on the third line; multiplications by constants are perfectly paired with multiplications by the same constants on the fourth line; squarings are perfectly paired with squarings on the sixth line; and multiplications by constants are imperfectly paired with multiplications by inputs on the seventh line. There is some loss of efficiency in, e.g., pairing the squaring with the multiplication, since this prohibits using faster squaring methods.
- Vectorizing across a batch of independent scalar-multiplication inputs, in applications where a suitably sized batch is available. This is relatively straightforward but increases cache traffic, often to problematic levels. In this paper we focus on the traditional case of a single input.

The second dimension of vectorizability is, as far as we know, a unique feature of HECC, and one that we heavily exploit for high performance.

For comparison, one can try to vectorize the well-known Montgomery ladder for ECC [42] across the field elements that represent a point, but (1) this provides only two-way vectorization (x and z), not four-way vectorization; and (2) many of the resulting pairings are imperfect. The Montgomery ladder for Curve25519 was vectorized by Costigan and Schwabe in [23] for the Cell, and then by Bernstein and Schwabe in [15] for the Cortex-A8, but both of those vectorizations had substantially higher overhead than our new vectorization of the HECC ladder.

3.1. Sandy Bridge Floating-Point Units. The only fast multiplier available on Intel's 32-bit platforms for many years, from the original Pentium twenty years ago through the Pentium M, was the floating-point multiplier. This was exploited by Bernstein for cryptographic computations in [8], [9], etc.

The conventional wisdom is that this use of floating-point arithmetic was rendered obsolete by the advent of 64-bit platforms: in particular, Intel now provides a reasonably fast 64-bit integer multiplier. However, floating-point units have also become more powerful; evidently Intel sees many applications that rely critically upon fast floating-point arithmetic. We therefore revisit Bernstein's approach, with the added challenge of vectorization.

We next describe the relevant features of the Sandy Bridge; see [25] for more information. Our optimization of HECC for the Sandy Bridge occupies the rest of Sections 3 and 4. The Ivy Bridge has the same features and should be expected to produce essentially identical performance for this type of code. The Haswell has important differences and is analyzed in Appendix B online; the Cortex-A8 is analyzed in Section 5.

Each Sandy Bridge core has several 256-bit vector units operating in parallel on vectors of 4 double-precision floating-point numbers:

- "Port 0" handles one vector multiplication each cycle, with latency 5.
- Port 1 handles one vector addition each cycle, with latency 3.
- Port 5 handles one permutation instruction each cycle. The selection of permutation instructions is limited and is analyzed in detail in Section 4.
- Ports 2, 3, and 4 handle vector loads and stores, with latency 4 from L1 cache and latency 3 to L1 cache. Load/store throughput is limited in various ways, never exceeding one 256-bit load per cycle.

Recall that a double-precision floating-point number occupies 64 bits, including a sign bit, a power of 2, and a "mantissa". Every integer between -2^{53} and 2^{53} can be represented exactly as a double-precision floating-point number. More generally, every real number of the form $2^e i$, where e is a small integer and i is an integer between -2^{53} and 2^{53}, can be represented exactly as a double-precision floating-point number. The computations discussed here do not approach the lower or upper limits on e, so we do not review the details of the limits.

Our final software uses fewer multiplications than additions, and fewer permutations than multiplications. This does not mean that we were free to use extra multiplications and permutations: if multiplications and permutations are not finished quickly enough then the addition unit will sit idle waiting for input. In many cases, noted below, we have the flexibility to convert multiplications to additions, reducing latency; we found that in some cases this saved time despite the obvious addition bottleneck.

3.2. Optimizing M (Field Multiplication). We decompose an integer f modulo $2^{127} - 1$ into six floating-point limbs in (non-integer) radix $2^{127/6}$. This means that we write f as $f_0 + f_1 + f_2 + f_3 + f_4 + f_5$ where f_0 is a small multiple of 2^0, f_1 is a small multiple of 2^{22}, f_2 is a small multiple of 2^{43}, f_3 is a small multiple of 2^{64}, f_4 is a small multiple of 2^{85}, and f_5 is a small multiple of 2^{106}. (The exact meaning of "small" is defined by a rather tedious, but verifiable, collection of bounds on the floating-point numbers appearing in each step of the program. It should be obvious that a simpler definition of "small" would compromise efficiency; for example, H cannot be efficient unless the bounds on H intermediate results and outputs are allowed to be larger than the bounds on H inputs.)

If g is another integer similarly decomposed as $g_0 + g_1 + g_2 + g_3 + g_4 + g_5$ then $f_0 g_0$ is a multiple of 2^0, $f_0 g_1 + f_1 g_0$ is a multiple of 2^{22}, $f_0 g_2 + f_1 g_1 + f_2 g_0$ is a multiple of 2^{43}, etc. Each of these sums is small enough to fit exactly in a double-precision floating-point number, and the total of these sums is exactly fg. What we actually compute are the sums

$$h_0 = f_0 g_0 + 2^{-127} f_1 g_5 + 2^{-127} f_2 g_4 + 2^{-127} f_3 g_3 + 2^{-127} f_4 g_2 + 2^{-127} f_5 g_1,$$
$$h_1 = f_0 g_1 + \quad f_1 g_0 + 2^{-127} f_2 g_5 + 2^{-127} f_3 g_4 + 2^{-127} f_4 g_3 + 2^{-127} f_5 g_2,$$
$$h_2 = f_0 g_2 + \quad f_1 g_1 + \quad f_2 g_0 + 2^{-127} f_3 g_5 + 2^{-127} f_4 g_4 + 2^{-127} f_5 g_3,$$
$$h_3 = f_0 g_3 + \quad f_1 g_2 + \quad f_2 g_1 + \quad f_3 g_0 + 2^{-127} f_4 g_5 + 2^{-127} f_5 g_4,$$
$$h_4 = f_0 g_4 + \quad f_1 g_3 + \quad f_2 g_2 + \quad f_3 g_1 + \quad f_4 g_0 + 2^{-127} f_5 g_5,$$
$$h_5 = f_0 g_5 + \quad f_1 g_4 + \quad f_2 g_3 + \quad f_3 g_2 + \quad f_4 g_1 + \quad f_5 g_0,$$

whose total h is congruent to fg modulo $2^{127} - 1$.

There are 36 multiplications $f_i g_j$ here, and 30 additions. (This operation count does not include carries; we analyze carries below.) One can collect the multiplications by 2^{-127} into 5 multiplications such as $2^{-127}(f_4 g_5 + f_5 g_4)$. We use another approach, precomputing $2^{-127} f_1, 2^{-127} f_2, 2^{-127} f_3, 2^{-127} f_4, 2^{-127} f_5$, for two reasons: first, this reduces the latency of each h_i computation, giving us more flexibility in scheduling; second, this gives us an opportunity to share precomputations when the input f is reused for another multiplication.

3.3. Optimizing S (Field Squaring) and m (Constant Field Multiplication).
For **S**, i.e., for $f = g$, we have

$$h_0 = \quad f_0 f_0 + \epsilon 2 f_1 f_5 + \epsilon 2 f_2 f_4 + \epsilon f_3 f_3, \quad h_1 = 2 f_0 f_1 + \epsilon 2 f_2 f_5 + \epsilon 2 f_3 f_4,$$
$$h_2 = 2 f_0 f_2 + \quad f_1 f_1 + \epsilon 2 f_3 f_5 + \epsilon f_4 f_4, \quad h_3 = 2 f_0 f_3 + \quad 2 f_1 f_2 + \epsilon 2 f_4 f_5,$$
$$h_4 = 2 f_0 f_4 + \quad 2 f_1 f_3 + \quad f_2 f_2 + \epsilon f_5 f_5, \quad h_5 = 2 f_0 f_5 + \quad 2 f_1 f_4 + \quad 2 f_2 f_3$$

where $\epsilon = 2^{-127}$. We precompute $2 f_1, 2 f_2, 2 f_3, 2 f_4, 2 f_5$ and $\epsilon f_3, \epsilon f_4, \epsilon f_5$; this costs 8 multiplications, where 5 of the multiplications can be freely replaced by additions. The rest of **S**, after this precomputation, takes 21 multiplications and 15 additions, plus the cost of carries.

For **m** we have simply $h_0 = c f_0$, $h_1 = c f_1$, etc., costing 6 multiplications plus the cost of carries. This does not work for arbitrary field constants, but it does work for the small constants stated in Section 2.6.

3.4. Carries.
The output limbs h_i from **M** are too large to be used in a subsequent multiplication. We carry $h_0 \to h_1$ by rounding $2^{-22} h_0$ to an integer c_0, adding $2^{22} c_0$ to h_1, and subtracting $2^{22} c_0$ from h_0. This takes 3 additions (the CPU has a rounding instruction, `vroundpd`, that costs just 1 addition) and 2 multiplications. The resulting h_0 is guaranteed to be between -2^{21} and 2^{21}.

We could similarly carry $h_1 \to h_2 \to h_3 \to h_4 \to h_5$, and carry $h_5 \to h_0$ as follows: round $2^{-127} h_5$ to an integer c_5, add c_5 to h_0, and subtract $2^{127} c_5$

from h_5. One final carry $h_0 \to h_1$, for a total of 7 carries (21 additions and 14 multiplications), would then guarantee that all of $h_0, h_1, h_2, h_3, h_4, h_5$ are small enough to be input to a subsequent multiplication.

The problem with this carry chain is that it has extremely high latency: 5 cycles for $2^{-22}h_0$, 3 more cycles for c_0, 5 more cycles for $2^{22}c_0$, and 3 more cycles to add to h_1, all repeated 7 times, for a total of 112 cycles, plus the latency of obtaining h_0 in the first place. The ladder step in Figure 2.4 has a serial chain of $H \to \mathbf{M} \to \mathbf{m} \to H \to \mathbf{S} \to \mathbf{M}$, for a total latency above 500 cycles, i.e., above 125500 cycles for a 251-bit ladder.

We do better in six ways. First, we use only 6 carries in $\mathbf{M}$ rather than 7, if the output will be used only for $\mathbf{m}$. Even if the output h_0 is several bits larger than 2^{22}, it will not overflow the small-constant multiplication, since our constants are all bounded by 2^{12}.

Second, pushing the same idea further, we do these 6 carries in parallel. First we round in parallel to obtain $c_0, c_1, c_2, c_3, c_4, c_5$, then we subtract in parallel, then we add in parallel, allowing all of $h_0, h_1, h_2, h_3, h_4, h_5$ to end up several bits larger than they would have been with full carries.

Third, we also use 6 parallel carries for a multiplication that *is* an $\mathbf{m}$. There is no need for a chain, since the initial $h_0, h_1, h_2, h_3, h_4, h_5$ cannot be very large.

Fourth, we also use 6 parallel carries for each $\mathbf{S}$. This allows the $\mathbf{S}$ output to be somewhat larger than the input, but this still does not create overflows in the subsequent $\mathbf{M}$. At this point the only remaining block of 7 carries is in the $\mathbf{M}^4$ by $(1 : x_1/y_1 : x_1/z_1 : x_1/t_1)$, where $\mathbf{M}^4$ means a vector of four field multiplications.

Fifth, for that $\mathbf{M}^4$, we run two carry chains in parallel, carrying $h_0 \to h_1$ and $h_3 \to h_4$, then $h_1 \to h_2$ and $h_4 \to h_5$, then $h_2 \to h_3$ and $h_5 \to h_0$, then $h_3 \to h_4$ and $h_0 \to h_1$. This costs 8 carries rather than 7 but chops latency in half.

Finally, for that $\mathbf{M}^4$, we use the carry approach from [8]: add the constant $\alpha_{22} = 2^{22}(2^{52}+2^{51})$ to h_0, and subtract α_{22} from the result, obtaining the closest multiple of 2^{22} to h_0; add this multiple to h_1 and subtract it from h_0. This costs 4 additions rather than 3, but reduces carry latency from 16 to 9, and also saves two multiplications.

4   Permutations: Vectorizing the Hadamard Transform

The Hadamard transform H in Section 2 is defined as follows: $H(x, y, z, t) = (x+y+z+t, x+y-z-t, x-y+z-t, x-y-z+t)$. Evaluating this as written would use 12 field additions (counting subtraction as addition), but a standard "fast Hadamard transform" reduces the 12 to 8.

Our representation of field elements for the Sandy Bridge (see Section 3) requires 6 limb additions for each field addition. There is no need to carry before the subsequent multiplications; this is the main reason that we use 6 limbs rather than 5.

In a ladder step there are 4 copies of H, each requiring 8 field additions, each requiring 6 limb additions, for a total of 192 limb additions. This operation count suggests that 48 vector instructions suffice. Sandy Bridge has a helpful vaddsubpd instruction that computes $(a - e, b + f, c - g, d + h)$ given (a, b, c, d) and (e, f, g, h), obviously useful inside H.

However, we cannot simply vectorize across x, y, z, t. In Section 3 we were multiplying one x by another, at the same time multiplying one y by another, etc., with no permutations required; in this section we need to add x to y, and this requires permutations.

The Sandy Bridge has a vector permutation unit acting in parallel with the adder and the multiplier, as noted in Section 3. But this does not mean that the cost of permutations can be ignored. A long sequence of permutations inside H will force the adder and the multiplier to remain idle, since only a small fraction of the work inside $\mathbf{M}$ can begin before H is complete.

Our original software used 48 vector additions and 144 vector permutations for the 4 copies of H. We then tackled the challenge of minimizing the number of permutations. We ended up reducing this number from 144 to just 36. This section presents the details; analyzes conditional swaps, which end up consuming further time in the permutation unit; and concludes by analyzing the total number of operations used in our Sandy Bridge software.

4.1. Limitations of the Sandy Bridge Permutations.
There is a latency-1 permutation instruction vpermilpd that computes (y, x, t, z) given (x, y, z, t). vaddsubpd then produces $(x - y, y + x, z - t, t + z)$, which for the moment we abbreviate as (e, f, g, h). At this point we seem to be halfway done: the desired output is simply $(f + h, f - h, e + g, e - g)$.

If we had (f, h, e, g) at this point, rather than (e, f, g, h), then we could apply vpermilpd and vaddsubpd again, obtaining $(f - h, h + f, e - g, g + e)$. One final vpermilpd would then produce the desired $(f + h, f - h, e + g, e - g)$. The remaining problem is the middle permutation of (e, f, g, h) into (f, h, e, g).

Unfortunately, Sandy Bridge has very few options for moving data between the left half of a vector, in this case (e, f), and the right half of a vector, in this case (g, h). There is a vperm2f128 instruction (1-cycle throughput but latency 2) that produces (g, h, e, f), but it cannot even produce (h, g, f, e), never mind a combination such as (f, h, e, g). (Haswell has more permutation instructions, but Ivy Bridge does not. This is not a surprising restriction: n-bit vector units are often designed as $n/2$-bit vector units operating on the left half of a vector in one cycle and the right half in the next cycle, but this means that any communication between left and right requires careful attention in the circuitry. A similar left-right separation is even more obvious for the Cortex-A8.) We could shift some permutation work to the load/store unit, but this would have very little benefit, since simulating a typical permutation requires quite a few loads and stores.

The vpermilpd instruction $(x, y, z, t) \mapsto (y, x, t, z)$ mentioned above is one of a family of 16 vpermilpd instructions that produce (x or y, x or y, z or t, z or t). There is an even more general family of 16 vshufpd instructions that produce (a or b, x or y, c or d, z or t) given (a, b, c, d) and (x, y, z, t). In the first versions of our software we applied vshufpd to (e, f, g, h) and (g, h, e, f), obtaining (f, h, g, e), and then applied vpermilpd to obtain (f, h, e, g).

Overall a single H handled in this way uses, for each limb, 2 `vaddsubpd` instructions and 6 permutation instructions, half of which are handling the permutation of (e, f, g, h) into (f, h, e, g). The total for all limbs is 12 additions and 36 permutations, and the large "bubble" of permutations ends up forcing many idle cycles for the addition unit. This occurs four times in each ladder step.

4.2. Changing the Input/Output Format.

There are two obvious sources of inefficiency in the computation described above. First, we need a final permutation to convert $(f - h, f + h, e - g, e + g)$ into $(f + h, f - h, e + g, e - g)$. Second, the middle permutation of (e, f, g, h) into (f, h, e, g) costs three permutation instructions, whereas (g, h, e, f) would cost only one.

The first problem arises from a tension between Intel's `vaddsubpd`, which always subtracts in the first position, and the definition of H, which always adds in the first position. A simple way to resolve this tension is to store (t, z, y, x) instead of (x, y, z, t) for the input, and (t', z', y', x') instead of (x', y', z', t') for the output; the final permutation then naturally disappears. It is easy to adjust the other permutations accordingly, along with constants such as $(1, a^2/b^2, a^2/c^2, a^2/d^2)$.

However, this does nothing to address the second problem. Different permutations of (x, y, z, t) as input and output end up requiring different middle permutations, but these middle permutations are never exactly the left-right swap provided by `vperm2f128`.

We do better by generalizing the input/output format to allow negations. For example, if we start with $(x, -y, z, t)$, permute into $(-y, x, t, z)$, and apply `vaddsubpd`, we obtain $(x + y, x - y, z - t, t + z)$. Observe that this is not the same as the $(x - y, x + y, z - t, t + z)$ that we obtained earlier: the first two entries have been exchanged.

It turns out to be best to negate z, i.e., to start from $(x, y, -z, t)$. Then `vpermilpd` gives $(y, x, t, -z)$, and `vaddsubpd` gives $(x - y, x + y, -z - t, t - z)$, which we now abbreviate as (e, f, g, h). Next `vperm2f128` gives (g, h, e, f), and independently `vpermilpd` gives (f, e, h, g). Finally, `vaddsubpd` gives $(f - g, h + e, h - e, f + g)$. This is exactly $(x', t', -z', y')$ where $(x', y', z', t') = H(x, y, z, t)$.

The output format here is not the same as the input format: the positions of t and y have been exchanged. Fortunately, Figure 2.4 is partitioned by the H rows into two separate universes, and there is no need for the universes to use the same format. We use the $(x, y, -z, t)$ format at the top and bottom, and the $(x, t, -z, y)$ format between the two H rows. It is easy to see that exactly the same sequence of instructions works for all the copies of H, either producing $(x, y, -z, t)$ format from $(x, t, -z, y)$ format or vice versa.

$\mathbf{S^4}$ and $\mathbf{M^4}$ do not preserve negations: in effect, they switch from $(x, t, -z, y)$ format to (x, t, z, y) format. This is not a big problem, since we can reinsert the negation at any moment using a single multiplication or low-latency logic instruction (floating-point numbers use a sign bit rather than twos-complement, so negation is simply xor with a 1 in the sign bit). Even better, in Figure 2.4(b), the problem disappears entirely: each $\mathbf{S^4}$ and $\mathbf{M^4}$ is followed immediately by a constant multiplication, and so we simply negate the appropriate constants. The resulting sequence of formats is summarized in Figure 4.3.

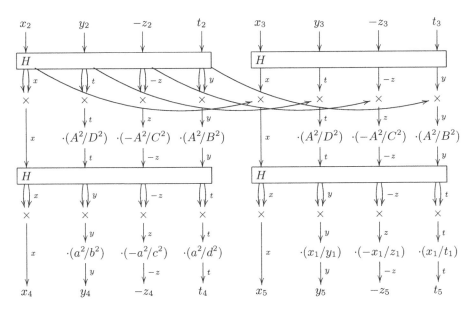

Fig. 4.3. Output format that we use for each operation in the right side of Figure 2.4 on Sandy Bridge, including permutations and negations to accelerate H

Each H now costs 12 additions and just 18 permutations. The number of non-addition cycles that need to be overlapped with operations before and after H has dropped from the original 24 to just 6.

4.4. Exploiting Double Precision. We gain a further factor of 2 by temporarily converting from radix $2^{127/6}$ to radix $2^{127/3}$ during the computation of H. This means that, just before starting H, we replace the six limbs $(h_0, h_1, h_2, h_3, h_4, h_5)$ representing $h_0 + h_1 + h_2 + h_3 + h_4 + h_5$ by three limbs $(h_0 + h_1, h_2 + h_3, h_4 + h_5)$. These three sums, and the intermediate H results, still fit into double-precision floating-point numbers.

It is essential to switch each output integer back to radix $2^{127/6}$ so that each output limb is small enough for the subsequent multiplication. Converting three limbs into six is slightly less expensive than three carries; in fact, converting from six to three and back to six uses exactly the same operations as three carries, although in a different order.

We further reduce the conversion cost by the following observation. Except for the $\mathbf{M^4}$ by $(1 : x_1/y_1 : x_1/z_1 : x_1/t_1)$, each of our multiplication results uses six carries, as explained in Section 3.4. However, if we are about to add h_0 to h_1 for input to H, then there is no reason to carry $h_0 \to h_1$, so we simply skip that carry; we similarly skip $h_2 \to h_3$ and $h_4 \to h_5$. These skipped carries exactly cancel the conversion cost.

For the $\mathbf{M^4}$ by $(1 : x_1/y_1 : x_1/z_1 : x_1/t_1)$ the analysis is different: h_0 is large enough to affect h_2, and if we skipped carrying $h_0 \to h_1 \to h_2$ then the output of H would no longer be safe as input to a subsequent multiplication. We thus carry $h_0 \to h_1$, $h_2 \to h_3$, and $h_4 \to h_5$ in parallel; and then $h_1 \to h_2$, $h_3 \to h_4$, and $h_5 \to h_0$ in parallel. In effect this $\mathbf{M^4}$ uses 9 carries, counting the cost of conversion, whereas in Section 3.4 it used only 8.

To summarize, all of these conversions for all four H cost just one extra carry, while reducing 48 additions and 72 permutations to 24 additions and 36 permutations.

4.5. Conditional Swaps. A ladder step starts from an input $(X(nP), X((n+1)P))$, which we abbreviate as $L(n)$, and produces $L(2n)$ as output. Swapping the two halves of the input, applying the same ladder step, and swapping the two halves of the output produces $L(2n + 1)$ instead; one way to see this is to observe that $L(-n - 1)$ is exactly the swap of $L(n)$.

Consequently one can reach $L(2n + \epsilon)$ for $\epsilon \in \{0, 1\}$ by starting from $L(n)$, conditionally swapping, applying the ladder step, and conditionally swapping again, where the condition bit is exactly ϵ. A standard ladder reaches $L(n)$ by applying this idea recursively. A standard *constant-time* ladder reaches $L(n)$ by applying this idea for exactly ℓ steps, starting from $L(0)$, where n is known in advance to be between 0 and $2^\ell - 1$. An alternate approach is to first add to n an appropriate multiple of the order of P, producing an integer known to be between (e.g.) $2^{\ell+1}$ and $2^{\ell+2} - 1$, and then start from $L(1)$. We use a standard optimization, merging the conditional swap after a ladder step into the conditional swap before the next ladder step, so that there are just $\ell + 1$ conditional swaps rather than 2ℓ.

One way to conditionally swap field elements x and x' using floating-point arithmetic is to replace (x, x') with $(x + b(x' - x), x' - b(x' - x))$ where b is the condition bit, either 0 or 1. This takes three additions and one multiplication (times 6 limbs, times 4 field elements to swap). It is better to use logic instructions: replace each addition with xor, replace each multiplication with and, and replace b with an all-1 or all-0 mask computed from b. On the Sandy Bridge, logic instructions have low latency and are handled by the permutation unit, which is much less of a bottleneck for us than the addition unit.

We further improve the performance of the conditional swap as follows. The $\mathbf{M^4}$ on the right side of Figure 4.3 is multiplying H of the left input by H of the right input. This is commutative: it does not depend on whether the inputs are swapped. We therefore put the conditional swap *after* the first row of H computations, and multiply the H outputs directly, rather than multiplying the swap outputs. This trick has several minor effects and one important effect.

A minor advantage is that this trick removes all use of the right half of the swap output; i.e., it replaces the conditional swap with a conditional move. This reduces the original 24 logic instructions to just 18.

Another minor advantage is as follows. The Sandy Bridge has a vectorized conditional-select instruction vblendvpd. This instruction occupies the permutation unit for 2 cycles, so it is no better than the 4 traditional logic instructions for a conditional swap: a conditional swap requires two conditional selects. However, this instruction *is* better than the 3 traditional logic instructions for a conditional move: a conditional move requires only one conditional select. This replaces the original logic instructions with 6 conditional-select instructions, consuming just 12 cycles.

A minor disadvantage is that the first $\mathbf{M^4}$ and $\mathbf{S^4}$ are no longer able to share precomputations of multiplications by 2^{-127}. This costs us 3 multiplication instructions.

The important effect is that this trick reduces latency, allowing the $\mathbf{M^4}$ to start much sooner. Adding this trick immediately produced a 5% reduction in our cycle counts.

4.6. Total Operations. We treat Figure 2.4(b) as $2\mathbf{M^4} + 3\mathbf{S^4} + 3\mathbf{m^4} + 4H$.

The main computations of h_i, not counting precomputations and carries, cost 30 additions and 36 multiplications for each $\mathbf{M^4}$, 15 additions and 21 multiplications for each $\mathbf{S^4}$, and 0 additions and 6 multiplications for each $\mathbf{m^4}$. The total here is 105 additions and 153 multiplications.

The $\mathbf{M^4}$ by $(1 : x_1/y_1 : x_1/z_1 : x_1/t_1)$ allows precomputations outside the loop. The other $\mathbf{M^4}$ consumes 5 multiplications for precomputations, and each $\mathbf{S^4}$ consumes 8 multiplications for precomputations; the total here is 29 multiplications. We had originally saved a few multiplications by sharing precomputations between the first $\mathbf{S^4}$ and the first $\mathbf{M^4}$, but this is incompatible with the more important trick described in Section 4.5.

There are a total of 24 additions in the four H, as explained in Section 4.4. There are also 51 carries (counting the conversions described in Section 4.4 as carries), each consuming 3 additions and 2 multiplications, for a total of 153 additions and 102 multiplications.

The grand total is 282 additions and 284 multiplications, evidently requiring at least 284 cycles for each iteration of the main loop. Recall that there are various options to trade multiplications for additions: each $\mathbf{S^4}$ has 5 precomputed doublings that can each be converted from 1 multiplication to 1 addition, and each carry can be converted from 3 additions and 2 multiplications to 4 additions and 0 multiplications (or 4 additions and 1 multiplication for $h_5 \to h_0$). We could use either of these options to eliminate one multiplication, reducing the 284-cycle lower bound to 283 cycles, but to reduce latency we ended up instead using the first option to eliminate 10 multiplications and the second option to eliminate 35 multiplications, obtaining a final total of 310 additions and 239 multiplications. These totals have been computer-verified.

We wrote functions in assembly for $\mathbf{M^4}$, $\mathbf{S^4}$, etc., but were still over 500 cycles. Given the Sandy Bridge floating-point latencies, and the requirement to keep *two* floating-point units constantly busy, we were already expecting instruction scheduling to be much more of an issue for this software than for typical integer-arithmetic software. We used various standard optimization techniques that were already used in several previous DH speed records: we merged the functions into a single loop, reorganized many computations to save registers, and eliminated many loads and stores. After building a new Sandy Bridge simulator and experimenting with different instruction schedules we ended up with our current loop, just 338 cycles, and a total of 88916 Sandy Bridge cycles for scalar multiplication. The main loop explains 84838 of these cycles; the remaining cycles are spent outside the ladder, mostly on converting $(x : y : z : t)$ to $(x/y : x/z : x/t)$ for output.

5   Cortex-A8

The low-power ARM Cortex-A8 core is the CPU core in the iPad 1, iPhone 4, Samsung Galaxy S, Motorola Droid X, Amazon Kindle 4, etc. Today a Cortex-A8 CPU, the Allwinner A10, costs just \$5 in bulk and is widely used in low-cost tablets, set-top boxes, etc. Like Sandy Bridge, Cortex-A8 is not the most recent microarchitecture, but its very wide deployment and use make it a sensible choice of platform for optimization and performance comparisons.

Bernstein and Schwabe in [15] (CHES 2012) analyzed the vector capabilities of the Cortex-A8 for various cryptographic primitives, and in particular set a new speed record for high-security DH, namely 460200 Cortex-A8 cycles. We do much better, just 274593 Cortex-A8 cycles, measured on a Freescale i.MX515. Our basic vectorization approach is the same for Cortex-A8 as for Sandy Bridge, and many techniques are reused, but there are also many differences. The rest of this section explains the details.

5.1. Cortex-A8 Vector Units. Each Cortex-A8 core has two 128-bit vector units operating in parallel on vectors of four 32-bit integers or two 64-bit integers:

- The arithmetic port takes one cycle for vector addition, with latency 2; or two cycles for vector multiplication (two 64-bit products ac, bd given 32-bit inputs a, b and c, d), with latency 7. Logic operations also use the arithmetic port.
- The load/store port handles loads, stores, and permutations. ARM's Cortex-A8 documentation [5] indicates that the load/store port can carry out one 128-bit load every cycle. Beware, however, that there are throughput limits on the L1 cache. We have found experimentally that the common TI Sitara Cortex-A8 CPU (used, e.g., in the Beaglebone Black development board) needs three cycles from one load until the next (this is what we call "Cortex-A8-slow"), while other Cortex-A8 CPUs ("Cortex-A8-fast") can handle seven consecutive cycles of loads without penalty.

There are three obvious reasons for Cortex-A8 cycle counts to be much larger than Sandy Bridge cycle counts: registers are only 128 bits, not 256 bits; there are only 2 ports, not 6; and multiplication throughput is 1 every 2 cycles, not 1 every cycle. However, there are also speedups on Cortex-A8. There is (as in Haswell's floating-point units—see Appendix B online) a vector multiply-accumulate instruction with the same throughput as vector multiplication. A sequence of m consecutive multiply-accumulate instructions that all accumulate into the same register executes in $2m$ cycles (unlike Haswell), effectively reducing multiplication latency from 7 to 1. Furthermore, Cortex-A8 multiplication produces 64-bit integer products, while Sandy Bridge gives only 53-bit-mantissa products.

5.2. Representation. We decompose an integer f modulo $2^{127} - 1$ into *five* integer pieces in radix $2^{127/5}$: i.e., we write f as $f_0 + 2^{26} f_1 + 2^{51} f_2 + 2^{77} f_3 + 2^{102} f_4$. Compared to Sandy Bridge, having 20% more room in 64-bit integers than in 53-bit floating-point mantissas allows us to reduce the number of limbs from 6 to 5. We require the small integers f_0, f_1, f_2, f_3, f_4 to be *unsigned* because this reduces carry cost from 4 integer instructions to 3.

We arrange four integers x, y, z, t modulo $2^{127} - 1$ in five 128-bit vectors: (x_0, y_0, x_1, y_1); (x_2, y_2, x_3, y_3); (x_4, y_4, z_4, t_4); (z_0, t_0, z_1, t_1); (z_2, t_2, z_3, t_3). This representation is designed to minimize permutations in **M**, **S**, and **H**. For example, computing $(x_0 + z_0, y_0 + t_0, x_1 + z_1, y_1 + t_1)$ takes just one addition without any permutations. The Cortex-A8 multiplications take two pairs of inputs at a time, rather than four as on Sandy Bridge, so there is little motivation to put (x_0, y_0, z_0, t_0) into a vector.

5.3. Optimizing M. Given an integer f as above and an integer $g = g_0 + 2^{26} g_1 + 2^{51} g_2 + 2^{77} g_3 + 2^{102} g_4$, the product fg modulo $2^{127} - 1$ is $h = h_0 + 2^{26} h_1 + 2^{51} h_2 + 2^{77} h_3 + 2^{102} h_4$, with

$$h_0 = f_0 g_0 + 2f_1 g_4 + 2f_2 g_3 + 2f_3 g_2 + 2f_4 g_1,$$
$$h_1 = f_0 g_1 + f_1 g_0 + f_2 g_4 + 2f_3 g_3 + f_4 g_2,$$
$$h_2 = f_0 g_2 + 2f_1 g_1 + f_2 g_0 + 2f_3 g_4 + 2f_4 g_3,$$
$$h_3 = f_0 g_3 + f_1 g_2 + f_2 g_1 + f_3 g_0 + f_4 g_4,$$
$$h_4 = f_0 g_4 + 2f_1 g_3 + f_2 g_2 + 2f_3 g_1 + f_4 g_0.$$

There are 25 multiplications $f_i g_j$; additions are free as part of multiply-accumulate instructions. We precompute $2f_1, 2f_2, 2f_3, 2f_4$ so that these values can be reused for another multiplication. These precomputations can be done by using either 4 shift or 4 addition instructions. Both shift and addition use 1 cycle per instruction, but addition has a lower latency. See Section 5.6 for the cost of carries.

5.4. Optimizing S. The idea of optimizing **S** in Cortex-A8 is quite similar to Sandy Bridge; for details see Section 3.3. We state here only the operation count. Besides precomputation and carry, we use 15 multiplication instructions; some of those are actually multiply-accumulate instructions.

5.5. Optimizing m. For **m** we compute only $h_0 = cf_0$, $h_1 = cf_1$, $h_2 = cf_2$, $h_3 = cf_3$, and $h_4 = cf_4$, again exploiting the small constants stated in Section 2.6.

Recall that we use *unsigned* representation. We always multiply absolute values, then negate results as necessary by subtracting from $2^{129} - 4$: $n_0 = 2^{28} - 4 - h_0$, $n_1 = 2^{27} - 4 - h_1$, $n_2 = 2^{28} - 4 - h_2$, $n_3 = 2^{27} - 4 - h_3$, $n_4 = 2^{27} - 4 - h_4$.

Negating any subsequence of x, y, z, t costs at most 5 vector subtractions. Negating only x or y, or both x and y, costs only 3 subtractions, because our representation keeps x, y within 3 vectors. The same comment applies to z and t. The specific **m** in Section 2.6 end up requiring a total of 13 subtractions with the same cost as 13 additions.

5.6. Carries. Each multiplication uses at worst 6 serial carries $h_1 \to h_2 \to h_3 \to h_4 \to h_0 \to h_1$, each costing 3 additions. Various carries are eliminated by the ideas of Section 3.4.

5.7. Hadamard Transform. See Appendix A online.

5.8. Total Arithmetic. We view Figure 2.4(b) as $4\mathbf{M}^2 + 6\mathbf{S}^2 + 6\mathbf{m}^2 + 4H$. Here we combine x multiplications and y multiplications into a vectorized $\mathbf{M}^2$, and similarly combine z multiplications and t multiplications; this fits well with the Cortex-A8 vector multiplication instruction, which outputs two products.

The main computations of h_i, not counting precomputations and carries, cost 0 additions and 25 multiplications for each $\mathbf{M}$, 0 additions and 15 multiplications for each $\mathbf{S}$, 0 additions and 5 multiplications for each $\mathbf{m}$, and 15 additions for each H block. The total here is 60 additions and 220 multiplications.

Each $\mathbf{M}$ costs 4 additions for precomputations, and each $\mathbf{S}$ also costs 4 additions for precomputations. Some precomputations can be reused. The cost of precomputations is 20 additions.

There are 10 carry blocks using 6 carries each, and 6 carry blocks using 5 carries each. Each carry consists of 1 shift, 1 addition, and 1 logical **and**. This cost is equivalent to 3 additions. There are another 13 additions needed to handle negation. Overall the carries cost 283 additions. Two conditional swaps, each costing 9 additions, sum up to 18 additions.

In total we have 381 additions and 220 multiplications in our inner loop. This means that the inner loop takes at least 821 cycles.

We scheduled instructions carefully but ended up with some overhead beyond arithmetic: even though the arithmetic and the load/store unit can operate in parallel, latencies and the limited number of registers leave the arithmetic unit idle for some cycles. Sobole's simulator at [48], which we found very helpful, reports 966 cycles. Actual measurements report 986 cycles; the 251 ladder steps thus account for 247486 of our 273349 cycles.

References

[5] ARM Limited: Cortex-A8 technical reference manual, revision r3p2 (2010), http://infocenter.arm.com/help/topic/com.arm.doc.ddi0344k/DDI0344K_cortex_a8_r3p2_trm.pdf

[8] Bernstein, D.J.: Floating-point arithmetic and message authentication (2004), http://cr.yp.to/papers.html#hash127

[9] Bernstein, D.J.: Curve25519: new Diffie-Hellman speed records. In: PKC 2006. LNCS, vol. 3958, pp. 207–228 (2006)

[10] Bernstein, D.J.: Elliptic vs. hyperelliptic, part 1 (2006), http://cr.yp.to/talks.html#2006.09.20

[11] Bernstein, D.J., Duif, N., Lange, T., Schwabe, P., Yang, B.-Y.: High-speed high-security signatures. In: CHES 2011. LNCS, vol. 6917 (2011); see also newer version [12]

[12] Bernstein, D.J., Duif, N., Lange, T., Schwabe, P., Yang, B.-Y.: High-speed high-security signatures. Journal of Cryptographic Engineering 2, 77–89 (2012); see also older version [11]

[13] Bernstein, D.J., Lange, T. (eds.): eBACS: ECRYPT Benchmarking of Crypto-graphic Systems, accessed 25 September 2014 (2014), http://bench.cr.yp.to
[15] Bernstein, D.J., Schwabe, P.: NEON crypto. In: CHES 2012. LNCS, vol. 7428, pp. 320–339 (2012)
[17] Bos, J.W., Costello, C., Hisil, H., Lauter, K.: Fast cryptography in genus 2. In: Eurocrypt 2013. LNCS, vol. 7881, pp. 194–210 (2013)
[18] Bos, J.W., Costello, C., Hisil, H., Lauter, K.: High-performance scalar multipli-cation using 8-dimensional GLV/GLS decomposition. In: CHES 2013. LNCS, vol. 8086, pp. 331–348 (2013)
[21] Chudnovsky, D.V., Chudnovsky, G.V.: Sequences of numbers generated by ad-dition in formal groups and new primality and factorization tests. Advances in Applied Mathematics 7, 385–434 (1986)
[22] Costello, C., Hisil, H., Smith, B.: Faster compact Diffie–Hellman: endomorphisms on the x-line. In: Eurocrypt 2014. LNCS, vol. 8441, pp. 183–200 (2014)
[23] Costigan, N., Schwabe, P.: Fast elliptic-curve cryptography on the Cell Broadband Engine. In: Africacrypt 2009. LNCS, vol. 5580, pp. 368–385 (2009)
[24] Faz-Hernández, A., Longa, P., Sánchez, A.H.: Efficient and secure algorithms for GLV-based scalar multiplication and their implementation on GLV-GLS curves. In: CT-RSA 2014. LNCS, vol. 8366, pp. 1–27 (2013)
[25] Fog, A.: Instruction tables: Lists of instruction latencies, throughputs and micro-operation breakdowns for Intel, AMD and VIA CPUs (2014), http://agner.org/optimize/
[26] Galbraith, S., Lin, X., Scott, M.: Endomorphisms for faster elliptic curve cryp-tography on a large class of curves. In: Eurocrypt 2009. LNCS, vol. 5479, pp. 518–535 (2009)
[27] Gallant, R.P., Lambert, R.J., Vanstone, S.A.: Faster point multiplication on el-liptic curves with efficient endomorphisms. In: Crypto 2001. LNCS, vol. 2139, pp. 190–200 (2001)
[30] Gaudry, P.: Variants of the Montgomery form based on Theta functions (2006); see also newer version [31], http://www.loria.fr/~gaudry/publis/toronto.pdf
[31] Gaudry, P.: Fast genus 2 arithmetic based on Theta functions. Journal of Math-ematical Cryptology 1, 243–265 (2007); see also older version [30]
[32] Gaudry, P., Lubicz, D.: The arithmetic of characteristic 2 Kummer surfaces and of elliptic Kummer lines. Finite Fields and Their Applications 15, 246–260 (2009)
[33] Gaudry, P., Schost, É: Genus 2 point counting over prime fields. Journal of Sym-bolic Computation 47, 368–400 (2012)
[34] Hamburg, M.: Fast and compact elliptic-curve cryptography (2012), https://eprint.iacr.org/2012/309
[40] Longa, P., Sica, F.: Four-dimensional Gallant–Lambert–Vanstone scalar multipli-cation. In: Asiacrypt 2012. LNCS, vol. 7658, pp. 718–739 (2012)
[42] Montgomery, P.L.: Speeding the Pollard and elliptic curve methods of factoriza-tion. Mathematics of Computation 48, 243–264 (1987)
[44] Oliveira, T., López, J., Aranha, D.F., Rodríguez-Henríquez, F.: Lambda coordi-nates for binary elliptic curves. In: CHES 2013. LNCS, vol. 8086, pp. 311–330 (2013)
[48] Sobole, É.: Calculateur de cycle pour le Cortex A8 (2012), http://pulsar.webshaker.net/ccc/index.php

Jacobian Coordinates on Genus 2 Curves

Huseyin Hisil[1] and Craig Costello[2]

[1] Yasar University, Izmir, Turkey
huseyin.hisil@yasar.edu.tr
[2] Microsoft Research, Redmond, USA
craigco@microsoft.com

Abstract. This paper presents a new projective coordinate system and new explicit algorithms which together boost the speed of arithmetic in the divisor class group of genus 2 curves. The proposed formulas generalise the use of Jacobian coordinates on elliptic curves, and their application improves the speed of performing cryptographic scalar multiplications in Jacobians of genus 2 curves over prime fields by an approximate factor of 1.25x. For example, on a single core of an Intel Core i7-3770M (Ivy Bridge), we show that replacing the previous best formulas with our new set improves the cost of generic scalar multiplications from 243,000 to 195,000 cycles, and drops the cost of specialised GLV-style scalar multiplications from 166,000 to 129,000 cycles.

Keywords: Genus 2, hyperelliptic curves, explicit formulas, Jacobian coordinates, scalar multiplication.

1 Introduction

Motivated by the popularity of low-genus curves in cryptography [29,22,23], we put forward a new system of projective coordinates that facilitates efficient group law computations in the Jacobians of hyperelliptic curves of genus 2. This paper combines several techniques to arrive at explicit formulas that are significantly faster than those in previous works [25,9]. The two main ingredients we use in the derivation are:

- The generalisation of Jacobian coordinates from the elliptic curve setting to the hyperelliptic curve setting: these coordinates essentially cast affine points into projective space according to the *weights* of x and y in the defining curve equation. While applying Jacobian coordinates to elliptic curves is straightforward, their application to hyperelliptic curves requires transferring the x-y weightings into weightings for the Mumford coordinates. As it does for the x-y coordinates in genus 1, this projection naturally balances the Mumford coordinates to facilitate substantial simplifications in the projective genus 2 group law formulas.
- The adaptation of Meloni's "co-Z" idea [28] to the genus 2 setting. Although originally proposed in the context of addition-only (e.g. Fibonacci-style) chains, this approach can also be used to gain performance in the more

P. Sarkar and T. Iwata (Eds.): ASIACRYPT 2014, PART I, LNCS 8873, pp. 338–357, 2014.

meaningful context of binary addition chains. Moreover, this idea is especially advantageous when used in conjunction with Jacobian coordinates.

The application of the above techniques, as well as some further optimisations discussed in the body of this paper, gives rise to the operation counts in Table 1 – the counts here include field multiplications ($\mathbf{M}$), squarings ($\mathbf{S}$), and multiplications by curve constants ($\mathbf{D}$). Here we make a brief comparison with the previous works in [25] and [9], by considering the two most common operations in the context of cryptographic scalar multiplications: a point doubling (denoted DBL), and a mixed-doubling-and-addition (denoted mDBLADD) between two points[1]. These two operations constitute the bottleneck of most state-of-the-art scalar multiplication routines, since the multiplication of a point in the Jacobian by an n-bit scalar typically requires α DBL operations and β mDBLADD operations, where $\alpha + \beta \approx n$. Thus, the improved operation counts in Table 1 give a rough idea of the speedups that we can expect when plugging these formulas into an existing genus 2 scalar multiplication routine that uses the formulas from [25] or [9]. (We give a better indication of the improvements over previous formulas by reporting concrete implementation numbers in Section 8.) As well as the reduction in field multiplications indicated in Table 1, the explicit formulas in this paper also require far fewer field additions than those in [25] and [9]. We note that the biggest relative difference occurs in the mDBLADD column: among other things, this difference results from the combination of the new coordinate system with the extension of Meloni's idea [28], which allows us to compute mDBLADD operations independently of the curve constants. On the other hand, when such curve constants are zero, certain operations in this paper become even faster (relatively speaking): for example, on the two special families exhibiting endomorphisms used in [7], the doubling formulas in [25] and [9] save $2\mathbf{D}$, while the new operation count reported for DBL in Table 1 saves $3\mathbf{S} + 2\mathbf{D}$ to drop down to $21\mathbf{M} + 9\mathbf{S}$.

Table 1. Field operation counts obtained in this work, versus two previous works, for the most common operations incurred during cryptographic scalar multiplications in Jacobians of genus 2 curves of the form $\mathcal{C}/K : y^2 = f(x)$, where $f(x)$ is of degree 5 and the characteristic of K is greater than 5

authors	DBL	mADD	mDBLADD
Lange [25]	$32\mathbf{M} + 7\mathbf{S} + 2\mathbf{D}$	$36\mathbf{M} + 5\mathbf{S}$	$68\mathbf{M} + 12\mathbf{S} + 2\mathbf{D}$
Costello-Lauter [9]	$30\mathbf{M} + 9\mathbf{S} + 2\mathbf{D}$	$36\mathbf{M} + 5\mathbf{S}$	$66\mathbf{M} + 14\mathbf{S} + 2\mathbf{D}$
This work	$21\mathbf{M} + 12\mathbf{S} + 2\mathbf{D}$	$29\mathbf{M} + 7\mathbf{S}$	$52\mathbf{M} + 11\mathbf{S}$

---

[1] For genus 2 scalar multiplications, it is usually advantageous to convert precomputed (lookup table) points to their affine representation using a shared inversion – see Section 7.2. This is why the double-and-add operations involve a "mixed" addition.

While the formulas in this paper target Jacobians of imaginary genus 2 curves, Gaudry showed in [16] that one can perform cryptographic scalar multiplications much more efficiently in the special case that the Jacobian of the curve C/K has K-rational two-torsion, by instead working on an associated Kummer surface. To illustrate the difference between working on the Kummer surface and working in the full Jacobian group, Gaudry's analogous operation counts are a blazingly fast $6\mathbf{M}+8\mathbf{S}$ for DBL and $16\mathbf{M}+9\mathbf{S}$ for mDBLADD. Referring back to Table 1, it is clear that raw scalar multiplications on the Kummer surface will remain unrivalled by those in the full Jacobian group. However, there are several cryptographic caveats related to the Kummer surface that justify the continued exploration of fast algorithms for traditional arithmetic in the Jacobian. Namely, Kummer surfaces do not support *generic* additions, so while they are extremely fast in the realm of key exchange (where such additions are not necessary), it is not yet known how to efficiently use the Kummer surface in a wider realm of cryptographic settings, e.g. for general digital signatures[2]. Furthermore, the absence of generic additions complicates the application of endomorphisms [7, §8.5], and from a more pragmatic standpoint, also prevents the use of standard precomputation techniques that exploit fixed system parameters (those of which give huge speedups in practice, even over the Kummer surface [7, §7.4]). Thus, all genus 2 implementations that either target signature schemes, use endomorphisms, or optimise the use of precomputation, are currently required to work in the full Jacobian group[3]; and in all of these cases, the formulas in this paper will now offer the most efficient route. The upshot is that in popular practical scenarios the most efficient genus 2 cryptography is likely to result from a hybrid combination of operations on the Kummer surface and in the full Jacobian group. We illustrate this in Section 8 by benchmarking genus 2 curves in the context of ephemeral elliptic curve Diffie-Hellman (ECDHE) with perfect forward secrecy: to exploit the best of both worlds, Alice's multiplications of the public generator P by each one of her ephemeral scalars a can make use of our new explicit formulas (and offline precomputations on P) in the full Jacobian, and her resulting ephemeral public keys $[a]P$ can then be mapped onto the corresponding Kummer surface, whose speed can be exploited by Bob in the computation of the shared secret $[b]([a]P)$.

A set of Magma [8] scripts verifying all of the explicit formulas and operation counts can be found in the full version [21], and is also publicly available at

http://research.microsoft.com/en-us/downloads/37730278-3e37-47eb-91d1-cf889373677a/   ;

and a complete mixed-assembly-and-C implementation of all explicit formulas and scalar multiplication routines is publicly available at

http://hhisil.yasar.edu.tr/files/hisil20140527jacobian.tar.gz   .

---

[2] At least one exception here, as Gaudry points out, is the hashed version of ElGamal signatures [16, §5.3].

[3] Lubicz and Robert [27] have recently broken through the "full addition restriction" on Kummer varieties, but it is not yet clear how competitive their *compatible addition* formula are in the context of raw scalar multiplications.

2   Preliminaries

For ease of exposition, we immediately restrict to the most cryptographically common case of genus 2 curves, where $\mathcal{C}$ is an imaginary hyperelliptic curve over a field K of characteristic greater than 5. (In terms of a general coverage of all genus 2 curves, we mention the interesting remaining scenarios in Section 9.) Every such curve can then be written as

$$\mathcal{C}/K : y^2 = f(x) := x^5 + a_3 x^3 + a_2 x^2 + a_1 x + a_0, \tag{1}$$

where we note the absence of an x^4 term in $f(x)$; it can always be removed via a trivial substitution thanks to $\mathrm{char}(K) \neq 5$.

Let $J_{\mathcal{C}}$ denote the Jacobian of $\mathcal{C}$. We assume that we are working with a general point $P \in J_{\mathcal{C}}(K)$, whose Mumford representation[4]

$$\begin{aligned} P \leftrightarrow (u(x), v(x)) &= \left(x^2 + qx + r, sx + t \right) \in K[x] \times K[x] \\ &\leftrightarrow (q, r, s, t) \in \mathbb{A}^4(K) \end{aligned} \tag{2}$$

encodes two affine points $(x_1, y_1), (x_2, y_2) \in \mathcal{C}(\overline{K})$, where we assume that $x_1 \neq x_2$ so that these two points are not the same, nor are they the hyperelliptic involution of one another. The Mumford coordinates (q, r, s, t) of P are uniquely determined according to $u(x_1) = u(x_2) = 0$, $v(x_1) = y_1$ and $v(x_2) = y_2$. That is,

$$q = -(x_1 + x_2), \quad r = x_1 x_2, \quad s = \frac{y_1 - y_2}{x_1 - x_2}, \quad t = \frac{x_1 y_2 - y_1 x_2}{x_1 - x_2}. \tag{3}$$

From (1), (2) and (3), it is readily seen that

$$v(x)^2 - f(x) = 0 \quad \text{in} \quad K[x]/\langle u(x) \rangle, \tag{4}$$

from which it follows that such general points P lie in the intersection of two hypersurfaces over K [9, §3], given as

$$\begin{aligned} \mathcal{S}_0 : &\quad r \left(s^2 + q^3 - (2r - a_3)q - a_2 \right) = t^2 - a_0, \\ \mathcal{S}_1 : &\quad q \left(s^2 + q^3 - (3r - a_3)q - a_2 \right) = 2st - r(r - a_3) - a_1. \end{aligned} \tag{5}$$

We note that a more simple relation is found by taking $r\mathcal{S}_1 - q\mathcal{S}_0$.

Our driving motivation for improving the explicit formulas for arithmetic in the Jacobian is the application of enhancing the fundamental operation in curve-based cryptosystems: the scalar multiplication $[k]P$ of an integer $k \in \mathbb{Z}$ by a general point P in $J_{\mathcal{C}}$. Such scalar multiplications are computed using a sequence of point doubling and addition operations, and so a common way of comparing different sets of addition formulas is to tally the number of field multiplications (**M**), field squarings (**S**), and field additions (denoted by **a**) that each point operation incurs. In cryptographic contexts, the input and output

---

[4] We adopted the notation (q, r, s, t) over (u_1, u_0, v_1, v_0) to avoid additional subscripts/superscripts when working with distinct elements in $J_{\mathcal{C}}$.

points are typically required to be in their unique affine form, whilst intermediate computations are carried out in projective space to avoid inversions. Thus, the most commonly reported operation counts include: DBL, which refers to the addition of a Jacobian point in projective form to itself; ADD, which refers to the addition between two distinct points in projective form; mADD, which refers to the *mixed* addition between a projective point and an affine point; and mDBLADD, which refers to the combined doubling of a projective point and subsequent addition of the result with an affine point.

As is done in [25, §5-6], in this paper we focus on deriving formulas for the most common cases of arithmetic in $J_\mathcal{C}$. This set of formulas is enough to perform and benchmark scalar multiplications in $J_\mathcal{C}$, since the possible input/output cases are extremely dense amongst all possible scenarios, i.e. for random input points P and scalars k, the cases not covered by these formulas have an exponentially small probability of being encountered in the scalar multiplication routine (see [25, §1.2] for a similar discussion). Nevertheless, the set of formulas we present are still far from a *complete* and cryptographically adequate coverage, so it is important to distinguish exactly which input/output cases they do apply to. We clarify this in Assumption 1 below, and return to this discussion in §7.3.

Assumption 1 (General Points and Operations in $J_\mathcal{C}$.). *Throughout this paper, we assume that all input and output points are "general" points in $J_\mathcal{C}$: we say that $P \in J_\mathcal{C}$ is general if the Mumford representation of P encodes two distinct affine points (x_1, y_1) and (x_2, y_2) on $\mathcal{C}$, where $x_1 \neq x_2$. Moreover, all operations in this paper are of the form $P_1 + P_2 = P_3$, where we assume that P_1, P_2 and P_3 are general points and that we are in one of two cases: (i) either $P_1 = P_2$, in which case we are computing the "doubling" $P_3 = [2]P_1$, where we further assume that neither of the two x-coordinates encoded by P_1 coincide with the two encoded by P_3, or (ii) that of the six points encoded by P_1, P_2 and P_3, no two share the same x-coordinate.*

3   Extending Jacobian Coordinates to Jacobians

Let λ be a nonzero element in K. Over fields of large characteristic, Jacobian coordinates have proven to be a natural and efficient way to work projectively on elliptic curves in short Weierstrass form $\mathcal{E}/K : y^2 = x^3 + ax + b$. Indeed, in cryptographic contexts, using the triple $(\lambda^2 X : \lambda^3 Y : \lambda Z) \in \mathbb{P}(2, 3, 1)(K)$ to represent the affine point $(X/Z^2, Y/Z^3) \in \mathbb{A}^2(K)$ on $\mathcal{E}$ was suggested by Miller in his seminal 1985 paper [29, p. 424], and his comment that this representation "appears best" still holds true after decades of further exploration: Jacobian coordinates (and extended variants) remain the most efficient way to work on such general Weierstrass curves [4]. Moreover, the weightings $\text{wt}(x) = 2$ and $\text{wt}(y) = 3$ are the orders of the poles of the functions x and y at the point at infinity on $\mathcal{E}$.

In the context of imaginary hyperelliptic curves of the form

$$\mathcal{C}/K : y^2 = x^5 + a_3 x^3 + a_2 x^2 + a_1 x + a_0,$$

the analogous weightings are

$$\mathrm{wt}(x) = 2, \qquad \text{and} \qquad \mathrm{wt}(y) = 5, \tag{6}$$

under which the affine point $(X/Z^2, Y/Z^5) \in \mathbb{A}^2(K)$ is represented by the triple $(\lambda^2 X : \lambda^5 Y : \lambda Z) \in \mathbb{P}(2, 5, 1)(K)$, which lies on

$$\mathcal{C}/K : Y^2 = X^5 + a_3 X^3 Z^4 + a_2 X^2 Z^6 + a_1 X Z^8 + a_0 Z^{10}. \tag{7}$$

Indeed, the weights $\mathrm{wt}(x) = 2$ and $\mathrm{wt}(y) = 5$ are the orders of the poles of x and y at the (unique) point at infinity on $\mathcal{C}$. Since we perform arithmetic using the Mumford coordinates in $J_{\mathcal{C}}$, rather than the x-y coordinates on $\mathcal{C}$, we transfer the above weightings across to the Mumford coordinates via Equation (3), which yields

$$\mathrm{wt}(q) = \mathrm{wt}(x), \quad \mathrm{wt}(r) = 2 \cdot \mathrm{wt}(x), \quad \mathrm{wt}(s) = \mathrm{wt}(y) - \mathrm{wt}(x), \quad \mathrm{wt}(t) = \mathrm{wt}(y). \tag{8}$$

Combining (6) and (8) then gives

$$\mathrm{wt}(q) = 2, \qquad \mathrm{wt}(r) = 4, \qquad \mathrm{wt}(s) = 3, \qquad \mathrm{wt}(t) = 5, \tag{9}$$

which suggests the use of $(\lambda^2 Q : \lambda^4 R : \lambda^3 S : \lambda^5 T : \lambda Z) \in \mathbb{P}(2, 4, 3, 5, 1)(K)$ to represent the affine point

$$(q, r, s, t) = \left(\frac{Q}{Z^2}, \frac{R}{Z^4}, \frac{S}{Z^3}, \frac{T}{Z^5} \right) \in \mathbb{A}^4(K). \tag{10}$$

Equation (10) is at the heart of this paper. We found these weightings to be highly advantageous for group law computations: the Mumford coordinates balance naturally under this projection, and significant simplifications occur regularly in the derivation of the corresponding explicit formulas. This coordinate system is referred to as *Jacobian coordinates* in this paper. We note that, in line with Assumption 1, we will not work with the full projective closure of the affine part in $\mathbb{P}(2, 4, 3, 5, 1)(K)$, but rather with the affine patch where $Z \neq 0$.

Just as in [25, §6], we found it useful to introduce an additional coordinate (independent of Z) in the denominator of the two coordinates corresponding to the v-polynomial in the Mumford representation. So, in addition to the Jacobian coordinate Z, we include the coordinate W and use the projective six-tuple $(\lambda^2 Q : \lambda^4 R : \lambda^3 \mu S : \lambda^5 \mu T : \lambda Z : \mu W)$ to represent the affine point

$$(q, r, s, t) = \left(\frac{Q}{Z^2}, \frac{R}{Z^4}, \frac{S}{Z^3 W}, \frac{T}{Z^5 W} \right) \in \mathbb{A}^4(K) \tag{11}$$

for some nonzero μ in K. This coordinate system is referred to as *auxiliary Jacobian coordinates* in this paper.

Remark 1. We note the distinction between the above coordinate weightings and the weightings used by Lange, which were also said to "generalise the concept of Jacobian coordinates ... from elliptic to hyperelliptic curves" [25, §6]. In terms of the first projective coordinate Z, Lange used $(q, r, s, t) = (Q/Z^2, R/Z^2, S/Z^3, T/Z^3)$. Although these weight the u- and v-polynomials of a point with the same (Jacobian) weightings as the x- and y-coordinates on an elliptic curve, the derivation of the weightings in (10) draws a closer analogy with the use of Jacobian coordinates in genus 1. This is why we dubbed the weightings used in this work as "Jacobian coordinates".

4   Adopting the "co-Z" Approach

With the aim of improving addition formulas on elliptic curves, Meloni [28] put forward a nice idea that is particularly suited to working in Jacobian coordinates. In the explicit addition of two elliptic curve points $(X_1 : Y_1 : Z_1)$ and $(X_2 : Y_2 : Z_2)$ in $\mathbb{P}(2, 3, 1)(K)$, which respectively correspond to the points $(X_1/Z_1^2, Y_1/Z_1^3)$ and $(X_2/Z_2^2, Y_2/Z_2^3)$ in $\mathbb{A}^2(K)$, Meloni observed that almost all expressions of the form $Z_1^i Z_2^j$ can completely vanish if $Z_1 = Z_2$. That is, the sum of the points $(X_1 : Y_1 : Z_1)$ and $(X_2 : Y_2 : Z_1)$ can be written as an expression of the form $(X_3 Z_1^6 : Y_3 Z_1^9 : Z_3 Z_1^3)$, which is projectively equivalent to $(X_3 : Y_3 : Z_3)$; here X_3 and Y_3 depend only on X_1, Y_1, X_2 and Y_2, so now it is only Z_3 that depends on Z_1. Since two projective points are unlikely to share the same Z-coordinate in general, the method starts by updating one or both of the input points to force this equivalence. The obvious way to do this is to respectively cross-multiply $(X_1 : Y_1 : Z_1)$ and $(X_2 : Y_2 : Z_2)$ into $(X_1 Z_2^2 : Y_1 Z_2^3 : Z_1 Z_2)$ and $(X_2 Z_1^2 : Y_2 Z_1^3 : Z_2 Z_1)$, but as it stands, performing this update would incur a significant overhead. The observation that is key to making this "co-Z" approach advantageous is that, in the context of scalar multiplications, these updated values (or the main subexpressions within them) are often already computed in the previous operation [28, p. 192], so this update can be performed either for free, or with a much smaller overhead.

Meloni did not apply his idea to classical "double-and-add" style addition chains, but subsequent papers [26,19] showed how his approach could be used to enhance performance in such binary chains. In genus 2 however, successful transferral of the "co-Z" idea has not yet been achieved: the work in [24] also uses non-binary addition chains, and crucially, it was performed without access to the hyperelliptic analogue of Jacobian coordinates (those which work in stronger synergy with Meloni's idea).

Our adaptation of the "co-Z" approach requires that both the Z and W coordinates are the same, for two different input points. The first projective formulas we derive in Section 6 are for the "co-ZW" addition between the two points $P_1 = (Q_1 : R_1 : S_1 : T_1 : Z_1 : W_1)$ and $P_2 = (Q_2 : R_2 : S_2 : T_2 : Z_1 : W_1)$, and this routine is then used as a subroutine for all subsequent operations (except for standalone doublings).

5 Arithmetic in Affine Coordinates with New Common Subexpressions

The explicit formulas for arithmetic in genus 2 Jacobians are significantly more complicated than their elliptic curve counterparts, so it is especially useful to start the derivation by looking for common subexpressions and advantageous orderings in the affine versions of the formulas (i.e., before the introduction of more coordinates complicates the situation further). Our derivation follows that of [9], but it is important to point out that the resulting affine formulas have been refined by grouping new subexpressions throughout; these groupings were strategically chosen to exploit the symmetries of the q and r coordinates, and especially for the application of Jacobian coordinates that follows in Section 6.

In what follows, we give the affine formulas for general point additions and general point doublings respectively. From Section 2, recall the abbreviated notation $(q, r, s, t) \in \mathbb{A}^4(K)$ for the point in J_C with Mumford representation $(x^2 + qx + r, sx + t)$.

Let $P_1 = (q_1, r_1, s_1, t_1)$, $P_2 = (q_2, r_2, s_2, t_2)$ and $P_1 + P_2 =: P_3 = (q_3, r_3, s_3, t_3)$ be points in J_C satisfying Assumption 1. The choice of the three subexpressions

$$
\begin{aligned}
A &:= (t_1 - t_2)(q_2(q_1 - q_2) - (r_1 - r_2)) - r_2(q_1 - q_2)(s_1 - s_2), \\
B &:= (r_1 - r_2)(q_2(q_1 - q_2) - (r_1 - r_2)) - r_2(q_1 - q_2)^2, \\
C &:= (q_1 - q_2)(t_1 - t_2) - (r_1 - r_2)(s_1 - s_2)
\end{aligned}
$$

is key to our refined derivation. The point P_3 is then given by

$$
\begin{aligned}
q_3 &= (q_1 - q_2) + 2\frac{A}{C} - \frac{B^2}{C^2}, \\
r_3 &= (q_1 - q_2)\frac{A}{C} + \frac{A^2}{C^2} + (q_1 + q_2)\frac{B^2}{C^2} - (s_1 + s_2)\frac{B}{C}, \\
s_3 &= (r_1 - r_3)\frac{C}{B} - q_3(q_1 - q_3)\frac{C}{B} + (q_1 - q_3)\frac{A}{B} - s_1, \\
t_3 &= (r_1 - r_3)\frac{A}{B} - r_3(q_1 - q_3)\frac{C}{B} - t_1.
\end{aligned}
\tag{12}
$$

These formulas are used to derive the projective co-ZW addition formulas in §6.1, those which form a basis for all of the other (non-doubling) formulas in this work.

Let $P_1 = (q_1, r_1, s_1, t_1)$ and $[2]P_1 =: P_3 = (q_3, r_3, s_3, t_3)$ be points in J_C satisfying Assumption 1. Again, it is particularly useful to make use of three subexpressions:

$$
\begin{aligned}
A &:= \left((q_1^2 - 4r_1 + a_3)q_1 - a_2 + s_1^2\right)(q_1 s_1 - t_1) + \left(3q_1^2 - 2r_1 + a_3\right)r_1 s_1, \\
B &:= 2(q_1 s_1 - t_1)t_1 - 2r_1 s_1^2, \\
C &:= \left((q_1^2 - 4r_1 + a_3)q_1 - a_2 + s_1^2\right)s_1 + \left(3q_1^2 - 2r_1 + a_3\right)t_1.
\end{aligned}
$$

The point P_3 is then given by

$$
\begin{aligned}
q_3 &= 2\frac{A}{C} - \frac{B^2}{C^2}, \\
r_3 &= \frac{A^2}{C^2} + 2q_1\frac{B^2}{C^2} - 2s_1\frac{B}{C}, \\
s_3 &= (r_1 - r_3)\frac{C}{B} - q_3(q_1 - q_3)\frac{C}{B} + (q_1 - q_3)\frac{A}{B} - s_1, \\
t_3 &= (r_1 - r_3)\frac{A}{B} - r_3(q_1 - q_3)\frac{C}{B} - t_1.
\end{aligned}
\tag{13}
$$

These formulas are used to derive projective doubling formulas in §6.5. The formulas in (12) and (13) agree with those of Costello-Lauter [9].

6 Projective Arithmetic in Extended Jacobian Coordinates

In this section we derive all of the explicit formulas that are needed for the scalar multiplication routines we describe in Section 7. The formulas are summarised in Table 2 below, where we immediately note the extension of auxiliary Jacobian coordinates discussed in Section 3 to include W^2; it is advantageous to carry this additional coordinate between consecutive operations because it is often computed *en route* to the output points already, and therefore comes for free as input into the following operation. We refer to this extended version of auxiliary Jacobian coordinates as *extended Jacobian coordinates*. Table 2 reports two sets of operation counts: the "plain" count, which corresponds to our deriving sets of formulas with the aim of minimising the total number of all field operations, and the "trade-offs" count.

If W^2 is dropped from the coordinate system, and we work only with auxiliary Jacobian coordinates, $(Q : R : S : T : Z : W)$, then we note that both DBL and DBLa2a3zero would require one extra squaring (in both the "plain" and "trade-off" formulas). The only other change resulting from this abbreviated coordinate system would be in the "trade-off" version of ADD, where a squaring would revert back to a multiplication. All other operation counts would remain unchanged.

Following on from the discussion in Section 4, in §6.1 we start the derivations by using the affine addition formulas in (12) to develop projective formulas for zwADD; these are then used in the derivation of the formulas for ADD in §6.2, for mADD in §6.3, and for mDBLADD in §6.4. Finally, we use the affine doubling formulas in (13) to develop projective formulas for DBL in §6.5.

6.1 Projective co-ZW Addition (zwADD)

Let $P_1 = (Q_1 : R_1 : S_1 : T_1 : Z_1 : W_1)$, $P_2 = (Q_2 : R_2 : S_2 : T_2 : Z_1 : W_1)$, and $P_1 + P_2 =: P_3 = (Q_3 : R_3 : S_3 : T_3 : Z_3 : W_3 : W_3^2)$ represent three points in $J_\mathcal{C}$ satisfying Assumption 1. We emphasize that P_1 and P_2 need not contain W_1^2, which is why both are given in auxiliary Jacobian coordinates. However, the output P_3 is in extended Jacobian coordinates. The projective form of (12)

Table 2. A summary of the explicit formulas derived in this section for various operations in the Jacobian, $J_\mathcal{C}$, of an imaginary hyperelliptic curve $\mathcal{C}/K$ of genus 2, with $\operatorname{char}(K) > 5$

operation in $J_\mathcal{C}$	description of operation	derived in	field operations "plain"	w. "trade-offs"
zwADD	$(Q_1 : R_1 : S_1 : T_1 : Z_1 : W_1)$ $+(Q_2 : R_2 : S_2 : T_2 : Z_1 : W_1)$	§6.1	$25M + 3S$ $+22a$	$23M + 4S$ $+40a$
ADD	$(Q_1 : R_1 : S_1 : T_1 : Z_1 : W_1 : W_1^2)$ $+(Q_2 : R_2 : S_2 : T_2 : Z_2 : W_2 : W_2^2)$	§6.2	$41M + 7S$ $+22a$	$35M + 12S$ $+56a$
mADD	$(Q_1 : R_1 : S_1 : T_1 : Z_1 : W_1 : W_1^2)$ $+(Q_2 : R_2 : S_2 : T_2 : 1 : 1 : 1)$	§6.3	$32M + 5S$ $+22a$	$29M + 7S$ $+44a$
mDBLADD	$[2](Q_1 : R_1 : S_1 : T_1 : Z_1 : W_1 : W_1^2)$ $+(Q_2 : R_2 : S_2 : T_2 : 1 : 1 : 1)$	§6.4	$57M + 8S$ $+42a$	$52M + 11S$ $+82a$
DBL	$[2](Q_1 : R_1 : S_1 : T_1 : Z_1 : W_1 : W_1^2)$	§6.5	$26M + 8S + 2D$ $+25a$	$21M + 12S + 2D$ $+52a$
DBL a2a3zero	$[2](Q_1 : R_1 : S_1 : T_1 : Z_1 : W_1 : W_1^2)$ (when $a_2a_3 = 0$)	§6.5	$25M + 6S$ $+22a$	$21M + 9S$ $+48a$

in extended Jacobian coordinates corresponds to the following. We define the subexpressions

$$A := (T_1 - T_2)(Q_2(Q_1 - Q_2) - (R_1 - R_2)) - R_2(Q_1 - Q_2)(S_1 - S_2),$$
$$B := (R_1 - R_2)(Q_2(Q_1 - Q_2) - (R_1 - R_2)) - R_2(Q_1 - Q_2)^2,$$
$$C := (Q_1 - Q_2)(T_1 - T_2) - (R_1 - R_2)(S_1 - S_2).$$

The point P_3 is then given by

$$
\begin{aligned}
W_3 &= W_1[B], \\
Q_3 &= \left(Q_1\left[C^2\right] - Q_2\left[C^2\right]\right) + 2AC - W_3^2, \\
R_3 &= \left(Q_1\left[C^2\right] - Q_2\left[C^2\right] + AC\right)AC+ \\
      &\quad \left(Q_1\left[C^2\right] + Q_2\left[C^2\right]\right)W_3^2 - S_1\left[C^3B\right] - S_2\left[C^3B\right], \\
S_3 &= \left(R_1\left[C^4\right] - R_3\right) + (AC - Q_3)\left(Q_1\left[C^2\right] - Q_3\right) - S_1\left[C^3B\right], \\
T_3 &= \left(R_1\left[C^4\right] - R_3\right)AC - R_3\left(Q_1\left[C^2\right] - Q_3\right) - T_1\left[C^5B\right], \\
Z_3 &= Z_1[C].
\end{aligned}
\tag{14}
$$

This operation, referred to as zwADD, not only computes P_3, but also produces the subexpressions $Q_1\left[C^2\right]$, $R_1\left[C^4\right]$, $S_1\left[C^3B\right]$, $T_1\left[C^5B\right]$, $Z_1[C]$, $W_1[B]$, $W_1^2\left[B^2\right]$; if desired, these can be used to update P_1 to be of the form

$$
\begin{aligned}
P_1 &= (Q_1 : R_1 : S_1 : T_1 : Z_1 : W_1 : W_1^2) \\
    &= (Q_1\left[C^2\right] : R_1\left[C^4\right] : S_1\left[C^3B\right] : T_1\left[C^5B\right] : Z_1[C] : W_1[B] : W_1^2\left[B^2\right]),
\end{aligned}
$$

so that it now has the same Z, W, and W^2 coordinates as P_3. The combination of the zwADD operation and this update will be denoted using the syntax

$$(P_3, P_1') := P_1 + P_2,$$

where P_1' is the updated (but projectively equivalent) version of P_1.

6.2 Projective Addition (ADD)

Rather than producing lengthy formulas for additions, we use a simple construction that exploits zwADD. Let $P_1 = (Q_1 : R_1 : S_1 : T_1 : Z_1 : W_1 : W_1^2)$, $P_2 = (Q_2 : R_2 : S_2 : T_2 : Z_2 : W_2 : W_2^2)$, and $P_1 + P_2 =: P_3 = (Q_3 : R_3 : S_3 : T_3 : Z_3 : W_3 : W_3^2)$ represent three points in J_C satisfying Assumption 1. We can then cross-multiply to define the points in auxiliary Jacobian coordinates

$$P_1' := \left(Q_1[Z_2^2] : R_1[Z_2^4] : S_1[Z_2^3 W_2] : T_1[Z_2^5 W_2] : Z_1[Z_2] : W_1[W_2] \right),$$
$$P_2' := \left(Q_2[Z_1^2] : R_2[Z_1^4] : S_2[Z_1^3 W_1] : T_2[Z_1^5 W_1] : Z_2[Z_1] : W_2[W_1] \right).$$

Observe that $P_1' = P_1$ and $P_2' = P_2$, but that P_1' and P_2' now share the same Z and W coordinates. This means that we can use the zwADD operation defined in §6.1 to compute $P_3 = P_1 + P_2$ as $(P_3, P_1'') := P_1' + P_2'$. Observe that $P_1'' = P_1$, and that P_1'' will share the same Z, W, and W^2 coordinates as P_3. We note that this update of P_1 into P_1'' can be useful in the generation of lookup tables [26], but is generally not useful during the main loop.

6.3 Projective Mixed Addition (mADD)

In a similar way, let $P_1 = (Q_1 : R_1 : S_1 : T_1 : Z_1 : W_1 : W_1^2)$, $P_2 = (Q_2 : R_2 : S_2 : T_2 : 1 : 1 : 1)$, and $P_1 + P_2 =: P_3 = (Q_3 : R_3 : S_3 : T_3 : Z_3 : W_3 : W_3^2)$ represent three points in J_C satisfying Assumption 1. This time we only need to update P_2 into P_2', which is performed in auxiliary Jacobian coordinates as

$$P_2' := \left(Q_2[Z_1^2] : R_2[Z_1^4] : S_2[Z_1^3 W_1] : T_2[Z_1^5 W_1] : [Z_1] : [W_1] \right),$$

where we observe that P_1 and P_2' now have the same Z and W coordinates. Subsequently, using the zwADD operation from §6.1 allows $P_3 = P_1 + P_2$ to be computed by $(P_3, P_1') := P_1 + P_2'$.

6.4 Projective Mixed Doubling-and-Addition (mDBLADD)

Let $P_1 = (Q_1 : R_1 : S_1 : T_1 : Z_1 : W_1 : W_1^2)$, $P_2 = (Q_2 : R_2 : S_2 : T_2 : 1 : 1 : 1)$, and $[2]P_1 + P_2 =: P_3 = (Q_3 : R_3 : S_3 : T_3 : Z_3 : W_3 : W_3^2)$, represent three points in J_C satisfying Assumption 1. To compute $[2]P_1 + P_2$, we schedule the higher level operations in the form $(P_1 + P_2) + P_1$ (see [11] and [26] for the same high level scheduling). This means that mDBLADD can be computed using an mADD operation before a zwADD operation. (Subsequently, we must also assume that P_1, the intermediate point $P_1 + P_2$, and the output point $[2]P_1 + P_2 =: P_3 = (Q_3 : R_3 : S_3 : T_3 : Z_3 : W_3 : W_3^2)$ represent three points in J_C satisfying Assumption 1.)

6.5 Projective Doubling (DBL)

Let $P_1 = (Q_1 : R_1 : S_1 : T_1 : Z_1 : W_1 : W_1^2)$ and $[2]P_1 =: P_3 = (Q_3 : R_3 : S_3 : T_3 : Z_3 : W_3 : W_3^2)$ represent two points in J_C satisfying Assumption 1.

The projective form of (13) in extended Jacobian coordinates corresponds to the following. We define the subexpressions

$$A := \left(\left(Q_1 \left(Q_1^2 - 4R_1 \right) + \left(Q_1 - \left(a_2/a_3 \right) Z_1^2 \right) a_3 Z_1^4 \right) W_1^2 + S_1^2 \right) \left(Q_1 S_1 - T_1 \right) + \\ \left(3Q_1^2 - 2R_1 + a_3 Z_1^4 \right) W_1^2 R_1 S_1,$$

$$B := 2 \left(Q_1 S_1 - T_1 \right) T_1 - 2R_1 S_1^2,$$

$$C := \left(\left(Q_1 \left(Q_1^2 - 4R_1 \right) + \left(Q_1 - \left(a_2/a_3 \right) Z_1^2 \right) a_3 Z_1^4 \right) W_1^2 + S_1^2 \right) S_1 + \\ \left(3Q_1^2 - 2R_1 + a_3 Z_1^4 \right) W_1^2 T_1.$$

We can then write P_3 as

$$
\begin{aligned}
W_3 &= W_1[B], \\
Q_3 &= 2AC - W_3^2, \\
R_3 &= (AC)^2 + 2Q_1 \left[C^2 \right] W_3^2 - 2S_1 \left[C^3 B \right], \\
S_3 &= \left(R_1 \left[C^4 \right] - R_3 \right) + (AC - Q_3) \left(Q_1 \left[C^2 \right] - Q_3 \right) - S_1 \left[C^3 B \right], \\
T_3 &= \left(R_1 \left[C^4 \right] - R_3 \right) AC - R_3 \left(Q_1 \left[C^2 \right] - Q_3 \right) - T_1 \left[C^5 B \right], \\
Z_3 &= Z_1[C].
\end{aligned}
\tag{15}
$$

The DBL operation not only computes P_3, but also produces the subexpressions $Q_1 \left[C^2 \right]$, $R_1 \left[C^4 \right]$, $S_1 \left[C^3 B \right]$, $T_1 \left[C^5 B \right]$, $Z_1[C]$, $W_1[B]$, $W_1^2 \left[B^2 \right]$; if desired, these can be used to update P_1 into

$$
\begin{aligned}
P_1 &= \left(Q_1 : R_1 : S_1 : T_1 : Z_1 : W_1 : W_1^2 \right) \\
&= \left(Q_1 \left[C^2 \right] : R_1 \left[C^4 \right] : S_1 \left[C^3 B \right] : T_1 \left[C^5 B \right] : Z_1[C] : W_1[B] : W_1^2 \left[B^2 \right] \right),
\end{aligned}
$$

in order to share the same Z, W, and W^2 coordinates with P_3. We define the operation DBLa2a3zero to be a doubling in the special case that the curve constants a_2 and a_3 are zero.

7   Implementation

We chose two different curves to showcase the explicit formulas derived in the previous section, both of which target the 128-bit security level.

The first curve was found in the colossal point counting effort undertaken by Gaudry and Schost [18]. From a security standpoint, it is both twist-secure and it is not considered to be special (e.g. it has a large discriminant); from a performance standpoint, it was chosen over the arithmetically advantageous field $\mathbb{F}_p$ with $p = 2^{127} - 1$, and with optimal cofactors such that the curve supports a Gaudry-style Kummer surface implementation [16]. This is the same Kummer surface that was used to set speed records in [7] and [3]. We chose the Jacobian of this curve to illustrate the performance that is gained when using our new formulas inside a general "double-and-add" scalar multiplication routine.

The second curve supports a 4-dimensional Gallant-Lambert-Vanstone (GLV) decomposition [15]. Over prime fields, requiring 4-dimensional GLV imposes that the Jacobian has complex multiplication (CM) by a special field – in this case it is $\mathbb{Q}(\zeta_5)$. This (specialness) means that we cannot hope to find a twist-secure curve over a particular prime, but rather that we must search over many primes. In the same vein as [7, §8.3], we also wanted this curve to support a rational Gaudry-style Kummer surface. This curve is defined over the prime field $p = 2^{128} - c$ with $c = 7689975$, which is the *smallest* $c > 0$ such that a curve with CM by $\mathbb{Q}(\zeta_5)$ over $\mathbb{F}_p$ is twist-secure with optimal cofactors[5]. This curve was chosen to exhibit the performance that is gained when using our new formulas inside a GLV-style multiexponentiation; in particular, each step of the multiexponentiation requires only an mDBLADD operation, and this is where our explicit formulas offer the largest relative speedup over the previous ones.

7.1 Working on the Gaudry-Schost Jacobian

Let $p = 2^{127} - 1$, and define the following constants in $\mathbb{F}_p$: $a := 11$, $b := -22$, $c := -19$, $d := -3$, $e := 1 + \sqrt{-833/363}$ and $f := 1 - \sqrt{-833/363}$. For the Rosenhain invariants $\lambda = \frac{ac}{bd}$, $\mu = \frac{ce}{df}$, $\nu = \frac{ae}{bf}$, the curve

$$\mathcal{C}_{Ros}/\mathbb{F}_p : y^2 = x(x-1)(x-\lambda)(x-\mu)(x-\nu)$$

is such that $\#J_{\mathcal{C}_{Ros}} = 2^4 \cdot r$ and $\#J_{\mathcal{C}'_{Ros}} = 2^4 \cdot r'$, where r and r' are 250- and 251-bit primes respectively [18], and where $\mathcal{C}'_{Ros}$ is the quadratic twist of $\mathcal{C}_{Ros}$. The coefficient of x^4 in $\mathcal{C}_{Ros}$ is $\alpha = -(1 + \lambda + \mu + \nu)$, and we choose to zero it under the transformation $\varphi : \mathcal{C}_{Ros} \to \tilde{\mathcal{C}}, (x, y) \mapsto (x - \alpha/5, y)$. The resulting curve, $\tilde{\mathcal{C}}$, has a coefficient of x^3 which is a fourth power in $\mathbb{F}_p$; let it be u^{-4}, where we chose $u = 1985974119227654614210545699131932$8298. We can then use the map $\psi : \tilde{\mathcal{C}} \to \mathcal{C}, (x, y) \mapsto (x \cdot u^2, y \cdot u^5)$ to work with the isomorphic curve $\mathcal{C}/\mathbb{F}_p : y^2 = x^5 + x^3 + a_2 x^2 + a_1 x + a_0$, where the coefficient of x^3 being 1 saves a multiplication inside every point doubling[6]. We use the name Jac1271 for the Jacobian $J_\mathcal{C}$, and use the name Kum1271 for the associated Kummer surface $\mathcal{K}$ – this is defined by the above constants a, b, c, d (see [16]).

In Section 8 we report two new sets of implementation numbers on Jac1271. First, we benchmark a generic scalar multiplication, using both the old and the new formulas, to illustrate the performance boost given by this work in the general case. In addition, we benchmark a *fixed-base* scalar multiplication, which uses the new formulas and takes advantage of precomputations on a public generator to give large speedups on Jac1271. In the context of ECDHE, this second benchmark corresponds to the "key_gen" phase, which compliments the performance numbers for the "shared_secret" scalar multiplications on Kum1271 in [7] and [3]. (We discuss some caveats related to this Jacobian/Kummer

---

[5] It is relatively straightforward to show that if $J_\mathcal{C}$ has CM by $\mathbb{Q}(\zeta_5)$ and full rational two-torsion, then either $J_\mathcal{C}$ or $J_{\mathcal{C}'}$ must contain a point of order 5; thus, the optimal cofactors are 16 and 80.

[6] If the coefficient of x^3 in $\tilde{\mathcal{C}}$ was not a fourth power, one could still use this form of transformation to achieve another "small" coefficient, or in this case, work on the twist instead.

combination in §7.3.) To tie these two sets of performance numbers together, we also benchmark the numbers for computing the map from Jac1271 onto Kum1271, which was made explicit in the AVIsogenies library [6], and for general points in $J_{\mathcal{C}_{Ros}}$ is given as

$$\Psi \: : \: J_{\mathcal{C}_{Ros}} \to \mathcal{K}, \qquad (x^2 + qx + r, sx + t) \mapsto (X : Y : Z : T),$$

where

$$X = a\left(r(\mu - r)(\lambda + q + \nu) - t^2\right), \quad Y = b\left(r(\nu\lambda - r)(1 + q + \mu) - t^2\right),$$
$$Z = c\left(r(\nu - r)(\lambda + q + \mu) - t^2\right), \quad T = d\left(r(\mu\lambda - r)(1 + q + \nu) - t^2\right). \quad (16)$$

For practical scenarios like ECDHE, it is fortunate that we only need the map in this direction, as the pullback map from $\mathcal{K}$ to $J_{\mathcal{C}_{Ros}}$ is much more complicated [16, §4.3]. Since we compute in $J_{\mathcal{C}}$ (rather than $J_{\mathcal{C}_{Ros}}$), we actually need to compute the composition of Ψ with $(\psi\varphi)^{-1}$, which when extended to general points in $J_{\mathcal{C}}$ is

$$(\psi\varphi)^{-1} \: : \: J_{\mathcal{C}} \to J_{\mathcal{C}_{Ros}} \quad , \quad (x^2 + qx + r, sx + t) \mapsto (x^2 + q'x + r', s'x + t'),$$

with $q' = u^{-2}q + 2\alpha/5$, $r' = u^{-4}r + \alpha/5q' - (\alpha/5)^2$, $s' = u^{-3}s$, $t' = u^{-5}t + \alpha/5s'$. Assuming that the input point in $J_{\mathcal{C}}$ is in extended Jacobian coordinates, the operation count for the full map $\Psi' = \Psi(\psi\varphi)^{-1}$ from $J_{\mathcal{C}}$ to $\mathcal{K}$ is $1\mathbf{I} + 31\mathbf{M} + 2\mathbf{S} + 19\mathbf{a}$; we benchmark it alongside the scalar multiplications in Section 8.

To draw a fair comparison against prior works, we inserted our formulas into the software made publicly available by Bos *et al.* [7], which itself employed the previous best formulas. (We tweaked both sets of formulas for Jac1271 to take advantage of the constant $a_3 = 1$.) This software computes the scalar multiplications on Jac1271 using an adaptation of the left-to-right signed sliding window recoding from [1] with a window size of $w = 5$, where the lookup table consists of 8 points and is constructed using the same approach as in [26, §4]. The timings are presented in Section 8.

7.2 Working on the Jacobian of a GLV curve

Let $p = 2^{128} - 7689975$ and define $\mathcal{C}/\mathbb{F}_p \: : \: y^2 = x^5 + 7^{10}$. The Jacobian groups $J_{\mathcal{C}}$ and $J_{\mathcal{C}'}$ have cardinalities $\#J_{\mathcal{C}} = 2^4 \cdot 5 \cdot r$ and $\#J_{\mathcal{C}'} = 2^4 \cdot r'$, where

$$r = (2^{252} + 37557692833123369178214679267779826721358413165176440 4159)/5,$$
$$r' = 2^{252} - 375576928331887882475846226038533397089218679777223482485$$

are both prime.

The implementation of a 4-dimensional GLV scalar multiplication in $J_{\mathcal{C}}$ follows that which is described in [7, §6]; again, we wrapped their GLV software around both their old and our new formulas for a fair comparison – we note that both instances were made to use the above curve, which we refer to as GLV128c.

Practically speaking, it does not make as much sense to benchmark GLV128c in the same ECDHE style as we discussed for Jac1271 and Kum1271. If there is enough storage to exploit a long-term public generator P, then the presence of endomorphisms is essentially redundant in the key_gen phase, since multiples of P can then be precomputed offline without using an endomorphism. On the shared_secret side, where *variable-base* scalar multiplications are performed on fresh inputs, our implementations show that a 4-dimensional decomposition on GLV128c is still slightly slower than a Kummer surface scalar multiplication, so in the case of ECDHE, it is likely to be faster on both sides to stick with the combination of Jac1271 and Kum1271. Nevertheless, there could be scenarios where it makes sense to use the endomorphism on GLV128c (e.g. for a signature verification), and still make use of the maps between the full Jacobian group and the associated Kummer surface. In this case, the map in (16) and the pullback map in [16, §4.3] can be exploited analogously to the case of Jac1271, keeping in mind that the maps would pass through the Jacobian of the Rosenhain form of $\mathcal{C}$.

Timings for a 4-dimensional GLV variable-base scalar multiplication on GLV128c using both the old and the new explicit formulas are given in Section 8.

We note that in all scalar multiplication routines, i.e. in both fixed- and variable-base scalar multiplications on Jac1271 and in 4-dimensional multiexponentiations on GLV128c, we always found it advantageous to convert the lookup table elements from extended Jacobian coordinates to affine coordinates using Montgomery's simultaneous inversion method [30]. This "decision" is generally made easier in genus 2, where the difference between mixed additions and full additions is greater, and the relative cost of a field inversion (compared to the rest of the scalar multiplication routine) is much less than it is in the elliptic curve case. Finally, we note that the single conversion of the output point from Jacobian to affine coordinates comes at a cost of $1\mathbf{I} + 10\mathbf{M} + 1\mathbf{S}$.

7.3   A Disclaimer: The Difficulties Facing Constant-Time, Exception-Free Scalar Multiplications in $J_\mathcal{C}$

We must point out that none of the scalar multiplications on Jac1271 or GLV128c that we report in this paper run in constant time, and that the difficulties of achieving such a routine in genus 2 Jacobians is closely related to Assumption 1. We note that these are not the same implementation-level difficulties pointed out in [3, §1.2]; indeed, while the Kummer surface implementations reported in [3] and [7] run in constant time, a truly constant-time genus 2 implementation that does not use the Kummer surface is yet to be documented in the literature.

More specifically, there are scalar recoding algorithms (cf. [20,12]) that make it possible to implement the Jac1271 or GLV128c routines such that scalar multiplications on random inputs will run in constant time with probability exponentially close to 1. However, in order to guard against active adversaries and to be considered *truly* constant-time, the routines should be guaranteed to execute identically and run correctly for *all* combinations of integer scalars and input points; this means the explicit formulas must be able to handle input combinations in $J_\mathcal{C}$ that are not "general" in the sense of Assumption 1. Although explicit formulas can be developed for each of these special cases, their culmination into an efficient and truly constant-time scalar multiplication algorithm remains an important open problem.

8   Results

In this section we present the timings of the routines described in the previous section. All of the benchmarks were performed on an Intel Core i7-3770M (Ivy Bridge) processor at 3.4 GHz with hyperthreading turned off and over-clocking ("turbo-boost") disabled, and all-but-one of the cores switched off in BIOS. The implementations were compiled with gcc 4.6.3 with the -O2 flag set and tested on a 64-bit Linux environment. Cycles were obtained using the SUPERCOP [5] toolkit and then rounded to the nearest 1,000 cycles.

The primary purpose of our benchmarks is to compare the performance of scalar multiplications in genus 2 Jacobians using both the old and new sets of explicit formulas. Table 3 reports that a generic scalar multiplication on Jac1271 using the explicit formulas in this paper gives a factor 1.25x improvement over one that uses the previous best formulas; this is the approximate speedup that one can expect when adopting extended Jacobian coordinates on any imaginary hyperelliptic curve of genus 2 over a large prime field. Table 4 reports that a 4-dimensional GLV multiexponentiation routine using the explicit formulas in this paper gives a factor 1.29x improvement over the same routine that calls the previous explicit formulas. We note that the benchmarked implementations of the new formulas always used the "plain" versions (see Table 2), since these proved to be more efficient than the "trade-off" versions in our implementations.

Table 3. Benchmarking the old and new explicit formulas in the context of a generic scalar multiplication on Jac1271

curve	coordinates	formulas from	cycles
Jac1271	homogeneous	[9,7]	243,000
Jac1271	ext. Jacobian	this work	195,000

Table 4. Benchmarking the old and new explicit formulas in the context of a 4-GLV scalar multiplication on GLV128c

curve	coordinates	formulas from	cycles
GLV128c	homogeneous	[9,7]	166,000
GLV128c	ext. Jacobian	this work	129,000

As a secondary set of benchmarks, in Table 5 we give summary performance numbers for the Gaudry-Schost curve in §7.1 in the context of ECDHE. Using extended Jacobian coordinates and precomputing a lookup table of size 256KB, each `key_gen` operation takes around 40,000 cycles in total. (Note that this cycle count excludes the cycles required to transfer the lookup table from main memory to the cache.) Together with the recent Kummer surface performance numbers of Bernstein *et al.* [3], this gives an idea of the performance that is possible when space permits a significant precomputation in genus 2 ECDHE. Note, however, that until an efficient remedy to the issues discussed in §7.3 is known, this style of `key_gen` in genus 2 is unprotected against side-channel attacks. We also benchmarked a fixed-base scalar multiplication with a much smaller 1KB lookup table, but it ran in 87,000, which when combined with the Ψ' map, is not faster than the scalar multiplication on `Kum1271` from [3].

Table 5. The performance of genus 2 in ECDHE on the Gaudry-Schost curve [18]

ECDHE operation	details	curve	implementation	cycles
`key_gen`	fixed-base scalar mul	`Jac1271`	this work	36,000
	Ψ' map	-	this work (and [6])	4,000
`shared_secret`	variable-base scalar mul	`Kum1271`	Bernstein *et al.* [3]	91,000

We reiterate that, to get the performance numbers in Tables 3 and 4, and those for `key_gen` in Table 5, we modified the software made publicly available by Bos *et al.* [7] to be able to call both sets of explicit formulas. This software already included routines for general scalar multiplications, 4-GLV scalar multiplications, and the fixed-base scenario. To complete the benchmarks in Table 5, we ran the publicly available software from [3] on our hardware.

9   Related Scenarios

We conclude by mentioning some related cases of interest, for which the analogue of (extended) Jacobian coordinates and/or the co-Z idea could also be applied. The takeaway message of this section is that, while we focussed on the most common instance of genus 2 curves, the ideas in this work have the potential to boost the speed of arithmetic in other scenarios too.

- **Real Hyperelliptic Curves.** In Section 2 we immediately specialised to the *imaginary* case, where C/K is hyperelliptic of degree 5 with one point at infinity. The other case in genus 2, where the curve is of degree 6 and has two points at infinity [13], has received less attention in papers pursuing high performance, since it is slightly slower than the imaginary case [10]. Moreover, it is often the case (at least among the scenarios of practical interest) that a degree 6 model contains a rational Weierstrass point and can therefore be transformed to a degree 5 model (e.g. the family in [17, §4.4]). On the other hand, there are some scenarios where this transformation is not always possible, so it is of interest to see how efficient projective arithmetic can be made in the real case, and whether analogues of the ideas in this work can be carried across successfully.

- **Pairings.** Genus 2 pairings are also likely to benefit from Jacobian coordinates. Roughly speaking, the explicit formulas in this paper inherently compute the additional components (i.e. the Miller functions) that are required in a pairing computation. However, the resulting savings would not be as drastic, as the operations in J_C are dominated by extension field operations in a pairing computation. In addition, genus 2 has not been as competitive in the realm of pairings as it has as a standard discrete logarithm primitive, largely because the construction of competitive ordinary, pairing-friendly hyperelliptic curves has been very limited. On the other hand, there are attractive constructions of supersingular genus 2 curves [14], which may be of interest in the "Type 1" setting, especially given that the fastest instantiations of such pairings are (in recent times) considered broken [2]. Interestingly, the construction in [14, §7] is one example of a scenario where the real model cannot be converted into an imaginary one in general.

- **Low Characteristic / Higher Genus.** The specialisation of Jacobian coordinates to low characteristic genus 2 curves and the extension to higher genus imaginary hyperelliptic curves follows analogously. However, the motivation in both directions is nowadays stunted by their respective security concerns. Nevertheless, it could be worthwhile to see how much faster the arithmetic in these cases can become when using Jacobian coordinates.

- **The RM Families.** We benchmarked the new explicit formulas in two scenarios; on a non-special "generic" curve, and on a curve with very special CM that subsequently comes equipped with an endomorphism. A third option comes from the families with explicit RM in [17], which perhaps achieves the best of both worlds in genus 2: they also come equipped with an endomorphism, but are much more general than the CM curve we used. This generality dispels any security concerns associated with special curves, and moreover allows them to be found over a fixed prime field. Thus, at the 128-bit security level, one could find such a curve over $p = 2^{127} - 1$ that facilitates both 2-dimensional GLV decomposition on its Jacobian and which supports a (twist-secure) Kummer surface. It would then be interesting to benchmark the new explicit formulas on one of these families, where the GLV routine would again make a higher relative frequency of calls to the fast mDBLADD routine.

Acknowledgements. We thank Joppe Bos, Michael Naehrig, Benjamin Smith, and Osmanbey Uzunkol for their useful comments on an early draft of this work. We also thank the anonymous referees for their valuable comments.

References

1. Avanzi, R.M.: A note on the signed sliding window integer recoding and a left-to-right analogue. In: Handschuh, H., Hasan, M.A. (eds.) SAC 2004. LNCS, vol. 3357, pp. 130–143. Springer, Heidelberg (2004)
2. Barbulescu, R., Gaudry, P., Joux, A., Thomé, E.: A heuristic quasi-polynomial algorithm for discrete logarithm in finite fields of small characteristic. In: Nguyen, P.Q., Oswald, E. (eds.) EUROCRYPT 2014. LNCS, vol. 8441, pp. 1–16. Springer, Heidelberg (2014)
3. Bernstein, D.J., Chuengsatiansup, C., Lange, T., Schwabe, P.: Kummer strikes back: new DH speed records. IACR Cryptology ePrint Archive, 2014:134 (2014)
4. Bernstein, D.J., Lange, T.: Explicit-formulas database, http://www.hyperelliptic.org/EFD/ (accessed January 2, 2014)
5. Bernstein, D.J., Lange, T.: eBACS: ECRYPT Benchmarking of Cryptographic Systems, http://bench.cr.yp.to (accessed September 28, 2013)
6. Bisson, G., Cosset, R., Robert, D.: AVIsogenies – a library for computing isogenies between abelian varieties (November 2012), http://avisogenies.gforge.inria.fr
7. Bos, J.W., Costello, C., Hisil, H., Lauter, K.: Fast cryptography in genus 2. In: Johansson, T., Nguyen, P.Q. (eds.) EUROCRYPT 2013. LNCS, vol. 7881, pp. 194–210. Springer, Heidelberg (2013), full version available at: http://eprint.iacr.org/2012/670
8. Bosma, W., Cannon, J., Playoust, C.: The Magma algebra system. I. The user language. J. Symbolic Comput. 24(3-4), 235–265 (1997) Computational algebra and number theory, London (1993)
9. Costello, C., Lauter, K.: Group law computations on jacobians of hyperelliptic curves. In: Miri, A., Vaudenay, S. (eds.) SAC 2011. LNCS, vol. 7118, pp. 92–117. Springer, Heidelberg (2012)
10. Erickson, S., Ho, T., Zemedkun, S.: Explicit projective formulas for real hyperelliptic curves of genus 2. Advances for Mathematics of Communications (to appear, 2014)
11. Fan, X., Gong, G.: Efficient explicit formulae for genus 2 hyperelliptic curves over prime fields and their implementations. In: Adams, C., Miri, A., Wiener, M. (eds.) SAC 2007. LNCS, vol. 4876, pp. 155–172. Springer, Heidelberg (2007)
12. Faz-Hernández, A., Longa, P., Sánchez, A.H.: Efficient and secure algorithms for GLV-based scalar multiplication and their implementation on GLV-GLS curves. In: Benaloh, J. (ed.) CT-RSA 2014. LNCS, vol. 8366, pp. 1–27. Springer, Heidelberg (2014)
13. Galbraith, S.D., Harrison, M., Mireles Morales, D.J.: Efficient hyperelliptic arithmetic using balanced representation for divisors. In: van der Poorten, A.J., Stein, A. (eds.) ANTS-VIII 2008. LNCS, vol. 5011, pp. 342–356. Springer, Heidelberg (2008)
14. Galbraith, S.D., Pujolàs, J., Ritzenthaler, C., Smith, B.A.: Distortion maps for supersingular genus two curves. J. Mathematical Cryptology 3(1), 1–18 (2009)
15. Gallant, R.P., Lambert, R.J., Vanstone, S.A.: Faster point multiplication on elliptic curves with efficient endomorphisms. In: Kilian, J. (ed.) CRYPTO 2001. LNCS, vol. 2139, pp. 190–200. Springer, Heidelberg (2001)
16. Gaudry, P.: Fast genus 2 arithmetic based on Theta functions. Journal of Mathematical Cryptology, JMC 1(3), 243–265 (2007)
17. Gaudry, P., Kohel, D.R., Smith, B.A.: Counting points on genus 2 curves with real multiplication. In: Lee, D.H., Wang, X. (eds.) ASIACRYPT 2011. LNCS, vol. 7073, pp. 504–519. Springer, Heidelberg (2011)

18. Gaudry, P., Schost, E.: Genus 2 point counting over prime fields. J. Symb. Comput. 47(4), 368–400 (2012)
19. Goundar, R.R., Joye, M., Miyaji, A., Rivain, M., Venelli, A.: Scalar multiplication on Weierstraß elliptic curves from Co-Z arithmetic. J. Cryptographic Engineering 1(2), 161–176 (2011)
20. Hamburg, M.: Fast and compact elliptic-curve cryptography. Cryptology ePrint Archive, Report 2012/309 (2012), http://eprint.iacr.org/
21. Hisil, H., Costello, C.: Jacobian coordinates on genus 2 curves. IACR Cryptology ePrint Archive, 2014:385 (2014)
22. Koblitz, N.: Elliptic curve cryptosystems. Mathematics of Computation 48(177), 203–209 (1987)
23. Koblitz, N.: Hyperelliptic cryptosystems. Journal of Cryptology 1(3), 139–150 (1989)
24. Kovtun, V., Kavun, S.: Co-Z divisor addition formulae in Jacobian of genus 2 hyperelliptic curves over prime fields. Cryptology ePrint Archive, Report 2010/498 (2010), http://eprint.iacr.org/
25. Lange, T.: Formulae for arithmetic on genus 2 hyperelliptic curves. Appl. Algebra Eng. Commun. Comput. 15(5), 295–328 (2005)
26. Longa, P., Miri, A.: New composite operations and precomputation scheme for elliptic curve cryptosystems over prime fields. In: Cramer, R. (ed.) PKC 2008. LNCS, vol. 4939, pp. 229–247. Springer, Heidelberg (2008)
27. Lubicz, D., Robert, D.: A generalisation of Miller's algorithm and applications to pairing computations on abelian varieties. Cryptology ePrint Archive, Report 2013/192 (2013), http://eprint.iacr.org/
28. Meloni, N.: New point addition formulae for ECC applications. In: Carlet, C., Sunar, B. (eds.) WAIFI 2007. LNCS, vol. 4547, pp. 189–201. Springer, Heidelberg (2007)
29. Miller, V.S.: Use of elliptic curves in cryptography. In: Williams, H.C. (ed.) CRYPTO 1985. LNCS, vol. 218, pp. 417–426. Springer, Heidelberg (1986)
30. Montgomery, P.L.: Speeding the Pollard and elliptic curve methods of factorization. Mathematics of Computation 48(177), 243–264 (1987)

Mersenne Factorization Factory

Thorsten Kleinjung[1], Joppe W. Bos[2,*], and Arjen K. Lenstra[1]

[1] EPFL IC LACAL, Station 14, CH-1015 Lausanne, Switzerland
[2] NXP Semiconductors, Leuven, Belgium

Abstract. We present work in progress to completely factor seventeen Mersenne numbers using a variant of the special number field sieve where sieving on the algebraic side is shared among the numbers. It is expected that it reduces the overall factoring effort by more than 50%. As far as we know this is the first practical application of Coppersmith's "factorization factory" idea. Most factorizations used a new double-product approach that led to additional savings in the matrix step.

Keywords: Mersenne numbers, factorization factory, special number field sieve, block Wiedemann algorithm.

1 Introduction

Despite its allegedly waning cryptanalytic importance, integer factorization is still an interesting subject and it remains relevant to test the practical value of promising approaches that have not been tried before. An example of the latter is Coppersmith's by now classical suggestion to amortize the cost of a precomputation over many factorizations [7]. The reason for the lack of practical validation of this method is obvious: achieving even a single "interesting" (i.e., record) factorization usually requires such an enormous effort [18] that an attempt to use Coppersmith's idea to obtain multiple interesting factorizations simultaneously would be prohibitively expensive, and meeting its storage requirements would be challenging.

But these arguments apply only to general numbers, such as RSA moduli [30], the context of Coppersmith's method. Given long-term projects such as [9,10,5] where many factoring-enthusiasts worldwide constantly busy themselves to factor many special numbers, such as for instance small-radix repunits, it makes sense to investigate whether factoring efforts that are eagerly pursued no matter what can be combined to save on the overall amount of work. This is what we set out to do here: we applied Coppersmith's factorization factory approach in order to simultaneously factor seventeen radix-2 repunits, so-called Mersenne numbers. Except for their appeal to makers of mathematical tables, such factorizations may be useful as well [16].

Let $S = \{1007, 1009, 1081, 1093, 1109, 1111, 1117, 1123, 1129, 1147, 1151, 1153, 1159, 1171, 1177, 1193, 1199\}$. For all $n \in S$ we have determined, or are in the

---

* Part of this work was conducted while this author was at Microsoft Research, One Microsoft Way, Redmond, WA 98052, USA.

P. Sarkar and T. Iwata (Eds.): ASIACRYPT 2014, PART I, LNCS 8873, pp. 358–377, 2014.

process of determining (for updates, see [19]), the full factorization of $2^n - 1$, using the method proposed in [7, Section 4] adapted to the number field sieve (SNFS, [22]). Furthermore, for two of the numbers a (new, but rather obvious) multi-SNFS approach was exploited as well. Most of our new factorizations will soundly beat the previous two SNFS records, the full factorizations of $2^{1039} - 1$ and $2^{1061} - 1$ reported in [1] and [6] respectively. Measuring individual (S)NFS-efforts, factoring $2^{1193} - 1$ would require about 20 times the effort of factoring $2^{1039} - 1$ or more than twice the effort of factoring the 768-bit RSA modulus from [18]. Summing the individual efforts for the seventeen numbers involved would amount to more than one hundred times the ($2^{1039} - 1$)-effort. Extrapolating our results so far, we expect that sharing the work à la Coppersmith will allow us to do it in about 50 times that effort. The practical implications of Coppersmith's method for general composites remain to be seen.

Although the factoring efforts reported here shared parts of the sieving tasks, each factorization still required its own separate matrix step. With seventeen numbers to be factored, and thus seventeen matrices to be dealt with, this gave us, and is still giving us, ample opportunity to experiment with a number of new algorithmic tricks in our block Wiedemann implementation, following up on the work reported in [1] and [18]. While the savings we obtained are relatively modest, given the overall matrix effort involved, they are substantial in absolute terms. Several of the matrices that we have dealt with, or will be dealing with, are considerably larger than the one from [18], the largest published comparable matrix problem before this work.

Section 2 gives background on the (S)NFS and Coppersmith's method as required for the paper. Section 3 introduces our two sets of target numbers to be factored, while sections 4 and 5 describe our work so far when applying the two main steps of the SNFS to these numbers. Section 6 provides evidence of the work completed so far and Section 7 presents a few concluding remarks.

All core years reported below are normalized to 2.2 GHz cores.

2 Background on (S)NFS and Coppersmith's Method

2.1 Number Field Sieve

To factor a composite integer N in the current range of interest using the number field sieve (NFS, [22]), a linear polynomial $g \in \mathbf{Z}[X]$ and a degree $d > 1$ polynomial $f \in \mathbf{Z}[X]$ are determined such that g and f have, modulo N, a root $m \approx N^{1/(d+1)}$ in common. For any m one may select $g(X) = X - m$ and $f(X) = \sum_{i=0}^{d} f_i X^i$ where $N = \sum_{i=0}^{d} f_i m^i$ and $0 \leq f_i < m$ (or $|f_i| \leq \frac{m}{2}$) for $0 \leq i \leq d$. Traditionally, everything related to the linear polynomial g is referred to as "rational" and everything related to the non-linear polynomial f as "algebraic".

Relations are pairs of coprime integers a, b with $b \geq 0$ such that $bg(a/b)$ and $b^d f(a/b)$ have only small factors, i.e., are *smooth*. Each relation corresponds to the vector consisting of the exponents of the small factors (omitting details

that are not relevant for the present description). Therefore, as soon as more
relations have been collected than there are small factors, the vectors are linearly
dependent and a matrix step can be used to determine an even sum of the vectors:
each of those has probability at least 50% to lead to a non-trivial factor of N.

Balancing the smoothness probability and the number of relations required
(which both grow with the number of small factors) the overall heuristic expected
NFS factoring time is $L((64/9)^{1/3}) \approx L(1.923)$ asymptotically for $N \to \infty$, where

$$L(c) = L[\frac{1}{3}, c] \quad \text{and} \quad L[\rho, c] = \exp((c + o(1))(\log(N))^{\rho}(\log(\log(N)))^{1-\rho})$$

for $0 \leq \rho \leq 1$ and the degree d is chosen as an integer close to $(\frac{3\log(N)}{\log(\log(N))})^{1/3}$.
A more careful selection of g and f than that suggested above (following for
instance [17]) can lead to a substantial overall speed-up but has no effect on the
asymptotic runtime expression.

For regular composites the f_i grow as $N^{1/(d+1)}$ which is only $N^{o(1)}$ for $N \to \infty$
but in general not $O(1)$. Composites for which the f_i are $O(1)$ are "special" and
the SNFS applies: its heuristic expected runtime is $L((32/9)^{1/3}) \approx L(1.526)$
asymptotically for $N \to \infty$, where the degree d is chosen as an integer close to
$(\frac{3\log(N)}{2\log(\log(N))})^{1/3}$. Both asymptotically and in practice the SNFS is much faster
than the NFS, with a slowly widening gap: for 1000-bit numbers the SNFS is
more than ten thousand times faster, for 1200-bit numbers it is more than 30
thousand times faster.

The function $L(c)$ satisfies various useful but unusual properties, due to the
$o(1)$ and $N \to \infty$: $L(c_1)L(c_2) = L(c_1 + c_2)$, $L(c_1) + L(c_2) = L(\max(c_1, c_2))$, and
for $c > 0$ and fixed k it is the case that $(\log(N))^k L(c) = L(c)/\log(L(c)) = L(c)$.

2.2   Relation Collection

We briefly discuss some aspects of the relation collection step that are relevant
for the remainder of the paper and that apply to both the NFS and the SNFS.
Let N be the composite to be factored, $c = (64/9)^{1/3}$ (but $c = (32/9)^{1/3}$ if
N is special), and assume the proper corresponding d as above. Heuristically
it is asymptotically optimal to choose $L(\frac{c}{2})$ as the upper bound for the small
factors in the polynomial values and to search for relations among the integer
pairs (a, b) with $|a| \leq L(\frac{c}{2})$ and $0 \leq b \leq L(\frac{c}{2})$. For the NFS the rational and
algebraic polynomial values then have heuristic probabilities $L(\frac{-c}{8})$ and $L(\frac{-3c}{8})$
to be smooth, respectively; for the SNFS both probabilities are $L(\frac{-c}{4})$. Either
way (i.e., NFS or SNFS) and assuming independence of the polynomial values,
the polynomial values are both smooth with probability $L(\frac{-c}{2})$. Over the entire
search space $L(c)L(\frac{-c}{2}) = L(\frac{c}{2})$ relations may thus be expected, which suffices.

Relation collection can be done using sieving because the search space is a
rectangle in $\mathbf{Z}^2$ and because polynomial values are considered. The latter implies
that if p divides $g(s)$ (or $f(s)$), then p divides $g(s + kp)$ (or $f(s + kp)$) for any
integer k, the former implies that given s all corresponding values $s + kp$ in the
search space are quickly located. Thus, for one of the polynomials, sieving is used

to locate all pairs in the search space for which the corresponding polynomial value has only factors bounded by $L(\frac{c}{2})$. This costs

$$\sum_{p \text{ prime}, \, p \leq L(\frac{c}{2})} \frac{L(c)}{p} = L(c)$$

(for $N \to \infty$, due to the $o(1)$ in $L(c)$) and leads to pairs for which the polynomial value is smooth. Next, in the same way and at the same cost, the pairs are located for which the other polynomial value is smooth. Intersecting the two sets leads to $L(\frac{c}{2})$ pairs for which both polynomial values are smooth.

Sieving twice, once for each polynomial, works asymptotically because $L(c) + L(c) = L(c)$. It may be less obvious that it is also a good approach in practice. After all, after the first sieve only pairs remain that are smooth with respect to the first polynomial, so processing those individually for the second polynomial could be more efficient than reconsidering the entire rectangular search space with another sieve. It will depend on the circumstances what method should be used. For the regular (S)NFS using two sieves is most effective, both asymptotically and in practice: sieving is done twice in a "quick and dirty" manner, relying on the intersection of the two sets to quickly reduce the number of remaining pairs, which are then inspected more closely to extract the relations. In Section 2.4, however, different considerations come into account and one cannot afford a second sieve – asymptotically or in practice – precisely because a second sieve would look at too many values.

As suggested in [28] the sieving task is split up into a large number of somewhat overlapping but sufficiently disjoint subtasks. Given a root z modulo a large prime q of one of the polynomials, a subtask consists of sieving only those pairs (a, b) for which $a/b \equiv z \bmod q$ and for which therefore the values of that polynomial are divisible by q. This implies that the original size $L(c)$ rectangular search space is intersected with an index-q sublattice of $\mathbf{Z}^2$, resulting in a size $L(c)/q$ search space. Sieving can still be used in the new smaller search space, but in a somewhat more complicated manner [28], as first done in [15] and later much better in [12]. Also, more liberal smoothness criteria allow several primes larger than $L(\frac{c}{2})$ in either polynomial value [11]. This complicates the decision of when enough relations have been collected and may increase the matrix size, but leads to a substantial overall speed-up. Another complication that arises is that duplicate relations will be found, i.e., by different subtasks, so the collection of relations must be made duplicate-free before further processing.

2.3   Matrix and Filtering

Assume that the numbers of distinct rational and algebraic small primes allowed in the smooth values during relation collection equal r_1 and r_2, respectively. With $r = r_1 + r_2$, each relation corresponds to an r-dimensional vector of exponents. With many distinct potential factors (i.e., large r_1 and r_2) of which only a few occur per smooth value, the exponent vectors are huge-dimensional (with r on the order of billions) and very sparse (on average about 20 non-zero entries).

As soon as $r+1$ relations have been collected, an even sum of the corresponding r-dimensional vectors (as required to derive a factorization) can in principle be found using linear algebra: with v one of the vectors and the others constituting the columns of an $r \times r$ matrix M_{raw}, an r-dimensional bit-vector x for which $M_{\text{raw}}x$ equals v modulo 2 provides the solution. Although a solution has at least a 50% chance to produce a non-trivial factorization, it may fail to do so, so in practice somewhat more relations are used and more than a single independent solution is derived.

The effort required to find solutions (cf. Section 5) grows with the product of the dimension r and the number of non-zero entries of M_{raw} (the *weight* of M_{raw}). A preprocessing *filtering* step is applied first to M_{raw} in order to reduce this product as much as is practically possible. It consists of a "best effort" to transform, using a sequence of transformation matrices, the initial huge-dimensional matrix M_{raw} of very low average column weight into a matrix M of much lower dimension but still sufficiently low weight. It is not uncommon to continue relation collection until a matrix M can be created in this way that is considered to be "doable" (usage of a second algebraic polynomial for some of our factorizations takes this idea a bit further than usual; cf. sections 3.2 and 4). Solutions for the original matrix M_{raw} easily follow from solutions for the resulting filtered matrix M.

2.4   Coppersmith's Factorization Factory

Coppersmith, in [7, Section 4], observed that a single linear polynomial g may be used for many different composites as long as their $(d+1)$st roots are not too far apart, with each composite still using its own algebraic polynomial. Thus smooth $bg(a/b)$-values can be precomputed in a sieving step and used for each of the different factorizations, while amortizing the precomputation cost. We sketch how this works, referring to [7, Section 4] for the details.

After sieving over a rectangular region of $L(2.007)$ rational polynomial values with smoothness bound $L(0.819)$ a total of $L(1.639)$ pairs can be expected (and must be stored for future use) for which the rational polynomial value is smooth. Using this stored table of $L(1.639)$ pairs corresponding to smooth rational polynomial values, any composite in the correct range can be factored at cost $L(1.639)$ per composite: the main costs per number are the algebraic smoothness detection, again with smoothness bound $L(0.819)$, and the matrix step. Factoring $\ell = L(\epsilon)$ such integers costs $L(\max(2.007, 1.639 + \epsilon))$, which is advantageous compared to ℓ-fold application of the regular NFS (at cost $L(1.923)$ per application) for $\ell \geq L(0.084)$. Thus, after a precomputation effort of $L(2.007)$, individual numbers can be factored at cost $L(1.639)$, compared to the individual factorization cost $L(1.923)$ using the regular NFS.

During the precomputation the $L(1.639)$ pairs for which the rational polynomial value is smooth are found by sieving $L(2.007)$ locations. This implies that, from an asymptotic runtime point of view, a sieve should not be used to test the resulting $L(1.639)$ pairs for algebraic smoothness (with respect to an applicable algebraic polynomial), because sieving would cost $L(2.007)$. As a result each individual factorization would cost more than the regular application of the NFS. Asymptotically, this issue is resolved by using the elliptic curve factoring method (ECM, [24]) for the algebraic smoothness test because, for smoothness bound $L(0.819)$, it processes each pair at cost $L(0)$, resulting in an overall algebraic smoothness detection cost of $L(1.639)$. In practice, if it ever comes that far, the ECM may indeed be the best choice, factorization trees ([3] and [14, Section 4]) may be used, or sieving may simply be the fastest option. Because the smooth rational polynomial values will be used by all factorizations, in practice the rational precomputation should probably include, after the sieving, the actual determination of all pairs for which the rational polynomial value is smooth: in the regular (S)NFS this closer inspection of the sieving results takes place only after completing *both* sieves.

These are asymptotic results, but the basic idea can be applied on a much smaller scale too. With a small number ℓ of sufficiently close composites to be factored and using the original NFS parameter choices (and thus a table of $L(1.683)$ as opposed to $L(1.639)$ pairs), the gain approaches 50% with growing ℓ (assuming the matrix cost is relatively minor and disregarding table-storage issues). It remains to be seen, however, if for such small ℓ individual processing is not better if each composite uses a carefully selected pair of polynomials as in [17], and if that effect can be countered by increasing the rational search space a bit while decreasing the smoothness bounds (as in the analysis from [7]).

We are not aware of practical experimentation with Coppersmith's method. To make it realistically doable (in an academic environment) a few suitable moduli could be concocted. The results would, however, hardly be convincing and deriving them would be mostly a waste of computer time – and electric power [20]. We opted for a different approach to gain practical experience with the factorization factory idea, as described below.

2.5   SNFS Factorization Factory

If we switch the roles of the rational and algebraic sides in Coppersmith's factorization factory, we get a method that can be used to factor numbers that share the same algebraic polynomial, while having different rational polynomials. Such numbers are readily available in the Cunningham project [9,10,5][1]. They have the additional advantage that obtaining their factorizations is deemed to be desirable, so an actual practical experiment may be considered a worthwhile effort. Our choice of target numbers is described in Section 3. First we present the theoretical analysis of the factorization factory with a fixed algebraic polynomial with $O(1)$ coefficients, i.e., the SNFS factorization factory.

Let $L(2\alpha)$ be the size of the sieving region for the fixed shared algebraic polynomial (with coefficient size $O(1)$), let $L(\beta)$ and $L(\gamma)$ be the algebraic and

---

[1] On an historical note, the desire to factor the ninth Fermat number $2^{2^9} + 1$, in 1988 the "most wanted" unfactored Cunningham number, inspired the invention of the SNFS, triggering the development of the NFS; the details are described in [22].

rational smoothness bounds, respectively. Assume the degree of the algebraic polynomial can be chosen as $\delta(\frac{\log(N)}{\log(\log(N))})^{1/3}$ for all numbers to be factored.

The algebraic polynomial values are of size $L[\frac{2}{3}, \alpha\delta]$ and are thus assumed to be smooth with probability $L(-\frac{\alpha\delta}{3\beta})$ (cf. [21, Section 3.16]). With the coefficients of the rational polynomials bounded by $L[\frac{2}{3}, \frac{1}{\delta}]$, the rational polynomial values are of size $L[\frac{2}{3}, \frac{1}{\delta}]$ and may be assumed to be smooth with probability $L(-\frac{1}{3\gamma\delta})$. To be able to find sufficiently many relations it must therefore be the case that

$$2\alpha - \frac{\alpha\delta}{3\beta} - \frac{1}{3\gamma\delta} \geq \max(\beta, \gamma). \tag{1}$$

The precomputation (algebraic sieving) costs $L(2\alpha)$ and produces $L(2\alpha - \frac{\alpha\delta}{3\beta})$ pairs for which the algebraic value is smooth. Per number to be factored, a total of $L(\max(\beta, \gamma) + \frac{1}{3\gamma\delta})$ of these pairs are tested for smoothness (with respect to $L(\gamma)$), resulting in an overall factoring cost

$$L(\max(2\beta, 2\gamma, \max(\beta, \gamma) + \frac{1}{3\gamma\delta}))$$

per number. If $\beta \neq \gamma$, then replacing the smaller of β and γ by the larger increases the left hand side of condition (1), leaves the right hand side unchanged, and does not increase the overall cost. Thus, for optimal parameters, it may be assumed that $\beta = \gamma$. This simplifies the cost to $L(\max(2\gamma, \gamma + \frac{1}{3\gamma\delta}))$ and condition (1) to $(2 - \frac{\delta}{3\gamma})\alpha \geq \gamma + \frac{1}{3\gamma\delta}$, which holds for some $\alpha \geq 0$ as long as $\delta < 6\gamma$. Fixing δ, the cost is minimized when $2\gamma = \gamma + \frac{1}{3\gamma\delta}$ or when $\gamma + \frac{1}{3\gamma\delta}$ attains its minimum; these two conditions are equivalent and the minimum is attained for $\gamma = (3\delta)^{-1/2}$. The condition $\delta < 6\gamma$ translates into $\delta < 12^{1/3}$ respectively $\gamma > 18^{-1/3}$. It follows that for δ approaching $12^{1/3}$ from below, the factoring cost per number approaches $L((4/9)^{1/3}) \approx L(0.763)$ from above, with a precomputation cost of $L(2\alpha), \alpha \to \infty$. These SNFS factorization factory costs should be compared to individual factorization cost $L((32/9)^{1/3}) \approx L(1.526)$ using the regular SNFS, and approximate individual factoring cost $L(1.639)$ after a precomputation at approximate cost $L(2.007)$ using Coppersmith's NFS factorization factory.

Assuming $\gamma = (3\delta)^{-1/2}$, the choices $\gamma = (2/9)^{1/3}$ and $\alpha = (128/343)^{1/3}$ lead to minimal precomputation cost $L((4/3)^{5/3}) \approx L(1.615)$, and individual factoring cost $L((4/3)^{2/3}) \approx L(1.211)$. This makes the approach advantageous if more than approximately $L(0.089)$ numbers must be factored (compare this to $L(0.084)$ for Coppersmith's factorization factory). However, with more numbers to be factored, another choice for γ (and thus larger α) may be advantageous (cf. the more complete analysis in [19]).

3   Targets for the SNFS Factorization Factory

3.1   Target Set

For our SNFS factorization factory experiment we chose to factor the Mersenne numbers $2^n - 1$ with $1000 \leq n \leq 1200$ that had not yet been fully factored, the seventeen numbers $2^n - 1$ with $n \in S$ as in the Introduction. We write $S = S_{\mathrm{I}} \cup S_{\mathrm{II}}$, where S_{I} is our first batch containing exponents that are $\pm 1 \bmod 8$ and S_{II} is the second batch with exponents that are $\pm 3 \bmod 8$. Thus

$$S_{\mathrm{I}} = \{1007, 1009, 1081, 1111, 1129, 1151, 1153, 1159, 1177, 1193, 1199\},$$

and

$$S_{\mathrm{II}} = \{1093, 1109, 1117, 1123, 1147, 1171\}.$$

Once these numbers have been factored, only one unfactored Mersenne number with $n \le 1200$ remains, namely $2^{991} - 1$. It can simply be dealt with using an individual SNFS effort, like the others with $n \le 1000$ that were still present when we started our project. Our approach would have been suboptimal for these relatively small n.

Around 2009, when we were gearing up for our project, there were several more exponents in the range $[1000, 1200]$. Before actually starting, we first used the ECM in an attempt to remove Mersenne numbers with relatively small factors and managed to fully factor five of them [4]: one with exponent $1 \bmod 8$ and four with exponents $\pm 3 \bmod 8$. Three, all with exponents $\pm 3 \bmod 8$, were later factored by Ryan Propper (using the ECM, [34]) and were thus removed from S_{II}. Some other exponents which were easier for the SNFS were taken care of by various contributors as well, after which the above seventeen remained.

3.2   Polynomial Selection for the Target Set

We used two different algebraic polynomials: $f_{\mathrm{I}} = X^8 - 2$ for $n = \pm 1 \bmod 8$ in S_{I} and $f_{\mathrm{II}} = X^8 - 8$ for $n = \pm 3 \bmod 8$ in S_{II}. This leads to the common roots m_n and rational polynomials g_n corresponding to n as listed in Table 1. Relations were collected using two sieves (one for f_{I} shared by eleven n-values, and one for f_{II} shared by six n-values) and seventeen factorization trees (one for each g_n), as further explained in Section 4. Furthermore, in an attempt to reduce the effort to process the resulting matrix, for $n \in \{1177, 1199\}$ additional relations were collected using the algebraic polynomial f_{I}', as specified in Table 1 along with the common roots m_n' and rational polynomials g_n'. Although $n = 1177$ and $n = 1199$ share f_{I}', to obtain the additional relations it turned out to be more convenient to use the vanilla all-sieving approach from [13] twice, cf. Section 4.4.

Another possibility would have been to select the single degree 6 polynomial $X^6 - 2$. Its relatively low degree and very small coefficients lead to a huge number of smooth algebraic values, all with a relatively large rational counterpart (again due to the low degree). Atypically, rational sieving could have been appropriate, whereas due to large cofactor sizes rational cofactoring would be relatively costly. Overall degree 8 can be expected to work faster, despite the fact that it requires two algebraic polynomials. Degree 7 would require three algebraic polynomials and may be even worse than degree 6 for our sets of numbers, but would have had the advantage that numbers of the form $2^n + 1$ could have been included too.

Table 1. The shared algebraic polynomials, roots, and rational polynomials for the $11 + 6 = 17$ Mersenne numbers $2^n - 1$ considered here

	$f_{\mathrm{I}} = X^8 - 2$				$f_{\mathrm{II}} = X^8 - 8$		
n	$n \bmod 8$	m_n	g_n	n	$n \bmod 8$	m_n	g_n
1007		2^{126}	$X - 2^{126}$	1093		2^{137}	$X - 2^{137}$
1111		2^{139}	$X - 2^{139}$	1109	-3	2^{139}	$X - 2^{139}$
1151	-1	2^{144}	$X - 2^{144}$	1117		2^{140}	$X - 2^{140}$
1159		2^{145}	$X - 2^{145}$				
1199		2^{150}	$X - 2^{150}$				
1009		2^{-126}	$2^{126}X - 1$	1123		2^{-140}	$2^{140}X - 1$
1081		2^{-135}	$2^{135}X - 1$	1147	3	2^{-143}	$2^{143}X - 1$
1129		2^{-141}	$2^{141}X - 1$	1171		2^{-146}	$2^{146}X - 1$
1153	1	2^{-144}	$2^{144}X - 1$				
1177		2^{-147}	$2^{147}X - 1$				
1193		2^{-149}	$2^{149}X - 1$				

$$f_{\mathrm{I}}' = X^5 + X^4 - 4X^3 - 3X^2 + 3X + 1$$

n	m_n'	g_n'
1177	$2^{107} + 2^{-107}$	$2^{107}X - (2^{214} + 1)$
1199	$2^{109} + 2^{-109}$	$2^{109}X - (2^{218} + 1)$

4   Relation Collection for the Target Set

4.1   Integrating the Precomputation

The first step of Coppersmith's factorization factory is the preparation and storage of a precomputed table of pairs corresponding to smooth rational polynomial values. With the parameters from [7] this table contains $L(1.639)$ pairs. Assuming composites of relevant sizes, this is too large to be practical. If we apply Coppersmith's idea as suggested in the second to last paragraph of Section 2.4 to a relatively small number of composites with the original NFS parameter choices, the table would contain $L(1.683)$ pairs, which is even worse.

Here we can avoid excessive storage requirements. First of all, with the original SNFS parameter choices the table would contain "only" $L(1.145)$ pairs corresponding to smooth algebraic polynomial values, because we are using the factorization factory for the SNFS with a shared algebraic polynomial. Though better, this is still impractically large. Another effect in our favor is that we are using degree 8 polynomials, which is a relatively large degree compared to what is suggested by the asymptotic runtime analysis: for our N-values the integer closest to $\left(\frac{3 \log(N)}{2 \log(\log(N))}\right)^{1/3}$ would be 6. A larger degree leads to larger algebraic values, fewer smooth values, and thus fewer values to be stored.

Most importantly, however, we know our set of target numbers in advance. This allows us to process precomputed pairs right after they have been generated, and to keep only those that lead to a smooth rational polynomial value as well. With ℓ numbers to be factored and $L(\frac{1.523}{2})$ as smoothness bound (cf. Section 2.2), this reduces the storage requirements from $L(1.523)L(\frac{-1.523}{4}) = L(1.145)$ to $\ell L(1.523)L(\frac{-1.523}{2}) = \ell L(0.763)$. For our target sets this is only on the order of TBs (less than six TBs for S_{II}.).

The precomputation and the further processing are described separately.

4.2   Algebraic Sieving

For the sieving of the polynomial $f_{\mathrm{I}} = X^8 - 2$ from Section 3.2 we used a search space of approximately 2^{66} pairs and varying smoothness bounds. At most two larger primes less than 2^{37} were allowed in the otherwise smooth f_{I}-values.

The sieving task is split up into a large number of subtasks: given a root z of f_{I} modulo a large prime number q, a subtask consists of finding pairs (a, b) for which $a/b \equiv z \bmod q$ (implying that q divides $b^8 f_{\mathrm{I}}(a/b)$) and such that the quotient $b^8 f_{\mathrm{I}}(a/b)/q$ is smooth (except for the large primes) with respect to the largest $h \cdot 10^8$ less than q, with $h \in \{3, 4, 6, 8, 12, 15, 20, 25, 30, 35\}$.

Pairs (a, b) for which $a/b \equiv z \bmod q$ form a two-dimensional lattice of index q in $\mathbf{Z}^2$ with basis $\binom{q}{0}, \binom{z}{1}$. After finding a reduced basis $u, v \in \mathbf{Z}^2$ for the lattice, the intersection of the original search space and the lattice is approximated as $\{\binom{a}{b} = iu + jv : i, j \in \mathbf{Z}, |i| < 2^I, 0 \le j < 2^J\}$. The bounds $I, J \in \mathbf{Z}_{>0}$ were (or, rather, "are ideally", as this is what we converged to in the course of our experiment) chosen such that $I + J + \log_2(q) \approx 65$ and such that $\max(|a|) \approx \max(b)$, thus taking the relative lengths of u and v into account. Sieving takes place in a size 2^{I+J+1} rectangular region of the (i, j)-plane while avoiding storage for the (even,even) locations, as described in [12]. After the sieving, all f_{I}-values corresponding to the reported locations are divided by q and trial-divided as also described in [12], allowing at most two prime factors between q and 2^{37}. Allowing three large primes turned out to be counterproductive with slightly more relations at considerably increased sieving time or many more relations at the expense of a skyrocketing cofactoring effort.

Each (a, b) with smooth algebraic polynomial value resulting from subtask (q, z) induces a pair $(-a, b)$ with smooth algebraic polynomial value for subtask $(q, -z)$. Subtasks thus come in pairs: it suffices to sieve for one subtask and to recover all smooth pairs for the other subtask. For $n \ge 1151$ we used most q-values with $4 \cdot 10^8 < q < 8 \cdot 10^9$ (almost 2^{33}), resulting in about 157 million pairs of subtasks. For the other n-values we used fewer pairs of subtasks: about 126 million for $n \in \{1007, 1009\}$ and about 143 million for the others.

Subtasks are processed in disjoint batches consisting of all (prime,root) pairs for a prime in an interval of length 2500 or 10 000. Larger intervals are used for larger q-values, because the latter are processed faster: their sieving region is smaller (cf. above), and their larger smoothness bounds require more memory and thus more cores. After completion of a batch, the resulting pairs are inspected for smoothness of their applicable rational polynomial values as further described below. Processing the batches, not counting the rational smoothness tests, required about 2367 core years. It resulted in $1.57 \cdot 10^{13}$ smooth algebraic values, and thus for each $n \in S_{\mathrm{I}}$ at most twice that many values to be inspected for rational smoothness. Storage of the $1.57 \cdot 10^{13}$ values (in binary format at five bytes per value) would have required 70 TB. As explained in Section 4.1 we avoided these considerable storage requirements by processing the smooth algebraic values almost on-the-fly; this also allowed the use of a more relaxed text format at about 20 bytes per value.

Sieving for $n \in S_{II}$ was done in the same way. For the polynomial $f_{II} = X^8 - 8$ and $n \in \{1147, 1171\}$ about 118 million pairs of subtasks were processed for most q-values with $3 \cdot 10^8 < q < 5.45 \cdot 10^9$. For the other n-values in S_{II} about 94% to 96% of that range of q-values sufficed. Overall, sieving for $n \in S_{II}$ required 1626 core years and resulted in $1.16 \cdot 10^{13}$ smooth algebraic values.

4.3   Rational Factorization Trees

Each time a batch of f_I-sieving subtasks is completed (cf. Section 4.2) the pairs (a, b) produced by it are partitioned over four initially empty queues $\mathcal{Q}_{34}, \mathcal{Q}_{35}, \mathcal{Q}_{36}$, and $\mathcal{Q}_{37}$: if the largest prime in the factorization of $b^8 f_I(a/b)$ has bitlength i for $i \in \{35, 36, 37\}$ then the pair is appended to $\mathcal{Q}_i$, all remaining pairs are appended to $\mathcal{Q}_{34}$.

After partitioning the new pairs among the queues, the following is done for each $n \in S_I$ (cf. Section 3.1). For all pairs (a, b) in $\cup_{i=34}^{\alpha(n)} \mathcal{Q}_i$, with $\alpha(n)$ as in Table 2, the rational polynomial value $bg_n(a/b)$ (with g_n as in Table 1) is tested for smoothness: if $bg_n(a/b)$ is smooth, then (a, b) is included in the collection of relations for the factorization of $2^n - 1$, else (a, b) is discarded. The smoothness test for the $bg_n(a/b)$-values is conducted simultaneously for all pairs $(a, b) \in \cup_{i=34}^{\alpha(n)} \mathcal{Q}_i$ using a factorization tree as in [14, Section 4] (see also [3]) with $\tau(n) \cdot 10^8$ and $2^{\beta(n)}$ as smoothness and cofactor bounds, respectively (with $\tau(n)$ and $\beta(n)$ as in Table 2). Here the cofactor bound limits the number and the size of the factors in $bg_n(a/b)$ that are larger than the smoothness bound.

For all $n \in S_I$, besides the runtimes Table 2 also lists the numbers of relations found, of free relations [23], of relations after duplicate removal (and inclusion of the free relations), and of prime ideals that occur in the relations before the first singleton removal (where the number of prime ideals is the actual dimension of the exponent vectors). All resulting raw matrices are over-square. For $n \in \{1193, 1199\}$ the over-squareness is relatively small. For $n = 1193$ we just dealt with the resulting rather large filtered matrix. For $n = 1199$, and for $n = 1177$ as well, additional sieving was done, as further discussed in the section below. The unusually high degree of over-squareness for the smaller n-values is a consequence of the large amount of data that had to be generated for the larger n-values, and that could be included for the smaller ones at little extra cost.

Completed batches of subtasks for f_{II}-sieving were processed in the same way. The results are listed in Table 2.

4.4   Additional Sieving

In an attempt to reduce the size of the (filtered) matrix we collected additional relations for $n \in \{1177, 1199\}$ using the degree 5 algebraic polynomial f_I' and

Table 2.

n	$\alpha(n)$	$\tau(n)$	$\beta(n)$	core years	relations found	free	total unique	occurring prime ideals
1007	34	5	99	26	6 157 265 485	47 681 523	4 083 240 054	1 488 688 670
1009	34	5	99	26	6 076 365 897	47 681 523	4 030 378 014	1 487 997 805
1081	35	5	103	48	7 704 145 069	92 508 436	5 484 250 026	2 828 752 381
1111	35	5	103	46	5 636 554 807	92 508 436	4 045 778 202	2 744 898 588
1129	35	5	103	47	4 860 167 788	92 508 436	3 447 412 400	2 690 405 347
1151	36	5	105	77	9 026 908 346	179 644 953	6 878 035 126	5 229 081 896
1153	36	5	105	78	8 919 329 699	179 644 953	6 798 580 785	5 219 976 433
1159	36	5	105	78	8 494 336 817	179 644 953	6 454 287 572	5 179 538 761
1177	37	20	138	140	15 844 796 536	349 149 710	12 687 801 912	10 098 132 272
1193	37	20	141	171	13 873 940 124	349 149 710	11 120 476 664	9 912 486 202
1199	37	20	141	169	13 201 986 116	349 149 710	10 600 157 337	9 795 656 570
core years for $n \in S_{\mathrm{I}}$:				906				
1093	35	5	103	37	5 380 284 567	92 508 436	3 777 018 420	2 736 825 054
1109	36	5	105	55	9 621 428 465	179 644 953	7 102 393 219	5 134 440 256
1117	36	5	105	55	8 930 755 992	179 644 953	6 762 813 242	5 220 018 492
1123	36	5	105	54	8 686 858 952	179 644 953	6 567 794 152	5 197 770 153
1147	37	20	138	122	15 404 494 545	349 149 710	12 096 909 112	9 967 719 536
1171	37	20	138	115	12 240 930 101	349 149 710	9 688 750 293	9 556 433 885
core years for $n \in S_{\mathrm{II}}$:				438				

the rational polynomials g'_n from Table 1. For various reasons these two n-values (though they share f'_{I}) were treated separately using the software from [13].

For $n = 1177$ we used on the rational side smoothness bound $3 \cdot 10^8$, cofactor bound 2^{109}, and large factor bound 2^{37}. On the algebraic side these numbers were $5 \cdot 10^8$, 2^{74}, and 2^{37}. Using large primes $q \in [3 \cdot 10^8, 3.51 \cdot 10^8]$ on the rational side (as opposed to the algebraic side above) we found 1 640 189 494 relations, of which 1 606 180 461 remained after duplicate removal. With 1 117 302 548 free relations this led to a total of 2 723 483 009 additional relations. With the 12 687 801 912 relations found earlier, this resulted in 15 411 284 921 relations in total, involving 15 926 778 561 prime ideals. Although this is not over-square (whereas the earlier relation set for $n = 1177$ from Section 4.3 was over-square), the new free relations contained many singleton prime ideals, so that after singleton removal the matrix was easily over-square. The resulting filtered matrix was small enough.

For $n = 1199$ the rational smoothness bound is $4 \cdot 10^8$. All other parameters are the same as for $n = 1177$. After processing the rational large primes $q \in [4 \cdot 10^8, 6.85 \cdot 10^8]$ we had 6 133 381 386 degree 5 relations (of which 5 674 876 905 unique) and 1 117 302 548 free relations. This led to 17 392 336 790 relations with 15 955 331 670 prime ideals and a small enough filtered matrix.

The overall reduction in the resulting filtered matrix sizes was modest, and we doubt that this additional sieving experiment, though interesting, led to an overall reduction in runtime. On the other hand, spending a few months (thus a few hundred core years) on additional sieving hardly takes any human effort, whereas processing (larger) matrices is (more) cumbersome. Another reason is that we have resources available that cannot be used for matrix jobs.

4.5 Equipment Used

Relation collection for $n \in S_I$ was done from May 22, 2010, until February 21, 2013, on clusters at EPFL as listed in Table 3: 82% on lacal_1 and lacal_2, 12% on pleiades, 3% on greedy, and 1.5% on callisto and vega each, spending 3273 $(2367 + 906)$ core years. Furthermore, 65 and 327 core years were spent on lacal_1 and lacal_2 for additional sieving for $n = 1177$ and $n = 1199$, respectively. Thus a total of 3665 core years was spent on relation collection for $n \in S_I$.

Relation collection for $n \in S_{II}$ was done from February 21, 2013, until September 11, 2014, on part of the XCG container cluster at Microsoft Research in Redmond, USA, and on clusters at EPFL: 46.5% on the XCG cluster, 45.5% on lacal_1 and lacal_2, 5% on castor, 2% on grid, and 1% on greedy, spending a total of 2064 $(1626 + 438)$ core years. It followed the approach described above for f_I, except that data were transported on a regular 500 GB hard disk drive that was sent back and forth between Redmond and Lausanne via regular mail.

Table 3. Description of available hardware. We have 100% access to the equipment at LACAL and to 134 nodes of the XCG container cluster (which contains many more nodes) and limited access to the other resources. A checkmark ($\checkmark$) indicates InfiniBand network. All nodes have 2 processors.

location	name	processor	nodes	cores per node	cores	GHz	GB RAM per node	core	TB disk space
EPFL	$\checkmark$ bellatrix	Sandy Bridge	424	16	6784	2.2	32	2	
	callisto	Harpertown	128	8	1024	3.0	32	4	
	castor	Ivy Bridge	52	16	832	2.6	$\begin{cases} 50: 64 \\ 2:256 \end{cases}$	$\begin{matrix} 4 \\ 16 \end{matrix}$	22
	greedy	≈ 1000 mixed cores, ≈ 1 GB RAM per core; 70% windows, 25% linux, 5% mac							
	vega	Harpertown	24	8	192	2.66	16	2	
LACAL	$\checkmark$ lacal_1	AMD	53	12	636	2.2	16	$1\frac{1}{3}$	
	$\checkmark$ lacal_2	AMD	28	24	672	1.9	32	$1\frac{1}{3}$	
	pleiades	Woodcrest	35	4	140	2.66	8	2	
	storage server	AMD	1	24	24	1.9	32	$1\frac{1}{3}$	58
Microsoft Research Switzerland	part of the XCG container cluster	AMD	134	8	1072	2.1	32	4	
	grid	several clusters at several Swiss institutes							

5 Processing the Matrices

Although relation collection could be shared among the numbers, the matrices must all be treated separately. Several of them required an effort that is considerably larger than the matrix effort reported in [18]. There a $192\,795\,550 \times 192\,796\,550$-matrix with on average 144 non-zeros per column (in this section all sizes and weights refer to matrices *after* filtering) was processed on a wide variety of closely coupled clusters in France, Japan, and Switzerland, requiring four months wall time and a tenth of the computational effort of the relation collection. So far it was the largest binary matrix effort that we are aware of, in the public domain. The largest matrix done here is about 4.5 times harder.

5.1   The Block Wiedemann Algorithm

Wiedemann's Algorithm. Given a sparse $r \times r$ matrix M over the binary field $\mathbf{F}_2$ and a binary r-dimensional vector v, we have to solve $Mx = v$ (cf. Section 2.3). The minimal polynomial F of M on the vector space spanned by $\{M^0v, M^1v, M^2v, \ldots\}$ has degree at most r. Denoting its coefficients by $F_i \in \mathbf{F}_2$ and assuming that $F_0 = 1$ we have $F(M)v = \sum_{i=0}^{r} F_i M^i v = 0$, so that x follows as $\sum_{i=1}^{r} F_i M^{i-1}v$. Wiedemann's method [32] determines x in three steps. For any j with $1 \leq j \leq r$ the j-th coordinates of the vectors $M^i v$ for $i = 0, 1, 2, \ldots$ satisfy the linear recurrence relation given by the F_i. Thus, once the first $2r+1$ of these j-th coordinates have been determined using $2r$ iterations of matrix×vector multiplications (Step 1), the F_i can be computed using the Berlekamp-Massey method [25] (Step 2), where it may be necessary to compute the least common multiple of the results of a few j-values. The solution x then follows using another r matrix×vector multiplications (Step 3).

Steps 1 and 3 run in time $\Theta(rw(M))$, where $w(M)$ denotes the number of non-zero entries of M. With Step 2 running in time $O(r^2)$ the effort of Wiedemann's method is dominated by steps 1 and 3.

Block Wiedemann. The efficiency of Wiedemann's conceptually simple method is considerably enhanced by processing several different vectors v simultaneously, as shown in [8,31]: on 64-bit machines, for instance, 64 binary vectors can be treated at the same time, at negligible loss compared to processing a single binary vector. Though this slightly complicates Step 2 and requires keeping the 64 first coordinates of each vector calculated per iteration in Step 1, it cuts the number of matrix×vector products in steps 1 and 3 by a factor of 64 and effectively makes Wiedemann's method 64 times faster. This *blocking factor* of 64 can, obviously, be replaced by $64t$ for any positive integer t. This calculation can be carried out by t independent threads (or on t independent clusters, [1]), each processing 64 binary vectors at a time while keeping the $64t$ first coordinates per multiplication in Step 1, and as long as the independent results of the t-fold parallelized first step are communicated to a central location for the Berlekamp-Massey step [1].

As explained in [8,18] a further speed-up in Step 1 may be obtained by keeping, for some integer $k > 1$, the first $64kt$ coordinates per iteration (for each of the t independent 64-bit wide threads). This reduces the number of Step 1 iterations from $2\frac{r}{64t}$ to $(\frac{1}{k}+1)\frac{r}{64t}$ while the number of Step 3 iterations remains unchanged at $\frac{r}{64t}$. However, it has a negative effect on Step 2 with time and space complexities growing as $(k+1)^\mu t^{\mu-1} r^{1+o(1)}$ and $(k+1)^2 tr$, respectively, for $r \to \infty$ and with μ the matrix multiplication exponent (we used $\mu = 3$).

Double Matrix Product. In all previous work that we are aware of a single filtered matrix M is processed by the block Wiedemann method. This matrix M replaces the original matrix M_{raw} consisting of the exponent vectors, and is calculated as $M = M_{\text{raw}} \times M_1 \times M_2$ for certain filtering matrices M_1 and M_2. For most matrices here, we adapted our filtering strategy, calculated $M_1' = M_{\text{raw}} \times M_1$, and applied the block Wiedemann method to the $r \times r$ matrix M

without actually calculating it but by using $M = M_1' \times M_2$. Because Mv can be calculated as $M_1'(M_2v)$ at (asymptotic) cost $w(M_2) + w(M_1')$ this is advantageous if $r(w(M_1') + w(M_2))$ is lower than the product of the dimension and weight resulting from traditional filtering. Details about the new filtering strategy will be provided once we have more experience with it.

Error Detection and Recovery. See [19] for the "folklore" methods we used.

5.2   Matrix Results

All matrix calculations were done at EPFL on the clusters with InfiniBand network (lacal_1, lacal_2, and bellatrix) and the storage server (cf. Table 3). Despite our limited access to bellatrix, it was our preferred cluster for steps 1 and 3 because its larger memory bandwidth (compared to lacal_1 and lacal_2) allowed us to optimally run on more cores at the same time while also cutting the number of core years by a factor of about two (compared to lacal_1). The matrix from [18], for instance, which would have required about 154 core years on lacal_1, would require less than 75 core years on bellatrix.

Table 4 lists most data for all matrices we processed, or are processing. Jobs were usually run on a small number of nodes (running up to five matrices at the same time), as that requires the least amount of communication and storage per matrix and minimizes the overall runtime. Extended wall times were and are of no concern. The Berlekamp-Massey step, for which there are no data in Table 4, was run on the storage server. Its runtime requirements varied from several days to two weeks, using just 8 of the 24 available cores, writing and reading intermediate results to and from disk to satisfy the considerable storage needs. For each of the numbers Step 2 thus took less than one core year.

6   Factorizations

For most n the matrix solutions were processed in the usual way [26,27,2] to find the unknown factors of $2^n - 1$. This required an insignificant amount of runtime. The software from [2] is, however, not set up to deal with more fields than the field of rational numbers and a single algebraic number field defined by a single algebraic polynomial (in our case f_I for $n \in S_I$ and f_{II} for $n \in S_{II}$). Using this software for $n \in \{1177, 1199\}$, the values for which additional sieving was done for the polynomials f_I' and g_n' from Table 1, would have required a substantial amount of programming. To save ourselves this non-trivial effort we opted for the naive old-fashioned approach used for the very first SNFS factorizations as described in [23, Section 3] of finding explicit generators for all first degree prime ideals in both number fields $\mathbf{Q}(\sqrt[8]{2})$ and $\mathbf{Q}(\zeta_{11} + \zeta_{11}^{-1})$ and up to the appropriate norms. Because both number fields have class number equal to one and the search for generators took, relatively speaking, an insignificant amount of time, this approach should have enabled us to quickly and conveniently deal with these two more complicated cases as well.

Table 4. Data about the matrices processed, as explained in Section 5.1, with M_1', M_2, and M matrices of sizes $r \times r'$, $r' \times (r+\delta)$, and $r \times (r+\delta)$, respectively, for a relatively small positive integer δ. Runtimes in italics are estimates for data that were not kept and runtimes between parentheses are extrapolations based on work completed. Starting from Step 3 for $n = 1151$ a different configuration was used, possibly including some changes in our code, and the programs ran more efficiently. Until $n = 1159$ a blocking factor of 128 was used (so t must be even), for $n \in \{1177, 1193, 1199\} \cup S_{\mathrm{II}}$ it was 64 in order to fit on 16 nodes. The green bars indicate the periods that the matrices were processed, on the green scale at the top. The red bars indicate the matrices that are currently being processed. Dates are in the format $yymmdd$.

$$\left[\begin{array}{l} 121207 \ldots \end{array}\right. \qquad\qquad \text{core years} \qquad \ldots 140914 \left.\begin{array}{l}\end{array}\right]$$

n	r, r', δ or r, δ (cf. above)	weight(s)	t	k	Step 1	Step 3	start - end
1007	$r = 38\,986\,666$, $r' = 61\,476\,801$, $\delta = 420$	$201.089r$, $31.518r'$	12	3	3.5	*2.6*	121207 - 130106 (30 days)
1009	$r = 39\,947\,548$, $r' = 64\,737\,522$, $\delta = 348$	$202.077r$, $36.958r'$	12	2	*3.9*	*2.6*	130424 - 130610 (47 days)
1081	$r = 79\,452\,919$, $r' = 122\,320\,052$, $\delta = 1624$	$183.296r$, $15.332r'$	16	2	20.3	13.5	130130 - 130311 (41 days)
1111	$r = 108\,305\,368$, $r' = 167\,428\,008$, $\delta = 1018$	$180.444r$, $13.887r'$	24	2	41.8	30.6	130109 - 130611 (154 days)
1129	$r = 132\,037\,278$, $r' = 204\,248\,960$, $\delta = 341$	$180.523r$, $13.434r'$	16	2	64.8	44.4	121231 - 130918 (262 days)
1151	$r = 164\,438\,818$, $r' = 253\,751\,725$, $\delta = 911$	$174.348r$, $11.810r'$	12	2	130.7	38.3	130316 - 131210 (270 days)
1153	$r = 168\,943\,024$, $r' = 260\,332\,296$, $\delta = 1830$	$169.419r$, $11.014r'$	8	2	75.4	43.3	130326 - 131026 (215 days)
1159	$r = 179\,461\,813$, $r' = 276\,906\,625$, $\delta = 1278$	$174.179r$, $11.688r'$	4	2	*87.0*	58.0	130808 - 140207 (184 days)
1177	$r = 192\,693\,549$, $r' = 297\,621\,101$, $\delta = 1043$	$216.442r$, $19.457r'$	4	3	89.3	74.1	140119 - 140525 (127 days)
1193	$r = 297\,605\,781$, $\delta = 1024$	$272.267r$	6	3	129.5	105.3	131029 - 140819 (295 days)
1199	$r = 270\,058\,949$, $\delta = 1064$	$217.638r$	6	3	104.8	(86.0)	started 140626 (≥ 51 days)
	core years for $n \in S_{\mathrm{I}}$:				751.0 +	498.7 =	1249.7
1093	$r = 90\,140\,482$, $r' = 138\,965\,105$, $\delta = 1854$	$204.151r$, $16.395r'$	8	3	13.4	10.1	140731 - 140912 (44 days)
1109	$r = 106\,999\,725$, $r' = 164\,731\,867$, $\delta = 1662$	$216.240r$, $15.976r'$	8	3	20.3	(16.7)	started 140801 (≥ 45 days)
1117	$r = 117\,501\,821$, $r' = 182\,813\,008$, $\delta = 1894$	$202.310r$, $15.638r'$	6	3	(25.5)	(20.9)	started 140805 (≥ 41 days)
1123	$r = 124\,181\,748$, $r' = 192\,010\,818$, $\delta = 3225$	$197.677r$, $14.222r'$	4	3	(29.4)	(24.1)	started 140819 (≥ 27 days)

For $n = 1177$, however, we ran into an unexpected glitch: the 244 congruences that were produced by the 256 matrix solutions (after dealing with small primes and units) were not correct modular identities involving squares of rational primes and first degree prime ideal generators. This means that the matrix step failed and produced incorrect solutions, or that incorrect columns (i.e., not corresponding to relations) were included in the matrix. Further inspection learned that the latter was the case. It turned out that due to a buggy adaptation to the dual number field case incorrect "relations" containing unfactored composites (due to the speed requirements unavoidably produced by sieving and cofactorization) were used as input to the filtering step. When we started counting the number of bad inputs, extrapolation of early counts suggested quite a few more than 244 bad entries, implying the possibility that the matrix step had to be redone because the 244 incorrect congruences may not suffice to produce correct congruences (combining incorrect congruences to remove the bad entries). We narrowly escaped because, due to circumstances beyond anyone's control [29], the count unexpectedly slowed down and only 189 bad entries were found. This then led to a total of 195 correct congruences, after which the factorization followed using the approach described above.

The factorizations that we obtained so far, ten for $n \in S_{\mathrm{I}}$ followed by a single one for $n \in S_{\mathrm{II}}$, are listed below: n, the lengths in binary and decimal of the unfactored part of $2^n - 1$, factorization date, the lengths of the smallest newly found prime factor, and the factor.

1007 : 843-bit $c254$, Jan 8 2013, 325-bit $p98$:
$\left\{ \begin{array}{l} 45664833523052628586495213371442511740075371951182478448819785894752 76 \\ 3553620148815526546415896369 \end{array} \right.$

1009 : 677-bit $c204$, Jun 12 2013, 295-bit $p89$:
$\left\{ \begin{array}{l} 32801629399316220386255938566077541078836238345868341181567256008155 63 \\ 8984594836583203447 \end{array} \right.$

1081 : 833-bit $c251$, Mar 11 2013, 380-bit $p115$:
$\left\{ \begin{array}{l} 14395810902323603067246527214972214758018935941043357067676291092775 02 \\ 59908332598995897457735306337226616870253764 1 \end{array} \right.$

1111 : 921-bit $c278$, Jun 13 2013, 432-bit $p130$:
$\left\{ \begin{array}{l} 94016992174261011260856274005378816886689234303060299026659472401120 85 \\ 572850557654128039535064932539432952669653208185411260693457 \end{array} \right.$

1129 : 1085-bit $c327$, Sep 20 2013, 460-bit $p139$:
$\left\{ \begin{array}{l} 26828635518494639415550122350613026061139195421171418141682190654697 41 \\ 0269731498119378612493808577720143084340172854729534287561205468229 11 \end{array} \right.$

1151 : 803-bit $c242$, Dec 12 2013, 342-bit $p103$:
$\left\{ \begin{array}{l} 83119194310395609642916349179778127659970015164447321362710006111747 75 \\ 26433792665734336910910066380404 7 \end{array} \right.$

1153 : 1099-bit $c331$, Oct 28 2013, 293-bit $p89$:
$\left\{ \begin{array}{l} 10122360961247873953624190885178888629606889980435179249683524293313 23 \\ 0115056983720103793 \end{array} \right.$

1159 : 1026-bit $c309$, Feb 9 2014, 315-bit $p95$:
$\left\{ \begin{array}{l} 62999265036082335900111964701462000438592932517815660818451881915621 15 \\ 434921003802703330934428 7 \end{array} \right.$

1177 : 847-bit $c255$, May 29 2014, 370-bit $p112$:
$\left\{ \begin{array}{l} 20156607875489234546625902056211238869700857614360215929428598475231 08 \\ 465523348455927947279783179798610711213193 \end{array} \right.$

1193 : 1177-bit $c355$, Aug 22 2014, 346-bit $p104$:
$\left\{ \begin{array}{l} 85227326201314361823893776605433636670217425388311906457714409016049 96 \\ 15075162304168221455997574624727 29 \end{array} \right.$

1093 : 976-bit $c294$, Sep 13 2014, 405-bit $p122$:
$\left\{ \begin{array}{l} 46116332943436452551540576315696985297990259869411311813222303231047 19 \\ 644416041896994679152055837869486391336398032829344 9 \end{array} \right.$

The total cost for the eleven factorizations for $n \in S_{\mathrm{I}}$ will be about 4915 core years, with relation collection estimated at 3665 core years, and all matrices in about 1250 core years. Relation collection for $n \in S_{\mathrm{II}}$ required 2064 core years, and three of the five remaining matrices are currently being processed. Individual factorization using the SNFS would have cost ten to fifteen thousand core years for all $n \in S_I$ and four to six thousand core years for all $n \in S_{II}$, so overall we expect a worthwhile saving. The completion date of the overall project depends on the resources that will be available to process the matrices. The online version [19] of this paper will be kept up-to-date with our progress.

With so far a smallest newly found factor of 89 decimal digits and a largest factor found using the ECM of 83 decimal digits [33], it may be argued that our ECM preprocessing did not "miss" anything yet.

7   Conclusion

We have shown that given a list of properly chosen special numbers their factorizations may be obtained using Coppersmith's factoring factory with considerable savings, in comparison to treating the numbers individually. Application of Coppersmith's idea to general numbers looks less straightforward. Taking the effects into account of rational versus algebraic precomputation (giving rise to many more smooth values) and of our relatively large algebraic degree (lowering our number of precomputed values), extrapolation of the 70 TB disk space estimate given at the end of Section 4.2 suggests that an EB of disk space may be required if a set S of 1024-bit RSA moduli to be factored is not known in advance. This is not infeasible, but not yet within reach of an academic effort. Of course, these excessive storage problems vanish if S is known in advance. But the relative efficiency of current implementations of sieving compared to factorization trees suggests that $|S|$ individual NFS efforts will outperform Coppersmith's factorization factory, unless the moduli get larger. This is compounded by the effect of advantageously chosen individual roots, versus a single shared root.

Regarding the SNFS factorization factory applied to Mersenne numbers, the length of an interval of n-values for which a certain fixed degree larger than our $d = 8$ is optimal, will be larger than our interval of n-values. And, as the corresponding Mersenne numbers $2^n - 1$ will be larger than the ones here, fewer will be factored by the ECM. Thus, we expect that future table-makers, who may wish to factor larger Mersenne numbers, can profit from the approach described in this paper to a larger extent than we have been able to – unless of course better factorization methods or devices have emerged. Obviously, the SNFS factorization factory can be applied to other Cunningham numbers, or Fibonacci numbers, or yet other special numbers. We do not elaborate.

Acknowledgements. We gratefully acknowledge the generous contribution by Microsoft Research to the relation collection effort. Specifically, we want to thank Kristin Lauter for arranging access to the XCG container lab, Lawrence LaVerne for technical support and Michael Naehrig for his assistance in the coordination of the relation collection. We thank Rob Granger for his comments. This work was supported by EPFL through the use of the facilities of its Scientific IT and Application Support Center: the SCITAS staff members have been tremendously helpful and forthcoming dealing with our attempts to process the seventeen matrix steps in a reasonable amount of time. Finally, this work was supported by the Swiss National Science Foundation under grant numbers 200021-119776, 200020-132160, and 206021-128727 and by the project SMSCG (Swiss Multi Science Compute Grid), with computational infrastructure and support, as part of the "AAA/SWITCH – e-infrastructure for e-science" program.

References

1. Aoki, K., Franke, J., Kleinjung, T., Lenstra, A.K., Osvik, D.A.: A kilobit special number field sieve factorization. In: Kurosawa, K. (ed.) ASIACRYPT 2007. LNCS, vol. 4833, pp. 1–12. Springer, Heidelberg (2007)
2. Bahr, F.: Liniensieben und Quadratwurzelberechnung für das Zahlkörpersieb, Diplomarbeit, University of Bonn (2005)
3. Bernstein, D.J.: How to find small factors of integers (June 2002), http://cr.yp.to/papers.html
4. Bos, J.W., Kleinjung, T., Lenstra, A.K., Montgomery, P.L.: Efficient SIMD arithmetic modulo a Mersenne number. In: IEEE Symposium on Computer Arithmetic – ARITH-20, pp. 213–221. IEEE Computer Society (2011)
5. Brillhart, J., Lehmer, D.H., Selfridge, J.L., Tuckerman, B., Wagstaff Jr., S.S.: Factorizations of $b^n \pm 1$, $b = 2, 3, 5, 6, 7, 10, 11, 12$ Up to High Powers, 1st edn. Contemporary Mathematics, vol. 22. American Mathematical Society (1983) (2nd edn. 1988), (3rd edn. 2002), Electronic book available at: http://homes.cerias.purdue.edu/~ssw/cun/index.html
6. Childers, G.: Factorization of a 1061-bit number by the special number field sieve. Cryptology ePrint Archive, Report 2012/444 (2012), http://eprint.iacr.org/
7. Coppersmith, D.: Modifications to the number field sieve. Journal of Cryptology 6(3), 169–180 (1993)
8. Coppersmith, D.: Solving homogeneous linear equations over GF(2) via block Wiedemann algorithm. Mathematics of Computation 62(205), 333–350 (1994)
9. Cunningham, A.J.C., Western, A.E.: On Fermat's numbers. Proceedings of the London Mathematical Society 2(1), 175 (1904)
10. Cunningham, A.J.C., Woodall, H.J.: Factorizations of $y^n \pm 1$, $y = 2, 3, 5, 6, 7, 10, 11, 12$ up to high powers. Frances Hodgson, London (1925)
11. Dodson, B., Lenstra, A.K.: NFS with four large primes: An explosive experiment. In: Coppersmith, D. (ed.) CRYPTO 1995. LNCS, vol. 963, pp. 372–385. Springer, Heidelberg (1995)
12. Franke, J., Kleinjung, T.: Continued fractions and lattice sieving. In: Special-purpose Hardware for Attacking Cryptographic Systems – SHARCS (2005), http://www.hyperelliptic.org/tanja/SHARCS/talks/FrankeKleinjung.pdf
13. Franke, J., Kleinjung, T.: GNFS for linux. Software (2012)
14. Franke, J., Kleinjung, T., Morain, F., Wirth, T.: Proving the primality of very large numbers with fastECPP. In: Buell, D.A. (ed.) ANTS 2004. LNCS, vol. 3076, pp. 194–207. Springer, Heidelberg (2004)
15. Golliver, R., Lenstra, A.K., McCurley, K.: Lattice sieving and trial division. In: Huang, M.-D.A., Adleman, L.M. (eds.) ANTS 1994. LNCS, vol. 877, pp. 18–27. Springer, Heidelberg (1994)
16. Harrison, J.: Isolating critical cases for reciprocals using integer factorization. In: IEEE Symposium on Computer Arithmetic – ARITH-16, pp. 148–157. IEEE Computer Society Press (2003)

17. Kleinjung, T.: On polynomial selection for the general number field sieve. Mathematics of Computation 75, 2037–2047 (2006)
18. Kleinjung, T., Aoki, K., Franke, J., Lenstra, A.K., Thomé, E., Bos, J.W., Gaudry, P., Kruppa, A., Montgomery, P.L., Osvik, D.A., te Riele, H., Timofeev, A., Zimmermann, P.: Factorization of a 768-bit RSA modulus. In: Rabin, T. (ed.) CRYPTO 2010. LNCS, vol. 6223, pp. 333–350. Springer, Heidelberg (2010)
19. Kleinjung, T., Bos, J.W., Lenstra, A.K.: Mersenne factorization factory. Cryptology ePrint Archive, Report 2014/653 (2014), http://eprint.iacr.org/
20. Lenstra, A.K., Kleinjung, T., Thomé, E.: Universal security. In: Fischlin, M., Katzenbeisser, S. (eds.) Buchmann Festschrift. LNCS, vol. 8260, pp. 121–124. Springer, Heidelberg (2013), http://eprint.iacr.org/2013/635
21. Lenstra, A.K., Lenstra Jr., H.W.: Algorithms in number theory. In: van Leeuwen, J. (ed.) Handbook of Theoretical Computer Science (Volume A: Algorithms and Complexity), pp. 673–715. Elsevier and MIT Press (1990)
22. Lenstra, A.K., Lenstra Jr., H.W.: The Development of the Number Field Sieve. LNM, vol. 1554. Springer (1993)
23. Lenstra, A.K., Lenstra Jr., H.W., Manasse, M.S., Pollard, J.M.: The number field sieve, pp. 11–42 in [22]
24. Lenstra Jr., H.W.: Factoring integers with elliptic curves. Annals of Mathematics 126(3), 649–673 (1987)
25. Massey, J.: Shift-register synthesis and BCH decoding. IEEE Transactions on Information Theory 15, 122–127 (1969)
26. Montgomery, P.: Square roots of products of algebraic numbers. In: Gautschi, W. (ed.) Mathematics of Computation 1943–1993: a Half-Century of Computational Mathematics, Proceedings of Symposia in Applied Mathematics, pp. 567–571. American Mathematical Society (1994)
27. Nguyen, P.Q.: A Montgomery-like square root for the number field sieve. In: Buhler, J.P. (ed.) ANTS 1998. LNCS, vol. 1423, pp. 151–168. Springer, Heidelberg (1998)
28. Pollard, J.M.: The lattice sieve, pp. 43–49 in [22]
29. Radford, B.: Why do people see guardian angels? (August 2013), http://news.discovery.com/human/psychology/why-people-see-guardian-angels-130813.htm
30. Rivest, R.L., Shamir, A., Adleman, L.: A method for obtaining digital signatures and public-key cryptosystems. Communications of the ACM 21, 120–126 (1978)
31. Thomé, E.: Subquadratic computation of vector generating polynomials and improvement of the block Wiedemann algorithm. Journal of Symbolic Computation 33(5), 757–775 (2002)
32. Wiedemann, D.: Solving sparse linear equations over finite fields. IEEE Transactions on Information Theory 32, 54–62 (1986)
33. Zimmermann, P.: 50 large factors found by ECM, http://www.loria.fr/~zimmerma/records/top50.html
34. Zimmermann, P.: Input file for Cunningham cofactors, http://www.loria.fr/~zimmerma/records/c120-355

Improving the Polynomial time Precomputation of Frobenius Representation Discrete Logarithm Algorithms

Simplified Setting for Small Characteristic Finite Fields

Antoine Joux[1,2] and Cécile Pierrot[2,3]

[1] CryptoExperts, France and Chaire de Cryptologie de la Fondation de l'UPMC, Paris
[2] Laboratoire d'Informatique de Paris 6, UPMC Sorbonnes Universités, Paris
[3] CNRS and Direction Générale de l'Armement, France
antoine.joux@m4x.org, Cecile.Pierrot@lip6.fr

Abstract. In this paper, we revisit the recent small characteristic discrete logarithm algorithms. We show that a simplified description of the algorithm, together with some additional ideas, permits to obtain an improved complexity for the polynomial time precomputation that arises during the discrete logarithm computation. With our new improvements, this is reduced to $O(q^6)$, where q is the cardinality of the basefield we are considering. This should be compared to the best currently documented complexity for this part, namely $O(q^7)$. With our simplified setting, the complexity of the precomputation in the general case becomes similar to the complexity known for Kummer (or twisted Kummer) extensions.

1 Introduction

Recently, the computation of discrete logarithms in small characteristic finite fields has been greatly improved [Jou14,GGMZ13a,BGJT14], with the introduction of a new family of Index Calculus algorithms for this case. In the sequel, we call the algorithms from this family: **Frobenius Representation algorithms**. Frobenius Representation algorithms can be seen as descendants of the pinpointing algorithm introduced in [Jou13a]. The first two Frobenius Representation algorithms appeared essentially simultaneously, one of them proposed by Joux in [Jou14] was first used in a discrete logarithm record in $\mathbb{F}_{2^{1778}}$ announced on Feb 11th 2013 on the NMBRTHRY mailing list, while the first draft of the article describing the $L(1/4)$ complexity analysis of the algorithm was posted as [Jou13b] on Feb 20th 2013. Between these two events, another Frobenius Representation algorithm with complexity $L(1/3)$ was proposed in [GGMZ13b] with a record in $\mathbb{F}_{2^{1971}}$ announced on Feb 19th 2013 on the same mailing list. From an asymptotic point of view, the best current Frobenius Representation algorithm is the quasi-polynomial time algorithm proposed in [BGJT14]. In practice, a lot of options are open depending on the exact finite field we want to

P. Sarkar and T. Iwata (Eds.): ASIACRYPT 2014, PART I, LNCS 8873, pp. 378–397, 2014.

address. However, there are currently many open questions about these algorithms. From a theoretical point of view, it would be extremely nice to remove the heuristic hypotheses that are used in the algorithms. A first step in this direction is proposed in [GKZ14b], with a simplified individual logarithms algorithm that only relies on the ability to descent finite field elements expressed by polynomials of even degree $2D$ to polynomials of degree D. Another theoretical question would be to get the complexity down to polynomial time instead of quasi-polynomial. From a practical point of view, the limiting step for setting records in the general case, as opposed to special cases such as Kummer extension, is usually the computation of the logarithm of the initial factor base elements. When working over a base field $\mathbb{F}_q$, the best documented complexity is $O(q^7)$ (see for example [AMORH14]). However, some authors mention an higher complexity, typically, for the computation performed in [GKZ14a], with $q = 2^6$, the authors explain that the dimension of the linear algebra is reduced from q^4 to $q^4/24 = q^4/\log_2(q^4)$. Asymptotically, with this approach the complexity would be $O(q^9/\log(q)^2)$. For specific cases such as Kummer extension, the complexity is lower of the order of $O(q^6)$.

In this paper, we give a new variation which achieves complexity $O(q^6)$ for the general case. Part of this work was already presented by the first-named author in several presentations during the development of our algorithm. It is presented here in writing for the first time. In these earlier talks, the variation was described as a simplified version with degraded performance, the main reason being that using polynomials of degree up to D over $\mathbb{F}_q$ seems essentially equivalent to using linear polynomials over $\mathbb{F}_{q^d}$, with $d = D$. However, instead of allowing us to compute logarithms in the field $\mathbb{F}_{q^{dk}}$ with k of the same order of magnitude as q, it only leads to logarithms in $\mathbb{F}_{q^k}$ and we lose the extra factor of d in the field exponent, which came for free with the standard approach (with a value of d usually between 2 and 4). Also note that a similar correspondance between low degree polynomials over a large field and higher degree polynomials over a smaller field also appears in [GKZ14b].

In order to make the algorithm efficient, D needs to be minimized. At first glance, it seems that we need to take at least $D = 3$ to bootstrap the computation. Our main contribution is that with this simplified approach, it is in fact possible, under a reasonable heuristic assumption, to reduce the degree of the polynomials in the initial factor base over $\mathbb{F}_q$ to $D = 2$. Once the initial factor base is computed, with a cost $O(q^5)$, we use it as a lever to obtain the logarithms of polynomials of degree $D = 3$ and $D = 4$ with a total cost $O(q^6)$. Using either the heuristic quasi-polynomial descent of [BGJT14] or the alternative version from [GKZ14b], it is possible to bring down arbitrary elements to $\mathbb{F}_{q^k}$ to this extended factor base formed of irreducible polynomials up to degree 4.

Outline of the Article. As any recent discrete logarithms algorithms for small characteristic finite fields, our simplified setting has several phases:

> ➤ The Preliminary phase, that finds a representation of the target finite field.
> ➤ The Relation Collection and Linear Algebra phases, that permit to recover the discrete logarithms of a small set of elements, the factor base.

➤ The Extension phase, specific to small characteristic finite fields, in which we obtain the discrete logarithms of a larger set containing the factor base. We call this new set the extended factor base.

➤ The Descent phase, that recovers the discrete logarithm of an arbitrary element of the finite field by rewriting it as products of elements of the extended factor base.

Following this common structure we introduce our simplified setting in Section 2. We present then in Section 3 the computation of the discrete logarithms of the factor base together with the Extension phase. Section 4 gives a short analysis of the total improved asymptotic complexity obtained. Finally, in Section 5, we illustrate the efficiency of the algorithm with a practical computation of discrete logarithms in the general case of a prime extension degree which does not divide[1] $q(q+1)(q-1)$. More precisely, we perform the computation of the logarithms in $\mathbb{F}_{q^k}$ with $q = 3^5$ and extension degree $k = 479$ (the largest prime smaller than $2q$).

2   Simplified Setting for Small Characteristic Finite Fields

When trying to compute discrete logarithms in a given finite field, let us say $\mathbb{F}_{q^k}$, the first step is to choose a convenient way to construct it. We first expose in Section 2.1 how Frobenius Representation algorithms represent the target field with the help of two polynomials h_0 and h_1. We present then an improved way to choose these two cornerstone polynomials in Section 2.2. Last but not least, we propose in Section 2.3 a simpler factor base. It is the combination of these two simplified choices that permits to obtain an improvement in the asymptotic complexity of the Relation Collection, Linear Algebra and Extension phases.

2.1   Frobenius Representation Algorithms

Like all Frobenius Representation algorithms, the algorithm we propose relies on two key elements. The first element is the well-known fact that over $\mathbb{F}_q[X]$, the following polynomial identity holds:

$$\prod_{\alpha \in \mathbb{F}_q} (X - \alpha) = X^q - X. \tag{1}$$

The second element is to define the target finite field $\mathbb{F}_{q^k}$, where we want to compute discrete logarithms, by determining two polynomials h_0 and h_1 of degree at most H and by requiring that there exists a monic irreducible polynomial $I(X)$ of degree k over $\mathbb{F}_q[X]$ such that:

$$I(X) \text{ divides } h_1(X)X^q - h_0(X). \tag{2}$$

---

[1] The known special cases which are very efficient for record being Kummer extensions of degree dividing $q - 1$, twisted Kummer extensions with degree dividing $q + 1$ and Artin-Schreier extensions.

If θ denotes a root of $I(X)$ in $\overline{\mathbb{F}}_q$, setting $\mathbb{F}_{q^k} = \mathbb{F}_q[X]/(I(X)) = \mathbb{F}_q(\theta)$ gives a representation of the finite field that satisfies $\theta^q = h_0(\theta)/h_1(\theta)$. Since the map that raises an element of $\overline{\mathbb{F}}_q$ to the power q is called the Frobenius map, this choice of representation explains the name of **Frobenius Representation** we use for this family of algorithms.

The Dual Frobenius Representation Variant. There is an alternative option proposed in [GKZ14a] for constructing the extension field where we require that:
$$I(X) \text{ divides } h_1(X^q)X - h_0(X^q).$$
The advantage of this option is to allow a wider range of possible extension degrees k for a given basefield $\mathbb{F}_q$. However, using this variation slightly complicates the description of the algorithm. With this variation, the finite field representation satisfies $\theta = h_0(\theta^q)/h_1(\theta^q)$. When referring to the variation by name, we will call it a **dual Frobenius Representation** or equivalently a **Verschiebung Representation**.

2.2 Improved Choice of h_0 and h_1

A Really Simple Construction. We recall that the usual choice is to take two quadratic polynomials to allows the possibility of representing, at least heuristically, a large range of finite fields. Since we know that using linear polynomials for h_0 and h_1 does not allow such a large range, we propose a slightly different choice. **We take for h_0 an affine polynomial and for h_1 a quadratic polynomial.** We assume furthermore that the constant term of h_1 is equal to 0. Note that, by factoring out a constant in the defining Equation (2), we can assume, without loss of generality, that h_1 is monic. For simplicity of the presentation, it is convenient to rewrite:
$$h_0(X) = rX + s \quad \text{and} \quad h_1(X) = X(X + t) \tag{3}$$

A Useful Variant. Another natural option is to take for h_0 a quadratic polynomial with a contant term equal to 0 and for h_1 an affine polynomial. In this case, it is convenient to rewrite:
$$h_0(X) = X(X + w) \quad \text{and} \quad h_1(X) = uX + v. \tag{4}$$

At first sight, nothing indicates that one of the two choices is better, and in fact, both are equivalent in term of complexity. However, as we show in Section 3, the first one leads in practice to a simpler description of the algorithm. As a mnemonic we can notice that $(\mathbf{r}, \mathbf{s}, t)$ are the coefficients of the really simple construction whereas $(\mathbf{u}, \mathbf{v}, w)$ are the one of the useful variant.

2.3 Seeking a Natural Factor Base

Once the representation of the target field is chosen, we need to fix the factor base. With the aim of simplifying the description of the algorithm, we propose to get rid of polynomials with coefficients in an extension field.

Irreducible Polynomials with Coefficients in the Basefield. We choose a parameter D and consider a factor base that contains all irreducible polynomials of degree $\leq D$ over $\mathbb{F}_q[X]$. This has to be compared with previous Frobenius Representation algorithms that consider irreducible polynomials with coefficients in an extension of $\mathbb{F}_q$. To generate equations, we let A and B be two polynomials of degree $\leq D$ and using Equations (1) and (2) we write:

$$B(\theta) \prod_{\alpha \in \mathbb{F}_q} (A(\theta) - \alpha B(\theta)) = B(\theta) A(\theta)^q - A(\theta) B(\theta)^q$$

$$= B(\theta) A(\theta^q) - A(\theta) B(\theta^q)$$

$$= B(\theta) A\left(\frac{h_0(\theta)}{h_1(\theta)}\right) - A(\theta) B\left(\frac{h_0(\theta)}{h_1(\theta)}\right).$$

For compactness, we match $B(\theta)$ with the point α at infinity on the projective line $\mathbb{P}_1(\mathbb{F}_q)$. This permits to rewrite throughout the sequel the first product as $\prod_{\alpha \in \mathbb{P}_1(\mathbb{F}_q)} (A(\theta) - \alpha B(\theta))$. We also introduce the following notation:

Definition 1. *Let D be an integer, and h_0, h_1, A, B be four polynomials such that A and B are of degree at most D. Then $[A, B]_D$ is called the D-bracket of A and B. It is defined as:*

$$[A, B]_D(X) = h_1(X)^D \left(B(X) A\left(\frac{h_0(X)}{h_1(X)}\right) - A(X) B\left(\frac{h_0(X)}{h_1(X)}\right) \right).$$

Proposition 1. *If h_0 and h_1 are polynomials of degree at most H and if A and B are polynomials of degree at most D then:*

> ➤ *$[A, B]_D$ is a polynomial of degree at most $(H + 1) \cdot D$.*
> ➤ *The map $[.,.]_D$ is bilinear and antisymmetric. In particular, $[A, A]_D = 0$.*

The proof of the two items of the proposition is straightforward. With these two notations, we rewrite the equality as:

$$\prod_{\alpha \in \mathbb{P}_1(\mathbb{F}_q)} (A(\theta) - \alpha B(\theta)) = \frac{[A, B]_D(\theta)}{h_1(\theta)^D}. \tag{5}$$

Since the numerator $[A, B]_D$ of the **right-hand** side of Equation (5) has a bounded degree, under a classical heuristic, the probability that it factors into irreducible polynomials of degree at most D can be lower bounded by a constant p_H. When using a dual Frobenius Representation, we similarly get:

$$\prod_{\alpha \in \mathbb{P}_1(\mathbb{F}_q)} (A(\theta) - \alpha B(\theta)) = \left(\frac{[A, B]_D(\theta)}{h_1(\theta)^D}\right)^q. \tag{6}$$

Degree of the Factor Base Polynomials. In order to choose the parameter D, we have to balance three ideas: to lower the complexity of the linear algebra

phase we require to have a small factor base, but, we also need to be able to generate enough good equations[2] and to descent larger polynomials to polynomials
of the factor base. The polynomial degree of the factor base must not be too
small in both cases, otherwise one at least of this two steps will not be possible.
Let us give more details about this degree.

The previous degree 3 barrier. When we consider the general case where h_0
and h_1 are polynomials of degree bounded by H, the analysis is as follows. The
number of equations that can be generated is obtained by counting the number
of pairs of polynomials (A, B) that remains once we take into account the fact
that the pairs are invariant under the action of $\mathrm{PGL}_2(\mathbb{F}_q)$. In other words, ignoring the cases where the degree is somehow reduced (see Appendix A for details)
in the **left-hand** side of Equation (5) we can assume that:

$$A(X) = X^D + a(X) \quad \text{and} \quad B(X) = X^{D-1} + b(X),$$

where $a(X)$ and $b(X)$ have degree at most $D - 2$. As a consequence, since polynomials of degree $D - 2$ have $D - 1$ coefficients, the number of good equations
that can be generated in this manner is of the order of $p_H \cdot q^{2D-2}$. Moreover,
the number of elements in the factor base, *i.e.* the number of irreducible of degree at most D is close to q^D/D. To get more equations than unknowns in the
linear algebra phase, i.e. to obtain $D \cdot p_H \cdot q^{D-2} \geqslant 1$, unless enlarging a lot the
probability p_H, we need that $D \geqslant 3$, as underlined in [GKZ14b].

As a consequence, the best hope we get for the complexity of computing
the logarithms of factor base elements is of the order of $(q^D)^2 \cdot q \geqslant q^7$. Note
that looking at the various existing record, this lower bound of q^7 is not always
attained, since some computations need to enlarge the factor base to $D = 4$, which
raises the complexity to $O(q^9)$. Typically, such an enlargement is performed
in [GKZ14a], even if, thanks to a judicious use of Galois invariance, they reduce
the cost of this enlargement compared to $O(q^9)$ by regrouping the degree 4
objects[3] into groups of 24 conjugates.

The reason for this enlargement is that the known techniques for descending
polynomials of degree larger than 4 to degree 4 do not work completely to descent degree 4 polynomials to degree 3, since in most cases, only a fraction of
degree 4 irreducible polynomials can be obtained in this manner. This is similar
to the situation reported in [AMORH14], where half of the quadratic polynomials over a cubic extension can be derived with the descent algorithm from linear
polynomials.

Breaking the barrier. Following the above argument, for $D = 2$ we expect about
$q^2/2$ irreducible polynomials and assuming that $H = 2$, one would expect a value
of p_H well below $1/2$. Thus, without any improvement on the probability, the
expected number of equations is too small compared to the number of unknowns
and it is not possible to derive the discrete logarithms of the small elements in
this manner... Yet, in our simplified setting **the factor base consists in all**

---

[2] We call **good equations** equations of the restricted form (5) where both right and
left-hand side can be written with polynomials of the factor base only.

[3] Those objects are in fact quadratic polynomials over a degree 2 extension.

the irreducible polynomial of degree 2 with coefficients in the base field. We explain in Section 3.1 how to get around this problem and to recover all the discrete logarithms of the factor base.

3   Improving Computations of the (Extended) Factor Base

In this section, we present two contributions which allow us to reduce the global cost of the polynomial part of discrete logarithm computations. The first contribution in Section 3.1 describes how we can adapt the use of Equation (5) to be able to perform an initial computation with a reduced initial factor base corresponding to $D = 2$ for a cost $O(q^5)$. We also show, in Section 3.2, that once this is done, the enlargement to $D = 3$ can be performed with a reduced cost $O(q^6)$, instead of the expected $O(q^7)$.

The second contribution presented in Section 3.3 is a new descent technique that only requires a small subset of degree 4 irreducible polynomials to be able to compute on the fly the logarithm of an overwhelming fraction of other degree 4 polynomials. If there is enough available memory, it is also possible using a adaptation of this technique to obtain the logarithms corresponding to an enlarged basis with $D = 4$. Both options can be performed with a time complexity $O(q^6)$.

3.1   A Reduced Degree 2 Factor Base

As previously said, if we choose a degree 2 factor base, it seems that we don't have enough good equations compared to the number of unknowns. We propose two approaches to get rid of this problem. First, we show that thanks to our smaller degree polynomials h_0 and h_1, we can improve p_H, the bound on the probability to obtain a good equation, by exhibiting systematic factors. In addition, we also use another source of equations to complete the system. A secondary advantage is that this second source leads to much sparser equations that the use of Equation (5).

Improving the Probability p_H Thanks to Systematic Factors. Once we have fixed $A(X) = X^D + a(X)$ and $B(X) = X^{D-1} + b(X)$, we see that both the left-hand side and the denominator of the right-hand side of Equation (5) or (6) can be written as products of elements of the factor base. So, we have to analysis the probability that the numerator of the right-hand side, namely the D-bracket of A and B, can be factorized in products of polynomials of degree at most 2.

The simple construction: h_0 affine and h_1 quadratic. Proposition 1 allows to upper-bound the degree of $[A, B]_D$ by $(H + 1) \cdot D$. As a consequence, for $H = 2$ and $D = 2$, this degree is lower than 6. The probability that a random polynomial of degree 6 factors into terms of degree less than 2 is well too small to permits to obtain enough equations. Though, as mentioned in [GKZ14b], we remark that a systematic term appears in the factorization of $[A, B]_D(X)$. To be more precise, we have the following result:

Lemma 1 (Systematic factor of a D-bracket). *Let A and B be two polynomials of degree at most D. Then $[A, B]_D(X)$ is divisible by $X h_1(X) - h_0(X)$.*

Proof. By bilinearity, if $A(X) = \sum_{i=0}^{D} a_i X^i$ and $B(X) = \sum_{i=0}^{D} b_i X^i$, we can write: $[A, B]_D = \sum_{i=0}^{D} \sum_{j=0}^{D} a_i b_j [X^i, X^j]_D$. Moreover, since $[.,.]_D$ is bilinear and anti-symmetric it is clear that $[X^i, X^j]_D = -[X^j, X^i]_D$ and $[X^i, X^i]_D = 0$. Thus, it suffices to consider the D-bracket of X^i and X^j where $i < j$. Lets us compute:

$$[X^i, X^j]_D = h_1^{D-j}(X) \left(X^j h_0(X)^i h_1(X)^{j-i} - X^i h_0(X)^j \right)$$
$$= h_1^{D-j}(X) X^i h_0(X)^i \left((X h_1(X))^{j-i} - h_0(X)^{j-i} \right)$$
$$= h_1^{D-j}(X) X^i h_0(X)^i (X h_1(X) - h_0(X)) \sum_{k=1}^{j-i} h_0(X)^{k-1} (X h_1(X))^{j-i-k}.$$

As a consequence $X h_1(X) - h_0(X)$ divides $[X^i, X^j]_D$ and the lemma follows. $\quad\square$

Thus, after dividing $[A, B]_D$ by this degree 3 systematic factor, the question is whereas a polynomial of degree 3 factors into terms of degree at most 2. Assuming that it behaves as a random polynomial in this respect, we can lower bound (see Appendix B) the probability by $2/3$. Since this is higher than $1/2$, we have now enough equations to compute the logarithms of the factor base.

The useful variant: h_0 quadratic and h_1 affine. We can check again that the numerator in the right-hand side of Equation (5) or (6) becomes systematically divisible by $\theta h_1(\theta) - h_0(\theta)$. Yet, in this variant, this systematic factor has degree 2 only. This partially improves the value of p_H, however, this is not sufficient to get enough equations.

To go further in reducing this degree, we have to remark that the bound on the degree of $[A, B]_D$ given in Proposition 1, which is $(H + 1) \cdot D$, can in fact be improved in the specific case where h_1 is affine. In truth, the degree is now upper-bounded by $(H + 1) \cdot D - 1$. For $H = 2$ and $D = 2$, this reduces for free the degree from 6 to 5. As a consequence, after dividing by the degree 2 systematic factor of Lemma 1, there remains as previously a polynomial of degree 3. Again the probability p_H is lower-bounded by $2/3 > 1/2$. In both cases, this probability would already suffice to produce enough equations.

Additional Equations. Despite the fact that the equations obtained with our improved choice of h_0 and h_1 in both the simple construction and the useful variant would suffice to solve the linear system with parameter $D = 2$, proposing a source of extra equations is also helpful. In this section, to produce additional equations, we simply consider a variation on the systematic equations that were introduced in [BMV85] and often used in the Function Field Sieve.

More precisely, let $f(X) = X^2 + f_1 X + f_0$ be an irreducible polynomial of degree 2 in $\mathbb{F}_q[X]$. We can write :

$$f(\theta)^q = f\left(\frac{h_0(\theta)}{h_1(\theta)}\right) = \frac{h_0(\theta)^2 + f_1 h_0(\theta) h_1(\theta) + f_0 h_1(\theta)^2}{h_1(\theta)^2}.$$

The numerator of the right-hand side is a polynomial of degree 4, since one of the two polynomials h_0 or h_1 is quadratic and the other one is affine. We remark that about half of these numerators are irreducible and the other half factor into a product of two degree 2 irreducible polynomials. For the case of a dual Frobenius Representation, the systematic equations are slightly different:

$$f(\theta) = f\left(\frac{h_0(\theta)}{h_1(\theta)}\right)^q = \left(\frac{h_0(\theta)^2 + f_1\, h_0(\theta)h_1(\theta) + f_0\, h_1(\theta)^2}{h_1(\theta)^2}\right)^q$$

but the principle remains identical. These systematic equations can easily be generalized to irreducible polynomials of arbitrary degree, with again a close to half/half repartition[4]:

Lemma 2. *Let h_0 and h_1 be two polynomials such that one is affine and the other quadratic. If f is a degree D monic irreducible polynomial in $\mathbb{F}_q[X]$, then $h_1(X)^{2D} f(h_0(X)/h_1(X))$ is a polynomial of degree $2D$ that has a probability equal to $1 - p$ to be irreducible and a probability equal to p to factor into two degree D irreducible polynomials, with:*

$$\frac{1}{q^D}\left(\frac{q^D - 1}{2} - \frac{q^{\lfloor D/2 \rfloor + 1} - q}{q - 1}\right) \le p \le \frac{q^D + 3}{2\, q^D}.$$

In particular, note that for irreducible polynomials of degree 1, which are part of the initial factor base for $D = 2$, we always obtain a systematic equation relating the given polynomial either to two other affine polynomials or to one quadratic polynomial. Note that we could also use the systematic equations for higher degree polynomials in Section 3.3 to ease the computation of the logarithm of degree 4 polynomials.

3.2   Enlarging the Factor Base to Degree 3

In order to be able to enlarge the factor base to degree 3 without performing linear algebra on a matrix of dimension q^3, we follow an approach quite similar to the one presented in [Jou14]. Namely, we divide first the set of irreducible polynomials of degree 3 into groups and search then for a way to generate enough equations involving only the polynomials within a group and polynomials of degree 1 or 2 whose logarithms are already known.

Groups of Degree 3 Polynomials for the Simple Construction. To define a group of degree 3 polynomials we start from an element g in the base field $\mathbb{F}_q$ and we consider $\mathcal{P}_g$ the corresponding group of degree 3 polynomials such that:

$$\mathcal{P}_g = \{(X^3 + g) + \alpha\, X^2 + \beta\, X | (\alpha, \beta) \in \mathbb{F}_q{}^2\}.$$

Clearly, if we generate a relation using Equation (5), or (6), with $A(X) = (X^3 + g) + \alpha\, X^2$ and $B(X) = (X^3 + g) + \beta\, X$, with a and b in $\mathbb{F}_q$, then all degree 3 polynomials that appear in the left-hand side belong to $\mathcal{P}_g$. The elements of $\mathcal{P}_g$ can be divided into two groups:

---

[4] We prove the following lemma in the extended version of this article.

➤ the reducible polynomials whose logarithms can be computed by taking the sum of the logarithms of their factors,
➤ and the irreducible polynomials which appear as unknowns. Note that the number of irreducible polynomials in a group $\mathcal{P}_g$ is approximately $q^2/3$.

For one fixed element g, by considering all possibilities for α and β, we find q^2 candidate relations. Yet, we keep only those whose right-hand side factors into terms of degree at most 2. The question is now whether we obtain enough equations to be able to solve the corresponding linear system.

For this, lets look into more details at the right-hand side. With our choice of h_0 and h_1 it is a polynomial of degree 9, as described in Proposition 1. Moreover, it follows from Lemma 1 that it is divisible by the degree 3 polynomial $\theta h_1(\theta) - h_0(\theta)$. As a consequence, we are left with a polynomial of degree 6 to factor in terms of degree at most 2. The probability to obtain a good relation is not yet higher than $1/3$. To improve on this probability, we first remark that with our specific choice of A and B the polynomial degree of the numerator of the right-hand side is in fact 8. Thus we are left with a polynomial of degree 5 to factor in terms of degree at most 2. Besides, we reveal a very simple systematic factor.

Lemma 3 (Systematic factor of particular 3-brackets in the simple construction). *Let h_0, h_1, A and B be four polynomials such that h_0 is affine, $h_1(X) = X(X+t)$, $A(X) = (X^3 + g) + \alpha X^2$ and $B(X) = (X^3 + g) + \beta X$, with t, g, α and β in $\mathbb{F}_q$. Then $[A, B]_3$ is a polynomial of degree at most 8 divisible by X.*

Proof. By bilinearity and antisymmetry we have $[A, B]_3 = \alpha [X^2, X^3 + g]_3 + \beta [X^3 + g, X]_3 + \alpha\beta [X^2, X]_3$. Let us compute the following 3-brackets:

$$\begin{aligned}
[X, X^2]_3 &= X^2 h_0 h_1^2 - X h_0^2 h_1 \\
[X^3 + g, X]_3 &= X(h_0^3 + g h_1^3) - (X^3 + g) h_0 h_1^2 \\
&= X[h_0^3 + g h_1^3 - (X^3 + g) h_0 X(X+t)^2] \\
[X^3 + g, X^2]_3 &= X^2(h_0^3 + g h_1^3) - (X^3 + g) h_0^2 h_1 \\
&= X[X(h_0^3 + g h_1^3) - (X^3 + g) h_0^2 (X+t)]
\end{aligned}$$

The result of the lemma comes from the fact that all the 3-brackets involved in the computation of $[A, B]_3$ are divisible by X. Moreover, considering the polynomials degrees of these elements we remark that $[X, X^2]_3$ has degree 6 whereas $[X^3 + g, X]_3$ has degree 7 and $[X^3 + g, X^2]_3$ has degree 8.   □

As a direct consequence, the remaining factor in the right-hand side when considering these groups is of degree 4. According to Appendix B the heuristic probability that it factors into terms of degree at most 2 is close to 41%. Since these is greater than $1/3$, we expect to find enough equations to compute all the discrete logarithms of the irreducible polynomials belonging to $\mathcal{P}_g$. Moreover, it is clear that any monic and irreducible polynomial of degree 3 belongs to one $\mathcal{P}_g$.

Groups of Degree 3 Polynomials for the Useful Variant. In this setting, computing discrete logarithms of degree 3 polynomials is a bit more tricky. To define a group in this case, we start from a triple (g_1, g_2, g_3) of elements in $\mathbb{F}_q$. The corresponding group of degree 3 polynomials is defined as:

$$\mathcal{P}_{g_1,g_2,g_3} = \{X^2(X - g_1) + \alpha\, X(X - g_2) + \beta\,(X - g_3)|(\alpha, \beta) \in \mathbb{F}_q^{\,2}\}.$$

Let us fix $(g_1, g_2, g_3) \in \mathbb{F}_q^{\,3}$. If we generate a relation using Equation (5) with $A(X) = X^2(X - g_1) + \alpha\,(X - g_3)$ and $B(X) = X(X - g_2) + \beta\,(X - g_3)$, with α and β in $\mathbb{F}_q$, then all degree 3 polynomials that appear in the left-hand side belong to the corresponding group $\mathcal{P}_{g_1,g_2,g_3}$. After keeping only the q^2 candidate relations whose right-hand side factors into terms of degree at most 2, the question is, again, whether we obtain enough equations to solve the linear system where the unknown are the $q^2/3$ irreducible polynomials of $\mathcal{P}_{g_1,g_2,g_3}$.

When h_0 is quadratic and h_1 affine, the right-hand side is still a polynomial of degree 8 divisible by $\theta h_1(\theta) - h_0(\theta)$. We are left with a polynomial of degree 6 to factor in terms of degree at most 2. Yet, without any further improvement, the probability of this remaining polynomial to factor into terms of degree at most 2 is still too small to obtain enough equations.

To overcome this obstacle, we no longer consider the general groups of this form. Our goal is to point out some groups in which we now that the right-hand sides have some extra systematic factors. Another argument for considering few special groups only comes when we remark that the number of degree 3 polynomials produced with all those general groups is way too large. Taking all q^3 groups of the form $\mathcal{P}_{g_1,g_2,g_3}$ is a clear overkill since they each contain q^2 elements whereas there are only q^3 monic polynomials of degree 3. In fact we expect that these polynomials could be mostly covered by q groups only. To put it in a nutshell, we restrict ourselves to the specific choice of g_1, g_2 and g_3 where we first choose a value $g_1 \in \mathbb{F}_q$ and compute then:

$$g_2 = G(g_1) \quad \text{and} \quad g_3 = G(g_2)$$

where $G : \mathbb{F}_q \mapsto \mathbb{F}_q$ is a particular map. We propose to consider:

$$G : g \mapsto \frac{v(v + w)}{(1 + u)(v + w - g)}. \tag{7}$$

We recall that u, v, w denote the coefficients of the polynomials h_0 and h_1, as given in (4). Assuming that both g_1 and g_2 are not equal to $v + w$ then all three values (g_1, g_2, g_3) are well-defined. With this specific choice, the right-hand side that now appear in Equation (5) or (6) gains a new systematic degree 2 factor $\theta h_1 + h_0 + (v + w)h_1 = (1 + u)\,\theta^2 + (1 + u)(v + w)\theta + vw + v^2$ as given in Lemma 4. Again, the remaining factor in the right-hand side when considering these groups is of degree 4. Since the probability of a degree 4 polynomial to factor in terms of degree at most 2 is higher than $1/3$, we can recover all the discrete logarithms of the irreducible polynomials of $\mathcal{P}_{g_1,G(g_1),G(G(g_1))}$.

Lemma 4 (Systematic factor of particular 3-brackets in the useful variant). *Let G denote the map of (7) and let h_0, h_1, A and B be four polynomials such that $h_0(X) = X(X + w)$, $h_1(X) = uX + v$, $A(X) = X^2(X - g) + a(X - G(G(g)))$ and $B(X) = X(X - G(g)) + b(X - G(G(g)))$, with u, v, w, a, b and g in $\mathbb{F}_q$. Then $[A, B]_3$ is divisible by $(1 + u)X^2 + (1 + u)(v + w)X + vw + v^2$.*

Proof. By bilinearity and antisymmetry: $[A, B]_3 = [X^2(X - g), X(X - G(g))]_3 + b[X^2(X - g), X - G(G(g))]_3 + a[X - G(G(g)), X(X - G(g)))]_3$. The result of the lemma comes from the computation of the 3 bracket of the three pairs of different elements made with $X^2(X - g)$, $X(X - G(g))$ and $X - G(G(g))$.   □

Fraction of Degree 3 Polynomials Covered by Our Groups. Since we can recover all the discrete logarithms of the irreducible polynomials that appear in a group, the question that remains is whether every polynomial belongs to one of these groups at least.

Valid groups. In the sequel we restrict ourselves to the case where $v + w \neq 0$. Yet, if $v + w = 0$ then G is the zero mapping. This case is studied in the extended version of our article. To study the properties of our group, it is convenient to remark that since G is an homography, we can transform it into a permutation of the projective line $\mathbb{P}_1(\mathbb{F}_q)$. As classically done, we add the two following values of G:

$$G(\infty) = 0 \quad \text{and} \quad G(v + w) = \infty.$$

With this additional definition, we see that the groups we consider are indexed by triple $(g, G(g), G(G(g)))$ which do not contain the value ∞. Since $v + w \neq 0$ then ∞ belongs to a cycle of length at least 3. Thus, there are $q - 2$ valid groups corresponding to the values of g in $\mathbb{F}_q - \{G^{-1}(\infty), G^{-1}(G^{-1}(\infty))\}$. With this description we reach at best $q^3 - 2q^2$ polynomials of degree 3.

Groups at infinity. To reach more polynomials we define three additional groups $\mathcal{P}_{g,G(g),G(G(g))}$ when $g, G(g)$ or $G(G(g))$ is equal to ∞. These groups are given by the following descriptions:

$$\mathcal{P}_{\infty,0,G(0)} = \left\{ X\left(X^2 + \frac{vw + v^2}{1 + u}\right) + \alpha X^2 + \beta(X - G(0)) | (\alpha, \beta) \in \mathbb{F}_q^2 \right\}.$$

$$\mathcal{P}_{\infty^{-1},\infty,0} = \left\{ X^2(X - \infty^{-1}) + \alpha X + \beta\left(X^2 + \frac{vw + v^2}{1 + u}\right) | (\alpha, \beta) \in \mathbb{F}_q^2 \right\}.$$

and $\quad \mathcal{P}_{\infty^{-2},\infty^{-1},\infty} = \{ X^2(X - \infty^{-2}) + \alpha X(X - \infty^{-1}) + \beta | (\alpha, \beta) \in \mathbb{F}_q^2 \}.$

where ∞^{-1} stands for $G^{-1}(\infty)$ and ∞^{-2} for $G^{-1}(G^{-1}(\infty))$. We remark that these three extra groups at infinity satisfy the same systematic divisibility properties as the usual groups. Moreover, we enlarge the number of available polynomials to $q^3 + q^2$, which is now enough to possibly cover all the monic degree 3 polynomials.

Covering every degree 3 *polynomials.* Let $P(X) = X^3 + a_2 X^2 + a_1 X + a_0$ be an arbitrary monic polynomial of degree 3. If P belongs to a valid group $\mathcal{P}_{g,G(g),G(G(g))}$, there exist α and β such that:

$$\alpha - g = a_2,$$
$$\beta - \alpha G(g) = a_1,$$
$$\text{and } -\beta G(G(g)) = a_0.$$

Substituting the equations into each other, we find that this implies:

$$a_0 = -(a_1 + (a_2 + g) G(g)) \cdot G(G(g)). \tag{8}$$

After simplification this becomes $H_{a_1,a_2}(g) = a_0$, where H_{a_1,a_2} is an homography whose coefficients depend on a_1 and a_2. If there is no degenerescence inside the coefficients of H_{a_1,a_2}, there is exactly one possible value for g. Let us write the homography $H_{a_1,a_2}(g) = \frac{\lambda + \mu g}{\lambda' + \mu' g}$ where $\lambda = -v(w + v)((1 + u)a_1 + va_2)$, $\mu = v((1+u)a_1 - v(v+w))$, $\lambda' = (1+u)(u(v+w)+w)$ and $\mu' = -(1+u)^2$. Thus, several cases appear:

- If $a_0 \neq \mu/\mu'$, then the homography is invertible.
  - As a consequence, as long as $g \neq \infty^{-1}$ and $g \neq \infty^{-2}$, the polynomial P belongs to the valid group generated by $g = H_{a_1,a_2}^{-1}(a_0), G(g)$ and $G(G(g))$, and only to this one. There are $q^3 - 3q^2$ such polynomials.
  - If $g = \infty^{-1}$ then $H_{a_1,a_2}(\infty^{-1}) = a_0$ becomes $a_0(\lambda' + \mu'(v+w)) = \lambda + \mu(v+w)$ and finally $a_0 = \infty^{-1}v(a_2 + \infty^{-1})/(1+u)$. Besides, P belongs to the group at infinity $\mathcal{P}_{\infty^{-1},\infty,0}$ if there exists α and β such that $\beta - \infty^{-1} = a_2$, $\alpha = a_1$, and $\beta v(v+w)/(1+u) = a_0$. Substituting the previous equations in β into each other, we find that this implies $a_0 = \infty^{-1}v(a_2 + \infty^{-1})/(1+u)$. Thus, the polynomial P belongs to $\mathcal{P}_{\infty^{-1},\infty,0}$. There are $q^2 - q$ such polynomials.
  - Similarly, if $g = \infty^{-2}$ then P belongs to the group at infinity $\mathcal{P}_{\infty^{-2},\infty^{-1},\infty}$ and, again, there are $q^2 - q$ such polynomials.
- If $a_0 = \mu/\mu'$ then Equation (8) is equivalent to $0 = g(a_0\mu' - \mu) = \lambda - \lambda'$. Moreover requiring $\lambda = \lambda'$ leads to $a_2 = \kappa(\kappa'a_1 + \kappa'')$ where $\kappa = (1+u)/(v^2(v+w))$, $\kappa' = v(v+w)$ and $\kappa'' = -u(v+w) - w$.
  - If $a_2 = \kappa(\kappa'a_1 + \kappa'')$ then P belongs to all the valid groups. There are q such polynomials.
  - If $a_2 \neq \kappa(\kappa'a_1 + \kappa'')$ the question is whether the $q^2 - q$ remaining polynomials belong to a group at infinity. Hopefully, if α denotes a_2 and β denotes $a_1 - v(v+w)/(1+u)$ then we have the following equality between polynomials: $X(X^2 + v(w+v)/(1+u)) + \alpha X^2 + \beta(X - G(0)) = X^3 + a_2 X^2 + a_1 X + v(v(v+w) - a_1(1+u))/(1+u)^2 = P(X)$. As a consequence, P belongs to the group at infinity $\mathcal{P}_{\infty,0,G(0)}$.

Remark 1. The previous proof does not interact with the restriction on a_2. Thus, the q polynomials satisfying $a_0 = \mu/\mu'$ and $a_2 = \kappa(\kappa'a_1 + \kappa'')$ belong also to the group at infinity $\mathcal{P}_{\infty,0,G(0)}$. Moreover, we notice that each intersection between two groups at infinity consists in q polynomials.

3.3 Discrete Logarithms of Degree 4 Polynomials

Previous Deadlocks. The natural approach for computing the logarithm of $I_4(\theta)$ where I_4 is an irreducible polynomial of degree 4 is to start from the two polynomials $A(X) = X^3 + a_1 X + a_0$ and $B(X) = X^2 + b_1 X + b_0$, construct a relation from Equation (5) and require that I_4 divides $[A, B]_3$. Rewriting this last condition as $[A, B]_3 = 0 \pmod{I_4}$, we obtain 4 bilinear equations in the 4 unknowns (a_0, a_1, b_0, b_1). Experimentally, as explained in [Jou14], this system is easy to solve using standard Gröbner basis algorithms. However, on average, the system has solutions only for half of the degree 4 polynomials. As a consequence, the other half polynomials are not accessible using this technique.

Another idea, already present in [AMORH14], is to use the additional relations from Section 3.1 to improve the probability of success. For an irreducible of degree 4 that failed to by expressed in terms of degree 3 polynomials, there is a 1/2 chance that its image by Frobenius, whose degree is 8, factors into 2 quartic polynomials. Each of them has a 1/2 chance to be expressed in terms of degree 3 polynomials. Thus, for a polynomial that failed, we have a 1/8 chance to compute its logarithms through this process. This increases the global probability of success for a degree 4 irreducible to 9/16. Repeating the process, we can further improve the success probability. Heuristically, we expect to have a probability of $p_0 = (4 - \sqrt{8})/2 \approx 0.586$. Unfortunately, this does not suffice to obtain all degree 4 polynomials. In order to bypass this problem, several techniques have been considered but none of them are sufficient in the general case. We propose here an approach that fits to the simple construction whereas the useful (but tricky) variant is detailed in the extended version of the article.

Improved Approach for Degree 4 Polynomials for the Simple Construction. The general approach we propose consists in dividing the degree 4 polynomials in groups of size q^3 and following an approach close to the case of the degree 3 polynomials presented in Section 3.2. We first compute all the discrete logarithms of a group $\mathcal{Q}_g$ of degree 4 polynomials of the form:

$$\mathcal{Q}_g = \{(X^4 + g) + \alpha X^3 + \beta X^2 + \gamma X | (\alpha, \beta, \gamma) \in \mathbb{F}_q^3\}. \tag{9}$$

To do so, we use a partition of this group $\mathcal{Q}_g = \cup_{g' \in \mathbb{F}_q} \mathcal{Q}_{g,g'}$ where:

$$\mathcal{Q}_{g,g'} = \{(X^4 + g) + \alpha X^3 + \beta X^2 + g'X | (\alpha, \beta) \in \mathbb{F}_q^2\}. \tag{10}$$

To build relations involving the polynomials from $\mathcal{Q}_{g,g'}$ we apply Equation (5) with polynomials of the form $A(X) = (X^4+g)+aX^2+g'X$ and $B(X) = X^3+bX^2$. With the simple construction, Lemma 5 shows that $[A, B]_4$ is of degree 11 and has a systematic factor of degree one. Together with the general degree 3 systematic factor coming from Lemma 1, we are left with a polynomial of degree 7. According to Appendix B the probability that it factors in terms of degree at most 3 is about 24%.

Besides, the number of irreducible polynomials in $\mathcal{Q}_{g,g'}$ is close to $q^2/4$. Combining with previous techniques, after removing the irreducibles whose logarithms can be obtained, we are left with approximately $(1-0.586) \cdot q^2/4 \approx 0.10\,q^2$

unknowns. Thus we obtain enough equations to solve the linear system. Finally, we recover the discrete logarithms of $\mathcal{Q}_g$ by computing the ones of its q subgroups.

Lemma 5 (Systematic factor of particular 4-brackets in the simple construction[5]). *Let h_0, h_1, A and B be four polynomials in $\mathbb{F}_q[X]$ such that h_0 is affine, $h_1(X) = X(X+t)$, $A(X) = (X^4+g) + \alpha X^2 + \alpha'X$ and $B(X) = X^3 + \beta X^2 + \beta'X$. Then $[A,B]_4$ is a polynomial of degree at most 11 divisible by X.*

Computing the remaining discrete logarithms. Let $I_4 \notin \mathcal{Q}_g$ be a degree 4 polynomial. We start again from $A(X) = (X^4+g) + aX^2 + a'X$ and $B(X) = X^3 + bX^2 + b'X$, and apply Equation (5) to construct a relation such that I_4 divides $[A,B]_4$. As in [Jou14], the heuristic probability to find a solution from the bilinear system is $1/2$. Extracting the degree one factor of Lemma 5 and the general degree 3 systematic factor of Lemma 1, and dividing then the degree 11 polynomial $[A,B]_4$ by our degree 4 polynomial I_4, we are left with a polynomial of degree 3, which logarithm is already known. Thus, with only one group of the form described in (9) we recover the discrete logarithms of approximately half[6] the irreducible missing polynomials of degree 4.

To obtain the remaining polynomials, we recursively apply this method to other groups of the form (9). We show in Section 4.3 that $O(\log(q))$ such groups suffice and that the cost of their computations is asymptotically dominated by the cost of the first one, which is $O(q^6)$, as announced.

4   Asymptotic Complexities

4.1   Recovering Discrete Logs of Degree 2 Irreducible Polynomials

We require to collect about q^2 equations in the Relation Collection phase. Since the probability to obtain a good relation is lower-bounded by $2/3$, this phase costs $O(q^2)$ operations. We perform then a sparse linear algebra phase on a matrix of size $O(q^2)$. We recall that due to the form of the relations that are created, the number of entries in each row is $O(q)$. The total cost to recover the discrete logarithms of degree 2 polynomials is so $O((q^2)^2 \cdot q) = O(q^5)$.

4.2   Recovering Discrete Logs of Degree 3 Irreducible Polynomials

With the really simple construction. Since each group $\mathcal{P}_g$ contains $O(q^2)$ unknowns and since the linear algebra is done with a matrix containing $O(q)$ entries per line, the cost of computing a single group is $O(q^5)$. There are q such groups and the global cost is, thus, $O(q^6)$.

---

[5] The proof of this lemma works as the one of Lemma 3.

[6] The probability to recover the logarithm of a missing polynomial is in fact higher than $1/2$, since we can use additional equations as presented in Section 3.1. Even there are very useful in practice, the $1/2$ probability already suffices for the analysis.

With the useful variant. We consider 3 groups at infinity and $q - 2$ valid groups with $O(q^2)$ unknowns each. Thus the global cost of this phase is $O(q^6)$.

4.3 Recovering Discrete Logs of Degree 4 Irreducible Polynomials

With the simple construction. We compute first the discrete logarithms of one group of the form (10). Since we have a system of dimension $O(q^2)$ with $O(q)$ entries per line, it can be solved for a cost of $O(q^5)$. To recover the logarithms of one group of the form (9), we need thus $O(q^6)$ operations.

Besides, the probability to recover the logarithm of an irreducible degree 4 polynomial from the first group of the form (9) is heuristically $1/2$. Considering that the probabilities are independent, with k such groups, the proportion of discrete logarithms that are left unknown is $1/2^k$. Clearly, as the number of available groups grows, this proportion quickly tends to 0. With $O(\log(q))$ such groups we expect to obtain all degree 4 polynomials. As a consequence, performing the computation of $O(\log(q))$ groups in this direct way, we would obtain a global complexity of $O(q^6 \log q)$. However, this overlooks the fact that for each new group that we wish to compute, the size of the corresponding linear system decreases and the rate of decrease follows a geometric progression[7]. As a consequence, the cost of computing the required $O(\log(q))$ groups is dominated by the computation of the first one.

Hence, the total complexity[8] of the precomputation phases becomes $O(q^6)$. This has to be compared with the previous $O(q^7)$ complexity for the same phases. However, we recall that the part of the algorithm that dominates the asymptotic complexity of each Frobenius Representation algorithm is the Descent phase, which is not under consideration in this article.

5 A Computational Example in Characteristic 3

To illustrate our algorithm, we have implemented our new ideas for a real-sized example in characteristic 3. Namely, we let $q = 3^5$ and define $\mathbb{F}_q = \mathbb{F}_3[\alpha]$, where α satisfies $\alpha^5 - \alpha + 1 = 0$. Choosing $h_0 = X^2 + \alpha^{111} X$ and $h_1 = \alpha X + 1$ we see that $X h_1(X^q) - h_0(X^q)$ has an irreducible factor of prime degree 479. We let U denote a root of this irreducible polynomial and construct $\mathbb{F}_{3^{5 \cdot 479}}$ as $\mathbb{F}_q[U]$.

The cardinality of the finite field we consider is a 3796-bit integer. A good point of comparison is the computation over $\mathbb{F}_{2^{12 \cdot 367}}$ performed in [GKZ14a].

---

[7] Another option is to continue the computation for all groups. Due to the geometric progression, the complexity of this part is the same. Yet, it yields a total runtime lower than the option of recomputing on the fly the missing degree 4 polynomials logarithms when required but as a side effect it raises the required amount of storage.

[8] We consider here algorithms of Wiedmann or Lanczos families, that has a complexity of $O(n^2)$ for a square matrix with n columns. Yet, using dense linear algebra with fast matrix multiplication instead of sparse linear algebra would lower the asymptotic complexity from $O(q^6)$ to $O(q^{5.746})$. We do not choose to consider these algorithms here since there are not at all competitive in practice.

Indeed, even if the bitsize of this computation was slightly larger than ours, being on 4404 bits, this total size included a factor of two in the exponent which comes for free when using the older Frobenius Representation algorithms. More precisely, the main drawback of our approach is that instead of computing logarithms in the field $\mathbb{F}_{q^{dk}}$ it only computes in $\mathbb{F}_{q^k}$. Many cryptographers have commented on this free factor, claiming that it is not really relevant in practice and that one should rather consider extension field of prime degree that can be embedded in the target field. For us, this is $\mathbb{F}_{3^{479}}$ a 760-bit field. This can also be compared to the largest computation of this form currently performed in the finite field $\mathbb{F}_{2^{809}}$ (see [BBD+13]).

With this example, computing all the discrete logarithms of the factor base with $D = 2$, containing 29 646 irreducible polynomials, required 16 sequential hours on a single core of an Intel Core i7 at 2.7 GHz. The equations themselves took 35 seconds to produce, the 16 hours being the cost of the linear algebra modulo:

$$M = \frac{3^{5 \cdot 479} - 1}{488246858}.$$

Enlarging the factor base to degree 3 polynomials was performed with 244 independent computations, each involving 19 602 unknowns in the corresponding linear system. On the same machine, the sequential cost of one such computation is 6.5 hours. Since these computations are independent, they are straightforward to parallelize.

For degree 4 polynomials, the first subset of 243 independent computations we considered contained on average 7 385 unknowns in each linear system. The largest system contained 7 571 unknowns and the smallest 7 212. Note that this used a suboptimal variation of the technique obtained in Section 3.3 and induced slightly larger system. Using the correct variation, we would expect a smaller number of unknowns per linear system (around 6100).

The second subset has on average 3 674 unknowns, the third 1 829, the fourth 909, the fifth 452. We see that as predicted, the rate of decrease is very steep, essentially a geometric series of ratio 1/2. As a consequence, the runtimes for these subsets rapidly becomes negligible compared to the main part of the computation consisting in tackling the degree 3 polynomials. Here again, our implementation is suboptimal, but this was not a critical part of the computation. In fact, for all subsets beyond the fifth, we only tried to the logarithms of the elements in terms of the first four subsets. Indeed, the resulting systems were so small (around 450 unknowns) and sparse that they could be solve with a straightforward Gaussian elimination. Thus for these subsets, the running time was dominated by the generation of the equations (around 2h for each subset) and it did not make sense to insist on reducing the size of the linear systems. In total, we computed 30 subsets and they were enough to express the logarithms of all the degree 4 elements encountered further during the computation.

For the descent phase, we followed the state of the art and were able to express the seeked discrete logarithm using a total of under 41 millions polynomials of degree 4 (and of course also polynomials of lower degree). For lack of space, we leave out the details, they will be reported in the extended version of this article. The total running time of the computation was under 8600 CPU-hours.

6   Conclusion

In this paper, we proposed an improved Frobenius Representation algorithm for the computation of discrete logarithms in small characteristic. Together with the aim of simplifying the description of previous algorithms, we reduce the complexity of the precomputation phase to $O(q^6)$ for general extension degree. Computations with such a cost were previously available only for special degrees such as Kummer extension.

References

AMORH14.   Adj, G., Menezes, A., Oliveira, T., Rodríguez-Henríquez, F.: Computing discrete logarithms in $\mathbb{F}_{3^{6 \cdot 137}}$ and $\mathbb{F}_{3^{6 \cdot 163}}$ using Magma. Cryptology ePrint Archive, Report 2014/057 (2014)

BBD$^+$13.   Barbulescu, R., Bouvier, C., Detrey, J., Gaudry, P., Jeljeli, H., Thomé, E., Videau, M., Zimmermann, P.: Discrete logarithm in $\mathbb{F}_{2^{809}}$ with ffs. Cryptology ePrint Archive, Report 2013/197 (2013)

BGJT14.   Barbulescu, R., Gaudry, P., Joux, A., Thomé, E.: A heuristic quasi-polynomial algorithm for discrete logarithm in finite fields of small characteristic. In: Nguyen, P.Q., Oswald, E. (eds.) EUROCRYPT 2014. LNCS, vol. 8441, pp. 1–16. Springer, Heidelberg (2014)

BMV85.   Blake, I.F., Mullin, R.C., Vanstone, S.A.: Computing logarithms in $\mathbb{F}_{2^n}$. In: Blakely, G.R., Chaum, D. (eds.) CRYPTO 1984. LNCS, vol. 196, pp. 73–82. Springer, Heidelberg (1985)

GGMZ13a.   Göloğlu, F., Granger, R., McGuire, G., Zumbrägel, J.: On the function field sieve and the impact of higher splitting probabilities - application to discrete logarithms in $\mathbb{F}_{2^{1971}}$ and $\mathbb{F}_{2^{3164}}$. In: Canetti, R., Garay, J.A. (eds.) CRYPTO 2013, Part II. LNCS, vol. 8043, pp. 109–128. Springer, Heidelberg (2013)

GGMZ13b.   Göloglu, F., Granger, R., McGuire, G., Zumbrägel, J.: On the function field sieve and the impact of higher splitting probabilities: Application to discrete logarithms in $\mathbb{F}_{2^{1971}}$. Cryptology ePrint Archive, Report 2013/074 (2013)

GKZ14a.   Granger, R., Kleinjung, T., Zumbrägel, J.: Breaking '128-bit secure' supersingular binary curves (or how to solve discrete logarithms in $\mathbb{F}_{2^{4 \cdot 1223}}$ and $\mathbb{F}_{2^{12 \cdot 367}}$). Cryptology ePrint Archive, Report 2014/119 (2014)

GKZ14b.   Granger, R., Kleinjung, T., Zumbrägel, J.: On the powers of 2. Cryptology ePrint Archive, Report 2014/300 (2014)

Jou13a.   Joux, A.: Faster index calculus for the medium prime case application to 1175-bit and 1425-bit finite fields. In: Johansson, T., Nguyen, P.Q. (eds.) EUROCRYPT 2013. LNCS, vol. 7881, pp. 177–193. Springer, Heidelberg (2013)

Jou13b.   Joux, A.: A new index calculus algorithm with complexity L(1/4 + o(1)) in very small characteristic. Cryptology ePrint Archive, Report 2013/095 (2013)

Jou14.   Joux, A.: A new index calculus algorithm with complexity L(1/4 + o(1)) in small characteristic. In: Lange, T., Lauter, K., Lisoněk, P. (eds.) SAC 2013. LNCS, vol. 8282, pp. 355–380. Springer, Heidelberg (2014)

A   Action of $\mathrm{PGL}_2(\mathbb{F}_q)$ on Polynomials

We detail here the reason why we can restrict ourselves to the case $A(X) = X^D + a(X)$ and $B(X) = X^{D-1} + b(X)$, with a and b polynomials of degree $D - 2$.

Assume that we are initially given an equation for two degree D polynomials A_0 and B_0. We may assume that these two polynomials are monic by multiplying Equation (5) by the inverse of the product of their leading coefficients. Moreover, thanks to Proposition 1 we have $[A_0, B_0]_D = [A_0, B_0 - A_0]_D$. Thus, we can replace B_0 by $B_1 = B_0 - A_0$. If there is no unexpected fall of degree (i.e. in the general case), B_1 has degree $D - 1$. We can again assume that it is monic. If the coefficient of X^{D-1} in A_0 is a_{D-1}, remarking that:

$$[A_0, B_1]_D = [A_0 - a_{D-1}B_1, B_1]_D,$$

we can replace A_0 by a polynomial A_1 whose coefficient of X^{D-1} is 0. Thus, the pair (A_1, B_1) generates the same equation as (A_0, B_0) and has the announced restricted form.

B   Estimating Probabilities of Factoring Polynomials

Throughout the paper, we need to estimate the probabilities that a polynomial of degree D factors into terms of degree at most d. This is often done by using the heuristic rule that the polynomial behaves in this respect like a random polynomial.

In this appendix, we analyze these probabilities for random polynomials. Let us start we a simple example and consider the probability that a random monic polynomial of degree D splits into linear factors. Over the finite field $\mathbb{F}_q$ there are q^D distinct monic polynomials of degree D. Among those it is easy to count the number of squarefree polynomials that split into linear terms, there are in correspondance with their D distinct roots in $\mathbb{F}_q$, thus there are precisely $\binom{q}{D} = \frac{q \cdot (q-1) \cdots (q-(D-1))}{D!}$ such polynomials. Hence, the fraction of polynomials that split is lower bounded by $\binom{q}{D} \cdot q^{-D}$, which tends to $1/D!$ as q tends to infinity.

To obtain an upper bound, we also need to count the polynomials that split and have multiple roots. The formula is more complex since we need to compute a sum over partitions of D into multiplicities. However, the number of terms in this sum is independent of q and each term is a multinomial that chooses the correct number of roots with each multiplicity. Since each term contains at most $D - 1$ roots, we can upper bound the contribution by $C(D) q^{D-1}$ where $C(D)$ does not depend on q. Thus, as q tends to infinity, the upper bound on the total fraction of polynomials that split tends to $1/D!$ too.

For more complex decomposition, this kind of analysis remains doable but messy for arbitrary fixed values of D and d. Thankfully, in the present paper, we are only considering values such that:

$$d + 1 > D/2.$$

Under this constraint the analysis becomes quite easy. Indeed, if a polynomial P of degree D does not factor into terms of degree at most d, it must have at least one factor F_k of large degree $k \geq d+1$. Since $k > D/2$, this factor is unique. Now, the probability that P can be written as $F_k \cdot Q$, with F_k an irreducible of degree k and Q an arbitrary polynomial of degree $D-k$ is precisely $(N_k \cdot q^{D-k})/q^D = N_k/q^k$, where N_k denotes the number of irreducible polynomials of degree k over $\mathbb{F}_q$. Thus, the probability is precisely the fraction of irreducibles among degree k polynomials and it is well-known that this tends to $1/k$ as q tends to infinity. As a consequence, as q tends to infinity the probability that a degree D polynomial factors into terms of degree at most d, when $d + 1 > D/2$ tends to:

$$1 - \sum_{k=d+1}^{D} \frac{1}{k}.$$

Using this we can easily estimate the probabilities required in the paper:

➤ For $D = 3$ and $d = 2$ the probability is $1 - \frac{1}{3} = \frac{2}{3}$.
➤ For $D = 4$ and $d = 2$ the probability is $1 - \frac{1}{3} - \frac{1}{4} = \frac{5}{12} \approx 0.4167$.
➤ For $D = 7$ and $d = 3$ the probability is $1 - \frac{1}{4} - \frac{1}{5} - \frac{1}{6} - \frac{1}{7} = \frac{101}{420} \approx 0.2405$.

Big Bias Hunting in Amazonia: Large-Scale Computation and Exploitation of RC4 Biases (Invited Paper)

Kenneth G. Paterson[1], Bertram Poettering[1], and Jacob C. N. Schuldt[2]

[1] Information Security Group, Royal Holloway, University of London, U.K.
[2] Research Institute for Secure Systems, AIST, Japan

Abstract. RC4 is (still) a very widely-used stream cipher. Previous work by AlFardan *et al.* (USENIX Security 2013) and Paterson *et al.* (FSE 2014) exploited the presence of biases in the RC4 keystreams to mount plaintext recovery attacks against TLS-RC4 and WPA/TKIP. We improve on the latter work by performing large-scale computations to obtain accurate estimates of the single-byte and double-byte distributions in the early portions of RC4 keystreams for the WPA/TKIP context and by then using these distributions in a novel variant of the previous plaintext recovery attacks. The distribution computations were conducted using the Amazon EC2 cloud computing infrastructure and involved the coordination of 2^{13} hyper-threaded cores running in parallel over a period of several days. We report on our experiences of computing at this scale using commercial cloud services. We also study Microsoft's Point-to-Point Encryption protocol and its use of RC4, showing that it is also vulnerable to our attack techniques.

Keywords: RC4, plaintext recovery attack, WPA, TKIP, MPPE.

1 Introduction

1.1 RC4 and Its Applications

The stream cipher RC4, originally designed by Ron Rivest, is a beautifully compact and fast algorithm. It became public in 1994 and has since been applied in a very wide variety of secure communications protocols, including SSL/TLS (as analysed in [1,7,13,16]); WEP [5] (where its particular usage led to devastating attacks including complete, efficient key recovery, see [20] for a summary and references); WPA [6] (as analysed in [21,20,22,15,18]); Microsoft's Point-to-Point Encryption protocol [14] (MPPE, as analysed here); and some Kerberos-related encryption modes [8]. A selection of additional, non-protocol specific analyses of RC4 can be found in [3,2,12,11,10,19].

Of particular relevance for this work are the results of AlFardan *et al.* [1]. They introduced a simple, Bayesian statistical method that recovers plaintexts that are repeatedly encrypted under RC4 by exploiting biases in RC4 keystreams. Their approach was successfully applied to RC4 in HTTPS (i.e., HTTP over

P. Sarkar and T. Iwata (Eds.): ASIACRYPT 2014, PART I, LNCS 8873, pp. 398–419, 2014.

SSL/TLS), where a fresh pseudorandom 128-bit key is used for each SSL/TLS connection, and where the repeated encryption of HTTP cookies can be arranged by having malicious JavaScript running in the target user's browser.

1.2   RC4 in WPA/TKIP

The work of AlFardan *et al.* motivated us to explore RC4's usage in other deployed protocols, in an attempt to determine whether similar weaknesses exist and are exploitable. Our first focus was the wireless network encryption protocol WPA/TKIP [6], with results presented in [15]. While WPA/TKIP was only ever intended as a stop-gap to replace WEP until stronger cryptography could be deployed, a recent survey [22] showed that it is still in widespread use.

In WPA/TKIP, fresh 16-byte (128-bit) RC4 keys are used for every frame transmitted on the wireless network, but the first three bytes of the key are determined by two bytes $\overline{TSC} = (TSC_0, TSC_1)$ of a public value, TSC, which increments on a frame-by-frame basis; the remaining 13 bytes of the per-frame key are generated pseudorandomly. As observed in [15] and independently in [18], the dependence of the RC4 key on $\overline{TSC}$ in turn induces large, $\overline{TSC}$-dependent, single-byte biases in the initial positions of RC4 keystreams. This suggests the attack proposed in [15]: bin the available ciphertexts into 2^{16} bins, one bin for each possible value of $\overline{TSC}$; perform a Bayesian analysis as per [1] for each bin; and then combine the results across all the bins to estimate the likelihood for each plaintext byte candidate. But this attack requires the computation of accurate single-byte distributions for RC4 keystreams for each of the 2^{16} values of $\overline{TSC}$. We estimated in [15] that the analysis of $2^{32} - 2^{40}$ RC4 keystreams per $\overline{TSC}$ would be needed to achieve sufficient accuracy, for a total of $2^{48} - 2^{56}$ RC4 keystreams. At that time, this was well beyond our computational capabilities. We resorted to working with 2^{24} keystream per $\overline{TSC}$ and using the sub-optimal procedure of examining the dependence of the RC4 keystream only on TSC_1, in effect aggregating over TSC_0 (since our intuition was that this byte would have a greater influence in determining the distribution than TSC_0).

Another avenue left unexplored in [15] for WPA/TKIP was the use of double-byte biases in plaintext recovery attacks. Such biases concern the distribution of adjacent pairs of keystream bytes. They were used in [1] in the preferred attack against SSL/TLS, because these biases are persistent throughout the RC4 keystream (whereas the single-byte biases disappear shortly after position 256) and, in the considered attack scenario, it was not possible to arrange for the target plaintext bytes (an HTTP cookie) to appear sufficiently early in the sequence of plaintext bytes. It's also possible that using a double-byte bias attack would improve plaintext recovery rates in the early positions. To extend the double-byte bias attack of [15] to the WPA/TKIP setting would then require the computation of the double-byte keystream distributions, ideally on a per-$\overline{TSC}$ basis. This would not only require enormous numbers of keystreams to obtain sufficient accuracy, but also significant storage: just to describe the double-byte distribution per position and $\overline{TSC}$ requires 2^{16} numbers, each typically 32 bits in size, leading to a total storage requirement of 8 Terabytes just to record the double-byte distributions for the first 512 keystream positions.

1.3   RC4 in MPPE

Microsoft's Point-to-Point Encryption (MPPE), as specified in [14,23], is a venerable security protocol that can be used on top of the Point-to-Point Tunnelling Protocol (PPTP). The latter is itself a general-purpose protocol encapsulation method that is commonly used for providing Virtual Private Networking services to devices running Microsoft operating systems, including Windows 8 and the Windows Server family of products.

MPPE uses RC4 with a non-standard method for selecting keys. For example, when a 40-bit key is used, MPPE starts with an 8-byte key $K = (K_0, \ldots, K_7)$ that is itself derived by hashing a user password, an authentication protocol challenge, and other public information. MPPE then sets $K_0 = \texttt{0xD1}, K_1 = \texttt{0x26}, K_2 = \texttt{0x9E}$. It is then natural to ask: does this method for selecting keys in MPPE lead to a different bias structure in its RC4 keystreams, and does this help or hinder plaintext recovery attacks akin to those of [1]?

1.4   Our Contributions and Paper Organisation

Section 2 provides further background on the RC4 stream cipher and its use in WPA and MPPE.

In Section 3, we report on our computations of more refined, per-$\overline{\text{TSC}}$, single-byte and double-byte RC4 keystream distributions for WPA/TKIP. In slightly more detail, we computed these distributions for the first 512 keystream bytes, based on 2^{48} keys for the single-byte case and 2^{46} keys for the double-byte case. We made use of the Amazon Elastic Compute Cloud (Amazon EC2)[1], which is part of Amazon Web Services, to perform the computations. We used approximately 30 virtual-core-years for the single-byte computation and 33 virtual core-years for the double-byte computation. Since, to us, a total of 63 virtual-core-years was quite a significant amount of computation (costing roughly US$41k[2]) and because we faced a number of obstacles in working at this scale, we report in some detail on our experiences of working with Amazon EC2. One notable feature revealed by our large-scale computations is the presence of $\overline{\text{TSC}}$-dependent, single-byte biases well beyond position 256 in the RC4 keystream.

Section 4 describes a plaintext recovery attack on WPA/TKIP that exploits our newly-computed and more accurate single-byte distributions for RC4 keystreams, comparing it to our previous results from [15].

Section 5 describes a novel plaintext recovery attack on WPA/TKIP that exploits per-$\overline{\text{TSC}}$, double-byte biases in RC4 keystreams. This attack combines the double-byte bias attack from [1] with the idea of binning that was developed for the case of single-byte biases in [15].

---

[1] http://aws.amazon.com/ec2/

[2] Here, and throughout, we quote prices exclusive of sales taxes at 20%.

Algorithm 1. RC4 KSA	Algorithm 2. RC4 PRGA

input : key K of l bytes
output: initial internal state st_0
begin
　for $i = 0$ to 255 **do**
　　$\lfloor$ $\mathcal{S}[i] \leftarrow i$
　$j \leftarrow 0$
　for $i = 0$ to 255 **do**
　　$j \leftarrow j + \mathcal{S}[i] + \mathrm{K}_{i \bmod l}$
　　$\mathrm{swap}(\mathcal{S}[i], \mathcal{S}[j])$
　$i, j \leftarrow 0$
　$st_0 \leftarrow (i, j, \mathcal{S})$
　return st_0
	input : internal state st_r
	output: keystream byte Z_{r+1}
	　　　　　updated internal state st_{r+1}
	begin
	　parse $(i, j, \mathcal{S}) \leftarrow st_r$
	　$i \leftarrow i + 1$
	　$j \leftarrow j + \mathcal{S}[i]$
	　$\mathrm{swap}(\mathcal{S}[i], \mathcal{S}[j])$
	　$Z_{r+1} \leftarrow \mathcal{S}[\mathcal{S}[i] + \mathcal{S}[j]]$
	　$st_{r+1} \leftarrow (i, j, \mathcal{S})$
	　return (Z_{r+1}, st_{r+1})

Fig. 1. Algorithms implementing the RC4 stream cipher. All additions are performed modulo 256.

In Section 6, we report on the single-byte keystream distributions for RC4 when it is keyed according to the MPPE specification. In short, we found the distributions to be highly skewed and amenable to exploitation using our attack techniques.

Finally, Section 7 presents our conclusions and remarks on open problems.

2 Further Background

2.1 The RC4 Stream Cipher

Technically, RC4 consists of two algorithms: a *key scheduling algorithm* (KSA) and a *pseudo-random generation algorithm* (PRGA), which are specified in Algorithms 1 and 2. The KSA takes as input a key K, typically a byte-array of length between 5 and 32 (i.e., 40 to 256 bits), and produces the initial internal state $st_0 = (i, j, \mathcal{S})$, where $\mathcal{S}$ is the canonical representation of a permutation on the set $[0, 255]$ as an array of bytes, and i, j are indices into this array. The PRGA will, given an internal state st_r, output 'the next' keystream byte Z_{r+1}, together with the updated internal state st_{r+1}.

2.2 WPA/TKIP

A detailed description of how RC4 is used in the WPA/TKIP context is given in [15]. In short, WPA/TKIP generates a fresh 128-bit key $K = (K_0, \ldots, K_{15})$ for RC4 for each frame that is transmitted; the key is a function of the temporal encryption key TK (128 bits), the TKIP sequence counter TSC (48 bits), and the transmitter address TA (48 bits). A single value of TK is used over many frames, while TSC increments from frame to frame; meanwhile TA is fixed. Very importantly, the function used to compute K adds a specific structure added to "preclude the use of known RC4 weak keys" [6]. More precisely, writing $\mathtt{TSC} = (\mathtt{TSC}_0, \mathtt{TSC}_1, \ldots, \mathtt{TSC}_5)$, we have

$$\mathtt{K}_0 = \mathtt{TSC}_1 \qquad \mathtt{K}_1 = (\mathtt{TSC}_1 \mid \mathtt{0x20}) \mathbin{\&} \mathtt{0x7f} \qquad \mathtt{K}_2 = \mathtt{TSC}_0 \qquad (1)$$

while $\mathtt{K}_3, \ldots, \mathtt{K}_{15}$ can be considered to be pseudorandom functions of $\mathtt{TK}$, $\mathtt{TSC}$ and $\mathtt{TA}$. Notably here, bytes $\mathtt{K}_0, \mathtt{K}_1, \mathtt{K}_2$ depend only on bytes $\mathtt{TSC}_0$ and $\mathtt{TSC}_1$ of $\mathtt{TSC}$. Moreover, the bits of $\mathtt{TSC}_1$ are used twice. So the bytes of $\mathtt{K}$ have more structure than they would if they were chosen with uniform distribution. The per-frame key $\mathtt{K}$ is then used to produce an RC4 keystream, following our description of RC4 above. The TKIP plaintext (consisting of the frame payload, a 64-bit MAC value $\mathtt{MIC}$, and a 32-bit Integrity Check Vector $\mathtt{ICV}$) is then XORed in a byte-by-byte fashion with the RC4 keystream.

2.3   MPPE

MPPE provides a confidentiality service over PPTP using the RC4 algorithm. Keys for the RC4 algorithm come from a separate authentication and key establishment protocol, such as MS-CHAPv1, MS-CHAPv2 or EAP-TLS; the first two of these were broken in [17] and [4], respectively, leading to the depre-cation of the first and the recommendation only to use the second with additional protection from PEAP[3].

RFC 3079 [23] describes in detail how the keys used in MPPE's instantiation of RC4 are derived from preceding authentication and key establishment protocols. Three different RC4 key lengths are supported, according to [23]: 40-bit, 56-bit and 128-bit. When a 40-bit key is used, MPPE starts with an 8-byte key $\mathtt{K} = (\mathtt{K}_0, \ldots, \mathtt{K}_7)$ that is itself derived by hashing the password, the authentication protocol challenge, and other public information. MPPE then overwrites $\mathtt{K}_0 = \mathtt{0xD1}, \mathtt{K}_1 = \mathtt{0x26}, \mathtt{K}_2 = \mathtt{0x9E}$. When a 56-bit key is used, the protocol starts with the same 8-byte key and then sets $\mathtt{K}_0 = \mathtt{0xD1}$; when a 128-bit key is used, a similar procedure involving password and challenge hashing is used to generate a 16-byte key $\mathtt{K}$, and no bytes of $\mathtt{K}$ are overwritten.

Furthermore, MPPE operates in two modes, with the mode in use being de-termined by a PPTP header field. In stateless mode, the RC4 key is refreshed and the cipher restarted for each PPTP packet sent. By contrast, in stateful mode, the RC4 key is refreshed only every 256 packets. In both cases, refreshing the key involves hashing the old key with the first key for the session (called $\mathtt{StartKey}$ in [14]) to generate a value $\mathtt{InterimKey}$, then an RC4 encryption step in which $\mathtt{InterimKey}$ is used to encrypt itself to generate a key $\mathtt{K}$ of either 8 or 16 bytes, and finally setting bytes as described above. See [14, Section 7] for details.

From the above description it may be remarked that, while the hashing and encryption steps used in deriving the RC4 keys may be intended to render them pseudorandom, in the 40-bit and 56-bit cases, they have additional structure

---

[3] https://technet.microsoft.com/library/security/2743314

that may be expected to lead to additional and/or different biases in the RC4 keystream as compared to the 128-bit case.Further, the use of stateless mode would mean a fresh RC4 key (with additional structure in the 40-bit and 56-bit cases) for every packet sent. These observations mean that MPPE in stateless mode can be expected to be vulnerable to plaintext recovery attacks similar to those developed in [1,15]. Since the protocol encapsulated by MPPE is likely to be IP, similar fields as those identified in [15] could be targeted. We note that while key lengths of 40 and 56 bits are small enough that a simple brute-force search might initially seem to be more efficient than mounting a bias analysis using our techniques, in the stateless case, such a brute-force search would only recover the key used for a single packet. Moreover, a basic analysis suggests that a 2^{64} attack would be needed to recover `StartKey` from which all keys in the session are derived. So our approach may be an attractive alternative if specific plaintext bytes are targeted for recovery.

3   Large-Scale Computation of RC4 Keystream Distributions for WPA/TKIP Keys

3.1   Computing Keystream Distributions and Finding New Biases

As noted in the introduction, in our previous work on WPA/TKIP in [15], we worked with a total of only 2^{40} keystreams and only with single-byte distributions for the first 256 positions. In an effort to further improve our attacks, we decided to perform larger-scale computations using, in addition to our own local resources, the Amazon EC2 cloud computing infrastructure to estimate both the single-byte and double-byte keystream distributions for the first 512 positions, on a per-$\overline{TSC}$ basis.

Because the double-byte biases are smaller than the single-byte ones (typically by a factor of roughly 2^8), many more keystreams would be needed to accurately estimate double-byte distributions than for single-byte ones. However, we chose to focus our effort on the single-byte case here, computing distributions based on 2^{32} keystreams per $\overline{TSC}$ in the single-byte case and based on 2^{30} keystreams per $\overline{TSC}$ in the double-byte case. The reasons for this focus are as follows. Using our local computational resources, we determined that it would be difficult to use the full per-$\overline{TSC}$ distributions in a double-byte attack akin to that of [1] because of the complexity of handling so much data when running attacks (for example, we would need to deal with 16GB of distribution data and perform 2^{48} multiplications of real numbers to analyse a single byte position). Rather, an aggregated approach seemed more likely to be feasible for the double-byte setting. Here our idea was to start with 2^{30} keystreams per $\overline{TSC}$ and combine the 2^8 distributions for each TSC_1 value (called TSC_0-aggregation in [15]) to obtain 2^8 different double-byte distributions, one per TSC_1-value, each distribution now based on 2^{38} keystreams. This not only boosts the number of keystreams per distribution estimate (good for accurately estimating biases), but also reduces the size of the distribution data and computation both by a factor of 256 (making simulation of attacks much more feasible).

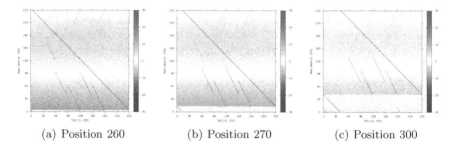

| (a) Position 260 | (b) Position 270 | (c) Position 300 |

Fig. 2. Pictorial representation of biases in RC4 keystreams for random TSC_0-aggregated WPA/TKIP keys at keystream positions 260, 270, and 300, for different TSC_1 values (x-axis) and byte values (y-axis). At each point we encode the bias in the keystream for the (TSC_1,value) combination as a colour; precisely, we encode the difference between the occurring probability and the (expected) probability $1/256$, scaled up by a factor of 2^{24}, capped to values in $[-30, +30]$.

Our computations went well beyond those of prior RC4 cryptanalyses in scale (e.g., [1,15]), and indeed we were rewarded by discovering new TSC_1-dependent single-byte biases in positions all the way up to 512 (see Figure 2 for examples at specific positions). The existence of these biases is surprising in view of the behaviour of single-byte biases observed in previous works and, in principle, would allow the recovery of plaintext using a single-byte attack like that presented in [15] and Section 4 below. It is an open problem to determine how far into the RC4 keystream these biases persist.

3.2   Reflections on Using Amazon EC2

The task of computing accurate estimates of RC4 keystream distributions is well-suited to distributed computation. In particular, in the case of WPA, the probability distribution for each $\overline{TSC}$ value can be estimated independently by generating keystreams using randomly chosen WPA keys for that $\overline{TSC}$ (having the structure described in Section 2.2). This makes performing the computation using cloud services such as Amazon EC2 seem appealing, on account of its virtually unlimited computing capacity being able to provide the computational resources required to complete the computation within an acceptable period of time.

For our computation of the per-$\overline{TSC}$ bias estimates, we used Amazon EC2 to create 256 virtual machines of the type 'c3.x8large', each providing 32 'virtual' cores. The underlying hardware of the virtual machines were servers equipped with Intel Xeon 2.8GHz processors. Note, however, that each of the cores of a virtual machine corresponds to a hyper-threaded core of the underlying CPU i.e., one 'c3.x8large' instance effectively corresponds to a machine with 16 physical cores. To manage the virtual machines, we utilized boto[4] which implements a Python interface to Amazon EC2. This provided a simple and straightforward

---

[4] http://github.com/boto/boto

way to automate management and access to the virtual machines, and made it relatively easy to set up the execution of the computation using a combination of Python and shell scripts. The virtual machines were all initialized with an Ubuntu 13.10 image obtained through the AWS Marketplace[5].

Each virtual machine was set up to compute the keystream distributions for all TSC_0 values given a fixed TSC_1 value, and to split this computation equally among the 32 available virtual cores. To make the WPA keystream generation efficient, we used the RC4 implementation in OpenSSL[6]. However, experiments showed that to reach the desired number of keystreams with our available budget, further optimizations were required. Additional experiments revealed that the amount of available cache in the underlying CPUs on which the virtual machines were running, and how this cache was utilized, played an important role in the performance of the keystream distribution generation. Specifically, the chance of cache misses occurring when updating the keystream distribution statistics was found to have a large influence on performance.

To address this, we used a combination of two different approaches to reduce the chance of cache misses occurring. Firstly, to fit the array storing the counters used to collect the statistics of the keystream distribution into the cache memory, we "packed" multiple small-width counters into single 64-bit integers and implemented logic for handling counter overflows. Secondly, instead of updating the keystream distribution statistics after each keystream has been generated, we stored multiple keystreams in memory before updating the statistics. This implies that multiple updates of the statistics for a single position can be done sequentially, which, assuming the appropriate memory layout, increases the chance of a cache hit. While these optimizations only provided small gains for the computation of single-byte biases, significant gains were achieved for the computation of double-byte biases.

Single-Byte Computations. Using the above setup, each virtual core was capable of generating and processing on average 294k length 512 WPA keystreams per second for single-byte distributions. Hence, computing the per-$\overline{TSC}$ single-byte distributions based on 2^{32} keystreams for each $\overline{TSC}$ value (i.e., 2^{48} keystreams in total), took 9.56×10^8 virtual core seconds in total, or approximately 30 virtual core years. Due to the large degree of parallelism in our setup, this corresponds to an actual running time of slightly more than 32 hours.

While each of the 256 virtual machine was set up identically, a single virtual machine ran significantly slower than the others, and was only capable of processing approximately 180k keystreams per second. We suspect that other virtual machines running on the same underlying hardware might have affected the performance of this virtual machine. Due to this issue, it took approximately 52 hours to complete the computation of the single-byte distributions.

At the time we did the experiments, the cost of running a single "c3.x8large" virtual machine instance was US$2.40 per hour, leading to a cost of US$614 per hour when running all 256 instances simultaneously.

---

[5] http://aws.amazon.com/marketplace/
[6] https://www.openssl.org/

To store the generated keystream distributions, we attached a separate Amazon Elastic Block Storage (EBS) volume to each virtual machine. This gave us the option of terminating a virtual machine without erasing the generated data, and furthermore allowed us to use a single virtual machine to inspect and process all generated data, by sequentially attaching the EBS volumes to this machine. The latter provided a more cost effective solution than running the virtual machines in parallel, and a faster solution than resuming each virtual machine sequentially. We stored the distribution for each $\overline{TSC}$ value as a sequence of binary encoded 32-bit integers, leading to a storage requirement of 512KB per distribution (128MB per virtual machine), or 32GB in total. However, since the minimum size of an EBS volume is 1GB, we allocated a total of 256GB of EBS storage (note that a single EBS volume cannot be mounted by multiple virtual machines simultaneously). The cost of EBS storage was US$0.05 per GB per month, leading to a cost of just US$12.60 a month to maintain the EBS volumes.

Double-Byte Computations. Working with double-byte keystream distributions introduced significant overheads compared to the single-byte case, both in terms of computation and storage. With the previously mentioned optimizations, each virtual core was capable of processing on average 67k WPA keystreams per second. Hence, computing the per-$\overline{TSC}$ double-byte keystream distributions based on 2^{30} keystreams for each $\overline{TSC}$ value (i.e., 2^{46} keystreams in total), took 1.05×10^9 virtual core seconds in total, or approximately 33 virtual core years. In our setup, this corresponds to an actual running time of slightly more than 34 hours, but due to the virtual machines being sequentially initialized and an issue with a single virtual machine, the time it took to complete the computation was approximately 48 hours. More specifically, the issue that arose was that the virtual machine in question was reset and rebooted during the computations, and hence did not complete its assigned task. We were unable to identify the cause of this event, and simply restarted the relevant computations manually.

As for the single-byte distributions, we created separate EBS volumes to store the double-byte distributions. However, each $\overline{TSC}$-specific double-byte distribution requires 128MB of storage when stored as a sequence of 32-bit integers, leading to a storage requirement of 32GB per virtual machine, or 8TB in total. This increased storage overhead not only led to an increased cost (US$410 per month), but also created additional practical issues which we had to handle. For example, since the EBS volumes are implemented via network attached storage (NAS), writing the distribution data to an EBS volume caused a significant delay in some instances. In particular, we observed that immediately after completion of the keystream distribution generation, detaching an EBS volume might not succeed, which in turn could interrupt the shutdown of a virtual machine. Furthermore, making all data available to a single machine at the same time, which is required to efficiently run attack simulations, was made more difficult by the 8TB size of the dataset. We decided to transfer the complete dataset to our local storage array both to run the attack simulations and to permanently store the data. For this purpose, we used bbcp[7], which is capable of transferring large amounts of data between network computers using multiple TCP streams and large transfer windows, and allowed us to obtain a transfer speed of approximately 50MB per second, leading to a total transfer time of slightly more than 48 hours. Note that data transfers out of Amazon EC2 were charged at US$0.12 per GB, resulting in a US$983 cost to move the complete 8TB dataset to our local storage.

Our experience of using Amazon EC2 to compute estimates of the per-$\overline{\mathsf{TSC}}$ biases suggests that Amazon provides a flexible platform which is well suited to perform this type of computation, and that the practical difficulties arising in the distribution of the computation can be overcome with moderate effort.

4   Plaintext Recovery Attacks against WPA/TKIP Based on Single-Byte Biases

4.1   The Attack of Paterson, Poettering and Schuldt[15]

The attack against WPA/TKIP in [15] builds on the single-byte bias attack (on TLS) of [1]. Both attacks work for the setting where the same plaintext is encrypted many times under different RC4 keys to obtain a set of ciphertexts. The key idea of both attacks is that, in any given position r of the ciphertext stream, a guess for the repeated plaintext byte in that position induces a distribution on the keystream in position r, via XORing the guess with byte r in each of the ciphertexts in turn. This induced distribution can be compared to the known distribution in keystream position r (which is obtained by sampling), and the choice of plaintext guess giving the "best fit" selected as the attack's output for position r. This is formalised as a Bayesian procedure, leading to the output in position r as being the plaintext candidate that maximises the probability of observing the induced keystream distribution in position r.

The innovation in [15] (and independently observed in [18]) was to recognise that in WPA/TKIP a different keystream distribution can – and should – be used for each value of the byte pair $\overline{\mathsf{TSC}} = (\mathsf{TSC}_0, \mathsf{TSC}_1)$ when estimating the probabilities of the induced keystream distributions. This leads to an algorithm that "bins" ciphertexts into 2^{16} groups, one group per $\overline{\mathsf{TSC}}$, computes the induced keystream probability for each group, and takes the product of these across the groups to compute the probabilities for each plaintext candidate. Since our new double-byte algorithm in Section 5 can be seen as an extension of our algorithm in [15], we explain the latter here in more detail.

We first obtain a detailed picture of the distributions of RC4 keystream bytes Z_r, for all positions r in some range, on a per $(\mathsf{TSC}_0, \mathsf{TSC}_1)$ pair basis, by gathering statistics from keystreams generated using a large number of random keys. That is, for all r in our selected range, we estimate

$$p_{\overline{\mathsf{TSC}}, r, k} := \Pr(Z_r = k), \quad \overline{\mathsf{TSC}} \in \mathcal{TscSp}, \quad k \in \mathcal{Byte},$$

---

[7] http://www.slac.stanford.edu/~abh/bbcp/

where here (and henceforth) $\mathcal{B}yte$ denotes the set $\{\texttt{0x00}, \ldots, \texttt{0xFF}\}$, $\mathcal{T}sc\mathcal{S}p$ denotes the set $\mathcal{B}yte \times \mathcal{B}yte$, and where the probability is taken over the random choice of the RC4 encryption key K, subject to the structure on K_0, K_1, K_2 induced by $\overline{\mathsf{TSC}}$.

Now suppose we have S ciphertexts $C_1, \ldots, C_S$ available for our attack. We partition these into 2^{16} groups according to the value of $\overline{\mathsf{TSC}}$ (recall that the $\overline{\mathsf{TSC}}$ value is public); for convenience, we assume the resulting bins of ciphertexts are all of equal size $T = S/2^{16}$. Let the bin of ciphertexts associated with a particular value of $\overline{\mathsf{TSC}}$ be denoted $\mathcal{S}_{\overline{\mathsf{TSC}}}$ and have members $C_{\overline{\mathsf{TSC}},j}$ for $j = 1, \ldots, T$; we denote the byte at position r of $C_{\overline{\mathsf{TSC}},j}$ by $C_{\overline{\mathsf{TSC}},j,r}$. For any position r and any candidate plaintext byte μ for that position, vector $\left(N_{\overline{\mathsf{TSC}},r,k}^{(\mu)}\right)_{k \in \mathcal{B}yte}$ with

$$N_{\overline{\mathsf{TSC}},r,k}^{(\mu)} = |\{j \in [1\,..\,T] \mid C_{\overline{\mathsf{TSC}},j,r} = k \oplus \mu\}| \qquad (\texttt{0x00} \leq k \leq \texttt{0xFF})$$

represents the distribution on Z_r required to obtain the observed ciphertext bytes $(C_{\overline{\mathsf{TSC}},j,r})_{1 \leq j \leq T}$ for bin $\mathcal{S}_{\overline{\mathsf{TSC}}}$ by encrypting μ. The probability $\lambda_{\overline{\mathsf{TSC}},r,\mu}$ that plaintext byte μ is encrypted to bytes $(C_{\overline{\mathsf{TSC}},j,r})_{1 \leq j \leq T}$ in bin $\mathcal{S}_{\overline{\mathsf{TSC}}}$ for position r now follows the distribution:

$$\lambda_{\overline{\mathsf{TSC}},r,\mu} = \prod_{k \in \mathcal{B}yte} \left(p_{\overline{\mathsf{TSC}},r,k}\right)^{N_{\overline{\mathsf{TSC}},r,k}^{(\mu)}}. \tag{2}$$

Note that this expression differs from that in [15] by the omission of factorial terms arising in the multinomial distribution. Those terms do not need to be included in the formal Bayesian procedure underlying the attack (since we are interested in the probability of a group of ciphertexts bytes as given in a particular sequence rather than in unordered form). Moreover, their removal makes the attack slightly easier to implement.

Now the probability that plaintext byte μ is encrypted to the vector of bytes $(C_{\overline{\mathsf{TSC}},j,r})_{1 \leq j \leq T}$ across all bins $\mathcal{S}_{\overline{\mathsf{TSC}}}$ in position r can be precisely calculated as

$$\lambda_{r,\mu} = \prod_{\overline{\mathsf{TSC}} \in \mathcal{T}sc\mathcal{S}p} \lambda_{\overline{\mathsf{TSC}},r,\mu}.$$

By computing $\lambda_{r,\mu}$ for all $\mu \in \mathcal{B}yte$, and identifying $P_r^* = \mu$ such that $\lambda_{r,\mu}$ is largest, we determine the maximum-likelihood plaintext byte value P_r^*.

Note that, for each position r and group of bytes $(C_{\overline{\mathsf{TSC}},j,r})_{1 \leq j \leq T}$, values $N_{\overline{\mathsf{TSC}},r,k}^{(\mu)}$ can be computed from values $N_{\overline{\mathsf{TSC}},r,k}^{(\mu')}$ by using the equation $N_{\overline{\mathsf{TSC}},r,k}^{(\mu)} = N_{\overline{\mathsf{TSC}},r,k \oplus \mu' \oplus \mu}^{(\mu')}$, for all k. Further, computing and comparing $\log(\lambda_{\overline{\mathsf{TSC}},r,\mu})$ and $\log(\lambda_{r,\mu})$ instead of $\lambda_{\overline{\mathsf{TSC}},r,\mu}$ and $\lambda_{r,\mu}$ makes the computation more efficient and accuracy easier to maintain. Adding these optimisations leads to the attack in Algorithm 3 (which differs from the corresponding attack in [15] only in the omission of a term $F_{\overline{\mathsf{TSC}}}$ corresponding to the factorial terms discussed above and some small notational changes).

Algorithm 3. Plaintext recovery attack using $\overline{\mathrm{TSC}}$ binning

---

input : $\{C_{\overline{\mathrm{TSC}},j}\}_{\overline{\mathrm{TSC}} \in \mathcal{T}sc\mathcal{S}p, 1 \leq j \leq T}$ – $S = 2^{16} \cdot T$ independent encryptions of fixed plaintext P

r – target byte position

$(p_{\overline{\mathrm{TSC}},r,k})_{\overline{\mathrm{TSC}} \in \mathcal{T}sc\mathcal{S}p, k \in \mathcal{B}yte}$ – keystream distributions for all $\overline{\mathrm{TSC}}$ at pos. r

output: P_r^* – estimate for plaintext byte P_r

begin

$N_{\overline{\mathrm{TSC}},k} \leftarrow 0$ for all $\overline{\mathrm{TSC}} \in \mathcal{T}sc\mathcal{S}p$, $k \in \mathcal{B}yte$

for $\overline{\mathrm{TSC}} = (\mathtt{0x00}, \mathtt{0x00})$ **to** $(\mathtt{0xFF}, \mathtt{0xFF})$ **do**

    for $j = 1$ **to** T **do**

        $k \leftarrow C_{\overline{\mathrm{TSC}},j,r}$

        $N_{\overline{\mathrm{TSC}},r,k} \leftarrow N_{\overline{\mathrm{TSC}},r,k} + 1$

for $\overline{\mathrm{TSC}} = (\mathtt{0x00}, \mathtt{0x00})$ **to** $(\mathtt{0xFF}, \mathtt{0xFF})$ **do**

    for $\mu = \mathtt{0x00}$ **to** $\mathtt{0xFF}$ **do**

        for $k = \mathtt{0x00}$ **to** $\mathtt{0xFF}$ **do**

            $N_{\overline{\mathrm{TSC}},r,k}^{(\mu)} \leftarrow N_{\overline{\mathrm{TSC}},r,k \oplus \mu}$

        $\lambda_{\overline{\mathrm{TSC}},r,\mu} \leftarrow \sum_{k \in \mathcal{B}yte} N_{\overline{\mathrm{TSC}},r,k}^{(\mu)} \log p_{\overline{\mathrm{TSC}},r,k}$

for $\mu = \mathtt{0x00}$ **to** $\mathtt{0xFF}$ **do**

    $\lambda_{r,\mu} \leftarrow \sum_{\overline{\mathrm{TSC}} \in \mathcal{T}sc\mathcal{S}p} \lambda_{\overline{\mathrm{TSC}},r,\mu}$

$P_r^* \leftarrow \arg\max_{\mu \in \mathcal{B}yte} \lambda_{r,\mu}$

return P_r^*

4.2 Attacks Based on Aggregation

One method of coping with noisy estimates for the probabilities $p_{\overline{\mathrm{TSC}},r,k}$ that was extensively explored in [15] was to consider aggregation of distributions over TSC_0 or over both TSC_0 and TSC_1 (effectively increasing the number of keys by factors of 2^8 and 2^{16}, respectively). It is not difficult to see how to modify Algorithm 3 to work with 2^8 bins, one for each value of TSC_1, instead of 2^{16} bins. The execution of the modified algorithm becomes in practice faster, since each estimate for a plaintext byte μ now only involves calculation of $\lambda_{\overline{\mathrm{TSC}},r,\mu}$ over 2^8 TSC_1 values instead of 2^{16} $(\mathrm{TSC}_0, \mathrm{TSC}_1)$ pair values. Similarly, one can modify the algorithm to work with just a single bin, one for all values of TSC_0 and TSC_1, in which case we recover the original algorithm of [1], albeit without the unnecessary factorial terms arising from the use of multinomial distributions and using WPA/TKIP-specific distributions in place of the original RC4 distributions reported in [1].

However, the cost of using aggregation is that it "throws away" statistical information that may be of use in improving the accuracy of the attack for a given number of ciphertexts S. Indeed, this is demonstrably the case: as we report below, using our new estimates for the probabilities $p_{\overline{\mathrm{TSC}},r,k}$ computed using a total of 2^{48} keystreams in a full binning (non-aggregated) attack leads to an improvement in accuracy.

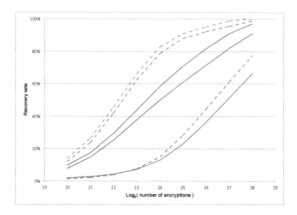

Fig. 3. Average success rates of non-aggregated (blue), TSC_0-aggregated (green), and fully aggregated (red) single-byte plaintext recovery attacks for byte positions 1 to 256 (based on 256 experiments). Punctured lines represent the average recovery rates for the odd byte positions.

4.3   Attack Simulation Results

We implemented the single-byte plaintext recovery attack of Algorithm 3 based on the keystream distribution estimates obtained from the Amazon EC2 computations described in Section 3. We furthermore implemented the TSC_0-aggregated and fully aggregated variants of the attack described in Section 4.2. To obtain bias estimates for the latter two attacks, we aggregated the Amazon EC2 data correspondingly, thereby obtaining estimates based on 2^{40} keystream per TSC_1 value, and 2^{48} keystreams, respectively.

The measured success rates of the attacks are shown in Figure 3. We observe that there is a significant difference in the recovery rates between the fully aggregated attack and the two other attacks, the non-aggregated attack being capable of achieving a similar success rate to the fully aggregated attack using almost 16 times fewer ciphertexts. Likewise, the non-aggregated attack clearly improves upon the TSC_0-aggregated attack, albeit not as significantly; the non-aggregated attack requires on average half as many ciphertexts to achieve a similar success rate to the TSC_0-aggregated attack.

In order to investigate the effect of our new and (presumably) more accurate single-byte keystream distributions, we also compared the performance of Algorithm 3 using keystream distribution estimates based on 2^{24} keystreams per $\overline{TSC}$ (as in our previous work [15]) and based on 2^{32} keystreams per $\overline{TSC}$ (obtained from the Amazon EC2 computations described in Section 3). Figure 4 shows the results, with the attacks using the two keystream distributions in combination with 2^{24} ciphertexts in each experiment. There is a clear boost to the success rate of the attack when moving to the refined keystream distribution estimates. The effect is particularly pronounced in the odd positions.

As noted in Section 3, we discovered significant TSC_1-dependent, single-byte biases in the RC4 keystreams for WPA/TKIP keys well beyond position 256.

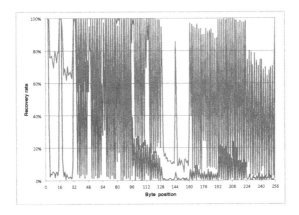

Fig. 4. Success rates of single-byte plaintext recovery attack against TKIP/WPA for positions 1 to 256 with 2^{24} ciphertexts, using keystream distribution estimates based on 2^{24} keystreams (red) and 2^{32} keystreams (blue) per $\overline{\text{TSC}}$ (success rates based on 256 experiments)

The biases are roughly comparable in size to the single-byte biases seen in RC4 keystreams at positions around 250 for random 128-bit keys (as used in TLS and reported in [1]). So we might expect to obtain reliable plaintext recovery with around $2^{30} - 2^{32}$ ciphertexts as in [1]. The full investigation of this avenue is left to future work.

5   Plaintext Recovery Attacks for WPA/TKIP Based on Double-Byte Biases

Our double-byte bias attack against WPA/TKIP builds on the attack in [1], and works in the same setting as the above described single-byte bias attack: the same plaintext is assumed to be encrypted many times under different RC4 keys, yielding a set of ciphertexts which is given as input to the attack algorithm. However, as opposed to the attack based on single-byte biases, the attack does not estimate the likelihoods of the individual plaintext bytes (or plaintext byte pairs). Instead, the basic idea of the attack is to estimate likelihoods of *sequences* of plaintext bytes by considering chains of overlapping plaintext byte pairs in combination with the double-byte biases in the keystream.

More precisely, the attack will construct likelihood estimates for sequences of plaintext bytes that are gradually increasing in length by extending already established sequences and their corresponding likelihood estimates. This is done as follows: consider a sequence of plaintext bytes with an already estimated likelihood, and a candidate for the next plaintext byte in the sequence. By XORing the pair consisting of the last plaintext byte of the existing sequence and the new candidate plaintext byte with the ciphertext byte pairs for the corresponding positions, an induced distribution on the keystream byte pairs is obtained.

By comparing this to the known double-byte keystream distribution, a likelihood estimate for the new candidate plaintext byte can be computed; combining this with the likelihood estimate for the initial plaintext sequence, a likelihood estimate for the extended sequence can be obtained.

Note that, using a naive algorithm, the complexity of computing the likelihood estimates for all possible plaintext sequences would grow exponentially in the length of the sequences. Furthermore, considering all possible candidates for the next plaintext byte, but only maintaining a small set of the most likely sequences after each extension, is not guaranteed to produce a plaintext byte sequence that maximises the value of the estimated likelihood. However, as highlighted in [1], by tracking which sequences produce the maximum value for the estimated likelihood for each possible value of the last byte in the sequence, the overall plaintext sequence which maximises the likelihood estimate is guaranteed to be found.

Compared to the algorithm from [1], the algorithm presented here provides two refinements made possible by the specific way RC4 is used in WPA/TKIP. Firstly, as in the attack described in Section 4, likelihood estimates are computed on a per-$\overline{\mathsf{TSC}}$ basis, and combined across all $\overline{\mathsf{TSC}}$ values to obtain improved overall likelihood estimates. Secondly, the attack not only exploits the per-$\overline{\mathsf{TSC}}$ double-byte biases in the WPA keystream, but also takes into account the single-byte biases in the computation of the likelihood estimates. A more detailed description of the algorithm is given next.

To run the algorithm, accurate estimates of both the single-byte and double-byte keystream distributions are required for all positions r the attack is targeting. By considering the statistics gathered by generating a large number of keystreams, we estimate

$$p_{\overline{\mathsf{TSC}},r,k} := \Pr(Z_r = k), \quad \text{and} \quad \tilde{p}_{\overline{\mathsf{TSC}},r,k_1,k_2} := \Pr(Z_r = k_1 \wedge Z_{r+1} = k_2)$$

where $\overline{\mathsf{TSC}} \in \mathcal{T}sc\mathcal{S}p$, $k, k_1, k_2 \in \mathcal{B}yte$, and the probability is taken over a random choice of RC4 key subject to the structure on $\mathsf{K}_0, \mathsf{K}_1, \mathsf{K}_2$ induced by $\overline{\mathsf{TSC}}$.

As in the single-byte bias attack, we suppose we have S ciphertexts available for our attack, and that, when grouped according to $\overline{\mathsf{TSC}}$ values, each group contains exactly $T = S/2^{16}$ ciphertexts. We likewise use the notation $C_{\overline{\mathsf{TSC}},j,r}$ to denote the ciphertext byte at position r in the jth member of the group of ciphertexts for the value $\overline{\mathsf{TSC}}$.

For a given position r, we can now use a similar approach to the single-byte bias attack to compute the likelihood of a candidate byte μ or a candidate byte pair (μ, μ') (at position $(r, r+1)$) corresponding to the encrypted plaintext byte or byte pair. More specifically, the vectors $\left(N_{\overline{\mathsf{TSC}},r,k_1}^{(\mu)}\right)_{k_1 \in \mathcal{B}yte}$ and $\left(\tilde{N}_{\overline{\mathsf{TSC}},r,k_1,k_2}^{(\mu,\mu')}\right)_{k_1,k_2 \in \mathcal{B}yte}$, where

$$N_{\overline{\mathsf{TSC}},r,k_1}^{(\mu)} = |\{j \in [1..T] \mid C_{\overline{\mathsf{TSC}},j,r} = k_1 \oplus \mu\}|$$

$$\tilde{N}_{\overline{\mathsf{TSC}},r,k_1,k_2}^{(\mu,\mu')} = |\{j \in [1..T] \mid (C_{\overline{\mathsf{TSC}},j,r}, C_{\overline{\mathsf{TSC}},j,r+1}) = (k_1 \oplus \mu, k_2 \oplus \mu')\}| \ ,$$

represent induced distributions on the keystream byte Z_r and keystream byte pair (Z_r, Z_{r+1}), respectively. Indeed, the probability that plaintext byte μ is encrypted at position r, which we will denote $\alpha_r^{(\mu)}$, and the probability that (μ, μ') is encrypted at position $(r, r+1)$, which we will denote $\beta_r^{(\mu,\mu')}$, can be computed as:

$$\alpha_r^{(\mu)} = \prod_{\overline{\mathrm{TSC}} \in \mathcal{TscSp}} \prod_{k_1 \in \mathcal{B}yte} \left(p_{\overline{\mathrm{TSC}},r,k_1} \right)^{N_{\overline{\mathrm{TSC}},r,k_1}^{(\mu)}},$$

$$\beta_r^{(\mu,\mu')} = \prod_{\overline{\mathrm{TSC}} \in \mathcal{TscSp}} \prod_{k_1,k_2 \in \mathcal{B}yte} \left(\tilde{p}_{\overline{\mathrm{TSC}},r,k_1,k_2} \right)^{\tilde{N}_{\overline{\mathrm{TSC}},r,k_1,k_2}^{(\mu,\mu')}}.$$

However, as highlighted earlier, instead of using the above probabilities for individual plaintext byte and byte pairs directly, we use these to construct likelihood estimates for longer sequences of plaintext bytes by considering chains of overlapping byte pairs. More specifically, consider a plaintext byte sequence $\mu_1 \parallel \cdots \parallel \mu_r$ for positions 1 to r with an already established likelihood estimate $\lambda_{\mu_1\parallel\cdots\parallel\mu_r}$, and a plaintext candidate byte μ_{r+1} for position $r+1$. Then we estimate the likelihood of the plaintext byte sequence $\mu_1 \parallel \cdots \parallel \mu_{r+1}$ as:

$$\lambda_{\mu_1\parallel\cdots\parallel\mu_{r+1}} = \delta_r^{(\mu_r,\mu_{r+1})} \cdot \lambda_{\mu_1\parallel\cdots\parallel\mu_r} \tag{3}$$

where $\delta_r^{(\mu_r,\mu_{r+1})}$ denotes the conditional probability that μ_{r+1} is the plaintext byte at position $r+1$ given that the plaintext byte at position r is μ_r. Note that, by the definition of conditional probability, we can compute $\delta_r^{(\mu_r,\mu_{r+1})}$ based on the estimates $\alpha_r^{(\mu_r)}$ and $\beta_r^{(\mu_r,\mu_{r+1})}$ as

$$\delta_r^{(\mu_r,\mu_{r+1})} = \beta_r^{(\mu_r,\mu_{r+1})} / \alpha_r^{(\mu_r)}.$$

In the description of the attack algorithm presented here, it is assumed that the plaintext byte P_1^* at position $r = 1$ is known. This serves as a starting point for the algorithm, i.e., the algorithm is initialized with a single plaintext sequence containing the byte value P_1^* for position $r = 1$ and with the estimated likelihood $\lambda_{P_1^*} = 1$. Now, using the above described method for extending a plaintext byte sequence and the corresponding likelihood estimate, the attack algorithm iterates over the range of considered positions as follows. For each position r, and for all possible values μ_{r+1} of the plaintext byte at position $r+1$, the extension with μ_{r+1} of each of the sequences from the previous iteration is considered, and, for each of the possible values of μ_{r+1}, the algorithm stores the "most likely" extended sequence having μ_{r+1} as the last byte value (that is, it stores the extended sequence which maximises the likelihood estimate expressed in equation (3)). When the attack algorithm reaches the last position, it simply returns the sequence with the highest likelihood estimate.

Note that this process is guaranteed to find the plaintext byte sequence with the highest likelihood estimate computed according to equation (3). However, we emphasise that this expression yields *only* an approximation to the actual plaintext likelihood, being based on the twin assumptions that plaintext bytes are independently and uniformly distributed and that keystream bytes have no dependencies beyond those in adjacent bytes as expressed in the double-byte distributions.

A full description of the attack algorithm is given in Algorithm 4 (on page 415). Note that the algorithm can easily be extended to work for the case where the plaintext byte at the initial position is unknown. In particular, by exploiting the single-byte biases, the likelihoods of all possible values of the initial plaintext byte can be estimated, and subsequently used as a starting point for the algorithm. Of course, the algorithm need not start at position $r = 1$ either.

Notice that the algorithm involves heavy nesting of loops, particularly in phase 2b, where for each position r we perform a computation over all possible values for the candidate plaintext byte pair (μ_{r-1}, μ_r), each such computation itself involving a sum over 2^{32} pairwise products of real numbers arising from the triple summation over $\overline{\text{TSC}}$, k_1 and k_2. Thus a direct implementation of this algorithm would require on the order of 2^{48} additions and products *per position*! This would be inconvenient, to say the least. For this reason, and because our double-byte, per-$\overline{\text{TSC}}$ keystream distributions are not particularly accurate (being based only on 2^{30} keystreams each), we would in preference use aggregated versions of the algorithm. Specifically, building on our experience in the single-byte case, we may consider a version of the algorithm that works with TSC_0-aggregated distributions and only works on a per-TSC_1 basis. It is not hard to see how to modify Algorithm 4 to operate in this way, saving a factor of 2^8 in its computational cost. The algorithm could be further modified to use fully aggregated distributions, saving another factor of 2^8 in computational cost, but now effectively ignoring any $\overline{\text{TSC}}$-related structure in the keystream distributions.

We have performed a very limited validation of our double-byte attack in its fully aggregated form. A complete evaluation of the algorithm and a comparison of its performance with the single-byte Algorithm 3 is deferred to the full version of the paper. We make one observation at this stage, however. Algorithm 4 makes use of ratios of probability expressions of the form $\beta_r^{(\mu_r, \mu_{r+1})} / \alpha_r^{(\mu_r)}$, where the numerator is a double-byte probability and the numerator is a single-byte probability. If the significant biases in the former probabilities actually arise from products of single-byte biases for adjacent positions, then such expressions can be simplified to just single-byte probability terms of the form $\alpha_{r+1}^{(\mu_{r+1})}$, in effect reducing our double-byte attack to our single-byte attack. Such behaviour can be expected in early byte positions, where single-byte biases are very large. Thus we do not expect our double-byte attack in Algorithm 4 to significantly out-perform our single-byte attack in the early positions. On the other hand, in regions where single-byte biases become smaller and fewer in number but double-byte biases still persist (as seems to be the case in later positions), then Algorithm 4 may be expected to perform better than our single-byte attack. Indeed, Algorithm 4 should be able to smoothly interpolate between regions where single-byte biases dominate and regions where they do not.

Algorithm 4. Double-byte bias attack

---

input : C – balanced vector of $2^{16} \cdot S$ encryptions of fixed plaintext P
$\quad\quad\quad$ ($C_{\overline{TSC},j,r}$ denotes r-th byte of j-th encryption of P for TSC-value $\overline{TSC}$)
$\quad\quad\quad$ L – length of P in bytes
$\quad\quad\quad$ m_1 and m_L – known first and last byte of P
$\quad\quad\quad$ $\{p_{\overline{TSC},r,k}\}_{\overline{TSC} \in \mathcal{TscSp},\, 1 \leq r \leq L,\, k \in \mathcal{Byte}}$ – single-byte key distribution
$\quad\quad\quad$ $\{\tilde{p}_{\overline{TSC},r,k_1,k_2}\}_{\overline{TSC} \in \mathcal{TscSp},\, 1 \leq r < L,\, k_1,k_2 \in \mathcal{Byte}}$ – double-byte key distribution
output: estimate P^* for plaintext P
begin
$\quad N_{\overline{TSC},r,k} \leftarrow 0 \quad$ for all $\overline{TSC} \in \mathcal{TscSp}$, $1 \leq r \leq L$, $k \in \mathcal{Byte}$
$\quad \tilde{N}_{\overline{TSC},r,k_1,k_2} \leftarrow 0 \quad$ for all $\overline{TSC} \in \mathcal{TscSp}$, $1 \leq r < L$, $k_1,k_2 \in \mathcal{Byte}$
$\quad$ initialise mappings $Q, Q' : \mathcal{Byte} \to \mathcal{Byte}^* \times \mathbb{R}$
$\quad$ // **Phase 1** $\quad$ (count occurrences of keystream bytes and byte pairs)
$\quad$ **for each** $\overline{TSC} \in \mathcal{TscSp}$ **do**
$\quad\quad$ **for** $j = 1$ **to** S **do**
$\quad\quad\quad$ **for** $r = 1$ **to** $L - 1$ **do**
$\quad\quad\quad\quad$ $N_{\overline{TSC},r,C_{\overline{TSC},j,r}} \leftarrow N_{\overline{TSC},r,C_{\overline{TSC},j,r}} + 1$
$\quad\quad\quad\quad$ $\tilde{N}_{\overline{TSC},r,C_{\overline{TSC},j,r},C_{\overline{TSC},j,r+1}} \leftarrow \tilde{N}_{\overline{TSC},r,C_{\overline{TSC},j,r},C_{\overline{TSC},j,r+1}} + 1$

$\quad$ // **Phase 2a** $\quad$ (derive likelihoods for plaintext byte at position 2)
$\quad$ **for** $\mu_2 = $ 0x00 **to** 0xFF **do**
$\quad\quad$ $\lambda_{m_1 \| \mu_2} \leftarrow + \displaystyle\sum_{\overline{TSC} \in \mathcal{TscSp}} \sum_{k_1,k_2 \in \mathcal{Byte}} \tilde{N}_{\overline{TSC},1,k_1 \oplus m_1,k_2 \oplus \mu_2} \log \tilde{p}_{\overline{TSC},1,k_1,k_2}$
$\quad\quad\quad\quad$ $- \displaystyle\sum_{\overline{TSC} \in \mathcal{TscSp}} \sum_{k \in \mathcal{Byte}} N_{\overline{TSC},1,k \oplus m_1} \log p_{\overline{TSC},1,k}$
$\quad\quad$ $Q[\mu_2] \leftarrow (\mu_2, \lambda_{m_1 \| \mu_2})$

$\quad$ // **Phase 2b** $\quad$ (derive likelihoods for plaintext bytes at positions $3 \ldots (L-1)$)
$\quad$ **for** $r = 3$ **to** $L - 1$ **do**
$\quad\quad$ **for** $\mu_r = $ 0x00 **to** 0xFF **do**
$\quad\quad\quad$ $L^* \leftarrow -\infty$
$\quad\quad\quad$ **for** $\mu_{r-1} = $ 0x00 **to** 0xFF **do**
$\quad\quad\quad\quad$ parse $Q[\mu_{r-1}]$ as $(P', \lambda_{P'})$
$\quad\quad\quad\quad$ $\lambda_{P' \| \mu_r} \leftarrow \lambda_{P'}$
$\quad\quad\quad\quad\quad$ $+ \displaystyle\sum_{\overline{TSC} \in \mathcal{TscSp}} \sum_{k_1,k_2 \in \mathcal{Byte}} \tilde{N}_{\overline{TSC},r-1,k_1 \oplus \mu_{r-1},k_2 \oplus \mu_r} \log \tilde{p}_{\overline{TSC},r-1,k_1,k_2}$
$\quad\quad\quad\quad\quad$ $- \displaystyle\sum_{\overline{TSC} \in \mathcal{TscSp}} \sum_{k \in \mathcal{Byte}} N_{\overline{TSC},r-1,k \oplus \mu_{r-1}} \log p_{\overline{TSC},r-1,k}$
$\quad\quad\quad\quad$ **if** $\lambda_{P' \| \mu_r} > L^*$ **then**
$\quad\quad\quad\quad\quad$ $(P^*, L^*) \leftarrow (P', \lambda_{P' \| \mu_r})$
$\quad\quad\quad$ $Q'[\mu_r] \leftarrow (P^* \| \mu_r, L^*)$
$\quad\quad$ $Q \leftarrow Q'$

$\quad$ // **Phase 3** $\quad$ (pick most likely plaintext out of candidate set)
$\quad$ $L^* \leftarrow -\infty$
$\quad$ **for** $\mu_{L-1} = $ 0x00 **to** 0xFF **do**
$\quad\quad$ parse $Q[\mu_{L-1}]$ as $(P', \lambda_{P'})$
$\quad\quad$ $\lambda_{P' \| m_L} \leftarrow \lambda_{P'}$
$\quad\quad\quad$ $+ \displaystyle\sum_{\overline{TSC} \in \mathcal{TscSp}} \sum_{k_1,k_2 \in \mathcal{Byte}} \tilde{N}_{\overline{TSC},L-1,k_1 \oplus \mu_{L-1},k_2 \oplus m_L} \log \tilde{p}_{\overline{TSC},L-1,k_1,k_2}$
$\quad\quad\quad$ $- \displaystyle\sum_{\overline{TSC} \in \mathcal{TscSp}} \sum_{k \in \mathcal{Byte}} N_{\overline{TSC},L-1,k \oplus \mu_{L-1}} \log p_{\overline{TSC},L-1,k}$
$\quad\quad$ **if** $\lambda_{P' \| m_L} > L^*$ **then**
$\quad\quad\quad$ $(P^*, L^*) \leftarrow (P', \lambda_{P' \| m_L})$
$\quad$ **return** $m_1 \| P^* \| m_L$

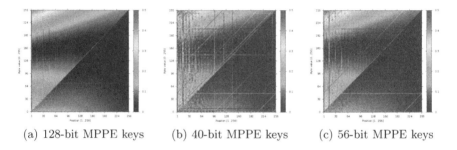

| (a) 128-bit MPPE keys | (b) 40-bit MPPE keys | (c) 56-bit MPPE keys |

Fig. 5. Pictorial representation of biases in RC4 keystreams for 128-bit, 40-bit and 56-bit MPPE keys, for different positions (x-axis) and byte values (y-axis). For each position we encode the bias in the keystream for the (position,value) combination as a colour; in each case, the colouring scheme encodes the absolute biases, i.e., the absolute difference between the occurring probabilities and the (expected) probability $1/256$, scaled up by a factor of 2^{16}, capped to a maximum of 0.5.

6 MPPE

6.1 Computing Keystream Distributions for MPPE Keys

We also computed the RC4 keystream distributions for the first 256 keystream bytes using MPPE keys having the structure described in Section 2.3. More specifically, we generated random 8-byte random keys $K = (K_0, \ldots, K_7)$ and then overwrote key bytes according to the MPPE specification for the 40-bit and 56-bit cases, while in the 128-bit case, we generated random 16-byte keys. We used more than 2^{39} keys in each case, with all computations being performed on our local computing facilities. Figure 5 compares the distributions obtained for random 128-bit RC4 keys (as used in 128-bit MPPE and in TLS) with those for 40-bit and 56-bit MPPE keys. As can be seen, the process of fixing certain key bytes to constant values produces many additional, strong biases in the corresponding keystreams.

6.2 Attack Simulation Results

We used the MPPE keystream distributions to simulate plaintext recovery attacks using the algorithm of [1], equivalent to the fully aggregated version of Algorithm 3. The results are depicted in Figure 6. As expected, the additional structure in RC4 keys introduced by MPPE in the 40-bit and 56-bit cases significantly aids plaintext recovery, with 40-bit keys leading to the highest success rate for a given number of ciphertexts. We also experimented with random 64-bit keys, finding success rates very close to the random 128-bit case. This indicates that it is not the reduction in key-size that makes the difference in MPPE, but rather the introduction of fixed key bytes.

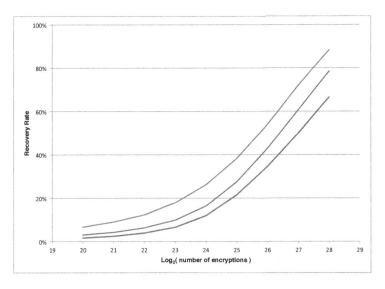

Fig. 6. Average success rates of single-byte plaintext recovery attacks against MPPE using 40-bit keys (blue), 56-bit keys (red), and 128-bit keys (green) over positions 1 to 256. The success rates are based on 256 experiments.

7   Conclusions

In this paper, we have explored the use of cloud computing facilities to perform large-scale computations in support of the cryptanalysis of WPA/TKIP. We expended 63 virtual-core-years of computational effort at a cost of US$43k to carry out two computations, one involving 2^{48} keystreams to estimate per-$\overline{\text{TSC}}$ single-byte distributions, the other involving 2^{46} keystreams to estimate per-$\overline{\text{TSC}}$ double-byte distributions. The total amount of computation was roughly one-twentieth of that used in the sieving stage for the factorisation of RSA-768[8]. The problems of developing efficient code for, and then managing, these computations were not insignificant but ultimately surmountable. This suggests that commercial cloud services can be used as a platform for this kind of work, instead of relying on owned infrastructure. Certainly, running 2^{13} hyper-threaded cores in parallel was an exhilarating, if expensive, way to explore the limits of commercial cloud computing capabilities.

The value of our keystream distribution computations for WPA/TKIP is aptly illustrated in Figure 4, which shows the marked improvement in success rate that accrues from moving from single-byte keystream distribution estimates based on 2^{24} keystreams per $\overline{\text{TSC}}$ to 2^{32} keystreams per $\overline{\text{TSC}}$. Our computations of RC4 keystream distributions in WPA/TKIP and MPPE also provide experimental data that may be useful in making hypotheses about keystream biases, and which may in turn lead to a better theoretical understanding of the operation

---

[8] Estimated at 1500 core-years for a single core 2.2 GHz AMD Opteron processor with 2GB RAM in [9].

of RC4 in these applications. Certainly, having an explanation for the long-lived TSC_1-specific single-byte biases that we observed experimentally would be very welcome. A similar project would investigate the effect of fixing key bytes in RC4 keys, and apply the results to provide a theoretical explanation for the observed biases in MPPE keystreams.

Our attack on WPA/TKIP based on double-byte biases requires further investigation: the time and budget available for this project has limited our experimentation with it and reduced our investment in its fine-tuning. Given the dominance of single-byte biases in early portions of the RC4 keystreams for WPA/TKIP, we expect this algorithm to come into its own when targeting repeated plaintext that is located later in WPA/TKIP frames (e.g. after position 256). Moreover, it provides a mechanism for smoothly transitioning attacks from the regime where single-byte biases dominate to the regime where these biases are no longer apparent but where double-byte biases are still present. It remains to investigate whether other forms of bias (such as the "ABSAB" biases from [11]) can be effectively integrated into a more general Bayesian approach, and how much impact this might have on overall attack performance.

Acknowledgements. The research of the authors was conducted in part while all authors were at Royal Holloway, University of London. The research was supported by an EPSRC Leadership Fellowship, EP/H005455/1 as well as a grant from the UK government. We thank Martin Albrecht, Jon Hart and Adrian Thomas at RHUL for their assistance with sourcing, building and maintaining our local computing infrastructure and for help in managing AWS. We thank Strombenzin for its generous donation of computing cycles. We thank the UK government for financing our adventures with Amazon's cloud computing infrastructure, and Mark Rowlands at Amazon Web Services for his assistance in maxing out the AWS US West data centre.

References

1. AlFardan, N.J., Bernstein, D.J., Paterson, K.G., Poettering, B., Schuldt, J.C.N.: On the security of RC4 in TLS. In: USENIX Security. USENIX Association (2013), https://www.usenix.org/conference/usenixsecurity13/security-rc4-tls
2. Fluhrer, S.R., Mantin, I., Shamir, A.: Weaknesses in the key scheduling algorithm of RC4. In: Vaudenay, S., Youssef, A.M. (eds.) SAC 2001. LNCS, vol. 2259, pp. 1–24. Springer, Heidelberg (2001)
3. Fluhrer, S.R., McGrew, D.A.: Statistical analysis of the alleged RC4 keystream generator. In: Schneier, B. (ed.) FSE 2000. LNCS, vol. 1978, pp. 19–30. Springer, Heidelberg (2001)
4. Hulton, D., Marlinspike, M.: Divide and conquer: Cracking MS-CHAPv2 with a 100% success rate (2012), https://www.cloudcracker.com/blog/2012/07/29/cracking-ms-chap-v2/
5. IEEE 802.11. Wireless LAN medium access control (MAC) and physical layer (PHY) specification (1997)
6. IEEE 802.11i. Wireless LAN medium access control (MAC) and physical layer (PHY) specification: Amendment 6: Medium access control (MAC) security enhancements (2004)

7. Isobe, T., Ohigashi, T., Watanabe, Y., Morii, M.: Full plaintext recovery attack on broadcast RC4. In: Moriai, S. (ed.) FSE 2013. LNCS, vol. 8424, pp. 179–202. Springer, Heidelberg (2014)

8. Jaganathan, K., Zhu, L., Brezak, J.: The RC4-HMAC Kerberos Encryption Types Used by Microsoft Windows. RFC 4757 (Informational) (December 2006), `http://www.ietf.org/rfc/rfc4757.txt`

9. Kleinjung, T., Aoki, K., Franke, J., Lenstra, A.K., Thomé, E., Bos, J.W., Gaudry, P., Kruppa, A., Montgomery, P.L., Osvik, D.A., te Riele, H., Timofeev, A., Zimmermann, P.: Factorization of a 768-bit RSA modulus. In: Rabin, T. (ed.) CRYPTO 2010. LNCS, vol. 6223, pp. 333–350. Springer, Heidelberg (2010)

10. Maitra, S., Paul, G., Sen Gupta, S.: Attack on broadcast RC4 revisited. In: Joux, A. (ed.) FSE 2011. LNCS, vol. 6733, pp. 199–217. Springer, Heidelberg (2011)

11. Mantin, I.: Predicting and distinguishing attacks on RC4 keystream generator. In: Cramer, R. (ed.) EUROCRYPT 2005. LNCS, vol. 3494, pp. 491–506. Springer, Heidelberg (2005)

12. Mantin, I., Shamir, A.: A practical attack on broadcast RC4. In: Matsui, M. (ed.) FSE 2001. LNCS, vol. 2355, pp. 152–164. Springer, Heidelberg (2002)

13. Ohigashi, T., Isobe, T., Watanabe, Y., Morii, M.: How to recover any byte of plaintext on RC4. In: Lange, T., Lauter, K., Lisoněk, P. (eds.) SAC 2013. LNCS, vol. 8282, pp. 155–173. Springer, Heidelberg (2013)

14. Pall, G., Zorn, G.: Microsoft Point-To-Point Encryption (MPPE) Protocol. RFC 3078 (Informational) (March 2001), `http://www.ietf.org/rfc/rfc3078.txt`

15. Paterson, K.G., Poettering, B., Schuldt, J.C.N.: Plaintext recovery attacks against WPA/TKIP. In: FSE, Lecture Notes in Computer Science. Springer (to appear, 2014)

16. Sarkar, S., Sen Gupta, S., Paul, G., Maitra, S.: Proving TLS-attack related open biases of RC4. Cryptology ePrint Archive, 2013/502, `https://eprint.iacr.org/2013/502`

17. Schneier, B.: Mudge. Cryptanalysis of Microsoft's Point-to-Point Tunneling Protocol (PPTP), `https://www.schneier.com/paper-pptp.pdf`

18. Sen Gupta, S., Maitra, S., Meier, W., Paul, G., Sarkar, S.: Dependence in IV-related bytes of RC4 key enhances vulnerabilities in WPA. In: FSE 2014. LNCS, Springer (to appear, 2014)

19. Sen Gupta, S., Maitra, S., Paul, G., Sarkar, S. (Non-) random sequences from (non-) random permutations – analysis of RC4 stream cipher. Journal of Cryptology 27(1), 67–108 (2014)

20. Sepehrdad, P., Vaudenay, S., Vuagnoux, M.: Statistical attack on RC4. In: Paterson, K.G. (ed.) EUROCRYPT 2011. LNCS, vol. 6632, pp. 343–363. Springer, Heidelberg (2011)

21. Tews, E., Beck, M.: Practical attacks against WEP and WPA. In: Basin, D.A., Capkun, S., Lee, W. (eds.) WISEC, pp. 79–86. ACM (2009)

22. Vanhoef, M., Piessens, F.: Practical verification of WPA-TKIP vulnerabilities. In: Chen, K., Xie, Q., Qiu, W., Li, N., Tzeng, W.-G. (eds.) ASIACCS, pp. 427–436. ACM (2013)

23. Zorn, G.: Deriving Keys for use with Microsoft Point-to-Point Encryption (MPPE). RFC 3079 (Informational) (March 2001), `http://www.ietf.org/rfc/rfc3079.txt`

Multi-user Collisions: Applications to Discrete Logarithm, Even-Mansour and PRINCE

Pierre-Alain Fouque[1], Antoine Joux[2], and Chrysanthi Mavromati[3]

[1] Université Rennes 1, France and Institut Universitaire de France, France
[2] CryptoExperts, France and Chaire de Cryptologie de la Fondation de l'UPMC
Laboratoire d'Informatique de Paris 6, UPMC Sorbonne Universités, France
[3] Sogeti/ESEC R&D Lab, France
Université de Versailles Saint-Quentin-en-Yvelines, France
`pierre-alain.fouque@univ-rennes1.fr, antoine.joux@m4x.org,`
`chrysanthi.mavromati@sogeti.com`

Abstract. In this paper, we investigate the multi-user setting both in public and in secret-key cryptanalytic applications. In this setting, the adversary tries to recover keys of many users in parallel more efficiently than with classical attacks, *i.e.*, the number of recovered keys multiplied by the time complexity to find a single key, by amortizing the cost among several users. One possible scenario is to recover a single key in a large set of users more efficiently than to recover a key in the classical model. Another possibility is, after some shared precomputation, to be able to learn individual keys very efficiently. This latter model is close to traditional time/memory tradeoff attacks with precomputation. With these goals in mind, we introduce two new algorithmic ideas to improve collision-based attacks in the multi-user setting. Both ideas are derived from the parallelizable collision search as proposed by van Oorschot and Wiener. This collision search uses precomputed chains obtained by iterating some basic function. In our cryptanalytic application, each pair of merging chains can be used to correlate the key of two distinct users. The first idea is to construct a graph, whose vertices are keys and whose edges are these correlations. When the graph becomes connected, we simultaneously recover all the keys. Thanks to random graph analysis techniques, we can show that the number of edges that are needed to make this event occurs is small enough to obtain some improved attacks. The second idea modifies the basic technique of van Oorschot and Wiener: instead of waiting for two chains to merge, we now require that they become *parallel*.

We first show that, using the first idea alone, we can recover the discrete logarithms of L users in a group of size N in time $\widetilde{O}(\sqrt{NL})$. We put these two ideas together and we show that in the multi-user Even-Mansour scheme, *all* the keys of $L = N^{1/3}$ users can be found with $N^{1/3+\epsilon}$ queries for each user (where N is the domain size). Finally, we consider the PRINCE block cipher (with 128-bit keys and 64-bit blocks) and find the keys of 2 users among a set of 2^{32} users in time 2^{65}. We also describe a new generic attack in the classical model for PRINCE.

P. Sarkar and T. Iwata (Eds.): ASIACRYPT 2014, PART I, LNCS 8873, pp. 420–438, 2014.

1   Introduction

The multi-user setting is a very interesting practical scenario, which is sometimes overlooked in cryptography. Indeed, cryptosystems are designed to be used by many users, and usually cryptographers prove the security of their schemes in a single-user model except in some cases such as key exchange, public-key encryption and signatures. At EUROCRYPT 2012, Menezes [20] gave an invited talk pointing out the discrepancy between security proofs for message authentication code in the single-user and in the multi-user setting. As it was already been pointed out in [10], he showed that there is a straightforward reduction between the security proof for one user and the security proof for L users with a success probability divided by L. Next, he recalled the key collision attack due to Biham [3] that matches this bound and that can be applied on various deterministic MACs (CMAC, SIV, OCB, EME, ...). In this attack, the adversary asks the MAC tag of a single message M for L different users; we call this the set of *secret* MACs. Then, for a subset W of size N/L of known keys (N is the key size), he computes $MAC(k, M)$ for all $k \in W$ and builds the set of *public* MACs. If a collision occurs between the public and secret sets, then we learn one of the L secret keys.[1] For MAC schemes with an 80-bit security level, it is possible with time/memory tradeoff to make this reasonably practical and derive a key recovery of a single key among a set of $L = 2^{20}$ users, using time and memory 2^{40}. Menezes thus insists that cryptographers have to consider this practical setting when devising or analyzing cryptosystems. For more results on multi-user attacks, the reader can also refer to [4].

In this paper, we are interested in collision-based attacks [24] in the multi-user setting. We rely on the distinguished point technique to propose new attacks on the generic discrete logarithm problem, on the Even-Mansour cipher and on PRINCE. Collision-based methods have been nicely improved by van Oorschot and Wiener to become parallelizable using the distinguished point technique of Rivest and Quisquater and Delescaille [22]. Here, we extend these methods and apply them to cryptanalysis in the multi-user setting.

Our Contributions. From a cryptanalytic point of view, there are many ways to perform attacks in the multi-user setting. In this paper, we are interested by several scenarios. The first option is to recover all the users' keys (or a large fraction thereof) in time less than the product of the number of users by the time complexity to recover one key. Another direction is to improve Biham's attack and recover a single key in the multi-user setting with a reduced memory cost. Finally, we consider time/memory attacks starting with a precomputation whose result can then be used later to recover individual keys much faster.

Giant connected component. The multi-user setting for the discrete logarithm problem has been studied by Kuhn and Struik in [17]. They show that it is

---

[1] Provided that the tag length is greater than the key length.

possible to adapt the parallel version of the Pollard rho technique with distinguished points to recover L keys in time $\sqrt{NL}$ where N is the size of the group as long as $L \ll \sqrt[4]{N}$. In the parallel version of Pollard rho method described by van Oorschot and Wiener, we run random walks in parallel, stop them once a distinguished point is reached and store this value for many starting points. We get a *public* set of distinguished points for the walks that begin at $y_a = g^a$ for which we know a and a *secret* set from a user public key y for starting points yg^b where b is known. Kuhn and Struik generalize this method by using many secret sets, one for each user. Once a distinguished point appears twice in the public and secret sets, the discrete logarithm of one user can be discovered, and consequently, we also know the discrete logarithm of all the distinguished points that were discovered during the random walks for this user. Therefore, as the number of "known" points increases, the probability of a collision between a secret point and a known one becomes higher. Similar results can be found in [19,1,2].

Here, we show another method that works without any restriction on L and keeps the symmetry between all read points. Indeed, we do not have to wait until the first collision between a public point and a secret one happens, but we also consider collisions between secret points. More precisely, as soon as a collision between the public walks and the secret walks happens, we learn many discrete logarithms, since when two secret chains collide, we learn the difference between the discrete logarithm. We can then construct a graph whose vertices are the users and we add an edge if we know the difference of the discrete logarithm between these users. At some point, when the number of edges becomes slightly larger than the users, a giant component emerges in our random graph and if the public user is in this component (with high probability in time $2L \ln L$), then the discrete logarithm of *all* users will be known.

Our method has an advantage towards the method proposed by Kuhn and Struik as we use parallelism extensively. However, a disadvantage is that in our case we do not learn any discrete logarithms until the very end, when a giant component appears in the graph. In contrast, Kuhn and Struik's algorithm is sequential and so they find each discrete logarithm one after the other. Overall, the main goal of section 2 is to provide an educational example of the graph connexity approach and show that it is much simpler to analyze.

Lambda Method for two different Even-Mansour style functions. We were also able to apply similar techniques on Even-Mansour with domain size N. Indeed, using some functions related to the encryption scheme, we show that we can learn the Xor between the keys of two users. The previous technique can also be used to recover the keys of all users. However, in this case, we get a new problem: the two functions we iterate are no longer the same. Consequently, contrary to the DL case, once a collision appears, the chains will no longer merge and we cannot use distinguished point technique. To solve this issue, we tweak the two functions and define related functions that will no longer merge but become parallel. We show that this parallel method is as efficient as the previous one. For instance, we show an attack that partially solves an open problem of Dunkelman *et al.* that asked to find a memoryless attack on Even-Mansour with D queries to the secret function and $T = N/D$ to the public function with $D \ll \sqrt{N}$. We propose an attack that matches these bounds $(D = N^{2/5}, T = N^{3/5})$ but where the memory is $N^{1/5}$ as an application of our lambda-method. Furthermore, we also describe a multi-user attack which allows to learn *all* the keys in a set of $N^{1/3}$ users in data complexity $N^{1/3+\epsilon}$ to each user and $T = N^{1/3+\epsilon}$ time complexity by combining the two algorithmic tools. This attack exhibits new tradeoff where the amortized data complexity per user times the time complexity is reduced to $N^{2/3+\epsilon}$ instead of N.

Application to PRINCE. PRINCE cipher [7] is a new block cipher recently introduced at ASIACRYPT 2012 with blocklength 64 bits and keylength 128 bits. Its design has a α-reflection property which is a related-key relation that transforms the decryption algorithm to the encryption process with a related-key. Here, we propose generic attacks on the full number of rounds. At FSE 2014 [8], an attack on 10 rounds of PRINCE has been presented, with time complexity $2^{60.7}$ and data complexity $2^{57.94}$. In [15], an attack with slightly less than 2^{128} allows to break all the rounds, but our attacks have a particular low time complexity. They are similar to the one on Even-Mansour but we have to take into account that in PRINCE, the internal permutation uses a secret key. They make use both of the α-reflection property and of the specific key scheduling of PRINCE, *i.e.* the relationship between the two whitening keys. The first attack allows to recover the keys of two users among a set of 2^{32} users in time 2^{65} and the second one allows to recover the keys of all users in time 2^{32} after a precomputation of time 2^{96} and 2^{64} in memory. Finally, we do not contradict the security bound showed in the original paper, but we show that different tradeoffs are possible.

Organization of the Paper. In section 2, we present our results on the discrete logarithm problem in the multi-user setting and we use the properties of random graph in this setting. Then, we present various results concerning the security of Even-Mansour: new time/memory/data (denoted T/M/D) tradeoffs, new time/memory (denoted T/M) attack solving the open problem of Dunkelman *et al.* and in the multi-user setting. In this part, we show how we can adapt the lambda-method when searching for collisions for two different functions based on the Even-Mansour idea. Finally, in the last section, we present various generic attacks on the PRINCE block cipher, one in the multi-user setting and the other in the classical model.

2   Discrete Logarithms in the Multi-user Setting

In this section, we present a new algorithmic idea for performing T/M attacks with distinguished points in the multi-users setting. Our technique allows to compute the discrete logarithms of L public keys $y_i = g^{x_i}$ for $i = 1, \ldots, L$ in time[2] $\widetilde{O}(\sqrt{NL})$ for any value of L where $N = |\langle g \rangle|$. Starting from the parallel

---

[2] The $\widetilde{O}$ notation hides logarithmic terms.

version of Pollard rho method [24], we compute $cL/2$ chains consisting of pseudo-random walks from y_i ($c/2$ chains for each user by randomizing the starting point) until we discover a distinguished point $d_i \in S_0$ where S_0 denotes the set of distinguished points[3]. Then, all distinguished points found are sorted and each collision between the distinguished points of different users d_i and d_j reveals a linear relation between x_i and x_j. We also compute a few chains starting from random points for which the discrete logarithm is known g^{x_0}. Finally, we construct the random graph where the vertices are the public keys and we add an edge between y_i and y_j if we have a collision between d_i and d_j (this process can be described more formally using a random graph process). This edge is labelled with the linear relation between x_i and x_j. Once we have computed a sufficient number of collisions, a small constant time the number of users, then a giant component will appear with high probability. More precisely, in a graph with L vertices and $cL/2$ randomly placed edges with $c > 1$, there is a giant component whose size is almost exactly $(1 - t(c))L$, (see [6]) where:

$$t(c) = \frac{1}{c} \sum_{k=1}^{\infty} \frac{k^{k-1}(ce^{-c})^k}{k!}.$$

For $c = 4$, we get $1 - t(c) = 0.98$. The discrete logarithm of all the points in the component of the x_0's are known. If we want to recover the discrete logarithm of *all* users with overwhelming probability, we need $2L \ln L$ edges to connect all connected components according to the coupon collectors problem and not $cL/2$, as it is recalled in Theorem 2.

Let ℓ the average length of the chains and S_0 the set of distinguished points. The average length of each chain is $\ell = N/|S_0|$. Assume we have computed i chains that do not collide, the probability that the $(i + 1)^{\text{th}}$ chain collides with one of the previous is $i\ell \times \ell/N$. Consequently, the expected number of collisions Coll is:

$$\mathbb{E}[\text{Coll}] = \sum_{i=1}^{L-1} \frac{i\ell^2}{N} \approx \frac{L^2}{2} \cdot \frac{\ell^2}{N} = \frac{L^2}{2} \cdot \frac{(N/|S_0|)^2}{N} = \frac{L^2 N}{2|S_0|^2}.$$

We want the number of collisions to be larger than $cL/2$, which implies $L^2 N/2|S_0|^2 \geq cL/2$, thus $|S_0| \leq \sqrt{LN/c}$. Consequently, the overall cost is dominated by the computation of the chains, *i.e.* $L \times N/|S_0|$ which is about $\sqrt{cLN}$ if $|S_0| = \sqrt{LN/c}$. Finally, in order to have $cL/2$ edges in our graph, each user has to compute a small number of chains using a small number of random input points of the form $g^{x_i+r_i}$ for known value of r_i. The overall complexity of our attack is $\tilde{O}(\sqrt{NL})$ for any value of L while Kuhn and Struik analysis achieves the value $\sqrt{2LN}$ for $L \ll \sqrt[4]{N}$.

Another possible approach to analyze known, unknown points and collisions between them would be to use a matrix. For this, we consider a symmetric matrix M where $M[i, j]$ represents the linear relation between the discrete logarithms of i and j. Then we apply a random variable in order to sparsify the matrix. More

---

[3] This algorithm can also be adapted to the Pollard-lambda algorithm [21].

precisely, we multiply the coefficient (i, j) of the matrix by 1 with probability p and by 0 with probability $(1 - p)$, where these probabilities are independent. When we multiply by 1, that means, that we know the differences between the discrete logarithms of i and j. The question then becomes how many rows (with 2 non-zero coefficients) do we need to achieve full column-rank, which naturally leads to the same results: $O(L * \log(L))$. However, when considering rows with $O(\log(L))$ non-zero coefficients, we only needs $O(L)$ rows. This would imply that for multi-user discrete logarithms the overall complexity can be reduced by a factor $\log(L)$ to $O(sqrt(L * N))$ by spending a factor $\log(L)$ more work in generating starting points of random combinations of $\log(L)$ known/unknown points (e.g., see [11]). We choose to analyze the complexity in the same form as Wiener and van Oorschot which is usually the case for crypto papers, *i.e* we do not care on the $\log N$ factors that arise in such birthday algorithms. Indeed, the Kuhn and Struik algorithm hides also a $\log(N)$ factor in order to get collisions with very high probability because a $1/2$ probability is not sufficient since we need many collisions of this type.

3   Even-Mansour in the Single and Multi-user Settings

3.1   Brief Description of Even-Mansour

At Asiacrypt 1991, Even and Mansour in [14] describe a very efficient design (called EM in the following) to construct a block cipher, i.e. a keyed permutation family Π_{K_1,K_2} from a large permutation π. The key K_1 is first xored with the plaintext, then the fixed permutation is applied and finally the key K_2 is xored to obtain the final value.

$$\Pi_{K_1,K_2}(P) = \pi(P \oplus K_1) \oplus K_2.$$

Their main result is a security proof that any attack that uses D on-line plaintext/ciphertext pairs (queries to Π) and T off-line computations (queries to π) must satisfy $DT = N$, where $N = 2^n$ with n the size of the plaintext and key and which will be called the EM curve. The important part of the proof is that it is a lower bound for all attacks including *known-plaintext* attacks. It appears that the use of two keys K_1 and K_2 does not add much more resistance to the scheme. This variant of using $K = K_1 = K_2$ has been proposed under the name *Single-Key Even-Mansour* and we denote it by Π_K. The security of this minimal version has been proved secure with the same bound as for the two-key version by Dunkelman *et al.* This minimal version is amazingly resistant and guarantees the same security bound, but it is not unexpected since usually the attacks look for the two keys independently and once the key K_1 is recovered, there is no security for K_2. In the following, we see that the two-key version does not improve the security since most of the attacks on the single-key can be levered to this version.

In this section, we describe new results concerning the security of the Even-Mansour scheme which has recently been the subject of many papers [13,18].

We recall the basic attacks and then, we present a basic T/M tradeoff for known plaintext attacks with better on-line complexity (Sect. 3.3) and a better T/M tradeoff for adaptive queries (Sect. 3.4). For this attack, we introduce our second algorithmic trick to discover collisions for two different functions based on the Even-Mansour construction. The main difficulty we have to solve is that when a lambda-like method is used to recover collisions, if two different functions are used, after the collision, the chain will no longer merge. To this end, we adapt the lambda-method to have parallel chains when the collision happens. Finally, we show that in the multi-user setting (Sect. 3.5) the precomputation cost can be amortized. It is possible to balance all the complexities to recover all the keys of $N^{1/3}$ users with $N^{1/3+\epsilon}$ adaptive queries to each user, a precomputation time of $N^{1/3+\epsilon}$ and the attack requires $N^{1/3+\epsilon}$ in memory and $N^{1/3+\epsilon}$ for the on-line time.

3.2   Previous Attacks on Even-Mansour

In [12], Daemen showed that the EM curve $TD = N$, is valid for a known plaintext attack at the point $(T = N/2, D = 2)$. He also gave a chosen-plaintext attack that matches the EM curve for any value of D and T and in particular at the point $(T = N^{1/2}, D = N^{1/2})$. Later, Biryukov and Wagner described a sliding attack that matches the EM curve for known-plaintext but only at the point $(T = N^{1/2}, D = N^{1/2})$. Recently, Dunkelman *et al.* introduce a new twist on the sliding attack whose complexities match the whole curve for any value of D and T using a known-plaintext attack which is exactly the result proved by Even and Mansour. Finally, Dunkelman *et al.* also provide a slidex attack on the two-key Even-Mansour scheme.

Simpler collision-based attack on the Single-Key Even-Mansour. In the single-key case a simpler attack achieves the same performance. The basic idea is to apply the Davies-Meyer construction to Π and to π. More precisely, write:

$$F_\Pi(x) = \Pi(x) \oplus x \quad \text{and } F_\pi(x) = \pi(x) \oplus x.$$

For any value of x, the equality $F_\Pi(x) = F_\pi(x \oplus K)$ is satisfied. Moreover, any collision between these two functions $F_\Pi(x) = F_\pi(y)$ indicates that $x \oplus y$ is a likely candidate for the key K.

  With this idea in mind, the problem of attacking the single key Even-Mansour scheme is reduced to the problem of finding a collision (or rather a few collisions) between F_Π and F_π. The simplest approach is simply to compute F_π on T distinct random values and F_Π on D distinct random values. When $DT \approx N$, one expects to find the required collisions.

  Moreover, this can be done in a more efficient way by using classical collision search algorithms with reduced memory. Indeed, it is possible to use Floyd's cycle finding algorithm to obtain such a solution for the special case $D = T = N^{1/2}$, without using memory. However, in this case the attack is no longer a known-plaintext attack and becomes an adaptively chosen plaintext attack.

Dunkelman, Keller and Shamir ask whether it is possible to generalize this and to find memoryless attacks using D queries to Π and N/D to π where $D \ll N^{1/2}$?

In this paper, we partially answer this question, proposing attacks that use less than $D \ll N^{1/2}$ data and memory lower than $\min(T, D)$ if we require the unkeyed queries to be precomputed. Without this requiring, we achieve a memoryless attack.

3.3   Extending the Simple Attack

Dealing with two keys Even-Mansour. A first important remark is that the simple attack on Single-Key EM can be extended to the two-key case. The idea is simply to replace the function $\pi(x) \oplus x$ by another function with similar properties. A first requirement is that the chosen function needs to be expressed by two different formulas, one based on π and the other on Π. The other requirement is that a collision on two evaluations, one of each type, should yield good candidates for the keys.

We now construct the required function and show that the simple attack on the single-key variant can be extended to two keys. We first choose a random non-zero constant δ and let:

$$F_\Pi(x) = \Pi(x) \oplus \Pi(x \oplus \delta) \text{ and } F_\pi(x) = \pi(x) \oplus \pi(x \oplus \delta).$$

We remark that $F_\Pi(x) = F_\pi(x \oplus K_1)$ and that $F_\Pi(x \oplus \delta) = F_\pi(x \oplus K_1)$ are both satisfied. As a consequence, every collision now suggests two distinct input keys $K_1 = x \oplus y$ and $K_1 = x \oplus y \oplus \delta$. Except for this detail, the attack remains unchanged. Note that once K_1 has been found, recovering K_2 is a trivial matter.

Reducing the on-line time complexity. In this section, we focus on *known-plaintext* attacks and we first show that the EM security model does not separate the on-line and off-line time complexities, as usually done in T/M/D tradeoff. It is then possible to use T/M/D tradeoff for this blockcipher design as suggested in [5] by Biryukov and Shamir.

Let us separate the on-line time denoted by T_{on} and the off-line time denoted by T_{off}. Clearly, the total time complexity T is $T_{on} + T_{off}$.

The main idea of this section is to use a different approach to find a collision between F_Π and F_π. More precisely, given a value of F_Π, we try to invert F_π on this value. If we succeed, we clearly obtain the desired collision. In order to inverse F_π, we rely on Hellman's algorithm. The T/M/D tradeoff is

$$T_{on} M^2 D^2 = N^2 \text{ and } D^2 \le T_{on} \le N.$$

In order to fully use Hellman tradeoff with multiple tables, we can use the δ in the definition of the function $F_\pi(x) = \pi(x) \oplus \pi(x \oplus \delta)$ to define different and independent functions for each table. These attacks achieve $T_{on} D \ll N$ while $TD = N$.

Using less data than memory. Despite its optimal efficiency in term of known-plaintext attack matching the EM curve, the Slidex attack presents an important drawback. Indeed, the public permutation π needs to be evaluated at points which depend on the result of the queries to the keyed Even-Mansour construction Π. As a consequence, with this attack, it is not possible to precompute the queries to π in order to improve the online time required to obtain the key to Π.

Our previous attack based on Hellman's tables no longer requires adaptive queries, however, it is less costly than the Slidex attack in term of on-line time complexity but more costly than the simple collision-based attack (which uses adaptive chosen plaintext). The goal of the next subsection is to present an attack on Π, which is based on classical collision search algorithms and works by using queries to π and Π without any cross-dependencies. However, the queries to Π are adaptive but this new attack is more flexible to perform T/M tradeoff.

3.4   Time/Memory/Data Tradeoff Attack on Even-Mansour

Attacking Even-Mansour using distinguished points methods. In order to attack Even-Mansour using a distinguished point method, we would like to construct a set of chains using the public permutation π and then find a collision with a chain obtained from the keyed permutation Π. One difficulty is that chains computing from π and from Π can never merge since they are based on different functions contrary to discrete logarithm section. We introduce here a new idea to solve this dilemma when the functions are based on the Even-Mansour construction. Let us define:

$$F_\Pi(x) = x \oplus \Pi(x) \oplus \Pi(x \oplus \delta) \text{ and } F_\pi(x) = x \oplus \pi(x) \oplus \pi(x \oplus \delta).$$

We remark that $F_\Pi(x \oplus K_1) = F_\pi(x) \oplus K_1$. As a consequence, two chains based on F_Π and F_π cannot merge, but they may become *parallel*. Indeed, using the equation $F_\Pi(x \oplus K_1) = F_\pi(x) \oplus K_1$ and let two points X and x such that $X = x \oplus K_1$, where X (resp. x) belongs to an F_Π chain (resp. x belongs to an F_π chain), the next element $Y = F_\Pi(X)$ in the F_Π chain and the next element $y = F_\pi(x)$ in the F_π chain will satisfy:

$$Y = F_\Pi(X) = F_\Pi(x \oplus K_1) = F_\pi(x) \oplus K_1 = y \oplus K_1.$$

So $Y = y \oplus K_1$, which means that Y and y satisfy the same relation as X and x, and so on. Therefore, as soon as by chance $X = x \oplus K_1$ where X is an element of an F_Π chain and x is an element of an F_π chain, the same relation remains with the subsequent points of the two chains, i.e. we get two parallel chains.

Moreover, the detection of this good event is compatible with the distinguished point method. Indeed, it suffices to define a distinguished point x as a point with a value of $\pi(x) \oplus \pi(x \oplus \delta)$ in S_0. Similarly, for chains constructed by using F_Π, we define a distinguished point X as a point with a value of $\Pi(X) \oplus \Pi(X \oplus \delta)$ in S_0. Now if $X = x \oplus K_1$ and x is a distinguished point in a π chain, then since

$$\Pi(X) \oplus \Pi(X \oplus \delta) = \pi(X \oplus K_1) \oplus \pi(X \oplus K_1 \oplus \delta) = \pi(x) \oplus \pi(x \oplus \delta),$$

the point X is also a distinguished point in the Π chain, and therefore $X \oplus x$ gives a candidate for K_1. Since the values $\pi(x) \oplus \pi(x \oplus \delta)$ and $\Pi(X) \oplus \Pi(X \oplus \delta)$ are needed to compute the next element in the chains, using this definition does not add any extra cost for distinguished point detection. The important point, is that for a parallel chain based on F_Π, a point $X = x \oplus K_1$ corresponds to a distinguished point x if and only if $\Pi(X) \oplus \Pi(X \oplus \delta)$ is in S_0.

An important difference compared to the classical search for collisions is that we do not need to backtrack to the beginning of the chains and identify where the chains merge. Indeed, seeing parallel distinguished points suffices to get candidates values for K_1.

Analysis of the attack with precomputation. Since there is a clear symmetry between the keyed and unkeyed queries, we may assume that the number of unkeyed queries T is larger than the number of keyed queries D. Let B_T the number of unkeyed chains to increase the probability of a collision between keyed and unkeyed chains. Moreover, this is the most reasonable scenario, since keyed queries are usually the most constrained resource. In this case, we need to choose the expected length ℓ of the chains we are going to construct and B_T that satisfy the following relations:

$$T = \ell \cdot B_T \quad \text{and} \quad N = B_T \ell^2.$$

Thus, $\ell = N/T$ and $B_T = T^2/N$. The required memory to store those chains is of size $O(B_T)$.

After terminating the computation of the unkeyed chains, we can turn to the keyed side. On this side, we want to perform about $D = N/T$ evaluations of the function. Since $D = \ell$, this means that we compute a single keyed chain and expect it to (parallel) collide with an unkeyed chain.

We are interested in values for M such that $M < D$. Consequently, as $M = T/D = N/D^2$, we have $N < D^3$. Let us consider $N^{1/3} < D = N^\alpha < N^{1/2}$. For example, if $D = N^{2/5}$ and $T = N^{3/5}$, then $M = N^{1/5}$ is much smaller than $N^{2/5}$. This attack requires a number of data $D \ll N^{1/2}$ and despite this attack is not memoryless (as in the open problem), the memory is less than the data.

Relaxing the precomputation requirement. Another alternative[4] is to perform the same attack while computing the keyed queries before the unkeyed ones. In this case, since there is a single keyed chain to be stored, we can achieve the attack using a constant amount of memory. Moreover, this variation works for any $D = N^\alpha \le N^{1/2}$ using $T = N/D$.

3.5   Attacks in the Multi-user Setting

In the multi-user setting, we assume that L different users are all using the Even-Mansour scheme based on the same public permutation π, with each user

---

[4] We thank an anonymous reviewer of Asiacrypt 2014 for pointing this out.

having its own key[5], chosen uniformly at random and independently from the keys of the other users.

Of course, the attack from Section 3.4 can be easily applied in this context. Depending on the exact goal of the cryptanalysis, we have two main options:

1. If the goal is to recover the key of all users, the previous attack can be applied by repeating the D key-dependent queries for each user, while amortizing the T unkeyed queries across users. A typical case is to consider $L = N^{1/3}$ users, to perform $T = N^{2/3+\epsilon}$ unkeyed queries ($N^{1/3+\epsilon}$ chains of $N^{1/3}$ queries, memory $N^{1/3}$). For each new user, we need $N^{1/3+\epsilon}$ key-dependent queries. As a consequence, the amortized cost per user (up to constant factors $c_0 = 20$) is $N^{1/3+\epsilon}$ queries of each type and the required memory also is $N^{1/3}$.
2. If the goal of the cryptanalyst is to obtain at least one user key among all the users, it suffices to split the D key-dependent queries arbitrarily across the users.

However, we present in this section a much more efficient tradeoff in the multi-user setting. This tradeoff becomes possible without precomputation in $N^{2/3}$, but by distributing the unkeyed queries among the users and by reusing the graph algorithmic idea of the section 2. For this, we construct a graph whose vertices are labelled by the users. Whenever we obtain a collision $F_\Pi^{(i)}(x) = F_\Pi^{(j)}(y)$ for users i and j, we add an edge between the corresponding vertices labelled with $x \oplus y$ which is expected equal to $K^{(i)} \oplus K^{(j)}$. Note that this indicates that we know the exclusive-or of the first keys of the two users.

If we have L vertices and $cL/2$ randomly edges with $c = 4$, there is a giant component whose size is 98% of the points, and with $cL \ln L$, all the points are in this component with overwhelming probability. Consequently, we obtain the exclusive-or of the first keys for an arbitrary pair of users. To conclude the attack, it suffices to find a single collision between any of the users functions F_Π of the large connected component and the unkeyed function F_π to reveal all the keys of these users.

Algorithm Description.

1. Create a constant number $c/2$ of chains for each user up to a distinguished point.
2. Sort the distinguished points.
3. Bring together the distinguished points into subsets, where we test whether the key candidate is really the good one. It is indeed easy to check with a few more queries if the xor of two keys is correct.
4. Construct the giant component and expect that the public user (the user with the unkeyed function), lies in this giant component. To this end, we initially begin with the set of reachable users containing only the public user. Then, we add to this set all the users that are in a group where a reachable user is present. At some point, the reachable set is stable and we stop.

---

[5] Or key-pair depending on whether we are considering the single or dual key scheme.

5. From the public user, we cross over the giant component and determine the keys of each user.

The first step requires $cL\ell/2$ data and time $O(c\ell)$ on average per user where ℓ is the average length of the chains. Then, the remaining steps are performed in time linear in the number of users L. Typical parameters are: for an arbitrary small positive constant c, we expect with $N^{1/3}$ users, $c \cdot N^{1/3}$ queries per user and $N^{1/3}$ unkeyed queries, to recover almost all the $N^{1/3}$ keys with overwhelming probability. If we want to recover *all* users, we need to have $L \ln L = cN^{1/3} \ln N = N^{1/3+\epsilon}$ edges (instead of $cL/2$) to connect all components according to the coupon collector's problem.

Analysis of the attack. We want to use results from graph theory to prove the correctness of our algorithm, this means that we have to prove that the assumptions of the giant component theorem are satisfied. We have to show that we construct of a random graph according to the Erdös-Rényi model of random graphs, in which each possible edge connecting pairs of a given set of L vertices is present, independently of the other edges, with probability p. In this case, we know that with this model of random graph, if the number of edges $c.L/2$ is larger than the number of vertices L, there is with high probability a single giant component, with all other components having size $O(\log L)$ according to [6].

Consequently, we need to prove that we construct a random graph and that the edges are added *independently* of each others. We will define an idealized version of the attack and we will show that the attack works in this version. Then, we will prove that the idealized version and the attack are equivalent using simulation argument.

In the idealized model, the simulator randomly chooses L keys $K_1, \ldots, K_L$ uniformly at random. Then it iterates the functions $F_\Pi^{(i)}(x) = K_i \oplus F_\pi(x \oplus K_i)$ until $x_\ell \oplus K_i \in S_0$, where S_0 is the set of pairs containing a distinguished point d_i and an identificator of this point $id(d_i)$. The identificators are unique, which means that we do not have collision on them. Finally, the simulator reveals the identificator of the point $x_\ell \oplus K_i$ and the point x_ℓ. The value K_i cannot be recovered from the information that the simulator returns.

To show that the attack works in this ideal model, we just have to see that if two users have the same identificator, then $x_\ell \oplus K_i = x_{\ell'} \oplus K_j$ and therefore $x_\ell \oplus x_{\ell'} = K_i \oplus K_j$ which is the same information as in the real attack.

Now, we will prove that the simulator does not need to know $F_\Pi^{(i)}$ and can simulate the information by only using the public random function F_π and that the distribution of its outputs is indistinguishable from the idealized model. The simulator generates at random L random keys for the EM scheme. For each key, we will show that the pairs distinguished point/identificator can be generated only using F_π. Indeed, x_ℓ the ℓth iteration of $F_\Pi^{(i)}$ with key K_i from the value x_0 is the value $K_i \oplus x_\ell$ and this value is also the result of the iteration of the public function F_π from the value $x_0 \oplus K_i$. Consequently, to generate the pairs (distinguished point, identificator), the simulator can compute $(x_\ell \oplus K_i, id(x_\ell))$ without interacting with the users. As in this last case, the pairs are generated at random without interacting and knowing the function and since the function F_π are random, the edges in the graph are added at random and independently of each others and so that the graph is a random graph according to the Erdös-Rényi graph model.

Experimental results. We implement the previous attacks on an Even-Mansour cryptosystem using the DES with a fixed key and $n = 64$. We simulate 2^{22} users and for each user we create 8 chains (80 for the public user). We use distinguished points containing 21 zeroes and so the expected length is 2^{21} on average. We bound the length of the chains to 2^{24}, this means that if we remove the chain if we have not seen a distinguished point after 2^{24} evaluations. In all, we have generated $33,543,077$ chains ($2^{25} = 33,554,432$, it misses the abandoned chains) and the number of groups containing at least two parallel chains is $4,109,961$. Experimentally, the size of the giant component contains $3,788,059$ users (among the $4,194,304$) and so we can deduce the keys of 90% of the users. This result is what is expected from theory since the number of vertices in this experiment is below the number of nodes. The 98% that is previously given as result in section 3.5, would require twice as many vertices.

The time to generate the chains is 1600 sec using 4096 cores in parallel and the analysis of the graph requires a few minutes on a standard PC.

4   Attacks on the PRINCE Cipher in the Multi-user and Classical Setting

PRINCE is a lightweight block cipher published at ASIACRYPT 2012 [7]. It is based on the FX construction [16] which is actually an Even-Mansour like construction. PRINCE has been the interest of many cryptanalysts [9,23,15] who attack either the full cipher, or its reduced version.

The designers of PRINCE claim that its security is ensured up to 2^{127-n} operations when an adversary acquires 2^n plaintext/ciphertext pairs. This bound has been reduced in [15] to 2^{126} operations with a single plaintext/ciphertext pair. After a brief presentation of PRINCE, we describe a generic attack in the multi-user setting that allow to recover the key of a pair of users in a set of 2^{32} users with complexity 2^{64} computations. The identification of the pair of users uses the idea similar to the attack on Even-Mansour. However, details are different since PRINCE is not an Even-Mansour scheme as the internal permutation uses a secret key. Finally, we present another generic attack in the classical model that after a precomputation of 2^{96} time and 2^{64} in memory, allows to recover the key of every single user in time 2^{32}. Both attacks work for all rounds of PRINCE.

4.1   Brief Description of PRINCE

PRINCE [7] uses a 64-bit block and a 128-bit key which is split into two equal parts of 64 bits, *i.e.* $k = k_0 \| k_1$. In order to extend the key to 192 bits it uses the mapping $k = (k_0 \| k_1) \rightarrow (k_0 \| k_0' \| k_1)$ where k_0' is derived from k_0 by using a linear function L':

$$L'(k_0) = (k_0 \ggg 1) \oplus (k_0 \gg 63),$$

where $\gg$ denotes the right shift and $\ggg$ the rotation of a 64-bit word. While subkeys k_0 and k_0' are used as input and output whitening keys, the 64-bit key k_1 is used for the 12-round internal block cipher which is called PRINCE_{core}. For simplicity, we refer to it as the core of PRINCE or simply the core function and we denote it by $Pcore$. So every plaintext P is transformed into the corresponding ciphertext C by using the function $E_k(P) = k_0' \oplus Pcore_{k_1}(P \oplus k_0)$ where $Pcore$ uses the key k_1 (see Fig.1).

Fig. 1. Structure of PRINCE

The core function consists of a key k_1 addition, a round constant (RC_0) addition, five forward rounds, a middle round, five backward rounds and finally a round constant (RC_{11}) and a key k_1 addition. The full schedule of the core is shown in Fig. 2.

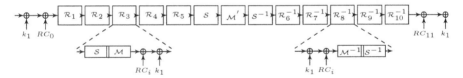

Fig. 2. Structure of the core of PRINCE

Each forward round of the core is composed by a 4-bit Sbox layer (S), a linear layer (64×64 matrix M), an addition of a round constant RC_i for $i \in \{1, \ldots, 5\}$ and the addition of the key k_1. The linear M layer is defined as $M = SR \circ M'$ where SR is the following permutation

| 0 | 1 | 2 | 3 | 4 | 5 | 6 | 7 | 8 | 9 | 10 | 11 | 12 | 13 | 14 | 15 | $\longrightarrow$ | 0 | 5 | 10 | 15 | 4 | 9 | 14 | 3 | 8 | 13 | 2 | 7 | 12 | 1 | 6 | 11 |

The M' layer, which is only used in the middle rounds, can be seen as a mirror in the middle of the core as the 5 backward rounds are defined as the inverse of the 5 forward rounds.

In every RC_i-add step, a 64-bit round constant is XORed with the state. It should be noted that $RC_i \oplus RC_{11-i} = \alpha = \texttt{0xc0ac29b7c97c50dd}$ for all $0 \le i \le 11$. From this, but also from the fact that the matrix M' is an involution, we can perform the decryption function of PRINCE by simply performing the encryption procedure with inverse order of keys k_0 and k_0' and by using the key $k_1 \oplus \alpha$ instead of k_1. That means, that for any key $(k_0 \| k_0' \| k_1)$, we have $D_{(k_0 \| k_0' \| k_1)}(\cdot) = E_{(k_0' \| k_0 \| k_1 \oplus \alpha)}(\cdot)$. This property is called the α-reflection property of PRINCE.

4.2   Attack on PRINCE in the Multi-user Setting

In the multi-user setting, we assume that we have L different users which are all using the block cipher PRINCE. Each user U_i with $0 \leq i < L$, chooses her key $k^{(i)} = k_0^{(i)} \| k_1^{(i)}$ at random and independently from all the other users. In order to attack PRINCE using the distinguished point method, we first construct a set of chains for every user using the function of PRINCE. For this, we use the function defined as follows:

$$F_{k_0^{(i)}, k_0'^{(i)}, k_1^{(i)}}(x) = x \oplus \text{PRINCE}_{k_0^{(i)}, k_0'^{(i)}, k_1^{(i)}}(x) \oplus \text{PRINCE}_{k_0^{(i)}, k_0'^{(i)}, k_1^{(i)}}(x \oplus \delta)$$

where δ is an arbitrary but fixed non zero constant. The key $k_0'^{(i)}$ vanishes from the equation and the function F thus takes the following form:

$$F_{k_1^{(i)}}(x) = x \oplus Pcore_{k_1^{(i)}}(x \oplus k_0^{(i)}) \oplus Pcore_{k_1^{(i)}}(x \oplus k_0^{(i)} \oplus \delta).$$

For every user U_i, we create one encryption $(\mathcal{E})$ chain and one decryption $(\mathcal{D})$ chain which are both based on the function F defined above. $\mathcal{E}$ uses the encryption function of PRINCE whereas $\mathcal{D}$ uses the decryption function. And so, for the user U_i, we define functions $\mathcal{E}$ and $\mathcal{D}$ as follows:

$$\mathcal{E}_{k_0^{(i)}, k_1^{(i)}}(x_j^{(i)}) = x_{j+1}^{(i)} = x_j^{(i)} \oplus Pcore_{k_1^{(i)}}(x_j^{(i)} \oplus k_0^{(i)}) \oplus Pcore_{k_1^{(i)}}(x_j^{(i)} \oplus k_0^{(i)} \oplus \delta)$$

$$\begin{aligned}
\mathcal{D}_{k_0'^{(i)}, k_1^{(i)} \oplus \alpha}(y_j^{(i)}) &= y_{j+1}^{(i)} \\
&= y_j^{(i)} \oplus Pcore_{k_1^{(i)} \oplus \alpha}(y_j^{(i)} \oplus k_0'^{(i)}) \\
&\oplus Pcore_{k_1^{(i)} \oplus \alpha}(y_j^{(i)} \oplus k_0'^{(i)} \oplus \delta).
\end{aligned}$$

Let us define:

$$f^{\mathcal{E}} = Pcore_{k_1^{(i)}}(x_j^{(i)} \oplus k_0^{(i)}) \oplus Pcore_{k_1^{(i)}}(x_j^{(i)} \oplus k_0^{(i)} \oplus \delta) \text{ and}$$

$$f^{\mathcal{D}} = Pcore_{k_1^{(i)} \oplus \alpha}(y_j^{(i)} \oplus k_0'^{(i)}) \oplus Pcore_{k_1^{(i)} \oplus \alpha}(y_j^{(i)} \oplus k_0'^{(i)} \oplus \delta).$$

We create encryption chains until $f^{\mathcal{E}}$ reaches a distinguished point (resp. decryption chains until $f^{\mathcal{D}}$ reaches a distinguished point). We search for a collision between the encryption and the decryption chain.

Let us consider two users, U_1 and U_2. Whenever the chains $\mathcal{E}_{k_0^{(1)}, k_1^{(1)}}(x^{(1)})$ and $\mathcal{D}_{k_0^{(2)}, k_1^{(2)} \oplus \alpha}(y^{(2)})$ arrive at the same distinguished point, we suspect that these two chains have become parallel. As the core of PRINCE is only parametrized by the key k_1, when we arrive at the same distinguished point we obtain a probable collision between keys $k_1^{(1)}$ and $k_1^{(2)} \oplus a$ used in $Pcore$. However, we must verify that this is a real collision and not just a random incident. For this, we verify that next points of $f^{\mathcal{E}}$ and $f^{\mathcal{D}}$ after reaching a distinguished point, continue to remain equal. If we obtained a real collision we know that:

$$k_1^{(1)} = k_1^{(2)} \oplus a.$$

This indicates that $x^{(1)} \oplus y^{(2)}$ is expected equal to $k_0^{(1)} \oplus k_0^{'(2)}$. It is obvious that since $k_1^{(1)} = k_1^{(2)} \oplus a$ we will also have $k_1^{(1)} \oplus a = k_1^{(2)}$. This indicates that we also know $k_0^{'(1)} \oplus k_0^{(2)}$.

Thus, we have:

$$k_0^{(1)} \oplus k_0^{'(2)} = A \quad \text{and} \quad k_0^{'(1)} \oplus k_0^{(2)} = B \quad (*).$$

Let $\{a_{63}, \ldots, a_0\}$ be the representation of the bits of $k_0^{(1)}$ and $\{b_{63}, \ldots, b_0\}$ the representation of bits of $k_0^{(2)}$. As, from the definition of PRINCE, $k_0' = (k_0 \ggg 1) \oplus (k_0 \gg 63)$, we have that:

$$k_0^{'(1)} = \{a_0, a_{63}, \ldots, a_2, a_1 \oplus a_{63}\} \quad \text{and} \quad k_0^{'(2)} = \{b_0, b_{63}, \ldots, b_2, b_1 \oplus b_{63}\}.$$

From $(*)$, we construct the system:

$$\{a_{63}, \ldots, a_0\} \oplus \{b_0, b_{63}, \ldots, b_2, b_1 \oplus b_{63}\} = \{A_{63}, \ldots, A_0\}$$
$$\{b_{63}, \ldots, b_0\} \oplus \{a_0, a_{63}, \ldots, a_2, a_1 \oplus a_{63}\} = \{B_{63}, \ldots, B_0\}$$

As this is an inversible linear system, we can easily find $k_0^{(1)}$ and $k_0^{(2)}$. Note that once $k_0^{(i)}$ has been found, recovering $k_1^{(i)}$ can be done with an exhaustive search whose cost is 2^{64}.

Analysis of the attack. Once the computation of a chain is finished we have to store $(x_{\ell-1}, d, d+1)$ where d is the distinguished point, $x_{\ell-1}$ is the point before the chain reaches a distinguished point and $d + 1$ is the point after the chain reached a distinguished point. We need to store $x_{\ell-1}$ as we have to test if the found collision is useful and we also need to store $d + 1$ to test if it is a real collision. If not, the search must continue.

As mentioned, PRINCE uses a 128-bit key which is split into two 64-bit parts, *i.e.* $k = k_0 \| k_1$. The attack consists in identifying and recovering all key material of a pair of users i and j for whom $k_1^{(i)} = k_1^{(j)} \oplus a$. We expect to find a collision $k_1^{(i)} = k_1^{(j)} \oplus a$ between two different users with high probability when the number of users will be at least 2^{32}. So the attack uses a set of 2^{32} users and for each one we create 2 chains (encryption and decryption chain). The cost per user is 2^{32} operations and the total cost for recovering the keys k_0 of 2 users is approximately 2^{64} operations. For recovering k_1, the cost of the exhaustive search is 2^{64}. So in total, we can deduce both k_0 and k_1 in 2^{65} operations.

4.3   Attack in the Classical Model

We show in this section that a classical attack that also uses the distinguished points technique can also be possible. For this, we will create encryption chains from the function $\mathcal{E}$ defined in section 4.2.

Precalculation. In the first phase of the attack, we aim to create encryption chains for every possible key $k_1^{(i)}$ with $0 \leq i < 2^{64}$. More specifically, for every possible $k_1^{(i)}$, we set $k_0^{(i)} = 0$ and we create for every (i) a chain $\mathcal{S}_i$ from the function $\mathcal{E}$ with length 2^{32}. We store all chains $\mathcal{S}_i$.

Attack. Now, our purpose is to find a collision with one of the chains created with the zero key $k_0^{(i)}$. For this, for a random starting point x_0 and for keys k_0 and k_1 we will calculate an encryption chain $\mathcal{T}$ from the function $\mathcal{E}$. The chain $\mathcal{T}$ will collide with high probability with one of the chains $\mathcal{S}_i$. As described in previous section 4.2, when we detect a collision between two distinguished points, we know that the chains had become parallel and so we obtain $k_0^{(i)} \oplus k_0$. As the key $k_0^{(i)} = 0$, we finally obtain the unknown k_0.

Analysis of the attack. For the precalculation phase, for every 2^{64} possible keys we calculate a chain with length 2^{32} and so our complexity is equal to 2^{96}. As we need to store all chains, the precalculation phase has also a cost of 2^{64} in memory. However, once the first phase is over, the attacker can perform the attack in only 2^{32} operations as she has to calculate only one chain. So, the total cost of the attack is 2^{96}. The proposed attack satisfies $DT = 2^{128}$ as $D = 2^{32}$ and $T = 2^{96}$. This attack does not improve the complexity of PRINCE given in [7] and [15]. However, in our case, T is not the on-line time complexity as it corresponds to the precalculation phase of the attack. Thus, in our attack, we have $DT_{on} = 2^{64}$.

5   Conclusion

In this paper, we have presented new tradeoffs for public-key and symmetric-key cryptosystems in the multi-user setting. We have introduced some algorithmic tools for collision-based attacks using the distinguished point technique. The first tool allows to look for the discrete logarithm of L users in parallel using only a $\widetilde{O}(\sqrt{L})$ penalty using random graph process behaviour. The second tool allows to achieve key-recovery of Even-Mansour and related ciphers and is a novel lambda technique to find collisions when two different functions are involved. For the Even-Mansour cipher, we show new tradeoffs that partially solve an open problem due to Dunkelman *et al.* and we propose an analysis in the multi-user setting. Finally, for the PRINCE cipher, we show generic attacks that improve the best published results in the sense that our time complexity corresponds to a precomputation phase and not to an on-line phase. This last result could also be adapted to similar ciphers such as DESX and would also improve on the best previous attack.

References

1. Bernstein, D.J., Lange, T.: Computing Small Discrete Logarithms Faster. In: Galbraith, S., Nandi, M. (eds.) INDOCRYPT 2012. LNCS, vol. 7668, pp. 317–338. Springer, Heidelberg (2012)
2. Bernstein, D.J., Lange, T.: Non-uniform Cracks in the Concrete: The Power of Free Precomputation. In: Sako, K., Sarkar, P. (eds.) ASIACRYPT 2013, Part II. LNCS, vol. 8270, pp. 321–340. Springer, Heidelberg (2013)
3. Biham, E.: How to decrypt or even substitute DES-encrypted messages in 2^{28} steps. Inf. Process. Lett. 84(3), 117–124 (2002)
4. Biryukov, A., Mukhopadhyay, S., Sarkar, P.: Improved Time-Memory Trade-Offs with Multiple Data. In: Preneel, B., Tavares, S. (eds.) SAC 2005. LNCS, vol. 3897, pp. 110–127. Springer, Heidelberg (2006), http://dx.doi.org/10.1007/11693383_8
5. Biryukov, A., Shamir, A.: Cryptanalytic Time/Memory/Data Tradeoffs for Stream Ciphers. In: Okamoto, T. (ed.) ASIACRYPT 2000. LNCS, vol. 1976, pp. 1–13. Springer, Heidelberg (2000)
6. Bollobás, B.: Random Graphs, 2nd edn. Cambridge studies in advanced mathematics (2001)
7. Borghoff, J., Canteaut, A., Güneysu, T., Kavun, E.B., Knezevic, M., Knudsen, L.R., Leander, G., Nikov, V., Paar, C., Rechberger, C., Rombouts, P., Thomsen, S.S., Yalçın, T.: PRINCE – A Low-Latency Block Cipher for Pervasive Computing Applications. In: Wang, X., Sako, K. (eds.) ASIACRYPT 2012. LNCS, vol. 7658, pp. 208–225. Springer, Heidelberg (2012)
8. Canteaut, A., Fuhr, T., Gilbert, H., Naya-Plasencia, M., Reinhard, J.-R.: Multiple Differential Cryptanalysis of Round-Reduced PRINCE (Full version). IACR Cryptology ePrint Archive, 2014:89 (2014)
9. Canteaut, A., Naya-Plasencia, M., Vayssière, B.: Sieve-in-the-middle: Improved MITM attacks. In: Canetti, R., Garay, J.A. (eds.) CRYPTO 2013, Part I. LNCS, vol. 8042, pp. 222–240. Springer, Heidelberg (2013)
10. Chatterjee, S., Menezes, A., Sarkar, P.: Another Look at Tightness. In: Miri, A., Vaudenay, S. (eds.) SAC 2011. LNCS, vol. 7118, pp. 293–319. Springer, Heidelberg (2012)
11. Costello, K.P., Vu, V.H.: The rank of random graphs. Random Structures & Algorithms 33(3), 269–285 (2008)
12. Daemen, J.: Limitations of the Even-Mansour Construction. In: Matsumoto, T., Imai, H., Rivest, R.L. (eds.) ASIACRYPT 1991. LNCS, vol. 739, pp. 495–498. Springer, Heidelberg (1993)
13. Dunkelman, O., Keller, N., Shamir, A.: Minimalism in Cryptography: The Even-Mansour Scheme Revisited. In: Pointcheval, D., Johansson, T. (eds.) EUROCRYPT 2012. LNCS, vol. 7237, pp. 336–354. Springer, Heidelberg (2012)
14. Even, S., Mansour, Y.: A construction of a cipher from a single pseudorandom permutation. In: Imai, H., Rivest, R., Matsumoto, T. (eds.) ASIACRYPT 1991. LNCS, vol. 739, pp. 210–224. Springer, Heidelberg (1993)
15. Jean, J., Nikolic, I., Peyrin, T., Wang, L.: S. Wu Security Analysis of PRINCE. In: Fast Software Encryption - 20th International Workshop, FSE 2013, Singapore, March 11-13, 2013. Revised Selected Papers, pp. 92–111 (2013)
16. Kilian, J., Rogaway, P.: How to Protect DES Against Exhaustive Key Search (an Analysis of DESX). J. Cryptology 14(1), 17–35 (2001)

17. Kuhn, F., Struik, R.: Random walks revisited: Extensions of pollard's rho algorithm for computing multiple discrete logarithms. In: Vaudenay, S., Youssef, A.M. (eds.) SAC 2001. LNCS, vol. 2259, pp. 212–229. Springer, Heidelberg (2001)
18. Lampe, R., Patarin, J., Seurin, Y.: An Asymptotically Tight Security Analysis of the Iterated Even-Mansour Cipher. In: Wang, X., Sako, K. (eds.) ASIACRYPT 2012. LNCS, vol. 7658, pp. 278–295. Springer, Heidelberg (2012)
19. Lee, H.T., Cheon, J.H., Hong, J.: Accelerating ID-based Encryption based on Trapdoor DL using Pre-computation. Cryptology ePrint Archive, Report 2011/187 (2011), http://eprint.iacr.org/
20. Menezes, A.: Another Look at Provable Security. In: Pointcheval, D., Johansson, T. (eds.) EUROCRYPT 2012. LNCS, vol. 7237, pp. 8–8. Springer, Heidelberg (2012)
21. Pollard, J.M.: Kangaroos, Monopoly and Discrete Logarithms. J. Cryptology 13(4), 437–447 (2000)
22. Quisquater, J.-J., Delescaille, J.-P.: How Easy Is Collision Search. New Results and Applications to DES. In: Brassard, G. (ed.) CRYPTO 1989. LNCS, vol. 435, pp. 408–413. Springer, Heidelberg (1990)
23. Soleimany, H., Blondeau, C., Yu, X., Wu, W., Nyberg, K., Zhang, H., Zhang, L., Wang, Y.: Reflection Cryptanalysis of PRINCE-Like Ciphers. In: Fast Software Encryption - 20th International Workshop, FSE 2013, Singapore, March 11-13, 2013. Revised Selected Papers, pp. 71–91 (2013)
24. van Oorschot, P.C., Wiener, M.J.: Parallel Collision Search with Cryptanalytic Applications. J. Cryptology 12(1), 1–28 (1999)

Cryptanalysis of Iterated Even-Mansour Schemes with Two Keys

Itai Dinur[1,*], Orr Dunkelman[2,3,**],
Nathan Keller[3,4,***], and Adi Shamir[4]

[1] Département d'Informatique, École Normale Supérieure, Paris, France
[2] Computer Science Department, University of Haifa, Israel
[3] Department of Mathematics, Bar-Ilan University, Israel
[4] Computer Science department, The Weizmann Institute, Rehovot, Israel

Abstract. The iterated Even-Mansour (EM) scheme is a generalization of the original 1-round construction proposed in 1991, and can use one key, two keys, or completely independent keys. In this paper, we methodically analyze the security of all the possible iterated Even-Mansour schemes with two n-bit keys and up to four rounds, and show that none of them provides more than n-bit security. Our attacks are based on a new cryptanalytic technique called *multibridge* which splits the cipher to different parts in a novel way, such that they can be analyzed independently, exploiting its self-similarity properties. After the analysis of the parts, the key suggestions are efficiently joined using a meet-in-the-middle procedure.

As a demonstration of the multibridge technique, we devise a new attack on 4 steps of the LED-128 block cipher, reducing the time complexity of the best known attack on this scheme from 2^{96} to 2^{64}. Furthermore, we show that our technique can be used as a generic key-recovery tool, when combined with some statistical distinguishers (like those recently constructed in reflection cryptanalysis of GOST and PRINCE).

Keywords: Cryptanalysis, meet-in-the-middle attacks, multibridge attack, iterated Even-Mansour, LED-128.

1 Introduction

Most block ciphers (such as the AES) have an iterated structure which alternately XOR's a secret key and applies some publicly known permutation (typically consisting of S-boxes and linear transformations) to the internal state. A generic way to describe such a scheme is to assume that the permutations are randomly chosen, with no weaknesses which can be exploited by the cryptanalyst.

* Some of the work presented in this paper was done while the first author was a postdoctoral researcher at the Weizmann Institute, Israel.
** The second author was supported in part by the German-Israeli Foundation for Scientific Research and Development through grant No. 2282-2222.6/2011.
*** The third author was supported by the Alon Fellowship.

P. Sarkar and T. Iwata (Eds.): ASIACRYPT 2014, PART I, LNCS 8873, pp. 439–457, 2014.

This approach has several advantages: First of all, this is a very clean construction with great theoretical appeal. In addition, we can use the randomness of the permutation in order to prove lower bounds on the complexity of all possible attacks, something we cannot hope to achieve when we instantiate the scheme with a particular choice of the permutation. Finally, any new generic attack on block ciphers with this general form can have broad practical applicability.

At Asiacrypt 1991 [11], Even and Mansour defined and analyzed the simplest example of such a block cipher, which consists of a single public permutation and two independently chosen secret keys XOR'ed before and after the permutation. We call such a scheme a 1-round 2-key Even-Mansour (EM) scheme. In their paper, Even and Mansour showed that in any attack on this scheme that succeeds with high probability, $TD \geq 2^n$. This implies that any attack on the scheme has overall complexity (i.e., the maximal complexity among the time,[1] memory and data complexities) of at least $2^{n/2}$. In such a case, we say that the *security* of the scheme is $2^{n/2}$, or $n/2$ bits.[2] At Eurocrypt 2012 [10], a matching upper bound in the known plaintext attack model was proved, and thus the security of this scheme is now fully understood.

Since the security provided by a 1-round 2-key EM is much smaller than the 2^{2n} time complexity of exhaustive key search, multiple papers published in the last couple of years had studied the security of iterated EM schemes with more than one round (e.g., [2,4,9,18,21,23]). These schemes differ not only in their number of rounds, but also in the number of keys they use and in the order in which these keys are used in the various rounds. This is somewhat analogous to the study of the security of generic Feistel structures with various numbers of rounds, which led to several fundamental results in theoretical cryptography in the last two decades (e.g., how to construct pseudo-random permutations from pseudo-random functions, and how many queries are required in order to distinguish them from truly random permutations [19,24]).

In this paper, we study the security of iterated EM constructions using two independent keys. As the security of the 1-round variant is already determined to be $2^{n/2}$, and as it is easy to see that a 2-round variant supplies security of at most 2^n, we analyze all 3-round and 4-round variants with two keys. We show that for any possible ordering of the two keys, all the r-round variants with $r \leq 4$ provide security of at most 2^n (compared to exhaustive key search which requires 2^{2n} time). Furthermore, for all such variants[3] we obtain a complete tradeoff curve of $DT = 2^{2n}$ in the known plaintext attack model.

---

[1] We define security in the computational model, which calculates the time complexity according to the number of operations that the attacker performs. This model is different from the information theoretical model (used, for example, in [4]), which only considers the number of queries to the internal permutations of the primitive.

[2] Note that, unlike the tradeoff attacks described in Hellman's paper [14], the overall complexity of an attack takes into account all attack stages. In particular, we do not allow any free preprocessing stage.

[3] Not including some weak variants, for which an attack of time complexity 2^n can be obtained given only 2 plaintext-ciphertext pairs (i.e., the unicity bound).

Since several concrete proposals for block ciphers use a relatively small number of fairly complex rounds, our theoretical analysis has immediate practical applications. For example, we can use our results in order to compare the best achievable security of schemes with various numbers of rounds and key schedules, and thus to guide the design of future schemes. More surprisingly, we can use our new generic attacks in order to improve by a large margin the running time of the best known attacks on the extensively studied lightweight block cipher LED-128, without even looking at its internal structure.

LED-128 [13] is a typical example of an iterated EM scheme. It is a 64-bit block cipher that uses two unrelated 64-bit keys, which are alternately XOR'ed in consecutive rounds. Since its publication at CHES 2011, reduced variants of LED-128 have been extensively analyzed, and in particular the 4-step[4] variant (reduced from the full 12) was analyzed in 3 consecutive papers at ACISP 2012 [16], Asiacrypt 2012 [21] and FSE 2013 [23], using a variety of cryptanalytic techniques (see Table 1).

Table 1. Attacks on 4-Step LED-128

Reference	Generic[†]	Data[††]	Time	Memory	**Security**
[16]	No	2^{16} CP	2^{112}	2^{16}	2^{112}
[21]	Yes	2^{64} KP	2^{96}	2^{64}	2^{96}
[23]	Yes	$D \leq 2^{32}$ KP	$2^{128}/D$	D	2^{96}
This paper	Yes	$D \leq 2^{64}$ KP	$2^{128}/D$	D	2^{64}

[†] "Generic" stands for an attack independent of the actual step function.
[††] The data complexity is given in chosen plaintexts (CP), or in known plaintexts (KP).

The first attack on 4-step LED-128 is described in [16]. The attack combines the splice-and-cut technique [3] with a meet-in-the-middle attack which is based on specific properties of the LED permutation. It has a time complexity of $T = 2^{112}$, and requires $D = 2^{16}$ chosen plaintext-ciphertext pairs. The second analysis of 4-step LED-128 is given in [21] and is applicable to all 4-round EM schemes with 2 alternating keys. When applied to 4-step LED-128, it has a reduced time complexity of $T = 2^{96}$ (compared to $T = 2^{112}$ of the attack of [16]), but it requires the full code-book of $D = 2^{64}$ plaintext-ciphertext pairs. The attack uses a technique related to Merkle and Hellman's attack on two-key triple-DES (2K3DES) [22], in combination with Daemen's chosen plaintext attack of EM [6]. Finally, the currently best known attack on 4-step LED-128 is a known plaintext attack given in [23]. The attack uses an extension of the SlideX attack [10] in order to obtain a flexible tradeoff curve of $TD = 2^{128}$ for any $D \leq N^{1/2}$.

---

[4] In the design of LED, the term "step" is used in order to describe what we refer to as a "round" of an iterated EM scheme.

By using our new generic attack on 4-round EM with alternating keys, we can extend the tradeoff curve all the way to $D = N$. We can thus reduce the time complexity of the best known attack on 4-step LED-128 by a large factor of 2^{32}, from the totally impractical $T = 2^{96}$ to a more practical $T = 2^{64}$. We note that when considering much smaller improvements over exhaustive search, attacks on up to 8 rounds of LED-128 have been published in [9].

In order to obtain our improved generic attacks, we had to develop a new cryptanalytic technique. The new technique stems from the dissection technique [7] and from the splice-and-cut technique [3], but has also additional features. Like the dissection technique, it divides the cipher into several parts treated independently by enumerating over an intermediate value, but unlike dissection, the parts are not consecutive but rather nested. In addition, as the splice-and-cut technique, the new attack takes advantage of "splicing" (or connecting) two ends of the cipher together. However, in the original splice-and-cut technique, the plaintexts and ciphertexts were "spliced" together, and as a result it was essentially a chosen plaintext attack. On the other hand, in our attack we bridge (or connect) together intermediate encryption values, and thus our attack does not have this constraint and can use known plaintexts. Once we connect a pair of intermediate encryption values using a bridge, we use a self-similarity property of the cipher in order to connect another pair of intermediate encryption values using another bridge. Thus, as our attack bridges between multiple parts of the cipher using multiple bridges, we call it the *multibridge* attack.

In addition to their application to iterated Even-Mansour ciphers with two keys, we notice that our techniques can also be combined with statistical distinguishers to give efficient key recovery attacks on certain block ciphers. These block ciphers have internal symmetric properties which allow us to connect (bridge) together intermediate encryption values at a relatively low cost. Such bridges are constructed in reflection cryptanalysis, a technique introduced by Kara in [17], and generalized more recently by Soleimany et al. in [28]. Thus, as an additional application of our multibridge attack, we show how to use it as a generic key-recovery tool in reflection cryptanalysis.

The self-similarity properties of the cipher that we exploit in multibridge attacks are similar to the ones exploited in the SlideX attack [10] on 1-round EM with one key and in later publications [9,23]. However, in the multibridge attack the connected parts are more complex, analyzed themselves using bridging techniques, and are joint using several meet-in-the-middle attacks.

The paper is organized as follows: in Section 2, we describe the notations and conventions used in this paper. In Section 3, we describe our new multibridge attack on the alternating key scheme, and its application to LED-128 and to reflection cryptanalysis. In Section 4, we classify all 4-round iterated EM schemes with two keys and summarize our attacks on them. We finish the analysis of 4-round iterated EM schemes in Section 5, and finally propose open problems and conclude the paper in Section 6.

2   Notations and Conventions

Notations. For a general r-round iterated EM scheme with a block size of n bits, we denote by F_i the public function of round i. We denote by K_{i-1} the round-key added at the beginning of round i (i.e., K_0 is added before round 1), while the last round-key is denoted by K_r (see Fig. 1). Given a plaintext-ciphertext pair (P, C), we denote the state after i encryption rounds by X_i (e.g., $X_0 = P$, X_1 is the state after one encryption round, etc.). In order to simplify our notation, we define $\hat{X}_i = X_i \oplus K_i$, and so $F_{i+1}(\hat{X}_i) = X_{i+1}$. In some of our attacks, we consider several parallel evaluations which are similarly denoted by $Y_{j\,|\,1} = F_{j+1}(\hat{Y}_j)$, $Z_{j+1} = F_{j+1}(\hat{Z}_j)$, etc.

Conventions. In this paper, we evaluate our attack algorithms in terms of the time complexity T, the data complexity D, and the memory complexity M, as a function of $N = 2^n$ where n is the block size. Note that this N is not necessarily the size of the key space, and exhaustive search of a 2-key EM scheme requires N^2 rather than N time. The complexities of our algorithms are generally exponential in n, and thus we can neglect multiplicative polynomial factors in n in our analysis.

We note that in all of our memory-consuming attacks, it is possible to use time-memory tradeoffs in order to reduce the amount of memory we use. However, in this paper we are mainly interested in tradeoffs between the data and time complexities of our attacks, and thus we simply assume that we have sufficient memory to execute the fastest possible version of the attack, i.e., given D known plaintext-ciphertext pairs, we always try to minimize T.

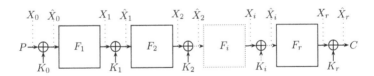

Fig. 1. Iterated Even-Mansour

3   A New Attack on 4-Round Iterated Even-Mansour with Two Alternating Keys

The currently best known attack on 4-round iterated EM scheme with 2 alternating keys (see Fig. 2) was proposed in [23] as part of the analysis of 4-step LED-128 (improving the previous attacks of [16,21]). The attack yields a tradeoff curve of $TD = N^2$, but is limited by an expensive outer loop that guesses one of the keys and performs computations on the entire data for each such guess. Therefore, the tradeoff $TD = N^2$ is restricted by the constraint $T \geq ND$ (or

$TD \geq ND^2$) and is valid only up to $D = N^{1/2}$. Consequently, the attack cannot efficiently exploit more than $D = N^{1/2}$ known plaintexts even when they are available. In this section, we describe a new attack, which can obtain the curve $TD = N^2$ for any amount of given data $D \leq N$. In order to provide sufficient background to our new attack, we start by describing the very simple variant of the SlideX attack (proposed in [10]) on 1-round EM with one key, and then describe the previous attack of [23] on 4-round iterated Even-Mansour with 2 alternating keys. After this background material, we describe the basic variant of our new attack on this scheme that applies in the case $D = N$, and then generalize the basic attack in order to obtain the complete curve $TD = N^2$. Finally, we apply the multibridge attack to 4-step LED-128, improving the running time of the best known attack on this well-studied scheme from 2^{96} to 2^{64}.

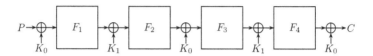

Fig. 2. 4-Round Iterated Even-Mansour with Alternating Keys

3.1 The SlideX Attack on 1-Round Even-Mansour with a Single Key

The SlideX attack [10] is an optimal known plaintext attack on 1-Round EM with one key. It is based on the observation that for each plaintext-ciphertext pair $(P, C) = (X_0, \hat{X}_1)$, by definition $P \oplus K = \hat{X}_0$ and $C \oplus K = X_1$, hence $P \oplus C = \hat{X}_0 \oplus X_1$ (see Fig. 3). As described in the attack below, this equality is exploited in order to match the plaintext-ciphertext pairs with independent evaluations of the public function F_1 by the attacker. Each such match yields a suggestion for the key, which we can easily test.

1. For each of the D plaintext-ciphertext pairs (P^i, C^i):
   (a) Calculate $P^i \oplus C^i$, and store it in a sorted list L, next to P^i.
2. For N/D arbitrary values $\hat{Y}_0^j$:
   (a) Compute $Y_1^j = F_1(\hat{Y}_0^j)$ and search $\hat{Y}_0^j \oplus Y_1^j$ in the list L.
   (b) For each match, obtain P^i and compute the suggestion $K = P^i \oplus \hat{Y}_0^j$.
   (c) Test the suggestion for K using a trial encryption, and if it succeeds, return it as the key.

As we have D plaintext-ciphertext pairs (P^i, C^i) and we evaluate N/D arbitrary values $\hat{Y}_0^j$, we have $D \cdot N/D = N$ pairs of the form (i, j). Thus, according to the birthday paradox, with high probability, there is a pair (i, j) such that $\hat{Y}_0^j = P^i \oplus K \triangleq \hat{X}_0^i$. This implies that $\hat{Y}_0^j \oplus Y_1^j = P^i \oplus C^i$, and thus we get a match in Step 2.(a), suggesting the correct key K. The time complexity of Step 1 is D. The time complexity of Step 2 is N/D, since for an arbitrary value of $\hat{Y}_0^j \oplus Y_1^j$, we expect a match in Step 2.(a) with probability D/N (and thus, on average, we perform only a constant number of operations for each value of $\hat{Y}_0^j$). Consequently, the time complexity of the attack is $max(D, N/D)$, i.e., the attack gives a tradeoff curve of $TD = N$, but only for $D \leq N^{1/2}$ (i.e., it cannot efficiently exploit more than $D = N^{1/2}$ known plaintexts).

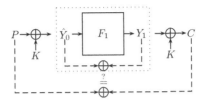

Fig. 3. The Slidex Attack on 1-Round Even-Mansour with 1 Key

3.2 The Best Previous Attack on 4-Round Iterated Even-Mansour with Two Alternating Keys [23]

The best previous attack [23] starts by guessing K_0. This guess makes it possible to eliminate the first and last XOR'ed keys and thus also the first and last permutations by partially encrypting (and decrypting) the plaintext (and ciphertext). In addition, guessing K_0 enables the attacker to combine the second and third applications of the permutations $F_3(F_2(x) \oplus K_0)$ into a single known permutation, $F'_{K_0}(x)$. This reduces the 4-round EM scheme into a single round EM scheme with a single key, which can be easily attacked by the SlideX technique (see Fig. 4). The details of this attack are described below.

1. For all values of K_0:
   (a) For each of the D plaintext-ciphertext pairs (P^i, C^i):
      i. Compute X_1^i and $\hat{X}_3^i$, and store $X_1^i \oplus \hat{X}_3^i$ in a sorted list L, next to X_1^i.
   (b) For N/D arbitrary values $\hat{Y}_1^j$:
      i. Compute Y_3^j and search $\hat{Y}_1^j \oplus Y_3^j$ in the list L.
      ii. For each match, obtain X_1^i and compute the suggestion $K_1 = X_1^i \oplus \hat{Y}_1^j$.
      iii. Test the suggestion for the full key (K_0, K_1) using a trial encryption, and if it succeeds, return it.

For the correct value of K_0, according to the birthday paradox, with high probability there is a pair (i, j) such that $\hat{Y}_1^j = \hat{X}_1^i$. This implies that $X_1^i \oplus \hat{X}_3^i = \hat{Y}_1^j \oplus Y_3^j$, and thus we get a match in Step 1.(b).i, suggesting the correct key (K_0, K_1). The time complexity of Step 1.(a) is D, and the complexity of Step 1.(b) is N/D (we do not expect more than one match in L in Step 1.(b).i for an arbitrary value of $\hat{Y}_1^j \oplus Y_3^j$). Thus, for each value of K_0 that we guess in Step 1, we perform $max(D, N/D)$ operations. Consequently, the attack gives a tradeoff curve of $TD = N^2$, but only for $D \leq N^{1/2}$, i.e., the time complexity must satisfy $T \geq N^{3/2}$. In particular, for $N = 2^{64}$, the best possible time complexity of this attack (for any available amount of data) is at least 2^{96}.

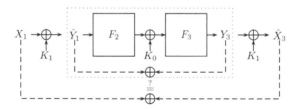

Fig. 4. The Best Previous Attack on 4-Round Iterated Even-Mansour with Two Alternating Keys

Applying a Generalized Version of the Attack to any 2-Key 4-Round Iterated Even-Mansour Scheme. Before describing our improved attack, we notice that in a general 4-round iterated EM scheme with 2 keys which can be used in any order, there is always a key that is added at most twice[5]. Thus, the attack of [23] can be easily generalized and applied with the same complexity to any 4-round iterated EM scheme with 2 keys. The generalized attack works by guessing the value of the most common key (i.e., the key that is added at least 3 times), partially encrypting (decrypting) the plaintexts (ciphertexts), and thus obtaining the inputs/outputs of a single-key EM scheme with a single permutation (which is fully known after guessing the most common key). However, as we show in the rest of this paper, when $D > N^{1/2}$, more efficient attacks exist on all 4-round 2-key EM schemes.

3.3   The Basic Version of Our New Multibridge Attack on 4-Round Iterated Even-Mansour with Two Alternating Keys

The approach of the previous attack was to guess K_0, and thus "peel off" the first and last rounds on the 4-round EM scheme with 2 alternating keys. Although this approach seems natural, it gives the tradeoff curve of $TD = N^2$ only for $D \leq N^{1/2}$, and thus its time complexity is at least $T \geq N^{3/2}$. We now present our new attack on this scheme which achieves the same tradeoff for any $D \leq N$, and thus enables us to reduce the time complexity to $T = N$.

Unlike the previous attack, which guessed the value of K_0, our attack guesses the value of some *internal state* for which a special self-similarity property holds. This property allows us to split the cipher into two parts which can be analyzed independently. While standard meet-in-the-middle attacks also split the cipher into two parts, in our attack the two parts of the cipher are nested (rather than concatenated), similarly to attacks based on the splice-and-cut technique [3].

---

[5] Schemes in which there is a key that is added only once are very weak (as shown in the full version of this paper [8]).

However, it is interesting to note that while splice-and-cut attacks consider the first and the last rounds of the cipher as consecutive rounds (i.e., the cipher is spliced using the plaintext-ciphertext pairs), here we connect (or bridge) the cipher internally and consider as consecutive rounds its two internal ends.

We begin by describing our multibridge attack for the specific case of $D = N$ (i.e., given the full code-book), for which the attack runs in time $T = N$. In this case, we look for some plaintext-ciphertext pair (P^i, C^i) with the internal fixed-point property $X_1^i = \hat{X}_3^i$ (i.e., we connect X_1^i and $\hat{X}_3^i$ using a "bridge"). Since XOR'ing the same key twice leaves the result unchanged, this self-similarity property also implies that $\hat{X}_1^i = X_3^i$ (i.e., $\hat{X}_1^i$ and X_3^i are now connected using another bridge, which we get "for free"), and this allows us to split the cipher into 2 nested parts[6], each independently suggesting a value for the key K_0. Finally, the suggestions are merged using a meet-in-the-middle technique. Note that for a specific plaintext-ciphertext pair, this internal fixed-point property occurs with probability $1/N$, and thus given $D = N$ data, with high probability, one of the plaintext-ciphertext pairs satisfies this property. The details of the basic multibridge attack are given below (see Fig. 5):

1. For each of the $D = N$ known plaintext-ciphertext pairs (P^i, C^i):
   (a) Calculate $P^i \oplus C^i$, and store it in a sorted list L_1, next to P^i.
2. For each of the N possible values of Y_1^j:
   (a) Compute $\hat{Y}_0^j = F_1^{-1}(Y_1^j)$.
   (b) Assume that $\hat{Y}_3^j = Y_1^j$, and compute $Y_4^j = F_4(\hat{Y}_3^j)$.
   (c) Compute $\hat{Y}_0^j \oplus Y_4^j$ and search for matches with this value in L_1.
   (d) For each match, obtain P^i, calculate a suggestion for $K_0 = P^i \oplus \hat{Y}_0^j$. Store all the suggestions in a sorted list L_2, next to Y_1^j. We expect L_2 to contain about N entries.
3. For each of the N possible values of $\hat{Z}_1^\ell$ (i.e., the intermediate encryption value obtained after applying 1 round and adding K_1):
   (a) Compute $Z_2^\ell = F_2(\hat{Z}_1^\ell)$.
   (b) Assume that $Z_3^\ell = \hat{Z}_1^\ell$, and compute $\hat{Z}_2^\ell = F_3^{-1}(Z_3^\ell)$.
   (c) Compute $K_0 = Z_2^\ell \oplus \hat{Z}_2^\ell$ and search for matches in L_2. We expect one match on average for a given value of K_0.
   (d) For each match, obtain Y_1^j, calculate a suggestion for $K_1 = Y_1^j \oplus \hat{Z}_1^\ell$.
   (e) Test the suggestion for the full key (K_0, K_1) using a trial encryption, and if it succeeds, return it.

The success of the attack is based on the observation above, namely, given $D = N$ plaintext-ciphertext pairs (P^i, C^i), then with high probability, there exists an i such that $X_1^i = \hat{X}_3^i$. Since we iterate over all possible values of Y_1^j in Step 2 of the attack, then for $Y_1^j = X_1^i$, we calculate $\hat{Y}_0^j \oplus Y_4^j = \hat{X}_0^i \oplus X_4^i = P^i \oplus C^i$ in step 2.(c). Thus, we get a match with the correct value of K_0 is Step 2.(d), and we store it next to $Y_1^j = X_1^i$ in the list L_2. Similarly, since we iterate over all

---

[6] In fact, as described in the detailed attack, the first part of the cipher is in itself also composed of 2 parts.

possible values of $\hat{Z}_1^\ell$, then for $\hat{Z}_1^\ell = \hat{X}_1^i$, we have $Z_3^\ell = \hat{Z}_1^\ell = \hat{X}_1^i = X_3^i$. Hence, we calculate the correct value of K_0 in Step 3.(c), obtain the match with L_2 such that $Y_1^j = X_1^i$, and obtain the correct $K_1 = Y_1^j \oplus \hat{Z}_1^\ell = X_1^i \oplus \hat{X}_1^i$. As a result, we encounter the correct suggestion for the full key in Step 3.(e) and return it.

The attack is composed of a sequential execution of 3 mains steps, each has a time complexity of N: in Step 1, we perform a simple XOR operation for each of the $D = N$ plaintext-ciphertext pairs, and allocate the list L_1, which is of size N. In Step 2, we iterate over N possible values of Y_1^j, and for each such value we expect a single match in L_1 in Step 2.(c), implying that the complexity of Step 2 is N. Finally, since the expected size of L_2 is N, for each suggestion of K_0 we expect a single match in Step 3.(c), and thus the time complexity of Step 3 is N, as claimed. In total, the analysis shows that the time complexity of the full attack is N, and its memory complexity is N as well.

Step 1: For all i

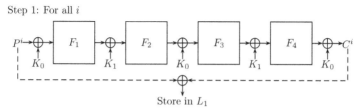

Step 2(a): For a given Δ^s for all j

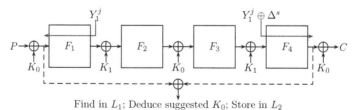

Step 2(b): For a given Δ^s for all ℓ

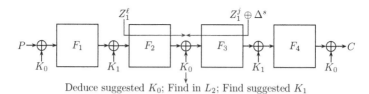

Fig. 5. The Multibridge Attack

3.4 Our Generalized Multibridge Attack on 4-Round Iterated Even-Mansour with Two Alternating Keys

Given $D < N$ data, we do not expect to have a plaintext-ciphertext pair that satisfies the internal fixed-point property. In order to generalize the attack for any $D \leq N$, we first notice that the internal fixed-point property $X_1^i = \hat{X}_3^i$ can be replaced by the more general "bridging" property $X_1^i = \hat{X}_3^i \oplus \Delta$, for any fixed known value of[7] Δ (the previously described fixed-point property is the special case of $\Delta = 0$). Thus, in Step 2.(b) we calculate $\hat{Y}_3^j = Y_1^j \oplus \Delta$, and similarly in Step 3.(b) we calculate $Z_3^\ell = \hat{Z}_1^\ell \oplus \Delta$.

When we fix one value of Δ, we expect to have a pair (P^i, C^i) such that $X_1^i = \hat{X}_3^i \oplus \Delta$ with probability of about D/N. Thus, in order to recover the key with high probability, we randomly choose N/D different values of Δ, indexed by Δ^s, and run a variant of the fixed-point multibridge attack independently for each value. This is a similar approach to the one used in [10] in order to extend the SlideX attack on 1-round 2-key EM to all $D \leq N^{1/2}$. The details of the generalized multibridge attack are given below:

1. For each of the D plaintext-ciphertext pairs (P^i, C^i):
   (a) Calculate $P^i \oplus C^i$, and store it in a sorted list L_1, next to P^i.
2. For N/D arbitrary values of Δ^s:
   (a) Apply a variant of the basic multibridge attack using Δ^s.

As we execute a variant of the fixed-point attack N/D times, the expected time complexity of the attack is N^2/D. The size of the list L_1 is D, implying that the size of L_2 (the second list allocated in the multibridge attack) is D as well, and thus the memory complexity of the attack is D.

We conclude by noting that this attack can also be applied directly to the attack of Merkle and Hellman against 2K3DES [14]. The resulting attack is essentially the known plaintext variant of van Oorschot and Wiener [29] to Merkle and Hellman's attack, i.e., an attack on 2K3DES with D known plaintexts and running time of N^2/D.

3.5   Application to 4-Step LED-128

LED is a 64-bit lightweight iterated EM block cipher, proposed at CHES 2011 [13]. The cipher has two main variants: a one-key version called LED-64, and a two-key version called LED-128. We concentrate on the 128-bit variant, which has 12 steps, in which the two keys are alternately used. The best previously known attack on 4-step LED-128 was described in [23] (and also described in Section 3.2 for a general 4-step EM cipher with alternating keys), and gives a tradeoff of $TD = 2^{128}$, but only for $D \leq 2^{32}$. We can directly apply our improved attack, described in Section 3.4, to 4-step LED-128, we obtain the tradeoff of $TD = 2^{128}$ for any $D \leq 2^{64}$. Thus, we improve the time complexity of the best known attack on this scheme from 2^{96} to 2^{64}.

We note that recently, up to 8 steps of the 2-key alternating EM scheme have been attacked faster than exhaustive search (see [9]). However, all the known

---

[7] Thus, we do not exploit the actual fixed-point in a strong way (such as in [1]), but merely some fixed linear relation between X_1^i and $\hat{X}_3^i$.

attacks on more than 4 steps are marginal in the sense that they improve the time complexity of exhaustive search only by a logarithmic factor in N, and thus our new attack on the 4-step version of LED-128 is currently the best non-marginal attack on this scheme.

3.6 Application to Reflection Cryptanalysis

Reflection cryptanalysis was introduced by Kara in [17] as a self-similarity attack on GOST and related block ciphers, and generalized to a statistical attack on a broader class of ciphers (called "PRINCE-like" ciphers) by Soleimany et al. in [28]. A PRINCE-like cipher is designed to have a specific symmetry property around its middle round, called α-reflection.[8] The definition and analysis of PRINCE-like ciphers in [28], was inspired by the block cipher PRINCE [5], that used the α-reflection property in order to realize decryption on top of encryption with a negligible additional cost.

In reflection cryptanalysis of PRINCE-like ciphers, we consider the encryption process of a single plaintext, and study the difference between its internal encryption values, which are symmetric with respect to the middle round of the cipher. The goal is to iteratively construct a *reflection distinguisher*, which is a strong non-random property, likely to be present in several rounds of PRINCE-like ciphers (as shown in [28]). In particular, a reflection distinguisher on r rounds of the cipher (denoted by E_K), gives a specific value of Δ for which $\Pr(X \oplus E_K(X) = \Delta) > 2^{-n}$ (where the probability is taken over the input X).

In this section, we present a variant of the multibridge attack as a generic key-recovery method for reflection cryptanalysis. This attack can be considered as the reflection cryptanalysis counterpart of the key-recovery attack of Daemen [6] for differential cryptanalysis of ciphers based on the Even-Mansour construction. The attack assumes that we have a reflection distinguisher with probability $p > 2^{-n}$ on r rounds of the cipher, and recovers the secret key for a total of $r + 2$ rounds, by adding one round at the beginning and one round at the end (i.e., the reflection distinguisher covers rounds $2, 3, \ldots, r + 1$). For the sake of simplicity, we first assume that the cipher is a single-key iterated Even-Mansour scheme, where the secret key is denoted by K. We now describe the attack, assuming that we obtain D plaintext-ciphertext pairs, such that $D > p^{-1}$.

1. For $2^n/(p \cdot D)$ arbitrary values of $\hat{Y}_0^j$:
   (a) Compute $Y_1^j = F_1(\hat{Y}_0^j)$.
   (b) Assume that $\hat{Y}_{r+1}^j = Y_1^j \oplus \Delta$ (where the value of Δ is given by the reflection distinguisher), and compute $Y_{r+2}^j = F_{r+2}(\hat{Y}_{r+1}^j)$.
   (c) Store $\hat{Y}_0^j \oplus Y_{r+2}^j$ in a sorted list L, next to $\hat{Y}_0^j$.
2. For each of the D plaintext-ciphertext pairs (P^i, C^i):
   (a) Compute $P^i \oplus C^i$, and search the list L for matches.

---

[8] If we denote by E_K the encryption of r rounds in the middle of the cipher under the key K, then the α-reflection property (for a fixed value of α) states that for any input X, $E_K(X) = E_{K \oplus \alpha}^{-1}(X)$.

(b) For each match obtain $\hat{Y}_0^j$, and calculate a suggestion for $K = P^i \oplus \hat{Y}_0^j$.
(c) Test the suggestion for the key K using a trial encryption, and if it succeeds, return it.

We have $D > p^{-1}$ plaintext-ciphertext pairs, out of which $p \cdot D > 1$ are expected to satisfy the reflection characteristic. As we evaluate $2^n/(p \cdot D)$ values of $\hat{Y}_0^j$ in Step 1 of the attack, according to the birthday paradox, we expect at least one match between $\hat{Y}_0^j$ and $P^i \oplus K$ such that (P^i, C^i) satisfies the reflection property. Once we obtain such a match (i.e., $\hat{Y}_0^j = P^i \oplus K$), we recover the correct key in Step 2.(c).

As we expect less than one match in L in Step 2.(a) for an arbitrary (P^i, C^i), the time complexity of the attack is $max(D, 2^n/(p \cdot D))$. The time complexity is minimized to $2^{n/2} \cdot p^{-1/2}$ by choosing $D = 2^{n/2} \cdot p^{-1/2}$ (note that it is not reasonable to exploit more than $2^{n/2} \cdot p^{-1/2}$ data). The memory complexity of the attack is $2^n/(p \cdot D)$, but can be easily reduced to D, by exchanging the order of steps 1 and 2 of the attack.

In order to apply the attack to more complex key schedules, the attacker can exploit the internal properties of the reflection distinguisher to recover more key material (perhaps using more data, or function evaluations in Step 1 of the attack). However, this extension is highly dependent on the internal properties of the cipher, and is thus out of the scope of this paper.

4   Classification and Summary of Our Attacks on All 4-Round 2-Key Iterated Even-Mansour Schemes

In the rest of the paper, we analyze all the remaining iterated EM schemes with 4 rounds and 2 keys, and show that the best attack on each one of them has a time complexity of N. We begin by noting that each such construction can be described by a sequence of 5 keys, which specifies the order in which the keys K_0 and K_1 are added (over $GF(2)$) to the internal state. For example, we denote the 4-round EM scheme with alternating keys (of Fig. 2) by $[K_0, K_1, K_0, K_1, K_0]$. Clearly, each such scheme has an equivalent representation which is obtained by renaming the keys K_0 and K_1 (e.g., $[K_0, K_0, K_1, K_1, K_0]$ is equivalent to $[K_1, K_1, K_0, K_0, K_1]$). In addition, since our attacks assume that the public permutations F_i (and F_i^{-1}) are chosen at random (i.e., we do not exploit any special properties of the public permutations), from a cryptanalytic point of view, the roles of encryption and decryption can be exchanged. Namely, if we reverse the order in which the keys are added, we get an equivalent scheme. For example, the scheme $[K_0, K_0, K_1, K_1, K_0]$ is equivalent to $[K_0, K_1, K_1, K_0, K_0]$, since any attack on $[K_0, K_0, K_1, K_1, K_0]$ can also be applied to $[K_0, K_1, K_1, K_0, K_0]$ (by reversing the roles of encryption and decryption), and vice-versa. Altogether, the scheme $[K_0, K_0, K_1, K_1, K_0]$ belongs to an equivalence class (EC) with 4 members, containing the 3 additional schemes $[K_1, K_1, K_0, K_0, K_1]$, $[K_0, K_1, K_1, K_0, K_0]$ and $[K_1, K_0, K_0, K_1, K_1]$. Since any attack on a member of an EC is applicable to its other members, we only need to describe an attack on a representative of the EC.

Table 2 lists the equivalence classes of all the 4-round 2-key iterated EM schemes, next to the complexities of our best attacks. For the sake of simplification, we will refer to each EC as a single scheme, using its ID as described in Table 2. For example, our attack on the schemes of the first EC is simply refereed to an attack on the "EC1 scheme", whose representative is $[K_0, K_1, K_1, K_1, K_1]$.

The attack on EC7, which is 4-round EM with alternating keys, was already described in Section 3.4. In the next section we present the most complex multi-bridge attacks on the classes EC8 and EC9. The simpler attacks on EC1–EC6 are presented in the full version of this paper [8].

Table 2. Classification and Attacks on Iterated Even-Mansour Schemes with Four Rounds and Two Keys

EC ID	EC Representative	Reference	Data	Time	Memory
EC1	$[K_0, K_1, K_1, K_1, K_1]$	[8]	$O(1)$	N	$O(1)$
EC2	$[K_0, K_1, K_0, K_0, K_0]$	[8]	$O(1)$	N	$O(1)$
EC3	$[K_0, K_0, K_1, K_0, K_0]$	[8]	$O(1)$	N	$O(1)$
EC4	$[K_0, K_0, K_1, K_1, K_1]$	[8]	$O(1)$	N	N
EC5	$[K_0, K_1, K_1, K_0, K_0]$	[8]	$O(1)$	N	N
EC6	$[K_0, K_1, K_1, K_1, K_0]$	[8]	$D \leq N$	N^2/D	D
EC7	$[K_0, K_1, K_0, K_1, K_0]$	Section 3.4	$D \leq N$	N^2/D	D
EC8	$[K_0, K_1, K_0, K_1, K_1]$	Section 5.1	$D \leq N$	N^2/D	D
EC9	$[K_0, K_1, K_0, K_0, K_1]$	Section 5.2	$D \leq N^{1/2}$	N^2/D	D
			$N^{1/2} < D \leq N$	N^2/D	N

Each EC (equivalence class) is described using an ID and a representative scheme.

Classification and Attacks on All 3-Round 2-Key Iterated Even-Mansour Schemes. We did not find any cryptanalytic techniques which are specifically applicable to 3-round 2-key EM schemes. However, for the sake of completeness, we also classify all 3-round 2-key iterated EM schemes and specify which variant of our 4-round attacks can be used to break it (with the same complexity parameters).

1. $[K_0, K_1, K_1, K_1]$ and $[K_0, K_1, K_0, K_0]$ can be broken with a variant of the attack on EC1.
2. $[K_0, K_1, K_1, K_0]$ can be broken with a variant of the attack on EC4.
3. $[K_0, K_1, K_0, K_1]$ can be broken with a variant of the attack on EC7.

5   Multibridge Attacks on EC8 and EC9

In this section we consider the schemes EC8 and EC9, and show that they can be attacked with complexity $DT = N^2$, for all $D \leq N$. The attacks on these schemes use the same general multibridge technique as our previous attack on EC7 in Section 3, namely, we use a generalized version of the internal fixed-point property in order to internally bridge different parts of the cipher. Finally, the suggestions for the key obtained from the two parts are merged using a meet-in-the-middle technique.

5.1 A Multibridge Attack on EC8

In order to attack the scheme $[K_0, K_1, K_0, K_1, K_1]$, we look for a plaintext-ciphertext pair (P^i, C^i) such that $\hat{X}_2^i = P^i \oplus \Delta^s$ (for arbitrary values of Δ^s). The details of the multibridge attack on EC8 are given below:

1. For N/D arbitrary values of Δ^s:
   (a) For each of the D plaintext-ciphertext pairs (P^i, C^i):
       i. Assume that $\hat{X}_2^i = P^i \oplus \Delta^s$ and compute $X_3^i = F_3(\hat{X}_2^i)$.
       ii. Compute $X_3^i \oplus C^i$ and store it in a sorted list L_1, next to C^i.
   (b) For each of the N possible values of $\hat{Y}_3^j$:
       i. Compute $Y_4^j = F_4(\hat{Y}_3^j)$.
       ii. Compute $\hat{Y}_3^j \oplus Y_4^j$, and search for matches in L_1.
       iii. For each match, obtain C^i, compute a suggestion for $K_1 = C^i \oplus Y_4^j$, and store the suggestion in a sorted list L_2, next to P^i.
   (c) For each of the N possible values of $\hat{Z}_0^\ell$:
       i. Compute $Z_1^\ell = F_1(\hat{Z}_0^\ell)$.
       ii. Assume that $Z_2^\ell = \hat{Z}_0^\ell \oplus \Delta^s$, and compute $\hat{Z}_1^\ell = F_2^{-1}(Z_2^\ell)$.
       iii. Compute a suggestion for $K_1 = Z_1^\ell \oplus \hat{Z}_1^\ell$ and search for it in the list L_2.
       iv. For each match, obtain P^i, compute a suggestion for $K_0 = P^i \oplus \hat{Z}_0^\ell$.
       v. Test the full key (K_0, K_1) using a trial encryption, and if it succeeds, return it.

The analysis of the attack is very similar to the analysis of our general multi-bridge attack in Section 3.4, and thus given $D \leq N$ known plaintext-ciphertext pairs, its time complexity is N^2/D and its memory complexity is D.

5.2 A Multibridge Attack on EC9

In order to attack the scheme $[K_0, K_1, K_0, K_0, K_1]$, we look for a plaintext-ciphertext pair (P^i, C^i) such that $X_1^i = C^i \oplus \Delta^s$ (for arbitrary values of Δ^s). The details of the multibridge attack on EC9 are given below:

1. For N/D arbitrary values of Δ^s:
   (a) For each of the D plaintext-ciphertext pairs (P^i, C^i):
       i. Assume that $X_1^i = C^i \oplus \Delta^s$ and compute $\hat{X}_0^i = F_1^{-1}(X_1^i)$.
       ii. Compute a suggestion for $K_0 = \hat{X}_0^i \oplus P^i$ and store it in a sorted list L_1, next to X_1^i.
   (b) For each of the N possible values of $\hat{Y}_1^j$:

     i. Compute $Y_2^j = F_2(\hat{Y}_1^j)$.

     ii. Assume that $Y_4^j = \hat{Y}_1^j \oplus \Delta^s$ and compute $\hat{Y}_3^j = F_4^{-1}(Y_4^j)$.

     iii. Compute $Y_2^j \oplus \hat{Y}_3^j$ and store this value on a sorted list L_2, next to $\hat{Y}_1^j$ and Y_2^j.

  (c) For each of the N possible values of $\hat{Z}_2^\ell$:

     i. Compute $Z_3^\ell = F_3(\hat{Z}_2^\ell)$.

     ii. Compute $\hat{Z}_2^\ell \oplus Z_3^\ell$ and search for it in the list L_2.

     iii. For each match, obtain Y_2^j (and $\hat{Y}_1^j$), compute a suggestion for $K_0 = Y_2^j \oplus \hat{Z}_2^\ell$, and search it in the sorted list L_1.

     iv. For each match, obtain X_1^i and compute a suggestion for $K_1 = X_1^i \oplus \hat{Y}_1^j$.

     v. Test the full key (K_0, K_1) using a trial encryption, and if it succeeds, return it.

Similarly to the multibridge attacks on EC7 and EC8, the time complexity of the attack is N^2/D for any $D \leq N$, as the time complexity of each of the Steps 1.(a), 1.(b) and 1.(c) is N. However, unlike the previous attacks which had a reduced memory complexity of D, the list L_2 contains N elements, and thus the memory complexity of this attack is N. As a result, when $D \leq N^{1/2}$, the most efficient attack on this scheme is the generalized version of the attack presented in Section 3.2, which has the same running time but requires less memory.

We note that in cases where $D > N^{1/2}$, but the available memory M satisfies $D \leq M < N$, it is possible obtain a tradeoff between the memory and time complexities of the attack. Although in this paper we mainly consider tradeoffs between data and time, an interesting open question is whether it is possible to reduce the memory complexity of the attack for $D > N^{1/2}$ without increasing its time complexity.

6   Conclusions and Open Problems

In this paper, we studied the security of iterated Even-Mansour schemes with two keys. We showed that all such schemes with at most 4 rounds provide security of at most 2^n (compared to the 2^{2n} complexity of exhaustive key search). Our theoretical results allowed us to reduce the complexity of the best known attack on 4-step LED-128 from 2^{96} to 2^{64}, and to develop a generic key-recovery tool for reflection cryptanalysis. In order to obtain these results, we developed the novel multibridge technique which combines the advantages of the dissection [7] and the splice-and-cut [3] techniques.

We conclude this paper with a list of several open problems and research directions which arise naturally from the results of our paper.

1. **Finding Better Attacks on 3-round EM with Two Keys.** Using our techniques, we could not find attacks on 3-round EM with alternating keys which are better than the attacks on 4-round EM with alternating keys. If such attacks indeed do not exist, then there is no security gain in adding

a round to the 3-round EM scheme. Such a situation is somewhat unusual, and hence, one may anticipate that better attacks exist on 3-round EM with alternating keys. We note that this is a similar scenario to cascade encryption, where the complexity of the best attack on 3-encryption is the same as the complexity of the best attack on 4-encryption [7]. However, in cascade encryption, the complexities are equal only for the specific attacks that minimize the time complexity, while in our case, the complexities are the same for all attacks on the tradeoff curve.

2. **Finding the Minimal number r for which r-round EM with Two Keys Provides $2n$-bit Security.** This is an interesting research direction whose equivalent has been extensively studied in the domain of Feistel constructions (see [20,25,26]). In the case of EM with two keys, we are not aware of any attacks on the 5-round alternating key scheme which improve over exhaustive search by a significant factor. On the other hand, when considering relatively small (polynomial in n) improvements over exhaustive search, up to 8 rounds can be broken (see [9]), but no attacks at all are known for $r \geq 9$ rounds. Clearly, this fundamental question can be generalized to more keys, namely, what is the minimal number of rounds for which mn-bit security can be achieved for n-bit iterated EM constructions with m independent keys?

3. **Other Attack Models.** In this paper, we concentrated on attacks in the most conservative model in which the adversary has access only to known plaintexts, and the complexity of the attack takes into consideration all operations (including a potential preprocessing stage). It would be interesting to see whether the complexities of the attacks can be reduced in other models, where chosen or even adaptively chosen plaintext queries are allowed, and perhaps precomputation is not counted in the overall complexity of the attack. We note that in a recent work of Joux and Fouque [12], such improved attacks were found for the 1-round EM construction with two keys, suggesting that similar results may be possible for iterated EM with two keys as well.

4. **Considering Memory Complexity.** As in all previous papers on iterated EM, we concentrated in this paper on tradeoffs between data and time complexities, assuming that we always have enough memory to apply the most efficient attack. It would be interesting to consider more general tradeoffs between data, memory and time complexities, and in particular, minimize the memory complexity for which the (presumably) optimal curve $DT = 2^{2n}$ can be obtained. We note that a similar question with respect to 1-round EM was asked in [10] and partially answered in [12].

5. **More Complex Key Schedules.** As stated in the introduction, iterated EM schemes can be considered with a wide variety of key schedules, generating an endless field of research. However, even when restricted to schemes with two keys as we do in this paper, one may consider more complex key schedules in which combinations of the keys $K0$ and $K1$ can be used as round keys. It seems that the attacks presented in this paper cannot target such key schedules, and for example, we could not find an attack of complexity 2^n on 4-round EM with the keys $[K0, K1, K0, K1, K0 \oplus K1]$. Hence, it will

be interesting to find new techniques that will be able to handle such key schedules, or to show lower bounds on the security of the respective iterated EM schemes.

References

1. Aerts, W., Biham, E., De Moitie, D., De Mulder, E., Dunkelman, O., Indesteege, S., Keller, N., Preneel, B., Vandenbosch, G.A.E., Verbauwhede, I.: A Practical Attack on KeeLoq. J. Cryptology 25(1), 136–157 (2012)
2. Andreeva, E., Bogdanov, A., Dodis, Y., Mennink, B., Steinberger, J.P.: On the Indifferentiability of Key-Alternating Ciphers. In: Canetti, R., Garay, J.A. (eds.) CRYPTO 2013, Part I. LNCS, vol. 8042, pp. 531–550. Springer, Heidelberg (2013)
3. Aoki, K., Sasaki, Y.: Preimage Attacks on One-Block MD4, 63-Step MD5 and More. In: Avanzi, R.M., Keliher, L., Sica, F. (eds.) SAC 2008. LNCS, vol. 5381, pp. 103–119. Springer, Heidelberg (2009)
4. Bogdanov, A., Knudsen, L.R., Leander, G., Standaert, F.-X., Steinberger, J.P., Tischhauser, E.: Key-Alternating Ciphers in a Provable Setting: Encryption Using a Small Number of Public Permutations - (Extended Abstract). In: Pointcheval, Johansson (eds.) [27], pp. 45–62
5. Borghoff, J., Canteaut, A., Güneysu, T., Kavun, E.B., Knezevic, M., Knudsen, L.R., Leander, G., Nikov, V., Paar, C., Rechberger, C., Rombouts, P., Thomsen, S.S., Yalçin, T.: PRINCE - A Low-Latency Block Cipher for Pervasive Computing Applications - Extended Abstract. In: Wang, Sako (eds.) [30], pp. 208–225
6. Daemen, J.: Limitations of the Even-Mansour Construction. In: Imai, et al. (eds.) [15], pp. 495–498
7. Dinur, I., Dunkelman, O., Keller, N., Shamir, A.: Efficient Dissection of Composite Problems, with Applications to Cryptanalysis, Knapsacks, and Combinatorial Search Problems. In: Safavi-Naini, R., Canetti, R. (eds.) CRYPTO 2012. LNCS, vol. 7417, pp. 719–740. Springer, Heidelberg (2012)
8. Dinur, I., Dunkelman, O., Keller, N., Shamir, A.: Cryptanalysis of iterated even-mansour schemes with two keys. Cryptology ePrint Archive, Report 2013/674 (2013), http://eprint.iacr.org/
9. Dinur, I., Dunkelman, O., Keller, N., Shamir, A.: Key Recovery Attacks on 3-round Even-Mansour, 8-step LED-128, and Full AES^2. In: Sako, K., Sarkar, P. (eds.) ASIACRYPT 2013, Part I. LNCS, vol. 8269, pp. 337–356. Springer, Heidelberg (2013)
10. Dunkelman, O., Keller, N., Shamir, A.: Minimalism in Cryptography: The Even-Mansour Scheme Revisited. In: Pointcheval, Johansson (eds.) [27], pp. 336–354
11. Even, S., Mansour, Y.: A construction of a cioher from a single pseudorandom permutation. In: Imai, et al. (eds.) [15], pp. 210–224
12. Fouque, P.-A., Joux, A., Mavromati, C.: Multi-user collisions: Applications to Discrete Logs, Even-Mansour and Prince. Cryptology ePrint Archive, Report 2013/761 (2013), http://eprint.iacr.org/
13. Guo, J., Peyrin, T., Poschmann, A., Robshaw, M.: The LED Block Cipher. In: Preneel, B., Takagi, T. (eds.) CHES 2011. LNCS, vol. 6917, pp. 326–341. Springer, Heidelberg (2011)
14. Hellman, M.E.: A Cryptanalytic Time-Memory Trade-Off. IEEE Transactions on Information Theory 26(4), 401–406 (1980)

15. Matsumoto, T., Imai, H., Rivest, R.L. (eds.): ASIACRYPT 1991. LNCS, vol. 739. Springer, Heidelberg (1993)

16. Isobe, T., Shibutani, K.: Security Analysis of the Lightweight Block Ciphers XTEA, LED and Piccolo. In: Susilo, W., Mu, Y., Seberry, J. (eds.) ACISP 2012. LNCS, vol. 7372, pp. 71–86. Springer, Heidelberg (2012)

17. Kara, O.: Reflection Cryptanalysis of Some Ciphers. In: Chowdhury, D.R., Rijmen, V., Das, A. (eds.) INDOCRYPT 2008. LNCS, vol. 5365, pp. 294–307. Springer, Heidelberg (2008)

18. Lampe, R., Patarin, J., Seurin, Y.: An Asymptotically Tight Security Analysis of the Iterated Even-Mansour Cipher. In: Wang, Sako (eds.) [30], pp. 278–295

19. Luby, M., Rackoff, C.: How to Construct Pseudorandom Permutations from Pscudorandom Functions. SIAM J. Comput. 17(2), 373–386 (1988)

20. Mandal, A., Patarin, J., Seurin, Y.: On the Public Indifferentiability and Correlation Intractability of the 6-Round Feistel Construction. In: Cramer, R. (ed.) TCC 2012. LNCS, vol. 7194, pp. 285–302. Springer, Heidelberg (2012)

21. Mendel, F., Rijmen, V., Toz, D., Varici, K.: Differential Analysis of the LED Block Cipher. In: Wang, Sako (eds.) [30], pp. 190–207

22. Merkle, R.C., Hellman, M.E.: On the Security of Multiple Encryption. Commun. ACM 24(7), 465–467 (1981)

23. Nikolić, I., Wang, L., Wu, S.: Cryptanalysis of Round-Reduced LED. In: Moriai, S. (ed.) FSE 2013. LNCS, vol. 8424, pp. 112–130. Springer, Heidelberg (2014)

24. Patarin, J.: Improved security bounds for pseudorandom permutations. In: Graveman, R., Janson, P.A., Neumann, C., Gong, L. (eds.) ACM Conference on Computer and Communications Security, pp. 142–150. ACM (1997)

25. Patarin, J.: Luby-Rackoff: 7 Rounds Are Enough for formula_image Security. In: Boneh, D. (ed.) CRYPTO 2003. LNCS, vol. 2729, pp. 513–529. Springer, Heidelberg (2003)

26. Patarin, J.: Security of Random Feistel Schemes with 5 or More Rounds. In: Franklin, M. (ed.) CRYPTO 2004. LNCS, vol. 3152, pp. 106–122. Springer, Heidelberg (2004)

27. Pointcheval, D., Johansson, T. (eds.): EUROCRYPT 2012. LNCS, vol. 7237. Springer, Heidelberg (2012)

28. Soleimany, H., Blondeau, C., Yu, X., Wu, W., Nyberg, K., Zhang, H., Zhang, L., Wang, Y.: Reflection Cryptanalysis of PRINCE-Like Ciphers. Journal of Cryptology, 1–27 (2013)

29. van Oorschot, P.C., Wiener, M.: A Known-Plaintext Attack on Two-Key Triple Encryption. In: Damgård, I.B. (ed.) EUROCRYPT 1990. LNCS, vol. 473, pp. 318–325. Springer, Heidelberg (1991)

30. Wang, X., Sako, K. (eds.): ASIACRYPT 2012. LNCS, vol. 7658, pp. 2012–2018. Springer, Heidelberg (2012)

Meet-in-the-Middle Attacks
on Generic Feistel Constructions

Jian Guo[1], Jérémy Jean[1], Ivica Nikolić[1], and Yu Sasaki[2]

[1] Nanyang Technological University, Singapore
[2] NTT Secure Platform Laboratories, Tokyo, Japan

ntu.guo@gmail.com, {JJean,INikolic}@ntu.edu.sg, sasaki.yu@lab.ntt.co.jp

Abstract. We show key recovery attacks on generic balanced Feistel ciphers. The analysis is based on the meet-in-the-middle technique and exploits truncated differentials that are present in the ciphers due to the Feistel construction. Depending on the type of round function, we differentiate and show attacks on two types of Feistels. For the first type, which is the most general Feistel, we show a 5-round distinguisher (based on a truncated differential), which allows to launch 6-round and 10-round attacks, for single-key and double-key sizes, respectively. For the second type, we assume the round function follows the SPN structure with a linear layer P that has a maximal branch number, and based on a 7-round distinguisher, we show attacks that reach up to 14 rounds. Our attacks outperform all the known attacks for any key sizes, have been experimentally verified (implemented on a regular PC), and provide new lower bounds on the number of rounds required to achieve a practical and a secure Feistel.

Keywords: Feistel, generic attack, key recovery, meet-in-the-middle.

1 Introduction

A Feistel network [13] is a scheme that builds n-bit permutations from smaller, usually $n/2$-bit permutations or functions. In ciphers based on the Feistel network, both the encryption and the decryption algorithms can be achieved with the use of a single scheme, thus such ciphers exhibit an obvious implementation advantage. The Feistel-based design approach is widely trusted and has a long history of usage in block ciphers. In particular, a number of current and former international or national block cipher standards such as DES [6], Triple-DES [19], Camellia [2], and CAST [5] are Feistels. In addition to the standard block ciphers, the Feistel construction is an attractive choice for many lightweight ciphers, for instance the recent NSA proposal SIMON [3], LBlock [26], Piccolo [24], etc. The application of the Feistel construction is not limited only to ciphers, and has been used to design other crypto primitives: the hash function SHAvite-3 [4], the CAESAR proposal for authentication scheme LAC [27] and others. The analysis of Feistel primitives and their provable security bounds depend on the type of the round function implemented. Luby and Rackoff [21] have

P. Sarkar and T. Iwata (Eds.): ASIACRYPT 2014, PART I, LNCS 8873, pp. 458–477, 2014.

shown that an n-bit pseudorandom permutation can be constructed from an $n/2$-bit pseudorandom function with 3-round Feistel network. In this construction, the round functions are chosen uniformly at random from a family of $2^{n/2 \cdot 2^{n/2}}$ functions – a set that can be enumerated with $n/2 \cdot 2^{n/2}$-bit keys. Later, Knudsen [20] considered a practical model, in which the round functions are chosen from a family of 2^k functions and showed a generic attack on up to 6 rounds. Knudsen's construction was coined as *Feistel-1* by Isobe and Shibutani in [18] to reflect the fact that it is the most general type of Feistels. They further introduced the term *Feistel-2* to denote ciphers in which the round functions are composed of an XOR of a subkey followed by an application of a public function or permutation. Generic attacks on Feistel-2 such as impossible differentials [20], all-subkey recovery [17,18], and integral-like attacks [25] penetrate up to 6 rounds when the key size equals the state size, and up to 9 rounds when the key is twice as large as the block. Better attacks have been published, but they are on so-called *Feistel-3* that has round functions based on substitution-permutation network (SPN), i.e. the rounds start with an XOR of a subkey, followed by a layer of S-Boxes and a linear diffusion layer. The attacks on Feistel-3 presented in [18] reach up to 7 rounds for equal key and state sizes, and 11 rounds for twice larger keys.

We present attacks on Feistel-2 and Feistel-3 ciphers based on the meet-in-the-middle cryptanalytic technique. Its most basic form corresponds to the textbook case of Double-DES [22] and in the past few years, a few improvements have been proposed to more specific cases, for instance, Dinur et al. [11] have generalized the attack on Double-DES when multiple encryption (more than two n-bit keys) is used. Besides the applications to preimage attacks on hash functions [1,16,23], a notable application of the meet-in-the-middle technique and a line of research that has been started by Demirci and Selçuk [8] are the attacks on the Advanced Encryption Standard (AES). They presented cryptanalysis of AES-192 and AES-256 reduced to 8 rounds by improving the collision attack due to Gilbert and Minier [14] and with the use of the meet-in-the-middle technique. Later, their strategy has been revisited by Dunkelman, Keller and Shamir [12], and most recently further improved by Derbez, Fouque and Jean [9,10]. In this advanced form, the attack combines both the classical differential attack and the meet-in-the-middle strategy. In the differential attack, a high-probability differential is used to detect statistical biases to deduce information on the last subkey used in a block cipher. The attacker detects correct subkey guesses by checking meet-in-the-middle equations during the encryption process. Namely, the attack starts with a precomputation phase which is used to fully tabulate the distinguishing behavior particular to the targeted cipher, e.g. AES, and later in the online phase, the attacker searches for messages verifying the distinguisher by checking the precomputed table.

Our Contributions. We show the best known generic attacks on Feistel-2 and Feistel-3 cipher constructions. Our analysis, and a preliminary step of the attacks, relies on a special differential behavior of several consecutive rounds that is inherited by the generic Feistel construction. This property can be seen as a distinguisher, and for Feistel-2 it extends to 5 rounds, while for Feistel-3 to 7 rounds. The attacks exploit the distinguishers, and by adding rounds before, in the middle, and after the distinguisher, they can penetrate higher number of rounds. The distinguisher allows the differential behavior of the Feistel rounds to be enumerated offline and without the knowledge of the actual subkeys. This in fact is the first step of our attacks: a precomputation phase used to create a large look-up table. The next step is the collection of a sufficient number of plaintext/ciphertext pairs, some of which will comply with the conditions of the distinguisher. Each such pair suggests candidates for the round subkeys, and the look-up table is used to filter the correct subkeys. This step is indeed the meet-in-the-middle part of the attack.

In the case of the Feistel-2 construction, the number of rounds that our attacks can reach depends on the ratio of key to state sizes k/n: the larger the ratio, the more rounds we can attack. Namely, $4s + 2$ rounds can be attacked for $k/n = (s+1)/2$, which translates to 6 rounds when $k = n$, 8 rounds for $k = 3n/2$, 10 for $k = 2n$, etc. As long as the ratio is increasing, the number of attacked rounds will grow. This property comes from the meet-in-the-middle nature of the attacks, i.e. when we increase the key by bit size equivalent to one Feistel branch (and thus allow the complexity of the attack to increase by this amount), then we can add one round to the distinguisher in the offline phase, and prepend one round in the online phase. Since the attack relies on the meet-in-the-middle strategy, the complexities of these two phases are not multiplied but simply added, hence the accumulative complexity remains below the trivial exhaustive key search. In the analysis of Feistel-2, regardless of the number of attacked rounds, we make no assumptions on the type of the round functions: they can be any invertible or one-way functions or permutations, unique for each round. What we assume, however, is that the round functions have standard differential behavior. That is, given a large set of input-output differences of these functions (which can be seen as a set of differentials), on average for each differential there is one solution that conforms to it.

For the Feistel-3 construction and a linear diffusion layer P with maximal branch number, we can attack up to 14 rounds of the ciphers when the key is twice as large as the state ($k = 2n$), while for smaller keys we have attacks on 12 and 10 rounds, for key sizes $k = 3n/2$ and $k = n$, respectively. The above generalization (the number of attacked rounds always increases when the key size increase) is no longer possible as the data complexity grows beyond the full codebook when key size is more than $2n$ bits. To reach more rounds compared to Feistel-2, we use the SPN structure of the round function in both the offline and online stages of the attack. The best such example given in the paper is the redefinition of the Feistel-3 by moving the linear layer from one round to the surrounding rounds: this allows to extend the attack by an additional round. Other improvements based on the SPN structure are better (in terms of number of rounds) distinguisher and key recovery. For the main Feistel-3 attacks, we assume that the P-layers of all rounds are the same, but in case they are different, we show that the attacks can be adapted on only one round less.

Table 1. Comparison of previous results and ours for n-bit block-length, k-bit key-length and c-bit S-Box length

Target	Round functions	#rounds and complexity						Reference
		$k = n$		$k = 3n/2$		$k = 2n$		
Feistel-2	bijective	5	$2^{3n/4}$	6	2^{n}	7	$2^{3n/2}$	[20]
	—	3	$2^{n/2}$	5	2^{n}	7	$2^{3n/2}$	[17]
	—	5	$2^{n/2}$	7	$2^{5n/4}$	9	$2^{3n/2}$	[18]
	bij., ident.	6	$2^{n/2}$	—	—	—	—	[25]
	—	**6**	$2^{3n/4}$	**8**	$2^{4n/3}$	**10**	$2^{11n/6}$	**Section 3**
Feistel-3	—	7	$2^{3n/4+c}$	9	2^{n+c}	11	$2^{7n/4+c}$	[18]
	—	**9**	$2^{n/2+4c}$	**11**	2^{n+4c}	**13**	$2^{3n/2+4c}$	**Section 4**
	identical	**10**	$2^{n/2+4c}$	**12**	2^{n+4c}	**14**	$2^{3n/2+4c}$	**Section 4**

Our analysis results in a recovery of the whole values (not only partial values or bytes) of certain subkeys. This is the main advantage of the attack, and by repeating it a few times, we can recover one by one all the subkeys and thus be able to encrypt and decrypt without the knowledge of the initial master key. Hence, the key schedule plays no role in the analysis and the attacks are in fact an all-subkey recovery. We have also experimentally confirmed the validity of our analysis on the case of small state Feistel-2[1]. The experiments ran on a regular PC supported the complexity evaluation and the correctness of the attacks. All of the results described in this paper are summarized in Table 1 and compared to the already-published generic analysis on Feistel-2 and Feistel-3.

Due to space constraints, in the sequel, we present only our main ideas that result in 6-round attack on Feistel-2 and 10-round attack on Feistel-3. The full version of the paper, including additional attacks, the technique to recover all the subkeys and the experimental results can be found in [15].

2   Preliminaries

Throughout the paper, we assume that the block size is n bits and the Feistel is balanced, thus the branch size is $n/2$ bits. The internal state value (the branch) is denoted by v_i and the n-bit plaintext is assigned to $v_0 \| v_{-1}$. We count the rounds starting from 0, and at round i, v_{i+1} is computed as $v_{i+1} \leftarrow$ RoundFunction(v_i, v_{i-1}, K_i). The round function depends on the class defined further, i.e. it is either Feistel-2 or Feistel-3. In the description of the attacks, we omit the network twist in the last round as it has not cryptographic significance.

Generic Feistel-2 Construction. A Feistel-2 round function consists of a subkey XOR and a subsequent public function as illustrated in Figure 1. Several

---

[1] The interested reader can find the implementations of our attacks at
http://www1.spms.ntu.edu.sg/~syllab/attacks/F2-6rounds.tar.gz and
http://www1.spms.ntu.edu.sg/~syllab/attacks/F2-8rounds.tar.gz.

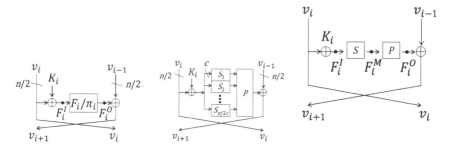

Fig. 1. Feistel-2   **Fig. 2.** Feistel-3   **Fig. 3.** Simplified Feistel-3

classes of public functions can be considered. Typical classifications are bijective or non-bijective, invertible or non-invertible, and different functions for different rounds or an identical function for all rounds.

Generic Feistel-3 Construction. A Feistel-3 round function consists of a subkey XOR, an S-layer, and a P-layer. The S-layer performs word-wise S-Boxes applications, while the P-layer performs a linear operation for mixing all words. Several classes of S-layers and P-layers can be considered. An example of the classification of the S-layer is different S-Boxes for different words or an identical S-Box for all words. The P-layers can be classified according to the branch number[2] of the linear transformation used in the layer. In our analysis, if c is the bit size of a word, then the internal state value has $n/2c$ words, and we assume that the branch number of the linear operation in the P-layer is $n/2c+1$, i.e. it is maximal. For example, a multiplication by an MDS matrix produces the maximal branch number of $n/2c + 1$. The Feistel-3 construction is shown in Figure 2. We often use the simplified description given in Figure 3.

Solutions of Differential Equations. In our analysis, we make the following assumption on the non-linear round functions F_i of the Feistel cipher. We assume that given a large set of fixed input and output differences of F_i, i.e. $(\Delta_{I_j}, \Delta_{O_j}), j = 1, 2, \ldots$, then on average there is one solution of each of the differential equations $F_i(X \oplus \Delta_{I_j}) \oplus F_i(X) = \Delta_{O_j}, j = 1, 2, \ldots$. That is, some of the equations may have many solutions and some none, however, we assume that on average (over a large set) the number of solution is one per equation. This requirement is sufficient for our analysis, as we solve the differential equations for a large number of (Δ_I, Δ_O), thus we can take the average case which is one solution per equation. Our computer simulations of the attacks confirmed this expectation and the complexity of the attacks was as predicted by our analysis, in part because the aforementioned assumption is true in the case of randomly chosen (Feistel-2 and Feistel-3) non-linear round functions. There are examples of round functions[3] where the assumption does not hold, for instance, linear

---

[2] The branch number of a linear transformation is the minimum number of active/non-zero input and output words over all inputs with at least one active/non-zero word.

[3] We do not claim attacks on Feistel-2 that have this type of round functions.

functions[4]. However, to the best of our knowledge, such round functions are either not used as building blocks of ciphers, or they can be attacked using other, more trivial attacks.

It is important to notice that although one solution is expected, it does not mean that it can be found trivially. To solve most of the equations, we use precomputation tables, i.e. we tabulate the functions, store their values, and later perform table lookups to solve the differential equations.

Definition 1 (δ-Set, [7]). *A δ-set for byte-oriented cipher is a set of 2^8 state values that are all different in 1 byte and are all equal in the remaining bytes.*

We introduce slightly modified definition (without byte-oriented sets).

Definition 2 (b-δ-Set). *A b-δ-set is a set of 2^b state values that are all different in b state bits (the active bits) and are all equal in the remaining state bits (the inactive bits).*

By this definition, the original Knudsen's δ-set from [7] can be seen as an 8-δ-set, since it takes all the values of a particular byte, which is an 8-bit value. To define b-δ-set, we have to specify not only the value of b, but also the position of the active bits. In some cases, however, the position is irrelevant and the analysis is applicable for any b active bits.

Given a state value v, we can construct a b-δ-set from v, by applying $2^b - 1$ differences to some b bits of the state v. Furthermore, we can take a function F, order all the possible $2^b - 1$ input differences, and obtain a sequence of output differences of F. An example of such sequence, when the active bits are the least significant bits, is $F(v) \oplus F(v \oplus 1), F(v) \oplus F(v \oplus 2), \ldots, F(v) \oplus F(v \oplus 2^b - 1)$.

The Attack Model. The key-recovery attacks presented in the paper follow the standard attack model. That is, the key of the block cipher is secret and chosen uniformly at random. The attacker can query both the encryption and the decryption functions of the block cipher. His task is to recover the secret key (or the subkeys produced from the key schedule) based on the queries. We explicitly state that the attacker has no information about the internal state values of the block cipher.

3   Key-Recovery Attacks against Feistel-2 Construction

In this section, we present a key-recovery attack on 6-round Feistel-2 ciphers for the case when the key and the state sizes are equal, i.e. $k = n$. The extensions of the attack to 8 rounds for $k = 3n/2$, 10 rounds for $k = 2n$, and in general to $(4 + 2s)$ rounds for $k = n(s + 1)/2$, can be found in the full version of the paper [15]. In our attack, the round functions can be either bijective or non-bijective, i.e. permutations or functions, and they can even be one-way. To make

---

[4] For linear function, the probability that a solution exist depends on the size of the large set.

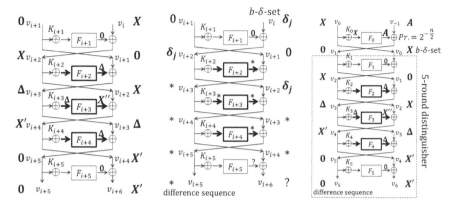

Fig. 4. 5-round differential characteristic

Fig. 5. b-δ-set construction

Fig. 6. 6-round key-recovery

the attack applicable to the most general type of constructions, in the sequel, we assume that the round functions are one-way and pairwise distinct.

We use F_i to denote the round function at round i of the construction. To refer to the input (resp. output) of F_i, we write $F_i^{\mathcal{I}}$ (resp. $F_i^{\mathcal{O}}$). Similarly, the input difference (resp. output difference) of F_i is denoted by $\Delta F_i^{\mathcal{I}}$ (resp. $\Delta F_i^{\mathcal{O}}$). Recall that the two branches, as well as the subkeys K_i, have $n/2$ bits each.

The 6-round key-recovery attack is based on a non-ideal behavior of 5 rounds of Feistel-2, which is described by the lemma and the proposition that follow. In the 6-round attack (refer to Figure 6), the last five rounds are the rounds where this distinguisher is used.

Lemma 1. *Let X and X', where $X \neq X'$, be two non-zero branch differences. If a 5-round Feistel-2 encrypts a pair of plaintexts (m, m') with difference $0 \| X$ to a pair of ciphertexts with difference $0 \| X'$, then the number of possible internal state values of the three middle rounds that correspond to the plaintext m is limited to $2^{n/2}$ on average.*

Proof. Note that $n/2$-bit round keys are added in each round, and hence the number of possible internal state values for the three middle rounds is limited by its size, $2^{3n/2}$. We show, however, that the bound can be tightened to $2^{n/2}$.

A 5-round differential characteristic, with input difference $0 \| X$ and output difference $0 \| X'$ is depicted in Figure 4 (the rounds are denoted from $i + 1$ to $i + 5$ to make this part of the analysis generic). From the figure, we can see that after the first round, the input difference $(0, X)$ must become a state difference $(X, 0)$. Similarly, after the inversion of the last round the output difference $(0, X')$ becomes $(0, X')$. This makes $\Delta F_{i+3}^{\mathcal{O}}$ to be $X'' \leftarrow X \oplus X'$. Since $X \neq X'$, it follows that $X'' \neq 0$ and thus $\Delta F_{i+3}^{\mathcal{I}} \neq 0$ – let us denote this difference with Δ. It means that both $\Delta F_{i+2}^{\mathcal{O}}$ and $\Delta F_{i+4}^{\mathcal{O}}$ also have the difference Δ. To summarize, we get that for each fixed Δ, the input and output differences of the round functions at rounds $i+2$, $i+3$, and $i+4$ are fixed. There exists one state value (one solution) that satisfies such input-output difference in each of the three rounds.

As Δ can take at most $2^{n/2}$ different values (one branch has $n/2$ bits), the states in rounds $i + 2$, $i + 3$, $i + 4$ can assume only $2^{n/2}$ different values. In Figure 4, the fixed value for each Δ is drawn by bold line. $\qquad\square$

We use Lemma 1 to prove the below proposition that will help us later to launch the attack on 6 rounds. To present the proposition, we need additional notations. Let $F : m \rightarrow F(m)$ be a 5-round Feistel-2 (we omit writing the key k as input) and let the function $F^\Delta : \{0,1\}^{\frac{3n}{2}} \rightarrow \{0,1\}^{\frac{n}{2}}$ be defined as $F^\Delta(m, \delta) = Trunc_{n/2}\Big(F(m) \oplus F(m \oplus (0\|\delta))\Big)$, where $Trunc_{n/2}$ denotes the truncation to the first $n/2$ bits. In other words, $F^\Delta(m, \delta)$ gives the output difference (of the left branch) in the pair of ciphertexts, produced by encryption of a pair of plaintexts $(m, m \oplus 0\|\delta)$ with the 5-round Feistel. Furthermore, instead of taking a single pair of plaintexts, let us create several pairs such that in each pair, the first element is always m, while the second is $m \oplus 0\|\delta_j$ where $\delta_j = 1, \ldots, 2^b - 1$ (the precise value of b is defined later in the section). In fact, we can see that the second elements of the pairs form a b-δ-sequence. The proposition given further claims that the sequence of differences in the ciphertexts pairs (that correspond to such plaintexts pairs) can take only $2^{n/2}$ values.

Proposition 1. *Let (m, m') be a pair of plaintexts that conforms to the 5-round differential characteristic given in Figure 4 and let $\delta_j = 1, \ldots, 2^b - 1, b \geq 1$ forms b-δ-sequence. Then, the sequence $F^\Delta(m, \delta_j)$, $\delta_j = 1, \ldots, 2^b - 1$ can assume only $2^{n/2}$ possible values.*

Remark 1. We note that the sequence can be constructed from any of the two plaintexts m or m' given in Proposition 1, as long as the pair (m, m') conforms to the differential characteristic.

Remark 2. From a theoretical point of view, Proposition 1 yields a distinguisher since the number of functions reached by the 5-round Feistel-2 construction is much less than the theoretical number of functions from a set of 2^b elements to a set of $2^{n/2}$ elements when $b \geq 1$. Indeed, for a fixed m, the latter equals $(2^{n/2})^{2^b} = 2^{2^b n/2}$, whereas it is only $2^{n/2}$ in the case of the 5-round Feistel-2 construction.

Proof. The initial pair of plaintexts (m, m') is only used to compute the state values of the three middle rounds that correspond to the plaintext m. We have seen from Lemma 1 that these three states can take only $2^{n/2}$ possible values (each of them corresponds to one of the values of Δ). We will show that if the values of these three states are fixed, then we can change the right half of the plaintext (instead of m, we take $m \oplus 0\|\delta_j$) and still be able to compute the output difference in the left half of the ciphertexts. In fact, we can change the value of the plaintext many times (i.e. we can produce many pairs of the form $(m, m \oplus 0\|\delta_j)$), and for each of them, we can easily compute the output difference in the right halves of the ciphertext. The number of plaintexts pairs adds no complexity in predicting the ciphertext difference – once the three middle states are fixed (and they can have only $2^{n/2}$ different values), the sequence of differences in the ciphertext pairs is uniquely determined.

Assume the difference Δ is fixed[5], and thus are fixed the three internal state values. Let $t_{i+2}, t_{i+3}, t_{i+4}$ be the input values to $F_{i+2}, F_{i+3}, F_{i+4}$ that correspond to the plaintext m, in which $t_{i+2}, t_{i+3}, t_{i+4}$ are determined depending on Δ. Let v_i be the values of the states that correspond to the plaintext m as shown in Fig. 5. Let us consider a new pair of plaintexts, $(m, m \oplus (0\|\delta_j))$, i.e. we introduce a difference δ_j to the right branch, i.e. $\Delta v_i = \delta_j$. Since the difference $\Delta F_{i+1}^{\mathcal{O}}$ is always zero, we obtain that $\Delta v_{i+2} = \Delta v_i = \delta_j$. In round $i+2$, the attacker knows the value of $F_{i+2}^{\mathcal{I}} = t_{i+2}$ and the difference $\Delta F_{i+2}^{\mathcal{I}} = \delta_j$. Hence, the new paired values of $F_{i+2}^{\mathcal{I}}$ are t_{i+2} and $t_{i+2} \oplus \delta_j$. Therefore, the new $\Delta F_{i+2}^{\mathcal{O}}$ can be obtained as $\Delta F_{i+2}^{\mathcal{O}} \leftarrow F_{i+2}(t_{i+2}) \oplus F_{i+2}(t_{i+2} \oplus \delta_j)$. In Figure 5, we represent this type of computable difference with '$*$'. The new difference for $\Delta F_{i+2}^{\mathcal{O}}$ is propagated forward to v_{i+3} and the same reasoning as in round $i+2$ is applied to round $i+3$. As we know the value of $F_{i+3}^{\mathcal{I}} = t_{i+3}$ and $\Delta F_{i+3}^{\mathcal{I}} = \Delta F_{i+2}^{\mathcal{O}}$, it follows that $(t_{i+3}, t_{i+3} \oplus \Delta F_{i+2}^{\mathcal{O}})$ are the paired values. The new $\Delta F_{i+3}^{\mathcal{O}}$ can therefore be computed as $\Delta F_{i+3}^{\mathcal{O}} \leftarrow F_{i+3}(t_{i+3}) \oplus F_{i+3}(t_{i+3} \oplus \Delta F_{i+2}^{\mathcal{O}})$. The knowledge of $\Delta F_{i+3}^{\mathcal{O}}$ gives the difference for v_{i+4} for the next round, namely: $\Delta v_{i+4} \leftarrow \Delta F_{i+3}^{\mathcal{O}} \oplus \delta_j$. The analysis continues the same way for round $i+4$. From the knowledge of the value of $F_{i+4}^{\mathcal{O}} = t_{i+4}$ and the new difference $\Delta F_{i+4}^{\mathcal{O}} = \Delta v_{i+4}$, the output difference of the round function $\Delta F_{i+4}^{\mathcal{O}}$ is computed, and finally Δv_{i+5} is computed as $\Delta F_{i+4}^{\mathcal{O}} \oplus \Delta v_{i+3} = \Delta F_{i+4}^{\mathcal{O}} \oplus \Delta F_{i+2}^{\mathcal{O}}$.

In summary, for an arbitrary δ_j, we can compute the output difference Δv_{i+5}, i.e., the mapping from δ_j to Δv_{i+5} becomes deterministic (as long as Δ is fixed). Therefore, for the ordered sequence of δ_j that takes the values $1, 2, \ldots, 2^{n/2} - 1$, we can determine the sequence of corresponding differences Δv_{i+5} (which indeed is the difference in the left half of the ciphertext). We emphasize that the mapping depends only on values of $t_{i+2}, t_{i+3}, t_{i+4}$, which in turn are determined from the value of Δ, X and X', and acts independently of the value of m. Since Δ takes at most $2^{n/2}$ values, the number of sequences of Δv_{i+5} is limited to $2^{n/2}$.   □

6-Round Key-Recovery Attack. We prepend one round to the 5-round distinguisher shown in Figure 4 and the resulting construction is illustrated in Figure 6. The attack consists of precomputation and online phases. The online phase is further divided into collecting pair and key recovery phases. In the precomputation phase, we choose many pairs (X, X'), where X is fixed while X' takes multiple values, and for each pair, we find all possible $2^{n/2}$ sequences of Δv_5 based on Proposition 1. We store all the sequences in a large table along with its corresponding internal state values. Next, in the online phase, we collect many pairs that satisfy one of the differential characteristics $(X, 0) \rightarrow (X', 0)$. Finally, for each of the obtained pairs, we compute Δv_5 sequences by guessing the first round key K_0. We then find a match of Δv_5 sequences between the precomputed table and the one computed online – this allows us to determine the internal states and to recover K_0. The meet-in-the-middle nature of our attack comes from the fact that the Δv_5 sequence is computed offline for the last

---

[5] Recall that this difference corresponds to an internal state difference for the plaintext pair (m, m').

five rounds and online for the first round, and the results are later matched in a meet-in-the-middle-like fashion.

Precomputation. From Proposition 1, the number of possible sequences of Δv_5 is $2^{n/2}$ for a fixed X and a fixed X'. We can achieve a time/memory tradeoff by relaxing the $n/2$-bit constraint of a fixed X' and allow $2^{x'}$ different possible differences for X', where $0 \leq x' \leq n/2$. Without loss of generality, assume that the values of X' differ in the last x' bits and are the same in the remaining $n/2-x'$ most significant bits (MSBs). In the sequel, we will determine the optimal value for x' to reach the best time/data/memory complexities for the attack.

First, we show how to compute all $2^{x'} \cdot 2^{n/2} = 2^{x'+n/2}$ sequences of 2^b differences as an offline precomputation in $2^{x'+n/2+b}$ time (encryptions), and $2^{x'+n/2+b}$ memory (blocks of $n/2$ bits). This offline precomputation results in a table T_δ, that contains all the sequences. Since the precomputation step is the same for all X' differences, further we show the procedure for a particular X' and assume that for the whole offline execution this procedure is repeated $2^{x'}$ times for the possible values of X' differences.

In rounds 2 and 4, the input differences to the round functions are fixed to X and X', respectively, while both of the output difference are Δ. To reduce the time complexity, we first tabulate completely the round functions F_2, F_3 and F_4 and thus we will have a constant-time access to paired values for some input or output differences. Namely, we construct precomputation tables T_2 and T_4, which take the difference Δ as input and return the paired values conforming to the differentials $X \rightarrow \Delta$ and $X' \rightarrow \Delta$ through F_2 and F_4, respectively. The strategy consists simply in iterating over all possible inputs, and storing the results indexed by output difference as described in Algorithm 1.

Similarly, in round 3 we want to construct the table T_3 that gives in constant time a paired-value input to F_3 resulting in the fixed output difference X''. However, since the function F_3 is assumed to be one-way and in the attack we need to invert it, we cannot compute F_3^{-1} to construct T_3. Thus, we first evaluate F_3 for all input values, store the values in a temporary table, and later consider the difference, as detailed in Algorithm 2. After this part of the precomputation phase, for an arbitrary fixed difference Δ (which is the difference $\Delta F_2^{\mathcal{O}} = \Delta F_3^{\mathcal{I}} = \Delta F_4^{\mathcal{O}}$), the corresponding state values in rounds 2, 3, and 4 can be looked up in tables T_2, T_3, and T_4 in constant time. Hence, we can compute the b-δ-set for all the $2^{n/2}$ possible choices of Δ and store the resulting sequences in the precomputation table T_δ, which later is used for the meet-in-the-middle check of the online phase. This step is described in Algorithm 3.

Finally, another table T_0 of size $2^{n/2}$ is generated to make more efficient the online phase and the recovery of the subkey K_0. That is, in round 0, for all values of $F_0^{\mathcal{I}}$, the corresponding $\Delta F_0^{\mathcal{O}}$ is computed. Namely, for $i = 0, 1, \ldots, 2^{n/2} - 1$, $F_0(i) \oplus F_0(i \oplus X)$ is computed and stored in T_0.

As stated previously, we repeat this procedure for $2^{x'}$ different choices of the difference X'. For the sake of simplicity, the resulting tables for each X' are all merged in the same table T_δ. For a fixed choice of X', building T_0, T_2, T_3 and T_4 requires $2^{n/2}$ round function computations each. Hence, constructing T_δ requires less[6] than $2^b \cdot 2^{n/2}$ encryptions. The entire analysis is iterated over $2^{x'}$ choices of X' so that the computational cost is less than $2^{x'+b+n/2}$ encryptions. The memory requirement to build T_0, T_2, T_3 and T_4 is $2^{n/2}$ blocks of $n/2$ bits, and is constant as we can reuse the memory across different X'. The size of T_δ increases with the iteration of $2^{x'}$ choices of X', namely, the memory requirement for the precomputation phase amounts to $2^b \cdot 2^{x'+n/2} = 2^{x'+n/2+b}$ blocks of $n/2$ bits.

Collecting Pairs. In the data collection phase, we query the encryption oracle with chosen plaintexts to get enough pairs such that one conforms to the whole 6-round differential characteristic. To do so, we construct a structure of $2^{n/2+1}$ plaintexts that consists of two lists of sizes $2^{n/2}$. All the elements of the first list are fixed to a constant random value v_0 on their left half, while the right halves are pairwise distinct. The second list is constructed similarly, except that the left half is fixed to $v_0 \oplus X$. As a result, we have 2^n pairs of plaintexts such that the difference in the left half equals X and the right half is nonzero.

For a single structure, the data complexity corresponds to encryption of $2^{n/2+1}$ chosen plaintexts, which can subsequently be sorted by their ciphertext values to detect the pairs that match on their left half ($n/2$ bits) and $n/2 - x'$ most significant bits of the right half. Consequently, we expect one structure of plaintexts to provide $2^n/2^{n/2+n/2-x'} = 2^{x'}$ pairs conforming to the truncated output difference, i.e. such that only the x' less significant bits of the right half are nonzero. To complete the attack, we need $2^{n/2}$ pairs, as the difference cancellation at the output of the first round holds with probability $2^{-n/2}$. Hence by repeating the data collection for $2^{n/2-x'}$ different values of v_0, we can expect one pair among the $2^{n/2}$ to follow the whole characteristic. Therefore, the data complexity amounts to $2^{n/2-x'} \times 2^{n/2+1} = 2^{n-x'+1}$ chosen plaintexts, requires the same amount of memory access as time complexity to be generated, and can be stored using only $2^{n/2}$ elements with the use of a hash table for the pairs that verify the truncated output difference. The whole procedure is described in Algorithm 4.

Recovery of K_0. The previous phase results in $2^{n/2}$ candidate pairs with a plaintext difference $(X, \Delta v_{-1})$ and an appropriate ciphertext difference. For each pair, we match against the precomputed table T_0 to find the corresponding value of $F_0^\mathcal{I}$, and thus determine uniquely a subkey candidate for K_0 by $K_0 \leftarrow v_0 \oplus F_0^\mathcal{I}$.

However, among these $2^{n/2}$ candidates for K_0, only one is correct while the remaining are false positives. To find the correct subkey, we use the results of Proposition 1 and the precomputation table T_δ, i.e. we construct a b-δ-set by modifying the active bits of v_0. For each modified plaintext, with the knowledge of K_0, we compute the corresponding $F_0^\mathcal{O}$ and modify v_{-1} so that the value of v_1 stays unchanged. Then, we query the plaintexts and observe the left half of the

---

[6] Less, as one evaluation of the round functions costs less than one encryption query.

Algorithm 1. Construction of the tables T_2 and T_4

1: **for** $i = 0, 1, \ldots, 2^{n/2} - 1$ **do**
2:     Compute $\Delta F_2^{\mathcal{O}} \leftarrow F_2(i) \oplus F_2(i \oplus X)$.
3:     Store $(i, \Delta F_2^{\mathcal{O}})$ in T_2 indexed by $\Delta F_2^{\mathcal{O}}$.
4:     Compute $\Delta F_4^{\mathcal{O}} \leftarrow F_4(i) \oplus F_4(i \oplus X')$
5:     Store $(i, \Delta F_4^{\mathcal{O}})$ in T_4 indexed by $\Delta F_4^{\mathcal{O}}$.

Algorithm 2. Construction of the table T_3

1: **for** $i = 0, 1, \ldots, 2^{n/2} - 1$ **do**
2:     Store $(i, F_3(i))$ in a temporal table `tmp` indexed by $F_3(i)$.
3: **for** $i = 0, 1, \ldots, 2^{n/2} - 1$ **do**
4:     Compute $F_3(i) \oplus X''$.
5:     Look up `tmp` to obtain j such that $F_3(j) = F_3(i) \oplus X''$.
6:     Store $(i, i \oplus j)$ in T_3 indexed by $i \oplus j$.

Algorithm 3. Construction of the sequences of Δv_5

1: **for** $\Delta = 1, \ldots, 2^{n/2} - 1$ **do**
2:     Obtain internal state values $F_2^{\mathcal{I}}$, $F_3^{\mathcal{I}}$ and $F_4^{\mathcal{I}}$ by looking up T_2, T_3 and T_4,
        respectively.
3:     **for** all b active bits of the b-δ-set **do**
4:         Modify Δv_0, and compute the corresponding Δv_5.
5:     Compute the sequence of Δv_5 and add it to T_δ.

Algorithm 4. Data collection phase of the 6-round attack

1: Choose $2^{x'}$ differences X' so that the $n/2 - x'$ MSBs of X' are 0 for all X'.
2: Choose a difference X such that $X \neq X'$.
3: **for** $2^{n/2-x'}$ different values of v_0 **do**
4:     **for** all $2^{n/2}$ choices of v_{-1} **do**
5:         Query (v_0, v_{-1}) and store it in L_0 sorted by the ciphertext value.
6:         Query $(v_0 \oplus X, v_{-1})$ and store it in L_1 sorted by the ciphertext value.
7:         Pick up the elements of $L_0 \times L_1$ whose ciphertexts match
            in the $n - x'$ most significant bits.

---

corresponding ciphertexts. Hence, we can compute the sequence of Δv_5. If this sequence is included in the precomputation table T_δ, K_0 is a correct guess with high probability, otherwise it is wrong. We note that this does not increase the data complexity, since the structures of plaintexts already includes the plaintexts for the b-δ-set evaluation.

Complexity Analysis.   In the online phase of the attack, we perform $2^{n/2}$ checks in the precomputed table T_δ that contains all the possible stored sequences of differences. If we do not store enough information in this table (if b is too small), many checks will wrongly yield to valid subkey candidates K_0. On the other hand, if we store too much information (if b is too large), the table will require higher time and memory complexity to be constructed. Thus, we need to select an optimal value of b. One check yields a false positive with probability

$2^{n/2}/2^{n2^b/2} = 2^{n(1-2^b)/2}$ as there are $2^{n/2}$ valid sequences of 2^b elements among the $2^{n2^b/2}$ theoretically possible ones. Therefore, we want $n(1-2^b)/2 + n/2 < 0$ so that among all the $2^{n/2}$ checks, only the correct K_0 results in a stored element, and thus $b \geq 2$.

In terms of tradeoff, adjusting the value x' balances the data, time and memory complexities. The data complexity is $2^{n-x'+1}$ chosen plaintexts, the time complexity is $2^{x'+n/2}$ encryptions to construct T_δ and $2^{n-x'+1}$ memory access to query the encryption oracle. The memory complexity is also $2^{x'+n/2}$ blocks of $n/2$ bits required to store T_δ. Consequently, the choice of $x' = n/4$ makes the data complexity to become about $2^{3n/4}$ chosen plaintexts, the time complexity equivalent to about $2^{3n/4}$ encryptions, and the memory complexity to $2^{3n/4}$ blocks of $n/2$ bits.

4   Key-Recovery Attacks against Feistel-3 Construction

In this section, we present a 10-round key-recovery attack on the Feistel-3 construction with $k = n$. In the attack, we assume that different S-Boxes are used for different words in a given round, but we consider they are the same across all of the rounds. Recall that all the S-Boxes operate on c-bit words, and thus there are $\frac{n}{2c}$ words per branch. We consider that the P-layer is identical for all rounds and it has the maximal branch number of $\frac{n}{2c} + 1$. The extensions of the attack to 12 and 14 rounds for key sizes of $k = 3n/2$ and $k = 2n$, respectively, and the analysis of a class of P-layers that not necessarily has a maximal branch number are given in the full version of the paper [15].

The 10-round key-recovery attack is based on a non-ideal behavior of 7 rounds of Feistel-3. We first present the 7-round distinguisher in the proposition below, and then use it to launch a key-recovery attack on a 10-round Feistel-3 primitive where the inner rounds are the ones from the distinguisher. To construct the distinguisher, we first apply an equivalent transformation to the 7-round primitive, as shown in Figure 7. Namely, the P-layer of round $i + 6$ is removed from this round, and linear transformations are added to three different positions in order to obtain a primitive that is computationally equivalent to the original one. Hereafter, v'_{i+7} represents the value of $P^{-1}(v_{i+7})$. We use the non-ideal behavior of the new representation to mount the 10-round key recovery attack by extending the 7-round differential by one round at the beginning and two rounds at the end. The newly-introduced P after v_{i+7} is later addressed in the key-recovery part.

As in the previous section, $F_i^{\mathcal{I}}$ and $\Delta F_i^{\mathcal{I}}$ denote the input value and input difference of the i-th round, respectively, that is the input to the S-layer in F_i. Similarly, $F_i^{\mathcal{M}}$ and $\Delta F_i^{\mathcal{M}}$ refer to the state value and state difference after the S-layer, that is between the S-layer and P-layer of F_i, and $F_i^{\mathcal{O}}$ and $\Delta F_i^{\mathcal{O}}$ denote the output value and output difference of the P-layer in F_i, respectively. For the branch-wise difference, we use $\mathbf{0}$ to refer to branch with no active words, $\mathbf{1}$ to the case when only a single pre-specified word is active, and $\mathcal{P}$ and $\mathcal{P}^{-1}$ for branch-wise differences obtained after $\mathbf{1}$ has been processed by P and P^{-1}, respectively.

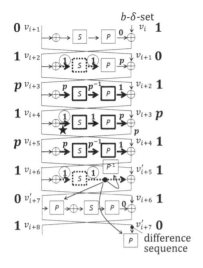

Fig. 7. 7-round differential

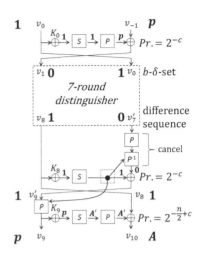

Fig. 8. 10-round key-recovery for $k = n$

Finally, $X[1]$ and $\Delta X[1]$, respectively, denote the pre-specified active-word value and difference of a branch-wise variable X.

The technique used to construct the 7-round distinguisher (described in the proposition below) is very similar to the technique we have used in the distinguisher on five rounds of Feistel-2. In other words, first we show that if a pair (m, m') of plaintexts follows a particular differential characteristic, then the number of possible internal state values that correspond to m is limited. Based on this, we can introduce a difference in the plaintext and predict the output difference in the ciphertext. Again, we introduce many pairs of plaintexts where each right half differs on δ_j (and thus get a b-δ-sequence) and observe that the pairs of ciphertexts have predictable difference. Unlike the proposition for Feistel-2 where we observed the difference in the left half of the ciphertext, for Feistel-3, we check the difference in one word of the right half in the ciphertext pairs (the position of this particular word plays no role in the analysis). That is why we have to redefine $F^\Delta(m, \delta_j)$. To avoid bulky notations, we define it informally as one-word difference in the right half of the ciphertext pair that are produced from the encryption of a plaintext pair $(m, m \oplus 0 \| \delta_j)$ through 7-round Feistel-3. In Figure 7, this is the ciphertext difference in the word v'_{i+7}.

Proposition 2. *Let (m, m') be a pair of plaintexts that conforms to the 7-round differential $(\mathbf{0}, \mathbf{1}) \overset{7R}{\to} (\mathbf{1}, \mathbf{0})$ shown in Figure 7 and let $\delta_j = 1, 2, \ldots 2^b - 1$ forms a b-δ-sequence. Then, the sequence $F^\Delta(m, \delta_j)$, $\delta_j = 1, \ldots, 2^b - 1$ can assume only $2^{n/2+4c}$ possible values.*

Proof. We show here that the number of internal state values for pairs satisfying the 7-round differential in Figure 7 is at most $2^{n/2+4c}$. Namely, we show they can be parameterized by five nonzero differences in five c-bit words (marked by circles in Figure 7), and by the values of $n/2 - c$ inactive bits of $F_{i+4}^{\mathcal{I}}$ (marked by a star '★' in Figure 7).

We first assume that the five word differences circled in Figure 7 are fixed, that is: $\Delta F_{i+2}^{\mathcal{I}}$, $\Delta F_{i+2}^{\mathcal{M}}$, $\Delta F_{i+4}^{\mathcal{I}}$, $\Delta F_{i+6}^{\mathcal{I}}$ and $\Delta F_{i+6}^{\mathcal{M}}$ are fixed to random nonzero values. When $\Delta F_{i+2}^{\mathcal{I}}$ and $\Delta F_{i+2}^{\mathcal{M}}$ are fixed, we expect one value on average to be determined for $F_{i+2}^{\mathcal{I}}[1]$. In Figure 7, the state in which the value is fixed only in one word is represented by dotted lines. Then, the corresponding $\Delta F_{i+2}^{\mathcal{O}} = \Delta v_{i+3} = \Delta F_{i+3}^{\mathcal{I}}$ can be fully computed linearly by $P(\Delta F_{i+2}^{\mathcal{M}})$. Since the branch number of P is $n/2c + 1$, $P(\Delta F_{i+2}^{\mathcal{M}})$ is fully active. Similarly, when $\Delta F_{i+6}^{\mathcal{I}}$ and $\Delta F_{i+6}^{\mathcal{M}}$ are fixed, one value on average can be determined for $F_{i+6}^{\mathcal{I}}[1]$, and the corresponding fully active difference $\Delta v_{i+5} = \Delta F_{i+5}^{\mathcal{I}}$ can also be computed linearly by $P(\Delta F_{i+6}^{\mathcal{M}})$. Then, $\Delta F_{i+4}^{\mathcal{O}}$ is computed by $\Delta v_{i+3} \oplus \Delta v_{i+5}$, where both Δv_{i+3} and Δv_{i+5} are of type $\mathcal{P}$. Since P is linear, $\Delta F_{i+4}^{\mathcal{O}}$ also has the form $\mathcal{P}$, which implies that the form of $\Delta F_{i+4}^{\mathcal{M}}$ is $P^{-1}(\mathcal{P}) = \mathbf{1}$. Then, the middle difference $\Delta F_{i+4}^{\mathcal{I}}$ is considered fixed. When $\Delta F_{i+4}^{\mathcal{I}} \neq \Delta F_{i+2}^{\mathcal{I}}$ and $\Delta F_{i+4}^{\mathcal{I}} \neq \Delta F_{i+6}^{\mathcal{I}}$, the corresponding differences $\Delta F_{i+3}^{\mathcal{O}}$ and $\Delta F_{i+5}^{\mathcal{O}}$ are computed by simply taking their XOR. Thus, both $\Delta F_{i+3}^{\mathcal{O}}$ and $\Delta F_{i+5}^{\mathcal{O}}$ are of type $\mathbf{1}$, which makes $\Delta F_{i+3}^{\mathcal{M}}$ and $\Delta F_{i+5}^{\mathcal{M}}$ fully active (denoted by $\mathcal{P}^{-1}$). Then, the values of $F_{i+3}^{\mathcal{I}}, F_{i+3}^{\mathcal{M}}, F_{i+3}^{\mathcal{O}}$ and $F_{i+5}^{\mathcal{I}}, F_{i+5}^{\mathcal{M}}, F_{i+5}^{\mathcal{O}}$ are uniquely determined, as well as the values for $F_{i+4}^{\mathcal{I}}[1], F_{i+4}^{\mathcal{M}}[1]$.

Finally, when we additionally consider the $n/2 - c$ inactive bits of $F_{i+4}^{\mathcal{I}}$ marked by a star in Figure 7 being fixed, along with the already-fixed c bits of the active word $\mathbf{1}$, the full $n/2$-bit values of $F_{i+4}^{\mathcal{M}}$ and $F_{i+4}^{\mathcal{O}}$ are determined. In summary, for each value of the five c-bit active differences circled in Figure 7 and the $n/2 - c$ inactive bits of $F_{i+4}^{\mathcal{I}}$, all the differences of the differential as well as one word values in rounds $i + 2$, $i + 6$, and all state values in rounds $i + 3$, $i + 4$, $i + 5$ are uniquely fixed.

For each of $5c + n/2 - c = n/2 + 4c$ word parameters, we can partially evaluate a b-δ-set v_i up to $\Delta v_{i+7}'[1]$. Namely, for one member of the pairs, $v_i[1]$ is modified so that $\Delta v_i[1]$ becomes δ_j. The modification changes the difference in subsequent rounds, but we can still compute the corresponding difference $\Delta v_{i+7}'[1]$ without requiring the knowledge of the subkey bits.

Indeed, in round $i + 1$, $\Delta F_{i+1}^{\mathcal{O}} = 0$, $\Delta v_{i+2} = \Delta F_{i+2}^{\mathcal{I}} = \delta_j$. In round $i + 2$, from the original active word value of $F_{i+2}^{\mathcal{I}}$ and updated difference $\Delta F_{i+2}^{\mathcal{I}} = \delta_j$, the updated $\Delta F_{i+2}^{\mathcal{O}}$ can be computed as $P \circ S(F_{i+2}^{\mathcal{I}}) \oplus P \circ S(F_{i+2}^{\mathcal{I}} \oplus \delta_j)$. This also derives the updated differences Δv_{i+3} and $\Delta F_{i+3}^{\mathcal{I}}$. Then, in round $i + 3$ to $i + 5$, from the original value and the updated difference of $F_x^{\mathcal{I}}$, the updated difference $\Delta F_x^{\mathcal{O}}$, and moreover the updated differences Δv_{x+1} and $\Delta F_{x+1}^{\mathcal{I}}$ can be computed for $x = i + 3, i + 4, i + 5$. Note that, in round $i + 4$, $\Delta F_{i+4}^{\mathcal{I}}$ originally has only one active word, while the updated difference is fully active. Because $n/2 - c$ inactive bits of $F_{i+4}^{\mathcal{I}}$ are parameters, and thus known to the attacker, $\Delta F_{i+4}^{\mathcal{M}}$ can be computed in all words. Finally, in round $i + 6$, the updated difference Δv_{i+6} is known in all words while the original value is known only in one active word. Since the position of the P-layer is moved, the attacker can still compute the 1-word updated difference $\Delta v_{i+7}'[1]$.

To conclude, for each of the $2^{n/2+4c}$ possible values of the parameters, the sequence of $\Delta v'_{i+7}[1]$ is uniquely obtained by computing $\Delta v'_{i+7}[1]$ for all δ_j in $\Delta v_i[1]$, which concludes the proof.   □

10-Round Key-Recovery Attack. Let us describe the 10-round key-recovery attack that uses the 7-round distinguisher. As shown in Figure 8, we extend the 7-round differential characteristic of the distinguisher by one round at the beginning and two rounds at the end (the analysis and complexity would be similar if we extend by two rounds at the beginning and one at the end). Recall that the additional P-layer after v'_7, introduced by the distinguisher, has to be addressed in the key-recovery part. We also note that the active word **1** in the branches can be located in any position, but the position has to be fixed beforehand to be able to conduct the attack. The P-layer in round 8 is moved to two different positions as shown in Figure 8. The newly-introduced P^{-1} transformation and the P transformation after v'_7 generated by the distinguisher cancel each other, we therefore ignore them. Similarly to the analysis for Feistel-2, the attack consists of three parts: the precomputation phase, followed by the data collection and finally the meet-in-the-middle check to detect correct subkey candidates.

Precomputation. Given the proof of Proposition 2, the precomputation phase is straightforward. For each of the $2^{n/2+4c}$ values of the parameters, and for any value of δ_j constructed at v_0, the corresponding $\Delta v'_7[1]$ can be computed easily as shown in Algorithm 5. As in the attack on Feistel-2, in this phase we construct the meet-in-the-middle table T_δ that contains all the sequences of differences in $\Delta v'_7[1]$ for $2^b < 2^c$ nonzero differences δ_j in v_0. The computational cost is about $2^{n/2+4c}$ encryptions as the b parameter is relatively small and we consider only a small fraction of all the rounds. Storing T_δ requires $2c/n \times 2^{n/2+4c+b}$ blocks of $n/2$ bits, as the sequences contains 2^b elements of c bits.

Collecting Pairs. To launch the attack, we need a pair that satisfies the 7-round differential characteristic in Figure 7, i.e. the plaintext difference $(\mathbf{1}, \mathcal{P})$ should propagate to the ciphertext difference $(\mathcal{P}, A)$, where A is a truncated difference. The probability that the plaintext difference $(\mathbf{1}, \mathcal{P})$ after the first round becomes $(\mathbf{0}, \mathbf{1})$ is 2^{-c}, while the probability that the ciphertext difference $(\mathcal{P}, A)$ after inversion of the last round becomes $(\mathbf{1}, \mathbf{1})$ is $2^{-n/2+c}$, and to become $(\mathbf{1}, \mathbf{0})$ after another inverse round is 2^{-c}. Therefore, a random pair verifying a plaintext difference $(\mathbf{1}, \mathcal{P})$ conforms to the inner 7-round differential with probability $2^{-n/2-c}$. Hence, we need to collect $2^{n/2+c}$ pairs satisfying the differential $(\mathbf{1}, \mathcal{P}) \overset{10R}{\to} (\mathcal{P}, A)$. Among all of them, one is expected to satisfy $(\Delta v_1, \Delta v_0) = (\mathbf{0}, \mathbf{1})$ and $(\Delta v_8, \Delta v_7) = (\mathbf{1}, \mathbf{0})$. The procedure is given in Algorithm 6.

For fixed values of the inactive bits in v_0 and v_{-1}, about 2^{4c} pairs can be generated, and we expect approximately $2^{4c} \cdot 2^{-n/2+c} = 2^{-n/2+5c}$ of them to verify the ciphertext truncated difference $(\mathcal{P}, A)$. By iterating the procedure for 2^{n-4c} different values, we obtain $2^{n-4c-n/2+5c} = 2^{n/2+c}$ pairs satisfying the desired $(\Delta v_0, \Delta v_{-1})$ and $(\Delta v_9, \Delta v_{10})$. The data complexity required to generate the $2^{n/2+c}$ pairs amounts to approximately $2^{2c+n-4c} = 2^{n-2c}$ chosen plaintexts,

---

Algorithm 5. Construction of the difference sequences of $\Delta v_7'[1]$ (precomputation)

1: **for** all $2^{n/2+4c}$ values of the parameters **do**
2:     Derive all differences of the differential.
3:     Derive 1-word state values in rounds 2 and 6.
4:     Derive all state values in rounds 3, 4 and 5.
5:     **for** 2^b different differences in v_0 **do**
6:         Modify $\Delta v_0[1]$, and update the corresponding sequence of $\Delta v_7'[1]$.
7:     Insert the sequence of $\Delta v_7'[1]$ in the table T_δ.

Algorithm 6. Data collection for the 10-round attack

1: Fix the $n/2 - c$ inactive bits of v_0 and v_{-1}.
2: **for** all 2^{2c} choices (v_0, v_{-1}) **do**
3:     Query (v_0, v_{-1}) to obtain (v_9, v_{10}).
4:     Store (v_9, v_{10}) in a hash table indexed by the wanted inactive bits in $P^{-1}(v_9)$.
5: Construct about $2^{4c}/2^{n/2-c} = 2^{-n/2+5c}$ pairs verifying the truncated ciphertext difference.
6: Iterate the analysis 2^{n-4c} times by changing the the inactive-bit value of v_0 and v_t.

---

the computational cost is equivalent to 2^{n-2c} memory accesses, and the memory requirement is about $2^{n/2+c}$ blocks of $n/2$ bits.

Detecting Subkeys. For each of the $2^{n/2+c}$ obtained pairs, we derive 2^c candidates for $n/2 + 2c$ bits of key material, namely $K_0[1]$, $K_8[1]$, and K_9. For each pair, we first guess the 1-word difference of $\Delta v_8[1]$. Then, we assume the differential characteristic is satisfied, i.e. $\Delta v_1 = 0$, $\Delta v_7' = 0$, and $\Delta v_8 = 1$. This fixes the input and output differences for the active words in rounds 0 and 8, and for all words in round 9. Then, the possible inputs for each of these S-Boxes can be reduced to a single value, and the corresponding subkeys $K_0[1]$, $K_8[1]$ and K_9 can be calculated.

Finally, we construct the b-δ-set by modifying $v_0[1]$. For each modified plaintext, with the knowledge of $K_0[1]$, we modify v_{-1} such that v_1 remains unchanged. From the corresponding ciphertexts, with the knowledge of K_9 and $K_8[1]$, we compute the sequence of 2^b differences $\Delta v_7'[1]$, and if it matches one of the entries in the precomputed table T_δ, then the guessed subkeys $K_0[1]$, $K_8[1]$, and K_9 are correct with high probability, otherwise they are wrong. When the values of c and n are in a particular range (see below), only the right guess will remain, thus the subkeys are recovered.

The computational cost of the key-recovery phase is the one for computing $\Delta v_7'[1]$ for $2^{n/2+c}$ pairs, 2^c guesses for $\Delta v_8[1]$, and 2^b choices of δ_j in the b-δ-set, which is upper bounded by $2^{n/2+3c}$ encryptions.

Complexity Analysis and Constraints on (n, c). As shown above, the data complexity requires 2^{n-2c} chosen plaintexts, the time complexity is equivalent to $2^{n-2c} + 2^{n/2+5c}$ encryptions and the memory complexity is $2^{n/2+5c}$ blocks of $n/2$

bits. We note that the overall complexity is balanced when $n/2c = 7$, i.e. when a branch includes 7 S-Boxes. It is possible to achieve a simple tradeoff where only a fraction $1/2^c$ of all the sequences are stored in T_δ, which decreases the memory complexity to $2^{n/2+4c}$ blocks of $n/2$ bits, but in turn makes the data complexity and the time complexity of the online phase increased by a factor 2^c as we have decreased the chance to hit one element in T_δ. With this tradeoff, the data complexity becomes 2^{n-c} chosen plaintexts, and the time complexity becomes about $2^{n-c} + 2^{n/2+4c}$ encryptions, which is balanced for $n/2c = 5$ S-Boxes per branch.

Moreover, to launch the attack, a branch must have at least 5 S-Boxes so that $n/2 + 4c < n$. Additionally, in the subkey detection phase, the number of remaining key candidates should be one or small enough. The number of sequences in T_δ is $2^{n/2+4c}$ and the number of candidates derived online is $2^{n/2+2c}$. Thus in total, 2^{n+6c} matches are examined, whether or not we use the tradeoff. In theory, there exists $2^{c \cdot 2^b}$ sequences from $b < c$ bits to c bits. Hence, the condition to extract only the correct subkey is $n + 6c - c \cdot 2^b < 0$, which gives $b > \log_2(6 + n/c)$. Since $2^b < 2^c$, by combining the two conditions, the valid range for (n, c) is $10c \le n < c(2^c - 6)$. For example, 128-bit block ciphers with 8-bit S-Boxes and 80-bit block ciphers with 5-bit S-Boxes can be attacked.

Another possible tradeoff is the one used to achieve the best attacks on reduced variants of the AES in [10]. If we add a second active word at the beginning of the differential characteristic, it allows to reduce the data complexity, while keeping the same overall complexity. This tradeoff is possible as long as there are at least 7 words per branch, i.e. $n/2c \ge 7$. The main advantage of adding an active word is to increase the size of the structures of plaintext from 2^{2c} to 2^{4c}, which allows to construct about 2^{8c} input pairs already verifying the input difference. The precomputation requires $2^{n/2+6c}$ encryptions and a memory of $2c/n \times 2^{n/2+6c+b}$ blocks of $n/2$ bits, the online phase requires more pairs, namely $2^{n/2+2c}$, but this is achieved with less data: only 2^{n-3c} chosen plaintexts. Therefore, the final time complexity is $2^{n-3c} + 2^{n/2+6c}$ for both the encryption of the data and the precomputation. This yields an attack as long as $n/2 + 6c < n$, which is true for $n/2c \ge 7$ S-Boxes. For example, with 8 S-Boxes per branch, the attack without the second active word requires $2^{14n/16}$ chosen plaintexts, $2^{14n/16}$ encryptions and the memory of about $2^{12n/16}$ blocks of $n/2$ bits, hence the overall complexity is $2^{14n/16}$. For the same primitive, but with an additional active word, the tradeoff gives an attack that requires the same overall time complexity while the data complexity is reduced to $2^{13n/16}$ chosen plaintexts.

5   Conclusion

With the use of the meet-in-the-middle technique, we have shown the best known generic attacks on balanced Feistel ciphers. As we imposed very small restrictions on the round functions, our attacks are applicable to almost all balanced Feistels. Such ciphers, with an arbitrary round function and a double key are insecure on up to 10 rounds. In the case when the round function is SPN, for a large class of linear P-layers, the attacks penetrate 14 rounds and recover all the subkeys. We have produced experimental verification of the attacks supporting our claims.

Our results give insights on the lower bound on the number of rounds a secure Feistel should have. They suggest that this number in the case of SPN round functions should be surprisingly high. Furthermore, from the attacks on Feistel-2, we show that as long as the ratio of key to state size is increasing, the number of rounds that can be attacked will grow, while the data complexity will always stay below the full codebook. Thus, we have shown that *a block cipher designer cannot fix a priori the number of rounds in a balanced Feistel and allow any (or very large) key size*, as for each increment of the key by amount of bits equivalent to the state size, we can attack four more rounds.

We have analyzed generic constructions and as such, we could not make any assumptions about the particular details of the ciphers, e.g. the key schedule, the permutation layer, etc. However, the attacks on the AES have shown that it is possible to take advantage of the cipher details in order to penetrate more rounds. Thus, we believe that our analysis can be used as a beginning step for attacks on larger number of rounds of specific Feistel ciphers.

Acknowledgments. The authors would like to thank the ASIACRYPT 2014 reviewers for their valuable comments. Jian Guo, Jérémy Jean and Ivica Nikolić are supported by the Singapore National Research Foundation Fellowship 2012 NRF-NRFF2012-06.

References

1. Aoki, K., Guo, J., Matusiewicz, K., Sasaki, Y., Wang, L.: Preimages for Step-Reduced SHA-2. In: Matsui, M. (ed.) ASIACRYPT 2009. LNCS, vol. 5912, pp. 578–597. Springer, Heidelberg (2009)
2. Aoki, K., Ichikawa, T., Kanda, M., Matsui, M., Moriai, S., Nakajima, J., Tokita, T.: *Camellia*: A 128-Bit Block Cipher Suitable for Multiple Platforms - Design and Analysis. In: Stinson, D.R., Tavares, S. (eds.) SAC 2000. LNCS, vol. 2012, pp. 39–56. Springer, Heidelberg (2001)
3. Beaulieu, R., Shors, D., Smith, J., Treatman-Clark, S., Weeks, B., Wingers, L.: The SIMON and SPECK Families of Lightweight Block Ciphers. Cryptology ePrint Archive, Report 2013/404 (2013)
4. Biham, E., Dunkelman, O.: The SHAvite-3 Hash Function. Submission to NIST, Round 2 (2009)
5. Communications Security Establishment Canada: Cryptographic algorithms approved for Canadian government use (2012)
6. Coppersmith, D.: The Data Encryption Standard (DES) and its Strength Against Attacks. IBM Journal of Research and Development 38(3), 243–250 (1994)
7. Daemen, J., Knudsen, L.R., Rijmen, V.: The Block Cipher SQUARE. In: Biham, E. (ed.) FSE 1997. LNCS, vol. 1267, pp. 149–165. Springer, Heidelberg (1997)
8. Demirci, H., Selçuk, A.A.: A Meet-in-the-Middle Attack on 8-Round AES. In: Nyberg, K. (ed.) FSE 2008. LNCS, vol. 5086, pp. 116–126. Springer, Heidelberg (2008)
9. Derbez, P., Fouque, P.A., Jean, J.: Improved Key Recovery Attacks on Reduced-Round AES in the Single-Key Setting. IACR Cryptology ePrint Archive, 477 (2012)

10. Derbez, P., Fouque, P.-A., Jean, J.: Improved Key Recovery Attacks on Reduced-Round AES in the Single-Key Setting. In: Johansson, T., Nguyen, P.Q. (eds.) EUROCRYPT 2013. LNCS, vol. 7881, pp. 371–387. Springer, Heidelberg (2013)

11. Dinur, I., Dunkelman, O., Keller, N., Shamir, A.: Efficient Dissection of Composite Problems, with Applications to Cryptanalysis, Knapsacks, and Combinatorial Search Problems. In: Safavi-Naini, R., Canetti, R. (eds.) CRYPTO 2012. LNCS, vol. 7417, pp. 719–740. Springer, Heidelberg (2012)

12. Dunkelman, O., Keller, N., Shamir, A.: Improved Single-Key Attacks on 8-Round AES-192 and AES-256. In: Abe, M. (ed.) ASIACRYPT 2010. LNCS, vol. 6477, pp. 158–176. Springer, Heidelberg (2010)

13. Feistel, H., Notz, W., Smith, J.: Some Cryptographic Techniques for Machine-to-Machine Data Communications. Proceedings of IEEE 63(11), 15545–1554 (1975)

14. Gilbert, H., Minier, M.: A Collision Attack on 7 Rounds of Rijndael. In: AES Candidate Conference, pp. 230–241 (2000)

15. Guo, J., Jean, J., Nikolić, I., Sasaki, Y.: Meet-in-the-Middle Attacks on Generic Feistel Constructions - Extended Abstract. Cryptology ePrint Archive, Temporary version (to appear, 2014),
http://www1.spms.ntu.edu.sg/~syllab/attacks/FeistelMitM.pdf

16. Guo, J., Ling, S., Rechberger, C., Wang, H.: Advanced Meet-in-the-Middle Preimage Attacks: First Results on Full Tiger, and Improved Results on MD4 and SHA-2. In: Abe, M. (ed.) ASIACRYPT 2010. LNCS, vol. 6477, pp. 56–75. Springer, Heidelberg (2010)

17. Isobe, T., Shibutani, K.: All Subkeys Recovery Attack on Block Ciphers: Extending Meet-in-the-Middle Approach. In: Knudsen, L.R., Wu, H. (eds.) SAC 2012. LNCS, vol. 7707, pp. 202–221. Springer, Heidelberg (2013)

18. Isobe, T., Shibutani, K.: Generic Key Recovery Attack on Feistel Scheme. In: Sako, K., Sarkar, P. (eds.) ASIACRYPT 2013, Part I. LNCS, vol. 8269, pp. 464–485. Springer, Heidelberg (2013)

19. ISO/IEC 18033-3:2010: Information technology–Security techniques–Encryption Algorithms–Part 3: Block ciphers (2010)

20. Knudsen, L.R.: The Security of Feistel Ciphers with Six Rounds or Less. J. Cryptology 15(3), 207–222 (2002)

21. Luby, M., Rackoff, C.: How to Construct Pseudorandom Permutations from Pseudorandom Functions. SIAM J. Comput. 17(2), 373–386 (1988)

22. Merkle, R.C., Hellman, M.E.: On the Security of Multiple Encryption. Commun. ACM 24(7), 465–467 (1981)

23. Sasaki, Y., Aoki, K.: Finding Preimages in Full MD5 Faster Than Exhaustive Search. In: Joux, A. (ed.) EUROCRYPT 2009. LNCS, vol. 5479, pp. 134–152. Springer, Heidelberg (2009)

24. Shibutani, K., Isobe, T., Hiwatari, H., Mitsuda, A., Akishita, T., Shirai, T.: *Piccolo*: An Ultra-Lightweight Blockcipher. In: Preneel, B., Takagi, T. (eds.) CHES 2011. LNCS, vol. 6917, pp. 342–357. Springer, Heidelberg (2011)

25. Todo, Y.: Upper Bounds for the Security of Several Feistel Networks. In: Boyd, C., Simpson, L. (eds.) ACISP. LNCS, vol. 7959, pp. 302–317. Springer, Heidelberg (2013)

26. Wu, W., Zhang, L.: LBlock: A Lightweight Block Cipher. In: Lopez, J., Tsudik, G. (eds.) ACNS 2011. LNCS, vol. 6715, pp. 327–344. Springer, Heidelberg (2011)

27. Zhang, L., Wu, W., Wang, Y., Wu, S., Zhang, J.: LAC: A Lightweight Authenticated Encryption Cipher. Submitted to the CAESAR competition (March 2014)

XLS is Not a Strong Pseudorandom Permutation

Mridul Nandi

Indian Statistical Institute, Kolkata, India
mridul@isical.ac.in

Abstract. In FSE 2007, Ristenpart and Rogaway had described a
generic method XLS to construct a length-preserving strong pseudorandom permutation (SPRP) over bit-strings of size at least n. It requires a
length-preserving permutation $\mathcal{E}$ over all bits of size multiple of n and a
blockcipher E with block size n. The SPRP security of XLS was proved
from the SPRP assumptions of both $\mathcal{E}$ and E. In this paper we disprove
the claim by demonstrating a SPRP distinguisher of XLS which makes
only three queries and has distinguishing advantage about $1/2$. XLS uses
a multi-permutation linear function, called mix2. In this paper, we also
show that if we replace mix2 by any invertible linear functions, the construction XLS still remains insecure. Thus the mode has inherit weakness.

Keywords: XLS, SPRP, Distinguishing Advantage, length-preserving
encryption.

1 Introduction

The notion of domain extension arises in many areas of cryptography such as
hash function, pseudorandom function or PRF, strong pseudorandom permutation or SPRP [12] etc. Usually, we design a building block defined for a small
and fixed bit size domain. Then, by applying the building block iteratively, we
obtain a similar kind of function defined over arbitrary domain. For example,
a blockcipher defined on n bits can be used to define an encryption algorithm
which can encrypt any message of size multiple of n. To define a ciphertext for
a message whose size is not a multiple of n, one can first apply some padding
rule to make the (padded) message of size multiple of n. This method can not
preserve length as it expands ciphertext length. A length-preserving encryption
is called an **enciphering scheme**. The length-preserving property makes our
task more difficult and restricted than length expanding encryptions. On the
other hand, designing enciphering schemes over all bit strings of size multiple
of block-size (i.e., n) seems to be easier than defining over arbitrary bit strings.
Many such enciphering schemes have been defined [10,9].

NON-GENERIC METHODS.   There are several known methods for turning a
blockcipher into an enciphering schemes over arbitrary bit strings. One can apply the underlying block cipher twice and use the intermediate output as an
one-time pad for partial block (used in EME [7], TET [8], HEH [16] etc.); The

P. Sarkar and T. Iwata (Eds.): ASIACRYPT 2014, PART I, LNCS 8873, pp. 478–490, 2014.

other constructions e.g., HCTR [17], HCH [3], XCB [13] use hash-then counter paradigm. A standard trick like ciphertext stealing can also be applied to specific constructions (e.g., AEZ [1]). However, all these approaches are not generic.

> We call a method **domain completion** (or generic domain completion) if it converts any enciphering scheme over bit strings of size multiple of n into an enciphering scheme over any bit strings (possibly of size at least n).

To our best knowledge, so far only two domain completions have been proposed.

1. A popular, efficient and neatly defined domain completion method is XLS (eXtended by Latin Square) designed by Ristenpart and Rogaway [15]. The design rational of XLS is similar to that of elastic blockcipher as both follow encrypt-then-mix paradigm.

2. Following hash-counter-hash paradigm, Nandi proposed a domain completion method in [14].

In addition to these, a heuristically described method, called **Elastic blockcipher** [4], was proposed by Cook, Yung and Keromytis. Later elastic blockcipher, defined over all bits of sizes in between n and $2n$, has been more formally defined in [5].

Applications of Domain Completion Method. While primarily interested in the theoretical question of how to obtain domain extension for ciphers, arbitrary-input-length enciphering is a problem with many applications. A well-known application is disk-sector encryption in which size of ciphertext and plaintext remain same as the sector size of the disk. In general, enciphering scheme is easy to define for input sizes of multiple of n (block size of the underlying blockcipher). Domain completion methods can be used to define the enciphering schemes for arbitrary bit strings. It is also used in other symmetric key algorithms such as authenticated encryption. For example, XLS is widely adapted in many authenticated encryptions, e.g. AES-COPA [2], Deoxys, Joltik, KIASU, Marble, SHELL etc. [1].

1.1   Our Contribution

In this paper we demonstrate a chosen plaintext-ciphertext adversary (CPCA) distinguisher against XLS (see Algorithm $\mathcal{A}_0$ in section 3.2). The attack makes only three encryption and decryption oracle queries in an interleaved manner and has distinguishing advantage about $1/2$. Thus, the security claim of XLS is wrong.

XLS uses a linear multi-permutation (very efficiently computable) mix2 which satisfies some property. Authors called any such linear permutation satisfying the property a *good mixing function*. It is natural to think a possible remedy of XLS to replace mix2 by other good mixing function or some other stronger linear permutations. Unfortunately, we show that these remedies would not work.

To establish our claim, we consider a generalized version of XLS (we call it GXLS) which applies any arbitrary linear permutations instead of mix2. Moreover, we consider keys of two invocations of the underlying blockcipher E to be independently chosen. We demonstrate similar CPCA-distinguishers (in section 4) for GXLS having advantage at least $1/4$. So we conclude that XLS has design flaws in its modes not in the choice of linear mixing functions.

2   XLS and Its General Form GXLS

Basics and Notation

1. An s-bit string X is denoted as $X = X[1]X[2]\cdots X[s]$ where $X[i] \in \{0,1\}$. We denote $X[i..j] = X[i]\cdots X[j]$ and $|X| = s$.
2. A length-preserving function f satisfies $|f(X)| = |X|$ for all X.
3. Any linear function from $\{0,1\}^s$ to $\{0,1\}^t$ can be represented by a $t \times s$ binary matrix. Let $\mathsf{rol}(X)$ represent left circular bit-rotation, that is, for any bit-string $X := X[1]X[2]\cdots X[s]$ of length s, let $\mathsf{rol}(X) = X[2]X[3]\cdots X[s]X[1]$. Note that rol is a linear invertible function and is represented by the following $s \times s$ invertible matrix:

$$\mathbf{R} = \begin{pmatrix} 0\,1\,0\cdots 0\,0 \\ 0\,0\,1\cdots 0\,0 \\ \vdots \\ 0\,0\,0\cdots 0\,1 \\ 1\,0\,0\cdots 0\,0 \end{pmatrix} = \begin{pmatrix} 0 & \mathbf{I}_{s-1} \\ 1 & 0 \end{pmatrix}.$$

Here, $\mathbf{I}_{s-1}$ represents the identity matrix of size $s-1$.
4. Throughout the paper, let n be a fixed integer representing the block-size of the underlying blockcipher E.

2.1   XLS and GXLS on $\{0,1\}^{2n-1}$

In this section we describe how XLS has been defined for bit strings of size $2n-1$. Later we show distinguishing attack of XLS by making queries of size $2n-1$ only. We refer readers to the original paper [15] for the definition of XLS over arbitrary bit strings. We first define a linear function $\mathsf{mix2} : \{0,1\}^{2n-2} \to \{0,1\}^{2n-2}$ as below

$$\mathsf{mix2}(AB) = (A \oplus \mathsf{rol}(A \oplus B), B \oplus \mathsf{rol}(A \oplus B)) = ((\mathbf{R}+\mathbf{I})\cdot A + \mathbf{R}\cdot B,\ \mathbf{R}\cdot A + (\mathbf{R}+\mathbf{I})\cdot B)$$

$$= \begin{pmatrix} \mathbf{R}+\mathbf{I} & \mathbf{R} \\ \mathbf{R} & \mathbf{R}+\mathbf{I} \end{pmatrix} \cdot \begin{pmatrix} A \\ B \end{pmatrix}$$

where $|A| = |B| = n-1$ and $\mathbf{I}$ is the identity matrix of size $n-1$. It is easy to see that the inverse of the linear map $\mathsf{mix2}$ is itself. Such a permutation is also called an involution. Now we describe the XLS algorithm [15] over the set of all $2n-1$

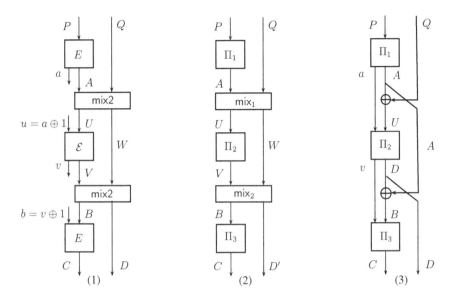

Fig. 2.1. Illustration of (1) XLS, (2) GXLS and (3) 3-round Elastic Blockcipher. The XLS and 3-round Elastic blockcipher are special cases of GXLS.

bit strings based on two n-bit (random) permutations E and $\mathcal{E}$ and the linear permutation mix2. We would like to note that we express the input, output and intermediate variables with different notations from those of [15] which would be used to describe our attack and analysis.

Algorithmic Definitions of XLS and GXLS. Now we describe the algorithms XLS and GXLS which are defined on $2n - 1$ bits.

Algorithm XLS$^{E,\mathcal{E}}$	Algorithm GXLS$[\Pi_1, \mathsf{mix}_1, \Pi_2, \mathsf{mix}_2, \Pi_3]$
Input: $(P,Q) \in \mathbb{F}_2^n \times \mathbb{F}_2^{n-1}$	**Input:** $(P,Q) \in \mathbb{F}_2^n \times \mathbb{F}_2^{n-1}$
Output: $(C,D) \in \mathbb{F}_2^n \times \mathbb{F}_2^{n-1}$.	**Output:** $(C,D) \in \mathbb{F}_2^n \times \mathbb{F}_2^{n-1}$.
01 $E(P) = a\|A, \quad a \in \mathbb{F}_2$.	01 $\Pi_1(P) = A$.
02 $u = a!, \quad (U,W) = \mathsf{mix2}(A,Q)$.	02 $(U,W) = \mathsf{mix}_1(A,Q)$.
03 $\mathcal{E}(u\|U) = v\|V$.	03 $\Pi_2(U) = V$.
04 $b = v!, \quad (B,D) = \mathsf{mix2}(V,W)$.	04 $(B,D) = \mathsf{mix}_2(V,W)$.
05 $E(b\|B) = C$.	05 $\Pi_3(B) = C$.
06 return (C,D).	06 return (C,D).

Here ! denotes bit complement. Here mix_1 and mix_2 are two invertible linear functions on $2n - 1$ bits and mix2 is a linear invertible function over $\{0,1\}^{2n-2}$ bits as described before. The Π_i's are independent uniform random (or pseudorandom) permutations whereas in XLS $\mathcal{E}$ and E are independent uniform random (or pseudorandom) permutations. We also denote the generalized-XLS as

$\mathsf{GXLS}[\Pi_1, \mathsf{mix}_1, \Pi_2, \mathsf{mix}_2, \Pi_3](P, Q) = (C, D)$ as above (in the right hand side of Fig. 2.1). Note that the XLS algorithm is nothing but $\mathsf{GXLS}[E, ! \| \mathsf{mix2}, \mathcal{E}, ! \| \mathsf{mix2}, E]$ where $(! \| f)(b, X) = b! \| f(X)$. In order for GXLS to be invertible, mix_1 and mix_2 should be invertible.

2.2   Elastic Blockcipher

The three round Elastic blockcipher can also be viewed as a $\mathsf{GXLS}[\Pi_1, \mathsf{mix3}, \Pi_2, \mathsf{mix3}, \Pi_3]$ where $\mathsf{mix3}(A, B) = ((A[i_1] \cdots A[i_s]) \oplus B, A[i_1] \cdots A[i_s])$, $|A| = n$, $|B| = s$ and $1 \le i_1 < \cdots < i_s \le n$ are some fixed integers (specific choices of these values depend on the underlying blockcipher). We illustrate this method in Fig 2.1 when $i_1 = n - s + 1, \ldots, i_s = n$. Basic mix function of it can be defined as $(X \| Y) \mapsto X \oplus Y \| X$ where $|X| = |Y| = s$. Similarly, four or higher number of rounds can be defined. So all of these follow the encrypt-mix paradigm iterated several rounds. We capture this paradigm for three iterations in GXLS. In the following sections, we prove that three rounds are not sufficient for having SPRP.

3   Insecurity of XLS

In this section we show that XLS is not SPRP (strong pseudorandom permutation). In fact we establish a distinguisher making only three oracle queries having distinguishing advantage about $1/2$. Moreover, if we repeat this attack independently, we can amplify the distinguishing advantage close to one. We first briefly define basics of security notions related to distinguishing advantages.

3.1   Security Definitions

Let R_i denote the uniform random function from $\{0, 1\}^i$ to $\{0, 1\}^i$, i.e., the uniform distribution on the set $\mathrm{Func}(\{0, 1\}^i, \{0, 1\}^i)$ of all functions from $\{0, 1\}^i$ to itself. Given a set $L \subseteq \mathbb{N} := \{1, 2, \cdots\}$, we denote R_L for the tuple $\langle \mathsf{R}_i \rangle_{i \in L}$ of random functions where R_i's are jointly independently distributed. We call the set L length set. We call R_L a *length-preserving uniform random function* on $\{0, 1\}^L := \cup_{i \in L} \{0, 1\}^i$. Similarly, let P_i denote the uniform random permutation on $\{0, 1\}^i$, i.e., the uniform distribution on the set $\mathrm{Perm}(\{0, 1\}^i, \{0, 1\}^i)$ of all permutations on $\{0, 1\}^i$. Note that the inverse random permutation, P_i^{-1}, is also an uniform random permutation. We similarly define length-preserving uniform random permutation P_L on $\{0, 1\}^L$ which is independent composition of P_i for all $i \in L$.

Now let $\mathcal{A}$ be an oracle algorithm which has access of two length-preserving oracles $\mathcal{O}_1$ and $\mathcal{O}_2$. Suppose $\mathcal{A}$ makes queries from the set $\{0, 1\}^L$ for both oracles. We define **SPRP-advantage** of $\mathcal{A}$ for a length-preserving random permutation F_L (not necessarily uniform) by

$$\mathbf{Adv}_{\mathsf{F}_L}^{\mathrm{sprp}}(\mathcal{A}) = \Pr[\mathcal{A}^{\mathsf{F}_L, \mathsf{F}_L^{-1}} = 1] - \Pr[\mathcal{A}^{\mathsf{P}_L, \mathsf{P}_L^{-1}} = 1].$$

In general, we can define advantage for two pairs of tuples of length-preserving random functions (F_L, F'_L) and (G_L, G'_L) as

$$\mathbf{Adv}_{\mathcal{A}}((F_L, F'_L), (G_L, G'_L)) = \Pr[\mathcal{A}^{F_L, F'_L} = 1] - \Pr[\mathcal{A}^{G_L, G'_L} = 1].$$

If $\mathcal{A}$ interacts with a length-preserving random permutation $\mathcal{O}_1$ and its inverse $\mathcal{O}_2$ then we can assume the following:

1. $\mathcal{A}$ is not making any repetition query.
2. If x_i is $\mathcal{O}_1$-query and y_i is its response then there is no $\mathcal{O}_2$-query $x_j = y_i$ for some $j > i$ and vice-versa.

We can assume these since the responses are determined for these types of queries. An adversary satisfying the above conditions is called an *allowed adversary*.

Theorem 1. *[11] Let* R_L *and* R'_L *be independently chosen length-preserving uniform random functions and let* P_L *be length-preserving uniform random permutation. Then for any allowed adversary* $\mathcal{A}$ *which makes at most* Q *queries, we have,*

$$\mathbf{Adv}_{\mathcal{A}}((P_L, P_L{}^{-1}), (R_L, R'_L)) \leq \frac{Q(Q-1)}{2^{m+1}}$$

where $m = \min\{\ell : \ell \in L\}$.

The above result says that an uniform length-preserving random permutation is very close to an uniform length-preserving random function. Thus if we want to prove that an enciphering scheme is not SPRP-secure by small number of queries then it would be enough to compute the distinguishing advantage from uniform random functions for an allowed adversary. For example, when $Q = 3$, if for length-preserving construction F_L, $\mathbf{Adv}_{\mathcal{A}}((P_L, P_L{}^{-1}), (F_L, F_L^{-1})) := c$ is significant for an allowed adversary then $\mathbf{Adv}_{F_L}^{\mathrm{sprp}}(\mathcal{A})$ is at least $c - 2^{-n+2}$ which is also significant.

Remark 1. The above is one side of the implication of the Theorem 1. The other application is to show a construction F_L SPRP by showing $\mathbf{Adv}_{\mathcal{A}}((F_L, F_L{}^{-1}), (R_L, R'_L))$ is negligible.

3.2 SPRP Distinguishing Algorithm

Distinguishing Algorithm $\mathcal{A}_0$ for XLS.

1. Make encryption query (P_1, Q_1) and obtains response (C_1, D_1).
2. Make decryption query $(C_2 := C_1, D_2 := D_1 + 1)$ and obtains response (P_2, Q_2).
3. Make encryption query $(P_3 = P_2, Q_3)$ and obtains response (C_3, D_3) where

$$Q_3 = Q_1 + (\mathbf{I} + \mathbf{R}^{-2}) \cdot (Q_1 + Q_2 + 1).$$

4. **return** 1 if $D_3 = Q_1 + Q_3 + D_1$, 0 otherwise.

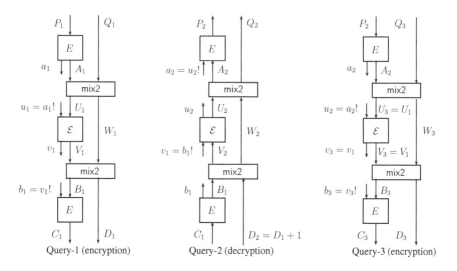

Fig. 3.1. $\mathcal{A}_0$ makes three queries and obtains collisions on U_1 and U_3 values with probability $1/2$ (due to the event that $a_1 = a_2$)

3.3   Analysis of Attack

To see why our attack works, let us first observe some useful relations among internal variables in the computations of XLS.

Lemma 1. *With the notations as described in the algorithm XLS, we have* $A + B = (\mathbf{R}^{-1} + \mathbf{I}) \cdot (Q + D)$.

Proof. Since mix2 is inverse of itself we have $(V, W) = \mathsf{mix2}(B, D)$. By equating W with line 02 of XLS algorithm, we have

$$\mathbf{R} \cdot B + (\mathbf{R} + \mathbf{I}) \cdot D = \mathbf{R} \cdot A + (\mathbf{R} + \mathbf{I}) \cdot Q.$$

Thus, $\mathbf{R} \cdot (A + B) = (\mathbf{R} + \mathbf{I}) \cdot (Q + D)$ and so the result follows.   □

Lemma 2. *With the notations as described in the algorithm XLS, we have* $U + V = \mathbf{R}^{-1} \cdot (Q + D)$.

Proof. Due to line 02 and 04, we have $\mathbf{R} \cdot U + (\mathbf{I} + \mathbf{R}) \cdot W = Q$ and $\mathbf{R} \cdot V + (\mathbf{I} + \mathbf{R}) \cdot W = D$. Thus, $\mathbf{R} \cdot (U + V) = (D + Q)$ and so the result follows.   □

The basic idea of our attack is to obtain an internal collision. Suppose we have two queries (P_i, Q_i) with responses (C_i, D_i), $i = 1, 2$ such that the U_i values remain the same. So the outputs V_i are also same. Due to above lemma, we have $Q_1 \oplus D_1 = Q_2 \oplus D_2$. For a uniform random permutations this event can occur with a probability of about 2^{-n+1}. Now we show that in query-1 and query-3, U values collide with probability $1/2$ and so we can distinguish XLS from uniform random permutation with advantage about $1/2$ (for large n, 2^{-n+1} is negligible).

Theorem 2. *The Algorithm of $\mathcal{A}_0$ has distinguishing advantage about $1/2$ against XLS.*

Proof. Note that $\mathcal{A}_0$ makes three encryption and decryption queries in an interleaved manner. Let us denote the intermediate variables of computations of ith query by using suffix i, $1 \leq i \leq 3$. Let us denote $\mathbf{G} = \mathbf{R}^{-1} + \mathbf{I}$. By Lemma 1, we have $A_1 + B_1 = \mathbf{G} \cdot (Q_1 + D_1)$ and $A_2 + B_1 = A_2 + B_2 = \mathbf{G} \cdot (Q_2 + D_2)$ and so $A_1 + A_3 = A_1 + A_2 = \mathbf{G} \cdot (Q_1 + D_1 + Q_2 + D_2) = \mathbf{G} \cdot (Q_1 + Q_2 + 1)$. Now we make our main claim:

Claim: $U_1 = U_3$.

$$
\begin{aligned}
U_1 + U_3 &= (\mathbf{R} + \mathbf{I}) \cdot (A_1 + A_3) + \mathbf{R} \cdot (Q_1 + Q_3) \\
&= (\mathbf{R} + \mathbf{I}) \cdot \mathbf{G}(Q_1 + Q_2 + 1) + \mathbf{R} \cdot (Q_1 + Q_3) \\
&= (\mathbf{R} + \mathbf{R}^{-1}) \cdot (Q_1 + Q_2 + 1) + \mathbf{R} \cdot (Q_1 + Q_3)
\end{aligned}
$$

Since $Q_3 = Q_1 + (\mathbf{I} + \mathbf{R}^{-2}) \cdot (Q_1 + Q_2 + 1)$, we have $U_1 = U_3$.   □

The rest of the proof is straightforward. As we have collision on U values, we have collision on V values, i.e., $V_1 = V_3$. But this can happen if the first bit of inputs of $\mathcal{E}$ in query 1 and 3 match which can happen with probability $1/2$. Assuming this, we can exploit the collision to make distinguishing attack as discussed before the theorem. We have $D_3 = Q_1 + Q_3 + D_1$. This can hold for a uniform random permutation $\mathcal{E}$ with probability about 2^{-n+1}. So the result follows.   □

Remark 2. The same attack works for any length of the form $kn - 1$ with same advantage. We only need the size of the partial block to be $n - 1$. Note that we need the first bit of output of E in query 1 and 3 should match which can happen with probability $1/2$. For other length inputs, the distinguishing advantage reduces as we need more bits collision. In general, if we want to distinguish XLS only on $kn + s$ bits inputs then we need collision on the first $n - s$ bits of outputs of E in query 1 and 3 which can happen with probability about 2^{s-n}. So the distinguishing advantage would be about $2^{s-n} - 2^{1-n}$. So if the partial block size s is small the distinguishing advantage of our attack reduces. This is very natural as most of the intermediate bits are processed through $\mathcal{E}$.

4 Distinguishing Attack on GXLS on $\{0,1\}^{2n-1}$

Now we demonstrate how we can modify the distinguishing attack for GXLS. This would suggest that any simple modification on XLS (such as changing mix functions with others) do not work. In other words, we show that the mode, not the mixing function, has inherent weakness. Behavior of this distinguishing attack depends on different cases of invertible mixing matrices mix_1 and mix_2. As we need to assume these as linear permutations, we can represent these by the following $(2n - 1) \times (2n - 1)$ invertible matrices.

$$\mathsf{mix}_1 = \begin{pmatrix} M[1]_{n \times n} & N[1]_{n \times (n-1)} \\ M[2]_{(n-1) \times n} & N[2]_{(n-1) \times (n-1)} \end{pmatrix},$$

$$\mathsf{mix}_2^{-1} = \begin{pmatrix} M'[1]_{n \times n} & N'[1]_{n \times (n-1)} \\ M'[2]_{(n-1) \times n} & N'[2]_{(n-1) \times (n-1)} \end{pmatrix}.$$

Before we demonstrate our attacks we state some notations and results which would be used.

Notations.   Given a $r \times s$ matrix $\mathbf{A}$ we denote $\mathcal{C}(\mathbf{A})$ the column space of the matrix.

Lemma 3.   *Let $M_{r \times s}$ and $N_{r \times t}$ be two matrices and $c_{r \times 1}$ is a vector such that $\mathcal{C}(N) \not\subseteq \mathcal{C}(M)$. For any two uniform random vectors $a \xleftarrow{\$} \{0,1\}^s$ and $q \xleftarrow{\$} \{0,1\}^t$ (not necessarily independent) $\mathsf{Pr}[M \cdot a = N \cdot q + c] \leq 1/2$.*

Proof. This is straightforward when M is of the form $\begin{pmatrix} \mathbf{I}_r & * \\ \mathbf{0} & \mathbf{0} \end{pmatrix}$ where r is the rank of M and $*$ means that the sub matrix could be anything. In this case there must exist $i > r$ such that i^{th} row of N is non-zero. As q_i is uniform on $\{0,1\}$, by equating the event on i^{th} bit we get probability at most $1/2$ to achieve the event.

For a general matrix M, we can find two non-singular square matrices S and T such that $S \cdot M \cdot T = \begin{pmatrix} \mathbf{I}_r & * \\ \mathbf{0} & \mathbf{0} \end{pmatrix}$. So the given probability p should be same as

$$\mathsf{Pr}[SMT \cdot (T^{-1}a) = SN \cdot q + S \cdot c].$$

Let us denote $M' = SMT$, $a' = T^{-1}a$, $c' = Sc$ and $N' = SN$. With this notation, we have $p = \mathsf{Pr}[M' \cdot a' = N' \cdot q + c']$. Now note that M' has the form as considered before. Due to invertible property of S and T, we have the property that $\mathcal{C}(N') \not\subseteq \mathcal{C}(M')$ and, a' and q' follow individually uniform distributions.   □

Now we describe our attacks for different cases of the sub matrices of the mix functions. Conventionally, we use suffix 1, 2 and 3 to denote intermediate values for the first, second and third queries respectively.

4.1   rank($M[2]$) $\leq n - 2$

In this case we first claim that the column space of $N[2]$ must contain a vector which does not belong to the column space of $M[2]$. Otherwise, the rank of $(n-1) \times (2n-1)$ matrix $(M[2]\ N[2])$ is less than $n - 2$ which contradicts that the matrix mix_1 is invertible.

Now we run the algorithm $\mathcal{A}_0$ only for the first two queries. As before, we have $M[2](A_1 + A_2) = N[2](Q_1 + Q_2) + N'[2] \cdot 1$. Using the lemma 3, we know that when the algorithm is interacting with uniform random permutation, the probability that $N[2](Q_1 + Q_2) + N'[2] \cdot 1$ belongs to the column space of $M[2]$ is less than $1/2$. However, for the case of GXLS it occurs with probability one. So we can distinguish with advantage at least $1/2$. We formally describe the attack algorithm by $\mathcal{A}_1$ below.

Distinguishing Algorithm $\mathcal{A}_1$.

1. Make encryption query (P_1, Q_1) and obtain response (C_1, D_1).
2. Make decryption query $(C_2 := C_1, D_2 := D_1+1)$ and obtain response (P_2, Q_2).
3. **return** 1 if $N[2](Q_1 + Q_2) + N'[2] \cdot 1 \in \mathcal{C}(M[2])$, 0 otherwise.

Note that given a vector v and a matrix $M[2]$, there is an efficient algorithm to check whether a vector v belongs to the column space of $M[2]$. For this we essentially need to solve the system of equations $M[2] \cdot x = v$ and whenever we arrive contradiction a solution does not exist equivalently v is not a member of the column space. Alternatively we can first find some invertible matrices S and T (by some standard elementary operations) such that

$$S \cdot M[2] \cdot T = \begin{pmatrix} \mathbf{I}_r & * \\ \mathbf{0} & \mathbf{0} \end{pmatrix}$$

where r denotes the rank of $M[2]$. So $M[2] \cdot x = v$ if and only if

$$\begin{pmatrix} \mathbf{I}_r & * \\ \mathbf{0} & \mathbf{0} \end{pmatrix}(T^{-1}x) = S \cdot v$$

which holds if and only if for all $i > r$, the ith entry of $S \cdot v$ is zero.

Remark 3. Similar attack works when $\text{rank}(M'[2]) \leq n - 2$. In this case we only need to interchange the role of encryption and decryption queries.

4.2 Case: $\text{rank}(M[2]) = \text{rank}(M'[2]) = n - 1$, $\text{rank}(N[1]) \leq n - 2$

As $N[1]$ does not have full rank, we can find $Q_1 \neq Q_2$ such that $N[1]Q_1 = N[1]Q_2$. So U values collide for two encryption queries (P, Q_1) and (P, Q_2). Now we write the relationship among intermediate variables. So $A_1 = A_2$ and due to choice of Q_1 and Q_2 we also have $U_1 = U_2$ and hence $V_1 = V_2$. Now, let us write mix_2 function as

$$\text{mix}_2 = \begin{pmatrix} M''[1]_{n \times n} & N''[1]_{n \times (n-1)} \\ M''[2]_{(n-1) \times n} & N''[2]_{(n-1) \times (n-1)} \end{pmatrix}.$$

By the applications of mix_1 and mix_2 functions for two queries, we have

1. $W_1 = M[2] \cdot A_1 + N[2] \cdot Q_1$, $W_2 = M[2] \cdot A_2 + N[2] \cdot Q_2$ and
2. $D_1 = M''[2] \cdot V_1 + N''[2] \cdot W_1$, $D_2 = M''[2] \cdot V_2 + N''[2] \cdot W_2$.

So $W_1 + W_2 = N[2] \cdot (Q_1 + Q_2)$ and $D_1 + D_2 = N''[2] \cdot (W_1 + W_2) = N''[2] \cdot N[2] \cdot (Q_1 + Q_2)$. Note that for a random function, we observe this with probability 2^{-n+1}. We formally describe the attack algorithm by $\mathcal{A}_2$ below.

Distinguishing Algorithm $\mathcal{A}_2$.

1. Let $N[1]Q_1 = N[1]Q_2$.
2. Make encryption query (P_1, Q_1) and obtains response (C_1, D_1).
3. Make encryption query (P_1, Q_2) and obtains response (C_2, D_2).
4. **return** 1 if $D_2 = D_1 + N''[2] \cdot N[2] \cdot (Q_1 + Q_2)$, 0 otherwise.

4.3   Case: $\mathrm{rank}(M[2]) = \mathrm{rank}(M'[2]) = n - 1$, $\mathrm{rank}(N[1]) = n - 1$

We make three queries same as $\mathcal{A}_0$ except the choice of Q_3 whose value is determined below. We have

1. $U_1 + U_3 = M[1](A_1 + A_2) + N[1]Q_1 + N[1]Q_3$.
2. $M[2](A_1 + A_2) = N[2](Q_1 + Q_2) + N'[2](D_1 + D_2)$ (from the computations of W_1 and W_2).

As the rank of $M[2]$ is $n - 1$ and the right hand side of item 2 above is known, we can guess $(A_1 + A_2)$ correctly with probability $1/2$ (since there are only two choices). So we can guess $M[1](A_1 + A_2)$ from $M[2](A_1 + A_2)$ with probability at least $1/2$. Let X be the guessed value of $M[1](A_1 + A_2)$. We now choose Q_3 such that $U_1 + U_3 = 0$ (i.e., $U_1 = U_3$). From the item 1 of above, we define $Q_3 = N[1]^{-1} \cdot X + Q_1$. Note that $N[1]$ is assumed to be invertible in this case. So $\Pr[U_1 = U_3] \geq 1/2$. This essentially leads to a similar distinguisher as in XLS. However, we need to compute the distinguishing event similar to the computation of the previous case. By the applications of mix_1, mix_2^{-1} and mix_2 functions for three queries, we have

1. $W_1 + W_2 = N'[2] \cdot (D_1 + D_2)$,
2. $W_2 + W_3 = N[2] \cdot (Q_2 + Q_3)$, and
3. $N''[2] \cdot (W_1 + W_3) = D_1 + D_3$.

So we have $D_3 = D_1 + N''[2] \cdot (N'[2] \cdot (D_1 + D_2) + N[2] \cdot (Q_1 + Q_3))$ which can be observed with probability 2^{-n+1} for a random function. We formally describe the attack algorithm by $\mathcal{A}_2$ below.

Distinguishing Algorithm $\mathcal{A}_3$.

1. Make encryption query (P_1, Q_1) and obtains response (C_1, D_1).
2. Make decryption query $(C_2 := C_1, D_2 := D_1 + 1)$ and obtains response (P_2, Q_2).
3. Guess $M[1](A_1 + A_2)$, denoted X, from $N[2](Q_1 + Q_2) + N'[2](D_1 + D_2)$
4. Choose Q_3 such that $N[1](Q_1 + Q_3) = X$.
5. Make encryption query $(P_3 = P_2, Q_3)$ and obtains response (C_3, D_3).
6. **return** 1 if $D_3 = D_1 + N''[2] \cdot (N'[2] \cdot (D_1 + D_2) + N[2] \cdot (Q_1 + Q_3))$,
7. **return** 0, otherwise.

---

5   Conclusion

In this paper we provide chosen plaintext and ciphertext distinguishing attack (i.e., SPRP distinguisher) of XLS. It makes three encryption and decryption calls and has distinguishing advantage about $1/2$. This attack can be further extended to a general form of XLS following mix-then-encrypt paradigm. We believe that it can not be repaired without introducing any non-linear functionality, e.g. an additional blockcipher call. So we need four blockcipher calls to make this types of design secure. Both Elastic blokcipher and Nandi's construction make four calls of non-linear functions. However, Nandi's construction could be potentially faster, as it requires two universal hash invocations (which can be achieved by applying four rounds of AES [6]) and one call of weak-PRF (optimistically, one can apply eight rounds of AES) in addition with a full blokcipher call (which is e.g., ten rounds of AES). So in total it requires 26 rounds of AES which is much faster than four full invocations of AES.

Acknowledgement. This work is supported by Centre of Excellence in Cryptology at Indian Statistical Institute, Kolkata.

References

1. CAESAR submissions (2014), http://competitions.cr.yp.to/caesar-submissions.html
2. Andreeva, E., Bogdanov, A., Luykx, A., Mennink, B., Tischhauser, E., Yasuda, K.: Parallelizable and authenticated online ciphers. In: Sako, K., Sarkar, P. (eds.) ASIACRYPT 2013, Part I. LNCS, vol. 8269, pp. 424–443. Springer, Heidelberg (2013)
3. Chakraborty, D., Sarkar, P.: HCH: A new tweakable enciphering scheme using the hash-encrypt-hash approach. In: Barua, R., Lange, T. (eds.) INDOCRYPT 2006. LNCS, vol. 4329, pp. 287–302. Springer, Heidelberg (2006)
4. Cook, D.L., Yung, M., Keromytis, A.D.: Elastic aes. IACR Cryptology ePrint Archive, 2004:141 (2004)
5. Cook, D.L., Yung, M., Keromytis, A.D.: Elastic block ciphers: method, security and instantiations. Int. J. Inf. Sec. 8(3), 211–231 (2009)
6. Daemen, J., Lamberger, M., Pramstaller, N., Rijmen, V., Vercauteren, F.: Computational aspects of the expected differential probability of 4-round aes and aes-like ciphers. Computing 85(1-2), 85–104 (2009)
7. Halevi, S.: EME*: Extending EME to handle arbitrary-length messages with associated data. In: Canteaut, A., Viswanathan, K. (eds.) INDOCRYPT 2004. LNCS, vol. 3348, pp. 315–327. Springer, Heidelberg (2004)
8. Halevi, S.: TET: A wide-block tweakable mode based on Naor-Reingold. Cryptology ePrint Archive, Report 2007/014 (2007), http://eprint.iacr.org/
9. Halevi, S., Rogaway, P.: A tweakable enciphering mode. In: Boneh, D. (ed.) CRYPTO 2003. LNCS, vol. 2729, pp. 482–499. Springer, Heidelberg (2003)
10. Halevi, S., Rogaway, P.: A parallelizable enciphering mode. In: Okamoto, T. (ed.) CT-RSA 2004. LNCS, vol. 2964, pp. 292–304. Springer, Heidelberg (2004)
11. Liskov, M., Rivest, R.L., Wagner, D.: Tweakable block ciphers. In: Yung, M. (ed.) CRYPTO 2002. LNCS, vol. 2442, pp. 31–46. Springer, Heidelberg (2002)

12. Luby, M., Rackoff, C.: How to construct pseudo-random permutations from pseudo-random functions. In: Williams, H.C. (ed.) CRYPTO 1985. LNCS, vol. 218, p. 447. Springer, Heidelberg (1986)
13. McGrew, D.A., Fluhrer, S.R.: The extended codebook (XCB) mode of operation. Cryptology ePrint Archive, Report 2004/278 (2004), http://eprint.iacr.org/
14. Nandi, M.: A generic method to extend message space of a strong pseudorandom permutation. Computación y Sistemas 12(3) (2009)
15. Ristenpart, T., Rogaway, P.: How to enrich the message space of a cipher. In: Biryukov, A. (ed.) FSE 2007. LNCS, vol. 4593, pp. 101–118. Springer, Heidelberg (2007)
16. Sarkar, P.: Improving upon the tet mode of operation. In: Nam, K.-H., Rhee, G. (eds.) ICISC 2007. LNCS, vol. 4817, pp. 180–192. Springer, Heidelberg (2007)
17. Wang, P., Feng, D., Wu, W.: HCTR: A variable-input-length enciphering mode. In: Feng, D., Lin, D., Yung, M. (eds.) CISC 2005. LNCS, vol. 3822, pp. 175–188. Springer, Heidelberg (2005)

Structure-Preserving Signatures on Equivalence Classes and Their Application to Anonymous Credentials

Christian Hanser and Daniel Slamanig

Institute for Applied Information Processing and Communications (IAIK),
Graz University of Technology (TUG), Inffeldgasse 16a, 8010 Graz, Austria
{christian.hanser,daniel.slamanig}@tugraz.at

Abstract. Structure-preserving signatures are a quite recent but important building block for many cryptographic protocols. In this paper, we introduce a new type of structure-preserving signatures, which allows to sign group element vectors and to consistently randomize signatures and messages without knowledge of any secret. More precisely, we consider messages to be (representatives of) equivalence classes on vectors of group elements (coming from a single prime order group), which are determined by the mutual ratios of the discrete logarithms of the representative's vector components. By multiplying each component with the same scalar, a different representative of the same equivalence class is obtained. We propose a definition of such a signature scheme, a security model and give an efficient construction, which is secure in the SXDH setting, where **EUF-CMA** security holds against generic forgers in the generic group model and the so called class hiding property holds under the DDH assumption.

As a second contribution, we use the proposed signature scheme to build an efficient multi-show attribute-based anonymous credential (ABC) system that allows to encode an arbitrary number of attributes. This is – to the best of our knowledge – the first ABC system that provides constant-size credentials and constant-size showings. To allow an efficient construction in combination with the proposed signature scheme, we also introduce a new, efficient, randomizable polynomial commitment scheme. Aside from these two building blocks, the credential system requires a very short and constant-size proof of knowledge to provide freshness in the showing protocol.

1 Introduction

Digital signatures are an important cryptographic primitive to provide a means for integrity protection, non-repudiation as well as authenticity of messages in a publicly verifiable way. In most signature schemes, the message space consists of integers in $\mathbb{Z}_{\mathrm{ord}(G)}$ for some group G or consists of arbitrary strings encoded either to integers in $\mathbb{Z}_{\mathrm{ord}(G)}$ or to elements of a group G using a suitable hash function. In the latter case, the hash function is usually required to be modeled as a random oracle (thus, one signs random group elements). In contrast,

P. Sarkar and T. Iwata (Eds.): ASIACRYPT 2014, PART I, LNCS 8873, pp. 491–511, 2014.
© International Association for Cryptologic Research 2014

structure-preserving signatures [33,6,1,2,21,5,4] can handle messages which are elements of two groups G_1 and G_2 equipped with a bilinear map, without requiring any prior encoding. Basically, in a structure-preserving signature scheme the public key, the messages and the signatures consist only of group elements and the verification algorithm evaluates a signature by deciding group membership of elements in the signature and by evaluating pairing product equations. Such signature schemes typically allow to sign vectors of group elements (from one of the two groups G_1 and G_2, or mixed) and also support some types of *randomization* (inner, sequential, etc., cf. [1,5]).

Randomization is one interesting feature of signatures, as a given signature can be randomized to another unlinkable version of the signature for the same message. Besides randomizable structure-preserving signatures, there are various other constructions of such signature schemes [24,25,18,43]. We emphasize that although these schemes are randomizable, they are still secure digital signatures in the standard sense (EUF-CMA security).

We are interested in constructions of structure-preserving signature schemes that do *not only* allow randomization of the signature, *but also* allow to randomize the signed message in particular ways. Such signature schemes are particularly interesting for applications in privacy-enhancing cryptographic protocols.

1.1   Contribution

This paper has three contributions: A novel type of structure-preserving signatures defined on equivalence classes on group element vectors, a novel randomizable polynomial commitment scheme, which allows to open factors of the polynomial committed to, and a new construction (type) of multi-show attribute-based anonymous credentials (ABCs), which is instantiated from the first two contributions.

Structure-Preserving Signature Scheme on Equivalence Classes. Inspired by *randomizable signatures*, we introduce a novel variant of structure-preserving signatures. Instead of signing particular message vectors as in other schemes, the scheme *produces signatures on classes of an equivalence relation* $\mathcal{R}$ defined on $(G_1^*)^\ell$ with $\ell > 1$ (where we use G_1^* to denote $G_1 \setminus \{0_{G_1}\}$). More precisely, we consider messages to be (representatives of) equivalence classes on $(G_1^*)^\ell$, which are determined by the mutual ratios of the discrete logarithms of the representative's vector components. By multiplying each component with the same scalar, a different representative of the same equivalence class is obtained. Initially, an equivalence class is signed by signing an arbitrary representative. Later, one can obtain a valid signature for every other representative of this class, without having access to the secret key. Furthermore, we require two representatives of the same class with corresponding signatures to be unlinkable, which we call *class hiding*. We present a definition of such a signature scheme along with game based notions of security and present an efficient construction, which produces short and constant-size signatures that are independent of the message vector length ℓ. In the full version [37], we prove the security of our construction in the generic group model against generic forgers and the DDH assumption, respectively.

Polynomial Commitments with Factor Openings. We propose a new, efficient, randomizable polynomial commitment scheme. It is computationally binding, unconditionally hiding, allows to commit to monic, reducible polynomials and is represented by an element of a bilinear group. It allows to open factors of committed polynomials and re-randomization (i.e., multiplication with a scalar) does not change the polynomial committed to, but requires only a consistent randomization of the witnesses involved in the factor openings. We present a definition as well as a construction of such a polynomial commitment scheme. In the full version [37], we give a security model in which we also prove the construction secure.

A Multi-Show Attribute-Based Anonymous Credential (ABC) System. We describe a new way to build multi-show ABCs (henceforth, we will only write ABCs) as an application of the first two contributions. From another perspective, the signature scheme allows to consistently randomize a vector of group elements and its signature. So, it seems natural to use this property to achieve unlinkability during the showings of an ABC system. To enable a compact attribute representation, which is compatible with the randomization property of the signature scheme, we encode the attributes to polynomials and commit to them using the introduced polynomial commitment scheme. During the issuing, the obtainer is, then, given a set of attributes and the credential, which is a message (vector) consisting of the polynomial commitment and the generator of the group plus the corresponding signature. During a showing, a subset of the issued attributes can be shown by opening the corresponding factors of the committed polynomial. The unlinkability of showings is achieved through the inherent re-randomization properties of the signature scheme and the polynomial commitment scheme, which are compatible to each other. Furthermore, to provide freshness during a showing, we require a very small, constant-size proof of knowledge. We emphasize that our approach to construct ABCs is very different from existing approaches, as we use *neither* zero-knowledge proofs for proving the possession of a signature *nor* for selectively disclosing attributes during showings. Recall that existing approaches rely on signature schemes that allow to sign vectors of attributes and use efficient zero-knowledge proofs to show possession of a signature and to prove relations about the signed attributes during a showing.

Interestingly, in our construction *the size of credentials as well as the size of the showings* are *independent of the number of attributes in the ABC system*, i.e., a small, constant number of group elements. This is, to the best of our knowledge, the first ABC system with this feature. The proposed ABC system is secure in a security model adapted from [23,8,26,27], where we refer the reader to the full version [37] for the proofs and the security model. Finally, we compare our system to other existing multi- and one-show ABC approaches. Although we are only dealing with multi-show credentials, for the sake of completeness, we also compare our approach to the one-show (i.e., linkable) anonymous credentials of Brands [20] (and, thus, also its provably secure generalization [12]).

1.2 Related Work

In [16], Blazy et al. present *signatures on randomizable ciphertexts* (based on linear encryption [18]) using a variant of Waters' signature scheme [43]. Basically, anyone given a signature on a ciphertext can *randomize the ciphertext* and adapt the signature accordingly, while maintaining public verifiability and neither knowing the signing key nor the encrypted message. However, as these signatures only allow to randomize the ciphertexts and not the underlying plaintexts, this approach is not useful for our purposes.

Another somewhat related approach is the proofless variant of the Chaum-Pedersen signature [31] which is used to build *self-blindable* certificates by Verheul in [42]. The resulting so called certificate as well as the initial message can be randomized using the same scalar, preserving the validity of the certificate. This approach works for the construction in [42], but it does not represent a secure signature scheme (as also observed in [42]) due to its homomorphic property and the possibility of efficient existential forgeries.

Homomorphic signatures for network coding [19] allow to sign any subspace of a vector space by producing a signature for every basis vector with respect to the same (file) identifier. Consequently, the message space consists of identifiers and vectors. These signatures are homomorphic, meaning that given a sequence of scalar and signature pairs $(\beta_i, \sigma_i)_{i=1}^{\ell}$ for vectors v_i, one can publicly compute a signature for the vector $v = \sum_{i=1}^{\ell} \beta_i v_i$ (this is called derive). If one was using a unique identifier per signed vector v, then such linearly homomorphic signatures would support a functionality similar to the one provided by our scheme, i.e., publicly compute signatures for vectors $v' = \beta v$ (although they are not structure-preserving). It is also known that various existing constructions, e.g., [19,10] are *strong context hiding*, meaning that original and derived signatures are unlinkable. Nevertheless, this does not help in our context, which is due to the following argument: If we do not restrict every single signed vector to a unique identifier, the signature schemes are homomorphic, which is not compatible with our unforgeability goal. If we apply this restriction, however, then we are not able to achieve *class hiding* as all signatures can be linked to the initial signature by the unique identifier. We note that the same arguments also apply to structure-preserving linearly homomorphic signatures [40].

The aforementioned context hiding property is also of interest in more general classes of homomorphic (also called malleable) signature schemes (defined in [7] and refined in [9]). In [29], the authors discuss malleable signatures that allow to derive a signature σ' on a message $m' = T(m)$ for an "allowable" transformation T, when given a signature σ for a message m. This can be considered as a generalization of signature schemes, such as quotable [10] or redactable signatures [38] with the additional property of being context hiding. The authors note that for messages being pseudonyms and transformations that transfer one pseudonym into another pseudonym, such malleable signatures can be used to construct anonymous credential systems. They also demonstrate how to build delegatable anonymous credential systems [15,14]. The general construction in [29] relies on malleable-ZKPs [28] and is not really efficient, even when instantiated with Groth-Sahai proofs [35]. Although it is conceptually totally different from our approach, we note that by viewing our scheme in a different way, our scheme fits into their definition of malleable signatures (such that their SigEval algorithm takes only a single message vector with corresponding signature and a single allowable transformation). However, firstly, our construction is far more efficient than their approach (and in particular really practical) and, secondly, [29] only focuses on transformations of single messages (pseudonyms) and does not consider multi-show attribute-based anonymous credentials at all (which is the main focus of our construction).

Signatures providing randomization features [24,25,18] along with efficient proofs of knowledge of committed values can be used to generically construct ABC systems. The most prominent approaches based on Σ-protocols are CL credentials [24,25]. With the advent of Groth-Sahai proofs, which allow (efficient) non-interactive proofs in the CRS model without random oracles, various constructions of so called delegatable (hierarchical) anonymous credentials have been proposed [15,14]. These provide per definition a non-interactive showing protocol, i.e., the show and verify algorithms do not interact when demonstrating the possession of a credential. In [34], Fuchsbauer presented the first delegatable anonymous credential system that also provides a non-interactive delegation protocol based on so called commuting signatures and verifiable encryption. We note that although such credential systems with non-interactive protocols extend the scope of applications of anonymous credentials, the most common use-case (i.e., authentication and authorization), essentially relies on interaction (to provide freshness/liveness). We emphasize that our goal is not to construct non-interactive anonymous credentials. Nevertheless, one could generically convert our proposed system to a non-interactive one: in the ROM using Fiat-Shamir or by replacing our single Σ-proof for freshness with a Groth-Sahai proof without random oracles, which is, however, out of scope of this paper.

2 Preliminaries

Definition 1 (Bilinear Map). Let G_1, G_2 and G_T be cyclic groups of prime order p, where G_1 and G_2 are additive and G_T is multiplicative. Let P and P' generate G_1 and G_2, respectively. We call $e : G_1 \times G_2 \to G_T$ *bilinear map* or *pairing* if it is efficiently computable and the following conditions hold:

Bilinearity: $e(aP, bP') = e(P, P')^{ab} = e(bP, aP')$ $\quad \forall a, b \in \mathbb{Z}_p$

Non-degeneracy: $e(P, P') \neq 1_{G_T}$, i.e., $e(P, P')$ generates G_T.

If $G_1 = G_2$, then e is called *symmetric* (Type-1) and *asymmetric* (Type-2 or Type-3) otherwise. For Type-2 pairings there is an efficiently computable isomorphism $\Psi : G_2 \to G_1$, whereas for Type-3 pairings no such efficient isomorphism is assumed to exist. Note that Type-3 pairings are currently the optimal choice [30], with respect to efficiency and security trade-off.

Definition 2 (Decisional Diffie Hellman Assumption (DDH)). Let p be a prime of bitlength κ, G be a group of prime order p generated by P and let $(P, aP, bP, cP) \in G^4$, where $a, b, c \in_R \mathbb{Z}_p^*$. Then, for every PPT adversary $\mathcal{A}$ distinguishing between $(P, aP, bP, abP) \in G^4$ and $(P, aP, bP, cP) \in G^4$ is infeasible, i.e., there is a negligible function $\epsilon(\cdot)$ such that

$$|\Pr[\text{true} \leftarrow \mathcal{A}(P, aP, bP, abP)] - \Pr[\text{true} \leftarrow \mathcal{A}(P, aP, bP, cP)]| \leq \epsilon(\kappa).$$

Definition 3 (Symmetric External DH Assumption (SXDH) [13]). Let G_1, G_2 and G_T be three distinct cyclic groups of prime order p and $e : G_1 \times G_2 \to G_T$ be a pairing. Then, the SXDH assumption states that in both groups G_1 and G_2 the DDH assumption holds.

Note that the SXDH assumption formalizes Type-3 pairings, i.e., the absence of an efficiently computable isomorphism between G_1 and G_2 as well as between G_2 and G_1.

Definition 4 (Bilinear Group Generator). Let BGGen be a PPT algorithm which takes a security parameter κ and generates a bilinear group BG $= (p, G_1, G_2, G_T, e, P, P')$ in the SXDH setting, where the common group order p of the groups G_1, G_2 and G_T is a prime of bitlength κ, e is a pairing and P as well as P' are generators of G_1 and G_2, respectively.

Definition 5 (t-Strong DH Assumption (t-SDH) [17]). Let p be a prime of bitlength κ, G be a group of prime order p generated by $P \in G$, $\alpha \in_R \mathbb{Z}_p^*$ and let $(\alpha^i P)_{i=0}^t \in G^{t+1}$ for some $t > 0$. Then, for every PPT adversary $\mathcal{A}$ there is a negligible function $\epsilon(\cdot)$ such that

$$\Pr\left[\left(c, \frac{1}{\alpha + c}P\right) \leftarrow \mathcal{A}((\alpha^i P)_{i=0}^t)\right] \leq \epsilon(\kappa) \quad \text{for some } c \in \mathbb{Z}_p \setminus \{-\alpha\}.$$

This assumption turns out to be very useful in bilinear groups (Type-1 or Type-2 setting). However, in a Type-3 setting (SXDH assumption), where the groups G_1 and G_2 are strictly separated, the presence of a pairing does not give any additional benefit. This is due to the fact that the problem instance is given either in G_1 or in G_2. As our constructions rely on the SXDH assumption, we introduce the following modified assumption, which can be seen as the natural counterpart for a Type-3 setting [30]:

Definition 6 (co-t-Strong DH Assumption (co-t-SDH$_i^*$)). Let G_1 and G_2 be two groups of prime order p (which has bitlength κ) generated by $P_1 \in G_1$ and $P_2 \in G_2$, respectively. Let $\alpha \in_R \mathbb{Z}_p^*$ and let $(\alpha^j P_1)_{j=0}^t \in G_1^{t+1}$ and $(\alpha^j P_2)_{j=0}^t \in G_2^{t+1}$ for some $t > 0$. Then, for every PPT adversary $\mathcal{A}$ there is a negligible function $\epsilon(\cdot)$ such that

$$\Pr\left[\left(c, \frac{1}{\alpha + c}P_i\right) \leftarrow \mathcal{A}((\alpha^j P_1)_{j=0}^t, (\alpha^j P_2)_{j=0}^t)\right] \leq \epsilon(\kappa) \quad \text{for some } c \in \mathbb{Z}_p \setminus \{-\alpha\}.$$

Note that for a compact representation, we make a slight abuse of notation, where it should be interpreted as $P_1 = P$ and $P_2 = P'$. Obviously, we have co-t-SDH$^*_i \leq_p t$-SDH in group G_i. The t-SDH assumption was originally proven to be secure in the generic group model in [17, Theorem 5.1] and further studied in [32]. The proof is done in a Type-2 pairing setting, where an efficiently computable isomorphism $\Psi : G_2 \to G_1$ exists. In the proof, the adversary is given the problem instance in group G_2 and is allowed to obtain encodings of elements in G_1 through isomorphism queries. As we are in a Type-3 setting, there is no such efficiently computable isomorphism. Thus, the problem instance given to the adversary must contain all corresponding elements in both groups G_1 and G_2. Then, the generic group model proof for the co-t-SDH$^*_i$ assumption can be done analogously to the proof in [17, proof of Theorem 5.1]. The main difference is that instead of querying the isomorphism, the adversary must compute the same sequence of computations performed in one group in the other group, in order to obtain an element containing the same discrete logarithm, which, however, preserves the asymptotic number of queries.

3   Structure-Preserving Signatures on Equivalence Classes

We are looking for an efficient, randomizable structure-preserving signature scheme for vectors with arbitrary numbers of group elements that allows to randomize messages and signatures consistently in the public. It seems natural to consider such messages as representatives of certain equivalence classes and to perform randomization via a change of representatives. Before we can introduce such a signature scheme and give an efficient construction, we detail these equivalence classes.

All elements of a vector $(M_i)^\ell_{i=1} \in (G^*_1)^\ell$ (for some prime order group G_1, where we write G^*_1 for $G_1 \setminus \{0_{G_1}\}$) share different mutual ratios. These ratios depend on their discrete logarithms and are invariant under the operation $\gamma : \mathbb{Z}^*_p \times (G^*_1)^\ell \to (G^*_1)^\ell$ with $(s, (M_i)^\ell_{i=1}) \mapsto s(M_i)^\ell_{i=1}$. Thus, we can use this invariance to partition the set $(G^*_1)^\ell$ into classes using the following equivalence relation:

$$\mathcal{R} = \{(M, N) \in (G^*_1)^\ell \times (G^*_1)^\ell : \exists s \in \mathbb{Z}^*_p \text{ such that } N = s \cdot M\} \subseteq (G^*_1)^{2\ell}.$$

It is easy to verify that $\mathcal{R}$ is indeed an equivalence relation given that G_1 has prime order. When signing an equivalence class $[M]_\mathcal{R}$ with our scheme, one actually signs an arbitrary representative $(M_i)^\ell_{i=1}$ of class $[M]_\mathcal{R}$. The scheme, then, allows to choose different representatives and to update corresponding signatures in the public, i.e., without any secret key. Thereby, one of our goals is to guarantee that two message-signature pairs on the same equivalence class cannot be linked. Note that such an approach only seems to work for structure-preserving signature schemes, where we have no direct access to scalars. Otherwise, if we wanted to sign vectors of elements of $\mathbb{Z}^*_p$, the direct access to the scalars would allow us to decide class membership efficiently. This is also the reason, why we subsequently define the class hiding property with respect to a random-message instead of a chosen-message attack.

3.1 Defining the Signature Scheme

Now, we formally define a signature scheme for the above equivalence relation and its required security properties.

Definition 7 (Structure-Preserving Signature Scheme for Equivalence Relation $\mathcal{R}$ (SPS-EQ-$\mathcal{R}$)). An SPS-EQ-$\mathcal{R}$ scheme consists of the following polynomial time algorithms:

$\mathsf{BGGen}_{\mathcal{R}}(\kappa)$: Is a probabilistic bilinear group generation algorithm, which on input a security parameter κ outputs a bilinear group BG.

$\mathsf{KeyGen}_{\mathcal{R}}(\mathsf{BG}, \ell)$: Is a probabilistic algorithm, which on input a bilinear group BG and a vector length $\ell > 1$, outputs a key pair $(\mathsf{sk}, \mathsf{pk})$.

$\mathsf{Sign}_{\mathcal{R}}(M, \mathsf{sk})$: Is a probabilistic algorithm, which on input a representative M of an equivalence class $[M]_{\mathcal{R}}$ and a secret key sk, outputs a signature σ for the equivalence class $[M]_{\mathcal{R}}$ (using randomness y).

$\mathsf{ChgRep}_{\mathcal{R}}(M, \sigma, \rho, \mathsf{pk})$: Is a probabilistic algorithm, which on input a representative M of an equivalence class $[M]_{\mathcal{R}}$, the corresponding signature σ, a scalar ρ and a public key pk, returns an updated message-signature pair $(\hat{M}, \hat{\sigma})$ (using randomness $\hat{y}$). Here, $\hat{M}$ is the new representative $\rho \cdot M$ and $\hat{\sigma}$ its updated signature.

$\mathsf{Verify}_{\mathcal{R}}(M, \sigma, \mathsf{pk})$: Is a deterministic algorithm, which given a representative M, a signature σ and a public key pk, outputs $\mathtt{true}$ if σ is a valid signature for the equivalence class $[M]_{\mathcal{R}}$ under pk and $\mathtt{false}$ otherwise.

When one does not care about which new representative is chosen, $\mathsf{ChgRep}_{\mathcal{R}}$ can be seen as consistent randomization of a signature and its message using randomizer ρ without invalidating the signature on the equivalence class. The goal is that the signature resulting from $\mathsf{ChgRep}_{\mathcal{R}}$ is indistinguishable from a newly issued signature for the new representative of the same class.

For security, we require the usual correctness property for signature schemes, but instead of single messages we consider the respective equivalence class and the correctness of $\mathsf{ChgRep}_{\mathcal{R}}$. More formally, we require:

Definition 8 (Correctness). An SPS-EQ-$\mathcal{R}$ scheme is called *correct*, if for all security parameters $\kappa \in \mathbb{N}$, for all $\ell > 1$, for all bilinear groups $\mathsf{BG} \leftarrow \mathsf{BGGen}_{\mathcal{R}}(\kappa)$, all key pairs $(\mathsf{sk}, \mathsf{pk}) \leftarrow \mathsf{KeyGen}_{\mathcal{R}}(\mathsf{BG}, \ell)$ and for all $M \in (G_1^*)^\ell$ it holds that

$$\mathsf{Verify}_{\mathcal{R}}(\mathsf{ChgRep}_{\mathcal{R}}(M, \mathsf{Sign}_{\mathcal{R}}(M, \mathsf{sk}), \rho, \mathsf{pk}), \mathsf{pk}) = \mathtt{true} \quad \forall \rho \in \mathbb{Z}_p^*.$$

Furthermore, we require a notion of EUF-CMA security. In contrast to the standard definition of EUF-CMA security, we consider a natural adaption, i.e., outputting a valid message-signature pair, corresponding to an unqueried equivalence class, is considered to be a forgery.

Definition 9 (EUF-CMA). An SPS-EQ-$\mathcal{R}$ scheme is called *existentially un-forgeable under adaptively chosen-message attacks*, if for all PPT algorithms $\mathcal{A}$ having access to a signing oracle $\mathcal{O}(\mathsf{sk}, M)$, there is a negligible function $\epsilon(\cdot)$ such that:

$$\Pr\left[\begin{array}{c} \mathsf{BG} \leftarrow \mathsf{BGGen}_{\mathcal{R}}(\kappa),\quad (\mathsf{sk}, \mathsf{pk}) \leftarrow \mathsf{KeyGen}_{\mathcal{R}}(\mathsf{BG}, \ell) \\ (M^*, \sigma^*) \leftarrow \mathcal{A}^{\mathcal{O}(\mathsf{sk}, \cdot)}(\mathsf{pk}) : \\ [M^*]_{\mathcal{R}} \neq [M]_{\mathcal{R}} \;\; \forall M \in Q \;\; \wedge \;\; \mathsf{Verify}_{\mathcal{R}}(M^*, \sigma^*, \mathsf{pk}) = \mathtt{true} \end{array}\right] \leq \epsilon(\kappa),$$

where Q is the set of queries which $\mathcal{A}$ has issued to the signing oracle $\mathcal{O}$.

Subsequently, we let Q be a list for keeping track of queried messages M and make use of the following oracles:

$\mathcal{O}^{RM}(\ell)$: A random-message oracle, which on input a message vector length ℓ, picks a message $M \xleftarrow{R} (G_1^*)^\ell$, appends M to Q and returns it.

$\mathcal{O}^{RoR}(\mathsf{sk}, \mathsf{pk}, b, M)$: A real-or-random oracle taking input a bit b and a message M. If $M \notin Q$, it returns $\perp$. On the first valid call, it chooses $R \xleftarrow{R} (G_1^*)^\ell$, computes $\mathcal{M} \leftarrow ((M, \mathsf{Sign}_{\mathcal{R}}(M, \mathsf{sk})), (R, \mathsf{Sign}_{\mathcal{R}}(R, \mathsf{sk})))$ and returns $\mathcal{M}[b]$. Any next call for $M' \neq M$ will return $\perp$ and $\mathsf{ChgRep}_{\mathcal{R}}(\mathcal{M}[b], \rho, \mathsf{pk})$ otherwise, where $\rho \xleftarrow{R} \mathbb{Z}_p^*$.

Definition 10 (Class Hiding). An SPS-EQ-$\mathcal{R}$ scheme on $(G_1^*)^\ell$ is called *class hiding*, if for every PPT adversary $\mathcal{A}$ with oracle access to $\mathcal{O}^{RM}$ and $\mathcal{O}^{RoR}$, there is a negligible function $\epsilon(\cdot)$ such that

$$\Pr\left[\begin{array}{c} \mathsf{BG} \leftarrow \mathsf{BGGen}_{\mathcal{R}}(\kappa),\ b \xleftarrow{R} \{0,1\},\ (\mathsf{state}, \mathsf{sk}, \mathsf{pk}) \leftarrow \mathcal{A}(\mathsf{BG}, \ell), \\ \mathcal{O} \leftarrow \{\mathcal{O}^{RM}(\ell), \mathcal{O}^{RoR}(\mathsf{sk}, \mathsf{pk}, b, \cdot)\},\ b^* \leftarrow \mathcal{A}^{\mathcal{O}}(\mathsf{state}, \mathsf{sk}, \mathsf{pk}) : \\ b^* = b \end{array}\right] - \frac{1}{2} \leq \epsilon(\kappa).$$

Here, the adversary is in the role of a signer, who issues signatures on random messages (in the sense of a random message attack) and can derive signatures for arbitrary representatives of queried classes. Observe that, if the adversary was able to pick messages on its own, e.g., knows the discrete logarithms of the group elements or puts identical group elements on different positions of the message vectors, it would trivially be able to distinguish the classes. Consequently, we define class hiding in a random message attack game and the random sampling of messages makes the probability of identical message elements at different positions negligible.

Definition 11 (Security). An SPS-EQ-$\mathcal{R}$ scheme is *secure*, if it is correct, EUF-CMA secure and class hiding.

3.2   Our Construction

In our construction, we sign vectors of $\ell > 1$ elements of G_1^*, where the public key only consists of elements in G_2 and we require the SXDH assumption to hold.

The signature consists of four group elements, where three elements are from G_1 and one element is from G_2. Two signature elements (Z_1, Z_2) are aggregates of the message vector under ℓ elements of the private key. In order to prevent an additive homomorphism on the signatures, we introduce a randomizer $y \in \mathbb{Z}_p^*$, multiply one aggregate with it and introduce two additional values $Y = yP$ and $Y' = yP'$. The latter elements (besides eliminating the homomorphic property) prevent simple forgeries, where Y' contains an aggregation of the public keys $X', X_1', \ldots, X_\ell'$ in G_2. This is achieved by verifying whether Y and Y' contain the same unknown discrete logarithms during verification. Our construction lets us switch to another representative $\hat{M} = \rho M$ of M by multiplying M and (Z_1, Z_2) with the respective scalar ρ. Furthermore, a consistent re-randomization of ρZ_2, Y and Y' with a scalar $\hat{y}$ yields a signature $\hat{\sigma}$ for $\hat{M}$ that is unlinkable to the signature σ of M. In Scheme 1, we present the detailed construction of the SPS-EQ-$\mathcal{R}$ scheme.

$\mathsf{BGGen}_\mathcal{R}(\kappa)$: Given a security parameter κ, output $\mathsf{BG} \leftarrow \mathsf{BGGen}(\kappa)$.

$\mathsf{KeyGen}_\mathcal{R}(\mathsf{BG}, \ell)$: Given a bilinear group description BG and vector length $\ell > 1$, choose $x \xleftarrow{R} \mathbb{Z}_p^*$ and $(x_i)_{i=1}^\ell \xleftarrow{R} (\mathbb{Z}_p^*)^\ell$, set the secret key as $\mathsf{sk} \leftarrow (x, (x_i)_{i=1}^\ell)$, compute the public key $\mathsf{pk} \leftarrow (X', (X_i')_{i=1}^\ell) = (xP', (x_i x P')_{i=1}^\ell)$ and output $(\mathsf{sk}, \mathsf{pk})$.

$\mathsf{Sign}_\mathcal{R}(M, \mathsf{sk})$: On input a representative $M = (M_i)_{i=1}^\ell \in (G_1^*)^\ell$ of equivalence class $[M]_\mathcal{R}$ and secret key $\mathsf{sk} = (x, (x_i)_{i=1}^\ell)$, choose $y \xleftarrow{R} \mathbb{Z}_p^*$ and compute

$$Z_1 \leftarrow x \sum_{i=1}^\ell x_i M_i, \quad Z_2 \leftarrow y \sum_{i=1}^\ell x_i M_i \quad \text{and} \quad (Y, Y') \leftarrow y \cdot (P, P').$$

Then, output $\sigma = (Z_1, Z_2, Y, Y')$ as signature for the equivalence class $[M]_\mathcal{R}$.

$\mathsf{ChgRep}_\mathcal{R}(M, \sigma, \rho, \mathsf{pk})$: On input a representative $M = (M_i)_{i=1}^\ell \in (G_1^*)^\ell$ of equivalence class $[M]_\mathcal{R}$, the corresponding signature $\sigma = (Z_1, Z_2, Y, Y')$, $\rho \in \mathbb{Z}_p^*$ and public key pk, this algorithm picks $\hat{y} \xleftarrow{R} \mathbb{Z}_p^*$ and returns $(\hat{M}, \hat{\sigma})$, where $\hat{\sigma} \leftarrow (\rho Z_1, \hat{y} \rho Z_2, \hat{y} Y, \hat{y} Y')$ is the update of signature σ for the new representative $\hat{M} \leftarrow \rho \cdot (M_i)_{i=1}^\ell$.

$\mathsf{Verify}_\mathcal{R}(M, \sigma, \mathsf{pk})$: Given a representative $M = (M_i)_{i=1}^\ell \in (G_1^*)^\ell$ of equivalence class $[M]_\mathcal{R}$, a signature $\sigma = (Z_1, Z_2, Y, Y')$ and public key $\mathsf{pk} = (X', (X_i')_{i=1}^\ell)$, check whether

$$\prod_{i=1}^\ell e(M_i, X_i') \stackrel{?}{=} e(Z_1, P') \quad \wedge \quad e(Z_1, Y') \stackrel{?}{=} e(Z_2, X') \quad \wedge \quad e(P, Y') \stackrel{?}{=} e(Y, P')$$

and if this holds output $\mathtt{true}$ and $\mathtt{false}$ otherwise.

Scheme 1. A Construction of an SPS-EQ-$\mathcal{R}$ Scheme

Note that a signature resulting from $\mathsf{ChgRep}_\mathcal{R}$ is indistinguishable from a new signature on the same class using the new representative (it can be viewed as issuing a signature with randomness $y \cdot \hat{y}$).

3.3   Security of Our Construction

In our construction, message vectors are elements of $(G_1^*)^\ell$, public keys are only available in G_2 and signatures are elements of G_1 and G_2. Furthermore, we rely on the SXDH assumption, and it seems very hard (to impossible) to analyze the EUF-CMA security of the scheme via a reductionist proof using accepted non-interactive assumptions. Abe et al. [3] show that for optimally short structure-preserving signatures, i.e., three-element signatures, such reductions using non-interactive assumptions cannot exist. But right now, it is not entirely clear how structure-preserving signatures for equivalence relation $\mathcal{R}$ fit into these results and if the lower bounds from [2] also apply. Independently of this, it appears that a reduction to a (non-interactive) assumption is not possible, since due to the class hiding property the winning condition cannot be checked efficiently (without substantially weakening the unforgeability notion). Therefore, we chose to prove the EUF-CMA security of our construction using a direct proof in the generic group model such as for instance the proof of Abe et al. [2, Lemma 1] (cf. [37] for the proof).

Now, we state the security of the signature scheme. The corresponding proofs can be found in the full version [37].

Theorem 1. *The SPS-EQ-$\mathcal{R}$ scheme in Scheme 1 is correct.*

Theorem 2. *In the generic group model for SXDH groups, Scheme 1 is an EUF-CMA secure SPS-EQ-$\mathcal{R}$ scheme.*

Theorem 3. *If the DDH assumption holds in G_1, Scheme 1 is a class hiding SPS-EQ-$\mathcal{R}$ scheme.*

Taking everything together, we obtain the following corollary:

Corollary 1. *The SPS-EQ-$\mathcal{R}$ scheme in Scheme 1 is secure.*

4   Polynomial Commitments with Factor Openings

In [39], Kate et al. introduced the notion of constant-size polynomial commitments. The authors present two distinct commitment schemes, where one is computationally hiding (PolyCommit$_{DL}$) and the other one is unconditionally hiding (PolyCommit$_{Ped}$). These constructions are very generic, as they allow to construct witnesses for opening arbitrary evaluations of committed polynomials.

Yet, we emphasize that in practical scenarios (and especially in our constructions) it is often sufficient to consider the roots of polynomials for encodings and to open factors of the polynomial instead of arbitrary evaluations. Moreover, we need a polynomial commitment scheme that is easily randomizable. Therefore, we introduce the subsequent commitment scheme for monic, reducible polynomials. Instead of opening evaluations, it allows to open factors of committed polynomials. Hence, we call this type of commitment *polynomial commitment with factor openings*. Our construction is unconditionally hiding, computationally binding and more efficient than the Pedersen polynomial commitment construction PolyCommit$_{Ped}$ of [39]. Now, we briefly present this construction, which we denote by PolyCommitFO.

$\mathsf{Setup}_{\mathsf{PC}}(\kappa, t)$: It takes input a security parameter $\kappa \in \mathbb{N}$ and a maximum polynomial degree $t \in \mathbb{N}$. It runs $\mathsf{BG} \leftarrow \mathsf{BGGen}(\kappa)$, picks $\alpha \overset{R}{\leftarrow} \mathbb{Z}_p^*$ and outputs
$\mathsf{sk} \leftarrow \alpha$ as well as $\mathsf{pp} \leftarrow (\mathsf{BG}, (\alpha^i P)_{i=1}^t, (\alpha^i P')_{i=1}^t)$.

$\mathsf{Commit}_{\mathsf{PC}}(\mathsf{pp}, f(X))$: It takes input the public parameters pp and a monic, reducible polynomial $f(X) \in \mathbb{Z}_p[X]$ with $\deg f \leq t$. It picks $\rho \overset{R}{\leftarrow} \mathbb{Z}_p^*$, computes
the commitment $\mathcal{C} \leftarrow \rho \cdot f(\alpha)P \in G_1$ and outputs $(\mathcal{C}, O)$ with opening
information $O \leftarrow (\rho, f(X))$. [1]

$\mathsf{Open}_{\mathsf{PC}}(\mathsf{pp}, \mathcal{C}, \rho, f(X))$: It takes input the public parameters pp, a polynomial
commitment $\mathcal{C}$, the randomizer ρ used for $\mathcal{C}$ and the committed polynomial
$f(X)$ and outputs $(\rho, f(X))$.

$\mathsf{Verify}_{\mathsf{PC}}(\mathsf{pp}, \mathcal{C}, \rho, f(X))$: It takes input the public parameters pp, a polynomial
commitment $\mathcal{C}$, the randomizer ρ used for $\mathcal{C}$ and the committed polynomial
$f(X)$. It verifies whether $\rho \overset{?}{\neq} 0 \ \wedge \ \mathcal{C} \overset{?}{=} \rho \cdot f(\alpha)P$ holds and outputs $\mathtt{true}$
on success and $\mathtt{false}$ otherwise.

$\mathsf{FactorOpen}_{\mathsf{PC}}(\mathsf{pp}, \mathcal{C}, f(X), g(X), \rho)$: It takes input the public parameters pp, a
polynomial commitment $\mathcal{C}$, the committed polynomial $f(X)$, a factor $g(X)$
of $f(X)$ and the randomizer ρ used for $\mathcal{C}$. It computes $h(X) \leftarrow \frac{f(X)}{g(X)}$, the
witness $\mathcal{C}_h \leftarrow \rho \cdot h(\alpha)P$ and outputs $(g(X), \mathcal{C}_h)$.

$\mathsf{VerifyFactor}_{\mathsf{PC}}(\mathsf{pp}, \mathcal{C}, g(X), \mathcal{C}_h)$: It takes input the public parameters pp, a polynomial commitment $\mathcal{C}$ to a polynomial $f(X)$, a polynomial $g(X)$ of positive
degree and a corresponding witness $\mathcal{C}_h$. It verifies that $g(X)$ is a factor of
$f(X)$ by checking whether $\mathcal{C}_h \overset{?}{\neq} 0_{G_1} \ \wedge \ e(\mathcal{C}_h, g(\alpha)P') \overset{?}{=} e(\mathcal{C}, P')$ holds. It
outputs $\mathtt{true}$ on success and $\mathtt{false}$ otherwise.

In analogy to the security notion in [39], a polynomial commitment scheme
with factor openings is *secure* if it is *correct, polynomial binding, factor binding,*
factor sound, witness sound and *hiding.* The above scheme can be proven secure
under the co-t-SDH_1^* assumption. For the security model and the formal proofs
of security, we refer the reader to the full version [37]. Note that one can also
define a scheme based on the co-t-SDH_2^* assumption with $\mathcal{C} \in G_1$ and $\mathcal{C}_h \in G_2$.
Although this would improve the performance of $\mathsf{VerifyFactor}_{\mathsf{PC}}$, we define it
differently to reduce the computational complexity of the prover in the ABC
system in Section 5.3. Also note that we use the co-t-SDH_1^* assumption in a
static way, as t is a system parameter and fixed a priori. Finally, observe that
$\mathsf{sk} = \alpha$ must remain unknown to the committer (and, thus, the setup has to be
run by a TTP), since it is a trapdoor commitment scheme otherwise.

5   Building an ABC System

In this section, we present an application of the signature scheme and the polynomial commitment scheme introduced in the two previous sections, by using

---

[1] Subsequently, we use $f(\alpha)P$ as short-hand notation for $\sum_{i=0}^{\deg f} f_i \cdot \alpha^i P$ even if α is
unknown.

them as basic building blocks for an ABC system. ABC systems are usually constructed in one of the following two ways. Firstly, they can be built from blind signatures: A user obtains a blind signature from some issuer on (commitments to) attributes and, then, shows the signature, provides the shown attributes and proves the knowledge of all unrevealed attributes [20,12]. The drawback of such a blind signature approach is that such credentials can only be shown once in an unlinkable fashion (*one-show*). Secondly, anonymous credentials supporting an arbitrary number of unlinkable showings (*multi-show*) can be obtained in a similar vein using different types of signatures: A user obtains a signature on (commitments to) attributes, then *randomizes* the signature (such that the resulting signature is unlinkable to the issued one) and proves in zero-knowledge the possession of a signature and the correspondence of this signature with the shown attributes as well as the undisclosed attributes [24,25]. Our approach also achieves multi-show ABCs, but differs from the latter significantly: We randomize the signature and the message and, thus, do not require costly zero-knowledge proofs (which are, otherwise, at least linear in the number of shown/encoded attributes) for the showing of a credential.

Subsequently, we start by discussing the model of ABCs. Then, we provide an intuition for our construction in Section 5.2 and present the scheme in Section 5.3. In Section 5.4, we discuss the security of the construction. Finally, we give a performance comparison with other existing approaches in Section 5.5.

5.1   Abstract Model of ABCs

In an ABC system there are different organizations issuing credentials to different users. Users can then anonymously demonstrate possession of these credentials to verifiers. Such a system is called multi-show ABC system when transactions (issuing and showings) carried out by the same user cannot be linked. A credential cred_i for user i is issued by an organization j for a set $\mathbb{A} = \{(\text{attr}_k, \text{attrV}_k)\}_{k=1}^n$ of attribute labels attr_k and values attrV_k. By $\#\mathbb{A}$ we mean the size of $\mathbb{A}$, which is defined to be the sum of cardinalities of all second components attrV_k of the tuples in $\mathbb{A}$. Moreover, we denote by $\mathbb{A}' \sqsubseteq \mathbb{A}$ a subset of the credential's attributes. In particular, for every k, $1 \le k \le n$, we have that either $(\text{attr}_k, \text{attrV}_k)$ is missing or $(\text{attr}_k, \text{attrV}'_k)$ with $\text{attrV}'_k \subseteq \text{attrV}_k$ is present. A showing with respect to $\mathbb{A}'$ only proves that a valid credential for $\mathbb{A}'$ has been issued, but reveals nothing beyond (selective disclosure).

We note that in some ABC system constructions, the entire key generation is executed by the Setup algorithm. However, we split these algorithms into three algorithms to make the presentation more flexible and convenient.

Definition 12 (Attribute-Based Anonymous Credential System). An *attribute-based anonymous credential (ABC) system* consists of the following polynomial time algorithms:

Setup: A probabilistic algorithm that gets a security parameter κ, an upper bound t for the size of attribute sets and returns the public parameters pp.

OrgKeyGen: A probabilistic algorithm that takes input the public parameters pp and $j \in \mathbb{N}$, produces and outputs a key pair $(\mathsf{osk}_j, \mathsf{opk}_j)$ for organization j.

UserKeyGen: A probabilistic algorithm that takes input the public parameters pp and $i \in \mathbb{N}$, produces and outputs a key pair $(\mathsf{usk}_i, \mathsf{upk}_i)$ for user i.

(Obtain, Issue): These (probabilistic) algorithms are run by user i and organization j, who interact during execution. Obtain takes input the public parameters pp, the user's secret key usk_i, an organization's public key opk_j and an attribute set $\mathbb{A}$ of size $\#\mathbb{A} \leq t$. Issue takes input the public parameters pp, the user's public key upk_i, an organization's secret key osk_j and an attribute set $\mathbb{A}$ of size $\#\mathbb{A} \leq t$. At the end of this protocol, Obtain outputs a credential cred_i for $\mathbb{A}$ for user i.

(Show, Verify): These (probabilistic) algorithms are run by user i and a verifier, who interact during execution. Show takes input public parameters pp, the user's secret key usk_i, the organization's public key opk_j, a credential cred_i for set $\mathbb{A}$ of size $\#\mathbb{A} \leq t$ and a second set $\mathbb{A}' \sqsubseteq \mathbb{A}$. Verify takes input pp, the public key opk_j and a set $\mathbb{A}'$. At the end of the protocol, Verify outputs `true` or `false` indicating whether the credential showing was accepted or not.

An ABC system is called *secure* if it is *correct*, *unforgeable* and *anonymous* (for formal definitions, we refer the reader to the full version [37]).

5.2   Intuition of Our Construction

Our construction of ABCs is based on the proposed signature scheme, on polynomial commitments with factor openings and on a *single* constant-size proof of knowledge (PoK) for guaranteeing freshness. In contrast to this, the number of proofs of knowledge in other ABC systems, like [23,20] and related approaches, is linear in the number of shown attributes. Nevertheless, aside from selective disclosure of attributes, they allow to prove statements about non-revealed attribute values, such as AND, OR and NOT, interval proofs, as well as conjunctions and disjunctions of the aforementioned. The expressiveness that we achieve with our construction, can be compared to existing alternative constructions of ABCs [26,27]. Namely, our construction supports selective disclosure as well as AND statements about attributes. Thereby, a user can either open some attributes and their corresponding values or solely prove that some attributes are encoded in the respective credential without revealing their concrete values. Furthermore, one may associate sets of values to attributes, such that one is not required to reveal the full attribute value, but only pre-defined "statements" about the attribute value such as {"01.01.1980"," > 16"," > 18"} for attribute `birthdate`. This allows us to emulate proving properties about attribute values and, thus, enhances the expressiveness of the system.

Credential Representation: In our construction, a credential cred_i of user i is a vector of two group elements (C_1, P) together with a signature under the proposed signature scheme (see Section 3.2). During a showing, the credential gets randomized, which is easily achieved by changing the representative. The meaning of its values will be discussed subsequently.

Attribute Representation: We use PolyCommitFO (cf. Section 4) to commit to a polynomial, which encodes a set of attributes $\mathbb{A} = \{(\mathtt{attr}_k, \mathtt{attrV}_k)\}_{k=1}^n$ (where the encoding is inspired from [36]). This commitment is represented by the credential value C_1.

Now, we show how we use polynomials to encode this set of attributes and values. Thereby, we use a collision-resistant hash function $H : \{0,1\}^* \to \mathbb{Z}_p^*$ and the following encoding function to generate the polynomials:

$$\mathsf{enc} : \mathbb{A} \mapsto \prod_{k=1}^n \prod_{M \in \mathtt{attrV_k}} (X - H(\mathtt{attr}_k \| M)).$$

This function is used to encode the set $\mathbb{A}$ in the issued credential, the shown attributes $\mathbb{A}'$ as well as its complement:

$$\overline{\mathbb{A}'} = \{(\mathtt{attr}, \mathtt{attrV} \setminus \mathtt{attrV}') : (\mathtt{attr}, \mathtt{attrV}) \in \mathbb{A}, (\mathtt{attr}, \mathtt{attrV}') \in \mathbb{A}'\} \cup$$
$$\{(\mathtt{attr}', \mathtt{attrV}) \in \mathbb{A} : (\mathtt{attr}', \cdot) \notin \mathbb{A}'\}$$

in every showing. The idea is that the credential includes a commitment to the encoding of $\mathbb{A}$ and that showings include a witness of the encoding of $\overline{\mathbb{A}'}$ (without opening it) as well as $\mathbb{A}'$ in plain for which the encoding is then recomputed by the verifier. To compute these values, we use the PolyCommitFO public parameters pp, which allow an evaluation of these polynomials in G_1 and G_2 at $\alpha \in \mathbb{Z}_p^*$ (without knowing the trapdoor α). Then, the verifier checks whether the multiplicative relationship $\mathsf{enc}(\mathbb{A}) = \mathsf{enc}(\mathbb{A}') \cdot \mathsf{enc}(\overline{\mathbb{A}'})$ between the polynomials is satisfied by checking the multiplicative relationship between the corresponding commitments and witnesses via a pairing equation. More precisely, the commitment to the encoding of $\mathbb{A}$ is computed as $C_1 = r_i \cdot \mathsf{enc}(\mathbb{A})(\alpha)P$ with r_i being the secret key of user i. We note that since no entity knows α, we must compute

$$C_1 \gets r_i \cdot \mathsf{enc}(\mathbb{A})(\alpha)P = r_i \cdot \sum_{i=0}^t e_i \alpha^i P, \quad \text{with } \mathsf{enc}(\mathbb{A}) = \sum_{i=0}^t e_i X^i \in \mathbb{Z}_p[X].$$

The verification of a credential, when showing $\mathbb{A}'$, requires checking whether the following holds:

$$\mathsf{VerifyFactor}_{\mathsf{PC}}(\mathsf{pp}, C_1, \mathsf{enc}(\mathbb{A}'), \mathcal{C}_{\overline{\mathbb{A}'}}) \stackrel{?}{=} \mathsf{true},$$

where $\mathcal{C}_{\overline{\mathbb{A}'}} = r_i \cdot \mathsf{enc}(\overline{\mathbb{A}'})(\alpha)P$ is part of the showing. A showing, then, simply amounts to randomizing C_1, opening a product of factors of the committed polynomial (representing the selective disclosure), providing a consistently randomized witness of the complementary polynomial and performing a small, constant-size PoK of the randomizer for freshness, as we will see soon.

Example. For the reader's convenience, we include an example of a set $\mathbb{A}$. We are given a user with the following set of attributes and values:

$$\mathbb{A} = \{(\mathtt{birthdate}, \{"01.01.1980", " > 18"\}), (\mathtt{drivinglicense}, \{\#, \mathtt{car}\})\}.$$

Note that # indicates an attribute value that allows to prove the possession of the attribute without revealing any concrete value. A showing could, for instance, involve the following attributes $\mathbb{A}'$ and its hidden complement $\overline{\mathbb{A}'}$:

$$\mathbb{A}' = \{(\texttt{drivinglicense}, \{\#\})\}$$
$$\overline{\mathbb{A}'} = \{(\texttt{birthdate}, \{"01.01.1980", " > 18"\}), (\texttt{drivinglicense}, \{\texttt{car}\})\}.$$

Freshness. We have to guarantee that no valid showing transcript can be replayed by someone not in possession of the credential and the user's secret key. To do so, we require the user to conduct a proof of knowledge $\mathsf{PoK}\{\gamma : C_2 = \gamma P\}$ of the discrete logarithm of the second component $C_2 = \rho P$ of a credential, i.e., the value ρ, in the showing protocol. This guarantees that we have a fresh challenge for every showing.

In order to prove the anonymity of the ABC system, we need a little trick. We modify the aforementioned PoK and require that the user delivers a proof of knowledge $\mathsf{PoK}\{\gamma : Q = \gamma P \vee C_2 = \gamma P\}$, where Q is an additional value in the public parameters pp with unknown discrete logarithm q. Consequently, the user needs to conduct the second part of the proof honestly, while simulating the one for Q. In the proof of anonymity, this allows us to let the challenger know q and simulate showings without knowledge of the discrete logarithm of C_2, which is required for our reduction to work. Due to the nature of the OR proof, this cannot be detected by the adversary.

5.3   The Construction of the ABC System

Now, we present our ABC system in Scheme 2, where we use the notation $X \leftarrow f(X)$ to indicate that the value of X is overwritten by the result of the evaluation of $f(X)$. Note that if a check does not yield true, the respective algorithm terminates with a failure and the algorithm Verify accepts only if VerifyFactor$_{PC}$ and Verify$_{\mathcal{R}}$ return true as well as PoK is valid. Also note that the first move in the showing protocol can be combined with the first move of the proof of knowledge. Therefore, the showing protocol consists of a total of three moves.

5.4   Security

In the full version [37], we introduce a security model for attribute-based anonymous credentials and we provide formal proofs for the following:

Theorem 4. *Scheme 2 is correct.*

Theorem 5. *If* PolyCommitFO *is factor-sound, H is a collision-resistant hash function, Scheme 1 is secure and the DLP is hard in* G_1*, then Scheme 2 is unforgeable.*

Theorem 6. *If Scheme 1 is class hiding, then Scheme 2 is anonymous.*

Taking everything together, we obtain the following corollary:

Setup: Given (κ, t), run $\mathsf{pp}' = (\mathsf{BG}, (\alpha^i P)_{i=1}^t, (\alpha^i P')_{i=1}^t) \leftarrow \mathsf{Setup}_{\mathsf{PC}}(\kappa, t)$ and let $H : \{0,1\}^* \rightarrow \mathbb{Z}_p^*$
be a collision-resistant hash function used inside $\mathsf{enc}(\cdot)$. Finally, choose $Q \xleftarrow{R} G_1$ and output
$\mathsf{pp} \leftarrow (H, \mathsf{enc}, Q, \mathsf{pp}')$.

OrgKeyGen: Given pp and $j \in \mathbb{N}$, return $(\mathsf{osk}_j, \mathsf{opk}_j) \leftarrow \mathsf{KeyGen}_{\mathcal{R}}(\mathsf{BG}, 2)$.

UserKeyGen: Given pp and $i \in \mathbb{N}$, pick $r_i \xleftarrow{R} \mathbb{Z}_p^*$, set $R_i \leftarrow r_i P$ and return $(\mathsf{usk}_i, \mathsf{upk}_i) \leftarrow (r_i, R_i)$.

(Obtain, Issue): Obtain and Issue interact in the following way:

Issue($\mathsf{pp}, \mathsf{upk}_i, \mathsf{osk}_j, \mathbb{A}$)		Obtain($\mathsf{pp}, \mathsf{usk}_i, \mathsf{opk}_j, \mathbb{A}$)
	$\xleftarrow{\;C_1\;}$	$C_1 \leftarrow r_i \cdot \mathsf{enc}(\mathbb{A})(\alpha)P$
$e(C_1, P') \stackrel{?}{=} e(R_i, \mathsf{enc}(\mathbb{A})(\alpha)P')$		
$\sigma \leftarrow \mathsf{Sign}_{\mathcal{R}}((C_1, P), \mathsf{osk}_j)$	$\xrightarrow{\;\sigma\;}$	$\mathsf{Verify}_{\mathcal{R}}((C_1, P), \sigma, \mathsf{opk}_j) \stackrel{?}{=} \mathsf{true}$
		$\mathsf{cred}_i \leftarrow ((C_1, P), \sigma)$

(Show, Verify): Show and Verify interact in the following way:

Verify($\mathsf{pp}, \mathsf{opk}_j, \mathbb{A}'$)		Show($\mathsf{pp}, \mathsf{usk}_i, \mathsf{opk}_j, (\mathbb{A}, \mathbb{A}'), \mathsf{cred}_i$)
		$\rho \xleftarrow{R} \mathbb{Z}_p^*$
		$\mathsf{cred}_i' \leftarrow \mathsf{ChgRep}_{\mathcal{R}}(\mathsf{cred}_i, \rho, \mathsf{opk}_j)$
$\Big[\mathsf{VerifyFactor}_{\mathsf{PC}}(\mathsf{pp}', C_1, \mathsf{enc}(\mathbb{A}'), C_{\overline{\mathbb{A}'}}) \wedge$	$\xleftarrow{\;\mathsf{cred}_i', C_{\overline{\mathbb{A}'}}\;}$	$C_{\overline{\mathbb{A}'}} \leftarrow (\rho \cdot \mathsf{usk}_i) \cdot \mathsf{enc}(\overline{\mathbb{A}'})(\alpha)P$
$\mathsf{Verify}_{\mathcal{R}}(\mathsf{cred}_i', \mathsf{opk}_j) \Big] \stackrel{?}{=} \mathsf{true}$	$\xleftrightarrow{\;\mathsf{PoK}\{\gamma: Q = \gamma P \vee C_2 = \gamma P\}\;}$	

where $\mathsf{cred}_i' = ((C_1, C_2), \sigma)$.

Scheme 2. A Multi-Show ABC System

Corollary 2. *Scheme 2 is a secure ABC system.*

Note that in the proof of Theorem 5, we can distinguish whether a forgery goes back to a signature forgery of Scheme 1 or not. The reason for this is that the knowledge extractor of the PoK gives us the possibility to extract the used credential, which allows us to determine whether a showing is based on a queried credential (and, in further consequence, on a queried signature) or not. Hence, we are able to efficiently check the winning condition of the EUF-CMA game.

5.5 Efficiency Analysis and Comparison

We provide a brief comparison with other ABC approaches and for completeness also include the most popular one-show approach. As other candidates for multi-show ABCs, we take the Camenisch-Lysyanskaya schemes [23,24,25] as well as schemes from BBS$^+$ signatures [18,11] which cover a broad class of ABC schemes from randomizable signature schemes with efficient proofs of knowledge. Furthermore, we take two alternative multi-show ABC constructions [26,27] as well as Brands' approach [20] (also covering the provable secure version [12]) for the sake of completeness, although latter only provides one-show ABCs. We omit other approaches such as [8] that only allow a single attribute per credential. We also omit approaches that achieve more efficient showings for existing ABC systems only in very special cases such as for attribute values that come from a very small set (and are, thus, hard to compare). For instance, the approach in

[22] for CL credentials in the strong RSA setting (encoding attributes as prime numbers) or in a pairing-based setting using BBS$^+$ credentials [41] (encoding attributes using accumulators), where the latter additionally requires very large public parameters (one BB signature [15] for every possible attribute value).

Table 1 gives an overview of these systems. Thereby, Type-1 and Type-2 refer to bilinear group settings with Type-1 and Type-2 pairings, respectively. In a stronger sense, XDH as well as SXDH stand for bilinear group settings, where the former requires the external Diffie-Hellman assumption and the latter requires the SXDH assumption to hold. Furthermore, G_q denotes a group of prime order q (e.g., a prime order subgroup of $\mathbb{Z}_p^*$ or of an elliptic curve group). By $|G|$, we mean the bitlength of the representation of an element from group G and the value c is a constant specified to be approximately 510 bits in [26]. We emphasize that, in contrast to other approaches, such as [25,27], our construction only requires a small and constant number of pairing evaluations in all protocol steps. Note that in the issuing step we always assume a computation of $O(L)$ for the user, as we assume that the user checks the validity of the obtained credential on issuing (most of the approaches, including ours, have cost $O(1)$ if this verification is omitted).

Table 1. Comparison of various approaches to ABC systems

		Parameter Size (L attributes)			Issuing			Showing (k-of-L attributes)		
	Setting	pp		Credential Size	Issuer	User	Com	Verifier	User	Com
[23,24]	sRSA	$O(L)$	$O(1)$	$3\lvert\mathbb{Z}_N\rvert$	$O(L)$	$O(L)$	$O(L)$	$O(L)$	$O(L)$	$O(L-k)$
[25]	Type-1	$O(L)$	$O(L)$	$(2L+2)\lvert G_1\rvert$	$O(L)$	$O(L)$	$O(L)$	$O(L)$	$O(L)$	$O(L)$
[18]	Type-2	$O(L)$	$O(1)$	$\lvert G_1\rvert+22\lvert\mathbb{Z}_q\rvert$	$O(L)$	$O(L)$	$O(1)$	$O(L)$	$O(L)$	$O(L)$
[26]	Type-2	$O(1)$	$O(L)$	$L\lvert G_1\rvert+c+\lvert G_2\rvert$	$O(L)$	$O(L)$	$O(L)$	$O(L)$	$O(1)$	$O(1)$
[27]	XDH	$O(L)$	$O(L)$	$(2L+2)(\lvert G_1\rvert+\lvert\mathbb{Z}_p\rvert)$	$O(L)$	$O(L)$	$O(L)$	$O(k)$	$O(k)$	$O(k)$
[20]	G_q	$O(L)$	$O(1)$	$2\lvert G_q\rvert+2\lvert\mathbb{Z}_q\rvert$	$O(L)$	$O(L)$	$O(1)$	$O(k)$	$O(k)$	$O(L-k)$
Our	SXDH	$O(L)$	$O(1)$	$4\lvert G_1\rvert+\lvert G_2\rvert$	$O(L)$	$O(L)$	$O(1)$	$O(k)$	$O(L-k)$	$O(1)$

6   Future Work

The proposed signature scheme seems to be powerful and there might be other applications that could benefit, like blind signatures or verifiably-encrypted signatures. We leave a detailed study and the analysis of such applications as future work. Future work also includes constructing revocable and delegatable anonymous credentials from this new approach to ABCs. Furthermore, it is an interesting question whether the proposed construction is already optimal, whether such signatures can be built for other interesting relations and whether it is possible to construct such signature schemes whose unforgeability can be proven under possible non-interactive assumptions or even to show that this is impossible.

Acknowledgments. This work has been supported by the European Commission through project FP7-FutureID, grant agreement number 318424. We thank the anonymous referees for their helpful comments.

References

1. Abe, M., Fuchsbauer, G., Groth, J., Haralambiev, K., Ohkubo, M.: Structure-Preserving Signatures and Commitments to Group Elements. In: Rabin, T. (ed.) CRYPTO 2010. LNCS, vol. 6223, pp. 209–236. Springer, Heidelberg (2010)
2. Abe, M., Groth, J., Haralambiev, K., Ohkubo, M.: Optimal Structure-Preserving Signatures in Asymmetric Bilinear Groups. In: Rogaway, P. (ed.) CRYPTO 2011. LNCS, vol. 6841, pp. 649–666. Springer, Heidelberg (2011)
3. Abe, M., Groth, J., Ohkubo, M.: Separating Short Structure-Preserving Signatures from Non-interactive Assumptions. In: Lee, D.H., Wang, X. (eds.) ASIACRYPT 2011. LNCS, vol. 7073, pp. 628–646. Springer, Heidelberg (2011)
4. Abe, M., Groth, J., Ohkubo, M., Tibouchi, M.: Structure-Preserving Signatures from Type II Pairings. In: Garay, J.A., Gennaro, R. (eds.) CRYPTO 2014, Part I. LNCS, vol. 8616, pp. 390–407. Springer, Heidelberg (2014)
5. Abe, M., Groth, J., Ohkubo, M., Tibouchi, M.: Unified, Minimal and Selectively Randomizable Structure-Preserving Signatures. In: Lindell, Y. (ed.) TCC 2014. LNCS, vol. 8349, pp. 688–712. Springer, Heidelberg (2014)
6. Abe, M., Haralambiev, K., Ohkubo, M.: Signing on Elements in Bilinear Groups for Modular Protocol Design. IACR Cryptology ePrint Archive (2010)
7. Ahn, J.H., Boneh, D., Camenisch, J., Hohenberger, S., Shelat, A., Waters, B.: Computing on Authenticated Data. In: Cramer, R. (ed.) TCC 2012. LNCS, vol. 7194, pp. 1–20. Springer, Heidelberg (2012)
8. Akagi, N., Manabe, Y., Okamoto, T.: An Efficient Anonymous Credential System. In: Tsudik, G. (ed.) FC 2008. LNCS, vol. 5143, pp. 272–286. Springer, Heidelberg (2008)
9. Attrapadung, N., Libert, B., Peters, T.: Computing on Authenticated Data: New Privacy Definitions and Constructions. In: Wang, X., Sako, K. (eds.) ASIACRYPT 2012. LNCS, vol. 7658, pp. 367–385. Springer, Heidelberg (2012)
10. Attrapadung, N., Libert, B., Peters, T.: Efficient completely context-hiding quotable and linearly homomorphic signatures. In: Kurosawa, K., Hanaoka, G. (eds.) PKC 2013. LNCS, vol. 7778, pp. 386–404. Springer, Heidelberg (2013)
11. Au, M.H., Susilo, W., Mu, Y.: Constant-Size Dynamic k-TAA. In: De Prisco, R., Yung, M. (eds.) SCN 2006. LNCS, vol. 4116, pp. 111–125. Springer, Heidelberg (2006)
12. Baldimtsi, F., Lysyanskaya, A.: Anonymous Credentials Light. In: CCS. ACM (2013)
13. Ballard, L., Green, M., de Medeiros, B., Monrose, F.: Correlation-Resistant Storage via Keyword-Searchable Encryption. IACR Cryptology ePrint Archive (2005)
14. Belenkiy, M., Camenisch, J., Chase, M., Kohlweiss, M., Lysyanskaya, A., Shacham, H.: Randomizable Proofs and Delegatable Anonymous Credentials. In: Halevi, S. (ed.) CRYPTO 2009. LNCS, vol. 5677, pp. 108–125. Springer, Heidelberg (2009)
15. Belenkiy, M., Chase, M., Kohlweiss, M., Lysyanskaya, A.: P-signatures and Non-interactive Anonymous Credentials. In: Canetti, R. (ed.) TCC 2008. LNCS, vol. 4948, pp. 356–374. Springer, Heidelberg (2008)
16. Blazy, O., Fuchsbauer, G., Pointcheval, D., Vergnaud, D.: Signatures on Randomizable Ciphertexts. In: Catalano, D., Fazio, N., Gennaro, R., Nicolosi, A. (eds.) PKC 2011. LNCS, vol. 6571, pp. 403–422. Springer, Heidelberg (2011)
17. Boneh, D., Boyen, X.: Short Signatures Without Random Oracles. In: Cachin, C., Camenisch, J.L. (eds.) EUROCRYPT 2004. LNCS, vol. 3027, pp. 56–73. Springer, Heidelberg (2004)

18. Boneh, D., Boyen, X., Shacham, H.: Short Group Signatures. In: Franklin, M. (ed.) CRYPTO 2004. LNCS, vol. 3152, pp. 41–55. Springer, Heidelberg (2004)

19. Boneh, D., Freeman, D., Katz, J., Waters, B.: Signing a Linear Subspace: Signature Schemes for Network Coding. In: Jarecki, S., Tsudik, G. (eds.) PKC 2009. LNCS, vol. 5443, pp. 68–87. Springer, Heidelberg (2009)

20. Brands, S.: Rethinking public-key Infrastructures and Digital Certificates: Building in Privacy. MIT Press (2000)

21. Camenisch, J., Dubovitskaya, M., Haralambiev, K.: Efficient Structure-Preserving Signature Scheme from Standard Assumptions. In: Visconti, I., De Prisco, R. (eds.) SCN 2012. LNCS, vol. 7485, pp. 76–94. Springer, Heidelberg (2012)

22. Camenisch, J., Groß, T.: Efficient Attributes for Anonymous Credentials. ACM Trans. Inf. Syst. Secur. 15(1), 4 (2012)

23. Camenisch, J.L., Lysyanskaya, A.: An Efficient System for Non-transferable Anonymous Credentials with Optional Anonymity Revocation. In: Pfitzmann, B. (ed.) EUROCRYPT 2001. LNCS, vol. 2045, pp. 93–118. Springer, Heidelberg (2001)

24. Camenisch, J.L., Lysyanskaya, A.: A Signature Scheme with Efficient Protocols. In: Cimato, S., Galdi, C., Persiano, G. (eds.) SCN 2002. LNCS, vol. 2576, pp. 268–289. Springer, Heidelberg (2003)

25. Camenisch, J.L., Lysyanskaya, A.: Signature Schemes and Anonymous Credentials from Bilinear Maps. In: Franklin, M. (ed.) CRYPTO 2004. LNCS, vol. 3152, pp. 56–72. Springer, Heidelberg (2004)

26. Canard, S., Lescuyer, R.: Anonymous credentials from (indexed) aggregate signatures. In: DIM, pp. 53–62. ACM (2011)

27. Canard, S., Lescuyer, R.: Protecting privacy by sanitizing personal data: a new approach to anonymous credentials. In: ASIACCS, pp. 381–392. ACM (2013)

28. Chase, M., Kohlweiss, M., Lysyanskaya, A., Meiklejohn, S.: Malleable Proof Systems and Applications. In: Pointcheval, D., Johansson, T. (eds.) EUROCRYPT 2012. LNCS, vol. 7237, pp. 281–300. Springer, Heidelberg (2012)

29. Chase, M., Kohlweiss, M., Lysyanskaya, A., Meiklejohn, S.: Malleable Signatures: Complex Unary Transformations and Delegatable Anonymous Credentials. IACR Cryptology ePrint Archive (2013)

30. Chatterjee, S., Menezes, A.: On cryptographic protocols employing asymmetric pairings - the role of ψ revisited. Discrete Applied Mathematics 159(13), 1311–1322 (2011)

31. Chaum, D., Pedersen, T.P.: Wallet Databases with Observers. In: Brickell, E.F. (ed.) CRYPTO 1992. LNCS, vol. 740, pp. 89–105. Springer, Heidelberg (1993)

32. Cheon, J.H.: Security analysis of the strong diffie-hellman problem. In: Vaudenay, S. (ed.) EUROCRYPT 2006. LNCS, vol. 4004, pp. 1–11. Springer, Heidelberg (2006)

33. Fuchsbauer, G.: Automorphic Signatures in Bilinear Groups and an Application to Round-Optimal Blind Signatures. IACR Cryptology ePrint Archive (2009)

34. Fuchsbauer, G.: Commuting Signatures and Verifiable Encryption. In: Paterson, K.G. (ed.) EUROCRYPT 2011. LNCS, vol. 6632, pp. 224–245. Springer, Heidelberg (2011)

35. Groth, J., Sahai, A.: Efficient Non-interactive Proof Systems for Bilinear Groups. In: Smart, N.P. (ed.) EUROCRYPT 2008. LNCS, vol. 4965, pp. 415–432. Springer, Heidelberg (2008)

36. Hanser, C., Slamanig, D.: Blank Digital Signatures. IACR Cryptology ePrint Archive, Report 2013/130 (2013)

37. Hanser, C., Slamanig, D.: Structure-Preserving Signatures on Equivalence Classes and their Application to Anonymous Credentials. Cryptology ePrint Archive, Report 2014/705 (2014)
38. Johnson, R., Molnar, D., Song, D., Wagner, D.: Homomorphic Signature Schemes. In: Preneel, B. (ed.) CT-RSA 2002. LNCS, vol. 2271, pp. 244–262. Springer, Heidelberg (2002)
39. Kate, A., Zaverucha, G.M., Goldberg, I.: Constant-Size Commitments to Polynomials and Their Applications. In: Abe, M. (ed.) ASIACRYPT 2010. LNCS, vol. 6477, pp. 177–194. Springer, Heidelberg (2010)
40. Libert, B., Peters, T., Joye, M., Yung, M.: Linearly Homomorphic Structure-Preserving Signatures and Their Applications. In: Canetti, R., Garay, J.A. (eds.) CRYPTO 2013, Part II. LNCS, vol. 8043, pp. 289–307. Springer, Heidelberg (2013)
41. Sudarsono, A., Nakanishi, T., Funabiki, N.: Efficient Proofs of Attributes in Pairing-Based Anonymous Credential System. In: Fischer-Hübner, S., Hopper, N. (eds.) PETS 2011. LNCS, vol. 6794, pp. 246–263. Springer, Heidelberg (2011)
42. Verheul, E.R.: Self-Blindable Credential Certificates from the Weil Pairing. In: Boyd, C. (ed.) ASIACRYPT 2001. LNCS, vol. 2248, pp. 533–551. Springer, Heidelberg (2001)
43. Waters, B.: Efficient Identity-Based Encryption Without Random Oracles. In: Cramer, R. (ed.) EUROCRYPT 2005. LNCS, vol. 3494, pp. 114–127. Springer, Heidelberg (2005)

On Tight Security Proofs for Schnorr Signatures

Nils Fleischhacker[1], Tibor Jager[2], and Dominique Schröder[1]

[1] Saarland University, Germany
[2] Horst Görtz Institute for IT Security
Ruhr-University Bochum, Germany

Abstract. The Schnorr signature scheme is the most efficient signature scheme based on the discrete logarithm problem and a long line of research investigates the existence of a *tight* security reduction for this scheme in the random oracle. Almost all recent works present lower tightness bounds and most recently Seurin (Eurocrypt 2012) showed that under certain assumptions the *non*-tight security proof for Schnorr signatures in the random oracle by Pointcheval and Stern (Eurocrypt 1996) is essentially optimal. All previous works in this direction rule out tight reductions from the (one-more) discrete logarithm problem. In this paper we introduce a new meta-reduction technique, which shows lower bounds for the large and very natural class of *generic* reductions. A generic reduction is independent of a particular representation of group elements and most reductions in state-of-the-art security proofs have this desirable property. Our approach shows *unconditionally* that there is no tight generic reduction from any *natural* computational problem Π defined over algebraic groups (including even interactive problems) to breaking Schnorr signatures, unless solving Π is easy.

Keywords: Schnorr signatures, black-box reductions, generic reductions, algebraic reductions, tightness.

1 Introduction

The security of a cryptosystem is nowadays usually confirmed by giving a security proof. Typically, such a proof describes a *reduction* from some (assumed-to-be-)hard computational problem to breaking a defined security property of the cryptosystem. A reduction is considered as *tight*, if the reduction solving the hard computational problem has essentially the same running time and success probability as the attacker on the cryptosystem. Essentially, a tight reduction means that a successful attacker can be turned into an efficient algorithm for the hard computational problem *without* any significant increase in the running time and/or significant loss in the success probability.[1] The tightness of a reduction thus determines the strength of the security guarantees provided by the security proof: a non-tight reduction gives weaker security guarantees than a tight one. Moreover, tightness of the reduction affects the efficiency of the cryptosystem when instantiated in practice: a tighter reduction allows to securely use smaller parameters (shorter moduli, a smaller group size, etc.). Therefore it is a very desirable property of a cryptosystem to have a tight security reduction.

---

[1] Usually even a polynomially-bounded increase/loss is considered as significant, if the polynomial may be large. An increase/loss by a small constant factor is not considered as significant.

P. Sarkar and T. Iwata (Eds.): ASIACRYPT 2014, PART I, LNCS 8873, pp. 512–531, 2014.

In the domain of digital signatures tight reductions are known for many fundamental schemes, like Rabin/Williams signatures (Bernstein, Eurocrypt 2008 [5]), many strong-RSA-based signatures (Schäge, Eurocrypt 2011 [25]), and RSA Full-Domain Hash (Kakvi and Kiltz, Eurocrypt 2012 [18]). The Schnorr signature scheme [26, 27] is one of the most fundamental public-key cryptosystems. Pointcheval and Stern have shown that Schnorr signatures are provably secure, assuming the hardness of the discrete logarithm (DL) problem [22], in the Random Oracle Model (ROM) [3]. However, the reduction of Pointcheval and Stern from DL to breaking Schnorr signatures is not tight: it loses a factor of q in the time-to-success ratio, where q is the number of random oracle queries performed by the forger.

A long line of research investigates the existence of tight security proofs for Schnorr signatures. At Asiacrypt 2005 Paillier and Vergnaud [21] gave a first lower bound showing that any algebraic reduction (even in the ROM) converting a forger for Schnorr signatures into an algorithm solving some computational problem Π must lose a factor of at least $q^{1/2}$. Their result is quite strong, as they rule out reductions even for adversaries that do not have access to a signing oracle and receive as input the message for which they must forge (UF-NM, see Section A for a formal definition). However, their result also has some limitations: It holds only under the interactive one-more discrete logarithm assumption, they only consider algebraic reductions, and they only rule out tight reductions from the (one-more) discrete logarithm problem. At Crypto 2008 Garg *et al.* [15] refined this result, by improving the bound from $q^{1/2}$ to $q^{2/3}$ with a new analysis and show that this bound is optimal if the meta-reduction follows a particular approach for simulating the forger. At Eurocrypt 2012 Seurin [28] finally closed the gap between the security proof of [22] and known impossibility results, by describing an elaborate simulation strategy for the forger and providing a new analysis. All previous works [21, 15, 28] on the existence of tight security proofs for Schnorr signatures have the following in common:

1. They only rule out the existence of tight reductions from certain strong computational problems, namely the (one-more) discrete logarithm problem [1]. Reduction from weaker problems like, e.g., the computational or decisional Diffie-Hellman problem (CDH/DDH) are not considered.

2. The impossibility results are themselves only valid under the very strong OMDL hardness assumption.

3. They hold only with respect to a limited (but natural) class of reductions, so-called *algebraic reductions*.

It is not unlikely that first the inexistence of a tight reduction from *strong* computational problems is proven, and later a tight reduction from some *weaker* problem is found. A concrete recent example in the domain of digital signatures where this has happened is RSA Full-Domain Hash (RSA-FDH) [4]. First, at Crypto 2000 Coron [8] described a non-tight reduction from solving the RSA-problem to breaking the security of RSA-FDH, and at Eurocrypt 2002 [9] showed that under certain conditions no tighter reduction from RSA can exist. Later, at Eurocrypt 2012, Kakvi and Kiltz [18] gave a tight reduction from solving a weaker problem, the so-called Phi-Hiding problem. The leverage used by Kakvi and Kiltz to circumvent the aforementioned impossibility results was to assume hardness of a weaker computational problem. As all previous works rule out only tight reductions from strong computational problems like DL and OMDL, this might happen again with Schnorr signatures and the following question was left open for 25 years:

Does a tight security proof for Schnorr signatures based on any weaker computational problem exist?

Our contribution. In this work we answer this question in the negative ruling out the existence of tight reductions in the random oracle model for virtually all natural computational problems defined over abstract algebraic groups. Like previous works, we consider universal unforgeability under no-message attacks (UF-NM-security). Moreover, our results hold unconditionally. In contrast to previous works, we consider *generic* reductions instead of algebraic reductions, but we believe that this restriction is marginal: The motivation of considering only algebraic reductions from [21] applies equally to generic reductions. In particular, to the best of our knowledge all known examples of algebraic reductions are generic.

Our main technical contribution is a new approach for the simulation of a forger in a meta-reduction, i.e., "a reduction against the reduction", which differs from previous works [21, 15, 28] and which allows us to show the following main result:

Theorem (Informal). *For almost any* natural *computational problem Π, there is no tight* generic *reduction from solving Π to breaking the universal unforgeability under no-message attacks of Schnorr signatures in the random oracle model.*

Technical approach. We begin with the hypothesis that there exists a tight generic reduction $\mathcal{R}$ from some hard (and possibly interactive) problem Π to the UF-NM-security of Schnorr signatures. Then we show that under this hypothesis there exists an efficient algorithm $\mathcal{M}$, a meta-reduction, which efficiently solves Π. This implies that the hypothesis is false. The meta-reduction $\mathcal{M} = \mathcal{M}^{\mathcal{R}}$ runs $\mathcal{R}$ as a subroutine, by efficiently *simulating* the forger $\mathcal{A}$ for $\mathcal{R}$.

All previous works in this direction [21, 15, 28] followed essentially the same approach. The difficulty with meta-reductions is that $\mathcal{M} = \mathcal{M}^{\mathcal{R}}$ must efficiently *simulate* the forger $\mathcal{A}$ for $\mathcal{R}$. Previous works resolved this by using a discrete logarithm oracle provided by the OMDL assumption, which allows to efficiently compute valid signatures in the simulation of forger $\mathcal{A}$. This is the reason why all previous results are only valid under the OMDL assumption, and were only able to rule out reductions from the discrete log or the OMDL problem. To overcome these limitations, a new simulation technique is necessary.

We revisit the simulation strategy of $\mathcal{A}$ applied in known meta-reductions, and put forward a new technique for proving impossibility results. It turns out that considering *generic* reductions provides a new leverage to simulate a successful forger efficiently, essentially by suitably re-programming the group representation to compute valid signatures. The technical challenge is to prove that the reduction does not notice that the meta-reduction changes the group representation during the simulation, except for some negligible probability. We show how to prove this by adopting the "low polynomial degree" proof technique of Shoup [30], which originally was introduced to analyze the complexity of certain algorithms for the discrete logarithm problem, to the setting considered in this paper.

This new approach turns out to be extremely powerful, as it allows to rule out reductions from any (even *interactive*) *representation-invariant* computational problem. Since almost all common hardness assumptions in algebraic groups (e.g., DL, CDH, DDH, OMDL, DLIN, etc.) are based on representation-invariant computational problems, we are able to rule out tight generic reductions from virtually any natural computational problem, without making any additional assumption. Even though we apply it specifically to Schnorr signatures, the overall approach is general. We expect that it is applicable to other cryptosystems as well.

Generic reductions vs. algebraic reductions. Similar to algebraic reductions, a generic reduction performs only group operations. The main difference is that the sequence of group operations performed by an algebraic reduction may (but, to our best knowledge, in all known examples does not) depend on a particular representation of group elements. A generic reduction, however, is required to work essentially identical for any representation of group elements. Generic reductions are by definition more restrictive than algebraic ones, however, we explain below why we do not consider this restriction as very significant.

An obvious question arising with our work is the relation between algebraic and generic reductions. Is a lower bound for generic reductions much less meaningful than a bound for algebraic reductions? We argue that the difference is not very significant. The restriction to algebraic reductions was motivated by the fact most reductions in known security proofs treat the group as a black-box, and thus are algebraic [21, 15, 28]. However, the same motivation applies to generic reductions as well, with exactly the same arguments. In particular, virtually all examples of algebraic reductions in the literature are also generic.

The vast majority of reductions in common security proofs for group-based cryptosystems treats the underlying group as a black-box (i.e., works for any representation of the group), and thus is generic. This is a very desirable feature, because then the cryptosystem can securely be instantiated with *any* group in which the underlying computational problem is hard. In contrast, representation-specific security proofs would require to re-prove security for any particular group representation the scheme is used with. Therefore considering generic reductions seems very reasonable.

Generic reductions vs. security proofs in the generic group model. One might wonder whether our result is implied by previous works (in particular by [28]), since we are considering generic reductions, because for generic algorithms most non-trivial computational problems in algebraic groups are equivalent to the discrete logarithm problem. The conclusion that therefore our result is implied by previous works is however not correct.

Note that a reduction does not solve the computational problem alone. It has access to an attacker $\mathcal{A}$. The algorithm which solves the computational problem is a composition $\mathcal{R}(\mathcal{A})$ of $\mathcal{R}$ and $\mathcal{A}$. If both $\mathcal{R}$ and $\mathcal{A}$ were generic algorithms, then the composition $\mathcal{R}(\mathcal{A})$ would also be a generic algorithm, and thus our results would indeed be trivial. But note that we do not require $\mathcal{A}$ to be generic. Therefore also the composition $\mathcal{R}(\mathcal{A})$

is not a generic algorithm, thus the generic equivalence of DLOG and other problems does not apply. See Section 2.4 and Figure 2 for further explanation.

Further related work. Dodis *et al.* [10] showed that it is impossible to reduce any computational problem to breaking the security of RSA-FDH in a model where the RSA-group $\mathbb{Z}_N^*$ is modeled as a generic group. This result extends [11]. Coron [9] considered the existence of tight security reductions for RSA-FDH signatures [4]. This result was generalized by Dodis and Reyzin [12] and later refined by Kiltz and Kakvi [18].

In the context of Schnorr signatures, Neven *et al.* [20] described necessary conditions the hash function must meet in order to provide existential unforgeability under chosen-message attacks (EUF-CM), and showed that these conditions are sufficient if the forger (not the reduction!) is modeled as a generic group algorithm.

In [13] Fischlin and Fleischhacker presented a result also about the security of Schnorr signatures which is orthogonal to our result. They show, again under the OMDL assumption, that a large class of reductions has to rely on re-programming the random oracle. Essentially they prove that in the *non-programmable* ROM [14] no reduction from the discrete logarithm problem can exist that invokes the adversary only ever on the same input. This class is limited, but encompasses all forking-lemma style reductions used to prove Schnorr signatures secure in the programmable ROM. As said before, the result is orthogonal to our main result, as it considers reductions in the *non-programmable* ROM.

2   Preliminaries

Notation. If S is a set, we write $s \leftarrow_\$ S$ to denote the action of sampling a uniformly random element s from S. If A is a probabilistic algorithm, we denote with $a \leftarrow_\$ A$ the action of computing a by running A. We denote with $\emptyset$ the empty string, the empty set, as well as the empty list, the meaning will always be clear from the context. We write $[n]$ to denote the set of integers from 1 to n, i.e., $[n] := \{1, \ldots, n\}$.

2.1   Schnorr Signatures

Let $\mathbb{G}$ be a group of order p with generator g, and let $H : \mathbb{G} \times \{0,1\}^k \to \mathbb{Z}_p$ be a hash function. The Schnorr signature scheme [26, 27] consists of the following efficient algorithms (Gen, Sign, Vrfy).

Gen(g): The key generation algorithm takes as input a generator g of $\mathbb{G}$. It chooses $x \leftarrow_\$ \mathbb{Z}_p$, computes $X := g^x$, and outputs (X, x).

Sign(x, m): The input of the signing algorithm is a private key x and a message $m \in \{0,1\}^k$. It chooses a random integer $r \leftarrow_\$ \mathbb{Z}_p$, sets $R := g^r$ as well as $c := H(R, m)$, and computes $y := x \cdot c + r \mod p$.

Vrfy($X, m, (R, y)$): The verification algorithm outputs the truth value of $g^y \stackrel{?}{=} X^c \cdot R$, where $c = H(R, m)$.

Remark 1. Note that the above description of Schnorr signatures deviates slightly from the original description in [26, 27], where a signature consists of (c, y) instead of (R, y), which reduces the length of signatures significantly. However, note that it is possible to compute R from (c, y) as $R := g^y \cdot X^{-c}$. Similarly, it is possible to compute c from (R, m) as $c := H(R, m)$. Thus both representations are equivalent. In particular, changing between these two representation does not affect our results.

2.2 Computational Problems

Let $\mathbb{G}$ be a cyclic group of order p and $g \in \mathbb{G}$ a generator of $\mathbb{G}$. We write $\mathrm{desc}(\mathbb{G}, g)$ to denote the list of group elements $\mathrm{desc}(\mathbb{G}, g) = (g, g^2, \ldots, g^p) \in \mathbb{G}^p$. We say that $\mathrm{desc}(\mathbb{G}, g)$ is the *enumerating description* of $\mathbb{G}$ with respect to g.

Definition 1. *A computational problem Π in $\mathbb{G}$ is specified by three (computationally unbounded) procedures $\Pi = (\mathcal{G}_\Pi, \mathcal{S}_\Pi, \mathcal{V}_\Pi)$, with the following syntax.*

$\mathcal{G}_\Pi(\mathrm{desc}(\mathbb{G}, g))$ *takes as input an enumerating description of $\mathbb{G}$, and outputs a state st and a problem instance (the challenge) $C = (C_1, \ldots, C_u, C') \in \mathbb{G}^u \times \{0, 1\}^*$. We assume in the sequel that at least C_1 is a generator of $\mathbb{G}$.*

$\mathcal{S}_\Pi(\mathrm{desc}(\mathbb{G}, g), st, Q)$ *takes as input $\mathrm{desc}(\mathbb{G}, g)$, a state st, and $Q = (Q_1, \ldots, Q_v, Q') \in \mathbb{G}^v \times \{0, 1\}^*$, and outputs (st', A) where st' is an updated state and $A = (A_1, \ldots, A_\nu, A') \in \mathbb{G}^\nu \times \{0, 1\}^*$.*

$\mathcal{V}_\Pi(\mathrm{desc}(\mathbb{G}, g), st, S, C)$ *takes as input $(\mathrm{desc}(\mathbb{G}, g), st, C)$ as defined above, and $S = (S_1, \ldots, S_w, S') \in \mathbb{G}^w \times \{0, 1\}^*$. It outputs 0 or 1.*

If $\mathcal{S}_\Pi$ always responds with $A = \emptyset$ (i.e., the empty string), then we say that Π is non-interactive. *Otherwise it is* interactive. *The exact description and distribution of st, C, Q, A, S depends on the considered computational problem.*

Definition 2. *An algorithm $\mathcal{A}$ (ϵ, t)-solves the computational problem Π if $\mathcal{A}$ has running time at most t and wins the following interactive game against a (computationally unbounded) challenger $\mathcal{C}$ with probability at most ϵ, where the game is defined as follows:*

1. *The challenger $\mathcal{C}$ generates an instance of the problem $(st, C) \leftarrow_\$ \mathcal{G}_\Pi(\mathrm{desc}(\mathbb{G}, g))$ and sends C to $\mathcal{A}$.*
2. *$\mathcal{A}$ is allowed to issue an arbitrary number of oracle queries to $\mathcal{C}$. To this end, $\mathcal{A}$ provides $\mathcal{C}$ with a query Q. $\mathcal{C}$ runs $(st', A) \leftarrow_\$ \mathcal{S}_\Pi(\mathrm{desc}(\mathbb{G}, g), st, Q)$, updates the state $st := st'$, and responds with A.*
3. *Finally, algorithm $\mathcal{A}$ outputs a candidate solution S. The algorithm $\mathcal{A}$ wins the game (i.e., solves the computational problem correctly) iff $\mathcal{V}_\Pi(\mathrm{desc}(\mathbb{G}, g), st, C, S) = 1$.*

Example 1. The *discrete logarithm problem* in $\mathbb{G}$ is specified by the following procedures. $\mathcal{G}_\Pi(\mathrm{desc}(\mathbb{G}, g))$ outputs (st, C) with $st = \emptyset$ and $C = (g, h)$, where $h \leftarrow_\$ \mathbb{G}$ is a random group element. $\mathcal{S}_\Pi(\mathrm{desc}(\mathbb{G}, g), st, Q)$ always outputs $(st', A) = (st, \emptyset)$. $\mathcal{V}_\Pi(\mathrm{desc}(\mathbb{G}, g), st, C, S)$ interprets $S = S' \in \{0, 1\}^*$ canonically as an integer in $\mathbb{Z}_p$, and outputs 1 iff $h = g^{S'}$.

Example 2. We describe the *u-one-more discrete logarithm problem* (u-OMDL) [2, 1] in $\mathbb{G}$ with the following algorithms. $\mathcal{G}_\Pi(\mathrm{desc}(\mathbb{G}, g))$ outputs (st, C) where $C = (C_1, \ldots, C_u) \leftarrow_\$ \mathbb{G}^u$ consists of u random group elements and $st = 0$. The algorithm

$\mathcal{S}_\Pi(\mathsf{desc}(\mathbb{G}, g), st, Q)$ takes as input state st and group element $Q \in \mathbb{G}$. It responds with $st' := st + 1$ and $A = A' \in \{0, 1\}^*$, where A' canonically interpreted as an integer in $\mathbb{Z}_p$ satisfies $g^{A'} = Q$. The verification algorithm $\mathcal{V}_\Pi(\mathsf{desc}(\mathbb{G}, g), st, C, S)$ interprets $S = (S_1', \ldots, S_u') \in \{0, 1\}^*$ canonically as a vector of u integers in $\mathbb{Z}_p$, and outputs 1 iff $st < u$ and $g_i = g^{S_i'}$ for all $i \subset [u]$.

Example 3. The UF-NM-forgery problem for Schnorr signatures in $\mathbb{G}$ with hash function H is specified by the following procedures. $\mathcal{G}_\Pi(\mathsf{desc}(\mathbb{G}, g))$ outputs (st, C) with $st = m$ and $C = (g, X, m) \in \mathbb{G}^2 \times \{0, 1\}^k$, where $X = g^x$ for $x \leftarrow_\$ \mathbb{Z}_p$ and $m \leftarrow_\$ \{0, 1\}^k$. $\mathcal{S}_\Pi(\mathsf{desc}(\mathbb{G}, g), st, Q)$ always outputs $(st', A) = (st, \emptyset)$. The verification algorithm $\mathcal{V}_\Pi(\mathsf{desc}(\mathbb{G}, g), st, C, S)$ parses S as $S = (R, y) \in \mathbb{G} \times \mathbb{Z}_p$, sets $c := H(R, st)$, and outputs 1 iff $X^c \cdot R = g^y$.

2.3 Representation-Invariant Computational Problems

In our impossibility results given below, we want to rule out the existence of a tight reduction from as large a class of computational problems as possible. Ideally, we want to rule out the existence of a tight reduction from any computational problem that meets Definition 1. However, it is easy to see that this is not achievable in this generality: as Example 3 shows, the problem of forging Schnorr signatures itself is a problem that meets Definition 1. However, of course there exists a trivial tight reduction from the problem of forging Schnorr signatures to the problem of forging Schnorr signatures! Therefore we need to restrict the class of considered computational problems to exclude such trivial, artificial problems.

We introduce the notion of *representation-invariant* computational problems. This class of problems captures virtually any reasonable computational problem defined over an abstract algebraic group, even interactive assumptions, except for a few extremely artificial problems. In particular, the problem of forging Schnorr signatures is *not* contained in this class (see Example 5 below).

Intuitively, a computational problem is *representation-invariant*, if a valid solution to a given problem instance remains valid even if the representation of group elements in challenges, oracle queries, and solutions is converted to a different representation of the same group. More formal is the following definition:

Definition 3. *Let* $\mathbb{G}, \hat{\mathbb{G}}$ *be groups such that there exists an isomorphism* $\phi : \mathbb{G} \to \hat{\mathbb{G}}$. *We say that* Π *is* representation-invariant, *if for all isomorphic groups* $\mathbb{G}, \hat{\mathbb{G}}$ *and for all generators* $g \in \mathbb{G}$, *all* $C = (C_1, \ldots, C_u, C') \leftarrow_\$ \mathcal{G}_\Pi(\mathsf{desc}(\mathbb{G}, g))$, *all* $st = (st_1, \ldots, st_t, st') \in \mathbb{G}^t \times \{0, 1\}^*$, *and all* $S = (S_1, \ldots, S_w, S') \in \mathbb{G}^w \times \{0, 1\}^*$ *holds that* $\mathcal{V}_\Pi(\mathsf{desc}(\mathbb{G}, g), st, C, S) = 1 \iff \mathcal{V}_\Pi(\mathsf{desc}(\hat{\mathbb{G}}, \hat{g}), \hat{st}, \hat{C}, \hat{S}) = 1$, *where* $\hat{g} = \phi(g) \in \mathbb{G}'$, $\hat{C} = (\phi(C_1), \ldots, \phi(C_u), C')$, $\hat{st} = (\phi(st_1), \ldots, \phi(st_t), st')$, *and* $\hat{S} = (\phi(S_1), \ldots, \phi(S_w), S')$.

Observe that this definition only demands the existence of an isomorphism $\phi : \mathbb{G} \to \hat{\mathbb{G}}$ and not that it is efficiently computable.

Example 4. The discrete logarithm problem is representation-invariant. Let $C = (g, h) \in \mathbb{G}^2$ be a discrete log challenge, with corresponding solution $S' \in \{0, 1\}^*$ such that S' canonically interpreted as an integer $S' \in \mathbb{Z}_p$ satisfies $g^{S'} = h \in \mathbb{G}$. Let $\phi : \mathbb{G} \to \hat{\mathbb{G}}$ be an isomorphism, and let $(\hat{g}, \hat{h}) := (\phi(g), \phi(h))$. Then it clearly holds that $\hat{g}^{\hat{S'}} = \hat{h}$, where $\hat{S'} = S'$.

Virtually all common hardness assumptions in algebraic groups are based on representation-invariant computational problems. Popular examples are, for instance, the discrete log problem (DL), computational Diffie-Hellman (CDH), decisional Diffie-Hellman (DDH), one-more discrete log (OMDL), decision linear (DLIN), and so on.

Example 5. The UF-NM-forgery problem for Schnorr signatures with hash function H is *not* representation-invariant for any hash function H. Let $C = (g, X, m) \leftarrow_\$ \mathcal{G}_\Pi(\text{desc}(\mathbb{G}, g))$ be a challenge with solution $S = (R, y) \in \mathbb{G} \times \mathbb{Z}_p$ satisfying $X^c \cdot R = g^y$, where $c := H(R, m)$.

Let $\hat{\mathbb{G}}$ be a group isomorphic to $\mathbb{G}$, such that $\mathbb{G} \cap \hat{\mathbb{G}} = \emptyset$ (that is, there exists no element of $\hat{\mathbb{G}}$ having the same representation as some element of $\mathbb{G}$).[2] Let $\mathbb{G} \to \hat{\mathbb{G}}$ denote the isomorphism. If there exists any R such that $H(R, m) \neq H(\phi(R), m)$ in $\mathbb{Z}_p$ (which holds in particular if H is collision resistant), then we have

$$g^y = X^{H(R,m)} \cdot R \quad \text{but} \quad \phi(g)^y \neq \phi(X)^{H(\phi(R),m)} \cdot \phi(R).$$

Thus, a solution to this problem is valid only with respect to a particular given representation of group elements.

The UF-NM-forgery problem of Schnorr signatures is not representation-invariant, because a solution to this problem involves the hash value $H(R, m)$ that depends on a concrete representation of group element R. We consider such complexity assumptions as rather unnatural, as they are usually very specific to certain constructions of cryptosystems.

2.4   Generic Reductions

In this section we recall the notion of *generic groups*, loosely following [30] (cf. also [19, 24], for instance), and define generic (i.e., representation independent) reductions.

Generic groups. Let $(\mathbb{G}, \cdot)$ be a group of order p and $E \subseteq \{0, 1\}^{\lceil \log p \rceil}$ be a set of size $|E| = |\mathbb{G}|$. If $g, h \in \mathbb{G}$ are two group elements, then we write $g \div h$ for $g \cdot h^{-1}$. Following [30] we define an *encoding function* as a random injective map $\phi : \mathbb{G} \to E$. We say that an element $e \in E$ is the *encoding* assigned to group element $h \in \mathbb{G}$, if $\phi(h) = e$.

A *generic group algorithm* is an algorithm $\mathcal{R}$ which takes as input $\hat{C} = (\phi(C_1), \ldots, \phi(C_u), C')$, where $\phi(C_i) \in E$ is an encoding of group element C_i for all $i \in [u]$, and $C' \in \{0, 1\}^*$ is a bit string. The algorithm outputs $\hat{S} = (\phi(S_1), \ldots, \phi(S_w), S')$, where $\phi(S_i) \in E$ is an encoding of group element S_i for all $i \in [w]$, and $S' \in \{0, 1\}^*$

---

[2] Such a group $\hat{\mathbb{G}}$ can trivially be obtained for any group $\mathbb{G}$, for instance by modifying the encoding by prepending a suitable fixed string to each group element, and changing the group law accordingly.

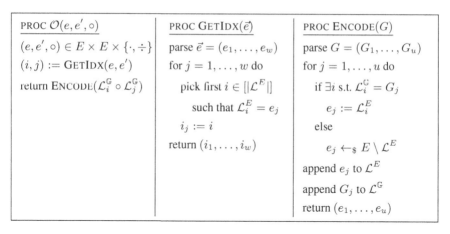

Fig. 1. Procedures implementing the generic group oracle

is a bit string. In order to perform computations on encoded group elements, algorithm $\mathcal{R} = \mathcal{R}^{\mathcal{O}}$ may query a *generic group oracle* (or "*group oracle*" for short). This oracle $\mathcal{O}$ takes as input two encodings $e = \phi(G)$, $e' = \phi(G')$ and a symbol $\circ \in \{\cdot, \div\}$, and returns $\phi(G \circ G')$. Note that $(E, \cdot_{\mathcal{O}})$, where $\cdot_{\mathcal{O}}$ denotes the group operation on E induced by oracle $\mathcal{O}$, forms a group which is isomorphic to $(\mathbb{G}, \cdot)$.

It will later be helpful to have a specific implementation of $\mathcal{O}$. We will therefore assume in the sequel that $\mathcal{O}$ internally maintains two lists $\mathcal{L}^{\mathbb{G}} \subseteq \mathbb{G}$ and $\mathcal{L}^{E} \subseteq E$. These lists define the encoding function ϕ as $\mathcal{L}_i^E = \phi(\mathcal{L}_i^{\mathbb{G}})$, where $\mathcal{L}_i^{\mathbb{G}}$ and $\mathcal{L}_i^E$ denote the i-th element of $\mathcal{L}^{\mathbb{G}}$ and $\mathcal{L}^E$, respectively, for all $i \in [|\mathcal{L}^{\mathbb{G}}|]$. Note that from the perspective of a generic group algorithm it makes no difference whether the encoding function is fixed at the beginning or lazily evaluated whenever a new group element occurs. We will assume that the oracle uses lazy evaluation to simplify our discussion and avoid unnecessary steps for achieving polynomial runtime of our meta-reductions.

Procedure ENCODE takes a list $G = (G_1, \ldots, G_u)$ of group elements as input. It checks for each $G_j \in L$ if an encoding has already been assigned to G_j, that is, if there exists an index i such that $\mathcal{L}_i^{\mathbb{G}} = G_j$. If this holds, ENCODE sets $e_j := \mathcal{L}_i^E$. Otherwise (if no encoding has been assigned to G_j so far), it chooses a fresh and random encoding $e_j \leftarrow_\$ E \setminus \mathcal{L}^E$. In either case G_j and e_j are appended to $\mathcal{L}^{\mathbb{G}}$ and $\mathcal{L}^E$, respectively, which gradually defines the map ϕ such that $\phi(G_j) = e_j$. Note also that the same group element and encoding may occur multiple times in the list. Finally, the procedure returns the list $(e_1, \ldots, e_u)$ of encodings.

Procedure GETIDX takes a list $(e_1, \ldots, e_w)$ of encodings as input. For each $j \in [w]$ it defines i_j as the smallest[3] index such that $e_j = \mathcal{L}_{i_j}^E$, and returns $(i_1, \ldots, i_w)$.[4]

---

[3] Recall that the same encoding may occur multiple times in $\mathcal{L}^E$.

[4] Note that GETIDX may receive only encodings $e_1, \ldots, e_w$ which are already contained in $\mathcal{L}^E$, as otherwise the behavior of GETIDX is undefined. We will make sure that this is always the case.

The lists $\mathcal{L}^G$, $\mathcal{L}^E$ are initially empty. Then $\mathcal{O}$ calls $(e_1, \ldots, e_u) \leftarrow_\$ \text{ENCODE}(G_1, \ldots, G_u)$ to determine encodings for all group elements $G_1, \ldots, G_u$ and starts the generic group algorithm on input $\mathcal{R}(e_1, \ldots, e_u, C')$.

$\mathcal{R}^{\mathcal{O}}$ may now submit queries of the form $(e, e', \circ) \in E \times E \times \{\cdot, \div\}$ to the generic group oracle $\mathcal{O}$. In the sequel we will restrict $\mathcal{R}$ to issue only queries $(e, e', \circ)$ to $\mathcal{O}$ such that $e, e' \in \mathcal{L}^E$. It determines the smallest indices i and j with $e = e_i$ and $e' = e_j$ by calling $(i, j) = \text{GETIDX}(e, e')$. Then it computes $\mathcal{L}_i^G \circ \mathcal{L}_j^G$ and returns the encoding $\text{ENCODE}(\mathcal{L}_i^G \circ \mathcal{L}_j^G)$. Furthemore, we require that $\mathcal{R}$ only outputs encodings $\phi(S_i)$ such that $\phi(S_i) \in \mathcal{L}^E$.

Remark 2. We note that the above restrictions are without loss of generality. To explain this, recall that the assignment between group elements and encodings is random. An alternative implementation $\mathcal{O}'$ of $\mathcal{O}$ could, given an encoding $e \notin \mathcal{L}^E$, assign a random group element $G \leftarrow_\$ \mathbb{G} \setminus \mathcal{L}^G$ to e by appending G to $\mathcal{L}^G$ and e to $\mathcal{L}^E$, in which case $\mathcal{R}$ would obtain an encoding of an independent, new group element. Of course $\mathcal{R}$ can simulate this behavior easily when interacting with $\mathcal{O}$, too.

Generic reductions. Recall that a (fully black-box [23]) reduction from problem Π to problem Σ is an efficient algorithm $\mathcal{R}$ that solves Π, having black-box access to an algorithm $\mathcal{A}$ solving Σ.

In the sequel we consider reductions $\mathcal{R}^{\mathcal{A}, \mathcal{O}}$ having black-box access to an algorithm $\mathcal{A}$ as well as to a generic group oracle $\mathcal{O}$. A generic reduction receives as input a challenge $C = (\phi(C_1), \ldots, \phi(C_\ell), C') \in \mathbb{G}^u \times \{0, 1\}^*$ consisting of u encoded group elements and a bit-string C'. $\mathcal{R}$ may perform computations on encoded group elements, by invoking a generic group oracle $\mathcal{O}$ as described above, and interacts with algorithm $\mathcal{A}$ to compute a solution $S = (\phi(S_1), \ldots, \phi(S_w), S') \in \mathbb{G}^w \times \{0, 1\}^*$, which again may consist of encoded group elements $\phi(S_1), \ldots, \phi(S_w)$ and a bit-string $S' \in \{0, 1\}^*$. Reductions from an *interactive* computational problem Π may additionally have access to an oracle $\mathcal{S}_\Pi$ corresponding to Π, we write $\mathcal{R}^{\mathcal{A}, \mathcal{O}, \mathcal{S}_\Pi}$.

We stress that the adversary $\mathcal{A}$ does not necessarily have to be a generic algorithm. It may not be immediately obvious that a generic reduction can make use of a non-generic adversary, considering that $\mathcal{A}$ might expect a particular encoding of the group elements. However, this is indeed possible. In particular, most reductions in security proofs for cryptosystems that are based on algebraic groups (like [22, 6, 31], to name a few well-known examples) are independent of a particular group representation, and thus generic.

Recall that $\mathcal{R}$ is fully blackbox, i.e., $\mathcal{A}$ is external to $\mathcal{R}$. Thus, the environment in which the reduction is run can easily translate between the two encodings. Consider as an example the reduction shown in Figure 2 that interacts with a non-generic adversary $\mathcal{A}$. Our notion of generic reductions merely formalizes that the reduction works identically for any group representation. This is illustrated in Figure 2 with an "environment" converting group elements received and output by the reduction from one group representation to another. Note also that essentially all security reductions (from a computational problem in an algebraic group) in the literature are generic. We stress that we model only the reduction $\mathcal{R}$ as a generic algorithm. We do not restrict the forger $\mathcal{A}$ in this way, as commonly done in security proofs in the generic group model.

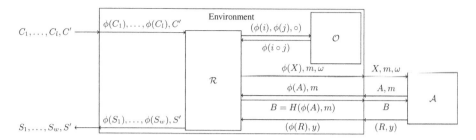

Fig. 2. An example of the interaction between a generic reduction $\mathcal{R}$ and a non-generic adversay $\mathcal{A}$ against the unforgeability of Schnorr signatures. All group elements – such as the challenge input, random oracle queries, and the signature output by $\mathcal{A}$ – are encoded by the environment before being passed to $\mathcal{R}$. In the other direction, encodings of group elements output by $\mathcal{R}$ – such as the public key that is the input of $\mathcal{A}$, random oracle responses, and the solution output by $\mathcal{R}$ – are decoded before being passed to the outside world.

It may not be obvious that this is possible, because $\mathcal{A}$ expects as input group elements in some specific encoding, while $\mathcal{R}$ can only specify them in the form of random encodings. However, the reduction only gets access to the adversary as a blackbox, which means that the adversary is external to the reduction, and the environment in which the reduction is run can easily translate between the encodings used by reduction and adversary. Further note, that while some reduction from a problem Π may be generic, the actual algorithm solving said problem is not $\mathcal{R}$ itself, but the composition of $\mathcal{R}$ and $\mathcal{A}$ which may be non-generic. In particular, this means that any results about equivalence of interesting problems in the generic group model do not apply to the reduction.

3   Unconditional Tightness Bound for Generic Reductions

In this section, we investigate the possibility of finding a tight *generic* reduction $\mathcal{R}$ that reduces a representation-invariant computational problem Π to breaking the UF-NM-security of the Schnorr signature scheme. Our results in this direction are negative, showing that it is impossible to find a generic reduction from any representation-invariant computational problem. This includes even interactive problems.

3.1   Single-Instance Reductions

We begin with considering a very simple class of reduction that we call *vanilla reductions*. A vanilla reduction is a reduction that runs the UF-NM forger $\mathcal{A}$ exactly once (without restarting or rewinding) in order to solve the problem Π. This allows us to explain and analyze the new simulation technique. Later we turn to reductions that may execute $\mathcal{A}$ repeatedly, like for instance the known security proof from [22] based on the Forking Lemma.

An Inefficient Adversary $\mathcal{A}$ In this section we describe an inefficient adversary $\mathcal{A}$ that breaks the UF-NM-security of the Schnorr signature scheme. Recall that a black-box reduction $\mathcal{R}$ must work for any attacker $\mathcal{A}$. Thus, algorithm $\mathcal{R}^{\mathcal{A}}$ will solve the challenge problem Π, given black-box access to $\mathcal{A}$. The meta-reduction will be able to simulate this attacker *efficiently* for any generic reduction $\mathcal{R}$. We describe this attacker for comprehensibility, in order to make our meta-reduction more accessible to the reader.

1. The input of $\mathcal{A}$ is a Schnorr public-key X, a message m, and random coins $\omega \in \{0, 1\}^{\kappa}$.
2. The forger $\mathcal{A}$ chooses q uniformly random group elements $R_1, \ldots, R_q \leftarrow_\$ \mathbb{G}$. (We make the assumption that $q \leq |\mathbb{G}|$.) Subsequently, the forger $\mathcal{A}$ queries the random oracle $\mathcal{H}$ on (R_i, m) for all $i \in [q]$. Let $c_i := \mathcal{H}(R_i, m) \in \mathbb{Z}_p$ be the corresponding answers.
3. Finally, the forger $\mathcal{A}$ chooses an index uniformly at random $\alpha \leftarrow_\$ [q]$, computes $y \in \mathbb{Z}_p$ which satisfies the equation $g^y = X^{c_\alpha} \cdot R_\alpha$, and outputs (R_α, y). For concreteness, we assume this computation is performed by exhaustive search over all $y \in \mathbb{Z}_p$ (recall that we consider an unbounded attacker here, we show later how to instantiate it efficiently).

Note that (R_α, y) is a valid signature for message m with respect to the public key X. Thus, the forger $\mathcal{A}$ breaks the UF-NM-security of the Schnorr signatures with probability 1.

Main Result for Vanilla Reductions Now we are ready to prove our main result for vanilla reductions.

Theorem 1. *Let $\Pi = (\mathcal{G}_\Pi, \mathcal{S}_\Pi, \mathcal{V}_\Pi)$ be a representation-invariant (possibly interactive) computational problem with a challenge consisting of u group elements and let p be the group order. Suppose there exists a generic vanilla reduction $\mathcal{R}$ that $(\epsilon_\mathcal{R}, t_\mathcal{R})$-solves Π, having one-time black-box access to an attacker $\mathcal{A}$ that $(\epsilon_\mathcal{A}, t_\mathcal{A})$-breaks the UF-NM-security of Schnorr signatures with success probability $\epsilon_\mathcal{A} = 1$ by asking q random oracle queries. Then there exists an algorithm $\mathcal{M}$ that (ϵ, t)-solves Π with*
$$\epsilon \geq \epsilon_\mathcal{R} - \frac{2(u+q+t_\mathcal{R})^2}{p} \text{ and } t \approx t_\mathcal{R}.$$

Remark 3. Observe that Theorem 1 rules out reductions from nearly arbitrary computational problems (even interactive). At a first glance this might look contradictory, for instance there always exists a trivial reduction from the problem of forging Schnorr signatures to solving the same problem. However, as explained in Example 5, forging Schnorr-signatures is not a representation-invariant computational problem, therefore this is not a contradiction.

Proof. Assume that there exists a generic vanilla reduction $\mathcal{R} := \mathcal{R}^{\mathcal{O}, \mathcal{S}'_\Pi, \mathcal{A}}$ that $(\epsilon_\mathcal{R}, t_\mathcal{R})$-solves Π, when given access to a generic group oracle $\mathcal{O}$, an oracle $\mathcal{S}'_\Pi$, and a forger $\mathcal{A}(\phi(X), m, \omega)$, where the inputs to the forger are chosen by $\mathcal{R}$. Furthermore, the reduction $\mathcal{R}$ simulates the random oracle $\mathcal{R}.\mathcal{H}$ for $\mathcal{A}$. We show how to build a meta-reduction $\mathcal{M}$ that has black-box access to $\mathcal{R}$ and to an oracle $\mathcal{S}_\Pi$ and that solves the representation-invariant problem Π directly.

We describe $\mathcal{M}$ in a sequence of games, beginning with an *inefficient* implementation $\mathcal{M}_0$ of $\mathcal{M}$ and we modify it gradually until we obtain an *efficient* implementation $\mathcal{M}_2$ of $\mathcal{M}$. We bound the probability with which any reduction $\mathcal{R}$ can distinguish each implementation $\mathcal{M}_i$ from $\mathcal{M}_{i-1}$ for all $i \in \{1,2\}$, which yields that $\mathcal{M}_2$ is an efficient algorithm that can use $\mathcal{R}$ to solve Π if $\mathcal{R}$ in tight.

In what follows let X_i denote the event that $\mathcal{R}$ outputs a valid solution to the given problem instance $\hat{C}$ of Π in Game i.

Game 0. Our meta-reduction $\mathcal{M}_0 := \mathcal{M}_0^{\mathcal{S}_\Pi}$ is an algorithm for solving a representation-invariant computational problem Π, as defined in Section 2.3. That is, $\mathcal{M}_0$ takes as input an instance $C = (C_1, \ldots, C_u, C') \in \mathbb{G}^u \times \{0,1\}^*$, of the representation-invariant computational problem Π, has access to oracle $\mathcal{S}_\Pi$ provided by Π, and outputs a candidate solution S. $\mathcal{R}$ is a generic reduction, i.e., a representation-independent algorithm for Π having black-box access to an attacker $\mathcal{A}$. Algorithm $\mathcal{M}_0$ runs reduction $\mathcal{R}$ as a subroutine, by simulating the generic group oracle $\mathcal{O}$, the $\mathcal{S}_\Pi$ oracle, and attacker $\mathcal{A}$ for $\mathcal{R}$. In order to provide the generic group oracle for $\mathcal{R}$, $\mathcal{M}_0$ implements the following procedures (cf. Figure 3).

PROC $\mathcal{M}_0(C)$	PROC $\mathcal{A}(\phi(X), m, \omega)$
# INITIALIZATION	for all $i \in [q]$
parse $C = (C_1, \ldots, C_u, C')$	$\quad c_i = \mathcal{R}.\mathcal{H}(\phi(R_i), m)$
$\mathcal{L}^{\mathbb{G}} := \emptyset \;\; ; \;\; \mathcal{L}^E := \emptyset$	$\alpha \leftarrow_\$ [q]$
$\vec{R} = (R_1, \ldots, R_q) \leftarrow_\$ \mathbb{G}^q$	$y := \log_g X^{c_\alpha} R_\alpha$
$\mathcal{I} := (C_1, \ldots, C_u, R_1, \ldots, R_q)$	return (R_α, y).
ENCODE($\mathcal{I}$)	
$\hat{C} := (\mathcal{L}_1^E, \ldots, \mathcal{L}_u^E, C')$	PROC $\mathcal{S}_\Pi{}'(Q)$
$\hat{S} \leftarrow_\$ \mathcal{R}^{\mathcal{O},\mathcal{A}}(\hat{C})$	parse $Q = (e_1, \ldots, e_v, Q')$
# FINALIZATION	$(i_1, \ldots, i_v) = $ GETIDX$(e_1, \ldots, e_v)$
parse $\hat{S} := (\hat{S}_1, \ldots, \hat{S}_w, S')$	$(A_1, \ldots, A_\nu, A') = \mathcal{S}_\Pi(\mathcal{L}_{i_1}, \ldots, \mathcal{L}_{i_\nu}, Q')$
$(i_1, \ldots, i_w) := $ GETIDX$(\hat{S}_1, \ldots, \hat{S}_w)$	$(f_1, \ldots, f_\nu) = $ ENCODE$(A_1, \ldots, A_\nu)$
return $(\mathcal{L}_{i_1}^{\mathbb{G}}, \ldots, \mathcal{L}_{i_w}^{\mathbb{G}}, S')$	return $(f_1, \ldots, f_\nu, A')$.

Fig. 3. Implementation of $\mathcal{M}_0$

INITIALIZATION OF $\mathcal{M}_0$: At the beginning of the game, $\mathcal{M}_0$ initializes two lists $\mathcal{L}^{\mathbb{G}} := \emptyset$ and $\mathcal{L}^E := \emptyset$, which are used to simulate the generic group oracle $\mathcal{O}$. Furthermore, $\mathcal{M}_0$ chooses $\vec{R} = (R_1, \ldots, R_q) \leftarrow_\$ \mathbb{G}^q$ at random (these values will later be used by the simulated attacker $\mathcal{A}$), sets $\mathcal{I} := (C_1, \ldots, C_u, R_1, \ldots, R_q)$, and runs ENCODE($\mathcal{I}$) to assign encodings to these group elements. Then $\mathcal{M}_0$ starts the reduction $\mathcal{R}$ on input $\hat{C} := (\mathcal{L}_1^E, \ldots, \mathcal{L}_u^E, C')$. Note that $\hat{C}$ is an encoded version of the challenge instance of Π received by $\mathcal{M}_0$. That is, we have $\hat{C} = (\phi(C_1), \ldots, \phi(C_u), C')$. Oracle queries of $\mathcal{R}$ are answered by $\mathcal{M}_0$ as follows:

GENERIC GROUP ORACLE $\mathcal{O}(e, e', \circ)$: To simulate the generic group oracle, $\mathcal{M}_0$ implements procedures ENCODE and GETIDX as described in Section 2.4. Whenever $\mathcal{R}$ submits a query $(e, e', \circ) \in E \times E \times \{\cdot, \div\}$ to the generic group oracle $\mathcal{O}$, the meta-reduction determines the smallest indices i and j such that $e = e_i$ and $e' = e_j$ by calling $(i, j) = \text{GETIDX}(e, e')$. Then it computes $\mathcal{L}_i^G \circ \mathcal{L}_j^G$ and returns $\text{ENCODE}(\mathcal{L}_i^G \circ \mathcal{L}_j^G)$.

ORACLE $\mathcal{S}'_\Pi(Q)$: This procedure handles queries issued by $\mathcal{R}$ to $\mathcal{S}'_\Pi$ by forwarding them to oracle $\mathcal{S}_\Pi$ provided by the challenger and returning the response. That is, whenever $\mathcal{R}$ submits a query $Q = (e_1, \ldots, e_v, Q') \in E^v \times \{0, 1\}^*$ to $\mathcal{S}'_\Pi$, the meta-reduction runs $(i_1, \ldots, i_v) := \text{GETIDX}(e_1, \ldots, e_v)$ and queries $\mathcal{S}_\Pi$ to compute $(A_1, \ldots, A_v, A') := \mathcal{S}_\Pi(\mathcal{L}_{i_1}, \ldots, \mathcal{L}_{i_v}, Q')$. Then $\mathcal{M}_0$ determines the corresponding encodings as $(f_1, \ldots, f_v) := \text{ENCODE}(A_1, \ldots, A_v)$ and returns $(f_1, \ldots, f_v, A')$ to $\mathcal{R}$.

THE FORGER $\mathcal{A}(\phi(X), m, \omega)$: This procedure implements a simulation of the inefficient attacker $\mathcal{A}$ described in Section 3.1. It proceeds as follows. When $\mathcal{R}$ outputs $(\phi(X), m, \omega)$ to invoke an instance of $\mathcal{A}$, $\mathcal{A}$ queries the random oracle $\mathcal{R}.\mathcal{H}$ provided by $\mathcal{R}$ on $(\phi(R_i), m)$ for all $i \in [q]$, to determine $c_i = \mathcal{H}(\phi(R_i), m)$. Afterwards, $\mathcal{M}_0$ chooses an index $\alpha \leftarrow_\$ [q]$ uniformly at random, computes the the discrete logarithm $y := \log_g X^{c_\alpha} R_\alpha$ by exhaustive search, and outputs (R_α, y). (This step is not efficient. We show in subsequent games how to implement this attacker efficiently.)

FINALIZATION OF $\mathcal{M}_0$: Eventually, the algorithm $\mathcal{R}$ outputs a solution $\hat{S} := (\hat{S}_1, \ldots, \hat{S}_w, S') \in E^w \times \{0, 1\}^*$. The algorithm $\mathcal{M}_0$ runs $(i_1, \ldots, i_w) := \text{GETIDX}(\hat{S}_1, \ldots, \hat{S}_w)$ to determine the indices of group elements $(\mathcal{L}_{i_1}^G, \ldots, \mathcal{L}_{i_w}^G)$ corresponding to encodings $(\hat{S}_1, \ldots, \hat{S}_w)$, and outputs $(\mathcal{L}_{i_1}^G, \ldots, \mathcal{L}_{i_w}^G, S')$.

Analysis of $\mathcal{M}_0$. Note that $\mathcal{M}_0$ provides a perfect simulation of the oracles $\mathcal{O}$ and $\mathcal{S}_\Pi$ and it also mimics the attacker from Section 3.1 perfectly. In particular, (R_α, y) is a valid forgery for message m and thus, $\mathcal{R}$ outputs a solution $\hat{S} = (\hat{S}_1, \ldots, \hat{S}_w, S')$ to $\hat{C}$ with probability $\Pr[X_0] = \epsilon_\mathcal{R}$. Since Π is assumed to be representation-invariant, $S := (S_1, \ldots, S_w, S')$ with $\hat{S}_i = \phi(S_i)$ for $i \in [w]$ is therefore a valid solution to C. Thus, $\mathcal{M}_0$ outputs a valid solution S to C with probability $\epsilon_\mathcal{R}$.

Game 1. In this game we introduce a meta-reduction $\mathcal{M}_1$, which essentially extends $\mathcal{M}_0$ with additional bookkeeping to record the sequence of group operations performed by $\mathcal{R}$. The purpose of this intermediate game is to simplify our analysis of the final implementation $\mathcal{M}_2$. Meta-reduction $\mathcal{M}_1$ proceeds identical to $\mathcal{M}_0$, except for a few differences (cf. Figure 4).

INITIALIZATION OF $\mathcal{M}_1$: The initialization is exactly like before, except that $\mathcal{M}_1$ maintains an additional list $\mathcal{L}^V$ of elements of $\mathbb{Z}_p^{u+q}$. Let $\mathcal{L}_i^V$ denote the i-th entry of $\mathcal{L}^V$.

List $\mathcal{L}^V$ is initialized with the $u + q$ canonical unit vectors in $\mathbb{Z}_p^{u+q}$. That is, let η_i denote the i-th canonical unit vector in $\mathbb{Z}_p^{u+q}$, i.e., $\eta_1 = (1, 0, \ldots, 0), \eta_2 = (0, 1, 0, \ldots, 0)$, $\ldots, \eta_{u+q} = (0, \ldots, 0, 1)$. Then $\mathcal{L}^V$ is initialized such that $\mathcal{L}_i^V := \eta_i$ for all $i \in [u + q]$.

GENERIC GROUP ORACLE $\mathcal{O}(e, e', \circ)$: In parallel to computing the group operation, the generic group oracle implemented by $\mathcal{M}_1$ also performs computations on vectors of $\mathcal{L}^V$.

Given a query $(e, e', \circ) \in E \times E \times \{\cdot, \div\}$, the oracle $\mathcal{O}$ determines the smallest indices i and j such that $e = e_i$ and $e' = e_j$ by calling GETIDX. It computes $a := \mathcal{L}_i^V \diamond \mathcal{L}_j^V \in \mathbb{Z}_p^{u+q}$, where $\diamond := +$ if $\circ = \cdot$ and $\diamond := -$ if $\circ = \div$, and appends a to $\mathcal{L}^V$. Finally it returns ENCODE$(\mathcal{L}_i^\mathbb{G} \circ \mathcal{L}_j^\mathbb{G})$.

Analysis of $\mathcal{M}_1$. Recall that the initial content $\mathcal{I}$ of $\mathcal{L}^\mathbb{G}$ is $\mathcal{I} = (C_1, \ldots, C_u, R_1, \ldots, R_q)$, and that $\mathcal{R}$ performs only group operations on $\mathcal{I}$. Thus, any group element $h \in \mathcal{L}^\mathbb{G}$ can be written as $h = \prod_{i=1}^u C_i^{a_i} \cdot \prod_{i=1}^q R_i^{a_{u+i}}$ where the vector $a = (a_1, \ldots, a_{u+q}) \in \mathbb{Z}_p^{u+q}$ is (essentially) determined by the sequence of queries issued by $\mathcal{R}$ to $\mathcal{O}$. For a vector $a \in \mathbb{Z}_p^{u+q}$ and a vector of group elements $V = (v_1, \ldots, v_{u+q}) \in \mathbb{G}^{u+q}$ let us write Eval(V, a) shorthand for Eval$(V, a) := \prod_{i=1}^{u+q} v_i^{a_i}$ in the sequel. In particular, it holds that Eval$(\mathcal{I}, a) = \prod_{i=1}^u C_i^{a_i} \cdot \prod_{i=1}^q R_i^{a_{u+i}}$. The key motivation for the changes introduced in Game 1 is that now (by construction of $\mathcal{M}_1$) it holds that $\mathcal{L}_i^\mathbb{G} =$ Eval$(\mathcal{I}, \mathcal{L}_i^V)$ for all $i \in [|\mathcal{L}^\mathbb{G}|]$. Thus, at any point in time during the execution of $\mathcal{R}$, the entire list $\mathcal{L}^\mathbb{G}$ of group elements can be recomputed from $\mathcal{L}^V$ and $\mathcal{I}$ by setting $\mathcal{L}_i^\mathbb{G} :=$ Eval$(\mathcal{I}, \mathcal{L}_i^V)$ for $i \in [|\mathcal{L}^V|]$. The reduction $\mathcal{R}$ is completely oblivious to this additional bookkeeping performed by $\mathcal{M}_1$, thus we have $\Pr[X_1] = \Pr[X_0]$.

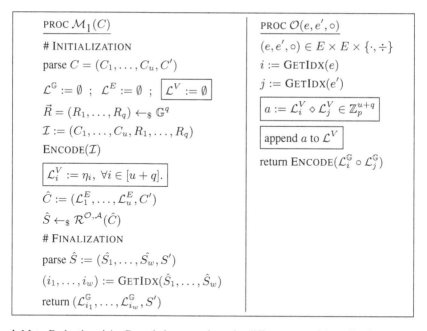

Fig. 4. Meta-Reduction $\mathcal{M}_1$. Boxed elements show the differences to $\mathcal{M}_0$. All other procedures are identical to $\mathcal{M}_0$ and thus omitted.

Game 2. Note that the meta-reductions described in previous games were not efficient, because the simulation of the attacker in procedure $\mathcal{A}$ needed to compute a discrete logarithm by exhaustive search. In this final game, we construct a meta-reduction $\mathcal{M}_2$

that simulates $\mathcal{A}$ efficiently. $\mathcal{M}_2$ proceeds exactly like $\mathcal{M}_1$, except for the following (cf. Figure 5).

THE FORGER $\mathcal{A}(\phi(X), m, \omega)$: When $\mathcal{R}$ outputs $(\phi(X), m, \omega)$ to invoke an instance of $\mathcal{A}$, $\mathcal{A}$ queries the random oracle $\mathcal{R}.\mathcal{H}$ provided by $\mathcal{R}$ on $(\phi(R_i), m)$ for all $i \in [q]$, to determine $c_i = \mathcal{H}(\phi(R_i), m)$. Then it chooses an index $\alpha \leftarrow_\$ [q]$ uniformly at random, samples an element y uniformly at random from $\mathbb{Z}_p$, computes $R_\alpha^* := g^y X^{-c_\alpha}$, and re-computes the entire list $\mathcal{L}^{\mathbb{G}}$ using R_α^* instead of R_α.

More precisely, let $\mathcal{I}^* := (C_1, \ldots, C_u, R_1, \ldots, R_{\alpha-1}, R_\alpha^*, R_{\alpha+1}, \ldots, R_q)$. Observe that the vector $\mathcal{I}^*$ is identical to the initial contents $\mathcal{I}$ of $\mathcal{L}^{\mathbb{G}}$, with the difference that R_α is replaced by R_α^*. The list $\mathcal{L}^{\mathbb{G}}$ is now recomputed from $\mathcal{L}^V$ and $\mathcal{I}^*$ by setting $\mathcal{L}_i^{\mathbb{G}} := \mathsf{Eval}(\mathcal{I}^*, \mathcal{L}_i^V)$ for all $i \in [|\mathcal{L}^V|]$. Finally, $\mathcal{M}_2$ returns $(\phi(R_\alpha^*), y)$ to $\mathcal{R}$ as the forgery.

Analysis of $\mathcal{M}_2$. First note that $(\phi(R_\alpha^*), y)$ is a valid signature, since $\phi(R_\alpha^*)$ is the encoding of group element R_α^* satisfying the verification equation $g^y = X^{c_\alpha} \cdot R_\alpha^*$, where $c_\alpha = \mathcal{H}(\phi(R_\alpha^*), m)$. Next we claim that $\mathcal{R}$ is not able to distinguish $\mathcal{M}_2$ from $\mathcal{M}_1$, except for a negligibly small probability. To show this, observe that Game 2 and Game 1 are perfectly indistinguishable, if for all pairs of vectors $\mathcal{L}_i^V, \mathcal{L}_j^V \in \mathcal{L}^V$ it holds that $\mathsf{Eval}(\mathcal{I}, \mathcal{L}_i^V) = \mathsf{Eval}(\mathcal{I}, \mathcal{L}_j^V) \iff \mathsf{Eval}(\mathcal{I}^*, \mathcal{L}_i^V) = \mathsf{Eval}(\mathcal{I}^*, \mathcal{L}_j^V)$, because in this case $\mathcal{M}_2$ chooses identical encodings for two group elements $\mathcal{L}_i^{\mathbb{G}}, \mathcal{L}_j^{\mathbb{G}} \in \mathcal{L}^{\mathbb{G}}$ *if and only if* $\mathcal{M}_1$ chooses identical encodings.

PROC $\mathcal{A}(\phi(X), m, \omega)$:

$\alpha \leftarrow_\$ [q]$

for all $i \in [q]$

    $c_i = \mathcal{R}.\mathcal{H}(\phi(R_i), m)$

$\boxed{y \leftarrow_\$ \mathbb{Z}_p \quad ; \quad R_\alpha^* := g^y X^{-c_\alpha}}$

$\boxed{\mathcal{I}^* := (C_1, \ldots, C_u, R_1, \ldots, R_{\alpha-1}, R_\alpha^*, R_{\alpha+1}, \ldots, R_q)}$

$\boxed{\text{for } j = 1, \ldots, |\mathcal{L}^{\mathbb{G}}| \text{ do}}$

    $\boxed{\mathcal{L}_i^{\mathbb{G}} := \mathsf{Eval}(\mathcal{I}^*, \mathcal{L}_i^V)}$

return $(y, \phi(R_\alpha^*))$

Fig. 5. Efficient simulation of attacker $\mathcal{A}$ by $\mathcal{M}_2$

Lemma 1. *Let F denote the event that $\mathcal{R}$ computes vectors $\mathcal{L}_i^V, \mathcal{L}_j^V \in \mathcal{L}^V$ such that*

$$\mathsf{Eval}(\mathcal{I}, \mathcal{L}_i^V) = \mathsf{Eval}(\mathcal{I}, \mathcal{L}_j^V) \quad \wedge \quad \mathsf{Eval}(\mathcal{I}^*, \mathcal{L}_i^V) \neq \mathsf{Eval}(\mathcal{I}^*, \mathcal{L}_j^V) \tag{1}$$

or

$$\mathsf{Eval}(\mathcal{I}, \mathcal{L}_i^V) \neq \mathsf{Eval}(\mathcal{I}, \mathcal{L}_j^V) \quad \wedge \quad \mathsf{Eval}(\mathcal{I}^*, \mathcal{L}_i^V) = \mathsf{Eval}(\mathcal{I}^*, \mathcal{L}_j^V). \tag{2}$$

Then

$$\Pr[F] \leq 2(u + q + t_{\mathcal{R}})^2/p.$$

The proof of Lemma 1 is deferred to the full version. We apply it to finish the proof of Theorem 1. By Lemma 1, the algorithm $\mathcal{M}_2$ fails to simulate $\mathcal{M}_1$ with probability at most $2(u + q + t_{\mathcal{R}})^2/p$. Thus, we have $\Pr[X_2] \geq \Pr[X_1] - 2(u + q + t_{\mathcal{R}})^2/p$.

Note also that $\mathcal{M}_2$ provides an efficient simulation of adversary $\mathcal{A}$. The total running time of $\mathcal{M}_2$ is essentially of the running time of $\mathcal{R}$ plus some minor additional computations and bookkeeping. Furthermore, if $\mathcal{R}$ is able to $(\epsilon_{\mathcal{R}}, t_{\mathcal{R}})$ solve Π, then $\mathcal{M}_2$ is able to (ϵ, t)-solve Π with probability at least

$$\epsilon \geq \Pr[X_2] \geq \epsilon_{\mathcal{R}} - \frac{2(u + q + t_{\mathcal{R}})^2}{p}.$$

Remark 4. Note that the simulated forger *re-computes* the entire list $\mathcal{L}^{\mathbb{G}}$ after replacing R_α with R_α^*. This ensures consistency of the attacker's view before and after replacing R_α with R_α^*, *if (and only if)* it holds that

$$\mathsf{Eval}(\mathcal{I}, \mathcal{L}_i^V) = \mathsf{Eval}(\mathcal{I}, \mathcal{L}_j^V) \iff \mathsf{Eval}(\mathcal{I}^*, \mathcal{L}_i^V) = \mathsf{Eval}(\mathcal{I}^*, \mathcal{L}_j^V) \tag{3}$$

Lemma 1 bounds the probability that 3 does not hold, thus it bounds the probability that an attacker is able to notice the re-programming by receiving different results before and after the re-programming.

4   Multi-instance Reductions

Now we turn to considering multi-instance reductions, which may run multiple sequential executions of the signature forger $\mathcal{A}$. This is the interesting case, in particular because the Forking-Lemma based security proof for Schnorr signatures by Pointcheval and Stern [22] is of this type.

The meta-reduction described in detail in the full version is heavily based on Seurin's meta-reduction [28]. Essentially, we show that our new simulation of forged signatures is compatible with Seurin's approach for simulating a sequence of Random Oracle queries. In combination this allows to prove that a generic reduction from *any* representation-invariant computational problem Π to breaking Schnorr signatures loses a factor of at least q, which essentially matches the upper bound of [22].

The description of the corresponding family of adversaries and the proof of the following theorem can be found in the full version.

Theorem 2. *Let Π be a representation-invariant computational problem. Suppose there exists a generic reduction $\mathcal{R}^{\mathcal{O}, \mathcal{S}'_\Pi, \mathcal{A}_{F,f}}$ that $(\epsilon_{\mathcal{R}}, t_{\mathcal{R}})$-solves Π, having n-time black-box access to an attacker $\mathcal{A}_{F,f}$ that $(\epsilon_{\mathcal{A}}, t_{\mathcal{A}}, q)$-breaks the UF-NM-security of Schnorr signatures with success probability $\epsilon_{\mathcal{A}} < 1$ in time $t_{\mathcal{A}} \approx q$. Then there exists an algorithm $\mathcal{M}$ that (ϵ, t)-solves Π with $t \approx t_{\mathcal{R}}$ and*

$$\epsilon \geq \epsilon_{\mathcal{R}} - 2n(u + nq + t_{\mathcal{R}})/p - n \ln\left((1 - \epsilon_{\mathcal{A}})^{-1}\right)/q(1 - p^{-1/4})$$

This bound allows essentially the same analysis as in [28] and thus we arrive (for $\epsilon_{\mathcal{A}} \approx 1 - (1 - 1/q)^q$) at a lower bound for ϵ of approximately $\epsilon_{\mathcal{R}} - \frac{n}{q}$. Therefore, $\mathcal{R}$ must necessarily lose a factor of almost $1/q$ if the discrete logarithm problem is indeed hard.

5   A Note on Tightly-Secure Schnorr-Type Signatures

There exist several variants of Schnorr signatures with *tight* security reductions from representation-invariant computational problems. This includes, for instance, the schemes of Goh and Jarecki [16] and Chevallier-Mames [7], which are based on the computational Diffie-Hellman problem, and the scheme of Shao [29].

It is natural to ask why our tightness bound, in particular our technique of re-programming the group representation, can not be applied to these schemes. Due to space limitations, we have to defer this discussion to the full version of this paper.

Acknowledgments. We thank the anonymous reviewers for valuable comments. Nils Fleischhacker and Dominique Schröder were supported by the German Federal Ministry of Education and Research (BMBF) through funding for the Center for IT-Security, Privacy, and Accountability (CISPA; see www.cispa-security.org). Dominique Schröder is also supported by an Intel Early Career Faculty Honor Program Award.

References

1. Bellare, M., Namprempre, C., Pointcheval, D., Semanko, M.: The one-more-RSA-inversion problems and the security of Chaum's blind signature scheme. Journal of Cryptology 16(3), 185–215 (2003)
2. Bellare, M., Palacio, A.: GQ and schnorr identification schemes: Proofs of security against impersonation under active and concurrent attacks. In: Yung, M. (ed.) CRYPTO 2002. LNCS, vol. 2442, p. 162. Springer, Heidelberg (2002)
3. Bellare, M., Rogaway, P.: Random oracles are practical: A paradigm for designing efficient protocols. In: Ashby, V. (ed.) Conference on Computer and Communications Security ACM CCS 1993, Fairfax, Virginia, USA, November 3–5, pp. 62–73. ACM Press (1993)
4. Bellare, M., Rogaway, P.: The exact security of digital signatures - how to sign with RSA and rabin. In: Maurer, U.M. (ed.) EUROCRYPT 1996. LNCS, vol. 1070, pp. 399–416. Springer, Heidelberg (1996)
5. Bernstein, D.J.: Proving tight security for rabin-williams signatures. In: Smart, N.P. (ed.) EUROCRYPT 2008. LNCS, vol. 4965, pp. 70–87. Springer, Heidelberg (2008)
6. Boneh, D., Boyen, X.: Secure identity based encryption without random oracles. In: Franklin, M. (ed.) CRYPTO 2004. LNCS, vol. 3152, pp. 443–459. Springer, Heidelberg (2004)
7. Chevallier-Mames, B.: An efficient CDH-based signature scheme with a tight security reduction. In: Shoup, V. (ed.) CRYPTO 2005. LNCS, vol. 3621, pp. 511–526. Springer, Heidelberg (2005)
8. Coron, J.-S.: On the exact security of full domain hash. In: Bellare, M. (ed.) CRYPTO 2000. LNCS, vol. 1880, pp. 229–235. Springer, Heidelberg (2000)
9. Coron, J.-S.: Optimal security proofs for PSS and other signature schemes. In: Knudsen, L.R. (ed.) EUROCRYPT 2002. LNCS, vol. 2332, pp. 272–287. Springer, Heidelberg (2002)
10. Dodis, Y., Haitner, I., Tentes, A.: On the instantiability of hash-and-sign RSA signatures. In: Cramer, R. (ed.) TCC 2012. LNCS, vol. 7194, pp. 112–132. Springer, Heidelberg (2012)
11. Dodis, Y., Oliveira, R., Pietrzak, K.: On the generic insecurity of the full domain hash. In: Shoup, V. (ed.) CRYPTO 2005. LNCS, vol. 3621, pp. 449–466. Springer, Heidelberg (2005)
12. Dodis, Y., Reyzin, L.: On the power of claw-free permutations. In: Cimato, S., Galdi, C., Persiano, G. (eds.) SCN 2002. LNCS, vol. 2576, pp. 55–73. Springer, Heidelberg (2003)

13. Fischlin, M., Fleischhacker, N.: Limitations of the meta-reduction technique: The case of schnorr signatures. In: Johansson, T., Nguyen, P.Q. (eds.) EUROCRYPT 2013. LNCS, vol. 7881, pp. 444–460. Springer, Heidelberg (2013)

14. Fischlin, M., Lehmann, A., Ristenpart, T., Shrimpton, T., Stam, M., Tessaro, S.: Random oracles with(out) programmability. In: Abe, M. (ed.) ASIACRYPT 2010. LNCS, vol. 6477, pp. 303–320. Springer, Heidelberg (2010)

15. Garg, S., Bhaskar, R., Lokam, S.V.: Improved bounds on security reductions for discrete log based signatures. In: Wagner, D. (ed.) CRYPTO 2008. LNCS, vol. 5157, pp. 93–107. Springer, Heidelberg (2008)

16. Goh, E.J., Jarecki, S.: A signature scheme as secure as the Diffie-Hellman problem. In: Biham, E. (ed.) EUROCRYPT 2003. LNCS, vol. 2656, pp. 401–415. Springer, Heidelberg (2003)

17. Goldwasser, S., Micali, S., Rivest, R.L.: A digital signature scheme secure against adaptive chosen-message attacks. SIAM Journal on Computing 17(2), 281–308 (1988)

18. Kakvi, S.A., Kiltz, E.: Optimal security proofs for full domain hash, revisited. In: Pointcheval, D., Johansson, T. (eds.) EUROCRYPT 2012. LNCS, vol. 7237, pp. 537–553. Springer, Heidelberg (2012)

19. Maurer, U.M.: Abstract models of computation in cryptography. In: Smart, N.P. (ed.) Cryptography and Coding 2005. LNCS, vol. 3796, pp. 1–12. Springer, Heidelberg (2005)

20. Neven, G., Smart, N.P., Warinschi, B.: Hash function requirements for schnorr signatures. J. Mathematical Cryptology 3(1), 69–87 (2009)

21. Paillier, P., Vergnaud, D.: Discrete-log-based signatures may not be equivalent to discrete log. In: Roy, B. (ed.) ASIACRYPT 2005. LNCS, vol. 3788, pp. 1–20. Springer, Heidelberg (2005)

22. Pointcheval, D., Stern, J.: Security proofs for signature schemes. In: Maurer, U.M. (ed.) EUROCRYPT 1996. LNCS, vol. 1070, pp. 387–398. Springer, Heidelberg (1996)

23. Reingold, O., Trevisan, L., Vadhan, S.P.: Notions of reducibility between cryptographic primitives. In: Naor, M. (ed.) TCC 2004. LNCS, vol. 2951, pp. 1–20. Springer, Heidelberg (2004)

24. Rupp, A., Leander, G., Bangerter, E., Dent, A.W., Sadeghi, A.-R.: Sufficient conditions for intractability over black-box groups: Generic lower bounds for generalized DL and DH problems. In: Pieprzyk, J. (ed.) ASIACRYPT 2008. LNCS, vol. 5350, pp. 489–505. Springer, Heidelberg (2008)

25. Schäge, S.: Tight proofs for signature schemes without random oracles. In: Paterson, K.G. (ed.) EUROCRYPT 2011. LNCS, vol. 6632, pp. 189–206. Springer, Heidelberg (2011)

26. Schnorr, C.-P.: Efficient identification and signatures for smart cards. In: Brassard, G. (ed.) CRYPTO 1989. LNCS, vol. 435, pp. 239–252. Springer, Heidelberg (1990)

27. Schnorr, C.P.: Efficient signature generation by smart cards. Journal of Cryptology 4(3), 161–174 (1991)

28. Seurin, Y.: On the exact security of schnorr-type signatures in the random oracle model. In: Pointcheval, D., Johansson, T. (eds.) EUROCRYPT 2012. LNCS, vol. 7237, pp. 554–571. Springer, Heidelberg (2012)

29. Shao, Z.: A provably secure short signature scheme based on discrete logarithms. Inf. Sci. 177(23), 5432–5440 (2007)

30. Shoup, V.: Lower bounds for discrete logarithms and related problems. In: Fumy, W. (ed.) EUROCRYPT 1997. LNCS, vol. 1233, pp. 256–266. Springer, Heidelberg (1997)

31. Waters, B.: Efficient identity-based encryption without random oracles. In: Cramer, R. (ed.) EUROCRYPT 2005. LNCS, vol. 3494, pp. 114–127. Springer, Heidelberg (2005)

A   Universal Unforgeability under No-Message Attacks

Consider the following security experiment involving a signature scheme $(\mathsf{Gen}, \mathsf{Sign}, \mathsf{Vrfy})$, an attacker $\mathcal{A}$, and a challenger $\mathcal{C}$.

1. The challenger $\mathcal{C}$ computes a key-pair $(X, x) \leftarrow_\$ \mathsf{Gen}(g)$ and chooses a message $m \leftarrow_\$ \{0, 1\}^k$ uniformly at random. It sends (X, m) to the adversary $\mathcal{A}$.
2. Eventually, $\mathcal{A}$ stops, outputting a signature σ.

Definition 4. *We say that $\mathcal{A}$ (ϵ, t)-breaks the* UF-NM-*security of* $(\mathsf{Gen}, \mathsf{Sign}, \mathsf{Vrfy})$, *if $\mathcal{A}$ runs in time at most t and* $\Pr[\mathcal{A}(X, m) = \sigma : \mathsf{Vrfy}(X, m, \sigma) = 1] \geq \epsilon.$

Note that UF-NM-security is a very weak security goal for digital signatures. Since we are going to prove a negative result, this is not a limitation, but makes our result only stronger. In fact, if we rule out reductions from some problem Π to forging signatures in the sense of UF-NM, then the impossibility clearly holds for stronger security goals, like existential unforgeability under adaptive chosen-message attacks [17], too.

Square Span Programs with Applications to Succinct NIZK Arguments*

George Danezis[1], Cédric Fournet[2], Jens Groth[1], and Markulf Kohlweiss[2]

[1] University College London, UK
[2] Microsoft Research

Abstract. We propose a new characterization of NP using square span programs (SSPs). We first characterize NP as affine map constraints on small vectors. We then relate this characterization to SSPs, which are similar but simpler than Quadratic Span Programs (QSPs) and Quadratic Arithmetic Programs (QAPs) since they use a single series of polynomials rather than 2 or 3.

We use SSPs to construct succinct non-interactive zero-knowledge arguments of knowledge. For performance, our proof system is defined over Type III bilinear groups; proofs consist of just 4 group elements, verified in just 6 pairings. Concretely, using the Pinocchio libraries, we estimate that proofs will consist of 160 bytes verified in less than 6 ms.

Keywords: Square span program, quadratic span program, SNARKs, non-interactive zero-knowledge arguments of knowledge.

1 Introduction

Gennaro, Gentry, Parno and Raykova [GGPR13] proposed a new, influential characterization of the complexity class **NP** using Quadratic Span Programs (QSPs), a natural extension of span programs defined by Karchmer and Wigderson [KW93]. Their main motivation was the construction of Succinct Non-interactive Arguments of Knowledge (SNARKs). Their work has lead to fast progress towards practical verifiable computations, whereby a resource-constrained client offloads the computation of an expensive function to a computationally endowed server or cloud, but still intends to verify the correctness of any returned results. For instance, using Quadratic Arithmetic Programs (QAPs), a generalization of QSPs for arithmetic circuits, Pinocchio [PHGR13] provides evidence that verified remote computation can be faster than local computation. At the same time, zero-knowledge variants of their construction enable the server to keep intermediate and additional values used in the computation private. Such constructions are at the forefront of privacy-friendly variants of Bitcoin, such as Pinocchio Coin [DFKP13] and Zerocash [BSCG+14].

---

* The research leading to these results has received funding from the European Research Council under the European Union's Seventh Framework Programme (FP/2007-2013) / ERC Grant Agreement n. 307937 and the Engineering and Physical Sciences Research Council grant EP/J009520/1.

P. Sarkar and T. Iwata (Eds.): ASIACRYPT 2014, PART I, LNCS 8873, pp. 532–550, 2014.

We introduce Square Span Programs (SSPs), a radical simplification of quadratic span programs, and we use them to build simpler and more efficient SNARKs and Non-Interactive Zero-Knowledge arguments (NIZKs) for the verified computation of binary circuits and the verification of SAT solving, two closely related problems. Thus, SSPs can be used to build NIZK arguments to support privacy properties while guaranteeing high integrity, at a minimal cost for the verifier.

Square span programs are based on the insight that every 2-input binary gate $g(a, b) = c$ can be specified using (1) an affine combination $\ell = \alpha a + \beta b + \gamma c + \delta$ of the gate's input and output wires that take exactly two values, $\ell = 0$ or $\ell = 2$, when the wires meet the gate's logical specification; and (2), equivalently, as a single 'square' constraint $(\ell - 1)^2 = 1$. Composing such constraints, a satisfying assignment for any binary circuit (or any SAT problem) can be specified first as a set of affine map constraints, then as a constraint on the span of a set of polynomials, defining the square span program for this circuit.

Due to their conceptual simplicity, SSPs offer several advantages over previous constructions for binary circuits. Their reduced number of constraints lead to smaller programs, and to lower sizes and degrees for the polynomials required to represent them, which in turn reduce the computation complexity required in proving or verifying NIZK arguments. Notably, their simpler 'square' form requires only a single polynomial to be evaluated for verification (instead of two for earlier QSPs, and three for Pinocchio [PHGR13]) leading to a simpler and more compact setup, smaller keys, and fewer operations required for proof and verification.

The resulting, SSP-based SNARKs may be the most compact constructions to date. For performance, our proof system is defined over Type III bilinear groups; to this end, we revisit and restate known assumptions for Type III bilinear groups. The communicated proofs consist of just 4 group elements (3 in the left group, and one in the right group); they can be verified in just 6 pairings, plus one multiplication for each (non-zero) bit of input, irrespective of the size of the circuit. Concretely, using the same groups as in the implementation of Pinocchio, we arrive at 160-byte proofs that we estimate can be verified in less than 6 ms, for circuits with millions of gates. For instance, our SNARKs would be entirely adequate to verify the solutions of large SAT problems offloaded to specialized servers and tools, such as those available in the annual SAT competition[1], without the need to communicate (or even reveal) their complete solutions.

2 Square Span Programs

In this section we will provide new characterizations of languages in NP. First, we show that circuit satisfiability can be recast as a set of constraints on affine maps over the integers. Next, we show in Section 2.2 that this leads to the NP-completeness of square span programs as defined below. The reader may find the example in Section 2.3 useful to illustrate the transformation from circuit

---

[1] http://satcompetition.org/

satisfiability to square span programs. We compare square span programs to quadratic span programs in Section 2.4.

Definition 1 (Square span program). *A square span program Q over the field $\mathbb{F}$ consists of $m + 1$ polynomials $v_0(x), v_1(x), \ldots, v_m(x)$ and a target polynomial $t(x)$ such that $\deg(v_i(x)) \leq \deg(t(x))$ for all $i = 0, \ldots, m$.*

We say that the square span program Q has size m and degree $d = \deg(t(x))$.

We say that Q accepts an input $(a_1, \ldots, a_\ell) \in \mathbb{F}^\ell$ if and only if there exist $a_{\ell+1}, \ldots, a_m \in \mathbb{F}$ satisfying

$$t(x) \quad divides \quad \left(v_0(x) + \sum_{i=1}^{m} a_i v_i(x)\right)^2 - 1.$$

We say that Q verifies a boolean function $f : \{0,1\}^\ell \rightarrow \{0,1\}$ if it accepts exactly those inputs $\boldsymbol{a} \in \mathbb{F}^\ell$ that satisfy $\boldsymbol{a} \in \{0,1\}^\ell$ and $f(\boldsymbol{a}) = 1$.

In the definition, we may see f as a binary circuit or, more abstractly, as a logical specification of a satisfiability problem. In our NIZK argument system in Section 3.3 we will split the ℓ inputs into ℓ_u public and ℓ_w private inputs. We remark that the public 'inputs' are considered from the viewpoint of the verifier: for an outsourced computation for instance, they may include both the inputs sent by the clients and the outputs returned by the server performing the computation together with its proof; for a SAT problem, they may provide a partial instantiation of the problem, or a part of its solution.

This treatment is strictly more general than classic Circuit-SAT which only cares about satisfiability and thus corresponds to the special case of $\ell_u = 0$, i.e., Q verifies a circuit C if it accepts exactly those w where $C(w) = 1$. Alternatively, if we want the same SSP Q to handle different circuits, it may be useful to let f be a universal circuit that takes as input an ℓ_u-bit description of a freely chosen circuit C and an ℓ_w-bit value w and returns 1 if and only if $C(w) = 1$.

2.1   The NP-completeness of Affine Map Constraints

In this section we will show that circuit satisfiability can be recast as a set of constraints on the image of an affine map $\boldsymbol{a} \mapsto \boldsymbol{a}V + \boldsymbol{b}$.

Groth, Ostrovsky and Sahai [GOS12] used that a NAND-gate with input wires a, b and output wire c can be "linearized". Given values $a, b, c \in \{0, 1\}$, with 0 meaning false and 1 meaning true, and writing $\bar{c}$ for $1 - c$, we have

$$c = \neg(a \wedge b) \quad \text{if and only if} \quad a + b - 2\bar{c} \in \{0, 1\}.$$

All logic gates with fan-in 2 can be linearized. We will without loss of generality ignore gates corresponding to $c = a$, $c = \bar{a}$, $c = b$, $c = \bar{b}$, $c = 0$ and $c = 1$ since they are trivial and can be eliminated from a circuit. This leaves us with 10 types of logic gates. Table 1 displays their truth tables and their linearizations.

Let C be a circuit with m wires and n fan-in 2 gates. We can use linearization of the logic gates to rewrite the circuit as a set of constraints on the output of an affine map.

Table 1. Linearization of logic gates with inputs a, b and output c. We omit the 6 remaining gates, which depend on at most one input and are not used in circuits.

AND

a	b	c
0	0	0
0	1	0
1	0	0
1	1	1

$a + b - 2c \in \{0, 1\}$

OR

a	b	c
0	0	0
0	1	1
1	0	1
1	1	1

$\bar{a} + \bar{b} - 2\bar{c} \in \{0, 1\}$

XOR

a	b	c
0	0	0
0	1	1
1	0	1
1	1	0

$a + b + c \in \{0, 2\}$

NAND

a	b	c
0	0	1
0	1	1
1	0	1
1	1	0

$a + b - 2\bar{c} \in \{0, 1\}$

NOR

a	b	c
0	0	1
0	1	0
1	0	0
1	1	0

$\bar{a} + \bar{b} - 2c \in \{0, 1\}$

XNOR

a	b	c
0	0	1
0	1	0
1	0	0
1	1	1

$a + b + \bar{c} \in \{0, 2\}$

$\bar{a} \wedge b$

a	b	c
0	0	0
0	1	1
1	0	0
1	1	0

$\bar{a} + b - 2c \in \{0, 1\}$

$\overline{\bar{a} \wedge b}$

a	b	c
0	0	1
0	1	0
1	0	1
1	1	1

$\bar{a} + b - 2\bar{c} \in \{0, 1\}$

$a \wedge \bar{b}$

a	b	c
0	0	0
0	1	0
1	0	1
1	1	0

$a + \bar{b} - 2c \in \{0, 1\}$

$\overline{a \wedge \bar{b}}$

a	b	c
0	0	1
0	1	1
1	0	0
1	1	1

$a + \bar{b} - 2\bar{c} \in \{0, 1\}$

Theorem 1. *For any circuit C with m wires and n fan-in 2 gates for a total size of $d = m + n$, there exists a matrix $V \in \mathbb{Z}^{m \times d}$ and a vector $\boldsymbol{b} \in \mathbb{Z}^d$ such that C is satisfiable if and only if there is a vector $\boldsymbol{a} \in \mathbb{Z}^m$ satisfying $\boldsymbol{a}V + \boldsymbol{b} \in \{0, 2\}^d$.*

The matrix V and the vector $\boldsymbol{b}$ can be constructed such that $\boldsymbol{a}V + \boldsymbol{b} \in \{0, 2\}^d$ implies $\boldsymbol{a} \in \{0, 1\}^m$ and $a_1, \ldots, a_m$ corresponds to the values on the wires in a satisfying assignment for C with the first ℓ bits being the input wires.

Proof. We represent an assignment to the wires as a vector $\boldsymbol{a} \in \mathbb{Z}^m$. The assignment is a satisfying witness for the circuit if and only if the entries belong to $\{0, 1\}$, the entries respect all gates, and the output wire is 1.

It is easy to impose the condition $\boldsymbol{a} \in \{0, 1\}^m$ by requiring $\boldsymbol{a}(2I) \in \{0, 2\}^m$. (Alternatively, whenever $a_i \in \{0, 1\}$ is clear from the context, for instance for the public inputs $a_1, \ldots, a_{\ell_u}$, this check can be omitted.)

Since $\bar{a} = 1 - a$, $\bar{b} = 1 - b$ and $\bar{c} = 1 - c$ and after scaling some of the gate equations from Table 1 by a factor 2, we can write all gate equations in the form $\alpha a + \beta b + \gamma c + \delta \in \{0, 2\}$. We want the circuit output wire c_{out} to have value 1. We do that by adding the condition $3\bar{c}_{\text{out}}$ to the linearization of the output gate, since if $c_{\text{out}} = 0$ this adds 3 to the linear equation and brings us outside $\{0, 2\}$ regardless of the type of logic gate.

Define $G \in \mathbb{Z}^{m \times n}$ and $\boldsymbol{\delta} \in \mathbb{Z}^n$ such that $\boldsymbol{a}G + \boldsymbol{\delta} \in \{0,2\}^n$ corresponds to the linearization of the gates as described above, and let

$$V = [\, 2I \mid G \,] \quad \text{and} \quad \boldsymbol{b} = (\, \mathbf{0} \mid \boldsymbol{\delta} \,).$$

The existence of $\boldsymbol{a}$ such that

$$\boldsymbol{a}V + \boldsymbol{b} \in \{0,2\}^d$$

is equivalent to a satisfying assignment to the wires in the circuit.   □

Note that V and $\boldsymbol{b}$ as we constructed them have some additional properties. The matrix V is sparse, since it only has $m+3n$ non-zero entries. The row vectors of V and $\boldsymbol{b}$ are all linearly independent. Furthermore, all entries in V and $\boldsymbol{b}$ are small integers. The small size of the integers gives us the following corollary.

Corollary 1. *For any circuit C with m wires and n fan-in 2 gates and for any $p \geq 8$ there exist a matrix $V \in \mathbb{Z}_p^{m \times d}$ (with $d = m + n$) and a vector $\boldsymbol{b} \in \mathbb{Z}_p^d$ (giving us $m+1$ linearly independent row vectors) such that C is satisfiable if and only if there exists a vector $\boldsymbol{a} \in \mathbb{Z}_p^m$ satisfying $\boldsymbol{a}V + \boldsymbol{b} \in \{0,2\}^d$. Furthermore, if $\boldsymbol{a}V + \boldsymbol{b} \in \{0,2\}^d$ then $\boldsymbol{a} \in \{0,1\}^m$ and $C(a_1,\ldots,a_\ell) = 1$.*

Relation to closest vector problem. There is an interesting connection between our construction of affine map constraints and the closest vector problem for integer lattices using the max-norm ℓ_∞. Consider a circuit made just from NAND-gates; then the affine map $\boldsymbol{a}V + \boldsymbol{b}$ constructed in the proof of Theorem 1 cannot take the value 1 for any index $i = 1,\ldots,d$, which means the circuit is satisfiable if and only if $\boldsymbol{a}V + \boldsymbol{b} - 1 \in \{-1,0,1\}^d$. This is equivalent to saying that the lattice generated by the rows of V has a vector $\boldsymbol{a}V$ with distance at most 1 from the target vector $\boldsymbol{t} = 1 - \boldsymbol{b}$, i.e., $\|\boldsymbol{a}V - \boldsymbol{t}\|_\infty \leq 1$, if and only if the circuit is satisfiable. Our construction therefore gives a very direct reduction of the closest vector problem in integer lattices to circuit satisfiability. The NP-hardness of the closest (nearest) vector problem was first demonstrated by van Emde Boas [vEB81] but using a more complicated reduction that relied on the partition problem.

2.2   The NP-completeness of Square Span Programs

We will now connect affine maps to square span programs, which gives a reduction of square span programs to circuit satisfiability.

Corollary 1 can be reformulated to say that, for any circuit C and $p \geq 8$, there exist V and $\boldsymbol{b}$ such that C is satisfiable if and only there exists $\boldsymbol{a} \in \mathbb{Z}_p^m$ satisfying

$$(\boldsymbol{a}V + \boldsymbol{b}) \circ (\boldsymbol{a}V + \boldsymbol{b} - 2) = 0,$$

where $\circ$ denotes the Hadamard product (entry-wise multiplication). We can rewrite this condition as

$$(\boldsymbol{a}V + \boldsymbol{b} - 1) \circ (\boldsymbol{a}V + \boldsymbol{b} - 1) = 1.$$

Let $r_1, \ldots, r_d$ be d distinct elements of $\mathbb{Z}_p$ for a prime $p \geq \max(d, 8)$. Define $v_0(x), v_1(x), \ldots, v_m(x)$ as the degree $d-1$ polynomials satisfying

$$v_0(r_j) = b_j - 1 \quad \text{and} \quad v_i(r_j) = V_{i,j}.$$

We can now reformulate Corollary 1 again. The circuit C is satisfiable if and only if there exists $\boldsymbol{a} \in \mathbb{Z}_p^m$ such that for all r_j

$$\left(v_0(r_j) + \sum_{i=1}^{m} a_i v_i(r_j) \right)^2 = 1.$$

Since the evaluations in $r_1, \ldots, r_d$ uniquely determine the polynomial $v(x) = v_0(x) + \sum_{i=1}^{m} a_i v_i(x)$ we can rewrite the condition as

$$\left(v_0(x) + \sum_{i=1}^{m} a_i v_i(x) \right)^2 \equiv 1 \mod \prod_{j=1}^{d}(x - r_j).$$

Theorem 2. *A circuit C with m wires and n fan-in 2 gates has for any prime $p \geq \max(n, 8)$ a square span program of size m and degree $d = m+n$ that verifies it over $\mathbb{Z}_p$.*

Proof. From the discussion above, we see that for any circuit C with m wires and n gates there exists polynomials $v_0(x), v_1(x), \ldots, v_m(x)$ and distinct roots $r_1, \ldots, r_d$ such that C is satisfiable if and only if

$$\prod_{j=1}^{d}(x - r_j) \quad \text{divides} \quad \left(v_0(x) + \sum_{i=1}^{m} a_i v_i(x) \right)^2 - 1.$$

Define $t(x) = \prod_{j=1}^{d}(x - r_j)$ to get an SSP $Q = (v_0(x), v_1(x), \ldots, v_m(x), t(x))$ that verifies C over $\mathbb{Z}_p$. $\qquad \square$

2.3 Example

As a small example of the process of generating a square span program, consider a circuit consisting of a single XOR-gate $a_3 = a_1 \oplus a_2$ (here $\ell = \ell_u + \ell_w = 2$ with $\ell_u = 0$ and $\ell_w = 2$). To guarantee $a_1, a_2, a_3 \in \{0, 1\}$ and the XOR-gate is respected we use the constraints $2a_i \in \{0, 2\}$ and $a_1 + a_2 + a_3 \in \{0, 2\}$. The output should be $a_3 = 1$, which we represent with the constraint $3\bar{a}_3 = 3(1 - a_3) = 0$. We add the latter constraint to the output wire's equation to get the combined $a_1 + a_2 - 2a_3 + 3 \in \{0, 2\}$, which at the same time guarantees $a_3 = a_1 \oplus a_2$ and $a_3 = 1$. We can represent the constraints as

$$\boldsymbol{a}V + \boldsymbol{b} = (a_1, a_2, a_3) \begin{pmatrix} 2 & 0 & 0 & 1 \\ 0 & 2 & 0 & 1 \\ 0 & 0 & 2 & -2 \end{pmatrix} + (0, 0, 0, 3) \in \{0, 2\}^4.$$

The satisfiability of the circuit can therefore be represented by 4 quadratic equations

$$(2a_1 - 1)^2 = 1 \quad (2a_2 - 1)^2 = 1 \quad (2a_3 - 1)^2 = 1 \quad (a_1 + a_2 - 2a_3 + 2)^2 = 1$$

corresponding to $(aV + b - 1) \circ (aV + b - 1) = 1$.

To get a square span program, let $p \geq 8$ be a prime and r_1, r_2, r_3, r_4 be four distinct elements in $\mathbb{Z}_p$. Pick degree 3 polynomials $v_0(x), v_1(x), v_2(x), v_3(x)$ such that

$$(v_0(r_1), v_0(r_2), v_0(r_3), v_0(r_4)) = b - 1 = (-1, -1, -1, 2)$$

and

$$\begin{pmatrix} v_1(r_1) & v_1(r_2) & v_1(r_3) & v_1(r_4) \\ v_2(r_1) & v_2(r_2) & v_2(r_3) & v_2(r_4) \\ v_3(r_1) & v_3(r_2) & v_3(r_3) & v_3(r_4) \end{pmatrix} = V = \begin{pmatrix} 2 & 0 & 0 & 1 \\ 0 & 2 & 0 & 1 \\ 0 & 0 & 2 & -2 \end{pmatrix}.$$

Let $t(x) = (x - r_1)(x - r_2)(x - r_3)(x - r_4)$ to get a square span program $(v_0(x), v_1(x), v_2(x), v_3(x), t(x))$ for the circuit such that

$$t(x) \quad \text{divides} \quad \left(v_0(x) + a_1 v_1(x) + a_2 v_2(x) + a_3 v_3(x) \right)^2 - 1$$

if and only if a_1, a_2, a_3 satisfy the circuit, i.e., $a_1, a_2 \in \{0, 1\}$, $a_3 = 1$ and $a_3 = a_1 \oplus a_2$.

2.4   Comparison to Quadratic Span Programs

Square span programs can be seen as a simplification of quadratic span programs. Below we recall the definition of quadratic span programs given by Gennaro, Gentry, Parno and Raykova [GGPR13].

Definition 2. *A quadratic span program over a field $\mathbb{F}$ contains two sets of polynomials $\mathcal{V} = \{v'_0(x), \ldots, v_m(x)\}$ and $\mathcal{W} = \{w'_0(x), \ldots, w_m(x)\}$ and a target polynomial $t(x)$. It also contains a partition of the indices $\mathcal{I} = \{1, \ldots, m\}$ into $\mathcal{I} = \mathcal{I}_{\text{labeled}} \cup \mathcal{I}_{\text{free}}$ and a further partition $\mathcal{I}_{\text{labeled}} = \cup_{i=1, j=0}^{\ell, 1} \mathcal{I}_{i,j}$.*

For input[2] $y \in \{0, 1\}^\ell$, let $\mathcal{I}_y = \mathcal{I}_{\text{free}} \cup_{i=1}^{\ell} \mathcal{I}_{i, y_i}$ be the set of indices that "belong" to y. The quadratic span program accepts an input $y \in \{0, 1\}^\ell$ if and only if there exist $a_i, b_i \in \mathbb{F}$ such that

$$t(x) \quad \text{divides} \quad \left(v'_0(x) + \sum_{i \in \mathcal{I}_y} a_i v_i(x) \right) \cdot \left(w'_0(x) + \sum_{i \in \mathcal{I}_y} b_i w_i(x) \right).$$

We say the quadratic span program verifies a boolean function $f : \{0, 1\}^\ell \to \{0, 1\}$ if it accepts exactly those inputs y where $f(y) = 1$. We say the size of the quadratic span program is m and the degree is $\deg(t(x))$.

---

[2] In the rest of the paper, we will be using inputs of the form $y = (u, w)$ where u of size ℓ_u is considered public and w is considered private.

Table 2. Costs compared with prior work (ℓ input wires, out of which ℓ_u are public, m wires in total and n gates). In a circuit with fan-in 2 gates $m \leq 2n + 1$, so we get rough bounds of size $2n$ and degree $3n$ when computed as a function of the number of gates n only (ignoring ℓ_u).

Size and degree of Span Programs

	Size	Degree
Quadratic span programs [GGPR13]	$36n$	$130n$
Quadratic span programs (Lipmaa) [Lip13]	$14n - 14\ell - 2$	$11n - 12\ell - 2$
Square span programs	m	$m + n - \ell_u$

A square span program uses the simpler condition

$$t(x) \quad \text{divides} \quad \left(v_0(x) + \sum_{i=1}^{m} a_i v_i(x) \right)^2 - 1,$$

which is equivalent to

$$t(x) \quad \text{divides} \quad \left(v_0(x) + 1 + \sum_{i=1}^{m} a_i v_i(x) \right) \cdot \left(v_0(x) - 1 + \sum_{i=1}^{m} a_i v_i(x) \right).$$

A square span program can therefore be seen as a particularly simple type of quadratic span program where $w_0'(x) = v_0'(x) - 2$ and $w_i(x) = v_i(x)$ and $a_i = b_i$. Furthermore, $\mathcal{I}_{\text{labeled}} = \{1, \ldots, \ell\}$ with $\mathcal{I}_{i,y_i} = \{i\}$ and $\mathcal{I}_{i,\bar{y}_i} = \emptyset$, and $\mathcal{I}_{\text{free}} = \{\ell + 1, \ldots, m\}$.

The compilation of a circuit into a quadratic span programs in [GGPR13] has a significant overhead. For a circuit with ℓ input wires and m wires in total and n gates, the size of the resulting quadratic span program is $36n$ and the degree is $130n$. Lipmaa [Lip13] gave a class of more efficient quadratic span programs. Included in this class is a quadratic span program of size $14n - 14\ell - 2$ and degree $11n - 12\ell - 2$. In comparison with these works our (square) quadratic span programs are much more compact with size $m - \ell_u$ and degree $m + n - \ell_u$ (assuming the verifier checks its inputs are all in $\{0, 1\}$.) These costs are summarised in Table 2.

A further advantage compared to the previous works is that we consider all types of logic gates, whereas they only consider NAND, AND and OR gates. We would expect that their constructions can be generalized to handle other logic gates but do not know whether this would increase the cost.

Remark. All three results prove that a circuit—fixed when the quadratic span program is generated—is satisfied for public input u and private input w. Universal circuits allow using a single program for all n' gate circuits at the cost of $n = n' \cdot 19 \log n'$ [Val76].

3   Succinct Non-interactive Arguments of Knowledge

We will now use square span programs to construct succinct non-interactive zero-knowledge arguments of knowledge using bilinear groups.

Notation. Given two functions $f, g : \mathbb{N} \to [0, 1]$ we write $f(\lambda) \approx g(\lambda)$ when $|f(\lambda) - g(\lambda)| = \lambda^{-\omega(1)}$. We say that f is *negligible* when $f(\lambda) \approx 0$ and that f is *overwhelming* when $f(\lambda) \approx 1$.

We write $y = A(x; r)$ when the algorithm A on input x and randomness r, outputs y. We write $y \leftarrow A(x)$ for the process of picking randomness r at random and setting $y = A(x; r)$. We also write $y \leftarrow S$ for sampling y uniformly at random from the set S. We will assume it is possible to sample uniformly at random from sets such as $\mathbb{Z}_p$.

Following Abe and Fehr [AF07] we write $(y; z) \leftarrow (\mathcal{A} \parallel \mathcal{X}_\mathcal{A})(x)$ when $\mathcal{A}$ on input x outputs y and $\mathcal{X}_\mathcal{A}$ on the same input (including random coins) outputs z.

3.1   Non-interactive Zero-Knowledge Arguments of Knowledge

Let $\{\mathcal{R}_\lambda\}_{\lambda \in \mathbb{N}}$ be a sequence of families of efficiently decidable binary relations R. For pairs $(u, w) \in R$ we call u the statement and w the witness. A non-interactive argument for $\{\mathcal{R}_\lambda\}_{\lambda \in \mathbb{N}}$ is a quadruple of efficient algorithms (Setup, Prove, Vfy, Sim) working as follows:

$(\sigma, \tau) \leftarrow \text{Setup}(1^\lambda, R)$: the setup algorithm takes as input a security parameter λ and a relation $R \in \mathcal{R}_\lambda$ and returns a common reference string σ and a simulation trapdoor τ for the relation R.

$\pi \leftarrow \text{Prove}(\sigma, u, w)$: the prover algorithm takes as input a common reference string σ for a relation R and $(u, w) \in R$ and returns an argument π.

$0/1 \leftarrow \text{Vfy}(\sigma, u, \pi)$: the verification algorithm takes as input a common reference string, a statement u and an argument π and returns 0 (reject) or 1 (accept).

$\pi \leftarrow \text{Sim}(\tau, u)$: the simulator takes as input a simulation trapdoor and a statement u and returns an argument π.

Definition 3. *We say* (Setup, Prove, Vfy, Sim) *is a perfect non-interactive zero-knowledge argument of knowledge for* $\{\mathcal{R}_\lambda\}_{\lambda \in \mathbb{N}}$ *if it has perfect completeness, perfect zero-knowledge and computational knowledge soundness as defined below.*

PERFECT COMPLETENESS. Completeness says that, given any true statement, an honest prover should be able to convince an honest verifier. For all $\lambda \in \mathbb{N}$, $R \in \mathcal{R}_\lambda$, $(u, w) \in R$

$$\Pr\left[(\sigma, \tau) \leftarrow \text{Setup}(1^\lambda, R); \pi \leftarrow \text{Prove}(\sigma, u, w) : \text{Vfy}(\sigma, u, \pi) = 1\right] = 1.$$

PERFECT ZERO-KNOWLEDGE. An argument is zero-knowledge if it does not leak any information besides the truth of the statement. We say (Setup, Prove, Vfy,

Sim) is perfect zero-knowledge if for all $\lambda \in \mathbb{N}, R \in \mathcal{R}_\lambda, (u, w) \in R$ and all adversaries $\mathcal{A}$, we have

$$\Pr\left[(\sigma, \tau) \leftarrow \text{Setup}(1^\lambda, R); \pi \leftarrow \text{Prove}(\sigma, u, w) : \mathcal{A}(\sigma, \tau, \pi) = 1\right]$$
$$= \Pr\left[(\sigma, \tau) \leftarrow \text{Setup}(1^\lambda, R); \pi \leftarrow \text{Sim}(\tau, u) : \mathcal{A}(\sigma, \tau, \pi) = 1\right].$$

COMPUTATIONAL KNOWLEDGE SOUNDNESS. We call (Setup, Prove, Vfy, Sim) an argument of knowledge if there is an extractor that can compute a witness whenever the adversary produces a valid argument. The extractor gets full access to the adversary's state, including any random coins.

Formally, we require that, for all sequences $(R_\lambda)_{\lambda \in \mathbb{N}}$ of polynomially bounded relations in $\{\mathcal{R}_\lambda\}_{\lambda \in \mathbb{N}}$ and non-uniform polynomial time adversaries $\mathcal{A}$, there exists a non-uniform polynomial time extractor $\mathcal{X}_\mathcal{A}$ such that

$$\Pr\left[\begin{array}{cc} (\sigma, \tau) \leftarrow \text{Setup}(1^\lambda, R_\lambda) & (u, w) \notin R_\lambda \\ ((u, \pi); w) \leftarrow (\mathcal{A} \parallel \mathcal{X}_\mathcal{A})(\sigma) & : \text{Vfy}(\sigma, u, \pi) = 1 \end{array}\right] \approx 0.$$

Remark. Our notion of knowledge soundness guarantees security against an *adaptive* adversary, cf. [AF07], that chooses the instance u depending on the CRS σ. However, to get adaptive security for circuit satisfiability, $\mathcal{R}_\lambda$ has to be universal, i.e., it has to check that a circuit u is satisfiable. For performance reasons, this is usually not what one wants, and adaptive soundness for a more restrictive $\mathcal{R}_\lambda$ is preferable. See Lipmaa [Lip14] for how to achieve adaptive soundness for some NP-complete languages, not including circuit satisfiability, while avoiding universal circuits.

3.2 Bilinear Groups

Let $\mathcal{G}$ be a bilinear group generator that, on security parameter λ, returns $(p, \mathbb{G}, \hat{\mathbb{G}}, \mathbb{G}_T, e) \leftarrow \mathcal{G}(1^\lambda)$ with the following properties:

- $\mathbb{G}, \hat{\mathbb{G}}, \mathbb{G}_T$ are groups of prime order p;
- $e : \mathbb{G} \times \hat{\mathbb{G}} \to \mathbb{G}_T$ is a bilinear map, that is, for all $U \in \mathbb{G}$, $V \in \hat{\mathbb{G}}$, $a, b \in \mathbb{Z}$, we have $e(U^a, V^b) = e(U, V)^{ab}$;
- if G is a generator for $\mathbb{G}$ and $\hat{G}$ is a generator for $\hat{\mathbb{G}}$ then $e(G, \hat{G})$ is a generator for $\mathbb{G}_T$; and
- there are efficient algorithms for computing group operations, evaluating the bilinear map, deciding membership of the groups, deciding equality of group elements and sampling generators of the groups.

There are many ways to set up bilinear groups both as symmetric bilinear groups where $\mathbb{G} = \hat{\mathbb{G}}$ and as asymmetric bilinear groups where $\mathbb{G} \neq \hat{\mathbb{G}}$. Our construction works for both symmetric and asymmetric bilinear groups. Currently, asymmetric bilinear groups are more efficient and therefore the most appropriate choice in practice [GPS08].

THE q-POWER KNOWLEDGE OF EXPONENT ASSUMPTION. The knowledge of exponent assumption (KEA) introduced by Damgård [Dam91] says that given $G, G' = G^\alpha$ it is infeasible to create V, V' such that $V' = V^\alpha$ without knowing a such that $V = G^a$ and $V' = G'^a$. Bellare and Palacio [BP04] extended this to the KEA3 assumption, which says that given G, G^s, G', G'^s it is infeasible to create $V, V' = V^\alpha$ without knowing a_0, a_1 such that $V = G^{a_0}(G^s)^{a_1}$. This assumption has been used also in symmetric bilinear groups by Abe and Fehr [AF07] who called it the extended knowledge of exponent assumption.

The q-power knowledge of exponent assumption is a generalization of these assumptions in bilinear groups. It says that given $(G, \hat{G}, G^s, \hat{G}^s, \ldots, G^{s^q}, \hat{G}^{s^q})$ it is infeasible to create $V, \hat{V}$ such that $e(V, \hat{G}) = e(G, \hat{V})$ without knowing $a_0, \ldots, a_q$ such that $V = \prod_{i=0}^{q}(G^{s^i})^{a_i}$. The q-power knowledge of exponent assumption was introduced in [Gro10] for symmetric bilinear groups using $\hat{G} = G^\alpha$ with α chosen at random. Here we adapt it with minor modifications to the general setting where it may be the case that $\mathbb{G} \neq \hat{\mathbb{G}}$ and $G, \hat{G}$ belong to different groups.

Definition 4 (q-PKE). *The $q(\lambda)$-power knowledge of exponent assumption holds relative to $\mathcal{G}$ for the class $\mathcal{Z}$ of auxiliary input generators if, for every non-uniform polynomial time auxiliary input generator $Z \in \mathcal{Z}$ and non-uniform polynomial time adversary $\mathcal{A}$, there exists a non-uniform polynomial time extractor $\mathcal{X}_\mathcal{A}$ such that*

$$\Pr\left[\begin{array}{l} gk := (p, \mathbb{G}, \hat{\mathbb{G}}, \mathbb{G}_T, e) \leftarrow \mathcal{G}(1^\lambda); G \leftarrow \mathbb{G}^* \\ s \leftarrow \mathbb{Z}_p^*; z \leftarrow Z(gk, G, \ldots, G^{s^q}); \hat{G} \leftarrow \hat{\mathbb{G}}^* \\ (V, \hat{V}; a_0, \ldots, a_q) \leftarrow (\mathcal{A} \parallel \mathcal{X}_\mathcal{A})(gk, G, \hat{G}, \ldots, G^{s^q}, \hat{G}^{s^q}, z) : \\ e(V, \hat{G}) = e(G, \hat{V}) \ \wedge \ V \neq G^{\sum_{i=0}^{q} a_i s^i} \end{array}\right] \approx 0.$$

An adaptation of the proof in Groth [Gro10] shows that the q-PKE assumption holds in the generic bilinear group model.

As demonstrated by Bitansky, Canetti, Paneth and Rosen [BCPR13], if indistinguishability obfuscators [BGI+12, GGH+13] exist, then there are auxiliary input generators for which the q-PKE assumption does not hold. However, their counterexample is specifically tailored to make extraction difficult and, as they explain, the q-PKE assumption may hold for "benign" auxiliary input generators. We will later use auxiliary input generators that generate group elements in $\mathbb{G}$ and $\hat{\mathbb{G}}$ in a specific manner according to the relations R_λ and we will conjecture that such auxiliary input generators are benign and that the q-PKE assumption holds with respect to them.

THE q-POWER DIFFIE-HELLMAN ASSUMPTION. The q-power Diffie-Hellman assumption says given $(G, \hat{G}, \ldots, G^{s^q}, \hat{G}^{s^q}, G^{s^{q+2}}, \hat{G}^{s^{q+2}}, \ldots, G^{s^{2q}}, \hat{G}^{s^{2q}})$ it is hard to compute the missing element $G^{s^{q+1}}$.

Definition 5 (q-PDH). *The $q(\lambda)$-power Diffie-Hellman assumption holds relative to $\mathcal{G}$ if for all non-uniform probabilistic polynomial time adversaries $\mathcal{A}$*

$$
\Pr\left[
\begin{array}{l}
gk := (p, \mathbb{G}, \hat{\mathbb{G}}, \mathbb{G}_T, e) \leftarrow \mathcal{G}(1^\lambda); G \leftarrow \mathbb{G}^*; \hat{G} \leftarrow \hat{\mathbb{G}}^*; s \leftarrow \mathbb{Z}_p^* \\
Y \leftarrow \mathcal{A}(gk, G, \hat{G}, \ldots, G^{s^q}, \hat{G}^{s^q}, G^{s^{q+2}}, \hat{G}^{s^{q+2}}, \ldots, G^{s^{2q}}, \hat{G}^{s^{2q}}) : \\
Y = G^{s^{q+1}}
\end{array}
\right] \approx 0.
$$

An adaptation of the proof in Groth [Gro10] shows that the q-PDH assumption holds in the generic bilinear group model.

THE q-TARGET GROUP STRONG DIFFIE-HELLMAN ASSUMPTION. We adapt the strong Diffie-Hellman assumption [BB08] in the target group [PHGR13] to the asymmetric setting. It says that given $(G, \hat{G}, \ldots, G^{s^q}, \hat{G}^{s^q})$ it is hard to find an $r \in \mathbb{Z}_p$ and compute $e(G, \hat{G})^{\frac{1}{s-r}}$.

Definition 6 (q-TSDH). *The $q(\lambda)$-target group strong Diffie-Hellman assumption holds relative to $\mathcal{G}$ if for all non-uniform probabilistic polynomial time adversaries $\mathcal{A}$*

$$
\Pr\left[
\begin{array}{l}
(p, \mathbb{G}, \hat{\mathbb{G}}, \mathbb{G}_T, e) \leftarrow \mathcal{G}(1^\lambda); G \leftarrow \mathbb{G}^*; \hat{G} \leftarrow \hat{\mathbb{G}}^*; s \leftarrow \mathbb{Z}_p^* \\
(r, Y) \leftarrow \mathcal{A}(p, \mathbb{G}, \hat{\mathbb{G}}, \mathbb{G}_T, e, G, \hat{G}, \ldots, G^{s^q}, \hat{G}^{s^q}) : \\
r \in \mathbb{Z}_p \setminus \{s\} \ \wedge \ Y = e(G, \hat{G})^{\frac{1}{s-r}}
\end{array}
\right] \approx 0.
$$

An adaptation of the proof in Boneh and Boyen [BB08] shows that the q-TSDH assumption holds in the generic bilinear group model.

3.3   Succinct Perfect NIZK Arguments

We will now construct succinct and perfect NIZK arguments of knowledge for any functions ℓ_u, ℓ_w and families $\{\mathcal{R}\}_\lambda$ of relations R of pairs $(u, w) \in \{0, 1\}^{\ell_u(\lambda)} \times \{0, 1\}^{\ell_w(\lambda)}$ that can be computed by polynomial size circuits with $m(\lambda)$ wires and $n(\lambda)$ gates for a total size of $d(\lambda) = m(\lambda) + n(\lambda)$.

$(\sigma, \tau) \leftarrow \text{Setup}(1^\lambda, R)$: Run $gk := (p, \mathbb{G}, \hat{\mathbb{G}}, \mathbb{G}_T, e) \leftarrow \mathcal{G}(1^\lambda)$. Parse R as a boolean circuit $C_R : \{0, 1\}^{\ell_u} \times \{0, 1\}^{\ell_w} \to \{0, 1\}$. Generate a square span program $Q = (v_0(x), \ldots, v_m(x), t(x))$ that verifies C_R over $\mathbb{Z}_p$. Pick $G \leftarrow \mathbb{G}^*$ and $\hat{G}, \tilde{G} \leftarrow \hat{\mathbb{G}}^*$ and $\beta, s \leftarrow \mathbb{Z}_p^*$ such that $t(s) \neq 0$. Return

$$
\begin{aligned}
\sigma &= (gk, G, \hat{G}, \ldots, G^{s^d}, \hat{G}^{s^d}, \{G^{\beta v_i(s)}\}_{i > \ell_u}, G^{\beta t(s)}, \tilde{G}, \tilde{G}^\beta, Q) \\
\tau &= (\sigma, \beta, s).
\end{aligned}
$$

$\pi \leftarrow \text{Prove}(\sigma, u, w)$: Parse u as $(a_1, \ldots, a_{\ell_u}) \in \{0, 1\}^{\ell_u}$ and use w to compute $a_{\ell_u+1}, \ldots, a_m$ such that $t(x)$ divides $\left(v_0(x) + \sum_{i=1}^m a_i v_i(x)\right)^2 - 1$. Pick $\delta \leftarrow \mathbb{Z}_p$ and let

$$
h(x) = \frac{\left(v_0(x) + \sum_{i=1}^m a_i v_i(x) + \delta t(x)\right)^2 - 1}{t(x)}.
$$

Use linear combinations of the elements in σ to compute

$$H = G^{h(s)} \qquad\qquad V_w = G^{\sum_{i>\ell_u}^m a_i v_i(s) + \delta t(s)}$$

$$B_w = G^{\beta\left(\sum_{i>\ell_u}^m a_i v_i(s) + \delta t(s)\right)} \qquad\qquad \hat{V} = \hat{G}^{v_0(s) + \sum_{i=1}^m a_i v_i(s) + \delta t(s)}$$

and return $\pi = (H, V_w, B_w, \hat{V})$.

$0/1 \leftarrow \mathrm{Vfy}(\sigma, u, \pi)$: Parse u as $(a_1, \ldots, a_{\ell_u}) \in \{0,1\}^{\ell_u}$ and π as $(H, V_w, B_w, \hat{V}) \in \mathbb{G}^3 \times \hat{\mathbb{G}}$. Compute $V = G^{v_0(s) + \sum_{i=1}^{\ell_u} a_i v_i(s)} V_w$ and return 1 if and only if

$$e(V, \hat{G}) = e(G, \hat{V}) \quad e(H, \hat{G}^{t(s)}) = e(V, \hat{V})e(G, \hat{G})^{-1} \quad e(V_w, \tilde{G}^\beta) = e(B_w, \tilde{G}).$$

$\pi \leftarrow \mathrm{Sim}(\tau, u)$: Parse u as $(a_1, \ldots, a_{\ell_u}) \in \{0,1\}^{\ell_u}$ and pick $\delta_w \leftarrow \mathbb{Z}_p$ at random. Let

$$h = \frac{\left(v_0(s) + \sum_{i=1}^{\ell_u} a_i v_i(s) + \delta_w\right)^2 - 1}{t(s)}$$

and return $\pi = (G^h, G^{\delta_w}, G^{\beta\delta_w}, \hat{G}^{v_0(s) + \sum_{i=1}^{\ell_u} a_i v_i(s) + \delta_w})$.

Let $\mathcal{Z}$ be a family of non-uniform polynomial time auxiliary input generators Z such that each of them corresponds to sequences of relations $(R_\lambda)_{\lambda \in \mathbb{N}}$ in a family of relations $\{\mathcal{R}_\lambda\}_{\lambda \in \mathbb{N}}$. They work such that Z corresponding to $(R_\lambda)_{\lambda \in \mathbb{N}}$ on input $(p, \mathbb{G}, \hat{\mathbb{G}}, \mathbb{G}_T, e, G, \ldots, G^{s^q})$ generates the final part of the common reference string, i.e., returns $z = (\{G^{\beta v_i(s)}\}_{i>\ell_u}, G^{\beta t(s)}, \tilde{G}, \tilde{G}^\beta, Q)$.

Theorem 3. *The construction above is a perfect NIZK argument for the family of relations $\{\mathcal{R}_\lambda\}_{\lambda \in \mathbb{N}}$ bounded by $d(\lambda)$ with computational knowledge soundness if the $d(\lambda)$-PKE, $d(\lambda)$-PDH and $d(\lambda)$-SDH assumptions hold relative to $\mathcal{G}$ and the family of auxiliary input generator $\mathcal{Z}$ defined above.*

Proof. Perfect completeness follows by direct verification.

Perfect zero-knowledge follows from observing that both a real argument and a simulated argument have a uniformly random V_w because $t(s) \neq 0$ and δ, δ_w are chosen uniformly at random. Once V_w has been fixed, the verification equations uniquely determine $B_w, \hat{V}$ and H. This means that for any $(u, w) \in R$ both the real arguments and the simulated arguments are chosen uniformly at random such that the verification equations will be satisfied.

We now describe the witness-extractor for computational knowledge soundness. The setup algorithm first generates a bilinear group $(p, \mathbb{G}, \hat{\mathbb{G}}, \mathbb{G}_T, e) \leftarrow \mathcal{G}(1^\lambda)$ and picks $G \leftarrow \mathbb{G}^*$ and $s \leftarrow \mathbb{Z}_p^*$, which are used to compute $G, \ldots, G^{s^d}$. This is exactly like the input given to the auxiliary input generator in a d-PKE challenge. The setup algorithm now generates a square span program Q over $\mathbb{Z}_p$ for the relation R_λ and elements $\{G^{\beta v_i(s)}\}_{i>\ell_u}$ and $\tilde{G}, \tilde{G}^\beta$. We can consider this as part of the auxiliary input z that Z outputs in the d-PKE definition. More precisely, let $\mathcal{A}'$ be the d-PKE adversary that, on $(p, \mathbb{G}, \hat{\mathbb{G}}, \mathbb{G}_T, e, G, \hat{G}, \ldots, G^{s^d}, \hat{G}^{s^d})$ and auxiliary input $z = (\{G^{\beta v_i(s)}\}_{i>\ell_u}, G^{\beta t(s)}, \tilde{G}, \tilde{G}^\beta, Q)$ runs $(u, H, V_w, B_w, \hat{V}) \leftarrow \mathcal{A}(\sigma)$ with $\sigma = (p, \ldots, \hat{G}^{s^d})$ and returns $(V, \hat{V})$ where

$V = G^{v_0(s)+\sum_{i=1}^{\ell_u} a_i v_i(s)} V_w$ when $u = (a_1, \ldots, a_{\ell_u}) \in \{0,1\}^{\ell_u}$. Let $\mathcal{X}_{\mathcal{A}'}$ be the corresponding extractor according to the d-PKE assumption that returns $c_0, \ldots, c_d$ such that $V = G^{\sum_{i=0}^{d} c_i s^i}$ when $e(V, \hat{G}) = e(G, \hat{V})$. Our witness-extractor $\mathcal{X}_{\mathcal{A}}$ given σ runs $(V, \hat{V}; c_0, \ldots, c_d) \leftarrow (\mathcal{A}' \parallel \mathcal{X}_{\mathcal{A}'})(p, \mathbb{G}, \hat{\mathbb{G}}, \mathbb{G}_T, e, G, \hat{G}, \ldots, G^{s^d}, \hat{G}^{s^d}, z)$, which defines a polynomial $\sum_{i=0}^{d} c_i x^i$. Define $\delta = c_d$ to get a degree $d-1$ polynomial $v(x) = \sum_{i=0}^{d} c_i x^i - \delta t(x)$. If it is possible to write the polynomial on the form $v(x) = v_0(x) + \sum_{i=1}^{m} a_i v_i(x)$ such that $(a_1, \ldots, a_m) \in \{0,1\}^m$ is a satisfying assignment for the circuit C_R with $u = (a_1, \ldots, a_{\ell_u})$ then the extractor returns $w = (a_{\ell_u+1}, \ldots, a_{\ell_u+\ell_w})$.

We will now show that with all but negligible probability the extracted polynomial $v(x)$ does indeed provide a valid witness $w \in \{0,1\}^{\ell_w}$ such that $(u, w) \in \mathcal{R}_\lambda$. Let Q be the square span program $(v_0(x), \ldots, v_m(x), t(x))$ specified in σ that verifies R_λ over $\mathbb{Z}_p$. We know by Theorem 2 that if $t(x)$ divides $v(x)^2 - 1$ and $v_{\mathrm{mid}}(x) = \sum_{i=0}^{d} c_i x^i - v_0(x) - \sum_{i=1}^{\ell_u} a_i v_i(x)$ belongs to the span of $\{v_i(x)\}_{i > \ell_u}$ then indeed $w \in \{0,1\}^{\ell_w}$ and $(u, w) \in \mathcal{R}_\lambda$. So we will in the following show that the two cases, $t(x)$ does not divide $v(x)^2 - 1$ or $v_{\mathrm{mid}}(x)$ is not in the appropriate span both happen with negligible probability breaking the d-TSDH assumption or the d-PDH assumption respectively.

Given a d-TSDH challenge $(p, \mathbb{G}, \hat{\mathbb{G}}, \mathbb{G}_T, e, G, \hat{G}, \ldots, G^{s^d}, \hat{G}^{s^d})$, we pick $\beta \leftarrow \mathbb{Z}_p^*$ and roots $r_1, \ldots, r_d$ in the same way the setup algorithm does and simulate a common reference string σ. Suppose the adversary and extractor return $u = (a_1, \ldots, a_{\ell_u}) \in \{0,1\}^{\ell_u}$, a valid proof $\pi = (H, V_w, B_w, \hat{V})$ and $c_0, \ldots, c_d$ such that $V = G^{v_0(s)+\sum_{i=1}^{\ell_u} a_i v_i(s)} V_w = G^{\sum_{i=0}^{d} c_i s^i}$. Let $v(x) = \sum_{i=0}^{d} c_i x^i - \delta t(x)$ with $\delta = c_d$ as before and define $p(x) = (v(x) + \delta t(x))^2 - 1$ and suppose $p(x)$ is not divisible by $t(x)$. Let r_i be a root of $t(x)$ such that $x - r_i$ does not divide $p(x)$. We can write $p(x) = a(x)(x - r_i) + b$, where $a(x)$ is a degree $2d - 1$ polynomial in $\mathbb{Z}_p[x]$ and $b \in \mathbb{Z}_p^*$. The verification equation $e(H, \hat{G}^{t(s)}) = e(V, \hat{V}) e(G, \hat{G})^{-1}$ gives us $e(H, \hat{G}^{\frac{t(s)}{s-r_i}}) = e(G, \hat{G})^{a(s)+\frac{b}{s-r_i}}$. The adversary can use generic group operations on the d-TSDH challenge to compute $\hat{G}^{\frac{t(s)}{s-r_i}}$ and $e(G, \hat{G})^{a(s)}$, which allows it to deduce $e(G, \hat{G})^{\frac{b}{s-r_i}}$. Rasing this to the power b^{-1} gives a solution $(r_i, e(G, \hat{G})^{\frac{1}{s-r_i}})$ to the d-TSDH challenge.

Given a d-PDH challenge $(p, \mathbb{G}, \hat{\mathbb{G}}, \mathbb{G}_T, e, G, \hat{G}, \ldots, G^{s^d}, \hat{G}^{s^d}, G^{s^{d+2}}, \hat{G}^{s^{d+2}}, \ldots, G^{s^{2d}}, \hat{G}^{s^{2d}})$ we pick a random degree d polynomial $a(x)$ such that $a(x)v_i(x)$ has coefficient 0 for x^d for all $\ell_u < i \leq m$ and $a(x)t(x)$ also has coefficient 0 for x^d. There are $d + \ell_u - m - 1 > 0$ degrees of freedom in choosing $a(x)$ so for a polynomial $v_{\mathrm{mid}}(x)$ outside the span of $\{v_i(x)\}_{i=\ell_u}^{m}$ and $t(x)$ the polynomial $a(x)v_{\mathrm{mid}}(x)$ has a random coefficient for x^d.

Now pick at random $b \leftarrow \mathbb{Z}_p$ and define $\beta(x) = a(x)x + b$ and let $\beta = \beta(s)$. Observe that $G^{\beta v_i(s)} = G^{(a(s)s+b)v_i(s)}$ can be constructed from our challenge without knowing $G^{s^{d+1}}$; and the same goes for $G^{\beta t(s)}$. Pick $\rho \leftarrow \mathbb{Z}_p^*$ at random and compute $\tilde{G} = \hat{G}^{\rho t(s)}$ and $\tilde{G}^{\beta} = \hat{G}^{\rho \beta t(s)}$. Give to the adversary a simulated common reference string

$$\sigma = (p, \mathbb{G}, \hat{\mathbb{G}}, \mathbb{G}_T, e, G, \hat{G}, \ldots, G^{s^d}, \hat{G}^{s^d}, \{G^{\beta v_i(s)}\}_{i > \ell_u}, G^{\beta t(s)}, \tilde{G}, \tilde{G}^\beta, Q).$$

Suppose the adversary and extractor return $u = (a_1, \ldots, a_{\ell_u}) \in \{0, 1\}^{\ell_u}$, a valid proof $\pi = (H, V_w, B_w, \hat{V})$ and $c_0, \ldots, c_d$ such that $V = G^{\sum_{i=0} c_i s^i}$. Define $v_{\mathrm{mid}}(x) = \sum_{i=0}^d c_i x^i - v_0(x) - \sum_{i=1}^{\ell_u} a_i v_i(x)$. Due to the random choice of b the value $\beta(s) = a(s)s + b$ does not reveal anything about $a(x)$, so if $v_{\mathrm{mid}}(x)$ is outside the span of $\{v_i(x)\}_{i > \ell_u}$ and $t(x)$ then $a(x)v_{\mathrm{mid}}(x)$ has a random coefficient for x^{d+1}. With probability $1 - \frac{1}{p}$ this means the adversary returns $B_w = G^{\beta(s)v_{\mathrm{mid}}(s)}$ where $\beta(x)v_{\mathrm{mid}}(x) = \sum_{i=0}^{2d} b_i x^i$ is a known polynomial with a non-trivial coefficient $b_{d+1} \neq 0$ for x^{d+1}. We can now take an appropriate linear combination of B_w and the elements $G, \ldots, G^{s^d}, G^{s^{d+2}}, \ldots, G^{s^{2d}}$ to compute $G^{s^{d+1}}$, which solves the d-PDH challenge. $\qquad\square$

The proof of Theorem 3 suffers a computational overhead in the reduction by using an extractor $\mathcal{X}_\mathcal{A}$ for $\mathcal{A}$. Except for this computational overhead, the security reduction for knowledge soundness is tight. It is possible to eliminate the q-TSDH assumption and rely solely on the q-PKE and q-PDH assumptions, but then the security reduction loses a factor q and is therefore not tight.

3.4   Efficiency

In this section, we will assume our NIZK argument is instantiated with the square span program that we constructed in Section 2.2. This choice of square span program enables a number of optimizations that makes the argument highly efficient.

The prover has to compute

$$
\begin{aligned}
V_w &= G^{\sum_{i > \ell_u}^m a_i v_i(s) + \delta t(s)} \\
B_w &= G^{\beta(\sum_{i > \ell_u}^m a_i v_i(s) + \delta t(s))} \\
\hat{V} &= \hat{G}^{v_0(s) + \sum_{i=1}^m a_i v_i(s) + \delta t(s)}.
\end{aligned}
$$

It is possible to compute the polynomials $\sum_{i > \ell_u}^m a_i v_i(x) + \delta t(x)$ and $v_0(x) + \sum_{i=1}^m a_i v_i(x) + \delta t(x)$ and then compute the appropriate exponentiations of the polynomials evaluated in s using the elements $G, \hat{G}, \ldots, G^{s^d}, \hat{G}^{s^d}, \{G^{\beta v_i(s)}\}_{i > \ell_u}$, and $G^{\beta t(s)}$ from the common reference string. However, this requires $O(d)$ exponentiations to the coefficients of the polynomials. Following [GGPR13] a significant saving can be made by precomputing $\{G^{v_i(s)}\}_{i > \ell_u}, G^{t(s)}$ and $\{\hat{G}^{v_i(s)}\}_{i=0}^m$, $\hat{G}^{t(s)}$. Since each $a_i \in \{0, 1\}$ this makes it possible to compute V_w, B_w and $\hat{V}$ using at most $3m + 1 - 2\ell_u$ multiplications and 3 exponentiations. (Pragmatically, taking advantage of our uniform support for all gates, we can profile the SSP and 'flip' internal values from a_i to $\bar{a}_i$ to ensure that a_i is more often equal to 0 than to 1, thereby on average performing less than half of those multiplications.)

The prover also has to compute $H = G^{h(s)}$, where $h(x) = \frac{(v(x) + \delta t(x))^2 - 1}{t(x)}$ with $v(x) = v_0(x) + \sum_{i=1}^m a_i v_i(x)$ and $t(x) = \prod_{i=1}^d (x - r_i)$. We can evaluate $h(x)$ in

d points $r'_1, \ldots, r'_d$ using two discrete Fourier transforms as follows. The degree $d-1$ polynomial $v(x)$ is uniquely determined by its evaluation in the d points $r_1, \ldots, r_d$. In our square span program the evaluations in the points $r_1, \ldots, r_d$ can be computed easily given the values of the wires in the circuit. Using an inverse discrete Fourier transform, we compute the coefficients of $v(x) = \sum_{i=0}^{d-1} c_i x^i$. Let $\gamma \in \mathbb{Z}_p^*$ be given such that $r'_1, \ldots, r'_d$ defined as $r'_i = \gamma^i r_i$ gives us $2d$ distinct values $r_1, \ldots, r_d, r'_1, \ldots, r'_d$. Compute $c'_i = \gamma^i c_i$ to get the coefficients of the polynomial $v'(x) = \sum_{i=0}^{d-1} c'_i x^i$ and use a discrete Fourier transform to evaluate $v'(x)$ in $r_1, \ldots, r_d$. This gives us evaluations of $v(x)$ in the points $r'_1, \ldots, r'_d$ since $v(r'_j) = v'(r_j)$. We have $h(x) = \frac{v(x)^2 - 1}{t(x)} + 2\delta v(x) + \delta^2 t(x)$. Assuming $t(r'_1)^{-1}, \ldots, t(r'_d)^{-1}$ have been precomputed, it only costs $3d$ multiplications in $\mathbb{Z}_p$ to evaluate $\frac{v(x)^2 - 1}{t(x)} + 2\delta v(x)$ in the d points $r'_1, \ldots, r'_d$. Using Lagrange interpolation in the exponent, this allows us to compute

$$G^{\frac{v(s)^2-1}{t(s)} + 2\delta v(s)} = \prod_{j=1}^{d} (G^{\ell'_j(s)})^{\frac{v(r'_j)^2-1}{t(r'_j)} + 2\delta v(r'_j)}$$

where $\ell'_j(x)$ is the Lagrange basis polynomial for r'_j. By multiplying with $(G^{t(s)})^{\delta^2}$ we then get $G^{h(s)}$.

To speed up the computation, we can set up a modified common reference string for the prover

$$\sigma_{\text{Prove}} = \left(\begin{array}{c} p, \mathbb{G}, \hat{\mathbb{G}}, \mathbb{G}_T, e, G, \hat{G}, \{G^{v_i(s)}\}_{i>\ell_u}, \{\hat{G}^{v_i(s)}\}_{i>\ell_u}, \{G^{\beta v_i(s)}\}_{i>\ell_u}, \\ G^{\beta t(s)}, \tilde{G}, \tilde{G}^\beta, \gamma, \{t(r'_j)^{-1}\}_{j=1}^{d}, \{G^{\ell'_j(s)}\}_{j=1}^{d}, G^{t(s)}, Q \end{array} \right).$$

The computational cost for the prover is dominated by d exponentiations in $\mathbb{G}$ and 2 discrete Fourier transforms in $\mathbb{Z}_p$. The two discrete Fourier transforms cost $O(d \log^2 d)$ multiplications in general but the computation can be reduced to $O(d \log d)$ multiplications when $\mathbb{Z}_p$ is of a form amenable to using the fast Fourier transform.

The verifier needs to compute $V = G^{v_0(s) + \sum_{i=1}^{\ell_u} a_i v_i(s)} V_w$ and evaluate three pairing product equations $e(V, \hat{G}) = e(G, \hat{V})$, $e(H, \hat{G}^{t(s)}) = e(V, \hat{V}) e(G, \hat{G})^{-1}$, and $e(V_w, \tilde{G}^\beta) = e(B_w, \tilde{G})$. The verifier does not need the full common reference string but can use a more compact common reference string

$$\sigma_{\text{Vfy}} = \left(p, \mathbb{G}, \hat{\mathbb{G}}, \mathbb{G}_T, e, G, \{G^{v_i(s)}\}_{i=0}^{\ell_u}, \hat{G}, \hat{G}^{t(s)}, \tilde{G}, \tilde{G}^\beta \right),$$

which only has $\ell_u + 6$ group elements.[3] Verification is also computationally efficient, in the worst case it requires $\ell_u + 1$ multiplications in $\mathbb{G}$, one multiplication in $\mathbb{G}_T$ and 6 pairings if we precompute $e(G, \hat{G})^{-1}$.

For a large circuit, the cost of verification can be much smaller than the cost of evaluating the circuit itself, even if the witness w is known to the verifier. This

---

[3] Using the binary representation of the public input u from [PHGR13] this can be further reduced to $\lceil \frac{\ell_u}{\lambda} \rceil + O(1)$ group elements.

Table 3. Size in number of group elements (either $\mathbb{G}$ or $\hat{\mathbb{G}}$), performance in terms of pairings (P) or multiplications in $\mathbb{G}$ or $\mathbb{G}_T$ respectively

Proof size and verification cost comparison with Pinocchio

	Proof Size (elements)	Verification cost
Pinocchio [PHGR13]	8	$14P + (\ell_u + 4)\mathbb{G} + 1\mathbb{G}_T$
This work	4	$6P + (\ell_u + 1)\mathbb{G} + 1\mathbb{G}_T$

makes the NIZK argument a succinct non-interactive argument of knowledge that is suitable for verifiable computation protocols.[4]

Partly due to the lack of benchmarks, it is hard to compare the performance of SNARK protocols quantitatively without carefully reimplementing them. Table 3 compares the proof sizes and operations performed by the verifier between our protocol and Pinocchio, arguably the state of the art in terms of proof size and verification speed for QAPs. On this basis and the numbers reported in [PHGR13], we conservatively estimate that an SSP implementation based on the Pinocchio library would offer 160-byte proofs verified in less than 6 ms.

4   Conclusion

We introduce a representation of logic circuits, or predicates on propositional formulae, using quadratic constraints on an affine map. The map is built using a linearization of each gate, and a set of constraints to ensure all values of wires are binary. This leads to a simple and elegant formulation of square span programs, and in turn to efficient, minimalistic constructions for NIZKs and SNARKs.

The simplifications are twofold: (i) our representation of boolean functions no longer requires wire checkers and (ii) square span programs consist of only a single set of polynomials that are summed and squared. The former improves prover efficiency, while the key advantage of the latter are SNARKs with an extremely compact proof, consisting of only four group elements, and an efficient verification procedure compared to more generic QSP characterisations of the same program.

As can be expected, binary programs such as SSPs remain less efficient than arithmetic programs for verifying computations on integers, involving e.g. 32-bit additions and multiplications. Those operations have to be encoded as binary adders and multipliers, leading to a significant blow-up in circuit size and computation costs for the prover. It remains an open problem how to extend the SSP approach with ideas from QAPs to verify such computations without sacrificing its conceptual simplicity and short proofs.

---

[4] In some cases, for instance when outsourcing computation, the verifier may be the one that sets up the common reference string. In that case the verifier may know β and s, which can further decrease the cost of verification.

References

[AF07]   Abe, M., Fehr, S.: Perfect NIZK with adaptive soundness. In: Vadhan, S.P.
            (ed.) TCC 2007. LNCS, vol. 4392, pp. 118–136. Springer, Heidelberg (2007)

[BB08]   Boneh, D., Boyen, X.: Short signatures without random oracles and the
            sdh assumption in bilinear groups. Journal of Cryptology 21(2), 149–177
            (2008)

[BCPR13]   Bitansky, N., Canetti, R., Paneth, O., Rosen, A.: Indistinguishability ob-
            fuscation vs. auxiliary-input extractable functions: One must fall. IACR
            Cryptology ePrint Archive, Report 2013/641 (2013)

[BGI+12]   Barak, B., Goldreich, O., Impagliazzo, R., Rudich, S., Sahai, A.,
            Vadhan, S.P., Yang, K.: On the (im)possibility of obfuscating programs.
            Journal of the ACM 59(2), 6 (2012)

[BP04]   Bellare, M., Palacio, A.: Towards plaintext-aware public-key encryption
            without random oracles. In: Lee, P.J. (ed.) ASIACRYPT 2004. LNCS,
            vol. 3329, pp. 48–62. Springer, Heidelberg (2004)

[BSCG+14]   Ben-Sasson, E., Chiesa, A., Garman, C., Green, M., Miers, I., Tromer,
            E., Virza, M.: Zerocash: Practical decentralized anonymous e-cash from
            bitcoin. In: Proceedings of the 2014 IEEE Symposium on Security and
            Privacy. IEEE (May 2014)

[Dam91]   Damgård, I.: Towards practical public key systems secure against cho-
            sen ciphertext attacks. In: Feigenbaum, J. (ed.) CRYPTO 1991. LNCS,
            vol. 576, pp. 445–456. Springer, Heidelberg (1992)

[DFKP13]   Danezis, G., Fournet, C., Kohlweiss, M., Parno, B.: Pinocchio coin: build-
            ing zerocoin from a succinct pairing-based proof system. In: Franz, M.,
            Holzer, A., Majumdar, R., Parno, B., Veith, H. (eds.) PETShop@CCS,
            pp. 27–30. ACM (2013)

[GGH+13]   Garg, S., Gentry, C., Halevi, S., Raykova, M., Sahai, A., Waters, B.: Can-
            didate indistinguishability obfuscation and functional encryption for all
            circuits. In: FOCS, pp. 40–49 (2013)

[GGPR13]   Gennaro, R., Gentry, C., Parno, B., Raykova, M.: Quadratic span pro-
            grams and succinct nizks without pcps. In: Johansson, T., Nguyen, P.Q.
            (eds.) EUROCRYPT 2013. LNCS, vol. 7881, pp. 626–645. Springer, Hei-
            delberg (2013)

[GOS12]   Groth, J., Ostrovsky, R., Sahai, A.: New techniques for noninteractive
            zero-knowledge. Journal of the ACM 59(3), 11:1–11:35 (2012)

[GPS08]   Galbraith, S.D., Paterson, K.G., Smart, N.P.: Pairings for cryptographers.
            Discrete Applied Mathematics 156(16), 3113–3121 (2008)

[Gro10]   Groth, J.: Short pairing-based non-interactive zero-knowledge arguments.
            In: Abe, M. (ed.) ASIACRYPT 2010. LNCS, vol. 6477, pp. 321–340.
            Springer, Heidelberg (2010)

[KW93]   Karchmer, M., Wigderson, A.: On span programs. In: Proc. of the 8th
            IEEE Structure in Complexity Theory, pp. 102–111. IEEE Computer So-
            ciety Press (1993)

[Lip13]   Lipmaa, H.: Succinct non-interactive zero knowledge arguments from span
            programs and linear error-correcting codes. In: Sako, K., Sarkar, P. (eds.)
            ASIACRYPT 2013, Part I. LNCS, vol. 8269, pp. 41–60. Springer, Heidel-
            berg (2013)

[Lip14]   Lipmaa, H.: Almost optimal short adaptive non-interactive zero knowl-
            edge. Cryptology ePrint Archive, Report 2014/396 (2014),
            http://eprint.iacr.org/

[PHGR13]   Parno, B., Howell, J., Gentry, C., Raykova, M.: Pinocchio: Nearly practical verifiable computation. In: IEEE Symposium on Security and Privacy, pp. 238–252 (2013)
[Val76]   Valiant, L.G.: Universal circuits (preliminary report). In: STOC, pp. 196–203 (1976)
[vEB81]   van Emde Boas, P.: Another NP-complete partition problem and the complexity of computing short vectors in a lattice. Technical report (1981), http://staff.science.uva.nl/~peter/vectors/mi8104c.html

Better Zero-Knowledge Proofs for Lattice Encryption and Their Application to Group Signatures[*]

Fabrice Benhamouda[1], Jan Camenisch[2], Stephan Krenn[2],
Vadim Lyubashevsky[3,1], and Gregory Neven[2]

[1] Département d'Informatique, École Normale Supérieure, Paris, France
fabrice.ben.hamouda@ens.fr, lyubash@di.ens.fr
[2] IBM Research Zurich – Rüschlikon, Switzerland
{jca,skr,nev}@zurich.ibm.com
[3] INRIA, France

Abstract. Lattice problems are an attractive basis for cryptographic systems because they seem to offer better security than discrete logarithm and factoring based problems. Efficient lattice-based constructions are known for signature and encryption schemes. However, the constructions known for more sophisticated schemes such as group signatures are still far from being practical. In this paper we make a number of steps towards efficient lattice-based constructions of more complex cryptographic protocols. First, we provide a more efficient way to prove knowledge of plaintexts for lattice-based encryption schemes. We then show how our new protocol can be combined with a proof of knowledge for Pedersen commitments in order to prove that the committed value is the same as the encrypted one. Finally, we make use of this to construct a new group signature scheme that is a "hybrid" in the sense that privacy holds under a lattice-based assumption while security is discrete-logarithm-based.

Keywords: Verifiable Encryption, Group Signatures, Zero-Knowledge Proofs for Lattices.

1 Introduction

There has been a remarkable increase of research in the field of lattice-based cryptography over the past few years. This renewed attention is largely due to a number of exciting results showing how cryptographic primitives such as fully homomorphic encryption [21] and multi-linear maps [20] can be built from lattices, while no such instantiations are known based on more traditional problems such as factoring or discrete logarithms. Lattice problems are also attractive to build standard primitives such as encryption and signature schemes, however, because of their strong security properties. In particular, their worst-case to average-case reductions as well as their apparent resistance against quantum computers set them apart from traditional cryptographic assumptions such as factoring or computing discrete logarithms, in particular in situations that require security many years or even decades into the future.

[*] The research leading to these results has received partial funding from the European Commission under the Seventh Framework Programme (CryptoCloud #339563, PERCY #321310, FutureID #318424) and from the French ANR-13-JS02-0003 CLE Project.

P. Sarkar and T. Iwata (Eds.): ASIACRYPT 2014, PART I, LNCS 8873, pp. 551–572, 2014.

Long-term integrity requirements, e.g., for digital signatures, can usually be fulfilled by re-signing documents when new, more secure signature schemes are proposed. The same approach does not work, however, for privacy requirements, e.g., for encryption or commitment schemes, because the adversary may capture ciphertexts or commitments now and store them until efficient attacks on the schemes are found.

Several lattice-based encryption schemes have been proposed in the literature, e.g., [22, 24, 30, 35], but many of their applications in more complex primitives require efficient zero-knowledge proofs of the encrypted plaintext. Examples include optimistic fair exchange [2], non-interactive zero-knowledge proofs [34], multiparty computation [18], and group signatures [12]. In this paper, we present a more efficient zero-knowledge proof for lattice-based encryption schemes. We then combine it with a non-lattice-based signature scheme to build a group signature scheme with privacy under lattice assumptions in the random-oracle model.

1.1 Improved Proofs of Plaintext Knowledge for Lattice Schemes

In a zero-knowledge proof of plaintext knowledge, the encryptor wants to prove in zero-knowledge that the ciphertext is of the correct form and that he knows the message. Efficient constructions of these primitives are known based on number-theoretic hardness assumptions such as discrete log, strong RSA, etc.

Encryptions in lattice-based schemes generally have the form $t = Ae \bmod q$, where A is some public matrix and e is the unique vector with small coefficients that satisfies the equation (in this general example, we are lumping the message with e). A proof that t is a valid ciphertext (and also a proof of plaintext knowledge), therefore, involves proving that one knows a short e such that $Ae = t$. It is currently known how to accomplish this task in two ways. The first uses a "Stern-type" protocol [37] in which every run has soundness error $2/3$ [26]. It does not seem possible to improve this protocol since some steps in it are inherently combinatorial and non-algebraic.

A second possible approach is to use the Fiat-Shamir approach for lattices using rejection sampling introduced in [27, 29]. But while the latter leads to fairly efficient Fiat-Shamir signatures, there are some barriers to obtaining a proof of knowledge. What one is able to extract from a prover are short vectors r', z' such that $Ae' = tc$ for some integer c, which implies that $Ae'c^{-1} = t$. Unfortunately, this does not imply that $e'c^{-1}$ is short unless $c = \pm 1$. This is the main way in which lattice-based Fiat-Shamir proofs differ from traditional schemes like the discrete-log based Schnorr protocol. In the latter, it is enough to extract any discrete log, whereas in lattice protocols, one must extract a *short* vector. Thus, the obvious approach at Fiat-Shamir proofs of knowledge for lattices (i.e., using binary challenge vectors) also leads to protocols with soundness error $1/2$.

Things do not improve for lattice-based proofs of knowledge even if one considers ideal lattices. Even if A and e are a matrix and a vector of polynomials in the ring $\mathbb{Z}_q[X]/(X^n + 1)$, and c is now a polynomial (as in the Ring-LWE based Fiat-Shamir schemes in [28, 29]), then one can again extract only a short e' such that $Ae'c^{-1} = t$. In this case, not only is c^{-1} not necessarily short, but it does not even necessarily exist (since the polynomial $X^n + 1$ can factor into up to n terms). At this point, we are not aware of any techniques that reduce the soundness error of protocols that prove plaintext knowledge of lattice encryptions, except by (parallel) repetition.

In a recent work, Damgård et al. [18] gave an improved *amortized* proof of plaintext knowledge for LWE encryptions. Their protocol allows one to prove knowledge of $O(k)$ plaintexts (where k is the security parameter) in essentially the same time as just one plaintext. The ideas behind their protocol seem do not reduce the time requirement for proving just one plaintext, nor do they apply to Ring-LWE based encryption schemes. In particular, Ring-LWE based schemes are able to encrypt $O(n)$ plaintext bits into one (or two) polynomial, which is often all that is needed. Yet, the techniques in [18] do not seem to be helpful here. The reason is that the challenge matrix required in [18] needs to be of a particular form and cannot simply be a ring element in $\mathbb{Z}_q[X]/(X^n + 1)$.

In this paper, we show that one can reduce the soundness error of lattice-based zero-knowledge proofs of knowledge for ciphertext validity from $1/2$ to $1/(2n)$, which in practice decreases the number of required iterations of the protocol by a factor of approximately 10. Interestingly, our techniques only work for ideal lattices, and we do not know how to adapt them to general ones. The key observation is that, when working over the ring $\mathbb{Z}[X]/(X^n + 1)$, the quantity $2/(X^i - X^j)$ for all $0 \leq i \neq j < n$ is a polynomial with coefficients in $\{-1, 0, 1\}$, cf. section 3.1.

This immediately allows us to prove that, given A and t, we know a vector of short polynomials e such that $Ae = 2t$. While this is not quite the same as proving that $Ae = t$, it is good enough for most applications, since it still allows us to prove knowledge of the plaintext. This result immediately gives improvements in all schemes that require such a proof of knowledge for Ring-LWE based encryption schemes such as the ring-version of the "dual" encryption scheme [22], the "two element" scheme of Lyubashevsky et al. [30], and NTRU [24, 35].

1.2   Linking Lattice-Based and Classical Primitives

A main step in applying our new proof protocol to construct a "hybrid" group signature scheme is to prove that two primitives, one based on classical cryptography and the other one on lattices, are committing to the same message (and that the prover knows that message). In our application, we will use the perfectly hiding Pedersen commitment scheme as the classical primitive, and a Ring-LWE encryption scheme as the lattice-based primitive.

While the Pedersen commitment and the lattice-based encryption scheme work over different rings, we show that we can still perform operations "in parallel" on the two. For example, if the message is $\mu_0, \mu_1, \ldots, \mu_{n-1}$, then it is encrypted in Ring-LWE schemes by encrypting the polynomial $\mu = \mu_0 + \mu_1 X + \ldots + \mu_{n-1} X^{n-1}$, and each μ_i is committed to individually using a Pedersen commitment. We will then want to prove that a Ring-LWE encryption of μ encrypts the same thing as n Pedersen commitments of the μ_i's. Even though the two computations are performed over different rings, we show that by mimicking polynomial multiplications over a polynomial ring by appropriate additions and multiplications of coefficients in exponents, we can use our previously mentioned proof of plaintext knowledge to both prove knowledge of μ and show that the Pedersen commitments are committing to the coefficients of the same μ. One reason enabling such a proof is that the terms dealing with μ (and μ_i) in the proof of knowledge are done "over the integers"—that is, no modular reduction needs to be performed on these terms. We describe this protocol in detail in section 4.

1.3 Applications to Group Signatures and Credentials

Group signatures [12] are schemes that allow members of a group to sign messages on behalf of the group without revealing their identity. In case of a dispute, the group manager can lift a signer's anonymity and reveal his identity. Currently known group signatures based on lattice assumptions are mainly proofs of concepts, rather than practically useful schemes. The schemes by Gordon et al. [23] and Camenisch et al. [10] have signature size linear in the number of group members. The scheme due to Laguil-laumie et al. [25] performs much better asymptotically with signature sizes logarithmic in the number of group members, but, as the authors admit, instantiating it with practical parameters would lead to very large keys and signatures. This is in contrast to classical number-theoretic solutions, where both the key and the signature size are constant for arbitrarily many group members.

One can argue that the privacy requirement for group signatures is a concern that is more long-term than traceability (i.e., unforgeability), because when traceability turns out to be broken, verifiers can simply stop accepting signatures for the broken scheme. When privacy is broken, however, an adversary can suddenly reveal the signers behind all previous signatures. Users may only be willing to use a group signature scheme if their anonymity is guaranteed for, say, fifty years in the future. It therefore makes sense to provide anonymity under lattice-based assumptions, while this is less crucial for traceability.

Following this observation, we propose a "hybrid" group signature scheme, where unforgeability holds under classical assumptions, while privacy is proved under lattice-based ones. This allows us to combine the flexible tools that are available in the classical framework with the strong privacy guarantees of lattice problems. Our group signature scheme has keys and signatures of size logarithmic in the number of group members; for practical choices of parameters and realistic numbers of group members, the sizes will even be independent of the number of users. Furthermore, by basing our scheme on ring-LWE and not standard LWE, we partially solve an open problem stated in [25].

Our construction follows a variant of a generic approach that we believe is folklore, as it underlies several direct constructions in the literature [3, 8] and was described explicitly by Chase and Lysyanskaya [11]. When joining the group, a user obtains a certificate from the group manager that is a signature on his identity under the group manager's public key. To sign a message, the user now encrypts his identity under the manager's public encryption key, and then issues a signature proof of knowledge that he possesses a valid signature on the encrypted plaintext. Our construction follows a variant of this general paradigm, with some modifications to better fit the specifics of our proof of plaintext knowledge for lattice encryption. To the best of our knowledge, however, the construction was never proved secure, so our proof can be seen as a contribution of independent interest.

2 Preliminaries

In this section, we informally introduce several notions. Formal definitions and proofs can be found in the full version.

2.1 Notation

We denote algorithms by sans-serif letters such as A, B. If S is a set, we write $s \xleftarrow{\$} S$ to denote that s was drawn uniformly at random from S. Similarly, we write $y \xleftarrow{\$} A(x)$ if y was computed by a randomized algorithm A on input x, and $d \xleftarrow{\$} D$ for a probability distribution D, if d was drawn according to D. When we make the random coins ρ of A explicit, we write $y \leftarrow A(x; \rho)$.

We write $\Pr[\mathcal{E} : \Omega]$ to denote the probability of event $\mathcal{E}$ over the probability space Ω. For instance, $\Pr[x = y : x, y \xleftarrow{\$} D]$ denotes the probability that $x = y$ if x, y were drawn according to a distribution D.

We identify the vectors $(a_0, \ldots, a_{n-1})$ with the polynomial $a_0 + a_1 X + \cdots + a_{n-1} X^{n-1}$. If v is a vector, we denote by $\|v\|$ its Euclidean norm, by $\|v\|_\infty$ its infinity norm, and by $v_{\ll l}$ an anti-cyclic shift of a vector v by l positions, corresponding to a multiplication by X^l in $R_q = \mathbb{Z}_q[X]/(X^n + 1)$. That is, $v_{\ll l} = (v_0, \ldots, v_{n-1})_{\ll l} = (-v_{n-l}, \ldots, -v_{n-1}, v_0, \ldots, v_{n-l-1})$.

Throughout the paper, λ denotes the main security parameter and ε denotes the empty string.

2.2 Commitment Schemes and Pedersen Commitments

Informally, a commitment scheme is a tuple (CSetup, Commit, COpen), where CSetup generates commitment parameters, which are then used to commit to a message m using Commit. A commitment cmt can then be verified using COpen. Informally, a commitment scheme needs to be binding and hiding. The former means that no cmt can be opened to two different messages, while the latter guarantees that cmt does not leak any information about the contained m.

The following commitment scheme was introduced by Pedersen [32]. Let be given a family of prime order groups $\{\mathbb{G}(\lambda)\}_{\lambda \in \mathbb{N}}$ such that the discrete logarithm problem is hard in $\mathbb{G}(\lambda)$ for security parameter λ, and let $\tilde{q} = \tilde{q}(\lambda)$ be the order of $\mathbb{G} = \mathbb{G}(\lambda)$.

To avoid confusion, all elements with order $\tilde{q}$ are denoted with a tilde in the following. To ease the presentation of our main result, we will write the group $\mathbb{G}(\lambda)$ additively.

CSetup. This algorithm chooses $\tilde{h} \xleftarrow{\$} \mathbb{G}$, $\tilde{g} \xleftarrow{\$} \langle \tilde{h} \rangle$, and outputs $cpars = (\tilde{g}, \tilde{h})$.

Commit. To commit to a message $m \in \mathcal{M} = \mathbb{Z}_{\tilde{q}}$, this algorithm first chooses $r \xleftarrow{\$} \mathbb{Z}_{\tilde{q}}$.
It then outputs the pair $(\widetilde{cmt}, o) = (m\tilde{g} + r\tilde{h}, r)$.

COpen. Given a commitment $\widetilde{cmt}$, an opening o, a public key $cpars$ and a message m, this algorithm outputs accept if and only if $\widetilde{cmt} \stackrel{?}{=} m\tilde{g} + o\tilde{h}$.

Theorem 2.1. *Under the discrete logarithm assumption for $\mathbb{G}$, the given commitment scheme is perfectly hiding and computationally binding.*

2.3 Semantically Secure Encryption and NTRU

A semantically secure (or IND-CPA secure) encryption scheme is a tuple (EncKG, Enc, Dec) of algorithms, where EncKG generates public/private key pair, Enc can be used to encrypt a message m under the public key, and the message can be recovered from the ciphertext by Dec using the secret key. Informally, while $\mathsf{Dec}(\mathsf{Enc}(m)) = m$ should always hold, only knowing the ciphertext and the public key should not leak any information about the contained message.

In this paper we present improved zero-knowledge proofs of plaintext knowledge for lattice-based encryption schemes, and show how to link messages being encrypted by these schemes to Pedersen commitments. Our improved proof of knowledge protocol will work for any Ring-LWE based scheme where the basic encryption operation consists of taking public key polynomial(s) a_i and computing the ciphertext(s) $b_i = a_i s + e_i$ where s and e_i are polynomials with small norms. Examples of such schemes include the ring-version of the "dual" encryption scheme [22], the "two element" scheme of Lyubashevsky et al. [30], and the NTRU encryption scheme [24, 35].

In this paper we will for simplicity only be working over the rings $R = \mathbb{Z}[X]/(X^n + 1)$ and $R_q = R/qR$, for some prime q. Also for simplicity, we will use NTRU as our encryption scheme because its ciphertext has only one element and is therefore simpler to describe in protocols. The NTRU scheme was first proposed by Hoffstein et al. [24], and we will be using a modification of it due to Stehlé and Steinfeld [35].

Definition 2.2. *The* discrete normal distribution *on $\mathbb{Z}^m$ centered at v with standard deviation σ is defined by the density function $D_{v,\sigma}^m(x) = \rho_{v,\sigma}^m(x)/\rho_\sigma(\mathbb{Z}^m)$, with $\rho_{v,\sigma}^m(x) = \left(\frac{1}{\sqrt{2\pi}\sigma}\right)^m e^{-\frac{\|x-v\|^2}{2\sigma^2}}$ being the continuous normal distribution on $\mathbb{R}^m$ and $\rho_\sigma(\mathbb{Z}^m) = \sum_{z \in \mathbb{Z}^m} \rho_{0,\sigma}^m(z)$ being the scaling factor required to obtain a probability distribution. When $v = 0$, we also write $D_\sigma^m = D_{0,\sigma}^m$.*

We will sometimes write $u \xleftarrow{\$} D_{v,\sigma}$ instead of $u \xleftarrow{\$} D_{v,\sigma}^n$ for a polynomial $u \in R_q$ if there is no risk of confusion.

In the following, let p be a prime less than q and $\sigma, \alpha \in \mathbb{R}$.

Message Space. The message space $\mathcal{M}$ is any subset of $\{y \in R : \|y\|_\infty < p\}$.

KeyGen. Sample f', g from D_σ, set $f = pf' + 1$, and resample, if $f \bmod q$ or $g \bmod q$ are not invertible. Output the public key $h = pg/f$ and the secret key f. Note here that h is invertible.

Encrypt. To encrypt a message $m \in \mathcal{M}$, set $s, e \xleftarrow{\$} D_\alpha$ and return the ciphertext $y = hs + pe + m \in R_q$.

Decrypt. To decrypt y with secret key f, compute $y' = fy \in R_q$ and output $m' = y' \bmod p$.

If the value of σ is large enough (approximately $\tilde{\mathcal{O}}(n\sqrt{q})$), then g/f is uniformly random in R_q [35], and the security of the above scheme is based on the Ring-LWE problem. For smaller values of σ, however, the scheme is more efficient and can be based on the assumption that $h = g/f$ is indistinguishable from uniform. This type of assumption, while not based on any worst-case lattice problem, has been around since the introduction of the original NTRU scheme over fifteen years ago. Our protocol works for either instantiation.

To obtain group signatures, we will need our encryption scheme to additionally be a commitment scheme. In other words, there should not be more than one way to obtain the same ciphertext. For the NTRU encryption scheme, this will require that we work over a modulus q such that the polynomial $X^n + 1$ splits into two irreducible polynomials of degree $n/2$, which can be shown to be always the case when n is a power of 2 and $q = 3 \bmod 8$ [36, Lemma 3].

Lemma 2.3. *Suppose that $q = 3 \mod 8$ and let $\#S, \#E$, and $\#M$ be the domain sizes of the parameters s, e, and m in the ciphertext $y = hs + pe + m$. Additionally suppose that for all $m \in M$, $\|m\|_\infty < p/2$. Then the probability that for a random h, there exists a ciphertext that can be obtained in two ways is at most $\frac{(2\#M+1)\cdot(2\#S+1)\cdot(2\#E+1)}{q^{n/2}}$.*

Note that the above lemma applies to NTRU public keys h that are uniformly random. If $h = pg/f$ is not random, then the ability to come up with two plaintexts for the same ciphertext would constitute a distinguisher for the assumed pseudorandomness of h.

2.4 Rejection Sampling

For a protocol to be zero-knowledge, the prover's responses must not depend on its secret inputs. However, in our protocols, the prover's response will be from a discrete normal distribution which is shifted depending on the secret key. To correct for this, we employ rejection sampling [28, 29], where a potential response is only output with a certain probability, and otherwise the protocol is aborted.

Informally, the following theorem states that for sufficiently large σ the rejection sampling procedure outputs results that are independent of the secret. The technique only requires a constant number of iterations before a value is output, and furthermore the output is also statistically close for every secret v with norm at most T. For concrete parameters we refer to the original work of Lyubashevsky [29, Theorem 4.6].

Theorem 2.4. *Let V be a subset of $\mathbb{Z}^\ell$ in which all elements have norms less than T, and let H be a probability distribution over V. Then, for any constant M, there exists a $\sigma = \tilde{\Theta}(T)$ such that the output distributions of the following algorithms A, F are statistically close:*

$\mathsf{A} : v \xleftarrow{\$} H;\; z \xleftarrow{\$} D^\ell_{v,\sigma};\; \text{output } (z,v) \text{ with probability } \min\left(D^\ell_\sigma(z)/(MD^\ell_{v,\sigma}(z)), 1\right)$

$\mathsf{F} : v \xleftarrow{\$} H;\; z \xleftarrow{\$} D^\ell_{0,\sigma};\; \text{output } (z,v) \text{ with probability } 1/M$

The probability that A outputs something is exponentially close to that of F, i.e., $1/M$.

2.5 Zero-Knowledge Proofs and Σ'-Protocols

On a high level, a zero-knowledge proof of knowledge (ZKPoK) is a two party protocol between a *prover* and a *verifier*, which allows the former to convince the latter that it knows some secret piece of information, without revealing anything about the secret apart from what the claim itself reveals. For a formal definition we refer to Bellare and Goldreich [4].

A language $\mathcal{L} \subseteq \{0,1\}^*$ has witness relationship $R \subseteq \{0,1\}^* \times \{0,1\}^*$ if $x \in \mathcal{L} \Leftrightarrow \exists (x, w) \in R$. We call w a witness for $x \in \mathcal{L}$. The ZKPoKs constructed in this paper will be instantiations of the following definition, which is a straightforward generalization of Σ-protocols [13, 15]:

Definition 2.5. *Let (P, V) be a two-party protocol, where V is PPT, and let $\mathcal{L}, \mathcal{L}' \subseteq \{0,1\}^*$ be languages with witness relations $\mathcal{R}, \mathcal{R}'$ such that $\mathcal{R} \subseteq \mathcal{R}'$. Then (P, V) is called a Σ'-protocol for $\mathcal{L}, \mathcal{L}'$ with completeness error α, challenge set $\mathcal{C}$, public input x and private input w, if and only if it satisfies the following conditions:*

- Three-move form: *The protocol is of the following form: The prover* P, *on input* (x, w), *computes a commitment* t *and sends it to* V. *The verifier* V, *on input* x, *then draws a challenge* c $\xleftarrow{\$} \mathcal{C}$ *and sends it to* P. *The prover sends a response* s *to the verifier. Depending on the protocol transcript* (t, c, s), *the verifier finally accepts or rejects the proof. The protocol transcript* (t, c, s) *is called* accepting, *if the verifier accepts the protocol run.*
- Completeness: *Whenever* $(x, w) \in \mathcal{R}$, *the verifier* V *accepts with probability at least* $1 - \alpha$.
- Special soundness: *There exists a PPT algorithm* E *(the* knowledge extractor*) which takes two accepting transcripts* $(t, c', s'), (t, c'', s'')$ *satisfying* $c' \neq c''$ *as inputs, and outputs* w' *such that* $(x, w') \in \mathcal{R}'$.
- Special honest-verifier zero-knowledge (HVZK): *There exists a PPT algorithm* S *(the* simulator*) taking* $x \in \mathcal{L}$ *and* $c \in \mathcal{C}$ *as inputs, that outputs* (t, s) *so that the triple* (t, c, s) *is indistinguishable from an accepting protocol transcript generated by a real protocol run.*

This definition differs from the standard definition of Σ-protocols in two ways. First, we allow the honest prover to fail in at most an α-fraction of all protocol runs, whereas the standard definition requires perfect completeness, i.e., $\alpha = 0$. However, this relaxation is crucial in our construction that is based on rejection sampling [28, 29], where the honest prover sometimes has to abort the protocol to achieve zero-knowledge. Second, we introduce a second language $\mathcal{L}'$ with witness relation $\mathcal{R}' \supseteq \mathcal{R}$, such that provers knowing a witness in $\mathcal{R}$ are guaranteed privacy, but the verifier is only ensured that the prover knows a witness for $\mathcal{R}'$. This has already been used in [1] and informally also in, e.g., [16, 19]. If the *soundness gap* between $\mathcal{R}$ and $\mathcal{R}'$ is sufficiently small, the implied security guarantees are often enough for higher-level applications. Note that the original definition of Σ-protocols is the special case that $\alpha = 0$ and $R = R'$.

We want to stress that previous results showing that a Σ-protocol is always also an honest-verifier ZKPoK with knowledge error $1/|\mathcal{C}|$ directly carry over to the modified definition whenever $1 - \alpha > 1/|\mathcal{C}|$. Zero-knowledge against arbitrary verifiers can be achieved by applying standard techniques such as Damgård et al. [14, 17].

Finally, it is a well known result that negligible knowledge and completeness errors in λ can be achieved, e.g., by running the protocol λ times in parallel and accepting if and only if at least $\lambda(1 - \alpha)/2$ transcripts were valid, if there exists a constant c such that $(1 - \alpha)/2 > 1/|\mathcal{C}| + c$.

Some of the Σ'-protocols presented in this paper will further satisfy the following useful properties:

- Quasi-unique responses: No PPT adversary A can output (y, t, c, s, s') with $s \neq s'$ such that $V(y, t, c, s) = V(y, t, c, s') = \texttt{accept}$.
- High-entropy commitments: For all $(y, w) \in \mathcal{R}$ and for all t, the probability that an honestly generated commitment by P takes on the value t is negligible.

3   Proving Knowledge of Ring-LWE Secrets

In the following we show how to efficiently prove knowledge of short $2s, 2e$ such that $2y = 2as + 2e$. This basic protocol can easily be adapted for proving more complex relations including more than one public image or more than two secret witnesses. Before presenting the protocol, we prove a technical lemma that is at the heart of the knowledge extractor thereof.

3.1 A Technical Lemma

The following lemma guarantees that certain binomials in $\mathbb{Z}[X]/(X^n + 1)$ can be inverted, and their inverses have only small coefficients.

Lemma 3.1. *Let n be a power of 2 and let $0 < i, j < 2n - 1$. Then $2(X^i - X^j)^{-1} \bmod (X^n + 1)$ only has coefficients in $\{-1, 0, 1\}$.*

Proof. Without loss of generality, assume that $j > i$. Using that $X^n = -1 \bmod (X^n + 1)$, we have that $2(X^i - X^j)^{-1} = -2X^{n-i}(1 - X^{j-i})^{-1}$. It is therefore sufficient to prove the claim for $i = 0$ only.

Now remark that, for every $k \geq 1$ it holds that: $(1 - X^j)(1 + X^j + X^{2j} + \ldots + X^{(k-1)j}) = 1 - X^{kj}$.

Let us write $j = 2^{j'} j''$, with j'' a positive odd integer and $0 \leq j' \leq \log_2(n)$, and let us choose $k = 2^{\log_2(n) - j'}$ (recall that n is a power of 2). We then have $jk = nj''$, and $X^{kj} = (-1)^{j''} = -1 \bmod (X^n + 1)$, hence $1 - X^{kj} = 2 \bmod (X^n + 1)$. Therefore, we have

$$2(1 - X^j)^{-1} = 1 + X^j + X^{2j} + \ldots + X^{(k-1)j} \bmod (X^n + 1)$$
$$= 1 \pm X^{j \bmod n} \pm X^{2j \bmod n} \pm \ldots \pm X^{(k-1)j \bmod n} \bmod (X^n + 1).$$

Finally, in this equation, no two exponents are equal, since otherwise that would mean that n divides jk' with $1 \leq k' < k$, which is impossible by definition of k. □

3.2 The Protocol

We next present our basic protocol. Let therefore be $y = as + e$, where the LWE-secrets $s, e \xleftarrow{\$} D_\alpha$ are chosen from a discrete Gaussian distribution with standard deviation α. Protocol 3.2 now allows a prover to convince a verifier that it knows s' and e' such that $2y = 2as' + 2e'$ with $2s'$ and $2e'$ being short (after reduction modulo q), i.e., the verifier is ensured that the prover knows short secrets for twice the public input. Here, by short we mean the following: An honest prover will always be able to convince the verifier whenever $\|s\|, \|e\| \leq \tilde{\mathcal{O}}(\sqrt{n}\alpha)$, which is the case with overwhelming probability if they were generated honestly. On the other hand, the verifier is guaranteed that the prover knows LWE-secrets with norm at most $\tilde{\mathcal{O}}(n^2\alpha)$. This soundness gap on the size of the witnesses is akin to those in, e.g., [1, 16].

To be able to simulate aborts when proving the zero-knowledge property of the protocol, we must not send the prover's first message in the plain, but commit to it and later open it in the last round of the Σ'-protocol. We therefore make use of an auxiliary commitment scheme (aCSetup, aCommit, aCOpen), and assume that honestly generated commitment parameters are given as common input to both parties. We do not make any assumptions on the auxiliary commitment scheme. However, if it is computationally binding, the resulting protocol is only sound under the respective assumption,

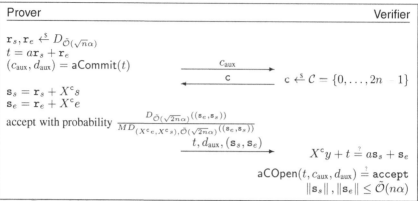

Protocol 3.2. Proof of knowledge of LWE-secrets s, e such that $y = as + e$

and similarly if it is computationally hiding. For simplicity, the reader may just think of the scheme as a random oracle.

Theorem 3.3. *Protocol 3.2 is an HVZK Σ'-protocol for the following relations:*

$$\mathcal{R} = \{((a,y),(s,e)) : y = as + e \quad \wedge \quad \|s\|, \|e\| \leq \tilde{\mathcal{O}}(\sqrt{n}\alpha)\}$$
$$\mathcal{R}' = \{((a,y),(s,e)) : 2y = 2as + 2e \quad \wedge \quad \|2s\|, \|2e\| \leq \tilde{\mathcal{O}}(n^2\alpha)\}$$

where $2s$ and $2e$ are reduced modulo q. The protocol has a knowledge error of $1/(2n)$, a completeness error of $1 - 1/M$, and high-entropy commitments.

We remark that in Protocol 3.2, the rejection sampling is applied on the whole vector (s_e, s_s) instead of applying it twice on s_e and on s_s. This yields better parameters (M or the "$\sigma = \tilde{\mathcal{O}}(T)$" in Theorem 2.4) by a factor of about $\sqrt{2}$, because of the use of the Euclidean norm.

Proof. We need to prove the properties from definition 2.5.
Completeness. First note that by theorem 2.4, the prover will respond with probability $1/M$. If the prover does not abort, we have that:

$$as_s + s_e = a(r_s + X^c s) + (r_e + X^c e) = X^c(as + e) + (ar_s + r_e) = X^c y + t.$$

For the norms we have that $\|s_s\| \leq \|r_s\| + \|s\| \leq \tilde{\mathcal{O}}(n\alpha)$ with overwhelming probability, as the standard deviations of r_s is $\tilde{\mathcal{O}}(\sqrt{n}\alpha)$, and similarly for s_e.
Honest-verifier zero-knowledge. Given a challenge value c, the simulator outputs the tuple (aCommit(0), c, $\perp$) with probability $1 - 1/M$. With probability $1/M$, the simulator S proceeds as follows: It chooses $s_s, s_e \xleftarrow{\$} D_{\tilde{\mathcal{O}}(\sqrt{n}\alpha)}$, and computes $t = as_s + s_e - X^c y$, and $(c_{\text{aux}}, d_{\text{aux}}) \xleftarrow{\$} \text{aCommit}(t)$. Finally, S outputs $(c_{\text{aux}}, c, (t, d_{\text{aux}}, (s_s, s_e)))$.

It follows from theorem 2.4 that if no abort occurs the distribution of s_e, s_s does not depend on s, e, and thus simulated and real protocol transcripts are indistinguishable.

In case that an abort occurs, the indistinguishability follows from the hiding property of aCommit and the fact that aborts are equally likely for every c.

Special soundness. Assume that we are given $(c_{\text{aux}}, c', (t', d'_{\text{aux}}, (s'_s, s'_e)))$ and $(c_{\text{aux}}, c'', (t'', d''_{\text{aux}}, (s''_s, s''_e)))$ passing the checks performed by the verifier. From the binding property of the auxiliary commitment scheme we get that $t' = t'' =: t$. Now, by subtracting the verification equations we get: $(X^{c'} - X^{c''})y = a(s'_s - s''_s) + (s'_e - s''_e)$. Multiplying by $2(X^{c'} - X^{c''})^{-1}$ yields:

$$2y = a\frac{2(s'_s - s''_s)}{X^{c'} - X^{c''}} + \frac{2(s'_e - s''_e)}{X^{c'} - X^{c''}} =: 2a\hat{s} + 2\hat{e}.$$

Furthermore, we get that $\|2\hat{s}\| \leq \|s'_e - s''_e\| \sqrt{n} \left\| \frac{2}{X^{c'} - X^{c''}} \right\| \leq \tilde{\mathcal{O}}(n^2\alpha)$, where in the second inequality we used lemma 3.1, and similarly for $\hat{e}$.

High-entropy commitments. This directly follows from the security of the auxiliary commitment scheme.   □

By section 2.5, both the completeness and the knowledge error can be made negligible if $n > M^2$.

4   Proving Equality among Classical and Lattice-Based Primitives

In the following we show how our basic protocol from section 3 can be used to link number-theory and lattice-based primitives via zero-knowledge proofs of knowledge. We exemplify this by showing how to prove that the messages contained in Pedersen commitments correspond to the plaintext in an encryption under the secure version of NTRU. We want to stress that in particular the choice of the encryption scheme is arbitrary, and it is easy to exchange it against other schemes, including standard NTRU [24] or Ring-LWE encryption [30].

Let $y = hs + pe + m \in R_q$ be the NTRU encryption of a message $m \in \{0, 1\}^n$, and let $p > 2n^2$ be coprime with q. Let further $\tilde{g}, \tilde{h}$ be a Pedersen commitment parameters, cf. section 2.2, and let $\widetilde{cmt}_i = m_i\tilde{g} + r_i\tilde{h}$ for $i = 0, \ldots, n-1$ be commitments to coefficients of m, where the order of $\tilde{g}$ and $\tilde{h}$ is $\tilde{q} > 2n^2$.

Then Protocol 4.1 can be used to prove, in zero-knowledge, that the commitments and the ciphertext are broadly well-formed and consistent, i.e., contain the same message. More precisely, the protocol guarantees the verifier that the prover knows the plaintext encrypted in $2y$, and that the coefficients of the respective message are all smaller than p. Furthermore, it shows that the messages are the same that are contained in $2\widetilde{cmt}_i$, i.e., $2y$ and the $2\widetilde{cmt}_i$ are consistent.

Prover	Verifier

$\mathbf{r}_s, \mathbf{r}_e \xleftarrow{\$} D_{\tilde{\mathcal{O}}(\sqrt{n}\alpha)}$

$\mathbf{r}_m \xleftarrow{\$} D_{\tilde{\mathcal{O}}(\sqrt{n})}$

$\mathbf{r}_{r,i} \xleftarrow{\$} \mathbb{Z}_{\tilde{q}}$ for $i = 0, \dots, n-1$

$\mathbf{t} = h\mathbf{r}_s + p\mathbf{r}_e + \mathbf{r}_m$

$\tilde{\mathbf{t}}_i = \mathbf{r}_{m,i}\tilde{g} + \mathbf{r}_{r,i}\tilde{h}$ for $i = 0, \dots, 2n-1$

$(c_{\text{aux}}, d_{\text{aux}}) = \mathsf{aCommit}(\mathbf{t}, (\tilde{\mathbf{t}}_i)_{i=0}^{n-1})$ $\xrightarrow{\quad c_{\text{aux}} \quad}$

$\xleftarrow{\quad c \quad}$ $\qquad c \xleftarrow{\$} \mathcal{C} = \{0, \dots, 2n-1\}$

$\mathbf{s}_s = \mathbf{r}_s + X^c s$

$\mathbf{s}_e = \mathbf{r}_e + X^c e$

$\mathbf{s}_m = \mathbf{r}_m + X^c m$

$\mathbf{s}_r = \mathbf{r}_r + (r_0, \dots, r_{n-1})_{\ll c}$

accept with probability $\dfrac{D_{\tilde{\mathcal{O}}(\sqrt{3n}\alpha)}((\mathbf{s}_e, \mathbf{s}_s, \mathbf{s}_m))}{M D_{(X^c e, X^c s, X^c m), \tilde{\mathcal{O}}(\sqrt{3n}\alpha)}(\mathbf{s}_e, \mathbf{s}_s, \mathbf{s}_m)}$

$\xrightarrow{\quad (\mathbf{t}, (\tilde{\mathbf{t}}_i)_{i=0}^{n-1}), d_{\text{aux}}, \quad}$
$\xrightarrow{\quad (\mathbf{s}_s, \mathbf{s}_e, \mathbf{s}_m, \mathbf{s}_r) \quad}$ $\qquad X^c y + \mathbf{t} \stackrel{?}{=} h\mathbf{s}_s + p\mathbf{s}_e + \mathbf{s}_m$

$(\widetilde{cmt}_0, \dots, \widetilde{cmt}_{n-1})_{\ll c} + (\tilde{\mathbf{t}}_0, \dots, \tilde{\mathbf{t}}_{n-1}) \stackrel{?}{=} \mathbf{s}_m\tilde{g} + \mathbf{s}_r\tilde{h}$

$\mathsf{aCOpen}((\mathbf{t}, \tilde{\mathbf{t}}_0, \dots, \tilde{\mathbf{t}}_{n-1}), c_{\text{aux}}, d_{\text{aux}}) \stackrel{?}{=} \text{accept}$

$\|\mathbf{s}_s\|, \|\mathbf{s}_e\| \leq \tilde{\mathcal{O}}(n\alpha)$

$\|\mathbf{s}_m\| \leq \tilde{\mathcal{O}}(n)$

Protocol 4.1. Proof that Pedersen commitments and NTRU encryption contain the same Plain Text

Theorem 4.2. *Protocol 4.1 is an HVZK Σ'-protocol for the following relations:*

$$\mathcal{R} = \Big\{ ((\tilde{g}, \tilde{h}, (\widetilde{cmt}_i)_{i=0}^{n-1}, h, p, y), (m, s, e, (r_i)_{i=0}^{n-1})) : y = hs + pe + m$$

$$\wedge \bigwedge_{i=0}^{n-1} \widetilde{cmt}_i = m_i\tilde{g} + r_i\tilde{h} \wedge \|m\|_\infty \leq 1 \wedge \|s\|, \|e\| \leq \tilde{\mathcal{O}}(\sqrt{n}\alpha) \Big\},$$

$$\mathcal{R}' = \Big\{ ((\tilde{g}, \tilde{h}, (\widetilde{cmt}_i)_{i=0}^{n-1}, h, p, y), (m, s, e, (r_i)_{i=0}^{n-1})) : 2y = 2hs + 2pe + 2m$$

$$\wedge \bigwedge_{i=0}^{n-1} 2\widetilde{cmt}_i = (2m \bmod q)_i\tilde{g} + 2r_i\tilde{h}$$

$$\wedge \|2m\|_\infty \leq 2n^2 \wedge \|2s\|, \|2e\| \leq \tilde{\mathcal{O}}(n^2\alpha) \Big\}.$$

where $(2m \bmod q)_i$ is the i-coefficient of $2m \in R_q$. The protocol has a knowledge error of $1/(2n)$, and a completeness error of $1 - 1/M$.

Furthermore, if for the auxiliary commitment scheme a commitment does not only bind the user to the message, but also to the opening information, the protocol has quasi-unique responses and high-entropy commitments.

A detailed proof is given in the full version. By the remark in section 2.5, both the completeness and the knowledge error can be made negligible if $n > M$.

5   Application to Group Signatures

We next show how Protocol 4.1 can be used to construct a group signature scheme with signature size logarithmic in the number of group members. The scheme is private under lattice assumptions, but traceable/unforgeable under non-lattice assumptions. As argued in the introduction, this may be realistic in applications where privacy needs to be guaranteed on the long term. For example, if group signatures are used to sign votes in electronic elections, unforgeability is mainly important when the votes are counted, but privacy needs to be preserved long after that.

Before presenting the actual signature scheme, we will prove secure a variation of a generic construction that we believe is folklore, as it underlies several direct schemes in the literature [3, 8] and was explicitly described by Chase and Lysyanskaya [11]. The resulting construction satisfies the following definition of group signatures providing full (CCA) anonymity put forth by Bellare et al. [5].

Definition 5.1. *A group signature scheme is a tuple* (GKG, GSign, GVerify, GOpen) *where:*

- *On input* $1^\lambda, 1^N$, *the key generation algorithm* GKG *outputs a group public key gpk, an opening key gok, and a vector of N signing keys* **gsk** *where* **gsk**$[i]$ *is given to user* $i \in \{1, \ldots, N\}$.
- *On input* $gsk = $ **gsk**$[i]$ *and message* $m \in \mathcal{M}$, *the signing algorithm* GSign *outputs a group signature* σ.
- *On input* gpk, m, σ, *the verification algorithm* GVerify *outputs* accept *or* reject.
- *On input* gok, m, σ, *the opening algorithm* GOpen *outputs the identity of the purported signer* $i \in \{1, \ldots, N\}$ *or* $\perp$ *to indicate failure.*

The algorithms satisfy the following properties:

- Correctness: *Verification accepts whenever keys and signatures are honestly generated, i.e., for all* $\lambda, N \in \mathbb{N}$, *all* $i \in \{1, \ldots, N\}$, *and all* $m \in \mathcal{M}$

$$\Pr\left[\begin{array}{c} \text{GVerify}(gpk, m, \sigma) = \text{accept} : \\ (gpk, gok, \textbf{gsk}) \xleftarrow{\$} \text{GKG}(1^\lambda, 1^N), \sigma \xleftarrow{\$} \text{GSign}(\textbf{gsk}[i], m) \end{array}\right] = 1 .$$

- Anonymity: *One cannot tell which signer generated a particular signature, even when given access to an opening oracle. Referring to Figure 1, for all PPT A there exists a negligible function* negl *such that*

$$\left| \Pr[\textbf{Exp}_A^{\text{anon}-0}(\lambda) = 1] - \Pr[\textbf{Exp}_A^{\text{anon}-1}(\lambda) = 1] \right| \leq \text{negl}(\lambda) .$$

- Traceability: *One cannot generate a signature that cannot be opened or that opens to an honest user. Referring to Figure 1, for all PPT A there exists a negligible function* negl *such that*

$$\Pr[\textbf{Exp}_A^{\text{trace}}(\lambda)] \leq \text{negl}(\lambda) .$$

Experiment $\mathbf{Exp}_A^{\text{anon}-b}(\lambda)$:
  $(gpk, gok, \mathbf{gsk}) \xleftarrow{\$} \mathsf{GKG}(1^\lambda, 1^N)$
  $(st, i_0^*, i_1^*, m^*) \xleftarrow{\$}$
     $A^{\mathsf{GOpen}(gok,\cdot,\cdot)}((gpk, \mathbf{gsk}), \varepsilon)$
  $\sigma^* \xleftarrow{\$} \mathsf{GSign}(\mathbf{gsk}[i_b], m^*)$
  $b' \xleftarrow{\$} A^{\mathsf{GOpen}(gok,\cdot,\cdot)}(\sigma^*, st)$
  If $(m^*, \sigma^*) \notin \mathcal{Q}_{\mathsf{GOpen}}$
    then return b' else return 0

Experiment $\mathbf{Exp}_A^{\text{trace}}(\lambda)$:
  $(gpk, gok, \mathbf{gsk}) \xleftarrow{\$} \mathsf{GKG}(1^\lambda, 1^N)$
  $(m, \sigma) \xleftarrow{\$}$
     $A^{\mathsf{GSign}(\mathbf{gsk}[\cdot],\cdot), \mathbf{gsk}[\cdot]}(gpk, gok)$
  $i \xleftarrow{\$} \mathsf{GOpen}(gok, m, \sigma)$
  If $\mathsf{GVerify}(gpk, m, \sigma) = 1 \;\wedge\; i \notin \mathcal{Q}_{\mathbf{gsk}}$
    $\wedge\; (i, m) \notin \mathcal{Q}_{\mathsf{GSign}}$
    then return 1 else return 0

Fig. 1. The anonymity (left) and traceability (right) experiments for group signatures. The sets $\mathcal{Q}_{\mathsf{GOpen}}, \mathcal{Q}_{\mathsf{GSign}}, \mathcal{Q}_{\mathbf{gsk}}$ contain all queries (m, σ), (i, m), and i that A submitted to its GOpen, GSign, and **gsk** oracles, respectively.

5.1 Building Blocks

The construction is based on weakly unforgeable standard signatures, and signature proofs of knowledge. In the following, we recap the respective definitions.

Informally, a signature scheme is a triple (SKG, SSign, SVerify), where SKG generates a signing/verification key pair (ssk, spk), SSign can be used to sign a message m using the signing key, and SVerify can be used to check the validity of a signature only using the public verification key. It should hold that honestly computed signatures are always valid, and that no adversary can come up with a valid signature on a new message after having received signatures on messages that he chose before obtaining spk. A formal definition can be found in the full version.

Concerning signature proofs of knowledge, we adapt the definitions of Chase and Lysyanskaya [11] to allow for signatures in the random-oracle model (ROM) that are simulated by programming the random oracle H and extracted through rewinding. We also generalize the definition to allow for a soundness gap: signing is performed using a witness from R for a language $\mathcal{L}$, while extraction only guarantees that the signer knows a witness from $R' \supseteq R$ for relation $\mathcal{L}'$. Finally, we add a definition of simulation soundness, meaning that an adversary cannot produce new signatures for false statements even after seeing simulated signatures on arbitrary statements.

Definition 5.2. *A signature of knowledge scheme for languages $\mathcal{L}, \mathcal{L}'$ with respective witness relations R, R' is a tuple* (SoKSetup, SoKSign, SoKVerify, SoKSim) *where:*

- *On input 1^λ, the setup algorithm* SoKSetup *outputs common parameters sokp.*
- *On input $sokp, x, w$ such that $(x, w) \in R$ and message $m \in \mathcal{M}$, the signing algorithm* SoKSign *outputs a signature of knowledge sok.*
- *On input $sokp, x, m, sok$, the verification algorithm* SoKVerify *outputs* accept *or* reject.
- *The stateful simulation algorithm* SoKSim *can be called in three modes. When called as $(sokp, st) \xleftarrow{\$}$* SoKSim(setup, $1^\lambda, \varepsilon$)*, it produces simulated parameters sokp, possibly keeping a trapdoor in its internal state st. When run as $(h, st') \xleftarrow{\$}$* SoKSim(ro, Q, st)*, it produces a response h for a random oracle query Q. When run as $(sok, st') \xleftarrow{\$}$* SoKSim(sign, x, m, st)*, it produces a simulated signature of knowledge sok without using a witness.*

For ease of notation, let $\mathsf{StpSim}(1^\lambda)$ *be the algorithm that returns the first part of* $\mathsf{SoKSim}(\texttt{setup}, 1^\lambda, st)$, *let* $\mathsf{ROSim}(Q)$ *be the algorithm that returns the first part of* $\mathsf{SoKSim}(\texttt{ro}, Q, st)$, *let* $\mathsf{SSim}(x, w, m)$ *be the algorithm that returns the first part of* $\mathsf{SoKSim}(\texttt{sign}, x, m, st)$ *if* $(x, w) \in R$ *and returns* $\perp$ *otherwise, and let* $\mathsf{SSim}'(x, m)$ *be the algorithm that returns the first part of* $\mathsf{SoKSim}(\texttt{sign}, x, m, st)$ *without checking language membership. The experiment keeps a single synchronized state for* SoKSim *across all invocations of these derived algorithms.*

The algorithms satisfy the following properties:

- Correctness*: Verification accepts whenever parameters and signatures are correctly generated, i.e., for all* $\lambda \in \mathbb{N}$, *all* $(x, w) \in R$, *and all* $m \in \mathcal{M}$, *there exists a negligible function* negl *such that*

$$\Pr\left[\begin{array}{c} \mathsf{SoKVerify}(sokp, x, m, sok) = \texttt{reject} : \\ sokp \xleftarrow{\$} \mathsf{SoKSetup}(1^\lambda),\ sok \xleftarrow{\$} \mathsf{SoKSign}(sokp, x, w, m) \end{array} \right] \leq \mathsf{negl}(\lambda)\ .$$

- Simulatability*: No adversary can distinguish whether it is interacting with a real random oracle and signing oracle, or with their simulated versions. Formally, for all PPT* A *there exists a negligible function* negl *such that*

$$\left| \Pr[b = 1 : sokp \xleftarrow{\$} \mathsf{SoKSetup}(1^\lambda), b \xleftarrow{\$} \mathsf{A}^{H(\cdot), \mathsf{SoKSign}(sokp, \cdot, \cdot, \cdot)}(sokp)] \right.$$
$$\left. - \Pr[b = 1 : sokp \xleftarrow{\$} \mathsf{StpSim}(1^\lambda), b \xleftarrow{\$} \mathsf{A}^{\mathsf{ROSim}(\cdot), \mathsf{SSim}(\cdot, \cdot, \cdot)}(sokp)] \right| \leq \mathsf{negl}(\lambda)\ .$$

- Extractability*: The only way to produce a valid signature of knowledge is by knowing a witness from* R'. *Formally, for all PPT* A *there exists an extractor* $\mathsf{SoKExt_A}$ *and a negligible function* negl *such that*

$$\Pr\left[\begin{array}{c} \mathsf{SoKVerify}(sokp, x, m, sok) = \texttt{accept} \\ \wedge (x, w, m) \notin \mathcal{Q} \wedge (x, w) \notin R' : \\ sokp \xleftarrow{\$} \mathsf{StpSim}(1^\lambda; \rho_S), \\ (x, m, sok) \xleftarrow{\$} \mathsf{A}^{\mathsf{ROSim}(\cdot), \mathsf{SSim}(\cdot, \cdot, \cdot)}(sokp; \rho_A), \\ w \xleftarrow{\$} \mathsf{SoKExt_A}(sokp, x, m, sok, \rho_S, \rho_A) \end{array} \right] \leq \mathsf{negl}(\lambda)\ ,$$

where $\mathcal{Q}$ *is the set of queries* (x, w, m) *that* A *submitted to its* SSim *oracle.*
- Simulation-soundness*: No adversary can produce a new signature on a false statement for* $\mathcal{L}'$, *even after seeing a signature on an arbitrary statement. Formally, for all PPT* A *there exists a negligible function* negl *such that*

$$\Pr\left[\begin{array}{c} \mathsf{SoKVerify}(sokp, x, m, sok) = \texttt{accept} \\ \wedge (x', m', sok') \neq (x, m, sok) \wedge x \notin \mathcal{L}' : \\ sokp \xleftarrow{\$} \mathsf{StpSim}(1^\lambda), (x, m, st) \xleftarrow{\$} \mathsf{A}^{\mathsf{ROSim}(\cdot)}(sokp), \\ sok \xleftarrow{\$} \mathsf{SSim}'(x, m), (x', m', sok') \xleftarrow{\$} \mathsf{A}^{\mathsf{ROSim}(\cdot)}(sok, st) \end{array} \right] \leq \mathsf{negl}(\lambda)\ .$$

5.2 Generic Construction

A folklore construction of group signatures is to have a user's signing key be a standard signature on his identity i, and to have a group signature on message m be an encryption of his identity together with a signature of knowledge on m that the encrypted identity is equal to the identity in his signing key. The construction appeared implicitly [3, 8] or explicitly [11] in the literature, but was never proved secure.

To obtain full anonymity, this generic construction would probably require CCA security from encryption scheme, but our NTRU variant is only semantically secure. We could apply a generic CCA-yielding transformation using random oracles or non-interactive zero-knowledge proofs of knowledge (NIZK), but this would make the signature of knowledge hopelessly inefficient. Instead, we take inspiration from the Naor-Yung construction [31, 33] by using a semantically secure scheme to encrypt the user's identity twice under two different public keys and letting the signature of knowledge prove that both ciphertexts encrypt the same plaintext. Moreover, our proof systems have a soundness gap: the adversary for the soundness game may use more noise in the ciphertexts than what the encryption algorithm Enc does, and may also encrypt plaintexts outside $\{0, 1\}^n$. We therefore give a generic construction that deviates slightly from the general idea, but that is sufficient and that we can efficiently instantiate with our protocol from section 3.

Let (EncKG, Enc, Dec) be an encryption scheme with message space $\mathcal{ID}$, let $\mathcal{ID}' \supseteq \mathcal{ID}$, and let Enc$'$ be an algorithm such that for all key pairs $(epk, esk) \xleftarrow{\$} \mathsf{EncKG}(1^\lambda)$, for all $i \in \mathcal{ID}$ and for all random tapes[1] ρ, ρ' and all $i \in \mathcal{ID}$, $i' \in \mathcal{ID}'$:

$$\mathsf{Enc}(epk, i; \rho) = \mathsf{Enc}'(epk, i; \rho) \text{ and } \mathsf{Dec}(esk, \mathsf{Enc}'(epk, i'; \rho')) = i' .$$

The algorithm Enc$'$ represents the way the adversary can generate the ciphertexts and still prove them valid. The above property ensures that completeness holds perfectly even with Enc$'$. The IND-CPA property still has to hold with Enc.

For our instantiation with the NTRU encryption scheme from Theorem 4.2, $\mathcal{ID} = \{0, 1\}^\ell$ which is identified with $\{0, \ldots, 2^\ell - 1\}$ (with $\ell \le n, \tilde{q}$), $\mathcal{ID}' = \mathbb{Z}_{\tilde{q}}$ and $\rho = (s, e)$. The algorithm Enc$'((h, p), i'; \rho')$ with $i' \in \mathcal{ID}'$ checks that either ρ' is a triple (s, e, i''), or $i' \in \mathcal{ID}$ and ρ' is a pair of vectors (s, e). In the latter case, i'' is just the binary vector in $\{0, 1\}^\ell$ corresponding to i'. In both cases, s and e must be such that $\|2s\|, \|2e\| \le \tilde{\mathcal{O}}(n^2 \alpha)$, $i'' \in R_q$, $2i' = \sum_{j=0}^{n-1} 2^j (2i'' \bmod q)_j \bmod \tilde{q}$, and $\|i''\|_\infty \le 2n^2 < p, \tilde{q}$. If all these requirements are met, Enc$'$ outputs $y \leftarrow hs + pe + i'$. We need to slightly change the algorithms EncKG and Enc to truncate the distribution of g, s and e, to ensure that $\|s\|, \|e\| \le \tilde{\mathcal{O}}(\sqrt{n}\alpha)$ and $\|g\|$ is small enough for the decryption below. We also change the algorithm Dec: to decrypt $C = y$ with secret key f, it computes $y' = 2fy \in R_q$, and outputs $i' = (\sum_{j=0}^{n-1} 2^j (y' \bmod p)_j)/2 \bmod \tilde{q}$. In other words, it decrypts $2C = 2y$ into $y' \bmod p$, and then recover the corresponding identity in $\mathcal{ID}' = \mathbb{Z}_{\tilde{q}}$. This does not touch security.

---

[1] To simplify notation in this section, we assume that the random tapes ρ, ρ' are not necessarily a uniform binary bitstrings as usual. Rather, we see ρ as the list of random values that Enc directly derives from the random tape, while ρ' can be seen as an auxiliary adversarial input to the Enc$'$ algorithm.

Let $(\mathsf{SKG}, \mathsf{SSign}, \mathsf{SVerify})$ be a signature scheme and let $(\mathsf{SoKSetup}, \mathsf{SoKSign}, \mathsf{SoKVerify}, \mathsf{SoKSim})$ be a signature of knowledge scheme for the languages $\mathcal{L}, \mathcal{L}'$ with witness relationships

$$R = \{((spk, epk_1, epk_2, C_1, C_2), (i, sig, \rho_1, \rho_2)) : \mathsf{SVerify}(spk, i, sig) = \texttt{accept}$$
$$\wedge\ C_1 = \mathsf{Enc}(epk_1, i; \rho_1) \wedge C_2 = \mathsf{Enc}(epk_2, i; \rho_2)\},$$
$$R' = \{((spk, epk_1, epk_2, C_1, C_2), (i', sig, \rho_1', \rho_2')) : \mathsf{SVerify}(spk, i', sig) = \texttt{accept}$$
$$\wedge\ C_1 = \mathsf{Enc}'(epk_1, i'; \rho_1') \wedge C_2 = \mathsf{Enc}'(epk_2, i'; \rho_2')\}.$$

Consider the following group signature scheme with user identities $i \in \mathcal{ID}$:

- $\mathsf{GKG}(1^\lambda, 1^N)$: The group manager generates signing keys $(spk, ssk) \xleftarrow{\$} \mathsf{SKG}(1^\lambda)$, encryption keys $(epk_1, esk_1) \xleftarrow{\$} \mathsf{EncKG}(1^\lambda)$, $(epk_2, esk_2) \xleftarrow{\$} \mathsf{EncKG}(1^\lambda)$, and parameters $sokp \xleftarrow{\$} \mathsf{SoKSetup}(1^\lambda)$. He computes $\mathbf{gsk}[i] \xleftarrow{\$} \mathsf{SSign}(ssk, i)$ for $i \in \mathcal{ID}$ and outputs $gpk = (spk, epk_1, epk_2, sokp)$, $gok = (gpk, esk_1)$, and $\mathbf{gsk}$.
- $\mathsf{GSign}(gsk, m)$: Signer i computes two ciphertexts $C_1 \leftarrow \mathsf{Enc}(epk_1, i; \rho_1)$ and $C_2 \leftarrow \mathsf{Enc}(epk_2, i; \rho_2)$, computes a signature of knowledge $sok \xleftarrow{\$} \mathsf{SoKSign}(sokp, (spk, epk_1, epk_2, C_1, C_2), (i, sig, \rho_1, \rho_2), m)$ and outputs the group signature $\sigma = (C_1, C_2, sok)$.
- $\mathsf{GVerify}(gpk, m, \sigma)$: To verify a group signature, one checks that $\mathsf{SoKVerify}(sokp, (spk, epk_1, epk_2, C_1, C_2), m, sok) = \texttt{accept}$.
- $\mathsf{GOpen}(gok, m, \sigma)$: The opener checks that $\mathsf{GVerify}(gpk, m, \sigma) = \texttt{accept}$, and returns $i \leftarrow \mathsf{Dec}(esk_1, C_1)$.

Theorem 5.3. *The group signature scheme sketched above is anonymous in the ROM if the encryption scheme is semantically secure and the signature of knowledge scheme is simulatable and simulation-sound.*

Theorem 5.4. *The group signature scheme is traceable in the ROM if the underlying signature scheme is weakly unforgeable and the signature of knowledge scheme is simulatable and extractable.*

The proofs of the last two theorems are omitted here and are given in the full version.

5.3 Signatures of Knowledge from Σ'-Protocols

We now show a construction of the required signatures of knowledge in the random-oracle model from a signature scheme and an encryption scheme with Σ'-protocol proofs. More particularly, we require that for the signature scheme one can prove knowledge of a signature on a committed message, while for the encryption scheme one can prove that an encrypted plaintext is equal to a committed message.

Let $(\mathsf{CSetup}, \mathsf{Commit}, \mathsf{COpen})$ be a commitment scheme, let $(\mathsf{EncKG}, \mathsf{Enc}, \mathsf{Dec})$ be an encryption scheme with message space $\mathcal{M}$ and let Enc' be an associated algorithm as described earlier. Let $(\mathsf{P_s}, \mathsf{V_s}, \mathsf{S_s})$ be a Σ-protocol for the language $\mathcal{L_s}$ with

$$R_\mathsf{s} = \{((spk, cpars, cmt), (sig, m, o)) :$$
$$\mathsf{SVerify}(spk, m, sig) = \texttt{accept} \wedge \mathsf{COpen}(cpars, m, cmt, o) = \texttt{accept}\}.$$

Let also $(\mathsf{P_e}, \mathsf{V_e}, \mathsf{S_e})$ be a Σ'-protocol for the languages $\mathcal{L}_e, \mathcal{L}'_e$ with

$$R_e = \{((epk, C, cpars, cmt), (m, \rho, o)) :$$
$$C = \mathsf{Enc}(epk, m; \rho) \wedge \mathsf{COpen}(cpars, m, cmt, o) = \mathtt{accept}\},$$
$$R'_e = \{((epk, C, cpars, cmt), (m, \rho', o)) :$$
$$C = \mathsf{Enc}'(epk, m; \rho') \wedge \mathsf{COpen}(cpars, m, cmt, o) = \mathtt{accept}\}.$$

Let $\mathcal{C}_s$ and $\mathcal{C}_e$ be the challenge spaces for these respective protocols, and let $H : \{0,1\}^* \to \mathcal{C}_s \times \mathcal{C}_e$. Consider the following construction of a signature of knowledge scheme for the languages $\mathcal{L}$ and $\mathcal{L}'$:

- SoKSetup(1^λ): Return $sokp = cpars \xleftarrow{\$} \mathsf{CSetup}(1^\lambda)$.
- SoKSign($sokp, x, w, m$): Create a commitment $(cmt, o) \xleftarrow{\$} \mathsf{Commit}(cpars, m)$. Compute the first round of the Σ'-protocols for a signature and two encryptions $(\mathsf{t_s}, st_s) \xleftarrow{\$} \mathsf{P_s}((spk, cpars, cmt), (sig, m, o))$ and $(\mathsf{t}_j, st_j) \xleftarrow{\$} \mathsf{P_e}((epk_j, C_j, cpars, cmt), (m, \rho_j, o))$ for $j = 1, 2$. Generate the challenges $(\mathsf{c_s}, \mathsf{c_e}) \leftarrow H(spk, cpars, cmt, epk_1, C_1, epk_2, C_2, \mathsf{t_s}, \mathsf{t_1}, \mathsf{t_2}, m)$. Compute responses $\mathsf{s_s} \leftarrow \mathsf{P_s}(\mathsf{c_s}, st_s)$ and $\mathsf{s}_j \leftarrow \mathsf{P_e}(\mathsf{c_e}, st_j)$ for $j = 1, 2$ and output the signature of knowledge $sok = (\mathsf{t_s}, \mathsf{t_1}, \mathsf{t_2}, \mathsf{s_s}, \mathsf{s_1}, \mathsf{s_2})$.
- SoKVerify($sokp, x, m, sok$): Recompute the challenges $(\mathsf{c_s}, \mathsf{c_e}) \leftarrow H(spk, cpars, cmt, epk_1, C_1, epk_2, C_2, \mathsf{t_s}, \mathsf{t_1}, \mathsf{t_2}, m)$. Return accept if $\mathsf{V_s}((spk, cpars, cmt), \mathsf{t_s}, \mathsf{c_s}, \mathsf{s_s}) = \mathtt{accept}$ and $\mathsf{V_e}((epk_j, C_j, cpars, cmt), \mathsf{t}_j, \mathsf{c_e}, \mathsf{s}_j) = \mathtt{accept}$ for $j = 1, 2$. Otherwise, return reject.
- SoKSim: The simulation algorithm keeps in its state its random tape, an initially empty table HT to keep track of previous random-oracle queries, and a counter ctr initialized to zero. The simulator's random tape ρ includes random-oracle responses $h_1, \ldots, h_{q_H + q_S} \xleftarrow{\$} \mathcal{C}_s \times \mathcal{C}_e$, where q_H and q_S are upper bounds on the number of random-oracle and signing queries that an adversary can make. When called as SoKSim($\mathtt{setup}, 1^\lambda, \varepsilon$), it generates commitment parameters $cpars \xleftarrow{\$} \mathsf{CSetup}(1^\lambda)$ and returns $(cpars, st = (\rho, HT, ctr, cpars))$. When run as SoKSim($\mathtt{ro}, Q, st$), it checks whether the query Q was made before. If so, it returns $h_{HT[q]}$. Otherwise, it increases the counter ctr, sets $HT[Q] \leftarrow ctr$, and returns h_{ctr}. When run as SoKSim($\mathtt{sign}, (spk, epk_1, epk_2, C_1, C_2), m, st$), the simulator first creates a commitment $(cmt, o) \xleftarrow{\$} \mathsf{Commit}(1, cpars)$. It then increases the counter ctr and parses h_{ctr} as $(\mathsf{c_s}, \mathsf{c_e})$. It runs the simulators $\mathsf{S_s}, \mathsf{S_e}$ to obtain simulated protocol transcripts $(\mathsf{t_s}, \mathsf{s_s}) \xleftarrow{\$} \mathsf{S_s}((spk, cpars, cmt), \mathsf{c_s})$ and $(\mathsf{t}_j, \mathsf{s}_j) \xleftarrow{\$} \mathsf{S_s}((epk_j, C_j, cpars, cmt), \mathsf{c_e})$ for $j = 1, 2$. If $HT[spk, cpars, cmt, epk_1, C_1, epk_2, C_2, \mathsf{t_s}, \mathsf{t_1}, \mathsf{t_2}, m]$ is not defined, then set it to h_{ctr}, else abort.

Theorem 5.5. *The above scheme is correct if the proof systems* $(\mathsf{P_s}, \mathsf{V_s})$ *and* $(\mathsf{P_e}, \mathsf{V_e})$ *have negligible completeness error.*

Theorem 5.6. *The above scheme is simulatable in the random-oracle model if the commitment scheme is hiding and the proof systems* $(\mathsf{P_s}, \mathsf{V_s})$ *and* $(\mathsf{P_e}, \mathsf{V_e})$ *are special HVZK and have high-entropy commitments.*

Theorem 5.7. *The above scheme is extractable in the random-oracle model if the commitment scheme is binding and the proof systems* (P_s, V_s) *and* (P_e, V_e) *are special-sound and have super-polynomial challenge spaces and negligible knowledge error.*

Theorem 5.8. *The above scheme is simulation-sound if the underlying commitment scheme is binding and the underlying* Σ'*-protocols* (P_s, V_s, S_s) *and* (P_e, V_e, S_e) *are special-sound, have quasi-unique responses, super-polynomial challenge spaces, and negligible knowledge error.*

Due to length limitations, the proofs of the previous theorems can be found in the full version.

5.4 Σ'-Protocols for Boneh-Boyen Signatures and the Group Signature Scheme

In the following we briefly recap the weakly unforgeable version of the Boneh-Boyen signature scheme [6, 7]. We assume that the reader is familiar with bilinear pairings and the strong Diffie-Hellman (SDH) assumption. The Boneh-Boyen signature scheme is defined as follows for a bilinear group generator BGGen:

SKG. This algorithm first computes $(\tilde{q}, \mathbb{G}_1, \mathbb{G}_2, \mathbb{G}_T, e) \xleftarrow{\$} \mathsf{BGGen}(1^\lambda)$. It chooses $\tilde{g}_1 \xleftarrow{\$} \mathbb{G}_1^\times$, $\tilde{g}_2 \xleftarrow{\$} \mathbb{G}_2^\times$, $x \xleftarrow{\$} \mathbb{Z}_{\tilde{q}}^\times$, and defines $\tilde{v} = x\tilde{g}_2$ and $\tilde{z} = e(\tilde{g}_1, \tilde{g}_2)$. It outputs $spk = ((\tilde{q}, \mathbb{G}_1, \mathbb{G}_2, \mathbb{G}_T, e), \tilde{g}_1, \tilde{g}_2, \tilde{v}, \tilde{z})$ and $ssk = x$.

SSign. To sign a message $m \in \mathbb{Z}_{\tilde{q}} \setminus \{-ssk\}$ with secret key $ssk = x$, this algorithm outputs the signature $\widetilde{sig} = \frac{1}{x+m}\tilde{g}_1$ if $x + m \neq 0$, and 0 otherwise.

SVerify. Given a signature public key spk, a message $m \in \mathbb{Z}_{\tilde{q}}$ and a signature $\widetilde{sig}$, this algorithm outputs accept if $\tilde{v} + m\tilde{g}_2 = 0$ in case $\widetilde{sig} = 0$, and if $e(\widetilde{sig}, \tilde{v} + m\tilde{g}_2) = \tilde{z}$ in case $\widetilde{sig} \neq 0$. In all other cases, it outputs reject.

Lemma 5.9. *If the SDH assumption holds for* BGGen, *then the above scheme is a weakly unforgeable signature scheme.*

We next show how a user can prove possession of a Boneh-Boyen signature on a message m, while keeping both, the message and the signature, private. In addition, the proof will additionally show that the m is also contained in a set of Pedersen commitments $\widetilde{cmt}_i = m_i\tilde{g} + r_i\tilde{h}$ such that $m = \sum_{i=0}^{n-1} 2^i m_i$, cf. section 2.2.

The idea underlying Protocol 5.10 is similar to that in Camenisch et al. [9]: The prover first re-randomizes the signature to obtain a value s, which it sends to the verifier. Subsequently, the prover and the verifier run a standard Schnorr proof for the resulting statement.

Theorem 5.11. *Protocol 5.10 is a perfectly HVZK Σ-proof of knowledge for the following relation:*

$$\mathcal{R} = \left\{ ((spk, (\widetilde{cmt}_i)_{i=0}^{n-1}), (\widetilde{sig}, m, r, (m_i, r_i)_{i=0}^{n-1})) : m = \sum_{i=0}^{n-1} 2^i m_i \wedge \right.$$

$$\left. \widetilde{cmt}_i = m_i\tilde{g} + r_i\tilde{h} \wedge \mathsf{SVerify}(spk, m, \widetilde{sig}) = \mathtt{accept} \right\}.$$

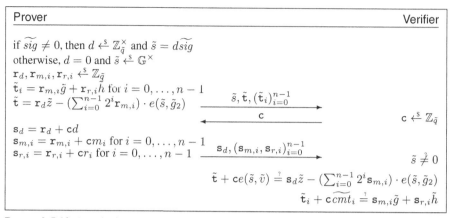

Protocol 5.10. Proof of possession of a signature on m, which is also contained in a set of Pedersen commitments

The protocol is perfectly complete, and has a knowledge error of $1/\tilde{q}$. Furthermore, the protocol has quasi unique responses (under the discrete logarithm assumption in $\mathbb{G}$) and high-entropy commitments.

The proof of this theorem is straightforward and can be found in the full version.

The Group Signature Scheme. Combining Protocols 4.1 and 5.10 now directly gives a group signature by the construction from section 5.3. The $\mathcal{ID}$ is given by $\{0,1\}^{\ell}$ (which can be identified with $\{0, \ldots, 2^{\ell} - 1\}$), where $\ell \leq n$ and $n/\tilde{q}$ is negligible, n is the dimension of the ring being used, and $\tilde{q}$ is the order of the groups of the commitment- and the signature schemes. The condition $n/\tilde{q}$ is just to ensure that with overwhelming probability, $ssk \notin \mathcal{ID}$, so that all signatures of an identity $i \in \mathcal{ID}$ is non-zero and can be used as a witness in Protocol 5.10. The commitment (CSetup, Commit, COpen) scheme from section 5.3, corresponds to the bit-by-bit Pedersen commitments $\widetilde{cmt}_i$.

References

1. Asharov, G., Jain, A., López-Alt, A., Tromer, E., Vaikuntanathan, V., Wichs, D.: Multiparty computation with low communication, computation and interaction via threshold FHE. In: Pointcheval, D., Johansson, T. (eds.) EUROCRYPT 2012. LNCS, vol. 7237, pp. 483–501. Springer, Heidelberg (2012)

2. Asokan, N., Shoup, V., Waidner, M.: Optimistic fair exchange of digital signatures. In: Nyberg, K. (ed.) EUROCRYPT 1998. LNCS, vol. 1403, pp. 591–606. Springer, Heidelberg (1998)

3. Ateniese, G., Camenisch, J., Joye, M., Tsudik, G.: A practical and provably secure coalition-resistant group signature scheme. In: Bellare, M. (ed.) CRYPTO 2000. LNCS, vol. 1880, pp. 255–270. Springer, Heidelberg (2000)

4. Bellare, M., Goldreich, O.: On defining proofs of knowledge. In: Brickell, E.F. (ed.) CRYPTO 1992. LNCS, vol. 740, pp. 390–420. Springer, Heidelberg (1993)

5. Bellare, M., Micciancio, D., Warinschi, B.: Foundations of group signatures: Formal definitions, simplified requirements, and a construction based on general assumptions. In: Biham, E. (ed.) EUROCRYPT 2003. LNCS, vol. 2656, pp. 614–629. Springer, Heidelberg (2003)

6. Boneh, D., Boyen, X.: Short signatures without random oracles. In: Cachin, C., Camenisch, J.L. (eds.) EUROCRYPT 2004. LNCS, vol. 3027, pp. 56–73. Springer, Heidelberg (2004)
7. Boneh, D., Boyen, X.: Short signatures without random oracles and the SDH assumption in bilinear groups. Journal of Cryptology 21(2), 149–177 (2008)
8. Boneh, D., Boyen, X., Shacham, H.: Short group signatures. In: Franklin, M. (ed.) CRYPTO 2004. LNCS, vol. 3152, pp. 41–55. Springer, Heidelberg (2004)
9. Camenisch, J., Dubovitskaya, M., Neven, G.: Oblivious transfer with access control. In: ACM Conference on Computer and Communications Security, pp. 131–140 (2009)
10. Camenisch, J., Neven, G., Rückert, M.: Fully anonymous attribute tokens from lattices. In: Visconti, I., De Prisco, R. (eds.) SCN 2012. LNCS, vol. 7485, pp. 57–75. Springer, Heidelberg (2012)
11. Chase, M., Lysyanskaya, A.: On signatures of knowledge. In: Dwork, C. (ed.) CRYPTO 2006. LNCS, vol. 4117, pp. 78–96. Springer, Heidelberg (2006)
12. Chaum, D., van Heyst, E.: Group signatures. In: Davies, D.W. (ed.) EUROCRYPT 1991. LNCS, vol. 547, pp. 257–265. Springer, Heidelberg (1991)
13. Cramer, R.: Modular Design of Secure yet Practical Cryptographic Protocols. Ph.D. thesis, CWI and University of Amsterdam (1997)
14. Damgård, I.B.: Efficient concurrent zero-knowledge in the auxiliary string model. In: Preneel, B. (ed.) EUROCRYPT 2000. LNCS, vol. 1807, pp. 418–430. Springer, Heidelberg (2000)
15. Damgård, I.: On Σ-Protocols. Lecture on Cryptologic Protocol Theory; Faculty of Science, University of Aarhus (2010)
16. Damgård, I.B., Fujisaki, E.: A statistically-hiding integer commitment scheme based on groups with hidden order. In: Zheng, Y. (ed.) ASIACRYPT 2002. LNCS, vol. 2501, pp. 125–142. Springer, Heidelberg (2002)
17. Damgård, I.B., Goldreich, O., Okamoto, T., Wigderson, A.: Honest verifier vs dishonest verifier in public coin zero-knowledge proofs. In: Coppersmith, D. (ed.) CRYPTO 1995. LNCS, vol. 963, pp. 325–338. Springer, Heidelberg (1995)
18. Damgård, I., Pastro, V., Smart, N., Zakarias, S.: Multiparty computation from somewhat homomorphic encryption. In: Safavi-Naini, R., Canetti, R. (eds.) CRYPTO 2012. LNCS, vol. 7417, pp. 643–662. Springer, Heidelberg (2012)
19. Fujisaki, E., Okamoto, T.: Statistical zero knowledge protocols to prove modular polynomial relations. In: Kaliski Jr., B.S. (ed.) CRYPTO 1997. LNCS, vol. 1294, pp. 16–30. Springer, Heidelberg (1997)
20. Garg, S., Gentry, C., Halevi, S.: Candidate multilinear maps from ideal lattices. In: Johansson, T., Nguyen, P.Q. (eds.) EUROCRYPT 2013. LNCS, vol. 7881, pp. 1–17. Springer, Heidelberg (2013)
21. Gentry, C.: Fully homomorphic encryption using ideal lattices. In: STOC, pp. 169–178 (2009)
22. Gentry, C., Peikert, C., Vaikuntanathan, V.: Trapdoors for hard lattices and new cryptographic constructions. In: STOC, pp. 197–206 (2008)
23. Gordon, S.D., Katz, J., Vaikuntanathan, V.: A group signature scheme from lattice assumptions. In: Abe, M. (ed.) ASIACRYPT 2010. LNCS, vol. 6477, pp. 395–412. Springer, Heidelberg (2010)
24. Hoffstein, J., Pipher, J., Silverman, J.H.: NTRU: A ring-based public key cryptosystem. In: Buhler, J.P. (ed.) ANTS 1998. LNCS, vol. 1423, pp. 267–288. Springer, Heidelberg (1998)
25. Laguillaumie, F., Langlois, A., Libert, B., Stehlé, D.: Lattice-based group signatures with logarithmic signature size. In: Sako, K., Sarkar, P. (eds.) ASIACRYPT 2013, Part II. LNCS, vol. 8270, pp. 41–61. Springer, Heidelberg (2013)

26. Ling, S., Nguyen, K., Stehlé, D., Wang, H.: Improved zero-knowledge proofs of knowledge for the ISIS problem, and applications. In: Kurosawa, K., Hanaoka, G. (eds.) PKC 2013. LNCS, vol. 7778, pp. 107–124. Springer, Heidelberg (2013)

27. Lyubashevsky, V.: Lattice-based identification schemes secure under active attacks. In: Cramer, R. (ed.) PKC 2008. LNCS, vol. 4939, pp. 162–179. Springer, Heidelberg (2008)

28. Lyubashevsky, V.: Fiat-shamir with aborts: Applications to lattice and factoring-based signatures. In: Matsui, M. (ed.) ASIACRYPT 2009. LNCS, vol. 5912, pp. 598–616. Springer, Heidelberg (2009)

29. Lyubashevsky, V.: Lattice signatures without trapdoors. In: Pointcheval, D., Johansson, T. (eds.) EUROCRYPT 2012. LNCS, vol. 7237, pp. 738–755. Springer, Heidelberg (2012)

30. Lyubashevsky, V., Peikert, C., Regev, O.: On ideal lattices and learning with errors over rings. J. ACM 60(6), 43 (2013), Preliminary version appeared in Gilbert, H. (ed.): EUROCRYPT 2010. LNCS, vol. 6110, pp. 1–23. Springer, Heidelberg (2010)

31. Naor, M., Yung, M.: Public-key cryptosystems provably secure against chosen ciphertext attacks. In: 22nd ACM STOC. pp. 427–437. ACM Press (May 1990)

32. Pedersen, T.P.: Non-interactive and Information-Theoretic Secure Verifiable Secret Sharing. In: Feigenbaum, J. (ed.) CRYPTO 1991. LNCS, vol. 576, pp. 129–140. Springer, Heidelberg (1992)

33. Sahai, A.: Non-malleable non-interactive zero knowledge and adaptive chosen-ciphertext security. In: 40th FOCS, pp. 543–553. IEEE Computer Society Press (October 1999)

34. Santis, A.D., Persiano, G.: Zero-knowledge proofs of knowledge without interaction (extended abstract). In: 33rd FOCS, pp. 427–436. IEEE Computer Society Press (October)

35. Stehlé, D., Steinfeld, R.: Making NTRU as secure as worst-case problems over ideal lattices. In: Paterson, K.G. (ed.) EUROCRYPT 2011. LNCS, vol. 6632, pp. 27–47. Springer, Heidelberg (2011)

36. Stehlé, D., Steinfeld, R., Tanaka, K., Xagawa, K.: Efficient public key encryption based on ideal lattices. In: Matsui, M. (ed.) ASIACRYPT 2009. LNCS, vol. 5912, pp. 617–635. Springer, Heidelberg (2009)

37. Stern, J.: A new identification scheme based on syndrome decoding. In: Stinson, D.R. (ed.) CRYPTO 1993. LNCS, vol. 773, pp. 13–21. Springer, Heidelberg (1994)

Author Index

Andreeva, Elena   I-105
Applebaum, Benny   II-162
Aranha, Diego F.   I-262

Belaïd, Sonia   II-306
Bellare, Mihir   II-102
Benhamouda, Fabrice   I-551
Bernstein, Daniel J.   I-317
Bilgin, Begül   II-326
Biryukov, Alex   I-63
Bogdanov, Andrey   I-105
Boneh, Dan   I-42
Bos, Joppe W.   I-358
Bouillaguet, Charles   I-63
Boura, Christina   I-179
Bruneau, Nicolas   II-344
Brzuska, Christina   II-122, II-142

Camenisch, Jan   I-551
Catalano, Dario   II-193
Chen, Yu   II-366
Chuengsatiansup, Chitchanok   I-317
Chung, Kai-Min   II-62
Cohen, Ran   II-466
Corrigan-Gibbs, Henry   I-42
Costello, Craig   I-338

Damgård, Ivan   II-213
Danezis, George   I-532
David, Bernardo   II-213
de Portzamparc, Frédéric   I-21
Dinur, Itai   I-439
Doche, Christophe   I-297
Ducas, Léo   II-22
Dunjko, Vedran   II-406
Dunkelman, Orr   I-439

Emami, Sareh   I-141

Faugère, Jean-Charles   I-21
Fitzsimons, Joseph F.   II-406
Fleischhacker, Nils   I-512
Forler, Christian   II-289
Fouque, Pierre-Alain   I-262, I-420,
  II-306

Fournet, Cédric   I-532
Fuchsbauer, Georg   II-82
Fujisaki, Eiichiro   II-426

Gérard, Benoît   I-262, I-282, II-306
Giacomelli, Irene   II-213
Gierlichs, Benedikt   II-326
Gilbert, Henri   I-200
Groth, Jens   I-532
Guilley, Sylvain   II-344
Guo, Jian   I-458
Guo, Qian   I-1
Guo, Yanfei   II-366

Hanser, Christian   I-491
Heuser, Annelie   II-344
Hirt, Martin   II-448
Hisil, Huseyin   I-338
Hu, Lei   I-158

Jager, Tibor   I-512
Jarecki, Stanislaw   II-233
Jean, Jérémy   I-458, II-274
Johansson, Thomas   I-1
Joo, Chihong   II-173
Joux, Antoine   I-378, I-420
Jovanovic, Philipp   I-85
Joye, Marc   II-1

Kammerer, Jean-Gabriel   I-262
Keller, Marcel   II-506
Keller, Nathan   I-439
Khovratovich, Dmitry   I-63
Khurana, Dakshita   II-386
Kiayias, Aggelos   II-233
Kleinjung, Thorsten   I-358
Kohlweiss, Markulf   I-532
Komargodski, Ilan   II-254
Konstantinov, Momchil   II-82
Krawczyk, Hugo   II-233
Krenn, Stephan   I-551

Lange, Tanja   I-317
Lenstra, Arjen K.   I-358
Libert, Benoît   II-1

Lindell, Yehuda   II-466
Ling, San   I-141
Liu, Zhenming   II-62
Löndahl, Carl   I-1
Longo, Jake   I-223
Lucks, Stefan   II-289
Luykx, Atul   I-85, I-105
Lyubashevsky, Vadim   I-551, II-22

Ma, Xiaoshuang   I-158
Maji, Hemanta K.   II-386
Malkin, Tal   II-42
Marcedone, Antonio   II-193
Martin, Daniel P.   I-223
Mather, Luke   I-243
Mavromati, Chrysanthi   I-420
Mennink, Bart   I-85, I-105
Mittelbach, Arno   II-122, II-142
Mohassel, Payman   II-486
Mouha, Nicky   I-105

Nandi, Mridul   I-126, I-478
Naor, Moni   II-254
Naya-Plasencia, María   I-179
Neven, Gregory   I-551
Nielsen, Jesper Buus   II-213
Nikolić, Ivica   I-141, I-458, II-274
Nikov, Ventzislav   II-326
Nikova, Svetla   II-326

Oswald, Elisabeth   I-223, I-243

Page, Daniel   I-223
Pass, Rafael   II-62
Paterson, Kenneth G.   I-398
Perret, Ludovic   I-21
Peters, Thomas   II-1
Peyrin, Thomas   II-274
Pieprzyk, Josef   I-141
Pierrot, Cécile   I-378
Pietrzak, Krzysztof   II-82
Poettering, Bertram   I-398
Portmann, Christopher   II-406
Prest, Thomas   II-22
Puglisi, Orazio   II-193

Qiao, Kexin   I-158

Rao, Vanishree   II-82
Raykov, Pavel   II-448
Renner, Renato   II-406
Rijmen, Vincent   II-326
Rioul, Olivier   II-344

Sadeghian, Saeed   II-486
Sahai, Amit   II-386
Sasaki, Yu   I-458
Scholl, Peter   II-506
Schröder, Dominique   I-512
Schuldt, Jacob C.N.   I-398
Schwabe, Peter   I-317
Shamir, Adi   I-439
Slamanig, Daniel   I-491
Smart, Nigel P.   II-486
Song, Ling   I-158
Stam, Martijin   I-223
Standaert, François-Xavier   I-282
Stepanovs, Igors   II-102
Suder, Valentin   I-179
Sun, Siwei   I-158

Teranishi, Isamu   II-42
Tessaro, Stefano   II-102
Tibouchi, Mehdi   I-262
Tunstall, Michael J.   I-223

Veyrat-Charvillon, Nicolas   I-282

Wang, Huaxiong   I-141
Wang, Peng   I-158
Wenzel, Jakob   II-289
Whitnall, Carolyn   I-243

Yasuda, Kan   I-105
Yogev, Eylon   II-254
Yun, Aaram   II-173
Yung, Moti   II-1, II-42

Zapalowicz, Jean-Christophe   I-262
Zhang, Jiang   II-366
Zhang, Zhenfeng   II-366
Zhang, Zongyang   II-366